Extracellular Matrix Remodeling

Extracellular Matrix Remodeling

Special Issue Editor

Nicoletta Gagliano

MDPI • Basel • Beijing • Wuhan • Barcelona • Belgrade

MDPI

Special Issue Editor
Nicoletta Gagliano
Università degli Studi di Milano
Italy

Editorial Office
MDPI
St. Alban-Anlage 66
4052 Basel, Switzerland

This is a reprint of articles from the Special Issue published online in the open access journal *Cells* (ISSN 2073-4409) from 2018 to 2019 (available at: https://www.mdpi.com/journal/cells/special_issues/ECM)

For citation purposes, cite each article independently as indicated on the article page online and as indicated below:

LastName, A.A.; LastName, B.B.; LastName, C.C. Article Title. *Journal Name* **Year**, *Article Number, Page Range.*

ISBN 978-3-03921-628-4 (Pbk)
ISBN 978-3-03921-629-1 (PDF)

Contents

About the Special Issue Editor

Nicoletta Gagliano, Ph.D., Associate Professor, graduated in Biological Sciences from the University of Milan in 1992, and in 1998, she obtained her Ph.D. degree in the Physiopathology of Aging. From 2002 to 2010, she was an Assistant Professor of Human Anatomy, and since 2010, she has been an Associate Professor of Histology at the Department of Biomedical Sciences for Health of the University of Milan, Italy. Nicoletta Gagliano is the director of the Extracellular Matrix Lab at the Department of Biomedical Sciences for Health. She is a molecular biologist with extensive experience in cell culture, gene and protein expression analysis, and morphologic methods. Her research is focused on the study of the expression of genes and proteins involved in extracellular matrix remodeling and collagen turnover in different conditions, such as fibrosis, tendon homeostasis, gingival overgrowth, and tumor invasion, as well as the role of extracellular matrix components on epithelial-to-mesenchymal transition mechanisms.

Preface to "Extracellular Matrix Remodeling"

All tissues and organs are composed of cells and extracellular matrix (ECM). ECM is more or less abundant in different tissues, influencing their anatomy and physiology. ECM is a combination of interacting components, such as collagen, elastic fibers, glycosaminoglycans and proteoglycans, and several glycoproteins, that surrounds cells. In most tissues, fibril-forming collagen type I is the major constituent of ECM.The ECM is where cells live. It plays a structure and mechanical role by providing structural support to cells and tissues, and at the same time, defining tissue and organ structure.

However, the function of the ECM goes beyond providing mechanical support to cells and tissues. As cells are embedded into the ECM and interact with its components through their surface receptors, the ECM acts as a key determinant of cell behavior and tissue homeostasis. In fact, cell–ECM interaction plays a key role in influencing different cell activities, such as cell proliferation and migration. The ECM also sequesters and releases growth factors affecting important cellular pathways.ECM is not static but is a highly dynamic structural network that under normal conditions continuously undergoes controlled remodeling to balance synthesis, secretion, and degradation. ECM degradation is mediated by matrix-degrading enzymes, including the matrix metalloproteinases (MMPs), cathepsins, heparanase, hyaluronidases, and other proteases. Tightly controlled ECM homeostasis is crucial for regulating cellular behavior. Any quantitative and/or qualitative deregulation of ECM remodeling and, especially, of collagen turnover, is responsible of the alteration of ECM composition and structure, associated with the development and progression of several pathological conditions, such as organ fibrosis (determined by the abnormal accumulation of ECM components) and tumor invasion.

This Special Issue, focused on ECM remodeling pathways and also including cell–ECM interactions, adds new knowledge about the diverse biological roles and properties of the ECM, and contributes to understanding the pivotal role of ECM remodeling in the development of new therapeutic tools for disease treatment.

Nicoletta Gagliano
Special Issue Editor

cells

MDPI

Article

Action of the Metalloproteinases in Gonadal Remodeling during Sex Reversal in the Sequential Hermaphroditism of the Teleostei Fish *Synbranchus marmoratus* (Synbranchiformes: Synbranchidae)

Talita Sarah Mazzoni [1,2], Fabiana Laura Lo Nostro [3], Fernanda Natália Antoneli [4] and Irani Quagio-Grassiotto [2,5,*]

[1] Department of Cell and Development Biology, Institute of Biomedical Sciences, Federal University of Alfenas (UNIFAL), Gabriel Monteiro da Silva 700, 37130-001 Alfenas-MG, Brazil; talitasarah@yahoo.com.br
[2] Department of Morphology, Botucatu Biosciences Institute, State University of São Paulo (UNESP), Prof. Dr. Antonio Celso Wagner Zanin 250, 18618-689 Botucatu-SP, Brazil
[3] Department of Biodiversity and Experimental Biology, Faculty of Exact and Natural Sciences, Ciudad Universitaria, Lab. 4 y 78; Piso 4to., Pabellón 2, Int. Güiraldes 2160, C1428EGA Buenos Aires, Argentina; fabi@bg.fcen.uba.ar
[4] Institute of Biology, UNICAMP, Bertrand Russel s/n, 13083-865 Campinas-SP, Brazil; antonelif@yahoo.com
[5] Aquaculture Center of UNESP (CAUNESP), Prof. Paulo Donato Castellane s/n, 14884-900 Jaboticabal-SP, Brazil
*   Correspondence: iraniqg@ibb.unesp.br; Tel.: +55-14-3880-0468

Received: 23 February 2018; Accepted: 19 April 2018; Published: 24 April 2018

Abstract: Teleostei present great plasticity regarding sex change. During sex reversal, the whole gonad including the germinal epithelium undergoes significant changes, remodeling, and neoformation. However, there is no information on the changes that occur within the interstitial compartment. Considering the lack of information, especially on the role played by metalloproteinases (MMPs) in fish gonadal remodeling, the aim of this study was to evaluate the action of MMPs on gonads of sex reversed females of *Synbranchus marmoratus*, a fresh water protogynic diandric fish. Gonads were processed for light microscopy and blood samples were used for the determination of plasma sex steroid levels. During sex reversal, degeneration of the ovaries occurred and were gradually replaced by the germinal tissue of the male. The action of the MMPs induces significant changes in the interstitial compartment, allowing the reorganization of germinal epithelium. Leydig cells also showed an important role in female to male reversion. The gonadal transition coincides with changes in circulating sex steroid levels throughout sex reversion. The action of the MMPs, in the gonadal remodeling, especially on the basement membrane, is essential for the establishment of a new functional germinal epithelium.

Keywords: extracellular matrix; collagenase; protogynous diandric fish; germinal epithelium; gonad

1. Introduction

Sex change, one of the most controversial and remarkable expressions of plasticity in sexual development, can be observed in a number of teleost orders and families. Functional hermaphroditism was confirmed in 27 teleost families in seven orders [1]. A species or population is considered to exhibit functional hermaphroditism if a proportion of individuals function as both sexes at some time in an individual's life history. This natural sex reversal process basically consists of the expression of both male and female reproductive functions in a single individual; the proliferation of secondary-sex gonadal tissue and the simultaneous degeneration of the primary-sex gonadal tissue [1–4].

There are two patterns recognized in fish sex reversal, simultaneous and sequential. Sequential hermaphroditism can vary in several ways, presenting as either protogynous or protandric. In cases of protogynous hermaphroditism, most frequently observed in teleosts, individuals first develop into females and later, have their functional ovaries gradually replaced by male tissue. Thus, a decrease in female gametogenesis and an increase in the male tissue activity are observed during sex reversal; then, protogynous species may show monandry, in which all males are secondary males arising from the sex reversed females, or diandry, in which there are two types of males: those that develop as primary males and those that are secondary males [1,5,6].

The swamp or marbled eel, *Synbranchus marmoratus* Bloch (1795), as *Monopterus albus* Zuiew (1793), the rice eel, also a Synbranchiformes [7], is a diandric protogynous fish that primarily—with an unrestricted lobular testis type—develops directly as males, while secondary males arise from the sex reversal of females [8–12]. They are widely distributed across Central and South America [13].

In this species, sex change is correlated to the length of the individual, where reversion takes place in those between 25.0 and 60.0 cm total length [9,12,14].

The sex reversal process in *Synbranchus marmoratus* begins in reproductive females and is characterized by a disorganization of the gonadal architecture; intense proliferation of myoid cells; appearance of new germline cysts located at the edges of the ovarian lamellae; massive degeneration of the female germ cells; intense phagocytic activity; increase of vascularization; and the presence of melanomacrophage centers [10,11,15].

During the gonadal development of teleosts, a constant remodeling of the interstitial tissue is required, according to the changes undergone by the germinal epithelium, either during reproductive cycles [16] or gonadal differentiation [17]. This remodeling of the connective tissue involves events of degradation and new synthesis of extracellular matrix components [18,19]. The degradation of the matrix components is effected by several proteolytic enzymes, with the matrix metalloproteinases (MMPs) being the main ones involved in this process [20].

Metalloproteinases (MMPs) are a group of structurally-related proteins (endopeptidases), calcium and zinc dependent, and active at physiological pH [21,22]. These enzymes act on the degradation of many components of the extracellular matrix during tissue remodeling and may be involved in the regulation of cell-cell and cell-matrix signaling [22,23]. In mammals, studies show that some of the major MMPs are involved in remodeling processes of the male germinal epithelium [24] and in the rupture of the ovarian follicle [25]. A recent study also shows the expression of proteinase genes and proteolytic enzymes in gonad development of the mouse [26]. However, in aspects regarding the reproductive biology in teleosts, there are few studies that relate gonadal remodeling to the action of the MMPs and none regarding their role during sex reversal.

Another important aspect during sex reversion is the role of the steroid hormones. In teleosts, sexual steroids are considered the main factor of gonadal sex development and reproduction and are involved on sex reversal in hermaphroditic species [1,3,12,15,27]. These hormones are produced by gonadal somatic cells: follicle and theca cells, in females; Sertoli and, especially, Leydig cells in males.

In most teleost fishes, as in other vertebrates, Leydig cells are usually located singly or in small clusters in the interstitial compartment of the testis [28]. Leydig cells have features of steroid-producing cells, such as steroid dehydrogenase enzymes (essential for the biosynthesis of most steroid hormones) and receptors for hormone/polypeptide growth factors [29,30]. In fact, different approaches and techniques have been used in teleost Leydig cells and have confirmed that these cells are the main source of testicular steroids [17,31–35].

Leydig cells, also in the swamp eel *Synbranchus marmoratus*, are known to have a steroidogenic function in both types of males [36]. Therefore, these cells may offer hormonal support to the development and proliferation of male structures during sex reversal. Consequently, in this protogynous hermaphroditic species, this cell type might be the earliest structure to arise in the ovary, thus considered the early transitional gonad.

Thus, in an attempt to determine the role of the matrix metalloproteinases (MMPs) in the gonadal tissue remodeling during sex reversal and examine whether Leydig cells are involved in the morphological and functional beginning of sex reversal in *Synbranchus marmoratus*, the transitional gonads were studied using histological high resolution light microscopy, polarizing light microscopy, and immunohistochemistry techniques for the detection of three types of metalloproteinases (MMP-2, MMP-9, and MMP-14) and the 3β-hydroxysteroid dehydrogenase (3β-HSD). Additionally, immunohistochemistry for the detection of proliferation of the Leydig cells and hormonal analysis of sex steroids were also performed.

2. Materials and Methods

2.1. Animals

Adult *Synbranchus marmoratus* specimens, from 25 to 54.9 cm total length, were collected in the Tietê river, in the Penápolis region, São Paulo State, Southeastern of Brazil ($21°17'$ S; $49°47'$ W). Fish were transferred to the laboratory and anesthetized by immersion in a solution of 0.1% benzocaine. Handling of animals was performed in compliance with international standards on animal welfare (Canadian Council on Animal Care, 2005—Ottawa, Canada), as well as being in compliance with the local Ethical Committee from Instituto de Biociências de Botucatu—Botucatu, SP—Brazil (n. 580-IBB—UNESP).

2.2. Sample Preparation for Light Microscopy

Animals were anesthetized and immediately following, were sacrificed by decapitation and the gonads were quickly removed. Samples were then fixed in glutaraldehyde 2% and paraformaldehyde 4% solution in Sorensen buffer (0.1 M at 7.2 pH) for at least 24 h at room temperature. After fixation, the samples were dehydrated in a crescent ethanol series and embedded in Historesin (Leica HistoResin®, Buffalo Grove, IL, USA). Cross sections (3 μm) were stained with Ferric Hematoxylin/Eosin (HE); Toluidine Blue (TB); Schiff Periodic Acid + Ferric Hematoxylin + Metanil Yellow (MY) [37]; and with the Reticulin Method that enhances basement membranes. The Reticulin stain [38] uses an oxidizing agent, potassium permanganate, to oxidize aldehyde groups. Subsequently, the oxidized aldehyde groups are detected by the deposition of positive silver ions followed by their reduction using formalin. The result is a black hue of the reticulin fibers. As reticulin fibers are part of basement membranes, the method clearly detects basement membranes.

The histological slides were also stained with Picrosirius Red and observed in polarized light for the localization of collagen fibers [39]. Gonadal tissues were evaluated by using a computerized image analyzer (Leica Qwin 2.5, Leica Microsystems, Buffalo Grove, IL, USA).

Immunohistochemistry for Metalloproteinases (MMPs), 3β-Hidroxysteroid Dehydrogenase(3β-HSD) Enzyme and Proliferating Cell Nuclear Antigen (PCNA)

For the detection of the MMPs (MMP14, MMP2, and MMP9), three β-HSD enzyme and PCNA samples were fixed in paraformoldehyde 4% for 1–3 h, embedded in Paraplast® (Sigma-Aldrich, St. Louis, MO, USA) and sectioned at 5 μm. Sections were deparaffinized, hydrated in TBS buffer (Tris+phosphate buffer, 5 mM, pH 7.6), and then treated with 3% hydrogen peroxide for 15 min to quench endogenous peroxidase activity. Antigen retrieval was performed in a steam pan with citrate buffer (0.01 M; pH 6.0) for 20 min. After buffer rinse, slides were treated with a protein blocker (1% of non-fat powdered milk in TBS) for 15 min. They were subsequently incubated with the primary antibody, specific to each antigen: Anti-MMP14 (anti-rabbit, ab53712, ABCam), Anti-MMP2 (anti-rabbit, ab37150, ABCam), and Anti-MMP9 (anti-rabbit, ab38898, ABCam) polyclonal Ab. (1:50); anti-3-β HSD monoclonal Ab. (1:100) (anti-mouse, SC-100466, Santa Cruz, CA, USA); and Anti-PCNA monoclonal Ab. (1:300) (NCL-L-PCNA, clone PC10, Novocastra) for 2 h at room temperature in a moisture chamber. Next, slides were washed and incubated with MR HRP-Polymer (MACH4 Universal HRP Polymer

Kit®) for 30 min. Immunostaining was visualized using 0.1% DAB (3′,3′-diamminobenzidine) in TBS buffer and 0.03% H_2O_2. Sections were lightly counterstained with Harris Hematoxylin, dehydrated, and mounted with DPX. Negative control sections were treated with TBS instead of primary antibody. Sections of ovaries from mouse specimens were used as the positive control.

2.3. Sex Steroid Levels

Blood samples of *Synbranchus marmoratus* were used for the quantification of plasma levels of steroids. Animals were anesthetized and peripheral blood was collected by puncture of the caudal vein with a heparin-coated needle, attached to a 3 mL syringe. Samples were centrifuged at $2500 \times g$ for 15 min at 4 °C to obtain plasma, which was stored at −20 °C.

The plasma profiles of 11-ketotestosterone (11-KT), testosterone (T), and 17β-estradiol (E2) were quantified by an enzyme-linked immunosorbent assay (ELISA) with commercial kits (for 11-KT: Cayman Chemical; for T and E2: Interkit). Minimum detectable concentrations in plasma samples were 45 pg mL^{-1} for E2, 57 pgmL^{-1} for 11-KT, and 59 pg mL^{-1} for T. The inter- and intra-assay coefficients of variation were 10.5% and 2.8 ± 0.2% for E2 ($n = 8$), 10.6% and 3.1 ± 0.2% for 11-KT ($n = 8$), and 9,6% and 4.4 ± 0.3% for T ($n = 7$), respectively. Main cross-reactivity (>1%; given by supplier) was detected with estradiol-3-glucoronide (17%) and estrone (4%) for the E2 antibody and with 5α-dihydrotestosterone (27.4%), 5β-dihydrotestosterone (18.9%), androstenedione (3.7%), and 11-KT (2.2%) for the T antibody. For 11-KT antiserum, cross-reactivity with other steroids was lower than 0.1%. Analyses were carried out following the manufacturer's instructions, samples were assayed in duplicate, the standard curve was run for each ELISA plate, and the absorbance (450 nm for E2, 421 nm for 11-KT and T) measurements were performed in a microplate reader (Expert Plus ®, BIOCHROM, Cambridge, UK).

2.4. Statistical Analysis

Plasma levels of each hormone were grouped in distinct individual phases during the sex reversal of *Synbranchus marmoratus*. Values were expressed as mean ± SEM (standard error of the mean) and then subjected to variance analysis (one way ANOVA) and to a Tukey Test (Statistica® 7.0—StatSoft, Palo Alto, CA, USA). Means were considered statistically different at $p < 0.05$.

3. Results

3.1. Gonodal Structure of Synbranchus marmoratus

Synbranchus marmoratus is a diandric protogynous hermaphrodite species, in which the population is composed of four types of sexual representatives (Figure 1): Females; Primary Males; Secondary Males; and Hermaphroditic Individuals. Primary males are those individuals that develop directly as male, while secondary males develop from a process of sex reversal of females, passing through a stage of intersexes.

The gonads are located dorsally to the digestive tract. They are elongated, cylindrical, and occupy 2/3 of the coelomic cavity, being connected to the urogenital papillae in the caudal portion. The female, male, or transitional gonads are attached to the dorsal wall of the coelomic cavity by a connective tissue membrane—the mesentery.

The ovaries (Figure 1A) constitute a single saculiform organ, surrounded by a thick capsule of connective tissue (the tunica albuginea). It presents a lumen, delimited by a germinal epithelium, which borders the ovigerous lamellae (Figure 1B). The germinal epithelium, supported by a basement membrane (Figure 1C), is composed of somatic and germ cells. In cross sections, ovigerous lamellae are projected into the organ (ovarian cavity) from the ventral region, being attached to the tunica albuginea by lateral supports.

Initially, the female gonads that undergo the sex change process (Figure 1D) have the same characteristics as an ovary, as described above. At the beginning of the transitional process, the epithelium becomes composed of a larger number of somatic and germ cells. The connective tissue (Figure 1E) becomes more developed. Thus, the germinal epithelium still supported by a basement membrane (Figure 1F) presents a greater number of germline cysts immersed in a connective tissue and surrounded by a basement membrane, which separates them from the interstitial compartment.

After the reversion, the "new" male gonad (Figure 1G), named secondary males, is drafted on a female gonad. Thus, the testis of the secondary male becomes a single structure, externally delimited by a connective tissue capsule (the former tunica albuginea of the ovary) (Figure 1G). The new testicular lobules, called testicular lamellae, develop on the former ovigerous lamellae, showing the same aspects as a primary male testes, that is, they are formed by a germinal epithelium supported by a basement membrane (Figure 1H,I). However, unlike primary males, testes that come from sex reversal have lateral supports (Figure 1G), which attach the old lamellae to the capsule of connective tissue. In addition, the lumen is maintained during the process of sex reversal. Thus, the testes of the secondary male present a non-functional coelomic cavity (pseudo-ovarian cavity) (Figure 1G).

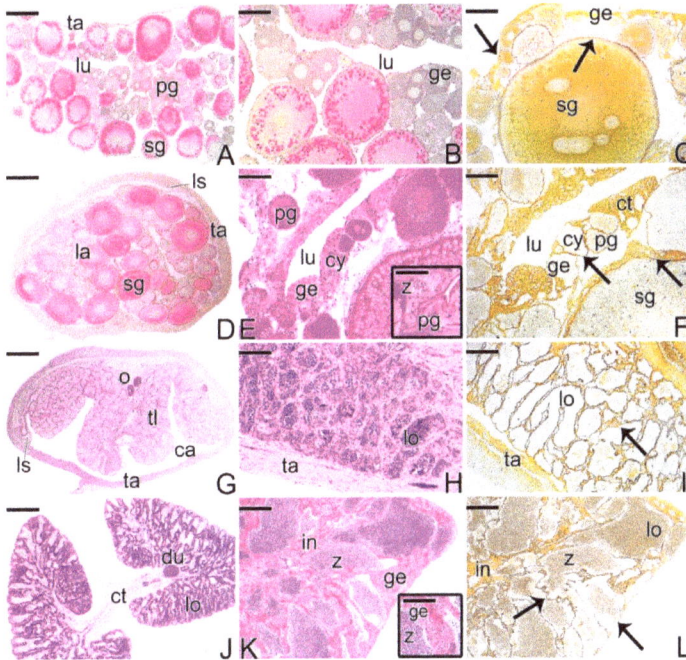

Figure 1. Four types of gonadal structures present in *S. marmoratus*: (**A–C**) Longitudinal section of ovary of a female. (**D–F**) Cross section of early transitional gonad, during the beginning of the sex reversal. (**G–I**) Cross section of testis of a secondary male. (**J–L**) Cross section of testis of a primary male. Note that the secondary male develops a single testis, whereas the primary male has a pair of testes. Note that the basement membrane, which supports the germinal epithelium, evidenced by the Reticulin Method, is present in all types of gonadal tissue organization (**C,F,I,L**). Tunica albuginea (ta), ovarian lumen (lu), primary growth oocyte (pg), secondary growth oocyte (sg), germinal epithelium (ge), lateral support (ls), ovigerous lamellae (la), germline cysts (cy), spermatozoa (z), connective tissue (ct), oocyte (o) testicular lamellae (tl), coelomic cavity (ca), testicular lobule (lo), testicular duct (du), interstitium (in), basement membrane (arrow). Staining: MY (**A,B,D,G**), HE (**E,H,J,K**), Reticulin Method (**C,F,I,L**). Bar: 300 μm (**A,D,G,J**), 50 μm (**B,C,E,E-inset, F,H,I,K,L**), 30 μm (**K-inset**).

The testes of primary males are paired organs, joined medially by connective tissue (Figure 1J). Histologically, they present a lobular unrestricted type organization (Figure 1K). The germinal epithelium of the testicular lobules, supported by a basement membrane (Figure 1L), is composed of somatic and male germ cells.

3.2. Female Gonad of Synbranchus marmoratus

As was described above, the ovaries of *Synbranchus marmoratus* present an ovarian cavity in which ovigerous lamellae are projected (Figure 2A). Along the ovigerous lamellae is the female germinal epithelium (Figure 2B), constituted by squamous somatic cells, with a basophilic nucleus andoogonia. These are located throughout the germinal epithelium, isolated or forming cell clusters, resulting from cell proliferation. Oogonia are surrounded by somatic cells forming germline cysts. In the cysts, the oogonia divide by mitosis and through the oogenesis process, enter into meiosis, giving rise to oocytes. In these niches of cell proliferation and differentiation, the formation of ovarian follicles occurs in a permanently active epithelium (Figure 2B–D).

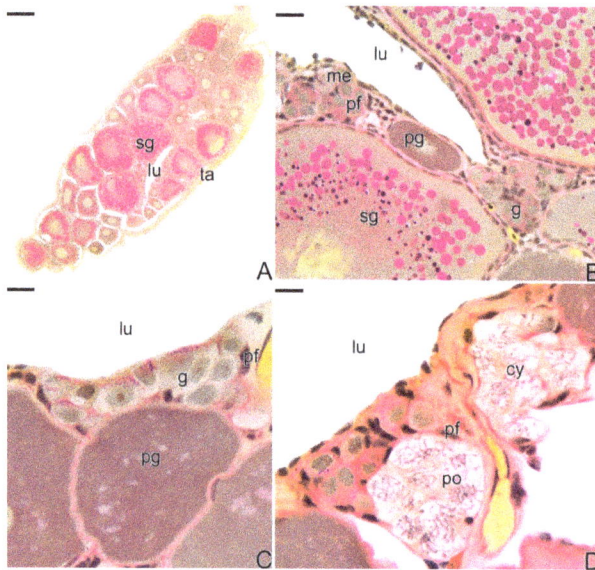

Figure 2. Histological section of ovary of *S. marmoratus*. (**A**) Cross section of ovary. (**B–D**) Details of A, showing the germinal epithelium, with germline cysts. Secondary growth oocyte (sg), ovarian lumen (lu), tunica albuginea (ta), metaphase (me), primary growth oocyte (pg), oogonia (g), pre-follicle cell (pf), pachytene oocyte (po), germline cyst (cy). Staining: MY. Bar: 200 μm (**A**), 20 μm (**B**), 10 μm (**C,D**).

3.3. The Gonadal Remodeling during the Sex Reversal—Formation of the Gonad of the Secondary Male

3.3.1. The Female Early Transitional Gonadal Tissue

The process of sexual reversion in *Synbranchus marmoratus* begins in a functional ovary, from the border of the ovarian cavity. This ovary shows an active female germinal epithelium, formed by oogonia, prophase oocytes, and primary and secondary growth oocytes.

The first signs that indicate the beginning of the transition are a structural disorganization of the female gonad, mainly consisting of the entrance into atresia of some secondary growth oocytes (Figure 3A,B). Concomitantly, there is a thickening of the connective tissue, adjacent to the epithelium (Figure 3C–F) that borders the ovigerous lamellae (Figure 3B,C). As a consequence of this thickening, there is the formation of invaginations from the coemolic cavity towards the epithelium (Figure 3D). In this region, a proliferation of gonia (Figure 3D) starts: entering into meiosis and increasing the number of cysts in the germinal epithelium (Figure 3E). Thus, the male gonadal tissue, initially scarce, gradually replaces the female germinal epithelium that initiates a constant regression. Just below the basement membrane, an increase of the collagen fibers occurs in the underlying interstitial tissue (Figure 3E,F).

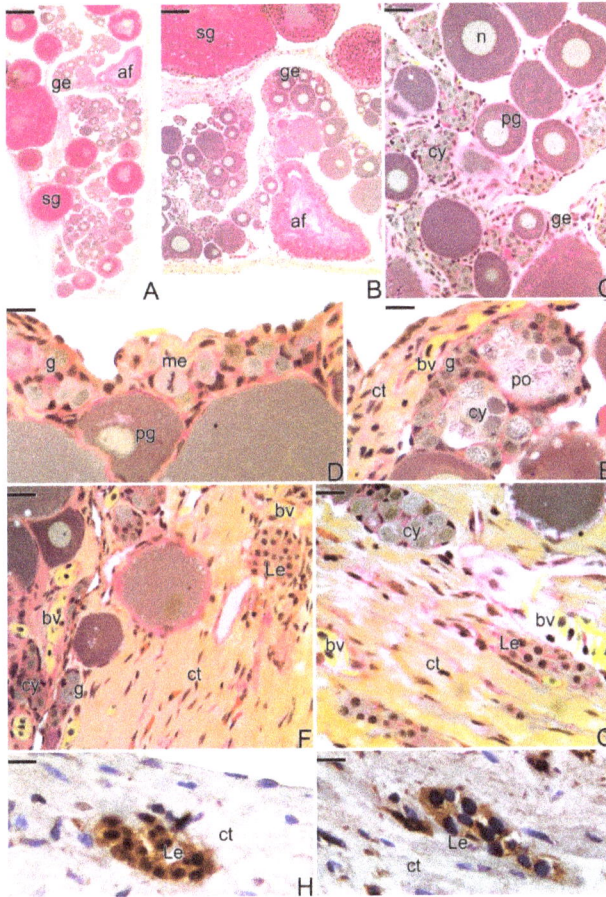

Figure 3. Histological sections of ovary at the beginning of gonadal transition of *S. marmoratus*. Note the presence of atretic follicles (**A,B**), a greater number of germline cysts, along the epithelium (**C–E**), increased blood vessels and connective tissue in the interstitial compartment, and the presence of Leydig cells, close to the blood vessels (**F–H**) Immunohistochemistry for detection of PCNA. The labeled cells that show positive response to PCNA (**H**) are Leydig cells. (**I**) Immunohistochemistry for detection of 3β-HSD enzyme, indicating that it is a steroidogenic cell. Germinal epithelium (ge), secondary growth oocyte (sg), atretic follicle (af), germline cyst (cy), nucleus (n), primary growth oocyte (pg), metaphase (me), gonia (g), connective tissue (ct), pachytene oocyte (po), blood vessel (bv), Leydig cell (Le). Staining: MY (**A–G**). Counterstaining: Harrys Hematoxylin (**H,I**). Bar: 200 µm (**A**), 100 µm (**B**), 20 µm (**C,F**), 10 µm (**D–I**).

Throughout the process of folliculogenesis and entrance of the oocyte into primary growth, the prophase oocytes, as well as the primary growth oocytes (Figure 3A–F), remain involved by somatic cells, as well as the pre-follicle and follicle cells, respectively, which synthesize a basement membrane, segregating them from the interstitial components. The basement membrane often becomes tortuous, next to the interstitial compartment, remaining rectilinear around the ovarian follicles.

At this stage of the sex reversal process, a proliferation of Leydig cells (Figure 3F,G) was observed in the interstitial tissue, which showed a positive response to immunohistochemistry for PCNA (Figure 3H) and for the detection of 3β-HSD (Figure 3I). Leydig cells have an oval shape and a small spherical nucleus with compacted chromatin (Figure 3F). They usually form cell clusters, either in the interstitial tissue or in the connective tissue capsule, often close to the blood vessels (Figure3G).

3.3.2. The Female Intermediary Transitional Gonad Tissue

As the gonadal remodeling progresses, increasing the structural disorganization of the gonad (Figure 4A), atretic follicles are frequently observed (Figure 4B), while the female portion of the gonad remains located in the peripheral region of the gonadal tissue, opposite to the germinal epithelium, which is being invaded by male gonadal tissue (Figure 4C,D). Hyperplasia of the connective tissue of the capsule (Figure 4A) and a strongly increased amount of blood vessels can be observed, both in the center of the gonad and in the region underlying the epithelium (Figure 4C,D).

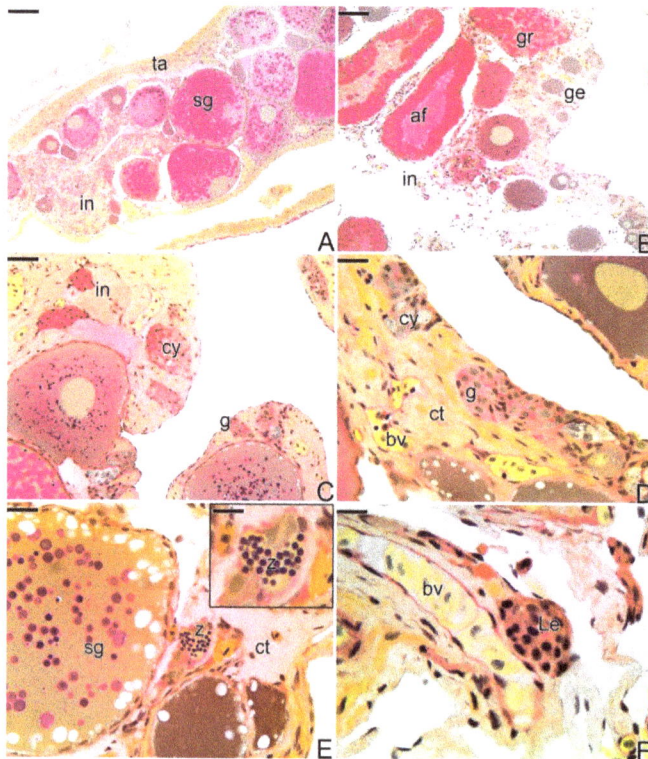

Figure 4. Histological sections of intermediary transitional gonad (ovary) of *S. marmoratus*. (**A**) Ovary with primary and secondary growth oocytes. (**B**) Atretic follicle and granulocytes in large quantity in gonadal tissue. (**C,D**) Male germinal epithelium developing gradually. (**E**) First spermatozoa produced. E-inset) Detail of the spermatozoa. (**F**) Leydig cells near blood vessels. Tunica albuginea (ta), interstitium (in), granulocyte (gr), secondary growth oocyte (sg), germinal epithelium (ge), atretic follicle (af), germline cyst (cy), connective tissue (ct), gonia (g), blood vessel (bv), Leydig cell (Le), spermatozoa (z). Staining: MY. Bar: 200 μm (**A**), 100 μm (**B**), 20 μm (**C–E**), 10 μm (**E-inset, F**).

At this stage, it is possible to observe the first spermatozoa in the lumen of small intra-tissue spaces newly formed (Figure 4E). These spaces constitute the primordium of the first testicular lobules, which will be established at the end of the process of sex reversion. The interstitial tissue develops progressively, being constituted by a large amount of fibroblasts, collagen fibers, myoid cells, and granulocytes. Near the blood vessels, the presence of groups of Leydig cells is often noted (Figure 4F).

3.3.3. The Final Transitional Gonad—The Intersex

At the final stage of sex reversal, the gonad becomes less thick due to the degeneration of the female gonadal tissue and predominance of the male gonadal tissue (Figure 5A). It is possible to observe some remaining oocytes in primary and secondary growth, but the male germline cysts occupy most of the gonad (Figure 5A–C). The distribution of the male gonadal tissue on the old ovigerous lamellae is observed (Figure 5A,B). The male gonadal tissue is formed by cysts containing spermatogonia and spermatocytes (Figure 5B,C). These are organized in cell clusters, such as acinar structures, in which there is not a fully defined lumen yet, except in some regions where spermatozoa are present in the lumen of a rudimentary testicular lobule (Figure 5F). Granulocytes infiltrate in large amounts within the interstitial tissue, being quite frequent near the atretic follicles and melanomacrophage centers (Figure 5D,E; Supplementary Figure S1).

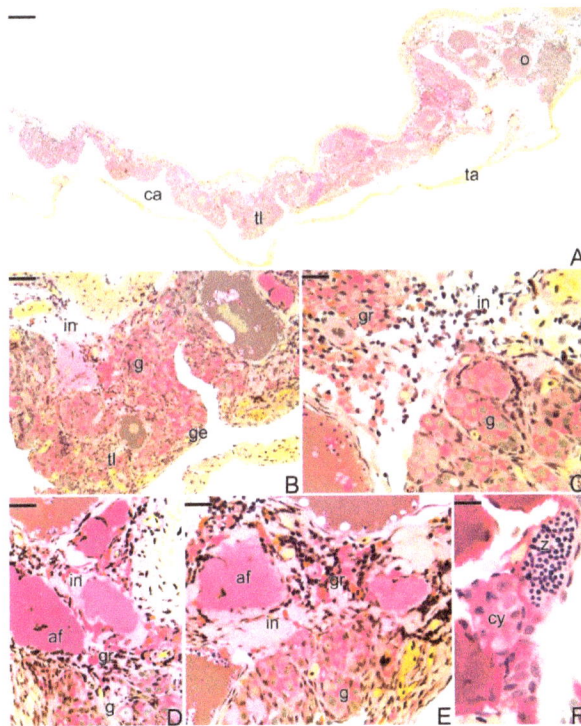

Figure 5. Histological section of final transitional gonad—the intersex of *S. marmoratus*. (**A**) General view of the gonad. (**B–F**) Details of A. Note the organization of the gonad in testicular lamellae (**B**) and the large amount of granulocytes that invade the interstitial tissue (**C–E**) close to atretic follicles (**D,E**). (**F**) Presence of the first spermatozoa. Coelomic cavity (ca), testicular lamellae (tl), tunica albuginea (ta), oocyte (o), interstitium (in), germinal epithelium (ge), gonia (g), granulocytes (gr), atretic follicle (af), spermatozoa (z). Staining: MY. Bar: 100 μm (**A**), 50 μm (**B**), 20 μm (**C–E**), 10 μm (**F**).

3.3.4. The Gonadal Tissue of the Secondary Male

At the end of the sex reversal process, the gonad does not presents intersex characteristics, showing predominantly male elements, such as the presence of testicular lobules (Figure 6A,B). The interstitial tissue underneath the male germinal epithelium is still disarranged. It is possible to observe some areas of necrosis with eosinophilic cells and a few remaining oocytes. The testicular lobules are formed from the acinar structures, constituted by germ cells. These germ cells, the spermatogonia, once encysted by the Sertoli cells, move away from one another, in the same cluster, forming small testicular lobules (Figure 6B). With the proliferation of the spermatogonia, the lobules grow, presenting a larger extension and a wider testicular lumen. Thus, the gonadal tissue is now completely remodeled into a male gonad, with male germinal epithelium completely established and identical to the germinal epithelium of the testis of a primary male. Anatomically, the testis of the secondary male remains quite similar to the anatomical structure of the ovary.

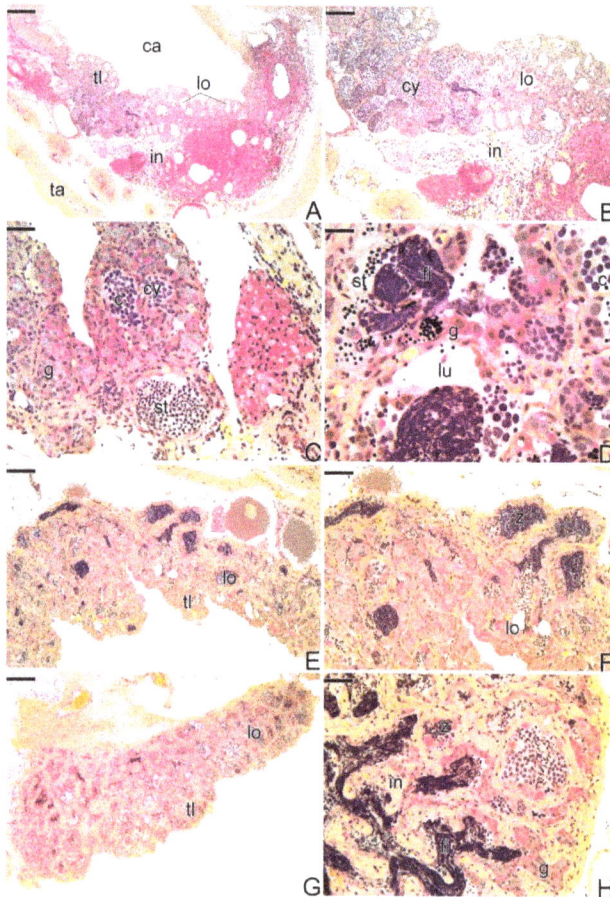

Figure 6. Histological sections of testes of secondary males of *S. marmoratus*. Note the progression of spermatogenesis, throughout the process of testicular development. Coelomic cavity (ca), interstitium (in), tunica albuginea (ta), testicular lamellae (tl), testicular lobules (lo), gemline cysts (cy), spermatocytes (c), spermatogonia (g), spermatids (st), basophilic filaments (fi), testicular lumen (lu). Staining: MY. Bar: 200 μm (**A,E,G**), 100 μm (**B**), 20 μm (**C**), 15 μm (**D**), 30 μm (**F**).

In the germinal epithelium of the lobules, the spermatogenesis begins and it is possible to observe all types of male germ cells (spermatogonia, spermatocytes, and spermatids) (Figure 6C,D). The germinal compartment, supported by a basement membrane, is composed of Sertoli and germ cells that give rise to sperm. At the end of the spermiogenesis, the lumen of the testicular lobules gradually becomes fully filled by spermatozoa (Figure 6E,F). It is still possible to observe a few remaining oocytes (Figure 6E).

Spermatogonia and pre-Sertoli cells present the same structural characteristics of oogonia and pre-follicle cells. The pre-Sertoli cells have a triangular nucleus, sparse cytoplasm, and cytoplasmic projections that interpose gradually between spermatogonia. Thus, spermatogonia are gradually and individually affected by cytoplasmic expansions of the now Sertoli cells, forming cysts delimited by cytoplasmic extensions of the Sertoli cells.

Now, the testis is fully formed and the secondary male is able to reproduce (Figure 6G,H). During the process, it is common to observe the presence of basophilic filaments distributed throughout the gonadal tissue (Figure 6H).

3.3.5. The Male Gonad of *Synbranchus marmoratus*—Primary Male

During all reproductive phases and/or the period of testicular differentiation of the primary male of *Synbranchus marmoratus*, the gonads remain paired, elongated, and cylindrical organs. At the beginning of the development, the gonadal tissue presents small testicular lobules, formed only by a germinal epithelium composed of cysts of spermatogonia, with frequent mitotic divisions. The luminal compartment of the lobes is still reduced and empty (Figure 7A,B).

Gradually, the spermatogonia continue to proliferate, enlarging the luminal compartment (Figure 7C,D). Spermatogonia enter into meiosis, initiating the spermatogenesis, which becomes continuous. With the increase of germ and somatic cells, the testis expands in width and length (Figure 7E). Spermatocytes in the early stages of the meiotic prophase become numerous in the testis (Figure 7F). Spermatozoa begin to be produced (Figure 7F).

With the progress of spermatogenesis, cysts of spermatid become numerous. As a consequence of the increase of germline cysts, there is an increase in the length of the lobules and the testis expands (Figure 7G). Cysts of spermatogonia are distributed throughout the testicular lobule among cysts of other germ cells, such as spermatocytes and spermatids, characterizing the testicular organization as an unrestricted lobular type (Figure 7H). The adjacent testicular lobules are separated from each other by highly developed interstitial tissue (Figure 7H). The production of spermatozoa increases and the testicular lumen, now more extensive, is filled by a large number of these cells (Figure 7H).

3.4. Detection of the Matrix Metalloproteinases MMP-9, MMP-2 and MMP-14 during the Sex Reversal

In females of *Synbranchus marmoratus*, the MMP-9 enzyme was detected through immunohistochemistry techniques in a few ovarian regions, especially in cells of the ovarian stroma and in some regions of the connective tissue capsule (tunica albuginea) (Figure 8). MMP-2 and MMP-14 were not detected in this ovarian stage.

During the beginning of the transitional process of the gonad, the labeling for the three types of metalloproteinases (MMPs) becomes intense and the immunolocalization coincides in all cases (Figure 9), predominating in ovarian stromal cells and around the primary and secondary growth oocytes, in the theca cells. Some mesenchymal cells were positively labeled for MMP-9 (Figure 9H).

During the intermediate transitional stage (Figure 10A–E), the gonad shows interstitial cells marked positively at the detection of MMP-2, but at a lower intensity compared to the previous stage (Figure 10A). Some gonia, theca cells, ovarian stromal cells, and granulocytes also showed a positive response to MMP-9 (Figure 10B–D). Also at this stage, the MMP-14 enzyme was detected only in theca cells (Figure 10E).

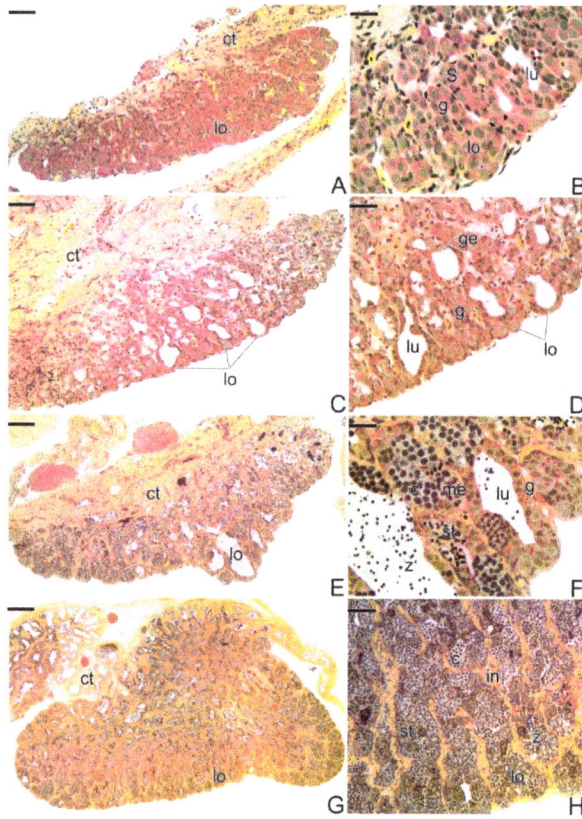

Figure 7. Histological sections of testes of primary males of *S. marmoratus*. Note the development of the testicular lobules, according to the progress of hte spermatogenesis. (**A,B**) Early testicular development. Germinal epithelium ir formed by Sertoli cells and spermatogonia. (**C,D**) Testicular lobules expand, increasing the testicular lumen. (**E,F**) Beginning of the spermatogenesis. (**G,H**) Testicular lobules are filled by spermatozoa. Connective tissue (ct), testicular lobules (lo), spermatogonia (g), Sertoli cell (S), testicular lumen (lu), germinal epithelium (ge), spermatocytes (c), metaphase (me), spermatids (st), spermatozoa (z). Staining: MY. Bar: 100 μm (**A,C,E**), 20 μm (**B,D,F**), 200 μm (**G**), 50 μm (**H**).

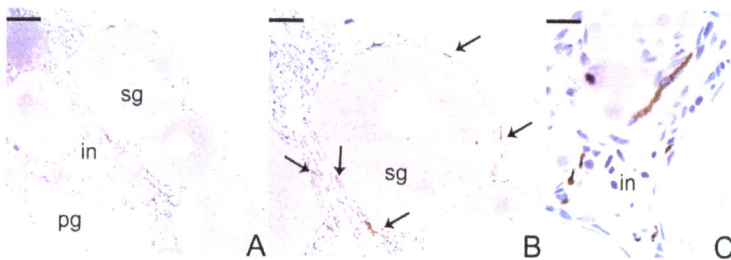

Figure 8. Immunohistochemistry for detection of MMP-9 metalloproteinase in ovary of *S. marmoratus*. Note the positive labeling in the interstitial cells (arrow). Primary growth oocyte (pg), secondary growth oocyte (sg), interstitium (in). Counterstaining: Harrys Hematoxylin. Bar: 80 μm (**A**), 30 μm (**B**), 15 μm (**C**).

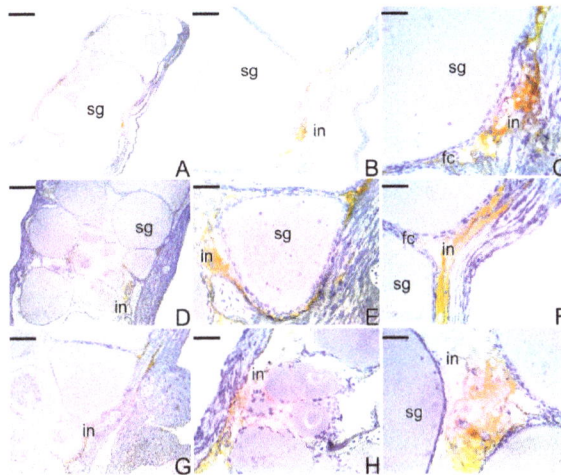

Figure 9. Immunohistochemistry for detection of MMP-14 (**A–C**), MMP-2 (**D–F**), and MMP-9 (**G–I**) metalloproteinases in an early transitional ovary of *S. marmoratus*. Secondary growth oocyte (sg), interstitium (in), follicle complex of the oocyte (fc). Counterstaining: Harrys Hematoxylin. Bar: 50 μm (**A**), 30 μm (**B,H**), 15 μm (**C,E,F,I**), 40 μm (**D,G**).

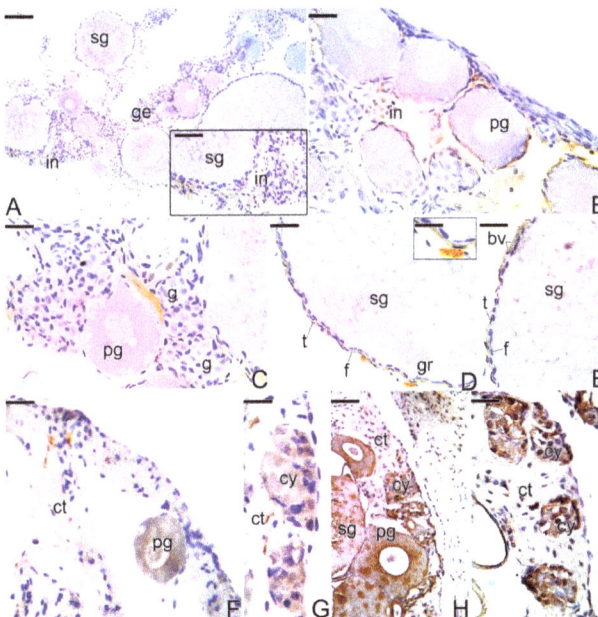

Figure 10. (**A–I**) Immunohistochemistry for detection of MMP-2 (**A**), MMP-9 (**B–D**), and MMP-14 (**E**) metalloproteinases in the intermediate transitional ovary of *S. marmoratus*. (**F–I**) Immunohistochemistry for detection of MMP-14 (**F–G**), MMP-9 (**H**), and MMP-2 (**I**) metalloproteinases in final transitional gonad (intersex) of *S. marmoratus*. Secondary growth oocyte (sg), interstitium (in), germinal epithelium (ge), primary growth oocyte (pg), gonia (g), granulocyte (gr), theca cell (t), follicle cells (f), germline cyst (cy), connective tissue (ct). Counterstaining: Harrys Hematoxylin (**A–I**). Bar: 100 μm (**A**), 40 μm (**A**-inset), 20 μm (**B–I**), 15 μm (**G**), 25 μm (**H**), 5 μm (**D**-inset).

During the intersex phase (Figure 10F–I), MMP-14 was poorly detected in some regions of connective tissue, as well as in primary growing oocytes and germline cysts of spermatogonia (Figure 10F,G). MMP-9 was detected in primary and secondary growth oocytes, as well as in germline cysts, showing a more intense marking when compared to the MMP-14 (Figure 10H). An intense marking of MMP-2 was also found in germline cysts (Figure 10I).

At the end of the sex reversal process, the gonad of the secondary male shows a weak reaction to the MMP-2, MMP-9, and MMP-14 proteins. MMP-9 was detected in somatic cells, in the dorsal region of the gonad, as well as in interstitial cells (Figure 11A–C). Granulocytes present in melanomacrophage centers also showed a positive response to MMP-9 detection, being absent in the cytoplasm of oocytes or in their follicle complexes (Figure 11D). The MMP-2 enzyme was not detected in the majority of the germline cysts (Figure 11E), but it showed a positive response in granulocytes located in the melanomacrophage centers (Figure 11F) and in some spermatogonia located in lobules full with sperm (Figure 11H). MMP-14 was expressed in granulocytes and in theca cells of follicle complexes (Figure 11G).

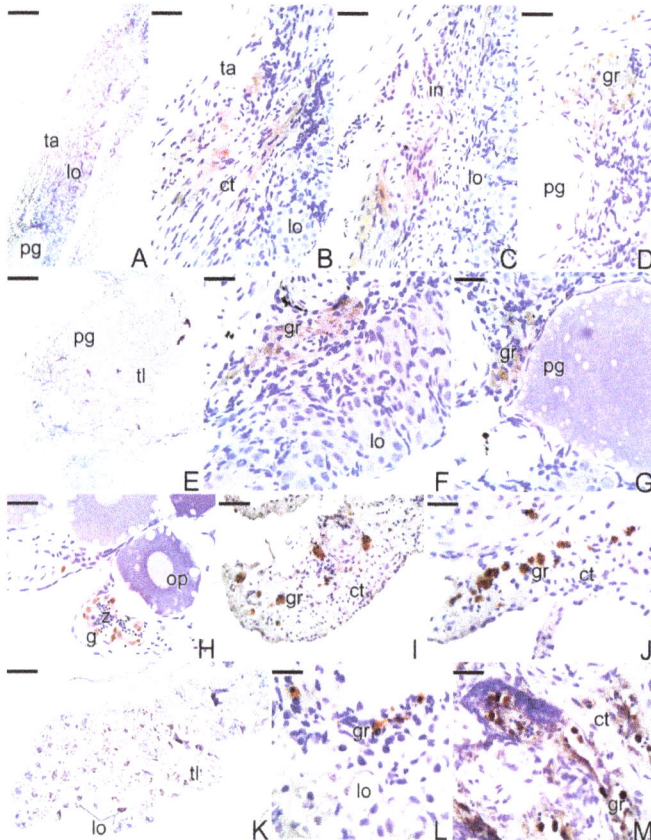

Figure 11. Immunohistochemistry for detection of MMP-9 (**A–D,J**), MMP-2 (**E,F,I,K–M**), and MMP-14 (**G**) metalloproteinases in testis of secondary males of *S. marmoratus*. Note the decline in the expression of the MMPs according to the end of the sex reversal process. Tunica albuginea (ta), testicular lobules (lo), connective tissue (ct), interstitium (in), primary growth oocyte (pg), granulocytes (gr), spermatogonia (g), testicular lamellae (tl), spermatozoa (z). Counterstaining: Harrys Hematoxylin. Bar: 80 μm (**A,E**), 40 μm (**B,C,I**), 30 μm (**D,F,G**), 20 μm (**H,J,L,M**).

After the beginning of the spermatogenesis, the testis of the secondary male (Figure 11K) shows a positive response to MMP-2 (Figure 11I) and MMP-9 (Figure 11J) in granulocytes from the connective tissue capsule. Granulocytes present in the interstitial compartment of the testis (Figure 11L) also show a positive response to MMP-2, which is also detected in interstitial tissue cells and granulocytes present inside the lumen of testes (Figure 11M).

In testes of the primary males, MMP-2 and MMP-9 (Figure 12C) were detected in the spermatogonia that constitute the germinal epithelium of the testicular lobules. The same positive result was observed for MMP-14 (Figure 12D,E). After the beginning of the spermatogenesis, none of the enzymes was detected, not even in granulocytes or in melanomacrophage centers (Figure 12F,G).

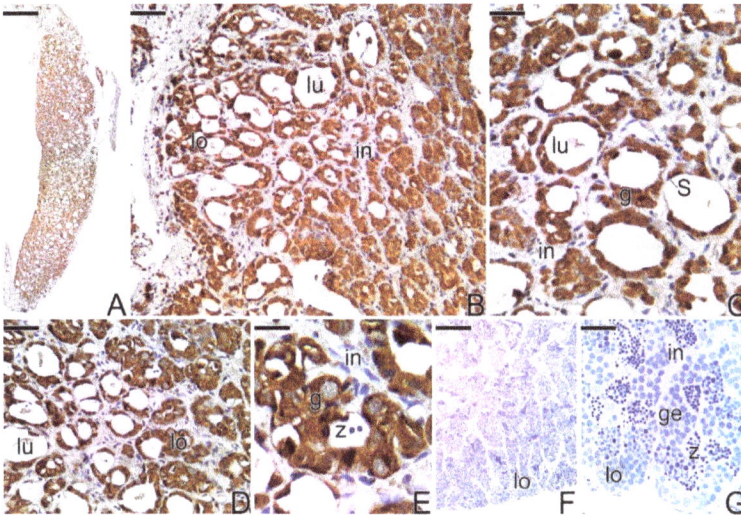

Figure 12. Immunohistochemistry for detection of metalloproteinases in testis of primary males of *S. marmoratus*. (**A**) General view of the testis. (**A**–**E**) Immunohistochemistry for detection of MMP-2 (**A**,**B**), MMP-9 (**C**), and MMP-14 (**D**,**E**). (**F**,**G**) In totally developed testes, there was no positive response to any of the MMPs. Testicular lobules (lo), testicular lumen (lu), interstitium (in), Sertoli cell (S), spermatogonia (g), spermatozoa (z), germinal epithelium (ge). Counterstaining: Harrys Hematoxylin. Bar: 100 μm (**A**,**F**), 40 μm (**B**,**D**), 20 μm (**C**), 10 μm (**E**), 30 μm (**G**).

The expression of the metalloproteinases (MMP-2, MMP-9 and MMP-14) was also confirmed by the negative control sections of the *Synbranchus marmoratus* treated with TBS instead of primary antibody (Supplementary Figure S2) and by the positive control in sections of ovaries from mouse specimens (Supplementary Figure S3).

The activity of the gelatinases (MMP-2 and MMP-9) was confirmed by in situ zymography in all types of gonads analyzed: ovary (Supplementary Figure S4A-H), transitional ovary during intersex (Supplementary Figure S4I-P), testis of secondary male (Supplementary Figure S5A-H) and testis of primary male (Supplementary Figure S5I-P).

3.5. Remodelation of the Collagen Fibers in Synbranchus marmoratus

Through the Picrosirius Red staining, a differential distribution of the collagen fibers was observed during the gonadal sex reversal of *Synbranchus marmoratus*.

In the females, the collagen fibers of the tunica albuginea are strongly birefringent, presenting reddish coloration (Figure 13A–D), corresponding to type I collagen.

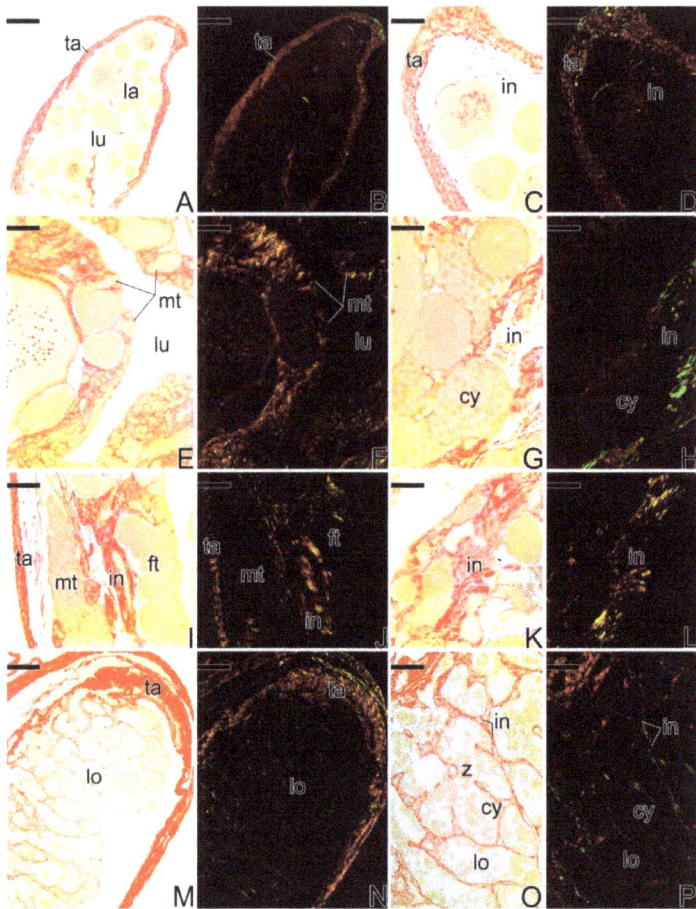

Figure 13. Histological sections of gonads of *S. marmoratus* showing gonadal remodeling during sex reversal. Picrosirius Red staining. (**A,C,E,G,I,K,M,O**) unpolarized light. (**B,D,F,H,J,L,N,P**) polarized light. (**A–D**). Ovary. Note the collagen fibers in the tunica albuginea are highly birefringent, while those of the interstitium are less birefringent. (**E–H**) Early transitional gonad. The collagen fibers are predominant next to the germinal epithelium, in that the remodeling begins. (**I–L**) Final transitional gonad. The birefringent collagen fibers are now found in the interstitium, in the central region of the gonad. (**M–P**) Testis of secondary male, showing the tunica albuginea defined as in the female. In the interstitial compartment, the organized collagen fibers show the definitive establishment of the germinal and interstitial compartment. Tunica albuginea (ta), ovigerous lamellae (la), ovarian lumen (lu), interstitium (in), male gonadal tissue (mt), female gonadal tissue (ft), testicular lobule (lo), germline cysts (cy), spermatozoa (z). Staining: Sirius red (red) and Picric Acid (yellow). Bar: 300 µm (**A,B**), 200 µm (**B,C,I,J,M,N**), 100 µm (**E,F**), 40 µm (**G,H,K,L,O,P**), 250 µm (**M,N**).

During the sex reversal, birefringent collagen fibers especially were observed next to the germinal epithelium that begins its remodeling. These fibers present yellowish and greenish tones corresponding to younger and older collagen fibers, respectively (Figure 13E–H). A similar pattern was observed at the end of the gonadal transition process (Figure 13I–L). In these transitional stages, there is a predominance of collagen fibers adjacent to the epithelium. However, the fibers within the gonadal tissue show low or no birefringence, such as in interstitial areas around oocytes or underlying the newly formed testicular lobules (Figure 13E–L).

In gonads of secondary male specimens, the tunica albuginea presents thicker collagen fibers than in the earlier transitional stages, with the prevalence of more mature collagen (Figure 13M,N). In this stage, there is a high birefringence of the less mature collagen (in green) around the testicular lobules, corresponding to the components of the basement membrane (Figure 13M–P).

During the development of of primary male testes (Figure 14), the collagen fibers of the tunica albuginea show greenish tones (less mature collagen) and later, reddish tones (more mature collagen). At the interstitium, the fibers become more birefringent as testicular development occurs, showing younger collagen fibers (in green) (Figure 14A–H) which become older in later stages of development (in red) (Figure 14I–L).

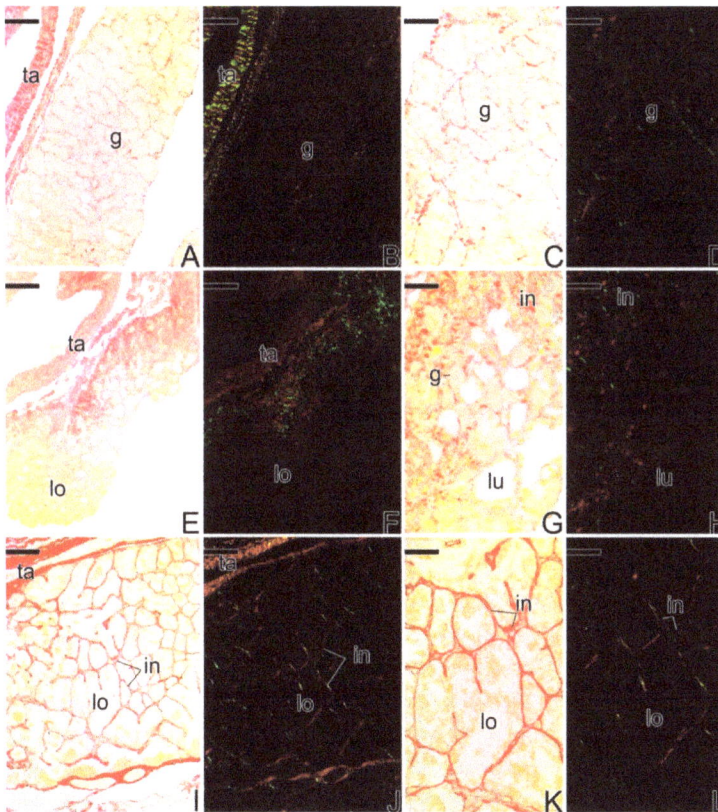

Figure 14. Histological sections of the stages of testicular development of primary males of *S. marmoratus*. Picrosirius Red staining. (**A,C,E,G,I,K**) unpolarized light. (**B,D,F,H,J,L**) polarized light. (**A–D**) Testis at the beginning of the development, with germinal epithelium formed only by spermatogonia. (**E–H**) Testis with germinal epithelium formed by cysts of spermatogonia, spermatocytes, and spermatids. There is no production of spermatozoa yet. The collagen fibers of the interstitium are more intense than the previous stage. (**I–L**) Testis of primary male capable of reproducing, showing the tunica albuginea defined. In the interstitial compartment, the collagen fibers are quite organized and defined. Tunica albuginea (ta), spermatogonia (g), interstitium (in), testicular lumen (lu), testicular lobule (lo). Staining: Sirius red (red) and Picric Acid (yellow). Bar: 100 µm (**A,B,I,J**), 50 µm (**C,D,G,H,K,L**), 150 µm (**E,F**).

3.6. Plasma Levels of Sexual Steroids in Synbranchus marmoratus

Plasma levels of 17β-estradiol (E2) showed no significant difference between classes of individuals (females, initial intersexes, mid intersexes, final intersexes, secondary male, and primary male) (Figure 15A). However, there was a clear decline from 0.47 ± 0.10 ng mL^{-1} (in females) to 0.19 ± 0.02 ng mL^{-1} (initial intersexes), remaining low until the end of sex reversal (secondary males). Levels of 17β-estradiol were very high in primary males (039 ± 0.05 ng mL^{-1}) compared to secondary males (0.18 ± 0.04 ng mL^{-1}).

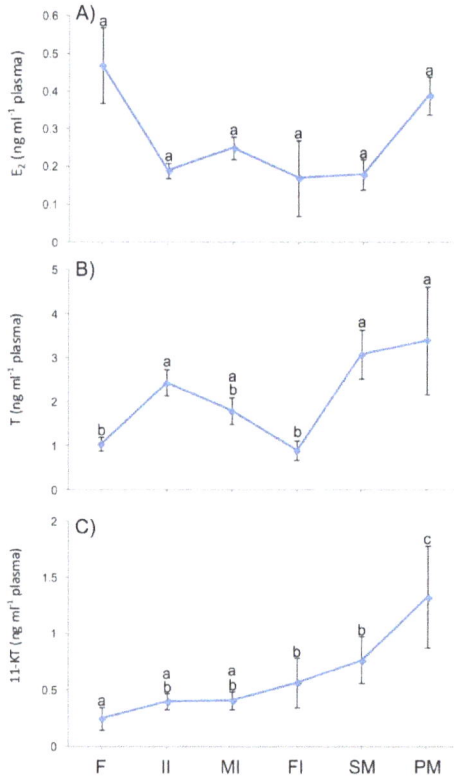

Figure 15. Plasma levels of: (**A**) 17β-Estradiol (E2); (**B**) Testosterone (T); and (**C**) 11-Ketotestosterone (11-KT) (mean $\pm$ S.E.M) in the different sex types of *S. marmoratus* individuals. F: females ($n = 49$), II: Initial intersexes ($n = 67$), MI: Mid intersexes ($n = 35$), FI: Final intersexes ($n = 4$), SM: secondary male ($n = 24$), PM: primary male ($n = 7$). Different letters denote significant differences ($p < 0.05$).

Plasma levels of testosterone (T) increased significantly from 1.04 ± 0.16 ng mL^{-1} (females) to 2.44 ± 0.30 ng mL^{-1} (initial intersexes), decreasing significantly and progressively until the end of the transitional process (final intersexes) (Figure 15B). The maximum plasma levels of testosterone were detected in primary and secondary males (3.41 ± 1.22 and 3.09 ± 0.56 ng mL^{-1}, respectively).

Regarding 11-ketotestosterone levels (11-KT), there was an increase from 0.40 ± 0.10 ng mL^{-1} (females) to 0.57 ± 0.07 ng mL^{-1} (final intersexes). The maximum plasma levels of 11-ketotestosterone were also detected in primary and secondary males (1.33 ± 0.45 and 0.77 ± 0.21 ng mL^{-1}) (Figure 15C).

4. Discussion

4.1. Gonadal Remodeling

The mechanism of sex reversal in *Synbranchus marmoratus* is characterized by the degeneration of the gonadal tissue corresponding to the first sex, followed by the substitution, development, and maturation of the germinal tissue of the opposite sex, having a period corresponding to an intersex phase, as seen in other species [5,10–12,17,40]. Moreover, in the morphological aspects, as herein reported, these events that result in the reversion of the sex in the adult animals are quite similar to those observed in the juvenile hermaphroditism [17].

This process shows the high bipotentiality of germ cells, as stem cells. Morphologically similar in both sexes, the primordial germ cells remain undifferentiated until they are exposed to external factors and hormonal influences that induce their differentiation into oogonia or spermatogonia [5]. In the present study, during the formation of secondary males, primordial germ cells initially differentiate into oogonia for development of an ovary. However, some of them may remain quiescent during gonadal development of the first sex and later, through various stimuli, become active, differentiating into spermatogonia and forming the male gonadal tissue. This same process occurs, for example, in juvenile individuals who have their sexual development altered by social or exogenous factors. In all cases, these animals exhibit enormous plasticity in their sexual patterns [1]. This is the case, for example, of adults of *Oryzias latipes* [41], which present undifferentiated gonia in the adult male gonadal tissue, which can be induced to differentiate into oogonia, giving rise to an ovary through the administration of steroids [42].

Therefore, regardless of the type of stimulus (natural or artificial), the formation of a new gonad during sequential hermaphroditism in adults is nothing more than a gonadal neodifferentiation that follows the same mechanisms found during the period of gonadal morphogenesis of juvenile individuals [17,40]. In addition to these factors, the infiltration of large quantities of granulocytes into the gonad to be remodeled, and the structural disorganization and the proliferation of Leydig cells coinciding with the beginning of the gonadal remodeling process in sequential hermaphroditism reinforces the idea of how similar these mechanisms are in a juvenile [17] and adult animal (present study). In *Sparus aurata*, for example, the infiltration of acidophilic granulocytes is essential for formation and remodelling of the testicular tissue during the gonadal differentiation [43]. In adults of *S. aurata*, an infiltration of these granulocytes is also observed at post-spawning. This event, associated with the role of MMPs, allows the regulation of the testicular physiology and the organization of the cysts during spermatogenesis and post-spawning [44].

4.2. The Basement Membrane in the Remodeling of the Germinal Epithelium

In *Synbranchus marmoratus*, the basement membrane was detected by the Reticulin Method at all stages of the sex reversal process, even during the intense gonadal remodeling. Considering *Gymnocorymbus ternetzi* [17], as the only example in which there is a study on the presence/absence of the basement membrane during hermaphroditism, *S. marmoratus* presents a marked difference, as regards the basement membrane around the ovarian follicles, even during the intersex. In the ovarian follicles of transitional gonads of *G. ternetzi*, the basement membrane is completely absent, even in remaining oocytes of completely developed testes [17]. However, these oocytes are stationary in primary growth, so the absence of the basement membrane should be responsible for preventing its development. Contrary data found in *S. marmoratus* may reinforce this idea that the basement membrane is decisive for the development of the ovarian follicle, since the oocytes of *S. marmoratus*, enveloped by a basement membrane always present, continue their development, entering secondary growth, and are capable of reaching final maturation.

The basement membrane is a thin sheet of extracellular matrix that separates the epithelial cells from the subjacent connective tissue. Among its many functions, it participates in the process of cell proliferation, establishes cellular polarity, induces cellular differentiation, and constitutes a structural barrier against the invasion of undesirable cells [45]. Therefore, it is responsible for the maintenance of tissue integrity and preventing inappropriate cell death [46].

When synthesized in the fish ovary, the basement membrane supports the epithelium and surrounds the ovarian follicles, allowing the development of the oocytes [47]. In the fish testis, the basement membrane is synthesized by Sertoli cells, and is thus intimately associated with the establishment of the blood testis barrier. In addition, the integrity of the basement membrane provides conditions for the nutrition, differentiation, and maintenance of the male germ cells, leading to the consequent development of the male germinal epithelium [48–51].

Despite the basement membrane being present in the transitional gonads of *Synbranchus marmotarus*, it becomes quite sinuous at this stage. Concomitantly, the oocytes enter into atresia and there is an invasion of a large amount of granulocytes in the gonadal tissue. The lack of a physical structure for the basement membrane of the ovary follicles occurs in the first events that signal the end of each reproductive cycle during the fish life [52]. Similarly, cell proliferation, differentiation, and migration, as well as different penetration by stroma, are crucial events for sexual differentiation of the gonads observed in *Xenopus laevis* during its development [53]. So, it is plausible to assume that this folding of the basement membrane results from the punctual lack of components of the extracellular matrix, caused by remodeling of the tissue, when the metalloproteinases become active.

4.3. Expression of the Matrix Metalloproteinases Enzymes during the Sex Reversion

Until now, there has been no data on the detailed formation of the basement membrane and of the interstitial compartment remodeling. Similarly, there is no information on the action of the matrix metalloproteinases in the sex reversal process in hermaphroditic species, as in the case of the animal model of this study, *Synbranchus marmoratus*.

The extracellular matrix regulates the basic processes in the tissues and their maintenance through the action of metalloproteinases (MMPs) that contribute to movement, growth, differentiation, and cellular surviving [54]. MMPs have an important role in the maintenance of the tissue of vertebrates, among them, the fish, as verified in this study, in the remodeling of the gonads. In vertebrates in normal conditions, lots of the metalloproteinases are considered a constitutive component of fibroblasts, epithelial, and endothelial cells, being expressed according to the need of the tissue in which they are [55].

In mammals, the MMP-9 and MMP-12 metalloproteinases, with gelatinolitic activity, participate as regulators of events that promote alterations in the male gonads, as well as being expressed in Sertoli and germ cells, participating in events of remodeling of the germinal epithelium [24].

In Teleostei, data about the action of metalloproteinases (MMPs) in the gonads are scarce; however, Chaves-Pozo and collaborators [44] found that those MMPs are expressed both by the germ cells and the somatic cells during the male and female reproductive cycle. Studies on the action of MMPs in males of Teleostei were also described by Santana and Quagio-Grassiotto [16]. These authors suggest that the fibroblasts may be the major cells responsible for the production and secretion of MMP-2 in the interstitial compartment of the catfish *Pimelodus maculatus* (Siluriformes: Pimelodidae), actively participating in the alterations that occur in the gonad during the reproductive cycle.

In this aspect, the changes in the interstitial compartment of the gonads of Teleostei are even more intense in animals that present sex reversal, in which there is an enormous remodeling of the gonadal tissue, especially in the interstitial components, which break a previous structure to reconstruct a new gonadal architecture, formed by the opposite sex gonad. In this study, this gonadal remodeling in adults of *Synbranchus marmoratus*, during the sequential hermaphroditism, was the result, at least partially, of the action of the MMP-2 and MMP-9 metalloproteinases, activated by a third transmembrane metalloproteinase, MMP-14 (also known as MT1-MMP). The three distinct types of MMPs were detected during the gonadal remodeling and produced by different cellular types and not only by fibroblasts.

In the initial stage of the process of gonadal remodeling in *Synbranchus marmoratus*, MMP-9 was detected, especially in the interstitial tissue, in gonads in the early transitional phase. However, during the intermediary transitional phase of the gonadal tissue, the oocytes significantly expressed the enzyme MMP-9. In females that did not undergo the gonadal remodeling process, the expression of MMP-9 was detected sporadically in stromal ovarian cells and never in the oocytes. This shows an active role of the oocytes in the gonad remodeling process. MMP-2 is expressed in male germline cysts, somatic cells, interstitium, and granulocytes. Its action decreases at the end of the sex reversal process, being more active during the intersex stage. After the total gonadal sex reversal, MMP-2 was only detected in granulocytes, similar to MMP-9. Given the above, it seems to be a relation between the activity of the MMPs and the momentary constitution of the gonadal tissue. That is, in the beginning of the remodeling process, when there is a prevalence of female gonadal tissue, MMP-9 is highly expressed, especially in the oocytes. At the end of the process, when the male gonadal tissue is predominant, MMP-2 is more active, especially in germline cysts and granulocytes. Other data obtained in this study, which sustains the hypothesis of the MMP-2 metalloproteinase having a significant participation in the remodeling of the male gonad, shows that this enzyme is intense during the development of the testicular lobules of primary males of *S. marmoratus*.

MMP-9 and MMP-14 metalloproteinases also presented intense expression in this stage, which is characterized by intense proliferation of spermatogonia, thus expanding the germinal epithelium, as well as widening the luminal compartments of the lobules. After the total development of the testis, none of these MMPs were detected in the stages analyzed. However, it is implicit that, if the male reproductive cycle is followed, similar data to those of Santana and Quagio-Grassiotto [16] will be obtained, since the granulocytes of fully developed testes of *Synbranchus marmoratus* showed a positive response to MMP-9. That is, its action and expression should vary according to different phases of the reproductive cycle.

The expression and co-localization of the MMP-14 metalloproteinase frequently coincided with the expression of the other MMPs studied here. Possibly, the correlation of this fact occurs because it is a transmembrane protein that activates the other MMPs, so that they can act on determined tissues [20].

Considering that the metalloproteinases (MMPs) are expressed in different cell types and in different phases of the gonadal remodeling, it is possible to think that, initially, the cells from the germinal epithelium start the production of MMPs, promoting the break-up of the basement membrane, leading to further infiltration of granulocytes, which in turn also participate actively in this process. The absence of the basement membrane can lead to cell death of part of the germinal epithelium, which would lose the support and nutrition by the basement membrane. In this way, several events occur by the action of MMPs: reorganization of the extracellular matrix; reestablishment of the germ and somatic cells, followed by the formation of a new basement membrane; and reconstruction of new gonadal tissue.

4.4. The Sexual Steroids and the Expression of the 3β-Hydroxyterid Dehydrognase (3β-HSD) during the Sex Reversal

The detection of the activity of enzyme 3β-hydroxysteroid dehydrogenase (3β-HSD) is used as a universal marker for Leydig cells [56]. The 3β-HSD enzyme is involved in several steps of the biosynthesis of steroid hormones, being responsible for the conversion of pregnenolone to progesterone; 17α-hydroxypregnenolone to 17α-hydroxyprogesterone; dehydroepiandrosterone to androstenedione; and androstenediol to testosterone [26,57]. Thus, it can be assumed that, at the time of its detection, plasma levels of testosterone should be high or at least present.

In this aspect, studies on gonadal differentiation in Teleostei show that the expression of 3β-HSD seems to be related to the onset of spermatogenesis [41]. However, they may be expressed early in male gonads with newly established germinal epithelium [58]. Thus, during gonadal differentiation, in gonads that undergo gonadal remodeling during sex reversal, 3β-HSD expression can be detected in interstitial cells, precursors of Leydig cells [17]. These cells also have a high proliferation rate, detected by immunohistochemistry for PCNA (proliferating cell nuclear antigen). These data are exactly the same as those found in the present study for adults of *Synbranchus marmoratus*, in which this cell type is also found forming cell clusters. Therefore, this data is clear evidence that these cells are involved in gonadal remodeling, helping in the process of construction of the new male gonad.

In relation to plasma levels, the lowest levels of testosterone (T) were found in females $(1.04 \text{ ng mL}^{-1})$ and in animals in the final transitional phase (0.9 ng mL^{-1}) of *Synbranchus marmoratus*. In both primary and secondary males, the levels are quite high (3.41 and 3.09 ng mL^{-1}, respectively). An interesting finding is a peak of2.44 ng mL^{-1} in animals at the beginning of transition, which showed a quite high positive response for detection of the 3β-HSD enzyme.

Plasma levels of 11-ketotestosterone (11-KT) in *Synbranchus marmoratus* always increased during sex reversal, from 0.4 ng mL^{-1} in the early transitional gonad, reaching 0.77 ng mL^{-1} in the secondary males. Thus, both testosterone and 11-KT showed increasing levels during sex reversal. In the Teleostei, both hormones are found in high levels in male adults, with 11-KT being responsible for the development of the testis, proliferation of the spermatogonia, final maturation of the gonad, and maintenance of the spermatogenesis [5,59,60]. These statements reinforce the events that occurred during the process of gonadal remodeling, with the female gonad in degeneration and development of new gonadal tissue, now a male gonad.

The variations in the plasma levels observed in the females in an early transitional gonad, with the increase of 17β-estradiol (E2) and decrease of testosterone (T) levels, can be associated with the aromatization of T, which leads to the synthesis of E2 [15,61].

5. Conclusions

The extracellular matrix components of the interstitial tissue of the gonads of the *Synbranchus marmoratus* are remodeled through the metalloproteinases (MMPs) according to morphophysiological alterations that occur during sex reversal. This renewal of extracellular matrix components is orchestrated by both cells from germinal and interstitial compartments. However, the breakdown of the basement membrane is the trigger to initiate the remodeling in both compartments and this is due to the action of MMPs, produced by different cell types. These data are confirmed by the constant renewal of the collagen fibers during the sex reversal process in *S. marmoratus*.

Helping in the sex reversal process, sex steroids begin to be synthesized as the proliferation and differentiation of Leydig cells occurs. At the same time, the remodeling of the gonadal tissue begins until it reaches total sex reversion. During the process, the metalloproteinases (MMPs) are expressed, promoting this remodeling, and consequently, the collagen fibers are replaced by new connective tissue elements. Thus, the synthesis of steroids coincides with gonadal remodeling, as the MMPs are synthesized by the cells of the germinal epithelium, as well as the interstitial tissue.

Using an excellent biological model (a hermaphroditic fish, which presents naturally sex reversion), this study will contribute significantly to the development of new techniques based on the inhibition and/or activation of the MMPs, providing new perspectives in experimental studies involving remodeling of the extracellular matrix in many different areas of the biology of the reproduction of fish, such as gonadal differentiation, the reproductive cycle, life history, and sex manipulation.

Supplementary Materials: The following are available online at: http://www.mdpi.com/2073-4409/7/5/34/s1. Figure S1. Cross section of the gonad of *S. marmoratus* showing melanomacrophage centers and granulocytes. Figure S2. Negative control of the immunohistochemistry for detection of the metalloproteinases in gonads of *S. marmoratus*. Figure S3. Positive control of the immunohistochemistry for detection of the metalloproteinases in ovary of mouse (Swiss). Figure S4. Activity of the gelatinases (MMP-2 and MMP-9) in the gonadal tissue of the *S. marmoratus* (ovary and transitional ovary during intersex). Figure S5. Activity of the gelatinases (MMP-2 and MMP-9) in the gonadal tissue of the *S. marmoratus* (testis of secondary and primary male).

Author Contributions: Talita Sarah Mazzoni performed the experiments, analyzed the data and wrote the paper; Fabiana Laura Lo Nostro contributed to the histological description of the gonad; Fernanda Natália Antoneli performed the experiments of enzyme-linked immunosorbent assay and transmission electron microscopy; and Irani Quagio-Grassiotto conceived and designed the experiments, and reviewed the paper.

Acknowledgments: We would like to thank: Brazilian Agencies CAPES/PROAP (Coordenação de Aperfeiçoamento de Pessoal de Nível Superior) and FAPESP (Fundação de Amparo à Pesquisa do Estado de São Paulo—n°processo: 2014/00868-3 and 2015/16358-7); Electron Microscopy Center/IBB-UNESP, Botucatu-SP for use of the facilities; The anonymous reviewer for the invaluable suggestions and insightful comments; and Piscicultura Santa Candida—Fish Farm for the fish used in the supplementary materials.

Conflicts of Interest: The authors declare no conflict of interest.

References

1. Sadovy de Mitcheson, Y.; Liu, M. Functional hermaphroditism in teleosts. *Fish Fish.* **2008**, *9*, 1–43. [CrossRef]
2. Francis, R.C. Sexual lability in teleosts: Developmental factors. *Q. Rev. Biol.* **1992**, *67*, 1–17. [CrossRef]
3. Frisch, A. Sex change and gonadal steroids in sequentially hermaphroditic teleost fish. *Rev. Fish Biol. Fish.* **2004**, *14*, 481–499. [CrossRef]
4. Wu, G.C.; Tomy, S.; Lee, M.F.; Lee, Y.H.; Yueh, W.S.; Lin, C.J.; Lau, E.L.; Chang, C.F. Sex differentiation and sex change in the protandrous black porgy, *Acanthopagrus schlegeli. Gen. Comp. Endocrinol.* **2010**, *167*, 417–421. [CrossRef] [PubMed]
5. Devlin, R.H.; Nagahama, Y. Sex determination and sex differentiation in fish: An overview of genetic, physiological and environmental influences. *Aquaculture* **2002**, *208*, 191–364. [CrossRef]
6. Asoh, K. Gonadal development and diandric protogyny in two populations of *Dascyllus reticulatus* from Madang, Papua New Guinea. *J. Fish Biol.* **2005**, *66*, 1127–1148. [CrossRef]
7. Liem, K.F. Sex Reversal as a Natural Process in the Synbranchiform Fish *Monopterus albus. Copeia* **1963**, *2*, 303–312. [CrossRef]
8. Liem, K.F. Geographical and taxonomic variation in the pattern of natural sex reversal in the teleost fish order Synbranchiformes. *J. Zool.* **1968**, *156*, 225–238. [CrossRef]
9. Lo Nostro, F.L.; Guerrero, G.A. Presence of primary and secondary males in a population of the protogynous *Synbranchus marmoratus. J. Fish Biol.* **1996**, *49*, 788–800. [CrossRef]
10. Lo Nostro, F.L.; Grier, H.J.; Andreone, L.; Guerrero, G.A. Involvement of the gonadal germinal epithelium during sex reversal and seasonal testicular cycling in the protogynous swamp eel, *Synbranchus marmoratus* Bloch 1795 (Teleostei, Synbranchidae). *J. Morphol.* **2003**, *257*, 107–126. [CrossRef] [PubMed]
11. Lo Nostro, F.L.; Grier, H.J.; Meijide, F.; Guerrero, G.A. Ultrastructure of the testis in *Synbranchus marmoratus* (Teleostei, Synbranchidae): The germinal compartment. *Tissue Cell* **2003**, *35*, 121–132. [CrossRef]
12. Antoneli, F.N. Perfil Morfo-Funcional da Inversão de Sexo em Synbranchidae (Teleostei, Synbranchiformes, Synbranchidae). Ph.D. Thesis, Biblioteca do Instituto de Biologia, UNICAMP, Campinas, Brazil, 2006.
13. Nelson, J.S.; Grande, T.C.; Wilson, M.V.H. *Fishes of the World*, 5th ed.; Wiley: New York, NY, USA, 2011; 752p, ISBN 978-0471250319.
14. Barros, N.H.C.; de Souza, A.A.; Peebles, E.B.; Chellappa, S. Dynamics of sex reversal in the marbled swamp eel (*Synbranchus marmoratus* Bloch, 1795), a diandric hermaphrodite from Marechal Dutra Reservoir, northeastern Brazil. *J. Appl. Ichthyol.* **2017**, *33*, 443–449. [CrossRef]
15. Ravaglia, M.; Lo Nostro, F.L.; Maggese, M.; Guerrero, G.; Somoza, G. Characterization of molecular variants of GnRH, induction of spermiation and sex reversal using sGnRH-A and domperidone in the protogynous diandric fish, *Synbranchus marmoratus* Bloch, (Teleostei, Synbranchidae). *Fish Physiol. Biochem.* **1997**, *5*, 425–436. [CrossRef]
16. Santana, J.C.; Quagio-Grassiotto, I. Extracellular matrix remodeling of the testes through the male reproductive cycle in Teleostei fish. *Fish Physiol. Biochem.* **2014**, *40*, 1863–1875. [CrossRef] [PubMed]
17. Mazzoni, T.S.; Grier, H.J.; Quagio-Grassiotto, I. The basement membrane and the sex establishment in the juvenile hermaphroditism during gonadal differentiation of the *Gymnocorymbus ternetzi* (Teleostei: Characiformes: Characidae). *Anat. Rec.* **2015**, *298*, 1984–2010. [CrossRef] [PubMed]
18. Hulboy, D.L.; Rudolph, L.A.; Matrisian, L.M. Matrix metalloproteinases as mediators of reproductive function. *Mol. Hum. Reprod.* **1997**, *3*, 27–45. [CrossRef] [PubMed]
19. Lu, P.; Takai, K.; Weaver, V.M.; Werb, Z. Extracellular matrix degradation and remodeling in development and disease. *Cold Spring Harb. Perspect. Biol.* **2011**, *3*, 50–58. [CrossRef] [PubMed]

20. Khokha, R.; Murthy, A.; Weiss, A. Metalloproteinases and their natural inhibitors in inflammation and immunity. *Nat. Rev. Immunol.* **2013**, *13*, 649–665. [CrossRef] [PubMed]
21. Birkedal-Hansen, H.; Moore, W.G.I.; Bodden, M.K. Matrix metalloproteinases: A review. *Crit. Rev. Oral Biol. Med.* **1993**, *4*, 197–250. [CrossRef] [PubMed]
22. Page-McCaw, A.; Ewald, A.J.; Werb, Z. Matrix metalloproteinases and the regulation of tissue remodelling. *Nat. Rev. Mol. Cell Biol.* **2007**, *8*, 221–233. [CrossRef] [PubMed]
23. Martins, V.L.; Caley, M.; O'Toole, E.A. Matrix metalloproteinases and epidermal wound repair. *Cell Tissue Res.* **2013**, *351*, 55–268. [CrossRef] [PubMed]
24. Longin, J.; Guillaumot, P.; Chauvin, M.A.; Morera, A.M.; Le Magueresse-Battistoni, B. MT1-MMP in rat testicular development and the control of Sertoli cell proMMP-2 activation. *J. Cell Sci.* **2001**, *114*, 2125–2134. [PubMed]
25. Hägglund, A.C.; Ny, A.; Leonardsson, G.; Ny, T. Regulation and localization of matrix metalloproteinases and tissue inhibitors of metalloproteinases in the mouse ovary during gonadotropin-induced ovulation. *Endocrinology* **1999**, *140*, 4351–4358. [CrossRef] [PubMed]
26. Piprek, R.P.; Kolasa, M.; Podkowa, D.; Kloc, M.; Kubiak, J.Z. Transcriptional profiling validates involvement of extracellular matrix and proteinases genes in mouse gonad development. *Mech. Dev.* **2018**, *149*, 9–19. [CrossRef] [PubMed]
27. Pandian, T.J. *Endocrine Sex Differentiation in Fish*; CRC Press: Boca Raton, FL, USA, 2013; 299p, ISBN 9781466575608.
28. Pudney, J. Leydig and Sertoli Cells Nonmmamalian. In *Encyclopedia of Reproduction*; Knobil, E., Skinner, M.A., Neill, J.D., Eds.; Academic Press: London, UK, 1998; pp. 1008–1021, ISBN 9780122270208.
29. Payne, A.H.; Hardy, M.P.; Russell, L.D. *The Leydig Cell*; Cache River Press: Vienna, Austria, 1996; 802p, ISBN 9781588297549.
30. Guraya, S.S. *Comparative Cellular and Molecular Biology of Testis in Vertebrates: Trends in Endocrine, Paracrine and Autocrine Regulation of Structure and Functions*; Science Publishers Inc.: Enfield, Australia, 2001; 91p, ISBN 1578081653.
31. Pudney, J. Comparative Cytology of the Leydig Cell. In *The Leydig Cell*; Payne, A.H., Hardy, M.P., Russel, L.D., Eds.; Cache River Press: Vienna, Austria, 1996; pp. 98–142, ISBN 9781588297549.
32. Loir, M. Trout steroidogenic testicular cells in primary culture. II Steroidogenic activity of interstitial cells, Sertoli cells and spermatozoa. *Gen. Comp. Endocrinol.* **1990**, *78*, 388–398. [CrossRef]
33. Kobayashi, T.; Nakamura, M.; Kajiura-Kobayashi, H.; Young, G.; Nagahama, Y. Immunolocalization of steroidogenic enzymes (P450scc, P450arom and 3β-HSD) in immature and mature testes of rainbow trout (*Oncorhynchus mykiss*). *Cell Tissue Res.* **1998**, *292*, 573–577. [CrossRef] [PubMed]
34. Ijiri, S.; Takei, N.; Kazeto, Y.; Todo, T.; Adachi, S.; Yamauchi, K. Changes in localization of cytochrome P450 cholesterol side-chain cleavage (P450scc) in Japanese eel testis and ovary during gonadal development. *Gen. Comp. Endocrinol.* **2006**, *145*, 75–83. [CrossRef] [PubMed]
35. Nóbrega, R.H.; Quagio-Grassiotto, I. Morphofunctional changes in Leydig cells throughout the continuous spermatogenesis of the freshwater teleost fish, *Serrasalmus spilopleura* (Characiformes, Characidae): An ultrastructural and enzyme study. *Cell Tissue Res.* **2007**, *329*, 339–349. [CrossRef] [PubMed]
36. Lo Nostro, F.L.; Antoneli, F.N.; Quagio-Grassiotto, I.; Guerrero, G.A. Testicular interstitial cells and steroidogenic detection in the protogynous fish, *Synbranchus marmoratus* (Teleostei, Synbranchidae). *Tissue Cell* **2004**, *36*, 221–231. [CrossRef] [PubMed]
37. Quintero-Hunter, I.; Grier, H.; Muscato, M. Enhancement of histological detail using Metanil Yellow as counterstain in periodic acid/Schiff's hematoxylin staining of glycol methacrylate tissue sections. *Biotech. Histochem.* **1991**, *66*, 169–172. [CrossRef] [PubMed]
38. Vidal, B.C. Histochemical and anisotropical properties characteristics of silver impregnation: The differentiation of reticulin fibers from the other interstitial collagens. *Zool. J. Anat.* **1988**, *117*, 485–494.
39. Junqueira, L.C.U.; Bignolas, G.; Brentani, R.R. Picrosirius staining plus polarization microscopy, a specific method for collagen detection in tissue sections. *Histochem. J.* **1979**, *11*, 447–455. [CrossRef] [PubMed]

40. Uchida, D.; Yamashita, M.; Kitano, T.; Iguchi, T. Oocyte apoptosis during the transition from ovary-like tissue to testes during sex differentiation of juvenile zebrafish. *J. Exp. Biol.* **2002**, *205*, 711–718. [PubMed]
41. Balch, G.C.; Mackenzie, C.A.; Metcalfe, C.D. Alterations to gonadal development and reproductive success in Japanese medaka (*Oryzias latipes*) exposed to 17alphaethinylestradiol. *Environ. Toxicol. Chem.* **2004**, *23*, 782–791. [CrossRef] [PubMed]
42. Dietrich, D.R.; Krieger, H.O. *Histological Analysis of Endocrine Disruptive Effects in Small Laboratory Fish*; John Wiley & Sons Inc.: Hoboken, NJ, USA, 2009; 319p, ISBN 9780471763581.
43. Chaves-Pozo, E.; Liarte, S.; Mulero, I.; Abellán, E.; Meseguer, J.; García-Ayala, A. Early presence of immune cells in the developing gonad of the gilthead seabream (*Sparus aurata* Linnaeus, 1758). *J. Reprod. Dev.* **2009**, *55*, 440–445. [CrossRef] [PubMed]
44. Chaves-Pozo, E.; Castillo-Briceño, P.; García-Alcázar, A.; Meseguer, J.; Mulero, V.; García-Ayala, A. A role for matrix metalloproteinases in granulocyte infiltration and testicular remodelation in a seasonal breeding teleost. *Mol. Immunol.* **2008**, *45*, 2820–2830. [CrossRef] [PubMed]
45. Crouch, E.C.; Martin, G.R.; Brody, J.S.; Laurie, G.W. Basement membrane. In *The Lung*; Crystal, R.G., West, J.B., Eds.; Lippincott-Raven: Philadelphia, PA, USA, 1997; Volume 1, pp. 769–791, ISBN 0397516320.
46. Domogatskaya, A.; Rodin, S.; Tryggvason, K. Functional diversity of laminins. *Annu. Rev. Cell. Dev. Biol.* **2012**, *28*, 523–553. [CrossRef] [PubMed]
47. Mazzoni, T.S.; Grier, H.J.; Quagio-Grassiotto, I. Germline cysts and the formation of the germinal epithelium during the female gonadal morphogenesis in *Cyprinus carpio* (Teleostei: Ostariophysi). *Anat. Rec.* **2010**, *293*, 1581–1606. [CrossRef] [PubMed]
48. Nagahama, Y. The functional morphology of teleost gonads. In *Fish Physiology*; Hoar, W.S., Randall, D.J., Donaldson, E.M., Eds.; Academic Press: New York, NY, USA, 1983; Volume 9, pp. 223–275, ISBN 9780080585338.
49. Pudney, J. Comparative cytology of the non-mammalian vertebrate Sertoli cell. In *The Sertoli Cell*; Russell, L.D., Griswold, M.D., Eds.; Cache River Press: Clearwater, FL, USA, 1993; pp. 612–657.
50. Meijide, F.J.; Lo Nostro, F.; Guerrero, G.A. Gonadal Development and Sex Differentiation in the Cichlid Fish *Cichlasoma dimerus* (Teleostei, Perciformes): A Light- and Electron-Microscopic Study. *J. Morphol.* **2005**, *264*, 191–210. [CrossRef] [PubMed]
51. Schulz, R.W.; França, L.R.; Lareyre, J.J.; LeGac, F.; Chiarini-Garcia, H.; Nóbrega, R.H.; Miura, T. Spermatogenesis in fish. *Gen. Comp. Endocrinol.* **2010**, *165*, 390–411. [CrossRef] [PubMed]
52. Grier, H.J.; Neidig, C.L.; Quagio-Grassiotto, I. Development and fate of the postovulatory follicle complex, postovulatory follicle and observations on folliculogenesis and oocyte atresia in ovulated common snook, *Centropomus undecimalis* (Bloch, 1792). *J. Morphol.* **2017**, *278*, 547–562. [CrossRef] [PubMed]
53. Piprek, R.P.; Kloc, M.; Tassane, J.P.; Kubiak, J.Z. Development of *Xenopus laevis* bipotential gonads into testis or ovary is driven by sex-specific cell-cell interactions, proliferation rate, cell migration and deposition of extracellular matrix. *Dev. Biol.* **2017**, *432*, 298–310. [CrossRef] [PubMed]
54. Lukashev, M.E.; Werb, Z. Ecmsignalling: Orchestrating cell behaviour and misbehaviour. *Trends Cell Biol.* **1998**, *8*, 437–441. [CrossRef]
55. Hipps, D.S.; Hembry, R.M.; Docherty, A.J.; Reynolds, J.J.; Murphy, G. Purification and characterization of human 72-kDa gelatinase (type IV collagenase). Use of immunolocalisation to demonstrate the non-coordinate regulation of the 72-kDa and 95-kDa gelatinases by human fibroblasts. *Biol. Chem. Hoppe Seyler* **1991**, *372*, 287–296. [CrossRef] [PubMed]
56. Chen, H.; Ge, R.S.; Zirkin, B.R. Leydig cells: From stem cells to aging. *Mol. Cell. Endocrinol.* **2009**, *306*, 9–16. [CrossRef] [PubMed]
57. Kobayashi, T.; Chang, X.T.; Nakamura, M.; Kajiura, H.; Nagahama, Y. Fish 3β-Hydroxysteroid Dehydrogenase/Δ5-Δ4Isomerase: Antibody Production and Their Use for the Immunohistochemical Detection of Fish Steroidogenic Tissues. *Zool. Sci.* **1996**, *13*, 909–914. [CrossRef]
58. Mazzoni, T.S.; Grier, H.J.; Quagio-Grassiotto, I. Male gonadal differentiation and the paedomorphic evolution of the testis in Teleostei. *Anat. Rec.* **2014**, *297*, 1137–1162. [CrossRef] [PubMed]
59. Nagahama, Y. Endocrine regulation of gametogenesis in fish. *Int. J. Dev. Biol.* **1994**, *38*, 217–229. [PubMed]

60. Schulz, R.W.; Miura, T. Spermatogenesis and its endocrine regulation. *Fish Physiol. Biochem.* **2002**, *26*, 43–56. [CrossRef]

61. Connaughton, M.A.; Aida, K. Female reproductive system fish. In *Encyclopedia of Reproduction*; Krobil, E., Neill, J.D., Eds.; Academic Press: San Diego, CA, USA, 1998; pp. 193–205, ISBN 9780122270208.

cells

MDPI

Brief Report

Sub-Cellular Localization of Metalloproteinases in Megakaryocytes

Alessandro Malara [1,2,†], **Daniela Ligi** [3,†], **Christian A. Di Buduo** [1,2], **Ferdinando Mannello** [3,‡] and **Alessandra Balduini** [1,2,4,*,‡]

[1]   Department of Molecular Medicine, University of Pavia, 27100 Pavia, Italy;
      alessandro.malara@unipv.it (A.M.); christian.dibuduo@unipv.it (C.A.D.B.)
[2]   Laboratory of Biotechnology, IRCCS San Matteo Foundation, 27100 Pavia, Italy
[3]   Section of Clinical Biochemistry and Molecular Genetics, Department of Biomolecular Sciences,
      University "Carlo Bo" of Urbino, 61029 Urbino, Italy; daniela.ligi@uniurb.it (D.L.);
      ferdinando.mannello@uniurb.it (F.M.)
[4]   Department of Biomedical Engineering, Tufts University, Medford, MA 02155, USA
*   Correspondence: alessandra.balduini@tufts.edu; Tel.: +3-90-382-502-968; Fax: +3-90-382-502-990
†   These authors equally contributed to this work.
‡   These authors are both co-senior authors.

Received: 13 June 2018; Accepted: 18 July 2018; Published: 20 July 2018

Abstract: Metalloproteinases (MMPs) are zinc-dependent endopeptidases that play essential roles as the mediator of matrix degradation and remodeling during organogenesis, wound healing and angiogenesis. Although MMPs were originally identified as matrixin proteases that act in the extracellular matrix, more recent research has identified members of the MMP family in unusual locations within the cells, exerting distinct functions in addition to their established role as extracellular proteases. During thrombopoiesis, megakaryocytes (Mks) sort MMPs to nascent platelets through pseudopodial-like structure known as proplatelets. Previous studies identified gelatinases, MMP-2 and MMP-9, as a novel regulator system of Mks and the platelet function. In this work we have exploited a sensitive immunoassay to detect and quantify multiple MMP proteins and their localization, in conditioned medium and sub-cellular fractions of primary human CD34[+]-derived Mks. We provide evidence that Mks express other MMPs in addition to gelatinases MMP-2 and MMP-9, peculiar isoforms of MMP-9 and MMPs with a novel nuclear compartmentalization.

Keywords: megakaryocyte; metalloproteinase; thrombopoiesis

1. Introduction

The bone marrow (BM) environment is composed of various types of cells surrounded by a meshwork of their secreted extracellular matrix (ECM) components [1]. The turnover of ECM is fundamental for the structural and functional homeostasis of BM hematopoiesis. The importance of ECM in physiologic hematopoiesis and its pathologic modifications in hematopoietic malignancies are becoming evident and are under extensive investigation [2]. Megakaryocytes (Mks) are rare cells in the BM and, besides releasing platelets, they participate in the establishment and maintenance of the BM cell niche in both physiologic and pathologic conditions [3–6]. Interestingly, Mks are involved in ECM deposition and remodeling [7], as demonstrated by their role in fibronectin (FN) fibrillogenesis [8] and the expression of ECM structure modifiers, such as lysyl oxidase and factor XIIIa, essential to the dynamic of Mk-ECM component interactions [8,9].

During thrombopoiesis, Mks sort metalloproteinases (MMPs) to nascent platelets through a pseudopodial-like structure known as proplatelets [10]. MMPs are zinc-dependent endopeptidases that play essential roles as the mediator of matrix degradation and remodeling during stem cell differentiation, organogenesis, wound healing and angiogenesis [11–13]. MMP family proteins are divided in five groups by their respective substrates or cellular localization: Stromelysins (e.g., MMP-3, MMP-10, MMP-11); matrilysins (e.g., MMP-7); collagenases (MMP-1, MMP-8, MMP-13, MMP-18); gelatinases (MMP-2 and MMP-9); membrane-type (MT-MMPs) and other MMPs (3).

Previous studies identified the gelatinases MMP-2 and MMP-9 as forming a novel regulator system of Mks and the platelet function. Induction of MMP-9 expression, but not MMP-2, by the chemotactic activity of Stromal Derived Factor-1 (SDF-1), is considered a key step in modulating Mk migration through the basement membrane of BM sinusoids and subsequent platelet release [14]. In resting platelets, MMP-2 is randomly distributed, not associated with platelet granules and is released upon platelet stimulation to regulate platelet activation and aggregation in physiological hemostasis or pathophysiological formation of occlusive thrombi [15–17]. The presence and activity of MMP-9 and its isoforms in platelets is still debated [18,19]. Aside from gelatinases, other MMPs have been detected in Mks and platelets by different groups using multiple techniques, including immunofluorescence, western blot, next generation RNA sequencing or PCR analysis [20,21]. Cecchetti et al. identified transcripts for MMP-1, 11, 14, 15, 17, 19, 24 and 25 by performing a RNAseq screen of MMP expression in primary human Mks, while expression of MMP-3 has still to be clarified [20,21].

Although MMPs were originally identified as proteases with a peculiar function in the extracellular matrix, more recent research has identified members of the MMP family in unusual locations within the cells, exerting distinct functions in addition to their established role as extracellular proteases [22]. To this regard, MMPs have been detected in the cytosol, organelles and extracellular compartments and more recently several types of MMPs were found in the nucleus [23–25] Nuclear MMPs are supposed to cleave nuclear matrix proteins, although other possible functions are beginning to emerge [26]. To this regard, MMP-1, MMP-2, MMP-9 and MMP-13 in cell nuclei of brain neurons, endothelial cells and cardiac myocytes are supposed to regulate the activity of proteins involved in DNA repair and apoptosis [27–29]. MMP-3 has an unprecedented role as a transcription factor that is independent of its enzymatic activity [30]. Further, new intracellular roles such as the cleavage of intracellular non-matrix proteins, activation/inactivation of intracellular substrates and signal transduction are increasing the functional plasticity of these enzymes.

Thus, following the emergence of these untraditional functions of MMPs in the extracellular space, as well as, in the cytosol and nucleus, we have performed a sensitive immunoassay to detect multiple MMP proteins and their localization, in conditioned medium and sub-cellular fractions of primary human Mks.

2. Material and Methods

2.1. Cell Culture

Human cord blood was collected following normal pregnancies and deliveries upon the informed consent of the parents. All samples were processed in accordance with the ethical committee of the IRCCS Policlinico San Matteo Foundation and the principles of the Declaration of Helsinki. CD34[+] hematopoietic were purified by immunomagnetic beads selection and cultured in StemSpan medium (Stem Cell Technologies, Vancouver, BC, Canada) supplemented with 1% L-glutamine, 1% penicillin-streptomycin, 10 ng/mL of human recombinant Thrombopoietin (TPO) and Interleukin-11 (IL-11) (PeproTech, London, UK), at 37 °C in a 5% CO_2 fully humidified atmosphere for 13 days as previously described [8]. Medium was changed at day 3, 7 and 10 of differentiation.

2.2. Flow Cytometry

Purity of Mk culture, at day 13 of differentiation, was analyzed by staining cells with FITC anti-human CD41 (clone HIP-8) and PE anti-human CD42b (clone HIP-1) antibodies (all from Biolegend, Milan, Italy). Samples were acquired with a Beckman Coulter FacsDiva flow cytometer (Beckman Coulter Inc., Milan, Italy). Relative isotype controls were used to set the correct analytical gating. FITC mouse IgG (clone MOPC-21) and PE-mouse IgG (clone MOPC-21), isotype controls were purchased from Biolegend (Milan, Italy). Off-line data analysis was performed using Beckman Coulter Kaluza® version software package.

2.3. Real Time PCR

Retrotranscription (RT) was performed using the iScriptTM cDNA Synthesis Kit according to manufacturer instructions (BioRad Laboratories Inc., Milan, Italy). For quantitative Real Time PCR, RT samples were diluted 1:3 with ddH2O and the resulting cDNA was amplified in triplicate in reaction mixture with 200 nM of each specific primer and SsoFast™ Evagreen® Supermix (Bio-rad Laboratories, Milan, Italy). The amplification reaction was performed in a CFX Real-time system (BioRad Laboratories Inc., Milan, Italy) with the following protocol: 95 °C for 5 min, followed by 35 cycles at 95 °C for 10 s, annealing at 60 °C for 15 s, extension at 72 °C for 20 s. Pre-designated KiCqStart™ primers for MMP1, MMP2, MMP3, MMP7, MMP8, MMP9, MMP10, MMP12, MMP13 and GAPDH genes were purchased from Sigma-Aldrich (Milan, Italy). The BioRad CFX Manager® software 3.0 was used for quantitative analysis (BioRad Laboratories Inc., Milan, Italy).

2.4. Zymography

Aliquots of all samples were analyzed by gelatin zymography carried out on 6.5% polyacrylamide gels copolymerized with 3 g/L 90 Bloom Type A gelatin from porcine skin (from Sigma-Aldrich, Milan, Italy). Samples were loaded native with the addition of SDS zymogram sample buffer (62.5 mM Tris-HCl, pH 6.8, 25% glycerol, 4% SDS, 0.01% bromophenol blue). SDS-PAGE gels were run using a Bio-Rad Mini-Protean III apparatus (Bio-Rad, Hercules, CA, USA) in SDS running buffer (25 mM Tris, 192 mM glycine, and 0.1% *w/v* SDS) at a constant voltage of 105 V. After electrophoresis, gels were incubated for 40 min at room temperature on a rotary shaker in Triton X-100 2.5%, to remove SDS. The gels were washed with distilled water and incubated for 24 h in enzyme incubation buffer (containing 50 mM Tris; 5 mM CaCl2; 100 mM NaCl; 1 mM ZnCl2; 0.3 mM NaN3, 0.2 g/L of Brij®-35; and 2.5% *v/v* of Triton X-100, pH 7.6) at 37 °C. Staining was performed using Coomassie Brilliant Blue R-250 (0.2% *w/v* Coomassie brilliant blue in 50% *v/v* methanol and 20% *v/v* acetic acid). Gels were destained with a destaining solution (50% *v/v* methanol and 20% *v/v* acetic acid) until clear gelatinolytic bands appeared against the uniform dark-blue background of undigested protein substrate. Gelatinase calibrators (as molecular weight standards) were prepared by diluting healthy capillary blood with 15 volumes of non-reducing Laemmli sample buffer. It is important to specify that whole capillary blood, used as a calibrator, presents only the zymogens of gelatinases: pro-MMP- 2 at 72 kDa, pro-MMP-9 at 92 kDa, and pro-MMP-9 complexes at 130 kDa (MMP-9/NGAL), and 225 kDa (MMP-9 multimeric form) as previously recognized by monoclonal anti-MMP-2 and anti-MMP-9 antibodies and characterized as latent pro-enzymes, activated by APMA and inhibited by both calcium and zinc chelators (EDTA and o-phenanthroline, respectively). Zymographic bands were densitometrically measured with the image analyzer LabImage 1D (Kapelan, Leipzig, Germany) [31].

2.5. Cell Fractionation

Megakaryocytes at day 13 of differentiation were collected, centrifuged and washed with PBS twice. At least 2×10^6 cells per experiments were used. The subcellular fractionation was performed according to the REAP method [32]. Briefly, the cells were collected in 1.5 mL microcentrifuge tubes in 1 mL of ice-cold PBS. After centrifugation (a popo-spin for 10 s in an Eppendorf table top microfuge), supernatants were removed from each sample and cell pellets were re-suspended in 900 μL of ice-cold 0.1% NP-40 detergent in PBS and triturated 5-times using a p1000 micropipette (whole lysate). An aliquot of 600 μL of the whole lysate was centrifuged for 10 s in 1.5 mL microcentrifuge tubes and then the supernatant was removed as the cytosolic fraction. The pellet was re-suspended in 0.5 mL of ice-cold 0.1% NP-40 detergent in PBS and centrifuged as above for 10 s and the supernatant was discarded. The pellet was re-suspended in 200 μL of native Laemmli sample buffer containing DNAse

I (0.01U/µL), triturated 5-times using a p200 micropipette, designated as nuclear fraction and used for both quantitative MMP assays and qualitative gelatinase zymography.

2.6. MMPs Multiplex Array

Quantification of MMPs was performed at day 13 of differentiation from 1×10^6 cells, prior to in viro platelets release, to avoid the release of intracellular proteins into conditioned supernatants as a consequence of the apoptotic-like process of proplatelet formation. Supernatants were collected, pre-cleared by centrifugation at 15,600 g at 4 °C for 20 min and stored at −80 °C or immediately analyzed. MMP concentrations in all samples (supernatants and subcellular fractions from at least seven cultures) were quantified through the Pro™ Human MMP 9-plex Assay (including: MMP-1, MMP-2, MMP-3, MMP-7, MMP-8, MMP-9, MMP-10, MMP-12, MMP-13). The assay was based on multiplex suspension immunomagnetic method using fluorescently dyed magnetic beads covalently conjugated with monoclonal antibodies (Bioplex, BioRad Labs, Hercules, CA, USA).

To avoid subcellular 'matrix' artifacts during assay caused by interfering substances in culture media, we serially diluted randomly selected serum-free media, reanalyzing them for the response linearity. The lower detection limit for all MMPs was 1.0 pg/mL, and the mean intra-assay variability was 10%.

Concentrations of all MMPs were determined using a Bio-Plex 200 system, based on Luminex X-Map Technology (BioRad Labs, Hercules, CA, USA). Data were analyzed using BioManager analysis software (version 6.1). The protein concentrations (expressed as pg/mL) were calculated through a standard curve [33].

2.7. Immunofluorescence

Megakaryocytes at day 13 of differentiation were cytospun onto Poly-L-Lysine-coated coverslips. Cells were fixed with PFA 4% and permeabilized with Triton 0.5%. Cells were stained with anti MMP-2 antibody (Abcam, Cambridge, UK, catalog number AB37150) diluted 1:200 overnight at 4 °C. Alexa 594-conjugated secondary antibody was purchased from Invitrogen (Milan, Italy). Nuclei were counterstained using Hoechst 33258 (100 ng/mL in PBS) at room temperature. Slides were then mounted with micro-cover glasses using Prolong Antifade Reagent (Invitrogen, Milan, Italy). Negative control was performed by omitting the primary antibody. Images were acquired with a TCS SP5 II confocal laser scanning microscope (Leica, Heidelberg, Germany).

3. Results and Discussion

To identify MMPs that are expressed during human thrombopoiesis, nine of 24 family members of MMPs were simultaneously analyzed (MMP-1, MMP-2, MMP-3, MMP-7, MMP-8, MMP-9, MMP-10, MMP-12, MMP-13) by means of a Multiplex Array. We focused on mature Mks at day 13 of differentiation to define the MMP repertoire that these cells harbor prior to platelet release. Mk purity was verified by FACS analysis after CD41 and CD42b staining. Only cell cultures that displayed a fraction of double positive CD41/CD42b cells higher than 90% by flow cytometry were then subjected to biochemical analysis of MMP content and zymography (Figure 1A). Analysis of fresh medium gave negative results for all the analytes (Data Not Shown).

Quantitative assessment of conditioned medium revealed that the MMP-9 protein was the most abundant in conditioned medium in agreement with its previous described role as the regulator of Mk function in the extracellular space (mean 4396 pg/mL from 1×10^6 cells) [14]. Significant amounts of MMP-2 (mean 1162 pg/mL), MMP-7 (mean 1045 pg/mL) and MMP-12 (mean 2287 pg/mL) were also detected (Figure 1B). Interestingly, the presence and activity of MMP-2 and MMP-9 can be readily appreciated by in gelatin zymography analysis of Mks conditioned medium at day 13 of culture (Figure 1C), and to the best of our knowledge, this is the first report that identifies all the isoforms of MMP-9. In particular, consistent gelatinolytic bands were easily identified at 82–86 kDa (as activated forms), and at 225 kDa (as complexed form). Interestingly, for the first time, we identified in the conditioned medium of Mks at day 13 of culture two further gelatinolytic bands at the range

160–180 kDa, probably dimers of MMP-9 activated forms (i.e., devoid of pro-domains) and unusual complexes of MMP9 with Neutrophil Gelatinase Associated-Lipocalin complexes (NGAL), which protects MMP-9 from proteolytic degradation and enhances its enzymatic activities [34]. Our evidence is in agreement with the crucial roles of MMPs (in particular of both gelatinases) in platelet functions, shedding further lights on the enzymatic activity of activated MMP-9 forms during the differentiation of Mks [35]. Notably, MMP-7 and MMP-12 may represent potential released products during thrombopoiesis. MMP-7 is capable of degrading a wide array of ECM components such as gelatin, fibronectin, laminin and elastin. Further, MMP-7 can cleave the pro-domain of the gelatinases MMP-2 and -9 [36], but its relevance under physiological conditions is still debated. To date, MMP-7 has been implicated in several physiological processes, such as wound healing, innate immunity and cancer [37,38]. MMP-12 however, is recognized as a macrophage-secreting proteinase. MMP-12 myeloid-restricted over-expression in mice has a significant impact on the development, differentiation and commitment of hematopoietic progenitor cells to myeloid lineage cells in the BM [39]. Thus, while MMP-12 may act as a pleiotrophic molecule with roles in hematopoiesis and myelopoiesis, involvement of MMP-7 in Mk biology is far from being fully understood. While MMP-7 has never been detected in platelets, Wang et al. recently reported on the expression of MMP-12 in platelets [40].

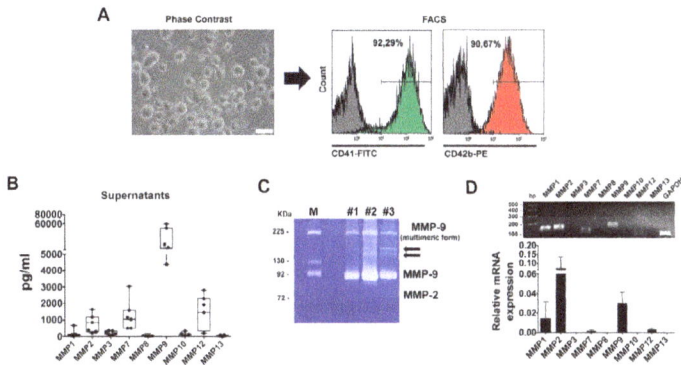

Figure 1. Expression profiling of MMPs in primary human Megakaryocytes. (**A**) Phase contrast image of Megakaryocyte (Mk) culture at day 13 of differentiation. Scale bar = 50 μm. Purity was analyzed by FACS, after staining with Mk markers CD41 (Green histogram) and CD42b (Red histogram). FITC Isotype IgG and PE Isotype IgG were used to set the analytical gate (Grey histograms). (**B**) Profiling and quantification of MMPs in supernatants from 1×10^6 megakaryocytes at day 13 of differentiation. N = 7. (**C**) Representative zymography of Mk supernatants at day 13 of culture. Three independent samples are shown. M = Molecular Marker. Arrows indicate potential dimers of MMP-9 activated forms (160–180 kDa). (**D**) Representative RT-PCR products of MMPs in Mks at day 13 of differentiation and relative quantification from at least three independent experiments. GAPDH expression was used for normalization.

On the contrary, very low levels of MMP-1, MMP-3, MMP-8, MMP-10 and MMP-13 were detected in culture conditioned medium (Figure 1B). RT-PCR was then applied to screen for MMP mRNAs that are expressed during human thrombopoiesis. As shown in Figure 1D, gene expression of MMPs, at day 13 of differentiation, confirmed a prevalence of MMP-1, MMP-2 and MMP-9 transcripts and to a lesser extent of MMP-7 and MMP-12. Moreover, other MMPs were not detected by this assay.

Next, to measure the intracellular localization of MMPs in cultured Mks, membrane/cytosolic and nuclear fractions from seven independent cultures were prepared. At day 13 of culture, 2×10^6 mature Mks were centrifuged, washed three times with PBS and subjected to cell fractionation. Quantification of MMPs by multiplex array revealed that cytosolic fractions displayed appreciable amounts of several MMPs, including MMP-1 (mean 299.18 pg/mL), MMP-2 (432 mean pg/mL), MMP-8 (mean 283 pg/mL),

MMP-9 (mean 527.09 pg/mL) and MMP-12 (mean 303 pg/mL) (Figure 2A). Negligible levels or almost complete absence of MMP-3, MMP-7 and MMP-13 were detected under our experimental conditions. More importantly, detection and quantification of MMPs in nuclear extracts revealed that MMP-1 (mean 568.57 pg/mL) and MMP-3 (mean 833.42 pg/mL) were the most abundant proteins with nuclear localization. To a lesser extent also MMP-2 (206.33 pg/mL) and MMP-9 (mean 232,72 pg/mL) were detected (Figure 2B). A schematic representation of intracellular distribution of individual MMPs and relative protein abundance is provided in Figure 2C.

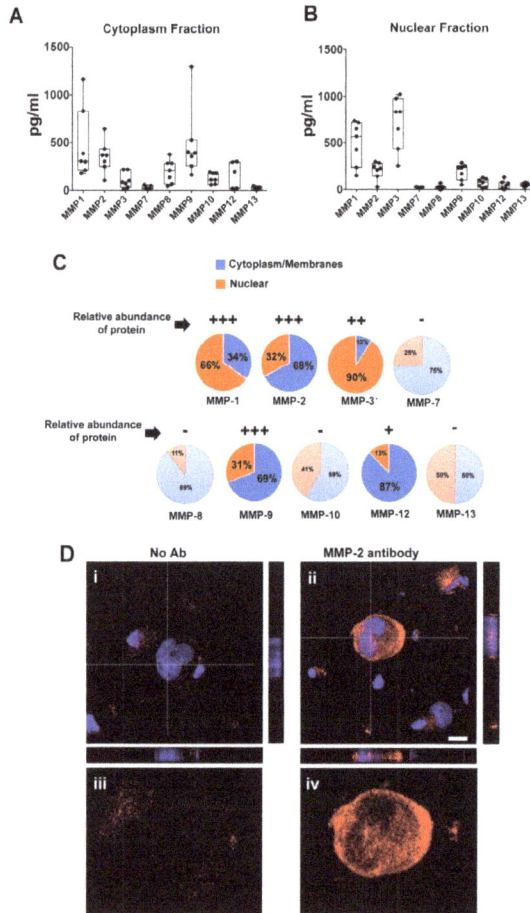

Figure 2. Subcellular localization of MMPs in primary human Megakaryocytes. (**A**,**B**) Profiling and quantification of MMPs in cytoplasm (**A**) and nuclear (**B**) fractions of 2×10^6 megakaryocytes at day 13 of differentiation. N = 7. (**C**) Schematic representation of intracellular distribution of individual MMP between the cytosolic/membrane and nuclear compartments. Relative intracellular abundance is also provided (−/absent, +/low level, ++/intermediate level, +++/high level). (**D**) Immunofluorescence analysis of intracellular localization of MMP-2 in megakaryocytes at day 13 of differentiation. In the panel (**i**), the primary antibody was omitted as negative control, while Hoechst was used to highlight nuclei. In the panel (**ii**), cells were stained with an anti MMP-2 antibody and Hoechst. Orthogonal cross-sections of representative z-stack from cell nuclei are also provided. In panels (**iii**,**iv**), higher magnification of MMP-2 staining in the absence or presence of the primary antibody is provided. Scale bar = 10 μm.

In addition, nuclear localization of MMPs was investigated by immunofluorescence procedures. Interestingly, analysis of MMP-2 localization, by immunofluorescence, showed a diffuse pattern in the cytoplasm as well as co-localization with Hoechst in nuclei of Mks (Figure 2D). Control cells, stained only with the secondary antibody, did not show a nuclear signal.

The few studies available addressing the function of nuclear MMPs suggest a role in apoptosis induction [26,27,29]. Nevertheless, of all the hematopoietic processes occurring in the BM, the production of Mks and, subsequently, platelets are the most complex and unusual. To this regard, the precise physiological role of the apoptotic process in Mks and platelets is yet to be established [41]. Apoptotic-like features are associated with Mk cytoplasm conversion into a mass of proplatelets, which are released from the cell. The remaining senescent and denuded Mk nuclei, after platelet release, are disposed by apoptosis and phagocytosis [42]. However, signals controlling these events are not known and, at the moment we can only speculate, about the potential involvement of nuclear MMPs in these processes.

Collectively, our data demonstrated that primary Mks expressed additional MMPs to the previously identified gelatinases, MMP-2 and MMP-9 peculiar isoforms, and revealed an unprecedented and novel/intriguing nuclear localization. To this regard, we are aware that nuclear localization of MMPs should be confirmed by means of several techniques, however the accuracy of the technology employed and the reproducibility of this novel compartmentalization of MMPs in Mks reveal novel regulatory functions. Further investigations will help in dissecting the role of the different MMPs in regulating platelet production.

Author Contributions: A.M., D.L. and C.A.D.B: designed and performed the experiments, analyzed the data and edited the manuscript. F.M. and A.B.: supervised the project, conceived the idea and wrote the manuscript.

Funding: This work was supported by Cariplo Foundation (2013-0717) to AB. The funders had no role in study design, data collection and analysis, decision to publish, or preparation of the manuscript.

Acknowledgments: We thank Cesare Perotti (IRCCS San Matteo Foundation, Pavia, Italy) for providing cord blood samples and Patrizia Vaghi ("Centro Grandi Strumenti" University of Pavia, Pavia, Italy) for technical assistance with confocal microscopy analysis.

Conflicts of Interest: The authors declare no conflict of interest.

References

1. Malara, A.; Currao, M.; Gruppi, C.; Celesti, G.; Viarengo, G.; Buracchi, C.; Laghi, L.; Kaplan, D.L.; Balduini, A. Megakaryocytes contribute to the bone marrow-matrix environment by expressing fibronectin, type IV collagen, and laminin. *Stem Cells* **2014**, *32*, 926–937. [CrossRef] [PubMed]
2. Malara, A.; Gruppi, C.; Celesti, G.; Abbonante, V.; Viarengo, G.; Laghi, L.; De Marco, L.; Muro, A.F.; Balduini, A. Alternatively spliced fibronectin extra domain a is required for hemangiogenic recovery upon bone marrow chemotherapy. *Haematologica* **2018**, *103*, e42–e45. [CrossRef] [PubMed]
3. Malara, A.; Abbonante, V.; Di Buduo, C.A.; Tozzi, L.; Currao, M.; Balduini, A. The secret life of a megakaryocyte: Emerging roles in bone marrow homeostasis control. *Cell Mol. Life Sci.* **2015**, *72*, 1517–1536. [CrossRef] [PubMed]
4. Gong, Y.; Zhao, M.; Yang, W.; Gao, A.; Yin, X.; Hu, L.; Wang, X.; Xu, J.; Hao, S.; Cheng, T.; et al. Megakaryocyte-derived excessive transforming growth factor β1 inhibits proliferation of normal hematopoietic stem cells in acute myeloid leukemia. *Exp. Hematol.* **2018**, *60*, 40–46. [CrossRef] [PubMed]
5. Bruns, I.; Lucas, D.; Pinho, S.; Ahmed, J.; Lambert, M.P.; Kunisaki, Y.; Scheiermann, C.; Schiff, L.; Poncz, M.; Bergman, A.; et al. Megakaryocytes regulate hematopoietic stem cell quiescence through cxcl4 secretion. *Nat. Med.* **2014**, *20*, 1315–1320. [CrossRef] [PubMed]
6. Zhao, M.; Perry, J.M.; Marshall, H.; Venkatraman, A.; Qian, P.; He, X.C.; Ahamed, J.; Li, L. Megakaryocytes maintain homeostatic quiescence and promote post-injury regeneration of hematopoietic stem cells. *Nat. Med.* **2014**, *20*, 1321–1326. [CrossRef] [PubMed]
7. Abbonante, V.; Di Buduo, C.A.; Gruppi, C.; Malara, A.; Gianelli, U.; Celesti, G.; Anselmo, A.; Laghi, L.; Vercellino, M.; Visai, L.; et al. Thrombopoietin/TGF-β1 loop regulates megakaryocyte extracellular matrix component synthesis. *Stem Cells* **2016**, *34*, 1123–1133. [CrossRef] [PubMed]

8. Malara, A.; Gruppi, C.; Rebuzzini, P.; Visai, L.; Perotti, C.; Moratti, R.; Balduini, C.; Tira, M.E.; Balduini, A. Megakaryocyte-matrix interaction within bone marrow: New roles for fibronectin and factor xiii-a. *Blood* **2011**, *117*, 2476–2483. [CrossRef] [PubMed]

9. Abbonante, V.; Chitalia, V.; Rosti, V.; Leiva, O.; Matsuura, S.; Balduini, A.; Ravid, K. Upregulation of lysyl oxidase and adhesion to collagen of human megakaryocytes and platelets in primary myelofibrosis. *Blood* **2017**, *130*, 829–831. [CrossRef] [PubMed]

10. Machlus, K.R.; Italiano, J.E., Jr. The incredible journey: From megakaryocyte development to platelet formation. *J. Cell Biol.* **2013**, *201*, 785–796. [CrossRef] [PubMed]

11. Krishnaswamy, V.R.; Mintz, D.; Sagi, I. Matrix metalloproteinases: The sculptors of chronic cutaneous wounds. *Biochim. Biophys. Acta* **2017**, *1864*, 2220–2227. [CrossRef] [PubMed]

12. Rundhaug, J.E. Matrix metalloproteinases and angiogenesis. *J. Cell Mol. Med.* **2005**, *9*, 267–285. [CrossRef] [PubMed]

13. Mannello, F.; Tonti, G.A.; Bagnara, G.P.; Papa, S. Role and function of matrix metalloproteinases in the differentiation and biological characterization of mesenchymal stem cells. *Stem Cells* **2006**, *24*, 475–481. [CrossRef] [PubMed]

14. Lane, W.J.; Dias, S.; Hattori, K.; Heissig, B.; Choy, M.; Rabbany, S.Y.; Wood, J.; Moore, M.A.; Rafii, S. Stromal-derived factor 1-induced megakaryocyte migration and platelet production is dependent on matrix metalloproteinases. *Blood* **2000**, *96*, 4152–4159. [PubMed]

15. Choi, W.S.; Jeon, O.H.; Kim, H.H.; Kim, D.S. Mmp-2 regulates human platelet activation by interacting with integrin alphaiibbeta3. *J. Thromb. Haemost.* **2008**, *6*, 517–523. [CrossRef] [PubMed]

16. Gresele, P.; Falcinelli, E.; Loffredo, F.; Cimmino, G.; Corazzi, T.; Forte, L.; Guglielmini, G.; Momi, S.; Golino, P. Platelets release matrix metalloproteinase-2 in the coronary circulation of patients with acute coronary syndromes: Possible role in sustained platelet activation. *Eur. Heart J.* **2011**, *32*, 316–325. [CrossRef] [PubMed]

17. Guglielmini, G.; Appolloni, V.; Momi, S.; De Groot, P.G.; Battiston, M.; De Marco, L.; Falcinelli, E.; Gresele, P. Matrix metalloproteinase-2 enhances platelet deposition on collagen under flow conditions. *J. Thromb. Haemost.* **2016**, *115*, 333–343.

18. Wrzyszcz, A.; Wozniak, M. On the origin of matrix metalloproteinase-2 and -9 in blood platelets. *Platelets* **2012**, *23*, 467–474. [CrossRef] [PubMed]

19. Mannello, F.; Medda, V. Differential expression of MMP-2 and MMP-9 activity in megakaryocytes and platelets. *Blood* **2011**, *118*, 6470–6471. [CrossRef] [PubMed]

20. Villeneuve, J.; Block, A.; Le Bousse-Kerdiles, M.C.; Lepreux, S.; Nurden, P.; Ripoche, J.; Nurden, A.T. Tissue inhibitors of matrix metalloproteinases in platelets and megakaryocytes: A novel organization for these secreted proteins. *Exp. Hematol.* **2009**, *37*, 849–856. [CrossRef] [PubMed]

21. Cecchetti, L.; Tolley, N.D.; Michetti, N.; Bury, L.; Weyrich, A.S.; Gresele, P. Megakaryocytes differentially sort mrnas for matrix metalloproteinases and their inhibitors into platelets: A mechanism for regulating synthetic events. *Blood* **2011**, *118*, 1903–1911. [CrossRef] [PubMed]

22. McCawley, L.J.; Matrisian, L.M. Matrix metalloproteinases: They're not just for matrix anymore! *Curr. Opin. Cell Biol.* **2001**, *13*, 534–540. [CrossRef]

23. Jobin, P.G.; Butler, G.S.; Overall, C.M. New intracellular activities of matrix metalloproteinases shine in the moonlight. *Biochim. Biophys. Acta* **2017**, *1864*, 2043–2055. [CrossRef] [PubMed]

24. Ip, Y.C.; Cheung, S.T.; Fan, S.T. Atypical localization of membrane type 1-matrix metalloproteinase in the nucleus is associated with aggressive features of hepatocellular carcinoma. *Mol. Carcinog.* **2007**, *46*, 225–230. [CrossRef] [PubMed]

25. Mohammad, G.; Kowluru, R.A. Novel role of mitochondrial matrix metalloproteinase-2 in the development of diabetic retinopathy. *Invest. Ophthalmol. Vis. Sci.* **2011**, *52*, 3832–3841. [CrossRef] [PubMed]

26. Mannello, F.; Medda, V. Nuclear localization of matrix metalloproteinases. *Prog. Histochem. Cytochem.* **2012**, *47*, 27–58. [CrossRef] [PubMed]

27. Aldonyte, R.; Brantly, M.; Block, E.; Patel, J.; Zhang, J. Nuclear localization of active matrix metalloproteinase-2 in cigarette smoke-exposed apoptotic endothelial cells. *Exp. Lung Res.* **2009**, *35*, 59–75. [PubMed]

28. Kwan, J.A.; Schulze, C.J.; Wang, W.; Leon, H.; Sariahmetoglu, M.; Sung, M.; Sawicka, J.; Sims, D.E.; Sawicki, G.; Schulz, R. Matrix metalloproteinase-2 (MMP-2) is present in the nucleus of cardiac myocytes and is capable of cleaving poly (adp-ribose) polymerase (parp) in vitro. *FASEB J.* **2004**, *18*, 690–692. [CrossRef] [PubMed]

29. Si-Tayeb, K.; Monvoisin, A.; Mazzocco, C.; Lepreux, S.; Decossas, M.; Cubel, G.; Taras, D.; Blanc, J.F.; Robinson, D.R.; Rosenbaum, J. Matrix metalloproteinase 3 is present in the cell nucleus and is involved in apoptosis. *Am. J. Pathol.* **2006**, *169*, 1390–1401. [CrossRef] [PubMed]

30. Eguchi, T.; Kubota, S.; Kawata, K.; Mukudai, Y.; Uehara, J.; Ohgawara, T.; Ibaragi, S.; Sasaki, A.; Kuboki, T.; Takigawa, M. Novel transcription-factor-like function of human matrix metalloproteinase 3 regulating the ctgf/ccn2 gene. *Mol. Cell. Biol.* **2008**, *28*, 2391–2413. [CrossRef] [PubMed]

31. Mannello, F.; Luchetti, F.; Canonico, B.; Falcieri, E.; Papa, S. Measurements, zymographic analysis, and characterization of matrix metalloproteinase-2 and -9 in healthy human umbilical cord blood. *Clin. Chem.* **2004**, *50*, 1715–1717. [CrossRef] [PubMed]

32. Suzuki, K.; Bose, P.; Leong-Quong, R.Y.; Fujita, D.J.; Riabowol, K. Reap: A two minute cell fractionation method. *BMC Res. Notes* **2010**, *3*, 294. [CrossRef] [PubMed]

33. Ligi, D.; Mosti, G.; Croce, L.; Raffetto, J.D.; Mannello, F. Chronic venous disease—Part II: Proteolytic biomarkers in wound healing. *Biochim. Biophys. Acta* **2016**, *1862*, 1900–1908. [CrossRef] [PubMed]

34. Tschesche, H.; Zölzer, V.; Triebel, S.; Bartsch, S. The human neutrophil lipocalin supports the allosteric activation of matrix metalloproteinases. *Eur. J. Biochem.* **2001**, *268*, 1918–1928. [CrossRef] [PubMed]

35. Shirvaikar, N.; Reca, R.; Jalili, A.; Marquez-Curtis, L.; Lee, S.F.; Ratajczak, M.Z.; Janowska-Wieczorek, A. Cfu-megakaryocytic progenitors expanded ex vivo from cord blood maintain their in vitro homing potential and express matrix metalloproteinases. *Cytotherapy* **2008**, *10*, 182–192. [CrossRef] [PubMed]

36. Wilson, C.L.; Matrisian, L.M. Matrilysin: An epithelial matrix metalloproteinase with potentially novel functions. *Int. J. Biochem. Cell Biol.* **1996**, *28*, 123–136. [CrossRef]

37. Dunsmore, S.E.; Saarialho-Kere, U.K.; Roby, J.D.; Wilson, C.L.; Matrisian, L.M.; Welgus, H.G.; Parks, W.C. Matrilysin expression and function in airway epithelium. *J. Clin. Invest.* **1998**, *102*, 1321–1331. [CrossRef] [PubMed]

38. Li, Q.; Park, P.W.; Wilson, C.L.; Parks, W.C. Matrilysin shedding of syndecan-1 regulates chemokine mobilization and transepithelial efflux of neutrophils in acute lung injury. *Cell* **2002**, *111*, 635–646. [CrossRef]

39. Qu, P.; Yan, C.; Du, H. Matrix metalloproteinase 12 overexpression in myeloid lineage cells plays a key role in modulating myelopoiesis, immune suppression, and lung tumorigenesis. *Blood* **2011**, *117*, 4476–4489. [CrossRef] [PubMed]

40. Wang, J.; Ye, Y.; Wei, G.; Hu, W.; Li, L.; Lu, S.; Meng, Z. Matrix metalloproteinase12 facilitated platelet activation by shedding carcinoembryonic antigen related cell adhesion molecule1. *Biochem. Biophys. Res. Commun.* **2017**, *486*, 1103–1109. [CrossRef] [PubMed]

41. Kile, B.T. The role of apoptosis in megakaryocytes and platelets. *Br. J. Haematol.* **2014**, *165*, 217–226. [CrossRef] [PubMed]

42. Patel, S.R.; Hartwig, J.H.; Italiano, J.E. The biogenesis of platelets from megakaryocyte proplatelets. *J. Clin. Investig.* **2005**, *115*, 3348–3354. [CrossRef] [PubMed]

cells

MDPI

Review

Coagulation, Microenvironment and Liver Fibrosis

Niccolò Bitto [1], Eleonora Liguori [1] and Vincenzo La Mura [2,3,4,]*

[1] Medicina Interna, Istituto di Ricovero e Cura a Carattere Scientifico (IRCCS) San Donato, Università Degli Studi di Milano, 20097 San Donato Milanese (MI), Italy; nicobitto@gmail.com (N.B.); eleonora.liguori1982@gmail.com (E.L.)

[2] Fondazione IRCCS Ca' Granda, Ospedale Maggiore Policlinico, UOC Medicina Generale-Emostasi e Trombosi, 20122 Milano, Italy

[3] Dipartimento di Scienze biomediche per la Salute, Università degli Studi di Milano, 20122 Milano, Italy

[4] A. M. and A. Migliavacca per lo studio delle Malattie del Fegato, 20122 Milano, Italy

[*] Correspondence: vincenzo.lamura@unimi.it; Tel.: +39-02-5503-5430

Received: 10 June 2018; Accepted: 20 July 2018; Published: 24 July 2018

Abstract: Fibrosis is the main consequence of any kind of chronic liver damage. Coagulation and thrombin generation are crucial in the physiological response to tissue injury; however, the inappropriate and uncontrolled activation of coagulation cascade may lead to fibrosis development due to the involvement of several cellular types and biochemical pathways in response to thrombin generation. In the liver, hepatic stellate cells and sinusoidal endothelial cells orchestrate fibrogenic response to chronic damage. Thrombin interacts with these cytotypes mainly through protease-activated receptors (PARs), which are expressed by endothelium, platelets and hepatic stellate cells. This review focuses on the impact of coagulation in liver fibrogenesis, describes receptors and pathways involved and explores the potential antifibrotic properties of drugs active in hemostasis in studies with cells, animal models of liver damage and humans.

Keywords: thrombin; protease-activated receptors; endothelial dysfunction; von Willebrand factor; hepatitis; cirrhosis; anticoagulation

1. Introduction

Fibrogenesis is a complex biochemical process that represents the hallmark of damage for the most common chronic diseases of the liver. The activation of hepatic stellate cells (HSC) is the key pathogenic mechanism for the initiation, progression, and regression of liver fibrosis. Several studies have gone into more depth on the complex and tightly regulated cross talk at the level of hepatic microcirculation owing to sinusoidal endothelial cells (SEC), Kuppfer cells (KC), and hepatocytes with HSC. This underlines the participation of several hepatic cellular types in fibrogenesis. Our manuscript offers an overview on the pathogenic role played by coagulation and thrombin generation in this complex cellular cross talk by considering fibrosis a wound healing process secondary to micro-thrombi in small hepatic and portal venules, sinusoidal ischemic injury and hepatocyte injury. In addition, thrombin may participate in fibrogenesis by interaction with HSC via protease-activated receptors (PAR-1 and PAR-4), promoting a myo-fibroblast phenotype, fibronectin fibril assembly, and may act as a chemoattractant for inflammatory cells. Altogether, these observations suggest that drugs interfering with the coagulation process have potential as antifibrotic drugs at any stage of chronic liver disease. The in vitro and in vivo studies on these aspects are the main focus of the review.

2. Coagulation in Fibrosis and Disease Progression

2.1. Hepatic Stellate Cells, Endothelium and Fibrosis: Role of PARs

During coagulation, the conversion of fibrinogen into fibrin is a key reaction catalyzed by thrombin, a serine protease which is generated on the surface of activated platelets in response to vascular or tissue injury [1]. Thrombin generation is a tightly regulated process, as it is the expression of the delicate balance between pro-coagulant and anti-coagulant factors. Besides its hemostatic function, thrombin orchestrates cell recruitment in response to any kind of tissue injury and activates endothelium [2–4]. Its interaction with inflammatory and mesenchymal cells is part of the wound healing process, in which hemostasis precedes and initiates tissue repair by fibrin deposition [5]. In 1991, the discovery of protease activated receptors (PARs) clarified the biological pathway of thrombin [6]. PARs are a family of receptors with proteolytic activity, which mediate thrombin (PAR 1, 3, 4)- or tryptase (PAR 2, 4)-induced cellular response. PARs are G-protein-coupled receptors and are activated by irreversible proteolytic cleavage of their N-terminal domain. They are expressed by several cellular types involved in fine regulation of vascular homeostasis and their signaling pathways are complex, as they are potentially coupled to G-proteins with different functions (Figure 1). As a result, they interact with a plethora of signaling transducers (e.g., Rho/Rho-kinase, c-Jun N terminal kinase, IP3, PI3K, JAK-STAT), with consequent pleiotropic effects [7,8]. Endothelium (via PAR1, PAR2) and platelets (PAR1, 4), are the main cells involved in the regulation of vasomotor function and hemostasis exerted by PARs [9]. At low concentrations, thrombin may induce a barrier protective response by endothelium, this effect is mediated by PAR-1 [10]. On the contrary, at high concentrations, thrombin induces a pro-inflammatory, pro-hemostatic and contracting phenotype of endothelium, as it increases the expression of TF, plasminogen activator inhibitor-1 expression (PAI-1), pro-inflammatory cytokines (IL6, IL8) and endothelin-1, among others [7]. This bi-modal effect of thrombin suggests that a disrupted regulation of thrombin generation, as occurs in pro-coagulant conditions, may overcome its physiological interaction with the endothelium and may induce significant tissue injury. Alongside endothelium, platelets are activated by PAR-1 and PAR-4, and inhibition of these receptors is a potent anti-platelet mechanism, confirming the important role played by these receptors on platelet function [11,12]. Thrombin is produced on the surface of activated platelets and its interaction with PARs may initiate and maintain the hemostatic process, leading to thrombus formation when anticoagulant factors are not able to counterbalance this process [1,7]. In recent years, the transcription factor Kruppel-like factor 2 (KFL2) has been recognized as a key regulator of endothelium homeostasis in response to inflammatory stimuli (e.g., tumor necrosis factor, TNFα, and interleukin 1) and hemodynamic forces like laminar shear stress [13,14]. Interestingly, Marrone et al. demonstrated that KLF2 overexpression in SEC and HSC proceeding from cirrhotic rats reduces HSC activation and ameliorates paracrine cross-talk with SEC [15,16]. This is in line with the reduction of fibrosis and portal pressure observed in animal studies in association with KFL2 expression [17]. In 2005, Li et al. demonstrated that KLF2 induction blunts the pro-inflammatory, pro-hemostatic transcriptional response of the endothelium exposed to noxious stimuli (e.g., TNFα), as it reduces tissue factor, Von Willebrand Factor (VWF) and PAI-1 [18]. Interestingly, ADAMTS-13, a metalloproteinase which regulates the pro-hemostatic function of VWF by its cleavage, is produced by HSC in physiological conditions and its activity declines alongside liver dysfunction [19,20]. Absolute deficiency of ADAMTS-13 leads to diffuse microvascular occlusion due to high molecular weight VWF multimers which promote platelet aggregation and microthrombi formation; therefore, low levels of ADAMTS-13 are of increasing interest in thrombotic-microangiopathies and those clinical conditions like sepsis in which liver failure, as well as other organ dysfunctions are frequently observed [21,22]. All these observations emphasize the role of HSCs-SEC interaction to maintain an anti-thrombotic phenotype at sinusoidal level in physiology, with a potential protective role of KFL2 due to its control of VWF and platelet aggregation. Moreover, KFL2 may also have a direct role in the direct control of hemostasis by the endothelium, since it inhibits PAR-1 expression on endothelial cells. This shows a direct link between KLF-2-induced regulation of endothelial physiology and the biological response of this cytotype to thrombin. Whereas in vascular medicine studies on PARs focus on platelets and endothelium, PARs expression by HSCs is central in liver fibrogenesis [17]. The progression of any chronic liver disease is characterized by the acquisition of a contractile and pro-fibrogenic phenotype by HSCs, along with an imbalance between vasoconstrictors and vasodilators produced by SECs. As a consequence, the liver parenchyma is distorted by the development of interstitial fibrosis and the constriction of sinusoids which, in turn, lead to the increase of portal pressure owing to the mechanical and functional increase of liver resistance to the portal blood flow [23]. Several studies have explored thrombin-PARs interaction on HSC in the process of liver fibrogenesis. They are summarized in Table 1. The action of thrombin on PARs (mainly PAR-1) induces fibrogenic response in the liver by reprogramming HSCs with the induction of a pro-fibrotic, activated phenotype [24–26]. Incremental doses of thrombin progressively transform HSCs into myofibroblasts, with increase of αSMA, pro-collagen, TGFβ-1 and other key cellular signals which are crucial in the wound healing response [24]. The uncontrolled persistence of a thrombin-related signaling through PARs, due to a pro-hemostatic milieu, is considered the main mechanism that binds hemostasis and fibrosis [27]. In line with this theory, experimental inhibition of PARs prevents the fibrogenic response of HSCs and the progression of fibrosis as demonstrated in pre-clinical studies with animal models of liver disease and cell cultures [26,28]. In addition to PAR-1, PAR-2 showed similar pro-fibrotic effects by inducing HSC contraction, collagen production and MMP-2 expression, this last promoting liver fibrosis due to extracellular matrix remodeling [25,29–31]. Furthermore, studies with PAR$^{-/-}$ transgenic mice confirmed the importance of this receptor in several models of liver fibrosis (xenobiotics, carbon-tetrachloride, CCL$_4$, and thioacetamide, TAA) [30,32,33] and, recently, even in a model fatty liver disease [34]. To our knowledge, just one study explored PAR-1 genotype and liver fibrosis in patients with chronic HCV infection. In this biopsy-proven study, a particular PAR-1 polymorphism (1426 C/T) correlated with increased liver fibrosis, thus confirming the above-mentioned results from pre-clinical studies [35]. Alongside PARs, tissue factor (TF) has been often investigated in liver fibrosis, since TF is a potent activator of hemostasis via factor VII (FVII) [36]. Interestingly, transgenic mice lacking of TF show a reduced rate of fibrosis development after exposure to various chronic damage stimuli, thus confirming a potential connection between the pro-hemostatic role of TF and liver fibrosis [31,34]. Recently, Ratou et al., in a study with mice after bile duct ligation (an animal model of liver fibrosis), demonstrated an increase of thrombin-antithrombin complexes, which are biomarkers of a pro-coagulant condition. This increase was prevented in mice lacking in TF. However, this anti-coagulant phenotype was not associated with a significant reduction of fibrosis [37], in contrast to other studies [30].

Table 1. Studies exploring the impact of coagulation on liver fibrosis.

Reference	Experimental Model	Pathway Explored	Methods	Results
Chambers 1998 [34]	human fetal lung fibroblasts	PAR-1	Exposure to incremental dose of thrombin; TRAPs (thrombin receptor-activating peptide) +/− inhibitors (hirudin/Phe-Pro-ArgCH2CL)	Thrombin ↑ α1-procollagen mRNA through PAR-1 activation
Gaça 2002 [35]	Cultured stellate HSEC	Thrombin, tryptase/PAR 1-2	PAR 1/2 mRNA RT-PCR analysis + northern blotting in lysate of HSEC. Use of PD98059 (kinase inhibitor)	↑ PAR-1/2 while fibroblast transforms in myofibroblast phenotype; ↑ HSC proliferation by PARs
Fiorucci et al. 2004 [26]	rat HSC cell line; BDL cirrhotic rat	Thrombin-PAR,	type I collagen mRNA expression; quantitative morphometric analysis; hepatic and urinary excretion of hydroxyproline	Thrombin triggers HSC activation and collagen deposition via PARs, prevented by PAR, antagonist
J Gillibert Duplantier et al. 2007 [38]	Human hepatic myofibroblasts	PAR-1; COX-2; Akt-1; platelet derived growth factor (PDGF)	Cell migration; RNA isolation and analysis for Prostaglandin E2 receptor; analysis of Akt-1 phosphorylation and PDGF-receptor phosphorylation.	Thrombin inhibits human hepatic myofibroblast migration via PAR-1; Thrombin inhibits PDGF induced migration (inhibition of PI3K)
Martinelli 2007 [35]	Patients with HCV (287 european, 90 brazilian)	PAR1	Cross-sectional study; fibrosis evaluated by liver biopsy; polymorphism of PAR-1 gene analysis (−1426 C/T; IVS-14, −506 I/D	↑ fibrosis in TT genotype of 1426 C/T polymorphism
Rullier 2008 [72]	PAR-1 −/− and +/− mice exposed to CCl4	PAR1	Histology; RT-PCR for type I collagen, MMP-2, PDGFβ-r, MP-1, mRNA	↓ fibrosis and activated fibrogenic cells ↓ type I collagen, MMP-2, PDGFβ-r mRNA ↓ T lymphocytes infiltration
B. P. Sullivan et al. 2010 [73]	Bile duct epithelial cells (BDECs); PAR1 −/−, TF +/− mice with low levels of human TF expression. All mice were fed with BDEC toxicant (ANIT); Human Liver Samples from patients with PBC/PSC	TF, PAR-1, αVβ6	Real-Time PCR of snap-frozen liver or adherent cells; immunofluorescence on liver frozen sections for αVβ6	TF and PAR-1 deficiency ↓ Liver Fibrosis/αVβ6 mRNA ↑ TGF-β1 related αVβ6 expression by PAR-1 αVβ6 inhibition ↓ fibrosis ↑ TF and PAR-1 mRNAs in livers from PBC/PSC patients
V. Knight et al. 2012 [51]	HSC cells; HSEC cells; (PAR-2 knockout mice; C57BL/6 mice; CCl4 cirrhotic mice	PARs	Hepatic hydroxyproline content in frozen liver tissue; PCR analysis of MMP-2, TIMP-1 and PAR-1/2; identification of α-SMA, F4/80 and CD68; TGF-β1 Production In Vitro; HSC Proliferation in Response to PAR Activation; Hepatic TGF-β1 Content	PAR-2 Deficiency ↓ Fibrosis/ procollagen mRNA/Hydroxyproline Content/ Stellate Cell Activation/ Hepatic TGF-β1 Expression/MMPs/ Activated Hepatic Macrophages; PAR 1/2 ↑ HSC Collagen Production/TGF-β1
R. Nault et al. 2016 [74]	PAR-1 −/− and +/− mice exposed to to TCDD (progression to NASH)	PAR-1;	Identification of Fibrin(ogen)	TCDD Exposure Activates the Coagulation Cascade; ↓ inflammation and collagen deposition in PAR-1 −/−
V. Knight et al. 2017 [50]	PAR-1 −/− mice; HSC cells; CCl4 treated mice	TF and PARs	Hepatic fibrosis assessment; Hepatic collagen content; Gene expression of TGF-β1, MMP-2, TIMP 1, PAR1 and 2; expression TGF-β1	↓ fibrosis/MMP2/activated macrophages in TF and PAR-1 −/-

In summary, hemostasis may drive a pro-fibrotic HSC phenotype via PARs. The cellular cross talk between HSCs and SECs and the expression of KLF2 may somehow reduce the fibrogenic process associated with a pro-coagulant imbalance under chronic conditions of liver damage.

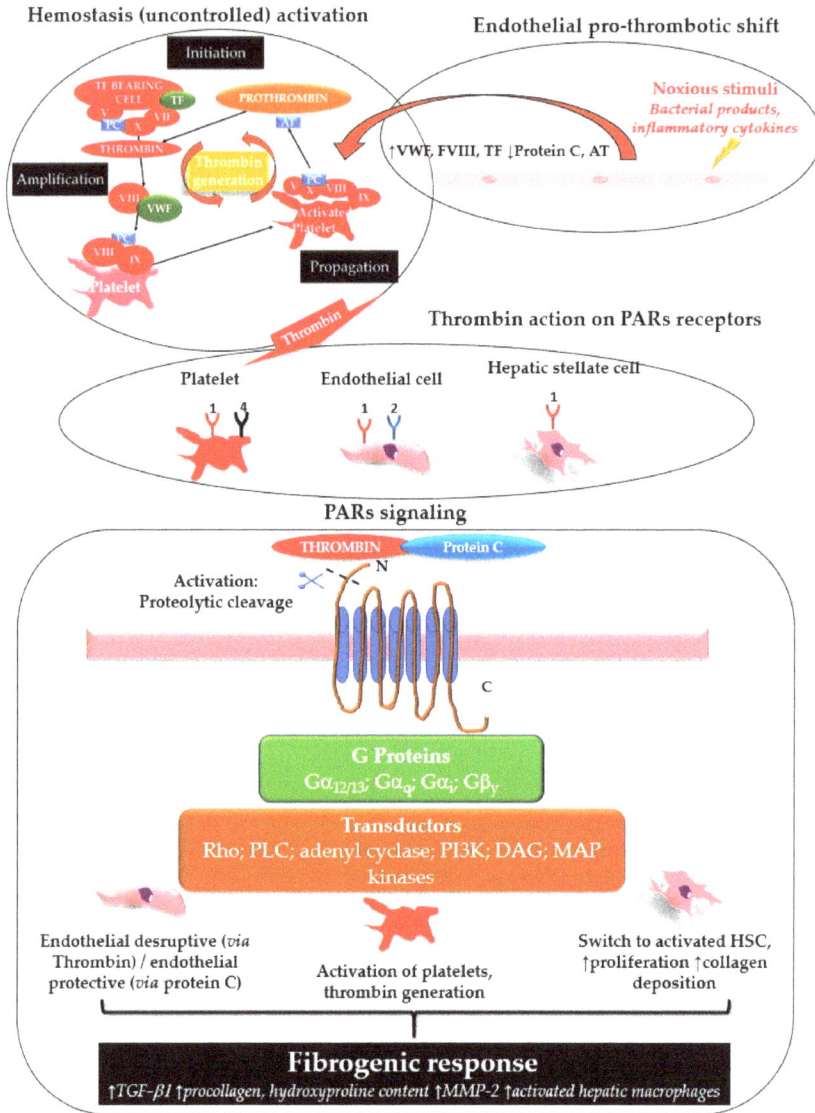

Figure 1. Schematic representation of PAR signaling.

2.2. Parenchymal Extinction: From Clot Generation to Liver Damage

An important step in the knowledge of coagulation as a mechanism of liver damage was the study by Wanless et al., who conducted a histological analysis by comparing 61 cirrhotic livers of any etiology removed at the time of transplantation with 24 livers from autopsy of normal subjects as controls [39]. The main purpose of the study was to confirm the previous observation that fibrosis co-localizes with vascular lesions of the hepatic venous system [40]. First, they distinguished origin (hepatic or portal), caliber (small, medium or large) and size (% of luminal narrowing) of vascular lesions. Second, they graded fibrosis with description by optical microscope and defined focal parenchymal extinction as a region of parenchymal loss filled by fibrosis. Hepatic and portal vein intimal fibrosis, highly suggestive of previous occlusion, were respectively evident in 70% and 36% of livers. In morphometric data on 534 hepatic veins pooled from 10 livers, hepatic vein occlusions were frequent in small veins and co-localized with a greater extent of fibrosis. The existence of a "post thrombotic syndrome" was also inferred by patchy distribution of fibrotic areas, multiple layers of fibrosis and the severe occlusion of the smallest veins. In another study, the same group analyzed 13 autopsy livers with congestive fibrosis, with another 12 livers as controls [41]. In this model, venous stasis was associated with thrombosis of sinusoids and terminal hepatic venules, with formation of fibrotic septa and sinusoidal fibrotic thickening. These changes were associated with the extension of thrombosis to larger veins, necrosis and parenchymal loss. Recently, Simonetto et al. confirmed these results in an animal model of congestive hepatopathy (partial inferior vena cava ligation), showing hepatic sinusoidal thrombosis with increase of liver stiffness and portal pressure [42]. Interestingly, fibrosis was accompanied by minimal inflammation, whereas mechanical forces seemed to prevail with stretch-induced fibronectin-fibrils assembly. Of note, tissue factor pathway inhibitor or warfarin treatment blunted sinusoidal thrombosis and fibrosis deposition, confirming the existence of a hemostasis-driven fibrogenesis in this model of liver congestion. They also analyzed liver specimens of patients with congenital heart failure due to chronic myocardial dysfunction or Fontan cardiac surgery, which is a set of surgical techniques causing venous hypertension after deriving the systemic venous flow directly into the pulmonary artery. In these patients, immunochemistry analysis revealed fibrin deposition within sinusoids, confirming the association between microthrombosis and fibrosis. The lack of inflammation in this study is apparently in line with the pre-clinical study by Cerini et al. who demonstrated a minimal anti-inflammatory effect of heparin, alongside a potent anti-fibrotic impact prevalently due to the anticoagulant properties of the drug. These results are in line with a recent work by Miyao et al., who demonstrated in a mice model of non-alcoholic fatty liver disease that sinusoidal endothelial injury may precede the activation of Kupffer cells, HSC, inflammation and fibrosis [43]. Therefore, despite inflammation being a cardinal element in development of a biological response to every kind of noxious stimuli, its link with hemostasis probably cannot explain alone the consequent fibrogenic response. Recent evidence, elsewhere reviewed by De Ridder et al. [44], focuses on the precise site of thrombin generation, identified in the intravascular or the interstitial anatomical space. Whereas intravascular activation is easily understood and studied in micro and macrovascular medicine, thrombin activity in the interstitial space is intriguing and often neglected [44]. However, liver fibrosis is, by definition, an interstitial process, and it is conceivable that thrombin exerts an important and complex action on fibrogenic response, for example by activating pro- [45] and anti-fibrotic [46]. metalloproteinases present in the extracellular matrix. The exact link, if it exists, between the intravascular and interstitial generation of thrombin during chronic hepatitis is certainly an open question, which is even more of interest for hepatologists, as recent studies have shown that anti-coagulation per se may favorably impact the natural history of cirrhosis [47,48].

In conclusion, parenchymal extinction theory represents the bridge between pre-clinical studies demonstrating a role of hemostasis on liver fibrogenesis and the pathological observations of liver parenchymal loss due to vascular occlusion, progressive necrosis and fibrosis replacement in humans. However, as thrombin explains its action also in the interstitium, further studies are warranted to confirm and define precisely the weight of microvascular and interstitial changes due to the activation of thrombin as consequence of a chronic liver damage.

2.3. Procoagulant Imbalance and Disease Progression: Clinical Observations

2.3.1. Common Inherited Pro-Hemostatic Genotype and Risk of Fibrosis Development

Unprovoked venous-thromboembolic events are often linked with pro-hemostatic mutations of clotting factors. FV Leiden and FII G20210 mutations are associated with thrombotic events in the general population, with relative common frequency (0.4–5% and 3%, respectively) [49–51]. FV Leiden missense mutation (ArG506Gln) leads to an intrinsic resistance to the anticoagulant action of protein C, whereas FII G20210 increases prothrombin levels and inhibits fibrinolysis by the reduction of the thrombin-activatable-fibrinolysis-inhibitor [52,53]. The potential impact on fibrosis development of a constitutional pro-thrombotic imbalance has been hypothesized and explored by several authors. In 2003 Wright et al. conducted a retrospective, biopsy-proven study aimed to describe the degree of association between the most common thrombophilic factors and the severity of liver fibrosis. In this study, FV Leiden, but not FII G20210 mutation was associated with accelerated fibrosis and cirrhosis development in patients with HCV infection [54]. In contrast, Maharshak et al. demonstrated an association between faster fibrosis and FII G20210 mutation with no evidence for FV Leiden [55]. These divergent results resemble subsequent data showing a potential [54–58] or doubtful impact of thrombophilia on the risk of liver fibrosis development [57,59,60]. A recent retrospective population study on 1055 patients demonstrated an association between FV Leiden and FII G20210 mutation and a significant increase in liver stiffness, which is a widely used non-invasive marker of liver fibrosis [56]. Moreover, in this study, non-0 blood group showed the highest liver stiffness in patients with pro-hemostatic mutations. These data are in line with a previous observation of an association of non-0 blood group and fibrosis severity in HCV-infected patients [61,62]. Interestingly, AB0 blood group is a major determinant of VWF and factor VIII (FVIII) levels in normal subjects, both potent pro-hemostatic factors, and non-0 blood group has been associated with increased levels of VWF and FVIII with increased risk of venous-thrombo-embolism [63,64]. In summary, a pro-hemostatic genotype may have a role in the development of fibrosis. However, evidence is limited to observational studies. The clinical question if thrombophilic inherited mutations may identify clusters of patients with high risk of fibrosis progression is appealing. Thus, it advocates proof of concept studies to clarify the impact and the magnitude of these mutations on fibrosis development.

2.3.2. Hemostatic Balance in Advanced Liver Disease

Every stage of liver disease results in a different degree of change in the hemostatic balance [65]. For years, the alteration of conventional coagulation tests (e.g., prothrombin time, partial activated thromboplastin time, bleeding time) disguised the coagulopathy of liver disease under a bleeding mask, represented by the assumption of spontaneous bleeding among patients with cirrhosis, the final grade of any chronic liver disease [66]. This is true in terms of spontaneous gastro-intestinal bleeding, but today we know that this is a consequence of portal hypertension and not of a disease-related reduction of plasma activating the coagulation cascade [48,67]. Indeed, in cirrhosis, the reduction of liver-dependent pro-hemostatic clotting factors (FII, V, VII, IX, X and XI) is counterbalanced by the reduction of anticoagulant factors and similar contrasting alterations in the fibrinolytic system [65]. As a result, the evaluation of the plasmatic hemostatic balance by the in vitro thrombin generation test, which takes into account both pro- and anti-coagulant factors, showed normal thrombin generation in these patients [68]. Therefore, the first seminal study by Tripodi et al. [68] allowed a shift from the old paradigm of an intrinsic bleeding tendency, to the concept of "re-balanced hemostasis" in patients with chronic liver disease [69,70]. Moreover, the same research group demonstrated a resistance to the action of thrombomodulin, a strong anticoagulant, which parallels disease severity and a progressive pro-coagulant imbalance of clotting factors [71,72]. The hypothesis of a pro-coagulant plasmatic milieu in cirrhosis is intriguing, as thrombotic events are common in this population [73]. Thrombosis of portal and splanchnic venous vessels ranges from 5 to 20%, and the highest rate is observed in the advanced stages of the disease [74,75]. Moreover, retrospective studies have shown that cirrhosis may represent a risk factor for venous-thrombo-embolic events in hospitalized patients [76–78]. The increase in FVIII, VWF and the resistance to the action of thrombomodulin due to protein C reduction are the best-described pro-hemostatic features, and worsen along with disease severity [72,79,80]. Interestingly, they were all independently associated with increased portal hypertension and worse prognosis, suggesting a potential impact on the pathogenesis of this clinical condition [81–85]. However, the design of these studies does not allow the uncovering of a cause-effect relationship between fibrosis and pro-hemostatic changes, despite an interesting role for VWF as a noninvasive marker of fibrosis in two studies [86,87]. Nevertheless, a potential impact of coagulation on fibrogenesis and parenchymal extinction is fascinating and is currently under investigation by several research groups. One potential limitation is the lack of a study investigating hemostasis in liver disease far from advanced stages or cirrhosis. Recently, two distinct leading groups in this field have published contrasting evidence on this topic in the clinical setting of non-alcoholic liver disease, which is expected to be the main increasing etiology of cirrhosis in next few years [88–91]. The recent debate that has risen on this topic [92,93] demonstrates the need of further investigations on the impact of hemostasis even in the earliest stages of any chronic disease of the liver.

3. Anticoagulation as Anti-Fibrotic Strategy

3.1. Heparin

In the era of etiological therapies, which will hopefully erase the burden of chronic viral hepatitis [94], powerful antifibrotic drugs are still lacking [95,96]. However, the increasing incidence of metabolic liver disease calls on such therapies, while etiologic treatments for NAFLD/NASH are not yet satisfactory [90]. Several studies have explored antifibrotic proprieties of drugs active on hemostasis (Table 2). Low molecular weight heparins (LMWH) inhibit factor X indirectly via antithrombin, thus lowering thrombin generation [97]. In a histological study in rats exposed to carbon-tetrachloride (CCL_4), LMWH reduced fibrosis and collagen deposition, while ultrastructural analysis on transmission electron microscope (TEM) showed reduced sinusoidal swelling and less distorted parenchymal architecture [98]. Dalteparin also showed fibrosis reduction in CCL_4 chronic damage, while increasing hepatic-growth factor and blunting pro-fibrotic expression of TGF-β1 and deactivating HSC (αSMA reduction). Interestingly, no effect on necrosis and inflammation was observed, with unchanged levels of TNF [99]. These results were confirmed in a study by Cerini et al., who explored enoxaparin in different rat-models of liver damage: CCL_4 (acute/short/long exposure) and TAA exposure [100]. Fibrosis and pro-fibrogenic stimuli were analyzed with histology, immunochemistry and HSC isolation. Additionally, portal pressure and hepatic vascular resistance were analyzed with isolation and perfusion of the liver. Enoxaparin markedly reduced fibrosis, with anti-fibrotic reprogramming of HSC with αSMA and pro-collagen I reduction. Moreover, it also reduced portal pressure without altering hepatic blood flow, thus reducing hepatic resistance in accordance to ohm's law (pressure = flow x resistance). These results were confirmed in both CCL_4 and TAA damage induction. Indeed, enoxaparin disclosed antifibrotic effects in chronic but not acute liver damage and this occurred without any anti-inflammatory action. Therefore, this solid biological background allows to promote LMWHs as potential antifibrotic strategy. Along these lines, relevant clinical data derive from the trial by Villa et al. [47] who randomized 70 patients with decompensated cirrhosis to receive, or not, enoxaparin in order to prevent de novo portal vein thrombosis. Surprisingly, the treatment arm prevented de novo portal vein thrombosis without any increase of the bleeding rate, and patients showed better clinical outcomes in term of new decompensating events (mainly ascites development) and survival. When treatment was interrupted, both arms turned to similar rates of clinical events and portal vein thrombosis development. This study was the first randomized clinical trial demonstrating the potential impact of anticoagulant on the natural history of cirrhosis, although as of today, no data exist to conclude that the beneficial effect of anticoagulation was mediated by the antifibrotic properties demonstrated in the above-mentioned pre-clinical studies.

Table 2. Main studies exploring anticoagulant-antifibrotic strategies.

Reference	Drug	Animal Model	Fibrosis/Cirrhosis Induction	Fibrosis Assessment	Results
Duplantier 2004 [25]	Wistars rat	Thrombin antagonist SSR182289	CCL4 (three or seven week exposure)	Histology; immunohistochemistry (IHC) for αSMA collagen type I, MMP-2, TIMP-1, and TIMP-2 mRNAs by RT-PCR	↓ 30% fibrosis (7 week CCL4 exposure); Early ↓αSMA positive cells/TIMP-1 mRNA
Abe 2007 [96]	Dalteparin	Female Wistars Rats	CCL4	Histology; IHC	↓ fibrosis, ↑HGF; ↓TGF-β1, COL1A1, αSMA; ↓ PDGF induced HSC proliferation
Anstee 2008 [101]	Warfarin	FV Leiden mutant mice, C57BL/6 control animals anticoagulated mice	CCL4	Histology; Liver Hidroxiproline content; αSMA mRNA expression	↑ fibrosis 80% in male FV mutant Warfarin effect: ↓ Hidroxiproline content ↓ fibrosis scores Effect blunted in FV mutant
Kassel 2012 [102]	Argatroban (via micro-osmotic pump)	LDLr−/− mice	Western diet	Histology; real time PCR hepatic mRNA expression of αSMA, COL1A1, PDGFβ, TIMP1/2, TGF-β1; IHC (anti CD68, F4/80, αSMA); MCP-1 Elisa	No change in collagen deposition ↓ αSMA, COL1A1, PDGFβ, TIMP1/2 No ↓TGF-β1 ↓inflammation (↓neutrophil/macrophage accumulation)
Cerini 2016 [100]	Enoxaparin	Male Wistars Rats	CCL4 (acute vs short vs long term exposure); TAA	Histology; IHC (anti FBN/αSMA/CD68); expression of procollagen I/ αSMA on isolated HSC	↓25–26% in short and long term CCL4 exposure; ↓ 41% in TAA ↓PP and HVR ↓αSMA, procollagen I in HSC No change on inflammation
Vilaseca 2017 [103]	Rivaroxaban	Cirrhotic wistar rats	CCL4; TAA	Histology; TEM analysis; Liver Hidroxiproline content; IHC (anti fibrinogen/ αSMA/CD68) and IF (anti FBN, anti VWF); real time PCR hepatic mRNA expression of αSMA, COL1A1, PDGFβ, TIMP1/2, TGF-β1; in vitro thrombin action on HSC	No ↓in CCL4, ↓25% TAA improved sinusoidal architecture ↓Hidroxiproline content/collagen/fibrin deposition ↓PP and HVR ↓HSC activity of profibrotic genes ↓VWF expression in vasculature No direct activity on HSC (in vitro studies)
Li 2017 [98]	Aspirin (low/high dose), enoxaparin	Sprague-Dawley rats	TAA	Histology (METAVIR score)	↓ in all treatment group (> for high dose aspirin)

3.2. Oral Anticoagulants: From Vitamin K Antagonists to Direct Oral Anticoagulants (DOACs)

Warfarin is an oral anticoagulant which inhibits the production of clotting factors, thus indirectly abolishing thrombin generation [104]. The laboratory testing of INR (a standardized measure derived from prothrombin time) is specifically designed to monitor the anticoagulant effect of vitamin k antagonists [105,106]. In 2008, Anstee et al. studied the effect of warfarin in mice with prothrombotic mutation of FV Leiden exposed to CCL$_4$ [101]. In this animal model, warfarin significantly reduced fibrosis progression and liver hydroxyproline content, while mice carrying FV mutation exhibited fibrosis progression with blunted effect of warfarin. In recent years, DOACs have radically changed management in hemostasis modulation [107]. This class of drugs directly inhibits the action of clotting factors (FX and FII), thus reducing thrombin generation [107]. The oral assumption and the lack of need of laboratory monitoring are progressively prompting the repeal of vit k antagonists in favor of this class of drugs, which is currently used in various thrombotic diseases [108]. In 2012, Kassel et al. studied the effect of argatroban, a direct inhibitor of FII, in LDLr$^{-/-}$ fed with a western diet [102]. Argatroban reduced hepatic mRNA expression of αSMA, COL1A1, PDGFβ, TIMP1/2, with no effect on TGF-β1 or collagen deposition. In this model of metabolic-induced damage, argatroban significantly reduced inflammation and neutrophil accumulation in the liver, globally showing early change to an anti-inflammatory, anti-fibrotic phenotype. In a recent study, Vilaseca et al. treated rats with chronic liver damage induced by CCL$_4$ and TAA with rivaroxaban, an FX direct inhibitor, which reduces thrombin generation [103]. In this study, rivaroxaban reduced portal pressure and hepatic vascular resistance, confirming the amelioration of liver microcirculation. In in vitro experiments on HSC, there was no clear thrombin-related activating effect. Otherwise, rivaroxaban treatment exerted an anti-fibrotic effect on mRNA expression of αSMA, COL1A1, PDGFβ, TIMP1/2 and TGF-β1. Moreover, rivaroxaban reduced fibrin deposition and ameliorated sinusoidal architecture, as seen in TEM analysis, thus suggesting a direct effect on microthrombosis.

In summary, preclinical studies suggest a direct anti-fibrotic effect of oral anticoagulants, which ameliorates liver microvascular perfusion, with an anti-fibrotic reprogramming on HSCs and, at last, reduced fibrin and collagen deposition. However, scant data exist on the use of direct oral coagulant in cirrhotic patients, and prescription is currently limited in this population, with few exceptions in patients with compensated disease [109]. Some registry-based studies are exploring the use of DOACs with promising results [110–112]; however, high-quality evidence in the form of clinical trials is eagerly awaited to confirm the safety profile of these drugs and, potentially, their impact on the natural history of the disease.

4. Future Directions: Hemostasis as Immune Response

The use of confocal microscopy recently shed a light on mechanisms of cell interactions in sterile or septic injury, due to the in vivo visualization allowed by the instrument [113,114]. The study of hepatic microcirculation, by in vivo visualization of sinusoids, confirmed a central role of the liver in the clearance of bacteria, as demonstrated after inoculation of S. aureus in a murine model [115–117]. Kupffer cells first gather in liver sinusoids after bacteria inoculation, and afterwards, neutrophils and platelets assemble and remain in the liver vasculature by VWF secretion and binding [116,118]. The platelet–neutrophil interaction leads to the organized destruction of neutrophils and the release of neutrophil extracellular traps (NETs), which are networks of neutrophil DNA and histones which entrap and kill bacteria gathered in the sinusoids [119]. This organized neutrophil death program is different from necrosis and apoptosis, and has been called NETosis [120]. While it is crucial in innate immune response, its uncontrolled activation may lead to tissue injury, and several experiments have demonstrated colocalization of NETs and subsequent necrosis. Hemostasis directly interact with NETosis by activated platelets and activation of coagulation in the site of the immune response [118,121,122]. Moreover, the demonstration of a VWF binding to histones, which precedes the discovery of NETs, suggests a continuum in hemostasis activation and tissue response to bacteria [123]. Therefore, in recent years, hemostasis has been revised as a direct effector of immune innate response, and in 2013, Engelmann used the term "immunothrombosis" to define thrombosis as an uncontrolled, deranged immune response to tissue injury [124]. Moreover, several studies have demonstrated an association between NET production and thrombosis [125–128]. Recently, McDonald et al. demonstrated an in vivo intravascular coagulation into sinusoids in response to sepsis (LPS administration and S. aureus inoculation in mice), which colocalize with NETs formation and tissue injury [129]. Interestingly, in this experiment, NET inhibition reduced thrombin activation and organ damage, while anticoagulation with argatroban alone did not reveal any effect on tissue injury. Collectively, these results confirm that the interaction between immune responses, platelets and coagulation is crucial in organ homeostasis in response to exogenous damage stimuli [122]. Therefore, immunothrombosis may represent a global mechanism which mediates tissue injury in response to acute and chronic damage and precedes fibrosis. In hepatology, increasing evidence advocate a pathogenic role for bacterial translocation from gut to general circulation [130,131]. Bacterial translocation is due to the increase of portal hypertension alongside liver disease severity, thus increasing gut permeability and disrupting the intestinal barrier [132]. This chronic exposition of enteric pathogens is associated with a progressively worsening inflammatory state, which has been recently presumed to be one of the main pathophysiological events in the development of cirrhosis-related complications [132–134]. Bacterial translocation is also associated with VWF, FVIII increase and platelet hyperactivation, thus confirming a pro-hemostatic role [79,135–138]. As immunothrombosis may originate from excessive response to bacteria in the liver vasculature, the existence of a chronic pathogen exposition may be crucial in sustaining inflammation, micro-thrombosis and consequent parenchymal extinction. Studies on the potential link with immune response, hemostasis activation and consequent fibrosis and disease progression are intriguing and highly anticipated.

5. Conclusions

Hemostasis has a non-negligible impact on liver fibrosis, as it induces a pro-fibrotic, activated HSC phenotype through thrombin–PARs interaction. Moreover, the increasing comprehension of liver immunology elucidates the crucial role of hemostasis in tissue injury mechanisms and may offer new potential druggable pathways by further defining this complex interplay. A pro-hemostatic milieu in liver microcirculation due to repetitive harmful stimuli may drive sinusoidal microthrombosis, which leads to parenchymal extinction and disease progression (Figure 2). As a result, an inherited or acquired pro-hemostatic imbalance is associated with fibrosis progression in pre-clinical and clinical studies. Moreover, anticoagulant drugs reduce fibrosis development, and may impact the natural history of liver disease, even in late stages of cirrhosis, which display a complex hemostatic balance. Therefore, the ever more precise understanding of the mechanisms that regulate hemostasis and its interactions with the pathophysiology of tissue damage will make it possible to better define new therapeutic targets in the clinical challenge of dampening liver fibrosis.

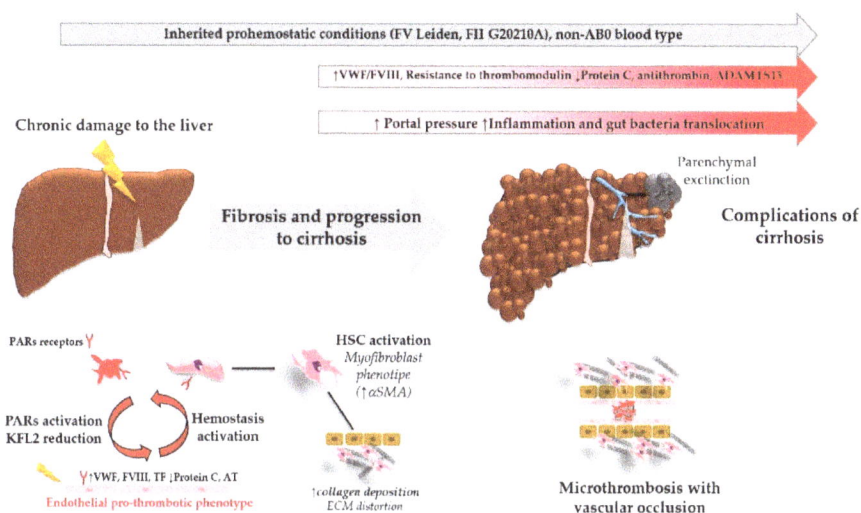

Figure 2. Hemostasis activation and liver disease progression.

Author Contributions: N.B. contributed to the systematic review of literature, and the design and writing of the manuscript; E.L. contributed to the systematic review of literature, and the writing of the manuscript; V.L.M. contributed to the supervision of the manuscript editing, and the revision of major intellectual content.

Funding: Authors received no external funding.

Conflicts of Interest: The authors declare no conflict of interest.

References

1. Hoffman, M.; Monroe, D.M. A cell-based model of hemostasis. *Thromb. Haemost.* **2001**, *85*, 958–965. [PubMed]
2. Weksler, B.B.; Ley, C.W.; Jaffe, E.A. Stimulation of Endothelial Cell Prostacyclin Production by Thrombin, Trypsin, and the Ionophore A 23187. *J. Clin. Investig.* **1978**, *62*, 923–930. [CrossRef] [PubMed]
3. Prescott, S.M.; Zimmerman, G.A.; McIntyre, T.M. Human endothelial cells in culture produce platelet-activating factor (1-alkyl-2-acetyl-sn-glycero-3-phosphocholine) when stimulated with thrombin. *Proc. Natl. Acad. Sci. USA* **1984**, *81*, 3534–3538. [CrossRef] [PubMed]
4. Sugama, Y.; Tiruppathi, C.; offakidevi, K.; Andersen, T.T.; Fenton, J.W.; Malik, A.B. Thrombin-induced expression of endothelial P-selectin and intercellular adhesion molecule-1: A mechanism for stabilizing neutrophil adhesion. *J. Cell Biol.* **1992**, *119*, 935–944. [CrossRef] [PubMed]
5. Laurens, N.; Koolwijk, P.; de Maat, M.P.M. Fibrin structure and wound healing. *J. Thromb. Haemost.* **2006**, *4*, 932–939. [CrossRef] [PubMed]
6. Vu, T.-K.H.; Hung, D.T.; Wheaton, V.I.; Coughlin, S.R. Molecular cloning of a functional thrombin receptor reveals a novel proteolytic mechanism of receptor activation. *Cell* **1991**, *64*, 1057–1068. [CrossRef]
7. Martorell, L.; Martínez-González, J.; Rodríguez, C.; Gentile, M.; Calvayrac, O.; Badimon, L. Thrombin and protease-activated receptors (PARs) in atherothrombosis. *Thromb. Haemost.* **2008**, *99*, 305–315. [CrossRef] [PubMed]
8. Kataoka, H.; Hamilton, J.R.; McKemy, D.D.; Camerer, E.; Zheng, Y.-W.; Cheng, A.; Griffin, C.; Coughlin, S.R. Protease-activated receptors 1 and 4 mediate thrombin signaling in endothelial cells. *Blood* **2003**, *102*, 3224–3231. [CrossRef] [PubMed]
9. Ramachandran, R.; Noorbakhsh, F.; Defea, K.; Hollenberg, M.D. Targeting proteinase-activated receptors: Therapeutic potential and challenges. *Nat. Rev. Drug Discov.* **2012**, *11*, 69–86. [CrossRef] [PubMed]

10. Bae, J.-S.; Kim, Y.; Park, M.-K.; Rezaie, A.R. Concentration dependent dual effect of thrombin in endothelial cells via PAR-1 and PI3 kinase. *J. Cell. Physiol.* **2009**, *219*, 744–751. [CrossRef] [PubMed]

11. Wiisanen, M.E.; Moliterno, D.J. Platelet protease-activated receptor antagonism in cardiovascular medicine. *Coron. Artery Dis.* **2012**, *23*, 375–379. [CrossRef] [PubMed]

12. Isermann, B. Homeostatic effects of coagulation protease-dependent signaling and protease activated receptors. *J. Thromb. Haemost.* **2017**, *15*, 1273–1284. [CrossRef] [PubMed]

13. SenBanerjee, S.; Lin, Z.; Atkins, G.B.; Greif, D.M.; Rao, R.M.; Kumar, A.; Feinberg, M.W.; Chen, Z.; Simon, D.I.; Luscinskas, F.W.; et al. KLF2 Is a Novel Transcriptional Regulator of Endothelial Proinflammatory Activation. *J. Exp. Med.* **2004**, *199*, 1305–1315. [CrossRef] [PubMed]

14. Nayak, L.; Lin, Z.; Jain, M.K. "Go with the flow": How Krüppel-like factor 2 regulates the vasoprotective effects of shear stress. *Antioxid. Redox Signal.* **2011**, *15*, 1449–1461. [CrossRef] [PubMed]

15. Marrone, G.; Russo, L.; Rosado, E.; Hide, D.; García-Cardeña, G.; García-Pagán, J.C.; Bosch, J.; Gracia-Sancho, J. The transcription factor KLF2 mediates hepatic endothelial protection and paracrine endothelial-stellate cell deactivation induced by statins. *J. Hepatol.* **2013**, *58*, 98–103. [CrossRef] [PubMed]

16. Marrone, G.; Maeso-Díaz, R.; García-Cardena, G.; Abraldes, J.G.; García-Pagán, J.C.; Bosch, J.; Gracia-Sancho, J. KLF2 exerts antifibrotic and vasoprotective effects in cirrhotic rat livers: Behind the molecular mechanisms of statins. *Gut* **2014**. [CrossRef] [PubMed]

17. Marrone, G.; Shah, V.H.; Gracia-Sancho, J. Sinusoidal communication in liver fibrosis and regeneration. *J. Hepatol.* **2016**, *65*, 608–617. [CrossRef] [PubMed]

18. Lin, Z.; Kumar, A.; SenBanerjee, S.; Staniszewski, K.; Parmar, K.; Vaughan, D.E.; Gimbrone, M.A.; Balasubramanian, V.; García-Cardeña, G.; Jain, M.K. Kruppel-like factor 2 (KLF2) regulates endothelial thrombotic function. *Circ. Res.* **2005**, *96*, e48–e57. [CrossRef] [PubMed]

19. Uemura, M.; Tatsumi, K.; Matsumoto, M.; Fujimoto, M.; Matsuyama, T.; Ishikawa, M.; Iwamoto, T.-A.; Mori, T.; Wanaka, A.; Fukui, H.; et al. Localization of ADAMTS13 to the stellate cells of human liver. *Blood* **2005**, *106*, 922–924. [CrossRef] [PubMed]

20. Uemura, M.; Fujimura, Y.; Matsumoto, M.; Ishizashi, H.; Kato, S.; Matsuyama, T.; Isonishi, A.; Ishikawa, M.; Yagita, M.; Morioka, C.; et al. Comprehensive analysis of ADAMTS13 in patients with liver cirrhosis. *Thromb. Haemost.* **2008**, *99*, 1019–1029. [CrossRef] [PubMed]

21. Levy, G.G.; Nichols, W.C.; Lian, E.C.; Foroud, T.; McClintick, J.N.; McGee, B.M.; Yang, A.Y.; Siemieniak, D.R.; Stark, K.R.; Gruppo, R.; et al. Mutations in a member of the ADAMTS gene family cause thrombotic thrombocytopenic purpura. *Nature* **2001**, *413*, 488–494. [CrossRef] [PubMed]

22. Levi, M.; Scully, M.; Singer, M. The role of ADAMTS-13 in the coagulopathy of sepsis. *J. Thromb. Haemost.* **2018**. [CrossRef] [PubMed]

23. García-Pagán, J.-C.; Gracia-Sancho, J.; Bosch, J. Functional aspects on the pathophysiology of portal hypertension in cirrhosis. *J. Hepatol.* **2012**, *57*, 458–461. [CrossRef] [PubMed]

24. Chambers, R.C.; Dabbagh, K.; McAnulty, R.J.; Gray, A.J.; Blanc-Brude, O.P.; Laurent, G.J. Thrombin stimulates fibroblast procollagen production via proteolytic activation of protease-activated receptor 1. *Biochem. J.* **1998**, *333*, 121–127. [CrossRef] [PubMed]

25. Gaça, M.D.A.; Zhou, X.; Benyon, R.C. Regulation of hepatic stellate cell proliferation and collagen synthesis by proteinase-activated receptors. *J. Hepatol.* **2002**, *36*, 362–369. [CrossRef]

26. Fiorucci, S.; Antonelli, E.; Distrutti, E.; Severino, B.; Fiorentina, R.; Baldoni, M.; Caliendo, G.; Santagada, V.; Morelli, A.; Cirino, G. PAR1 antagonism protects against experimental liver fibrosis. Role of proteinase receptors in stellate cell activation. *Hepatology* **2004**, *39*, 365–375. [CrossRef] [PubMed]

27. Anstee, Q.M.; Dhar, A.; Thursz, M.R. The role of hypercoagulability in liver fibrogenesis. *Clin. Res. Hepatol. Gastroenterol.* **2011**, *35*, 526–533. [CrossRef] [PubMed]

28. Duplantier, J.G.; Dubuisson, L.; Senant, N.; Freyburger, G.; Laurendeau, I.; Herbert, J.-M.; Desmoulière, A.; Rosenbaum, J. A role for thrombin in liver fibrosis. *Gut* **2004**, *53*, 1682–1687. [CrossRef] [PubMed]

29. Takahara, T.; Furui, K.; Funaki, J.; Nakayama, Y.; Itoh, H.; Miyabayashi, C.; Sato, H.; Seiki, M.; Ooshima, A.; Watanabe, A. Increased expression of matrix metalloproteinase-II in experimental liver fibrosis in rats. *Hepatology* **1995**, *21*, 787–795. [PubMed]

30. Knight, V.; Lourensz, D.; Tchongue, J.; Correia, J.; Tipping, P.; Sievert, W. Cytoplasmic domain of tissue factor promotes liver fibrosis in mice. *World J. Gastroenterol.* **2017**, *23*, 5692–5699. [CrossRef] [PubMed]

31. Knight, V.; Tchongue, J.; Lourensz, D.; Tipping, P.; Sievert, W. Protease-activated receptor 2 promotes experimental liver fibrosis in mice and activates human hepatic stellate cells. *Hepatology* **2012**, *55*, 879–887. [CrossRef] [PubMed]

32. Rullier, A.; Gillibert-Duplantier, J.; Costet, P.; Cubel, G.; Haurie, V.; Petibois, C.; Taras, D.; Dugot-Senant, N.; Deleris, G.; Bioulac-Sage, P.; et al. Protease-activated receptor 1 knockout reduces experimentally induced liver fibrosis. *Am. J. Physiol. Gastrointest. Liver Physiol.* **2008**, *294*, G226–G235. [CrossRef] [PubMed]

33. Sullivan, B.P.; Weinreb, P.H.; Violette, S.M.; Luyendyk, J.P. The Coagulation System Contributes to αVβ6 Integrin Expression and Liver Fibrosis Induced by Cholestasis. *Am. J. Pathol.* **2010**, *177*, 2837–2849. [CrossRef] [PubMed]

34. Nault, R.; Fader, K.A.; Kopec, A.K.; Harkema, J.R.; Zacharewski, T.R.; Luyendyk, J.P. From the Cover: Coagulation-Driven Hepatic Fibrosis Requires Protease Activated Receptor-1 (PAR-1) in a Mouse Model of TCDD-Elicited Steatohepatitis. *Toxicol. Sci.* **2016**, *154*, 381–391. [CrossRef] [PubMed]

35. Martinelli, A.; Knapp, S.; Anstee, Q.; Worku, M.; Tommasi, A.; Zucoloto, S.; Goldin, R.; Thursz, M. Effect of a thrombin receptor (protease-activated receptor 1, PAR-1) gene polymorphism in chronic hepatitis C liver fibrosis. *J. Gastroenterol. Hepatol.* **2008**, *23*, 1403–1409. [CrossRef] [PubMed]

36. Mackman, N. Role of Tissue Factor in Hemostasis, Thrombosis, and Vascular Development. *Arterioscler. Thromb. Vasc. Biol.* **2004**, *24*, 1015–1022. [CrossRef] [PubMed]

37. Rautou, P.-E.; Tatsumi, K.; Antoniak, S.; Owens, A.P.; Sparkenbaugh, E.; Holle, L.A.; Wolberg, A.S.; Kopec, A.K.; Pawlinski, R.; Luyendyk, J.P.; et al. Hepatocyte Tissue Factor Contributes to the Hypercoagulable State in a Mouse Model of Chronic Liver Injury. *J. Hepatol.* **2015**. [CrossRef] [PubMed]

38. Gillibert-Duplantier, J.; Neaud, V.; Blanc, J.-F.; Bioulac-Sage, P.; Rosenbaum, J. Thrombin inhibits migration of human hepatic myofibroblasts. *Am. J. Physiol. Gastrointest. Liver Physiol.* **2007**, *293*, G128–G136. [CrossRef] [PubMed]

39. Wanless, I.R.; Wong, F.; Blendis, L.M.; Greig, P.; Heathcote, E.J.; Levy, G. Hepatic and portal vein thrombosis in cirrhosis: Possible role in development of parenchymal extinction and portal hypertension. *Hepatology* **1995**, *21*, 1238–1247. [PubMed]

40. Hou, P.C.; Mcfadzean, A.J. Thrombosis and Intimal Thickening in the Portal System in Cirrhosis of the Liver. *J. Pathol. Bacteriol.* **1965**, *89*, 473–480. [PubMed]

41. Wanless, I.R.; Liu, J.J.; Butany, J. Role of thrombosis in the pathogenesis of congestive hepatic fibrosis (cardiac cirrhosis). *Hepatology* **1995**, *21*, 1232–1237. [CrossRef] [PubMed]

42. Simonetto, D.A.; Yang, H.; Yin, M.; de Assuncao, T.M.; Kwon, J.H.; Hilscher, M.; Pan, S.; Yang, L.; Bi, Y.; Beyder, A.; et al. Chronic passive venous congestion drives hepatic fibrogenesis via sinusoidal thrombosis and mechanical forces. *Hepatology* **2015**, *61*, 648–659. [CrossRef] [PubMed]

43. Miyao, M.; Kotani, H.; Ishida, T.; Kawai, C.; Manabe, S.; Abiru, H.; Tamaki, K. Pivotal role of liver sinusoidal endothelial cells in NAFLD/NASH progression. *Lab. Investig.* **2015**, *95*, 1130–1144. [CrossRef] [PubMed]

44. de Ridder, G.G.; Lundblad, R.L.; Pizzo, S.V. Actions of thrombin in the interstitium. *J. Thromb. Haemost.* **2016**, *14*, 40–47. [CrossRef] [PubMed]

45. Koo, B.-H.; Han, J.H.; Yeom, Y.I.; Kim, D.-S. Thrombin-dependent MMP-2 activity is regulated by heparan sulfate. *J. Biol. Chem.* **2010**, *285*, 41270–41279. [CrossRef] [PubMed]

46. Cabrera, S.; Gaxiola, M.; Arreola, J.L.; Ramírez, R.; Jara, P.; D'Armiento, J.; Richards, T.; Selman, M.; Pardo, A. Overexpression of MMP9 in macrophages attenuates pulmonary fibrosis induced by bleomycin. *Int. J. Biochem. Cell Biol.* **2007**, *39*, 2324–2338. [CrossRef] [PubMed]

47. Villa, E.; Cammà, C.; Marietta, M.; Luongo, M.; Critelli, R.; Colopi, S.; Tata, C.; Zecchini, R.; Gitto, S.; Petta, S.; et al. Enoxaparin prevents portal vein thrombosis and liver decompensation in patients with advanced cirrhosis. *Gastroenterology* **2012**, *143*, 1253–1260.e4. [CrossRef] [PubMed]

48. La Mura, V.; Braham, S.; Tosetti, G.; Branchi, F.; Bitto, N.; Moia, M.; Fracanzani, A.L.; Colombo, M.; Tripodi, A.; Primignani, M. Harmful and Beneficial Effects of Anticoagulants in Patients With Cirrhosis and Portal Vein Thrombosis. *Clin. Gastroenterol. Hepatol.* **2018**, *16*, 1146–1152. [CrossRef] [PubMed]

49. Ho, W.K.; Hankey, G.J.; Quinlan, D.J.; Eikelboom, J.W. Risk of recurrent venous thromboembolism in patients with common thrombophilia: A systematic review. *Arch. Intern. Med.* **2006**, *166*, 729–736. [CrossRef] [PubMed]

50. Marchiori, A.; Mosena, L.; Prins, M.H.; Prandoni, P. The risk of recurrent venous thromboembolism among heterozygous carriers of factor V Leiden or prothrombin G20210A mutation. A systematic review of prospective studies. *Haematologica* **2007**, *92*, 1107–1114. [CrossRef] [PubMed]

51. Connors, J.M. Thrombophilia Testing and Venous Thrombosis. *N. Engl. J. Med.* **2017**, *377*, 1177–1187. [CrossRef] [PubMed]

52. Colucci, M.; Binetti, B.M.; Tripodi, A.; Chantarangkul, V.; Semeraro, N. Hyperprothrombinemia associated with prothrombin G20210A mutation inhibits plasma fibrinolysis through a TAFI-mediated mechanism. *Blood* **2004**, *103*, 2157–2161. [CrossRef] [PubMed]

53. Van Cott, E.M.; Khor, B.; Zehnder, J.L. Factor V Leiden. *Am. J. Hematol.* **2016**, *91*, 46–49. [CrossRef] [PubMed]

54. Wright, M.; Goldin, R.; Hellier, S.; Knapp, S.; Frodsham, A.; Hennig, B.; Hill, A.; Apple, R.; Cheng, S.; Thomas, H.; et al. Factor V Leiden polymorphism and the rate of fibrosis development in chronic hepatitis C virus infection. *Gut* **2003**, *52*, 1206–1210. [CrossRef] [PubMed]

55. Maharshak, N.; Halfon, P.; Deutsch, V.; Peretz, H.; Berliner, S.; Fishman, S.; Zelber-Sagi, S.; Rozovski, U.; Leshno, M.; Oren, R. Increased fibrosis progression rates in hepatitis C patients carrying the prothrombin G20210A mutation. *World J. Gastroenterol.* **2011**, *17*, 5007–5013. [CrossRef] [PubMed]

56. Plompen, E.P.C.; Murad, S.D.; Hansen, B.E.; Loth, D.W.; Schouten, J.N.L.; Taimr, P.; Hofman, A.; Uitterlinden, A.G.; Stricker, B.H.; Janssen, H.L.A.; et al. Prothrombotic genetic risk factors are associated with an increased risk of liver fibrosis in the general population: The Rotterdam Study. *J. Hepatol.* **2015**, *63*, 1459–1465. [CrossRef] [PubMed]

57. Assy, N.; Bekirov, I.; Mejritsky, Y.; Solomon, L.; Szvalb, S.; Hussein, O. Association between thrombotic risk factors and extent of fibrosis in patients with non-alcoholic fatty liver diseases. *World J. Gastroenterol.* **2005**, *11*, 5834–5839. [CrossRef] [PubMed]

58. D'Amico, M.; Pasta, F.; Pasta, L. Thrombophilic genetic factors PAI-1 4G-4G and MTHFR 677TT as risk factors of alcohol, cryptogenic liver cirrhosis and portal vein thrombosis, in a Caucasian population. *Gene* **2015**, *568*, 85–88. [CrossRef] [PubMed]

59. Poujol-Robert, A.; Rosmorduc, O.; Serfaty, L.; Coulet, F.; Poupon, R.; Robert, A. Genetic and acquired thrombotic factors in chronic hepatitis C. *Am. J. Gastroenterol.* **2004**, *99*, 527–531. [CrossRef] [PubMed]

60. Goulding, C.; O'Brien, C.; Egan, H.; Hegarty, J.E.; McDonald, G.; O'Farrelly, C.; White, B.; Kelleher, D.; Norris, S. The impact of inherited prothrombotic risk factors on individuals chronically infected with hepatitis C virus from a single source. *J. Viral Hepat.* **2007**, *14*, 255–259. [CrossRef] [PubMed]

61. Poujol-Robert, A.; Boëlle, P.-Y.; Wendum, D.; Poupon, R.; Robert, A. Association between ABO blood group and fibrosis severity in chronic hepatitis C infection. *Dig. Dis. Sci.* **2006**, *51*, 1633–1636. [CrossRef] [PubMed]

62. Shavakhi, A.; Hajalikhani, M.; Minakari, M.; Norian, A.; Riahi, R.; Azarnia, M.; Liaghat, L. The association of non-O blood group and severity of liver fibrosis in patients with chronic hepatitis C infection. *J. Res. Med. Sci.* **2012**, *17*, 466–469. [PubMed]

63. Koster, T.; Vandenbroucke, J.P.; Rosendaal, F.R.; Briët, E.; Rosendaal, F.R.; Blann, A.D. Role of clotting factor VIII in effect of von Willebrand factor on occurrence of deep-vein thrombosis. *Lancet* **1995**, *345*, 152–155. [CrossRef]

64. O'Donnell, J.; Laffan, M.A. The relationship between ABO histo-blood group, factor VIII and von Willebrand factor. *Transfus. Med.* **2001**, *11*, 343–351. [CrossRef] [PubMed]

65. Tripodi, A.; Mannucci, P.M. The coagulopathy of chronic liver disease. *N. Engl. J. Med.* **2011**, *365*, 147–156. [CrossRef] [PubMed]

66. Ratnoff, O.D.; Patek, A.J. The Natural History of Laennec's Cirrhosis of the Liver an Analysis of 386 Cases. *Medicine* **1942**, *21*, 207–268. [CrossRef]

67. La Mura, V.; Nicolini, A.; Tosetti, G.; Primignani, M. Cirrhosis and portal hypertension: The importance of risk stratification, the role of hepatic venous pressure gradient measurement. *World J. Hepatol.* **2015**, *7*, 688–695. [CrossRef] [PubMed]

68. Tripodi, A.; Salerno, F.; Chantarangkul, V.; Clerici, M.; Cazzaniga, M.; Primignani, M.; Mannuccio Mannucci, P. Evidence of normal thrombin generation in cirrhosis despite abnormal conventional coagulation tests. *Hepatology* **2005**, *41*, 553–558. [CrossRef] [PubMed]

69. Lisman, T.; Porte, R.J. Rebalanced hemostasis in patients with liver disease: Evidence and clinical consequences. *Blood* **2010**, *116*, 878–885. [CrossRef] [PubMed]

70. Tripodi, A.; Primignani, M.; Mannucci, P.M.; Caldwell, S.H. Changing Concepts of Cirrhotic Coagulopathy. *Am. J. Gastroenterol.* **2017**, *112*, 274–281. [CrossRef] [PubMed]

71. Tripodi, A.; Primignani, M.; Chantarangkul, V.; Dell'Era, A.; Clerici, M.; de Franchis, R.; Colombo, M.; Mannucci, P.M. An imbalance of pro- vs. anti-coagulation factors in plasma from patients with cirrhosis. *Gastroenterology* **2009**, *137*, 2105–2111. [CrossRef] [PubMed]

72. Tripodi, A.; Primignani, M.; Lemma, L.; Chantarangkul, V.; Mannucci, P.M. Evidence that low protein C contributes to the procoagulant imbalance in cirrhosis. *J. Hepatol.* **2013**, *59*, 265–270. [CrossRef] [PubMed]

73. Ambrosino, P.; Tarantino, L.; Minno, G.D.; Paternoster, M.; Graziano, V.; Petitto, M.; Nasto, A.; Minno, M.N.D.D. The risk of venous thromboembolism in patients with cirrhosis. *Thromb. Haemost.* **2017**, *117*, 139–148. [CrossRef] [PubMed]

74. Nonami, T.; Yokoyama, I.; Iwatsuki, S.; Starzl, T.E. The Incidence of Portal Vein Thrombosis at Liver Transplantation. *Hepatology* **1992**, *16*, 1195–1198. [CrossRef] [PubMed]

75. Tsochatzis, E.A.; Senzolo, M.; Germani, G.; Gatt, A.; Burroughs, A.K. Systematic review: Portal vein thrombosis in cirrhosis. *Aliment. Pharmacol. Ther.* **2010**, *31*, 366–374. [CrossRef] [PubMed]

76. Northup, P.G.; McMahon, M.M.; Ruhl, A.P.; Altschuler, S.E.; Volk-Bednarz, A.; Caldwell, S.H.; Berg, C.L. Coagulopathy does not fully protect hospitalized cirrhosis patients from peripheral venous thromboembolism. *Am. J. Gastroenterol.* **2006**, *101*, 1524–1528. [CrossRef] [PubMed]

77. Søgaard, K.K.; Horváth-Puhó, E.; Grønbaek, H.; Jepsen, P.; Vilstrup, H.; Sørensen, H.T. Risk of venous thromboembolism in patients with liver disease: A nationwide population-based case-control study. *Am. J. Gastroenterol.* **2009**, *104*, 96–101. [CrossRef] [PubMed]

78. Wu, H.; Nguyen, G.C. Liver cirrhosis is associated with venous thromboembolism among hospitalized patients in a nationwide US study. *Clin. Gastroenterol. Hepatol.* **2010**, *8*, 800–805. [CrossRef] [PubMed]

79. Ferro, D.; Quintarelli, C.; Lattuada, A.; Leo, R.; Alessandroni, M.; Mannucci, P.M.; Violi, F. High plasma levels of von Willebrand factor as a marker of endothelial perturbation in cirrhosis: Relationship to endotoxemia. *Hepatology* **1996**, *23*, 1377–1383. [CrossRef] [PubMed]

80. Albornoz, L.; Alvarez, D.; Otaso, J.C.; Gadano, A.; Salviú, J.; Gerona, S.; Sorroche, P.; Villamil, A.; Mastai, R. Von Willebrand factor could be an index of endothelial dysfunction in patients with cirrhosis: Relationship to degree of liver failure and nitric oxide levels. *J. Hepatol.* **1999**, *30*, 451–455. [CrossRef]

81. La Mura, V.; Reverter, J.C.; Flores-Arroyo, A.; Raffa, S.; Reverter, E.; Seijo, S.; Abraldes, J.G.; Bosch, J.; García-Pagán, J.C. Von Willebrand factor levels predict clinical outcome in patients with cirrhosis and portal hypertension. *Gut* **2011**, *60*, 1133–1138. [CrossRef] [PubMed]

82. Ferlitsch, M.; Reiberger, T.; Hoke, M.; Salzl, P.; Schwengerer, B.; Ulbrich, G.; Payer, B.A.; Trauner, M.; Peck-Radosavljevic, M.; Ferlitsch, A. von Willebrand factor as new noninvasive predictor of portal hypertension, decompensation and mortality in patients with liver cirrhosis. *Hepatology* **2012**, *56*, 1439–1447. [CrossRef] [PubMed]

83. La Mura, V.; Tripodi, A.; Tosetti, G.; Cavallaro, F.; Chantarangkul, V.; Colombo, M.; Primignani, M. Resistance to thrombomodulin is associated with de novo portal vein thrombosis and low survival in patients with cirrhosis. *Liver Int.* **2016**, *36*, 1322–1330. [CrossRef] [PubMed]

84. Kalambokis, G.N.; Oikonomou, A.; Baltayiannis, G.; Christou, L.; Kolaitis, N.I.; Tsianos, E.V. Thrombin generation measured as thrombin-antithrombin complexes predicts clinical outcomes in patients with cirrhosis. *Hepatol. Res.* **2015**. [CrossRef] [PubMed]

85. Kalambokis, G.N.; Oikonomou, A.; Christou, L.; Kolaitis, N.I.; Tsianos, E.V.; Christodoulou, D.; Baltayiannis, G. von Willebrand factor and procoagulant imbalance predict outcome in patients with cirrhosis and thrombocytopenia. *J. Hepatol.* **2016**, *65*, 921–928. [CrossRef] [PubMed]

86. Maieron, A.; Salzl, P.; Peck-Radosavljevic, M.; Trauner, M.; Hametner, S.; Schöfl, R.; Ferenci, P.; Ferlitsch, M. Von Willebrand Factor as a new marker for non-invasive assessment of liver fibrosis and cirrhosis in patients with chronic hepatitis C. *Aliment. Pharmacol. Ther.* **2014**, *39*, 331–338. [CrossRef] [PubMed]

87. Hametner, S.; Ferlitsch, A.; Ferlitsch, M.; Etschmaier, A.; Schöfl, R.; Ziachehabi, A.; Maieron, A. The VITRO Score (Von Willebrand Factor Antigen/Thrombocyte Ratio) as a New Marker for Clinically Significant Portal Hypertension in Comparison to Other Non-Invasive Parameters of Fibrosis Including ELF Test. *PLoS ONE* **2016**, *11*, e0149230. [CrossRef] [PubMed]

88. Tripodi, A.; Fracanzani, A.L.; Primignani, M.; Chantarangkul, V.; Clerici, M.; Mannucci, P.M.; Peyvandi, F.; Bertelli, C.; Valenti, L.; Fargion, S. Procoagulant imbalance in patients with non-alcoholic fatty liver disease. *J. Hepatol.* **2014**, *61*, 148–154. [CrossRef] [PubMed]

89. Wong, R.J.; Aguilar, M.; Cheung, R.; Perumpail, R.B.; Harrison, S.A.; Younossi, Z.M.; Ahmed, A. Nonalcoholic steatohepatitis is the second leading etiology of liver disease among adults awaiting liver transplantation in the United States. *Gastroenterology* **2015**, *148*, 547–555. [CrossRef] [PubMed]

90. Younossi, Z.M.; Koenig, A.B.; Abdelatif, D.; Fazel, Y.; Henry, L.; Wymer, M. Global Epidemiology of Non-Alcoholic Fatty Liver Disease–Meta-Analytic Assessment of Prevalence, Incidence and Outcomes. *Hepatology* **2015**. [CrossRef]

91. Potze, W.; Siddiqui, M.S.; Boyett, S.L.; Adelmeijer, J.; Daita, K.; Sanyal, A.J.; Lisman, T. Preserved hemostatic status in patients with non-alcoholic fatty liver disease. *J. Hepatol.* **2016**, *65*, 980–987. [CrossRef] [PubMed]

92. Tripodi, A.; Fracanzani, A.L.; Chantarangkul, V.; Primignani, M.; Fargion, S. Procoagulant imbalance in patients with non-alcoholic fatty liver disease. *J. Hepatol.* **2017**, *66*, 248–250. [CrossRef] [PubMed]

93. Potze, W.; Sanyal, A.J.; Lisman, T. Reply to: "Procoagulant imbalance in patients with non-alcoholic fatty liver disease.". *J. Hepatol.* **2017**, *66*, 250–251. [CrossRef] [PubMed]

94. Bruno, S.; Di Marco, V.; Iavarone, M.; Roffi, L.; Crosignani, A.; Calvaruso, V.; Aghemo, A.; Cabibbo, G.; Viganò, M.; Boccaccio, V.; et al. Survival of patients with HCV cirrhosis and sustained virologic response is similar to the general population. *J. Hepatol.* **2016**, *64*, 1217–1223. [CrossRef] [PubMed]

95. Schuppan, D.; Pinzani, M. Anti-fibrotic therapy: Lost in translation? *J. Hepatol.* **2012**, *56* (Suppl. 1), S66–S74. [CrossRef]

96. Schuppan, D.; Ashfaq-Khan, M.; Yang, A.T.; Kim, Y.O. Liver fibrosis: Direct antifibrotic agents and targeted therapies. *Matrix Biol.* **2018**. [CrossRef] [PubMed]

97. Hirsh, J.; Levine, M.N. Low molecular weight heparin. *Blood* **1992**, *79*, 1–17. [CrossRef] [PubMed]

98. Li, C.-J.; Yang, Z.-H.; Shi, X.-L.; Liu, D.-L. Effects of aspirin and enoxaparin in a rat model of liver fibrosis. *World J. Gastroenterol.* **2017**, *23*, 6412–6419. [CrossRef] [PubMed]

99. Abe, W.; Ikejima, K.; Lang, T.; Okumura, K.; Enomoto, N.; Kitamura, T.; Takei, Y.; Sato, N. Low molecular weight heparin prevents hepatic fibrogenesis caused by carbon tetrachloride in the rat. *J. Hepatol.* **2007**, *46*, 286–294. [CrossRef] [PubMed]

100. Cerini, F.; Vilaseca, M.; Lafoz, E.; García-Irigoyen, O.; García-Calderó, H.; Tripathi, D.M.; Avila, M.; Reverter, J.C.; Bosch, J.; Gracia-Sancho, J.; et al. Enoxaparin reduces hepatic vascular resistance and portal pressure in cirrhotic rats. *J. Hepatol.* **2016**, *64*, 834–842. [CrossRef] [PubMed]

101. Anstee, Q.M.; Goldin, R.D.; Wright, M.; Martinelli, A.; Cox, R.; Thursz, M.R. Coagulation status modulates murine hepatic fibrogenesis: Implications for the development of novel therapies. *J. Thromb. Haemost.* **2008**, *6*, 1336–1343. [CrossRef] [PubMed]

102. Kassel, K.M.; Sullivan, B.P.; Cui, W.; Copple, B.L.; Luyendyk, J.P. Therapeutic administration of the direct thrombin inhibitor argatroban reduces hepatic inflammation in mice with established fatty liver disease. *Am. J. Pathol.* **2012**, *181*, 1287–1295. [CrossRef] [PubMed]

103. Vilaseca, M.; García-Calderó, H.; Lafoz, E.; García-Irigoyen, O.; Avila, M.A.; Reverter, J.C.; Bosch, J.; Hernández-Gea, V.; Gracia-Sancho, J.; García-Pagán, J.C. The anticoagulant rivaroxaban lowers portal hypertension in cirrhotic rats mainly by deactivating hepatic stellate cells. *Hepatology* **2017**, *65*, 2031–2044. [CrossRef] [PubMed]

104. Bell, R.G.; Sadowski, J.A.; Matschiner, J.T. Mechanism of action of warfarin. Warfarin and metabolism of vitamin K1. *Biochemistry* **1972**, *11*, 1959–1961. [CrossRef] [PubMed]

105. Kirkwood, T.B. Calibration of reference thromboplastins and standardisation of the prothrombin time ratio. *Thromb. Haemost.* **1983**, *49*, 238–244. [CrossRef] [PubMed]

106. Ansell, J.; Hirsh, J.; Hylek, E.; Jacobson, A.; Crowther, M.; Palareti, G. Pharmacology and management of the vitamin K antagonists: American College of Chest Physicians Evidence-Based Clinical Practice Guidelines (8th Edition). *Chest* **2008**, *133*, 160S–198S. [CrossRef] [PubMed]

107. Barnes, G.D.; Kurtz, B. Direct oral anticoagulants: Unique properties and practical approaches to management. *Heart* **2016**, *102*, 1620–1626. [CrossRef] [PubMed]

108. Barnes, G.D.; Lucas, E.; Alexander, G.C.; Goldberger, Z.D. National Trends in Ambulatory Oral Anticoagulant Use. *Am. J. Med.* **2015**, *128*, 1300–1305.e2. [CrossRef] [PubMed]

109. Intagliata, N.M.; Maitland, H.; Caldwell, S.H. Direct Oral Anticoagulants in Cirrhosis. *Curr. Treat. Opt. Gastroenterol.* **2016**, *14*, 247–256. [CrossRef] [PubMed]
110. Intagliata, N.M.; Henry, Z.H.; Maitland, H.; Shah, N.L.; Argo, C.K.; Northup, P.G.; Caldwell, S.H. Direct Oral Anticoagulants in Cirrhosis Patients Pose Similar Risks of Bleeding When Compared to Traditional Anticoagulation. *Dig. Dis. Sci.* **2016**. [CrossRef] [PubMed]
111. Hum, J.; Shatzel, J.J.; Jou, J.H.; Deloughery, T.G. The efficacy and safety of direct oral anticoagulants vs. traditional anticoagulants in cirrhosis. *Eur. J. Haematol.* **2017**, *98*, 393–397. [CrossRef] [PubMed]
112. De Gottardi, A.; Trebicka, J.; Klinger, C.; Plessier, A.; Seijo, S.; Terziroli, B.; Magenta, L.; Semela, D.; Buscarini, E.; Langlet, P.; et al. Antithrombotic treatment with direct-acting oral anticoagulants in patients with splanchnic vein thrombosis and cirrhosis. *Liver Int.* **2017**, *37*, 694–699. [CrossRef] [PubMed]
113. Marques, P.E.; Antunes, M.M.; David, B.A.; Pereira, R.V.; Teixeira, M.M.; Menezes, G.B. Imaging liver biology in vivo using conventional confocal microscopy. *Nat. Protoc.* **2015**, *10*, 258–268. [CrossRef] [PubMed]
114. Wang, J.; Hossain, M.; Thanabalasuriar, A.; Gunzer, M.; Meininger, C.; Kubes, P. Visualizing the function and fate of neutrophils in sterile injury and repair. *Science* **2017**, *358*, 111–116. [CrossRef] [PubMed]
115. McDonald, B.; Jenne, C.N.; Zhuo, L.; Kimata, K.; Kubes, P. Kupffer cells and activation of endothelial TLR4 coordinate neutrophil adhesion within liver sinusoids during endotoxemia. *Am. J. Physiol. Gastrointest. Liver Physiol.* **2013**, *305*, G797–G806. [CrossRef] [PubMed]
116. Kolaczkowska, E.; Jenne, C.N.; Surewaard, B.G.J.; Thanabalasuriar, A.; Lee, W.-Y.; Sanz, M.-J.; Mowen, K.; Opdenakker, G.; Kubes, P. Molecular mechanisms of NET formation and degradation revealed by intravital imaging in the liver vasculature. *Nat. Commun.* **2015**, *6*, 6673. [CrossRef] [PubMed]
117. Weber, C. Liver: Neutrophil extracellular traps mediate bacterial liver damage. *Nat. Rev. Gastroenterol. Hepatol.* **2015**, *12*, 251. [CrossRef] [PubMed]
118. Andrews, R.K.; Arthur, J.F.; Gardiner, E.E. Neutrophil extracellular traps (NETs) and the role of platelets in infection. *Thromb. Haemost.* **2014**, *112*, 659–665. [CrossRef] [PubMed]
119. Brinkmann, V.; Reichard, U.; Goosmann, C.; Fauler, B.; Uhlemann, Y.; Weiss, D.S.; Weinrauch, Y.; Zychlinsky, A. Neutrophil extracellular traps kill bacteria. *Science* **2004**, *303*, 1532–1535. [CrossRef] [PubMed]
120. Yipp, B.G.; Kubes, P. NETosis: How vital is it? *Blood* **2013**, *122*, 2784–2794. [CrossRef] [PubMed]
121. Healy, L.D.; Puy, C.; Itakura, A.; Chu, T.; Robinson, D.K.; Bylund, A.; Phillips, K.G.; Gardiner, E.E.; McCarty, O.J.T. Colocalization of neutrophils, extracellular DNA and coagulation factors during NETosis: Development and utility of an immunofluorescence-based microscopy platform. *J. Immunol. Methods* **2016**. [CrossRef] [PubMed]
122. Deppermann, C.; Kubes, P. Platelets and infection. *Semin. Immunol.* **2016**. [CrossRef] [PubMed]
123. Ward, C.M.; Tetaz, T.J.; Andrews, R.K.; Berndt, M.C. Binding of the von Willebrand factor A1 domain to histone. *Thromb. Res.* **1997**, *86*, 469–477. [CrossRef]
124. Engelmann, B.; Massberg, S. Thrombosis as an intravascular effector of innate immunity. *Nat. Rev. Immunol.* **2013**, *13*, 34–45. [CrossRef] [PubMed]
125. Diaz, J.A.; Fuchs, T.A.; Jackson, T.O.; Kremer Hovinga, J.A.; Lämmle, B.; Henke, P.K.; Myers, D.D.; Wagner, D.D.; Wakefield, T.W.; Michigan Research Venous Group. Plasma DNA is Elevated in Patients with Deep Vein Thrombosis. *J. Vasc. Surg. Venous Lymphat Disord.* **2013**, *1*. [CrossRef]
126. Liaw, P.C.; Ito, T.; Iba, T.; Thachil, J.; Zeerleder, S. DAMP and DIC: The role of extracellular DNA and DNA-binding proteins in the pathogenesis of DIC. *Blood Rev.* **2015**. [CrossRef] [PubMed]
127. Yang, C.; Sun, W.; Cui, W.; Li, X.; Yao, J.; Jia, X.; Li, C.; Wu, H.; Hu, Z.; Zou, X. Procoagulant role of neutrophil extracellular traps in patients with gastric cancer. *Int. J. Clin. Exp. Pathol.* **2015**, *8*, 14075–14086. [PubMed]
128. Michels, A.; Albánez, S.; Mewburn, J.; Nesbitt, K.; Gould, T.J.; Liaw, P.C.; James, P.D.; Swystun, L.L.; Lillicrap, D. Histones link inflammation and thrombosis through the induction of Weibel-Palade Body exocytosis. *J. Thromb. Haemost.* **2016**. [CrossRef] [PubMed]
129. McDonald, B.; Davis, R.P.; Kim, S.-J.; Tse, M.; Esmon, C.T.; Kolaczkowska, E.; Jenne, C.N. Platelets and neutrophil extracellular traps collaborate to promote intravascular coagulation during sepsis in mice. *Blood* **2017**. [CrossRef] [PubMed]
130. Cirera, I.; Bauer, T.M.; Navasa, M.; Vila, J.; Grande, L.; Taurá, P.; Fuster, J.; García-Valdecasas, J.C.; Lacy, A.; Suárez, M.J.; et al. Bacterial translocation of enteric organisms in patients with cirrhosis. *J. Hepatol.* **2001**, *34*, 32–37. [CrossRef]

131. Wiest, R.; Lawson, M.; Geuking, M. Pathological bacterial translocation in liver cirrhosis. *J. Hepatol.* **2014**, *60*, 197–209. [CrossRef] [PubMed]
132. Bellot, P.; Francés, R.; Such, J. Pathological bacterial translocation in cirrhosis: Pathophysiology, diagnosis and clinical implications. *Liver Int.* **2013**, *33*, 31–39. [CrossRef] [PubMed]
133. Bellot, P.; García-Pagán, J.C.; Francés, R.; Abraldes, J.G.; Navasa, M.; Pérez-Mateo, M.; Such, J.; Bosch, J. Bacterial DNA translocation is associated with systemic circulatory abnormalities and intrahepatic endothelial dysfunction in patients with cirrhosis. *Hepatology* **2010**, *52*, 2044–2052. [CrossRef] [PubMed]
134. Bernardi, M.; Moreau, R.; Angeli, P.; Schnabl, B.; Arroyo, V. Mechanisms of decompensation and organ failure in cirrhosis: From peripheral arterial vasodilation to systemic inflammation hypothesis. *J. Hepatol.* **2015**, *63*, 1272–1284. [CrossRef] [PubMed]
135. Violi, F.; Ferro, D.; Basili, S.; Saliola, M.; Quintarelli, C.; Alessandri, C.; Cordova, C. Association between low-grade disseminated intravascular coagulation and endotoxemia in patients with liver cirrhosis. *Gastroenterology* **1995**, *109*, 531–539. [CrossRef]
136. Ferro, D.; Basili, S.; Lattuada, A.; Mantovani, B.; Bellomo, A.; Mannucci, P.M.; Violi, F. Systemic clotting activation by low-grade endotoxaemia in liver cirrhosis: A potential role for endothelial procoagulant activation. *Ital. J. Gastroenterol. Hepatol.* **1997**, *29*, 434–440. [PubMed]
137. Raparelli, V.; Basili, S.; Carnevale, R.; Napoleone, L.; Del Ben, M.; Nocella, C.; Bartimoccia, S.; Lucidi, C.; Talerico, G.; Riggio, O.; et al. Low-grade endotoxemia and platelet activation in cirrhosis. *Hepatology* **2016**. [CrossRef] [PubMed]
138. Carnevale, R.; Raparelli, V.; Nocella, C.; Bartimoccia, S.; Novo, M.; Severino, A.; De Falco, E.; Cammisotto, V.; Pasquale, C.; Crescioli, C.; et al. Gut-derived endotoxin stimulates factor VIII secretion from endothelial cells. Implications for hypercoagulability in cirrhosis. *J. Hepatol.* **2017**, *67*, 950–956. [CrossRef] [PubMed]

cells

MDPI

Article

New Insights into the Occurrence of Matrix Metalloproteases -2 and -9 in a Cohort of Breast Cancer Patients and Proteomic Correlations

Gianluca Di Cara [1], Maria Rita Marabeti [1], Rosa Musso [1], Ignazio Riili [2], Patrizia Cancemi [3,*] and Ida Pucci Minafra [1,*]

[1] Centro di Oncobiologia Sperimentale, Università di Palermo, 90146 Palermo, Italy; lucadicar@gmail.com (G.D.C.); marabetimariarita@yahoo.it (M.R.M.); rosi82.m@libero.it (R.M.)
[2] La Maddalena Hospital, 90146 Palermo, Italy; urp@lamaddalenanet.it
[3] Dipartimento di Scienze e Tecnologie Biologiche Chimiche e Farmaceutiche, Università di Palermo, 90100 Palermo, Italy
*   Correspondence: patrizia.cancemi@unipa.it (P.C.); pucci.ida@gmail.com (I.P.M.)

Received: 8 June 2018; Accepted: 24 July 2018; Published: 28 July 2018

Abstract: Matrix metalloproteases (MMPs) are a family of well-known enzymes which operate prevalently in the extracellular domain, where they fulfil the function of remodeling the extracellular matrix (ECM). Within the 26 family members, encoded by 24 genes in humans, MMP-2 and MMP-9 have been regarded as primarily responsible for the basement membrane and peri-cellular ECM rearrangement. In cases of infiltrating carcinomas, which arise from the epithelial tissues of a gland or of an internal organ, a marked alteration of the expression and the activity levels of both MMPs is known to occur. The present investigation represents the continuation and upgrading of our previous studies, now focusing on the occurrence and intensity levels of MMP-2 and -9 and their proteomic correlations in a cohort of 80 breast cancer surgical tissues.

Keywords: matrix metalloproteases; breast cancer; proteomics

1. Introduction

Breast cancer (BC) is one of the most common types of cancer in women, resulting in more than one million cases annually with disparities in incidence and mortality worldwide [1,2]. BC is a heterogeneous disease, both at the inter- and intra-tumoral levels, leading to different prognostic implications: indeed, some carcinoma subtypes have a benign course, while others behave very aggressively and are potentially metastatic, causing treatment failure and probable mortality [3]. Usually, the prognosis for patients becomes worse in cases of late diagnosis [4]. Thus, periodical screening is of great importance for the detection of early-stage tumors. Concurrently, a fundamental goal for both the diagnosis and therapy of patients is to increase the knowledge of cancer biology.

Modern cancer research aims to provide information on the mechanisms leading to cell transformation and tumor progression to be used for the development of new markers and new therapies. Several traditional prognostic markers for breast carcinoma, including tumor size, nodal involvement, histological tumor type, and differentiation, as well as the age of the patients at diagnosis, are currently in use. Furthermore, biological markers which are predictive for treatments, such as the hormone-dependency of the tumor and its HER-2 expression, are of increasing importance [5,6]. However, new markers are still necessary to better identify and to distinguish patients who need the highly intensive treatments from those who could be spared the most toxic treatments.

Metastasis is the leading cause of death in patients with cancer. Thus, a crucial aspect of tumor progression is the ability of cancer cells to cross tissue barriers and spread to distant anatomical sites.

The extracellular matrix (ECM) is the first barrier against cell invasion. Therefore, the proteolytic degradation of the ECM is a key aspect of tumor progression. Matrix metalloproteinases (MMPs) are a family of highly homologous, zinc- and calcium-dependent extracellular enzymes [7]. They are classified into five subgroups (collagenases, gelatinases, stromelysins, matrilysins and membrane metalloproteinases) based on substrate specificity, protein domain structure, sequence homology and the ability/inability to be secreted. Therefore, MMPs may play a critical role in the conversion of an "in situ" breast cancer to an invasive lesion. In particular, MMP-2 and -9 can break down several collagenous components of the basement membranes that surround the tissue confines [8,9]. Several studies in the past decade have shown that these enzymes are involved in breast cancer initiation and growth through the interaction with tumor suppressor genes involved in the early stage of tumorigenesis [10]. Several studies correlated increased levels of the gelatinases with higher metastatic spread and reduced survival [11,12].

However, the complex molecular networks that supervise the "entry into play" of these lytic enzymes is not yet fully understood.

The purpose of this study was to examine the activity levels of MMP-2 and MMP-9 in a cohort of 80 surgical samples of breast cancer patients, and to uncover their putative correlations with the proteomic assembly of the same patients.

2. Materials and Methods

2.1. Patients and Tissue Samples

In this study, we examined the activity levels of gelatinases in tissue samples from 80 breast cancer (BC) patients, diagnosed as ductal infiltrating carcinomas with histological grading G2/G3. No other clinical markers were included in this study, because they were not required for the aim of the present research. The 30 adjacent non-tumor tissues were located at least 5 cm away from the primary tumor.

The study was carried out after fulfilling all required ethical standards with the informed consent of patients. The tissue samples were obtained following surgical interventions at "La Maddalena" Hospital of Palermo and were immediately cryo-preserved at $-80\,^{\circ}\text{C}$ until use. All samples were intended to be discarded after the completion of the histopathological examinations.

The patients in this study did not receive any cytotoxic/endocrine treatment prior to surgery.

2.2. Tissue Processing

The frozen samples were weighed, cut into small pieces, homogenized in an ice bath with an extracting buffer (50 mM Tris, 0.003% penicillin, 0.005% streptomycin, pH 7.5) and mixed overnight at $4\,^{\circ}\text{C}$. After incubation, tissue lysates were centrifuged several times at 10,000 rpm for 20 min at $4\,^{\circ}\text{C}$ in order to remove debris. Supernatants containing proteins were collected and dialyzed against Millipore water for 24 h at $4\,^{\circ}\text{C}$, with several changes. Samples were then freeze-dried and stored at $-80\,^{\circ}\text{C}$. For the analyses, samples were solubilized in 50 mM Tris pH 7.5, and the protein content was quantified by Bradford assay [13].

2.3. Gelatin Zymography

Protein samples (18 µg) in Tris pH 7.5 were solubilized in Laemmli sample buffer under non-reducing conditions and loaded onto SDS-PAGE on 7.5% polyacrylamide gels co-polymerized with 0.1% gelatin at 150 V in a Tris-glycine buffer. Purified MMP-2 and MMP-9 [14] were used as references.

After electrophoresis, gels were washed in a washing buffer (2.5% Triton X-100, 50 mM Tris, pH 7.5) for 1 h to remove SDS and allow a partial renaturation of the protein. Gels were then incubated overnight at $37\,^{\circ}\text{C}$ in an incubation buffer (0.15 M NaCl, 10 mM CaCl$_2$, 50 mM Tris, pH 7.5) that allows the activation of the metalloproteinases. Subsequently, gels were fixed and stained with 0.2% Coomassie Brilliant Blue R-250 in 40% methanol and 10% acetic acid, and destained in 7% methanol and 5% acetic acid. The location of gelatinolytic activity was detectable as clear bands against the blue background of stained gelatin.

2.4. Quantification of Enzymatic Activity

Following zymography, the degree of gelatin digestion was quantified using ImageMaster 2D Platinum software, Version 5.0 (Amersham, Little Chalfont, UK). The image was digitally inverted, so that the integration of bands was reported as positive values. The pixel density was determined after background subtraction and used to calculate the integrated density of a selected band. Values of integrated density with area were reported in volume units, as calculated by the Image Master algorithms.

2.5. Two-Dimensional Gel Electrophoresis

The protein extracts were dissolved as reported previously [15]. Protein samples (45 μg for the analytical gels, or 1.5 mg for preparative gels) were rehydrated in a solution containing 8 M urea, 2% CHAPS, 10 mM DTE and 0.5% carrier ampholytes (Resolyte 3.5–10; Amersham, Little Chalfont, UK), and applied to the strips for isoelectrofocusing (IEF) (18 cm long, pH range 3.0–10, Bio-Rad, Hercules, CA, USA). After the IEF, the strips were incubated in a solution composed of 50 mM Tris-HCl pH 6.8, 6 M urea, 0.5% SDS, 30% glycerol, 130 mM DTE and 135 mM iodoacetamide (Sigma-Aldrich, St. Louis, MO, USA).

The strips containing the focused proteins were then loaded onto 9–16% linear gradient polyacrylamide gels (SDS-PAGE) and the electrophoresis run was performed by applying a constant current of 20 mA/gel at 10 °C. The protein spots were revealed by ammoniacal silver staining [14]. Silver-stained gels were analyzed with ImageMaster 2D Platinum software with the support of the ExPaSy molecular biology server.

2.6. Protein Identification and Functional Association

The spots of interest were submitted to peptide mass fingerprinting using the Voyager DE-PRO mass spectrometer (MALDI-TOF, AbSciex, Framingham, MA, USA) as described [16]. The in-gel digestion of the selected protein spots was performed with sequencing-grade trypsin (Promega, Madison, WI, USA). The peptides were re-dissolved in 0.1% trifluoroacetic acid (TFA) and spotted in a HCCA (R-cyano-4-hydroxycinnamic acid) matrix (Sigma-Aldrich). The mass spectra were recorded in the 500–5000 Da range, using a minimum of 150 shots of laser per spectrum. Internal calibration was performed using trypsin autolysis fragments (m/z 842.5100, 1045.5642, and 2211.1046 Da). Peptide mass fingerprinting was compared to the theoretical masses from the Swiss-Prot databases using the Mascot algorithms [17]. Search parameters were: 50 ppm of mass tolerance, carbamidomethylation of cysteine residues, oxidation of methionine residues and one missed enzymatic cleavage for trypsin. A minimum of four peptide mass hits was required for a match.

Protein-protein interactions were deduced by the informatic platform of the STRING database [18].

3. Results

3.1. Activity Levels of MMP-2 and MMP-9 in Breast Cancer Tissues and Their Adjacent Non-Tumoral Tissues

Firstly, we investigated the collagenolytic activities in tumor tissues compared to their healthy counterparts adjacent to the tumor. Supplementary Table S1 reports the current clinical parameters of the selected patients.

For this purpose, we performed gelatin zymographies in 30 pairs of tumoral and non-tumoral tissues run in parallel with purified MMP-2 and MMP-9 as a standard. As shown in Figure 1, the majority of the tumoral tissues, contrary to their non-tumoral counterparts, were positive for the lytic activities corresponding to the latent and activated form of MMP-2 (72 kDa and 62 kDa, respectively) and the latent and activated form of MMP-9 (92 kDa and 82 kDa, respectively).

Occasionally, lytic bands of higher molecular weight also occurred in some patients. These bands may represent homodimeric forms of MMP-9 and complexes of pro-MMP-9/TIMP-1, respectively [14].

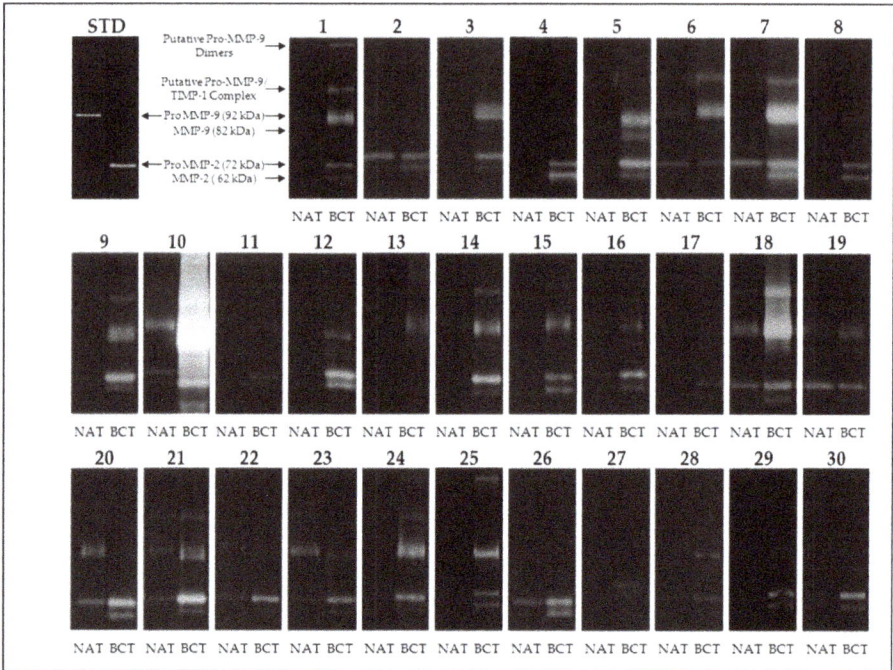

Figure 1. Gelatin zymography of 30 selected breast cancer tissues (BCT) and their non-tumoral adjacent tissues (NAT). Samples were run in parallel with purified MMP-2 and MMP-9 as a standard (lanes STD). The higher molecular weight (MW) of the lytic bands were previously identified as MMP-9 dimers and MMP-9/TIMP-1 complexes [14].

Gelatin zymograms were subjected to densitometric analysis to quantify the activity levels as relative volumes, using the Image Master software. The activity levels of pro-MMP-2, MMP-2, pro-MMP-9, and MMP-9 in tumor tissues were compared with the corresponding activities of the adjacent non-tumoral tissues when positive for lytic activity. As shown in the graphs in Figure 2, the activity levels of MMP-2 and MMP-9 were much higher in the tumoral tissues, compared to the positive non-tumoral tissues.

3.2. Gelatinolytic Activities of MMP-2 and MMP-9 in Breast Cancer Patients

We further extended the analysis to 80 BC surgical tissues to evaluate the activity levels of gelatinases in a significant number of cases. Supplementary Table S2 reports the current clinical parameters of the selected patients.

As shown in Figure 3, the gelatinolytic activity was detected in almost all breast cancer tissues analyzed, even though some variations concerning the presence and intensity levels of the zymographic bands were seen. The diagrams in Figure 4 also show this heterogeneity among the selected cases.

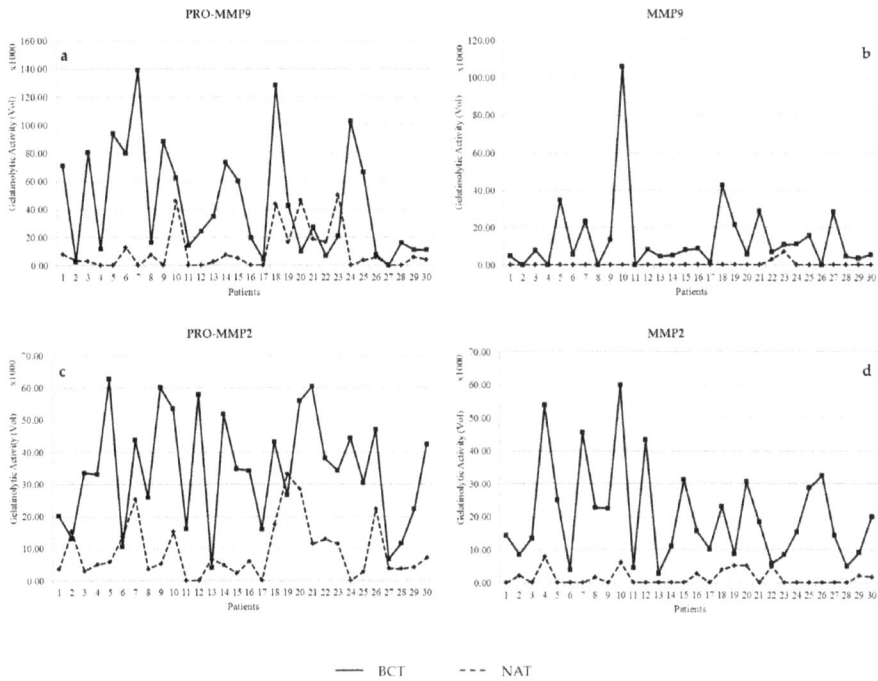

Figure 2. Densitometric profiles of the gelatinolytic bands corresponding to pro-MMP-9 (**a**); MMP-9 (**b**); pro-MMP-2 (**c**); MMP-2 (**d**) in the 30 breast cancer tissues compared to the respective non-tumoral adjacent tissues. Note that the active forms of both the MMPs are absent or rarely present in the non-tumoral tissues.

In addition, we compared the activity levels of the collagenolytic bands within the group of 80 patients in order to find possible cross-correlations between them. The volumes derived from the densitometric analyses of each lytic band, as described in "Materials and Methods", were used for the measurement of the Pearson's correlation coefficient (*r*) by using the "Graphpad" system (PRISM4 Demo software).

The results showed a weak linear correlation between the pro-MMP2 and pro-MMP9, a significant correlation between the pro-enzymatic forms and their respective activated forms, and finally a moderate correlation between the two activated enzymes (Table 1).

Table 1. Statistical correlations (Pearson *r*) between the activity levels of pro-MMP-9/pro-MMP-2; pro-MMP-9/MMP-9; pro-MMP-2/MMP-2 and MMP-9/MMP-2, detected in the breast cancer extracts from the 80 patients selected for this study. The number of asterisks indicates the level of significance.

	PRO-MMP-9/PRO-MMP-2	PRO-MMP-9/MMP-9	PRO-MMP-2/MMP-2	MMP-9/MMP-2
Number of XY Pairs	80	80	80	80
Pearson *r*	0.249	0.6805	0.6244	0.292
95% confidence interval	0.03091 to 0.4444	0.5418 to 0.7831	0.4690 to 0.7423	0.07722 to 0.4809
p value (two-tailed)	0.0259	$p < 0.0001$	$p < 0.0001$	0.0086
p value summary	*	***	***	**
Is the correlation significant? (alpha = 0.05)	Yes	Yes	Yes	Yes
R squared	0.06199	0.4631	0.3899	0.08529

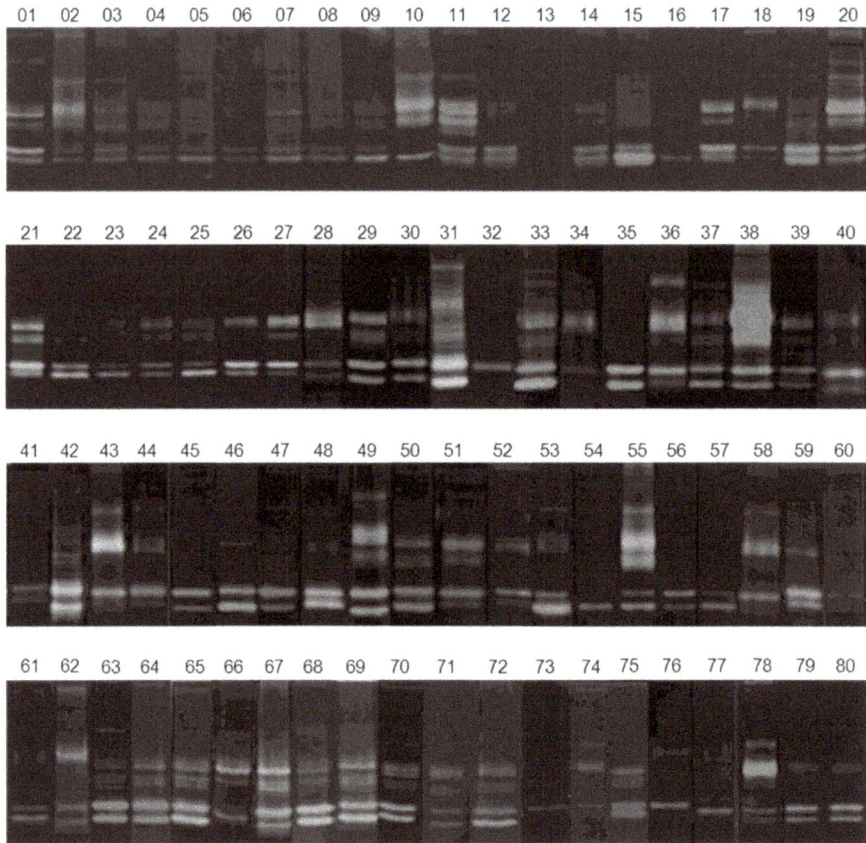

Figure 3. Gelatin zymographies of tissue extratcs from 80 breast cancer patients, showing a heterogeous level of the lytic bands. Gels were loaded with 18 ug of protein extracts and developed overnight at 37 °C, before staining with Coomassie-blue. Note the heterogeneous intensity level of the lytic bands among the patients.

3.3. MMP Levels and Proteomic Correlations

Surgical tissue fragments from the group of 80 selected patients were submitted for proteomic analysis according to the methodology described [15]. Collectively, the spots identified with UniProt Accession number (AC), were 458 proteins (including isoforms) corresponding to 274 AC entries. This list (see Supplementary Table S3) was subjected to bio-informatic analyses. Qualitative correlations were deduced by the "STRING" functional protein association networks, through known/predicted evidence. This platform traced the network between the tissue proteins of the tumoral samples with MMP-2 and MMP-9, respectively, as illustrated by the diagrams in Figure 5. In detail, by entering the list of the 274 AC identities with MMP-9, the system generated a network of interactions, 20 of which were directly linked, while many others occurred through other intermediate interactors. The same analysis with MMP-2 revealed 11 direct interactions. As shown in the Venn diagram in Figure 6, six interactors were common between the two enzymes. Each protein of the network can interplay with a variable number of secondary interactors, as reported in brackets.

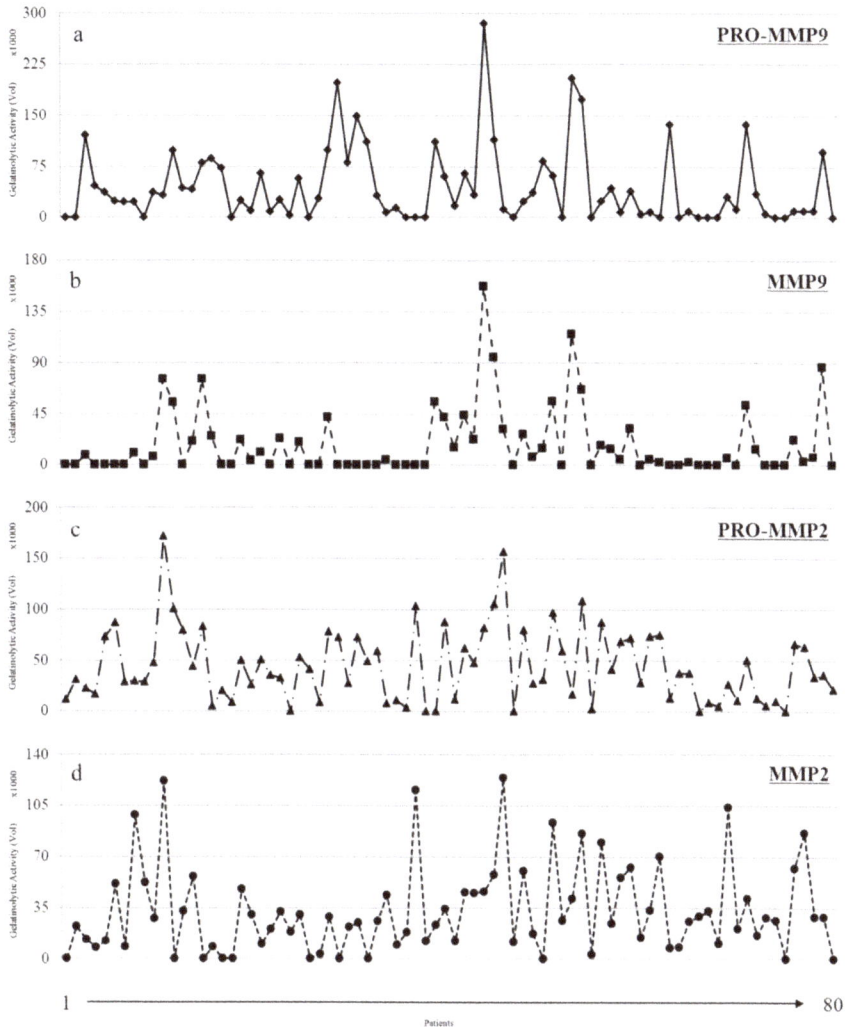

Figure 4. Densitometric profiles of the gelatinolytic bands identified as pro-MMP-9 (**a**), MMP-9 (**b**), pro-MMP-2 (**c**), MMP-2 (**d**) in the 80 breast cancer tissues.

Collectively, these have been classified into the following categories according to the current gene ontologies: cell proliferation, anti-apoptosis, metabolic pathways, cytoskeleton organization and cell motility, cell surface projections and response to stress and extracellular activities.

Due to the multifunctional roles of the majority of the detected proteins, they were included in several functional classes, as shown in Figure 7a–c, which highlights the interactors of MMP-9 (**b**) and MMP-2 (**c**) generated by and extracted from the STRING platform (Figure 5).

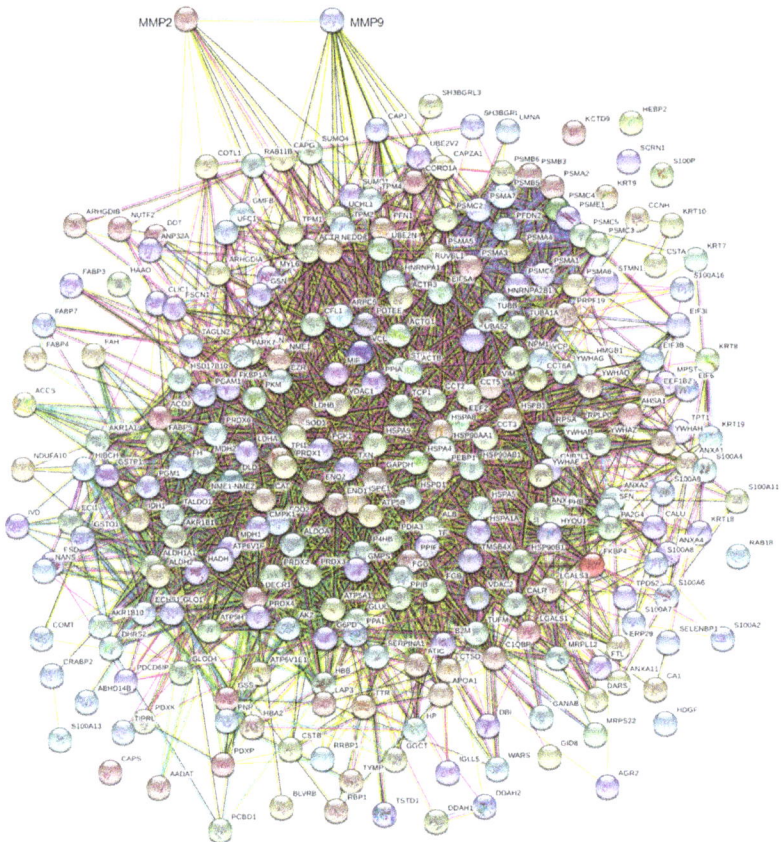

Figure 5. Interactome of the 274 genes coding for the 458 protein spots identified on the proteomic maps of the 80 breast cancer tissues selected for the study. The interactome was obtained by the software STRING, available online [18]. MMP-2 and MMP-9 were added to the list of 274 genes, in order to trace their predicted interactions with the proteomic profiles of the selected breast tissues.

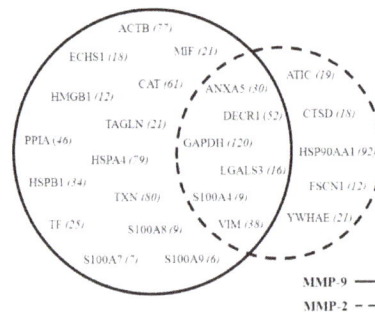

Figure 6. Venn diagram representing the predicted direct interactions of MMP-9 and MMP-2 with protein identities within the 274 entries identified in the proteomic profiles of the 80 selected breast cancer tissues. The intersection of the two circles includes six identities shared by the two MMPs. In brackets are the reported second-level interactors for each putative partner.

A

MMP9	MMP2	CELL PROLIFERATION	ANTI-APOPTOSIS	METABOLIC PATHWAYS	CYTOSKELETON ORGANIZATION AND CELL MOTILITY	CELL SURFACE PROJECTIONS	RESPONSE TO STRESS	EXTRACELLULAR ACTIVITY
ACTB					ACTB		ACTB	
ANXA5	ANXA5		ANXA5			ANXA5	ANXA5	ANXA5
	ATIC							
CAT			CAT	CAT				
	CTSD							CTSD
DECR1	DECR1			DECR1				
ECHS1				ECHS1				
	FSCN1					FSCN1		
GAPDH	GAPDH	GAPDH		GAPDH				
HMGB1					HMGB1		HMGB1	
HSPA4							HSPA4	
HSPB1			HSPB1		HSPB1		HSPB1	HSPB1
	HSP90AA1	HSP90AA1	HSP90AA1				HSP90AA1	
LGALS3	LGALS3		LGALS3		LGALS3		LGALS3	
MIF		MIF	MIF				MIF	
PPIA					PPIA		PPIA	
S100A4	S100A4				S100A4			S100A4
S100A7							S100A7	
S100A8					S100A8			
S100A9					S100A9			
TAGLN					TAGLN			
TF		TF						TF
TXN			TXN		TXN		TXN	
VIM	VIM				VIM	VIM		
	YWHAE						YWHAE	

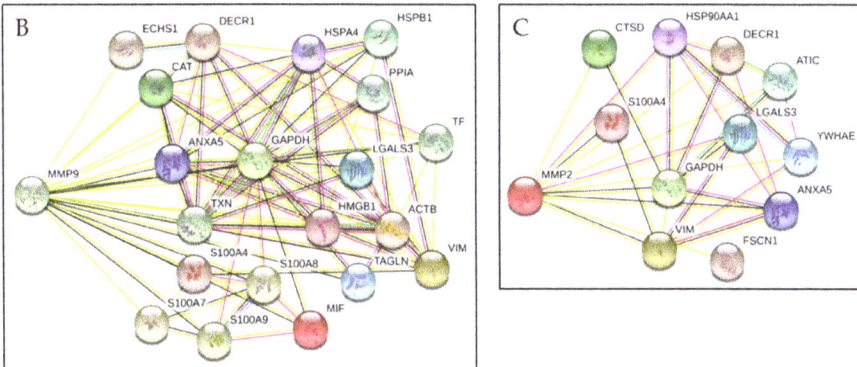

B

C

Figure 7. (**A**) Lists of the direct interactions of MMP-9 and MMP-2 with their putative partners at the primary level. The interacting proteins are distributed in the functional categories given by the databases. The lower panels show the interactors of MMP-9 (**B**) and MMP-2 (**C**), generated by and extracted from the STRING platform as shown in Figure 5.

4. Discussion

The potentially important role of matrix-degrading enzymes in breast cancer has been stated for many years, in particular in relation to the activities of MMP-2 and MMP-9 [14]. Indeed, both collagenolytic enzymes have been individually correlated with breast cancer progression [19] and tumor vascularization, invasion and metastasis, and differentiation and proliferation [20]. We initially aimed to verify the occurrence of gelatinase activity in 30 tissue samples of breast carcinomas vs. the adjacent non-affected tissues. The results demonstrated that both MMP-9 and MMP-2 were present, as pro- and active forms in the majority of the tumoral tissues, and absent or detectable at very low levels in the adjacent non-affected paired tissues.

This result encouraged us to continue investigating the expression of the lytic activities of the pro- and active forms of MMP-2 and MMP-9 on a larger number of cases, consisting of 80 surgical tissue fragments.

These two collagenolytic enzymes have been traditionally considered to be the most important promoters of tumor invasion, because of their ability to degrade the first tissue barrier, i.e., the basal lamina. This, as is well known, has the role of maintaining the right tissue architecture and to prevent cell mobility of stationary tissues, such as epithelial tissue.

The first question of our approach was to understand if the two enzymatic activities were interrelated. The results showed a significant Pearson correlation between the pro-enzymatic forms and their corresponding activated forms; while a moderate correlation was found between the two active enzymes. These observations suggest that the two enzymes can act in an independent manner, but in some way complementary.

In the belief that genes and proteins never operate alone, we wanted to search for potential interactors of the two enzymes. For this purpose, we used the STRING platform that builds functional protein association networks based on literature evidence.

An impressive result was the observation of a noticeably high number of hypothetical functional interactions at several levels. The first level concerns the primary interactions that each enzyme can establish with its putative partners. The second level concerns the secondary interactions, and beyond, and presumes the presence of intermediaries among the functional interactions.

4.1. Common Interactors for MMP-2 And MMP-9

The first-level interactions were 20 contacts for MMP-9, and 11 for MMP-2. A point of interest was the observation that six contacts are shared between the two enzymes, namely: GAPDH, ANXA5, LGAL3, S100A4, DECR1 and VIM.

Glyceraldehyde-3-phosphate dehydrogenase, G3P (GAPDH). Among the shared interactors, the presence of the G3P, which is generally over-expressed in breast cancer, as already reported in our previous studies [21], deserves particular attention. The G3P, besides being a key enzyme of glycolysis, under some circumstances, can translocate into the nucleus and act as a co-promoter for cell proliferation [22,23]. Its putative relationship with the MMPs has never reported before.

Annexin A5, ANXA5 (ANX5A). This protein is a member of a multigene family consisting of 13 members of Ca^{2+} and phospholipid binding proteins with peripheral membrane location. In our previous proteomics study, we found a ubiquitous and over-expressed presence of ANXA5, which is known to play an anti-apoptotic role. Similarly to other members of the family, it is thought to be involved at different levels in the tumor development, progression and invasivity [24–26]. Moreover, Annexin A5 is known to exert an anticoagulant effect, acting as an indirect inhibitor of the thromboplastin-specific complex [27].

Galectin, LEG3 (LGALS3). Galectin is a galactose-specific lectin associated with the cell membrane and involved in several processes of membrane trafficking by interacting with other surface proteins [28]. Intracellularly, it participates in the cytoskeleton organization and cell motility by interacting with ACTB and other associated proteins (COF1, EZRI, MIF, and CAPG). Extracellularly, it interacts with a variety of cell surface glycoproteins, and in cancer, it seems to participate in the dynamics of cell migration and in escaping the T cell-mediated immune response [29].

S100 proteins. The S100 proteins represent a multigene family of calcium-binding proteins of the EF-hand type, encoded by 21 genes in humans [30]. Members of the S100 family are differentially expressed in normal tissues and are frequently upregulated in cancer [31]. They may perform a large variety of functions, either intracellularly or extracellularly, in a cytokine-like manner through the receptor for "advanced glycation end products" (RAGE) [32]. In particular, the S100A4 protein is known to be secreted by tumor and/or stromal cells to support tumorigenesis by stimulating angiogenesis and promoting endothelial cell migration [33].

In addition, it has been reported that S100A4 may support MMP-9 and MMP-13 gene expression [34], and also that it may enhance the activity of some MMPs, causing higher cell dissociation and cancer metastasis [35,36].

Vimentin, VIME (VIM). Another interesting partner of the MMPs appears to be VIM. Vimentins are class-III intermediate filaments found in various non-epithelial cells, especially mesenchymal cells. Its high expression in breast cancer testifies to the occurrence of the epithelial-mesenchyme transition, already postulated at the proteomic level [21]. Its partnership with MMP-2 and -9 is suggestive for joined mechanisms of matrix degradation and cell migration.

2,4-Dienoyl-CoA reductase, DECR (DECR1). Somewhat ambiguous is the involvement of DECR1 2,4-dienoyl-CoA reductase, a mitochondrial auxiliary enzyme of beta-oxidation, for which no reports have been produced in relation to cancerogenesis, or with the MMPs in particular.

4.2. MMP-9 Direct Interactors

Fourteen out of the 20 putative interactors were exclusive for MMP-9. They were the following:

Transgelin, TAGL (TAGLN) is an actin-binding protein involved in calcium interactions and contractile properties of the cell. Using an expression cloning strategy, Nair et al. identified transgelin as a novel suppressor of MMP-9 expression [37]. This finding attributes the unusual role of tumor-suppressor to the MMP-9, which deserves further attention.

Macrophage Migration Inhibitory Factor, MIF (MIF) is an interesting small protein (approximately 12.5 kDa) involved in several biological activities, including the stimulation of the production of cytokines, chemokines, growth factors and angiogenic factors that may favor tumor growth and metastatic spreading. The overexpression of MIF in breast cancer cells, and its reported interaction with HSP90 and CXCR-4, is known to induce resistance to apoptosis and stimulation of proliferation via the AKT pathway. This opens new scenery regarding the possible correlations between matrix degradation and cell proliferation. Moreover, MIF is involved in the innate immune response and in regulating the function of macrophages in host defense [38–40].

Heat Shock 70 kDa Protein 4, HSP74 (HSPA4) and *Heat Shock 27kDa Protein, (HSP27) (HSPB1)* are two significant members of the multigenic heat shock protein family [41]. The HSPA4 (heat shock 70 kDa protein 4) is a major component of the HSP chaperone family involved in the folding of nascent proteins and in the degradation of misfolded polypeptides [42].

The other subset of chaperones consists of HSPs of a molecular weight of less than 30 kDa (sHSPs). Among some of their functions, sHPSs participate in cell survival, cytoskeletal motility, and disruption of protein aggregation [43]. Moreover, it has been reported that the HSPs display elevated expression levels in cancer, where they may perform anti-apoptotic activities, both spontaneous and generated by therapy [44]. In particular, the high expression of HSPB1 (HSP27) has been associated with poor prognosis in several carcinomas and osteosarcomas [45]. Their putative correlation with MMP-9 suggests a synergy between the two key mechanisms of cancer progression: matrix degradation and anti-apoptotic effects.

Peptidylprolyl isomerase A (cyclophilin A), PPIA (PPIA) accelerates the folding of proteins and catalyzes the cis-trans isomerization of proline imidic peptide bonds in oligopeptides. Recently, a key role of PPIA in tumor biology has been proposed [46]. The postulated interaction with MMP-9 remains to be clarified.

Other presumed interactors are listed below.

Catalase, CATA (CAT) occurs in almost all aerobic organisms and serves to protect cells from the toxic effects of hydrogen peroxide (H_2O_2). Moreover, it can reduce the activity of MMPs and promote cellular growth of many cell types, including T-cells, B-cells, myeloid leukemia cells, transformed fibroblast cells and others [47,48].

Enoyl-CoA hydratase, ECHM (ECHS1) is a key enzyme involved in the oxidation of fatty acids and branched-chain amino acids. Moreover, it has been reported to be associated with the progression of a variety of tumors [49], including gastric cancer [50]. However, no correlation has been described so far with the matrix metalloproteases.

Thioredoxin, THIO (TXN) is a multifunctional cellular factor with thiol-mediated redox activity. It plays pivotal roles in the regulation of many cellular processes, including proliferation, apoptosis, and gene expression, both in normal and tumoral cells [44]. Its frequent high expression has been reported in many cancers [51–53].

Transferrin, TRFE (TF) is a transport protein which can bind two Fe (3+) ions in association with the binding of an anion, usually bicarbonate. It is responsible for the transport of iron from sites of absorption and heme degradation to those of storage and utilization. Serum transferrin may also have a further role in stimulating cell proliferation [54].

High mobility group protein B1, HMGB1 (HMGB1) is a multifunctional redox-sensitive protein with various roles in different cellular compartments. In the nucleus, it is one of the major non-histone/chromatin-associated proteins and acts as a DNA chaperone involved in replication, transcription, DNA repair and genome stability. It has been reported that it promotes host inflammatory reactions to external signals and immune responses. In the cytoplasm, it functions as a damage-associated molecule inducing inflammatory mediator release [55].

S100 proteins, such as the group of S100-A7, -A8 and -A9 calcium binding proteins, are known to exert several functions in normal and tumoral tissues, where they may be expressed at high levels (as has been reported by our group [56] and other authors [57–60]). Their correlation with MMP-9 has never been proposed before.

Actin, (ACTB) ACTB is the primary component of the cytoskeleton in eukaryotic cells. The actin cytoskeleton is therefore responsible for the integrity and shape-maintenance of cells. The cytoskeleton reorganization is a physiological event during cell growth, differentiation, and senescence of stationary cells, and it is responsible for the cell motility of migratory cells [61].

Under pathological conditions, like cancer, the actin cytoskeleton may undergo deregulated fragmentation. It has been recently reported that induced alteration of actin cytoskeletal integrity in human trabecular meshwork cells (HTMC) is associated with MMP-2 activation, presumably through the upregulation of its activator, MT1-MMP [62].

4.3. MMP-2 Direct Interactors

The functional interactors attributed exclusively to MMP-2 are the following:

Bifunctional purine biosynthesis protein PURH, PUR9 (ATIC) is a bifunctional enzyme that catalyzes two steps in purine biosynthesis. Moreover, it has been reported that it may promote auto-phosphorylation and internalization of the insulin receptor (INSR) [63]. No information is currently found on potential functional interactions with MMP-2 and other MMPs

Cathepsin D, CATD (CTSD) is an acid protease that is active in the breakdown of intracellular proteins, which has been reported to be involved in the pathogenesis of several diseases, such as breast cancer and possibly Alzheimer's disease via non-proteolytic pathways [64,65]. The postulated direct correlation between CTSD and MMP-2 is an interesting observation which needs further insight.

Fascin, FSCN1 (FSCN1) is one of the organizers of the actin filaments into bundles. Therefore, it plays a role in the formation of microspikes, membrane ruffles, and stress fibers, as well as other cell projections such as filopodia, which are essential for cell motility and migration. Its frequent overexpression in cancer has been related to the promotion of an actin-independent cell migration [66]. Its correlation with MMP-2 could be instrumental in the increased aptitude of cells to migrate during the invasive phase of the breast cancer,

Heat shock protein HSP 90-alpha, HS90A (HSP90AA1) is a molecular chaperone that has been recognized as one of those responsible for the structural maintenance of cells, and also for cell cycle control and signal transduction pathways [67]. It has been reported that HSP90, when translocated into the nucleus, may influence the activity of many transcription factors [68].

14-3-3 Protein epsilon, 1433E (YWHAE) is a member of a multigenic protein family. In mammals, it consists of seven members (β, ε, η, γ, τ, ζ and σ), which act as adapter proteins involved in the regulation of a large number of signaling pathways [69,70] and in the maintenance of epithelial cell polarity [71]. It is interesting to note that this protein is frequently overexpressed in breast cancer [21]. However, no correlation with collagenolytic activities has been reported before.

In conclusion, the present investigation (based on a double line of approach, experimental and "in silico") highlights for the first time the complexity of the interactive networks that MMP-2 and MMP-9, through a series of interactors (partly common and partly exclusive) may accomplish within the cell. These complex interactive molecular circuits, where the two collagenolytic enzymes appear to be included, suggest their potential involvement in other important cellular activities, besides that of remodeling the extracellular matrix. These interactors are deduced by the direct inquiry between MMP-2 and MMP-9 with the proteomic platform of the breast tissues, and are not inclusive of the MMP inhibitors (TIMP) and activator (MMP14). The scenery becomes even more complex when analyzing the members of the second level of interactors, or beyond. In our opinion, this is the first comprehensive description of potential activities where MMP-2 and MMP-9 can be involved in the complicated scenario in which the mechanisms of tumor progression are correlated with unfavorable prognosis.

Supplementary Materials: The following are available online at http://www.mdpi.com/2073-4409/7/8/89/s1, Table S1: Basic clinical parameters corresponding to 30 selected patients, Table S2: Basic clinical parameters corresponding to 80 selected patients, Table S3: List of 274 genes encoding for 458 protein spots identified on our reference proteomic map of Breast Cancer Tissue.

Author Contributions: Conceptualization, I.P.M. and G.D.; Methodology, G.D. and M.R.M.; Software, G.D. and R.M.; Validation, I.P.M. and P.C.; Resources, COBS; Data Curation, I.R.; Writing—Original Draft Preparation, I.P.M.; Writing—Review & Editing, I.P.M.

Funding: The present work was mainly supported by a State-made contribution (5 × 1000) to COBS as distinguished no-profit research institution.

Acknowledgments: Guido Filosto, President of La Maddalena Hospital, is gratefully acknowledged for continuous support to the Centro di Oncobiologia Sperimentale (COBS).

Conflicts of Interest: The authors declare no conflict of interest.

References

1. Siegel, R.L.; Miller, K.D.; Jemal, A. Cancer statistics. 2018. *CA Cancer J. Clin.* **2018**, *68*, 7–30. [CrossRef] [PubMed]
2. Hashim, D.; Boffetta, P.; La Vecchia, C.; Rota, M.; Bertuccio, P.; Malvezzi, M.; Negri, E. The global decrease in cancer mortality: Trends and disparities. *Ann. Oncol.* **2016**, *27*, 926–933. [CrossRef] [PubMed]
3. Turkoz, F.P.; Solak, M.; Petekkaya, I.; Keskin, O.; Kertmen, N.; Sarici, F.; Arik, Z.; Babacan, T.; Ozisik, Y.; Altundag, K. Association between common risk factors and molecular subtypes in breast cancer patients. *Breast* **2013**, *22*, 344–350. [CrossRef] [PubMed]
4. Caplan, L. Delay in breast cancer: Implications for stage at diagnosis and survival. *Front. Public Health* **2014**, *2*, 87. [CrossRef] [PubMed]
5. Pondé, N.; Brandão, M.; El-Hachem, G.; Werbrouck, E.; Piccart, M. Treatment of advanced HER2-positive breast cancer: 2018 and beyond. *Cancer Treat Rev.* **2018**, *67*, 10–20. [CrossRef] [PubMed]
6. Keegan, N.M.; Gleeson, J.P.; Hennessy, B.T.; Morris, P.G. PI3K inhibition to overcome endocrine resistance in breast cancer. *Expert Opin. Investig. Drugs* **2018**, *27*, 1–15. [CrossRef] [PubMed]
7. Rawlings, N.D.; Barrett, A.J.; Thomas, P.D.; Huang, X.; Bateman, A.; Finn, R.D. The MEROPS database of proteolytic enzymes, their substrates and inhibitors in 2017 and a comparison with peptidases in the PANTHER database. *Nucleic Acids Res.* **2018**, *46*, D624–D632. [CrossRef] [PubMed]
8. Holanda, A.O.; Oliveira, A.R.; Cruz, K.J.; Severo, J.S.; Morais, J.B.; Silva, B.B.; Marreiro, D.D. Zinc and metalloproteinases 2 and 9: What is their relation with breast cancer? *Rev. Assoc. Med. Bras.* **2017**, *63*, 78–84. [CrossRef] [PubMed]

9. Daniele, A.; Abbate, I.; Oakley, C.; Casamassima, P.; Savino, E.; Casamassima, A.; Sciortino, G.; Fazio, V.; Gadaleta-Caldarola, G.; Catino, A.; et al. Clinical and prognostic role of matrix metalloproteinase-2, -9 and their inhibitors in breast cancer and liver diseases: A review. *Int. J. Biochem. Cell Biol.* **2016**, *77*, 91–101. [CrossRef] [PubMed]
10. Skerenova, M.; Mikulova, V.; Capoun, O.; Zima, T.; Tesarova, P. Circulating tumor cells and serum levels of MMP-2, MMP-9 and VEGF as markers of the metastatic process in patients with high risk of metastatic progression. *Biomed. Pap. Med. Fac. Univ. Palacky Olomouc Czech Repub.* **2017**, *161*, 272–280. [CrossRef] [PubMed]
11. Nikkola, J.; Vihinen, P.; Vuoristo, M.S.; Kellokumpu-Lehtinen, P.; Kähäri, V.M.; Pyrhönen, S. High serum levels of matrix metalloproteinase-9 and matrix metalloproteinase-1 are associated with rapid progression in patients with metastatic melanoma. *Clin. Cancer Res.* **2005**, *11*, 5158–5166. [CrossRef] [PubMed]
12. Leppä, S.; Saarto, T.; Vehmanen, L.; Blomqvist, C.; Elomaa, I. A high serum matrix metalloproteinase-2 level is associated with an adverse prognosis in node-positive breast carcinoma. *Clin. Cancer Res.* **2004**, *10*, 1057–1063. [CrossRef] [PubMed]
13. Bradford, M.M. A rapid and sensitive method for the quantitation of microgram quantities of protein utilizing the principle of protein-dye binding. *Anal. Biochem.* **1976**, *72*, 248–254. [CrossRef]
14. Pucci-Minafra, I.; Minafra, S.; La Rocca, G.; Barranca, M.; Fontana, S.; Alaimo, G.; Okada, Y. Zymographic analysis of circulating and tissue forms of colon carcinoma gelatinase A (MMP-2) and B (MMP-9) separated by mono- and two-dimensional electrophoresis. *Matrix Biol.* **2001**, *20*, 419–427. [CrossRef]
15. Pucci-Minafra, I.; Cancemi, P.; Marabeti, M.R.; Albanese, N.N.; Di Cara, G.; Taormina, P.; Marrazzo, A. Proteomic profiling of 13 paired ductal infiltrating breast carcinomas and non-tumoral adjacent counterparts. *Proteom. Clin. Appl.* **2007**, *1*, 118–129. [CrossRef] [PubMed]
16. Pucci Minafra, I.; Di Cara, G.; Musso, R.; Peri, G.; Valentino, B.; D'Arienzo, M.; Martini, D.; Raspanti, M.; Minafra, S. Proteomic profiling of In Vitro bone-conditioned skbr3 breast cancer cells. *J. Proteom. Bioinf.* **2016**, *9*, 75–83.
17. Perkins, D.N.; Pappin, D.J.; Creasy, D.M.; Cottrell, J.S. Probability-based protein identification by searching sequence databases using mass spectrometry data. *Electrophoresis* **1999**, *20*, 3551–3567. [CrossRef]
18. Szklarczyk, D.; Morris, J.H.; Cook, H.; Kuhn, M.; Wyder, S.; Simonovic, M.; Santos, A.; Doncheva, N.T.; Roth, A.; Bork, P.; et al. The STRING database in 2017: Quality-controlled protein-protein association networks; made broadly accessible. *Nucleic Acids Res.* **2017**, *45*, D362–D368. [CrossRef] [PubMed]
19. Jezierska, A.; Motyl, T. Matrix metalloproteinase-2 involvement in breast cancer progression: A mini-review. *Med. Sci. Monit.* **2009**, *2*, 32–40.
20. Mehner, C.; Hockla, A.; Miller, E.; Ran, S.; Radisky, D.C.; Radisky, E.S. Tumor cell-produced matrix metalloproteinase 9 (MMP-9) drives malignant progression and metastasis of basal-like triple negative breast cancer. *Oncotarget* **2014**, *5*, 2736–2749. [CrossRef] [PubMed]
21. Pucci-Minafra, I.; Di Cara, G.; Musso, R.; Cancemi, P.; Albanese, N.N.; Roz, E.; Minafra, S. Retrospective Proteomic Screening of 100 Breast Cancer Tissues. *Proteomes* **2017**, *5*, 15. [CrossRef] [PubMed]
22. Tristan, C.; Shahani, N.; Sedlak, T.W.; Sawa, A. The diverse functions of GAPDH: Views from different subcellular compartments. *Cell Signal.* **2011**, *23*, 317–323. [CrossRef] [PubMed]
23. Sirover, M.A. Subcellular dynamics of multifunctional protein regulation: Mechanisms of GAPDH intracellular translocation. *J. Cell. Biochem.* **2012**, *113*, 2193–2200. [CrossRef] [PubMed]
24. Mussunoor, S.; Murray, G.I. The role of annexins in tumour development and progression. *J. Pathol.* **2008**, *216*, 131–140. [CrossRef] [PubMed]
25. Wehder, L.; Arndt, S.; Murzik, U.; Bosserhoff, A.K.; Kob, R.; von Eggeling, F.; Melle, C. Annexin A5 is involved in migration and invasion of oral carcinoma. *Cell Cycle* **2009**, *8*, 1552–1558. [CrossRef] [PubMed]
26. Peng, B.; Guo, C.; Guan, H.; Liu, S.; Sun, M.Z. Annexin A5 as a potential marker in tumors. *Clin. Chim. Acta* **2014**, *427*, 42–48. [CrossRef] [PubMed]
27. Sun, J.; Bird, P.; Salem, H.H. Effects of annexin V on the activity of the anticoagulant proteins C and S. *Thromb. Res.* **1993**, *69*, 279–287. [CrossRef]
28. Shetty, P.; Bargale, A.; Patil, B.R.; Mohan, R.; Dinesh, U.S.; Vishwanatha, J.K.; Gai, P.B.; Patil, V.S.; Amsavardani, T.S. Cell surface interaction of annexin A2 and galectin-3 modulates epidermal growth factor receptor signaling in Her-2 negative breast cancer cells. *Mol. Cell. Biochem.* **2016**, *411*, 221–233. [CrossRef] [PubMed]

29. Fukumori, T.; Takenaka, Y.; Yoshii, T.; Kim, H.R.; Hogan, V.; Inohara, H.; Kagawa, S.; Raz, A. CD29 and CD7 mediate galectin-3-induced type II T-cell apoptosis. *Cancer Res.* **2003**, *63*, 8302–8311. [PubMed]

30. HUGO Gene Nomenclature Committee (HGNC). Available online: http://www.genenames.org/cgi-bin/genefamilies/set/459 (accessed on 17 March 2017).

31. Salama, I.; Malone, P.S.; Mihaimeed, F.; Jones, J.L. A review of the S100 proteins in cancer. *Eur. J. Surg. Oncol.* **2008**, *34*, 357–364. [CrossRef] [PubMed]

32. Leclerc, E.; Fritz, G.; Vetter, S.W.; Heizmann, C.W. Binding of S100 proteins to RAGE: An update. *Biochim. Biophys. Acta* **2009**, *1793*, 993–1007. [CrossRef] [PubMed]

33. Xu, H.; Li, M.; Zhou, Y.; Wang, F.; Li, X.; Wang, L.; Fan, Q. S100A4 participates in epithelial-mesenchymal transition in breast cancer via targeting MMP2. *Tumour Biol.* **2016**, *37*, 2925–2932. [CrossRef] [PubMed]

34. Ismail, T.M.; Bennett, D.; Platt-Higgins, A.M.; Al-Medhity, M.; Barraclough, R.; Rudland, P.S. S100A4 Elevation Empowers Expression of Metastasis Effector Molecules in Human Breast Cancer. *Cancer Res.* **2017**, *77*, 780–789. [CrossRef] [PubMed]

35. Buetti-Dinh, A.; Pivkin, I.V.; Friedman, R. S100A4 and its role in metastasis—Computational integration of data on biological networks. *Mol. Biosyst.* **2015**, *11*, 2238–2246. [CrossRef] [PubMed]

36. Buetti-Dinh, A.; Pivkin, I.V.; Friedman, R. S100A4 and its role in metastasis–simulations of knockout and amplification of epithelial growth factor receptor and matrix metalloproteinases. *Mol. Biosyst.* **2015**, *11*, 2247–2254. [CrossRef] [PubMed]

37. Nair, R.R.; Solway, J.; Boyd, D.D. Expression cloning identifies transgelin (SM22) as a novel repressor of 92-kDa type IV collagenase (MMP-9) expression. *J. Biol. Chem.* **2006**, *281*, 26424–26436. [CrossRef] [PubMed]

38. Conroy, H.; Mawhinney, L.; Donnelly, S.C. Inflammation and cancer: Macrophage migration inhibitory factor (MIF)—The potential missing link. *Q. J. Med.* **2010**, *103*, 831–836. [CrossRef] [PubMed]

39. Nobre, C.; de Araújo, J.M.; Fernandes, T.A.; Cobucci, R.N.; Lanza, D.C.; Andrade, V.S.; Fernandes, J.V. Macrophage migration inhibitory factor (MIF): Biological activities and relation with cancer. *Pathol. Oncol. Res.* **2016**, *23*, 235–244. [CrossRef] [PubMed]

40. Richard, V.; Kindt, N.; Saussez, S. Macrophage migration inhibitory factor involvement in breast cancer (Review). *Int. J. Oncol.* **2015**, *47*, 1627–1633. [CrossRef] [PubMed]

41. Li, Z.; Srivastava, P. Heat-Shock Proteins. *Curr. Protoc. Immunol.* **2004**. [CrossRef]

42. Abisambra, J.F.; Jinwal, U.K.; Jones, J.R.; Blair, L.J.; Koren, J.; Dickey, C.A. Exploiting the diversity of the heat-shock protein family for primary and secondary tauopathy therapeutics. *Curr. Neuropharmacol.* **2011**, *9*, 623–631. [CrossRef] [PubMed]

43. Bakthisaran, R.; Tangirala, R.; Rao, C.M. Small heat shock proteins Role in cellular functions and pathology. *Biochim. Biophys. Acta (BBA) Proteins Proteom.* **2015**, *1854*, 291–319. [CrossRef] [PubMed]

44. Wu, J.; Liu, T.; Rios, Z.; Mei, Q.; Lin, X.; Cao, S. Heat shock proteins and cancer. *Trends Pharmacol. Sci.* **2017**, *38*, 226–256.

45. Ciocca, D.R.; Calderwood, S.K. Heat shock proteins in cancer: Diagnostic, prognostic, predictive, and treatment implications. *Cell Stress Chaperones* **2005**, *10*, 86–103. [CrossRef] [PubMed]

46. Nath, P.R. Peptidyl-prolyl isomerase (PPIase): An emerging area in tumor biology. *Cancer Res. Front.* **2017**, *3*, 126–143. [CrossRef]

47. Tehrani, H.S.; Moosavi-Movahedi, A.A. Catalase and its mysteries. *Prog. Biophys. Mol. Biol.* **2018**, *9*, 30293–30296. [CrossRef]

48. Takeuchi, A.; Miyamoto, T.; Yamaji, K.; Masuho, Y.; Hayashi, M.; Hayashi, H.; Onozaki, K. A human erythrocyte-derived growth-promoting factor with a wide target cell spectrum: Identification as catalase. *Cancer Res.* **1995**, *55*, 1586–1589. [PubMed]

49. Zhang, J.; Ibrahim, M.M.; Sun, M.; Tang, J. Enoyl-coenzyme A hydratase in cancer. *Clin. Chim. Acta* **2015**, *448*, 13–17. [CrossRef] [PubMed]

50. Zhu, X.S.; Gao, P.; Dai, Y.C.; Xie, J.P.; Zeng, W.; Lian, Q.N. Attenuation of enoyl coenzyme A hydratase short chain 1 expression in gastric cancer cells inhibits cell proliferation and migration in vitro. *Cell. Mol. Biol. Lett.* **2014**, *19*, 576–589. [CrossRef] [PubMed]

51. Raninga, P.V.; Trapani, G.D.; Tonissen, K.F. Cross Talk between Two Antioxidant Systems, Thioredoxin and DJ-1: Consequences for Cancer. *Oncoscience* **2014**, *1*, 95–110. [CrossRef] [PubMed]

52. Fan, J.; Yu, H.; Lu, Y.; Yin, L. Diagnostic and prognostic value of serum thioredoxin and DJ-1 in non-small cell lung carcinoma patients. *Tumour Biol.* **2016**, *37*, 1949–1958. [CrossRef] [PubMed]

53. Karlenius, T.C.; Tonissen, K.F. Thioredoxin and Cancer: A Role for Thioredoxin in all States of Tumor Oxygenation. *Cancers (Basel)* **2010**, *2*, 209–232. [CrossRef] [PubMed]

54. Laskey, J.; Webb, I.; Schulman, H.M.; Ponka, P. Evidence that transferrin supports cell proliferation by supplying iron for DNA synthesis. *Exp. Cell Res.* **1988**, *176*, 87–95. [CrossRef]

55. Lee, H.J.; Kim, A.; Song, I.H.; Park, I.A.; Yu, J.H.; Ahn, J.H.; Gong, G. Cytoplasmic expression of high mobility group B1 (HMGB1) is associated with tumor-infiltrating lymphocytes (TILs) in breast cancer. *Pathol. Int.* **2016**, *66*, 202–209. [CrossRef] [PubMed]

56. Cancemi, P.; Di Cara, G.; Albanese, N.N.; Costantini, F.; Marabeti, M.R.; Musso, R.; Riili, I.; Lupo, C.; Roz, E.; Pucci-Minafra, I. Differential occurrence of S100A7 in breast cancer tissues: A proteomic-based investigation. *Proteom. Clin. Appl.* **2012**, *6*, 364–373. [CrossRef] [PubMed]

57. Mandal, S.; Curtis, L.; Pind, M.; Murphy, L.C.; Watson, P.H. S100A7 (psoriasin) influences immune response genes in human breast cancer. *Exp. Cell Res.* **2007**, *313*, 3016–3025. [CrossRef] [PubMed]

58. West, N.R.; Watson, P.H. S100A7 (psoriasin) oncostatin-M and interleukin-6 in human breast cancer. *Oncogene* **2010**, *29*, 2083–2092. [CrossRef] [PubMed]

59. Nasser, M.W.; Qamri, Z.; Deol, Y.S.; Ravi, J.; Powell, C.A.; Trikha, P.; Schwendener, R.A.; Bai, X.F.; Shilo, K.; Zou, X.; et al. S100A7 enhances mammary tumorigenesis through upregulation of inflammatory pathways. *Cancer Res.* **2012**, *72*, 604–615. [CrossRef] [PubMed]

60. Chen, H.; Xu, C.; Jin, Q.; Liu, Z. S100 protein family in human cancer. *Am. J. Cancer Res.* **2014**, *4*, 89–115. [PubMed]

61. Bildyug, N. Matrix metalloproteinases: An emerging role in regulation of actin microfilament system. *Biomol. Concepts* **2016**, *7*, 321–329. [CrossRef] [PubMed]

62. Sanka, K.; Maddala, R.; Epstein, D.L.; Rao, P.V. Influence of actin cytoskeletal integrity on matrix metalloproteinase-2 activation in cultured human trabecular meshwork cells. *Investig. Ophthalmol. Vis. Sci.* **2007**, *48*, 2105–2114. [CrossRef] [PubMed]

63. Boutchueng-Djidjou, M.; Collard-Simard, G.; Fortier, S.; Hébert, S.S.; Kelly, I.; Landry, C.R.; Faure, R.L. The last enzyme of the de novo purine synthesis pathway 5-aminoimidazole-4-carboxamide ribonucleotide formyltransferase/IMP cyclohydrolase (ATIC) plays a central role in insulin signaling and the Golgi/endosomes protein network. *Mol. Cell. Proteom.* **2015**, *14*, 1079–1092. [CrossRef] [PubMed]

64. Garcia, M.; Platet, N.; Liaudet, E.; Laurent, V.; Derocq, D.; Brouillet, J.P.; Rochefort, H. Biological and clinical significance of cathepsin D in breast cancer metastasis. *Stem Cells* **1996**, *14*, 642–650. [CrossRef] [PubMed]

65. Pranjol, M.Z.I.; Gutowski, N.J.; Hannemann, M.; Whatmore, J.L. Cathepsin D non-proteolytically induces proliferation and migration in human omental microvascular endothelial cells via activation of the ERK1/2 and PI3K/AKT pathways. *Biochim. Biophys. Acta* **2018**, *1865*, 25–33. [CrossRef] [PubMed]

66. Lee, H.J.; An, H.J.; Kim, T.H.; Kim, G.; Kang, H.; Heo, J.H.; Kwon, A.Y.; Kim, S. Fascin expression is inversely correlated with breast cancer metastasis suppressor 1 and predicts a worse survival outcome in node-negative breast cancer patients. *J. Cancer* **2017**, *8*, 3122–3129. [CrossRef] [PubMed]

67. Jarosz, D. Hsp90: A global regulator of the genotype-to-phenotype map in cancers. *Adv. Cancer Res.* **2016**, *129*, 225–247. [PubMed]

68. Calderwood, S.K.; Neckers, L. Hsp90 in Cancer: Transcriptional roles in the nucleus. *Adv. Cancer Res.* **2016**, *129*, 89–106. [PubMed]

69. Aghazadeh, Y.; Papadopoulos, V. The role of the 14-3-3 protein family in health, disease, and drug development. *Drug Discov. Today* **2016**, *21*, 278–287. [CrossRef] [PubMed]

70. Ling, C.; Zuo, D.; Xue, B.; Muthuswamy, S.; Muller, W.J. A novel role for 14-3-3 sigma in regulating epithelial cell polarity. *Genes Dev.* **2010**, *24*, 947–956. [CrossRef] [PubMed]

71. Freeman, A.K.; Morrison, D.K. 14-3-3 Proteins: Diverse functions in cell proliferation and cancer progression. *Semin. Cell Dev. Biol.* **2011**, *22*, 681–687. [CrossRef] [PubMed]

cells

MDPI

Review
MicroRNAs and Osteoarthritis

Charles J. Malemud

Department of Medicine, Division of Rheumatic Diseases, University Hospitals Cleveland Medical Center, Foley Medical Building, 2061 Cornell Road, Cleveland, OH 44106-5076, USA; cjm4@cwru.edu; Tel.: +1-(216)-536-1945; Fax: +1-(216)-844-2288

Received: 11 July 2018; Accepted: 29 July 2018; Published: 1 August 2018

Abstract: An imbalance in gene expressional events skewing chondrocyte anabolic and catabolic pathways toward the latter causes an aberrant turnover and loss of extracellular matrix proteins in osteoarthritic (OA) articular cartilage. Thus, catabolism results in the elevated loss of extracellular matrix proteins. There is also evidence of an increase in the frequency of chondrocyte apoptosis that compromises the capacity of articular cartilage to undergo repair. Although much of the fundamental OA studies over the past 20 years identified and characterized many genes relevant to pro-inflammatory cytokines, apoptosis, and matrix metalloproteinases (MMPs)/a disintegrin and metalloproteinase with thrombospondin motif (ADAMTS), more recent studies focused on epigenetic mechanisms and the associated role of microRNAs (miRs) in regulating gene expression in OA cartilage. Thus, several miRs were identified as regulators of chondrocyte signaling pathways, apoptosis, and proteinase gene expression. For example, the reduced expression of miR-146a was found to be coupled to reduced type II collagen (COL2) in OA cartilage, whereas MMP-13 levels were increased, suggesting an association between *MMP-13* gene expression and *COL2A1* gene expression. Results of these studies imply that microRNAs could become useful in the search for diagnostic biomarkers, as well as providing novel therapeutic targets for intervention in OA.

Keywords: apoptosis; articular cartilage; autophagy; chondrocytes; extracellular matrix; microRNA

---

1. Introduction

The main focus of basic studies designed to unravel the pathology of osteoarthritis (OA) in humans recently employed non-surgical animal models with genetically modified mice as replicas of the human disease [1,2]. The results of these studies provided many important insights into the pathogenesis of human OA which were based mainly on compelling evidence that chondrocyte metabolic pathways that regulate anabolic and catabolic events were skewed toward the latter [3]. Thus, at its most fundamental level, human OA is characterized by chronic inflammation, progressive destruction of articular cartilage, and subchondral bone sclerosis [4].

In addition to the significance of molecular events that drive the progression of OA, there is now ample evidence that "chondroscenesence", a term used to describe chondrocytes in the older individual, was linked to inflammation and was characterized by an abnormal interplay between autophagy and the inflammasome [5]. Of note, it was also previously postulated using a comparison-type analysis that OA-like cartilage lesions in "elderly" rats responded more favorably to moderate physical activity and normal mechanical loading compared with rats not subjected to the exercise program [6]. This suggested that physical activity could potentially blunt the progression of OA even in older individuals, which may be associated with improved lubrication of the articular cartilage.

Recent advances also increased our understanding of how the loss of articular cartilage extracellular matrix (ECM) proteins characteristic of OA synovial joints results from the action of pro-inflammatory cytokines (e.g., interleukin-1β (IL-1β), IL-6, IL-15, IL-17, IL-21) and several other interleukin family members [7–9]. In addition to several of these interleukins, tumor necrosis factor-α

(TNF-α) mediates the upregulation of matrix metalloproteinases (MMPs), and elevated levels of a family of enzymes termed a disintegrin and metalloproteinase with thrombospondin motif (e.g., ADAMTS-4,-5) with selective aggrecanase (i.e., aggrecanase-1 and aggrecanase-2) activities [10–12]. Additional soluble mediators of inflammation, including nitric oxide and the transcription nuclear factor kappa b (NF-κB), as well as aberrant mechanical stressors, also appear to be involved in cartilage ECM protein degradation in OA [4]. However, the medical therapy of OA remains fixated on primarily relieving symptomatic synovial joint pain [12]. This appears to be in keeping with the identification of various pain pathways which are implicated in the progression of OA [13–15].

Basic studies also implicated programmed cell death, also known as apoptosis, or in the chondrocyte context, "chondropotosis", in the OA pathological process [16]. The increased frequency of apoptotic chondrocytes in human OA leads to reduced chondrocyte vitality, which is accompanied by a feeble response of articular cartilage for repairing OA lesions [17–19]. In this context, Terminal deoxynucleotidyl transferase dUTP nick end labeling.

(TUNEL)/Hoechst 33258 staining revealed an elevated frequency of apoptosis in freshly isolated chondrocytes from OA cartilage [20], as well as altered patterns of pro-apoptosis and anti-apoptosis factors, exemplified by an analysis of the extrinsic apoptosis pathway which was determined by measurements of Bcl-2-associated X protein(Bax), B-cell lymphoma-2 (Bcl-2), TNF-related apoptosis-inducing ligand (TRAIL), death receptor-5 (DR5), and caspase-3 [21].

In addition to those genes which effectively reduce chondrocyte viability, other specific genes and signaling pathways, include genes that encode bone morphogenetic proteins, the WNT/β-catenin proteins, leukemia inhibitory factor, hypoxia-inducible factor-1-α/2α (HIF-1α/2α) [22,23], as well as *GREM1, FRZB, DKK1, VEGF, EGF* [23–25], *GDF5* [26] genes, and the NOTCH/NF-κB pathway [27]. Each of these were linked to various stages in the OA cartilage degenerative process.

Significant advances also occurred in defining the role that signal transduction pathways play in OA. For example, pro-inflammatory cytokine-mediated activation of mitogen-activated protein kinase (MAPK) and Janus kinase/signal transducers and activators of transcription (JAK-STAT) pathways demonstrated how these signaling pathways control gene expression associated with pro-apoptosis protein synthesis and MMP synthesis, both of which were found to be dysregulated in OA articular chondrocytes [28–33].

2. A Role for Epigenetics in OA

Epigenetics is a field of genetics wherein variations in cell phenotype are considered to result from external or environmental factors that affect how gene activity is regulated. This mechanism of gene regulation exists in contradistinction to genes regulated by changes in DNA sequence. Thus, in this way, epigenetic factors can regulate the dynamic potential of gene transcription.

Results of recent studies implicated epigenetics as an important contributor to the pathology and progression of cartilage ECM protein turnover and damage in OA. In that regard, genome-wide association studies (GWAS) identified several critical genes, including *IL-1β, IL-6*, and *TNF-α*, which were subject to epigenetic modifications during the OA process [34]. In fact, these genes and others were implicated in inflammatory processes characteristic of OA progression, as well as how these genes could be responsible for producing the chondrocyte hypertrophic phenotype characteristic of altered cartilage structure in OA [35]. Of note, the repertoire of altered methylation patterns in fibroblast-like synoviocytes that defines the inflammatory process in rheumatoid arthritis (RA) was employed to distinguish primary inflammatory episodes in RA from those found in OA synovium. Furthermore, an integrative analysis of the methylome transcriptome and proteome from OA chondrocytes defined several components which could be amenable to therapeutic intervention [36]. In addition, an epigenetic analysis shed new light on the role of methylation in regulating the gene expression of transcription factors, cytokines, ECM proteins, and MMPs pertinent to OA [37]. Just as critical to this analysis were the results stemming from a review of GWAS in OA [38]. This review concluded that there were several types of factors that contributed to the development of OA [38]. Thus, GWAS deemed systemic and genetic factors (e.g., *GDF5, ASTN2,*

DOT1L genes), intrinsic factors (e.g., congenital defects, past surgeries, infection), as well as extrinsic biomechanical factors (e.g., physical activity, high body mass index (BMI), past joint injury) as particularly germane to the development of OA.

3. MicroRNAs and OA

Overview

Microribonucleic acids (miRNAs; miRs) are ~22-nucleotide non-coding RNAs which interact with cognate messenger RNAs (mRNAs) [39]. Thus, it appears that the principal role of miRs is to modify posttranscriptional regulation of genes either through enhancement of degradation, by suppressing translation, or via other mechanisms.

Several groups of investigators [40–44] proposed that miRs could play a key role in OA by virtue of the fact that mechanical loading was shown to affect miR expression. In support of this contention, previously published evidence indicated that miRs were regulators of hundreds of genes which were relevant to cartilage development, homeostasis, and OA pathology [45]. The results of many studies reviewed by Wu et al. [46] revealed more than 25 miRs which were implicated in cartilage development and OA, and, in particular, miRs which were associated with regulating chondrocyte hypertrophy, and proteolytic enzyme synthesis. Of note, OA cartilage signaling pathways, exemplified by those involving transforming growth factor-ß (TGF-β)/SMA and Drosophila MADs (SMADs)/Bone Morphogenetic Proteins (BMPs), MMPs, ADAMTS, inducible nitric acid synthase (iNOS) IL-1, and TNF-α, were found, in part, to be regulated by miRs [47]. Furthermore, general pathologic phenomena associated with cartilage degeneration and OA, including inflammation, obesity, apoptosis, and defective autophagy were also found to be related to the specific activity of miRs [19]. With respect to autophagy, miRs were reported to most likely to play an important role in the aberrant autophagic response of OA chondrocytes by virtue of their capacity to regulate apoptosis [48,49] and reactive oxygen species [49].

Specifically, several miRs were identified as playing a role in OA pathology by virtue of their abnormal expression in OA. These included miR-9, miR-27, miR-34a, miR-140, miR-146a, miR-558, and miR-602 [50]. In fact, some of these miRs were identified as regulators of interleukin-mediated expression of MMP-13 (i.e., collagenase-3), an MMP critical in the degradation of cartilage type II collagen. Of note, Cong et al. [50], in reviewing the published literature, indicated that many miRs were differentially expressed in OA, whereby upregulated miRs primarily targeted events occurring in the nucleus, and downregulated miRs primarily targeted transcription.

4. Specific miRs: Role in Chondrocyte Signaling Pathways, Apoptosis, and Proteinase Production

4.1. Specific miRs: Chondrocyte Signaling

One approach to recognizing the extent to which miRs may be involved in altering articular cartilage homeostasis was conducted by determining which miRs alter various aspects of chondrocyte signaling, apoptosis, and proteinase production [51]. In that regard, the results of several recent studies focused on how miRs regulate those events which are germane to OA pathophysiology. For example, Zhang et al. [52] showed that miR-210 targeted the 3′-untranslated region (UTR) of death receptor-6 (DR6) in cultured chondrocytes which resulted in the inhibition of DR6 gene expression. The inhibition of DR6, in turn, inhibited NF-κB-mediated signaling. A similar finding was reported in vivo in rats using the anterior cruciate ligament transection model of OA whereby mIR-210 expression was also reduced. Of note, cytokine production, and NF-κB and DR6 expression were all inhibited following miR-210 lentivirus administration to the animals, suggesting that miR-210 was involved in altering articular cartilage homeostasis in this model of experimental OA. However, miR-210 is not the only miR involved in NF-κB pathway signaling; miR-26a and miR-26b suppressed karyopherin subunit alpha-3 (*KPNA3*) gene expression, the latter identified as a mediator affecting the capacity of the NF-κB p65 subunit to translocate from the cytoplasm to the nucleus [45]. Importantly, levels of MMP-3, -9, and -13, as well as of cyclooxygenase-2 (COX-2), were upregulated after chondrocytes were transfected with an inhibitor of miR-26a or miR-26b [53]. In addition, the levels of miR-26a/miR-26b, as well as of miR-138 and miR-140, were reduced in cartilage from OA patients [53]. Of note, Wei et al. [51] also showed that miR-138 was greatly reduced in OA cartilage compared to normal cartilage. Moreover, miR-138 was reduced following incubation of both OA and normal human chondrocytes with TNF-α, whereas miR-138 overexpression suppressed p65, COX-2, and IL-6 in human OA chondrocytes and in the SW1353 line of chondrogenic cells. Additional miRs were also implicated in altering chondrocyte signaling pathways and gene targets (Table 1).

Table 1. Chondrocyte signaling pathways and genes regulated by microRNAs (miRs).

miR	Signaling Target/Other Targets	Reference
miR-139	IGF1R [1]/EIF4G2 [2]	[54]
miR-140	SMAD1 [3]	[55]
miR-29a/miR-140	IL-1β/MMP13 [4] ↓/TIMP1 [5] ↑	[56]
miR-130a	Tumor necrosis factor-α ↓	[57]
miR-27a–3p	MAPK; NF-κB/ADAMTS5 [6]	[58]
	SMAD3 [7]	[59]
miR-634	PI3K [8]/PIK3R1 [9]	[60]
miR-449a	IL-1β/SIRT1 [10]	[61]
miR-9	L-1β/IL-6; (MCP1) [11]	[62]
miR-29	SMAD; NF-κB; WNT-related genes, FZD3 [12], FZD5 [13], DVL3 [14], FRAT2 [15], CK2A2 [16]	[63]
miR-92a-3p	Histone deacetylase 2↓	[64]
miR-381	Histone deacetylase 4↓	[65]
miR-370/miR-373	SHMT2 [17]/MECP2 [18]	[66]
miR-21	GDF5 [19]	[67]
miR-29b	COL2A1 [20]↓; COL1A1 [21] ↑	[68]
miR-146a	Camk2d [22]/Ppp32r [23]	[69]
miR-146b	SOX5 [24]	[70]
miR-155	Autophagy-Related Genes: Ulk1; FoxO3; Atg14; Atg5; Atg3; Gabarappl1; Map1lc3	[71]
miR-30b	BECN1 [25]/Atg5	[72,73]

[1] Insulin-like growth factor-1 receptor; [2] eukaryotic translation initiation factor 4 gamma 2; [3] mothers against decapentaplegic homolog 1; [4] Matrix Metalloproteinase-13; [5] Tissue Inhibitor of Metalloproteinase-1. [6] a disintegrin and metalloproteinase with thrombospondin motif-5; [7] mothers against decapentaplegic homolog 3; [8] phosphatidylinositol 3-kinase; [9] regulatory subunit 1 of class 1 PI3K (p85α); [10] NAD-dependent deacetylase sirtuin-1; [11] monocyte chemoattractant protein-induced protein 1; [12] frizzled class receptor 3; [13] frizzled class receptor 5; [14] disheveled segment polarity protein 3; [15] frequently rearranged in advanced T-cell lymphomas 2; [16] CK2A2 encodes a subunit of protein kinase CK2; [17] serine hydroxymethyl transferase-2; [18] methyl Cpg binding protein 2; [19] growth/differentiation factor 5; [20] Collagen Type II; [21] Collagen Type I; [22] calcium/calmodulin-dependent protein kinase II delta; [23] phosphatase 2 regulatory subunit B; [24] SRY-related HMG-box; [25] beclin-1.

Most of the targets shown in Table 1 were cited from evidence presented in papers from the PubMed database (https://www.ncbi.nlm.nih.gov/pubmed). This evidence demonstrated that many of the miRs (e.g., miR-29, miR-130a, miR-138, miR-139, miR-210, miR-26a, miR-26b, miR-140, miR-130a, and miR-27a–3p) were implicated in targeting those genes that are involved in dysregulated signal transduction (e.g., MAPK, IGFR1, PI3K, SMAD1, NF-κB), ECM protein turnover (e.g., COL2A1,

COL1A1), pro-inflammatory cytokines (e.g., *TNFα*, *IL1ß*, *IL6*), methylation/demethylation (e.g., histone deacetylase-2, -4), MMP and ADAMTS (e.g., *MMP13*, *TIMP1*, *ADAMTS5*), and autophagy (e.g., *Atg3*, *Atg5*, *Atg14*), many of which were shown to be critical in altering the chondrocyte phenotype characteristic of OA cartilage [51–59].

4.2. Specific miRs: Apoptosis

Apoptosis is regulated by the balanced expression of pro-apoptotic and anti-apoptotic proteins, including the influential effects of a class of proteins termed cellular inhibitor of apoptosis proteins (c-IAPs) [32]. Evidence from animal models of OA and from measurements in human OA cartilage showed that the frequency of apoptotic chondrocytes was significantly increased in human OA cartilage [4,17,19]. The results of a study indicated a relationship between miR-146a and additional biomarkers of inflammation, including mothers against decapentaplegic homolog 4 (SMAD4) and vascular endothelial growth factor (VEGF) [74]. Importantly, upregulation of miR-146a increased apoptosis in human chondrocytes in association with increased levels of *VEGF* and decreased *SMAD4* gene expression. Consequently, downregulation of miR-146a reduced human chondrocyte apoptosis, whereas upregulation of VEGF induced by miR-146a was shown to correlate with SMAD4 in chondrocytes stimulated by mechanical injury [75]. More recent studies also indicated that over-expression of miR-34a induced chondrocyte apoptosis while also inhibiting proliferation which involved sirtuin-1 (SIRT1)/p53 signaling. Thus, an anti-miR-34a sequence administered to rats who had developed OA-like cartilage lesions following anterior cruciate ligament transection reduced the progression of cartilage degradation in this rodent OA model [76]. Wu et al. [77] showed that miR-181, which targets phosphatase and tensin homolog deleted on chromosome 10 (PTEN), a regulator of the phosphatidylinositol 3-kinase (PI3K)/Akt/Mechanistic Target of Rapamycin (mTOR) signaling pathway [78], upregulated the expression of caspase-3, poly-ADP-ribose polymerase (PARP), MMP-2, and MMP-9. These changes inhibited chondrocyte proliferation while also inducing apoptosis. Taken together, the results of these experimental studies indicated that targeting genes with miRs could ultimately become an effective way of thwarting the elevated frequency of apoptotic chondrocytes characteristic of human cartilage in OA synovial joints.

4.3. Specific miRs: MMPs and ADAMTS

A host of miRs were discovered to regulate the expression of several MMPs and ADAMTS which are relevant to the regulation of chondrocyte homeostasis, and to the pathogenesis and progression of OA. Thus, it was instructive in establishing a role for miRs in altering human OA cartilage ECM when Yamasaki et al. [74] showed that expression of miR-146a and *COL2A1* gene expression (type II collagen) were decreased in OA cartilage with low-grade OA, whereas MMP-13 expression was increased at a similar cartilage grade, suggesting that decreased miR-146a was associated with elevated MMP-13 levels. The results of this study [74] also showed that miR-146a increased following the incubation of human chondrocytes with IL-1β. In another study, Song et al. [79] showed that miR-21 regulated the expression of the growth arrest-specific 5 gene (*GAS5*) in normal and OA regions of human cartilage removed at total knee replacement. In addition, overexpression of *GAS5* increased the expression of MMPs, including *MMP2*, *MMP3*, *MMP9*, *MMP13*, and *ADAMTS4*, while also stimulating apoptosis and suppressing autophagy. Furthermore, miR-21 was also shown to regulate GAS5, as evidenced by its reduced level in OA and the capacity of miR-21 to significantly increase cartilage destruction in "OA mice". Song et al. [80] also showed that miR-222 regulated MMP-13 via *histone deacetylase 4* (*HDAC4*). Thus, miR-222 was significantly reduced in OA chondrocytes, and overexpression of miR-222 suppressed apoptosis while also reducing *HDAC4* and *MMP13* gene expression. The biomarker of senescence, also known as p16INK4a, was shown to be a target for miR-24 [81], and when *p16INK4a* was over-expressed in chondrocytes, the levels of *MMP1* and *MMP13* increased. Thus, miR-24 was identified as a negative regulator of p16INK4a, as well as of cartilage development and OA. A few years ago, Meng et al. [82] also identified miR-320 as a regulator of *MMP13*, where NF-κB and MAPK activation was shown to downregulate miR-320 expression. This conclusion was later extended by Jin et al. [83], who showed that overexpression of miR-320a, which targets the pre-B-cell leukemia transcription factor 3 (*PBX3*) gene, increased MMP-13, whereas *PBX3* and the proteasome inhibitor, MG132, suppressed the effects of miR-320a on the expression of *COL2A1, ACAN* (i.e., the aggrecan gene), sulfated glycosaminoglycans, and MMP-13. Moreover, miR-105 was also found to be important for the fibroblast growth factor-2 (FGF2)-induced gene expression of *ADAMTS4, 5, 7*, and *12* [84]. These are enzymes with activity toward type II collagen and aggrecan in OA. In addition, miR-105 was reduced in OA and was inversely correlated with *Runx2* expression. Thus, the FGF2/p65/miR-105/ADAMTS axis appears to be important in OA pathogenesis. Mao et al. [85] also reported that miR-92a–3p regulated *ADAMTS4/ADAMTS-5* expression, which also involved NF-κB and MAPK activation. Additionally, miR-30a was also identified as a regulator of ADAMTS-5 [86], whereby IL-1β suppressed miR-30a expression via activation of the activator protein-1 (AP-1) complex consisting of c-jun/c-fos. Thus, AP1/miR-30a appears to be essential for IL-1β-induced increase in *ADAMTS5* gene expression.

5. Conclusions and Future Perspectives

Although persuasive evidence confirmed the critical role of pro-inflammatory cytokines as potent inducers of chronic inflammation, apoptosis, and proteinase gene expression in OA, the results of recent studies extended this analysis to demonstrate how various miRs regulate the effects of these cytokines, the balance between pro-apoptotic and anti-apoptotic proteins [87], and *MMP/ADAMTS* gene expression [88], all of which contribute to altered articular cartilage structure in OA. However, the complexity inherent in these responses, which resulted from identification of the numerous miRs involved in the pathogenesis and progression of human OA, created uncertainty as to how employing exogenous miRs could alter articular cartilage ECM aberrant turnover and ECM degradation, all of which are consistent characteristics of OA progression to joint failure. Perhaps more germane to their potential as translational medicines is that miRs were employed to identify additional targets for intervention in OA, including those involved in autophagy and proteasome-mediated protein degradation [89]. Furthermore, results of experimental studies performed in well-validated animal models of OA in which miRs were administered to these animals confirmed that miRs were active in vivo *T*. Thus, the results of these pre-clinical studies provide a compelling platform for considering the use of miRs in OA clinical safety trials going forward.

Funding: This research received no external funding.

Acknowledgments: I thank my many colleagues for helpful discussions.

Conflicts of Interest: The author declares no conflicts of interest.

References

1. Glasson, S.S. In vivo osteoarthritis target validation utilizing genetically-modified mice. *Curr. Drug Targets* **2007**, *8*, 367–376. [CrossRef] [PubMed]
2. Veronesi, F.; Della Bella, E.; Cepollaro, S.; Brogini, S.; Martin, L.; Fini, M. Novel therapeutic targets in osteoarthritis: Narrative review on knock-out genes involved in disease development in mouse animal models. *Cytotherapy* **2016**, *18*, 593–612. [CrossRef] [PubMed]
3. Rahmati, M.; Nalesso, G.; Mobasheri, A.; Mozafari, M. Aging and osteoarthritis: Central role of the extracellular matrix. *Ageing Res. Rev.* **2017**, *40*, 20–30. [CrossRef] [PubMed]
4. Malemud, C.J. Biological basis of osteoarthritis: State of the evidence. *Curr. Opin. Rheumatol.* **2015**, *27*, 289–294. [CrossRef] [PubMed]
5. Mobasheri, A.; Matta, C.; Zákány, R.; Musumeci, G. Chondrosenescence: Definition, hallmarks and potential role in the pathogenesis of osteoarthritis. *Maturitas* **2015**, *80*, 237–244. [CrossRef] [PubMed]

6. Musumeci, G.; Castrogiovanni, P.; Trovato, M.; Imbesi, R.; Giunta, S.; Szychlinska, M.A.; Loreto, C.; Castorina, S.; Mobasheri, A. Physical activity ameliorates cartilage degeneration in a rat model of aging: A study on lubricin expression. *Scand. J. Med. Sci. Sports* **2015**, *25*, e222–e230. [CrossRef] [PubMed]

7. Gargiulo, S.; Gamba, P.; Poli, G.; Leonarduzzi, G. Metalloproteinases and metalloproteinase inhibitors in age-related diseases. *Curr. Pharm. Des.* **2014**, *20*, 2993–3018. [CrossRef] [PubMed]

8. Malemud, C.J. *Role of Proteases in Cellular Dysfunction;* Springer Science: New York, NY, USA, 2013; pp. 63–77.

9. Kapor, M.; Martel-Pelletier, J.; Lajeunesse, D.; Pelletier, J.-P.; Fahmi, H. Role of proinflammatory cytokines in the pathophysiology of osteoarthritis. *Nat. Rev. Rheumatol.* **2011**, *7*, 33–42. [CrossRef] [PubMed]

10. Wang, H.; Shen, J.; Jin, H.; Im, J.H.; Sandy, J.; Chen, D. Recent progress in understanding molecular mechanisms of cartilage degeneration during osteoarthritis. *Ann. N. Y. Acad. Sci.* **2011**, *1240*, 61–69. [CrossRef] [PubMed]

11. Meszaros, E.; Malemud, C.J. Prospects for treating osteoarthritis: Enzyme-protein interactions regulating matrix metalloproteinase activity. *Adv. Ther. Chronic Dis.* **2012**, *3*, 219–229. [CrossRef] [PubMed]

12. Malemud, C.J. The medical therapy of osteoarthritis: "Thinking outside the box". *J. Osteoarthr.* **2016**, *1*, e101.

13. Mease, P.J.; Hanna, S.; Frakes, E.P.; Altman, R.D. Pain mechanisms in osteoarthritis: Understanding the role of central pain and current approaches to its treatment. *J. Rheumatol.* **2011**, *38*, 1546–1551. [CrossRef] [PubMed]

14. Perrot, S. Osteoarthritis pain. *Best Pract. Res. Clin. Rheumatol.* **2015**, *29*, 90–97. [CrossRef] [PubMed]

15. Miller, R.E.; Block, J.A.; Malfait, A.M. What is new in pain modification in osteoarthritis? *Rheumatol. (Oxf. UK)* **2018**, *57*, iv99–iv107. [CrossRef] [PubMed]

16. Schulte, M.E.; Malemud, C.J. Is there a final common pathway for arthritis? *Future Rheumatol.* **2008**, *3*, 253–268.

17. Malemud, C.J.; Gillespie, H.J. The role of apoptosis in arthritis. *Curr. Rheum. Rev.* **2005**, *1*, 131–142. [CrossRef]

18. Charlier, E.; Relic, B.; Deroyer, C.; Malaise, O.; Neuville, S.; Collée, J.; Malaise, M.G.; De Seny, D. Insights on molecular mechanisms of chondrocyte death in osteoarthritis. *Int. J. Mol. Sci.* **2016**, *17*, 2146. [CrossRef] [PubMed]

19. Musumeci, G.; Castrogiovanni, P.; Trovato, F.M.; Weinberg, A.M.; Al-Wasiyah, M.K.; Algahtani, M.H.; Mobasheri, A. Biomarkers of chondrocyte apoptosis and autophagy in osteoarthritis. *Int. J. Mol. Sci.* **2015**, *16*, 20560–20575. [CrossRef] [PubMed]

20. Musumeci, G.; Loreto, C.; Carnazza, M.L.; Strehin, I.; Elisseeff, J. OA cartilage derived chondrocytes encapsulated in poly (ethylene glycol) diacrylate (PEGDA) for the evaluation of cartilage restoration and apoptosis in an in vitro model. *Histol. Histopathol.* **2011**, *26*, 1265–1278. [PubMed]

21. Musumeci, G.; Loreto, C.; Carnazza, M.L.; Martinez, G. Characterization of apoptosis in articular cartilage derived from knee joints of patients with osteoarthritis. *Knee Surg. Sports Traumatol. Arthrosc.* **2011**, *19*, 307–313. [CrossRef] [PubMed]

22. Wu, L.; Huang, X.; Li, L.; Xu, R.; Luyten, W. Insights on biology and pathology of HIF-1α/2α, TGFß/BMP, Wnt/ß-catenin, and NF-κB pathways in osteoarthritis. *Curr. Pharm. Des.* **2012**, *18*, 3293–3312. [CrossRef] [PubMed]

23. Jiang, Y.; Xiao, Q.; Hu, Z.; Pu, B.; Shu, J.; Yang, Q.; Lao, H.; Hao, J. Tissue levels of leukemia inhibitory factor vary by osteoarthritis grade. *Orthopedics* **2014**, *37*, e460–e464. [CrossRef] [PubMed]

24. Leitjen, J.C.; Bos, S.D.; Landman, E.B.; Georgi, N.; Jahr, H.; Meulenbelt, I.; Post, J.N.; van Bilterswiik, C.A.; Karperien, M. GREM1, FRZB, and DKK1 mRNA levels correlate with osteoarthritis and are regulated by osteoarthritis-related factors. *Arthritis Res. Ther.* **2013**, *15*, R126.

25. Melas, N.; Chairakaki, A.D.; Chatzopoulou, E.I.; Messinis, D.E.; Katopodi, T.; Pliaka, V.; Samara, S.; Mitsos, A.; Dailiana, Z.; Kollia, P.; et al. Modeling of signaling pathways in osteoarthritis based on phosphoproteomics and cytokine release data. *Osteoarthr. Cartil.* **2014**, *22*, 509–518. [CrossRef] [PubMed]

26. Muoh, O.; Malemud, C.J.; Askari, A.D. Clinical significance and implications of genetic and genomic studies in patients with osteoarthritis. *Adv. Genom. Genet.* **2014**, *4*, 193–206.

27. Saito, T.; Tanaka, S. Molecular mechanisms underlying osteoarthritis development: Notch and NF-κB. *Arthritis Res. Ther.* **2017**, *19*, 94. [CrossRef] [PubMed]

28. Islam, N.; Haqqi, T.M.; Jepsen, K.J.; Kraay, M.; Welter, J.F.; Goldberg, V.M.; Malemud, C.J. Hydrostatic pressure induces apoptosis in human chondrocytes from osteoarthritic cartilage through up-regulation of tumor necrosis factor-α, inducible nitric oxide synthase, p53, c-myc and bax-α and suppression of bcl-2. *J. Cell Biochem.* **2002**, *87*, 266–278. [CrossRef] [PubMed]

29. Malemud, C.J. Protein kinases in chondrocyte signaling and osteoarthritis. *Clin. Orthop. Relat. Res.* **2004**, *427*, S145–S151. [CrossRef]

30. Ivashkiv, L.B.; Hu, X. Signaling by STATs. *Arthritis Res. Ther.* **2004**, *6*, 159–168. [CrossRef] [PubMed]

31. Stark, G.R.; Darnell, J.E., Jr. The JAK-STAT pathway at 20. *Immunity* **2012**, *36*, 503–514. [CrossRef] [PubMed]

32. Malemud, C.J. Negative regulators of JAK/STAT signaling in rheumatoid arthritis and osteoarthritis. *Int. J. Mol. Sci.* **2017**, *18*, 484. [CrossRef] [PubMed]

33. Akeson, G.; Malemud, C.J. A role for soluble IL-6 receptor in osteoarthritis. *J. Funct. Morphol. Kinesiol.* **2017**, *2*, 27. [CrossRef] [PubMed]

34. Raman, S.; FitzGerald, U.; Murphy, J.M. Interplay of inflammatory mediators with epigenetics and cartilage modifications in osteoarthritis. *Front. Bioeng. Biotechnol.* **2018**, *6*, 22. [CrossRef] [PubMed]

35. Ripmeester, E.G.J.; Timur, U.T.; Caron, M.M.J.; Welting, T.M.J. Recent insights into the contribution of the changing hypertrophic chondrocyte phenotype in the development and progression of osteoarthritis. *Front. Bioeng. Biotechnol.* **2018**, *6*, 18. [CrossRef] [PubMed]

36. Hammaker, D.; Firestein, G.S. Epigenetics of inflammatory arthritis. *Curr. Opin. Rheumatol.* **2018**, *30*, 188–196. [CrossRef] [PubMed]

37. Zhang, M.; Wang, J. Epigenetics and osteoarthritis. *Genes Dis.* **2015**, *2*, 69–75. [CrossRef] [PubMed]

38. Warner, S.C.; Valdes, A.M. The genetics of osteoarthritis: A. review. *J. Funct. Morphol. Kinesiol.* **2016**, *1*, 140–153. [CrossRef]

39. D'Adamo, S.; Cetrullo, S.; Minquzzi, M.; Silvestri, Y.; Borzi, R.M.; Flamigni, F. MicroRNAs and autophagy: Fine players in the control of chondrocyte homeostatic activities in osteoarthritis. *Oxid. Med. Cell Longev.* **2017**, *2017*, 3720128. [CrossRef] [PubMed]

40. Yu, C.; Chen, W.P.; Wang, X.H. MicroRNA in osteoarthritis. *J. Int. Med. Res.* **2011**, *39*, 1–9. [CrossRef] [PubMed]

41. Miyaki, S.; Asahara, H. Macro view of microRNA function in osteoarthritis. *Nat. Rev. Rheumatol.* **2012**, *8*, 543–552. [CrossRef] [PubMed]

42. Sondag, G.R.; Haqqi, T.M. The role of microRNAs and their targets in osteoarthritis. *Curr. Rheumatol. Rep.* **2016**, *18*, 56. [CrossRef] [PubMed]

43. De Palma, A.; Cheleschi, S.; Pascarelli, N.A.; Tenti, S.; Galeazzi, M.; Fioravanti, A. Do microRNAs have a key epigenetic role in osteoarthritis and in mechanotransduction? *Clin. Exp. Rheumatol.* **2017**, *35*, 518–526. [PubMed]

44. Zhang, M.; Lygrisse, K.; Wang, L. Role of microRNA in osteoarthritis. *J. Arthritis* **2017**, *6*, 239. [CrossRef] [PubMed]

45. Mirzamohammad, F.; Papaioannou, G.; Kobayashi, T. MicroRNAs in cartilage development, homeostasis and disease. *Curr. Osteoporos. Rep.* **2014**, *12*, 410–419. [CrossRef] [PubMed]

46. Wu, C.; Tian, B.; Qu, X.; Liu, F.; Tang, T.; Qin, A.; Zhu, Z.; Dai, K. MicroRNAs play a role in chondrogenesis and osteoarthritis (review). *Int. J. Mol. Med.* **2014**, *34*, 13–23. [CrossRef] [PubMed]

47. Li, Y.P.; Wei, X.C.; Li, P.C.; Chen, C.W.; Wang, X.H.; Jiao, Q.; Wang, D.M.; Wei, F.Y.; Zhang, J.; Wei, L. The role of microRNAs in cartilage homeostasis. *Curr. Genom.* **2015**, *16*, 393–404. [CrossRef] [PubMed]

48. Li, Y.S.; Zhang, F.J.; Zeng, C.; Luo, W.; Xiao, W.F.; Gao, S.G.; Lei, G.H. Autophagy in osteoarthritis. *Jt. Bone Spine* **2016**, *83*, 143–148. [CrossRef] [PubMed]

49. Nugent, M. MicroRNAs: Exploring new horizons in osteoarthritis. *Osteoarthr. Cartil.* **2016**, *24*, 573–580. [CrossRef] [PubMed]

50. Cong, L.; Zhu, Y.; Tu, G. A bioinformatics analysis of microRNAs role in osteoarthritis. *Osteoarthr. Cartil.* **2017**, *25*, 1362–1371. [CrossRef] [PubMed]

51. Wei, Z.J.; Liu, J.; Qin, J. miR-138 suppressed the progression of osteoarthritis mainly through targeting p65. *Eur. Rev. Med. Pharmacol. Sci.* **2017**, *21*, 2177–2184. [PubMed]

52. Zhang, D.; Cao, X.; Li, J.; Zhao, G. MiR-210 inhibits NF-κB signaling pathway by targeting DR6 in osteoarthritis. *Sci. Rep.* **2015**, *5*, 12775. [CrossRef] [PubMed]

53. Yin, X.; Wang, J.Q.; Yan, S.Y. Reduced miR-26a and miR-26b expression contributes to the pathogenesis of osteoarthritis via the promotion of p65 translocation. *Mol. Med. Rep.* **2017**, *15*, 551–558. [CrossRef] [PubMed]

54. Hu, W.; Zhang, W.; Li, F.; Guo, F.; Chen, A. miR-139 is up-regulated in osteoarthritis and inhibits chondrocyte proliferation and migration possibly via suppressing EIF4G2 and IGF1R. *Biochem. Biophys. Res. Commun.* **2016**, *474*, 296–302. [CrossRef] [PubMed]

55. Li, C.; Hu, Q.; Chen, Z.; Shen, B.; Yang, J.; Kang, P.; Zhou, Z.; Pei, F. MicroRNA-140 suppresses human chondrocyte hypertrophy by targeting SMAD1 and controlling bone morphogenetic protein pathway in osteoarthritis. *Am. J. Med. Sci.* **2018**, *355*, 477–487. [CrossRef] [PubMed]

56. Li, X.; Zhen, Z.; Tang, G.; Zheng, C.; Yang, G. MiR-29 and MiR-140 protect chondrocytes against the anti-proliferation and cell matrix signaling changes by IL-1ß. *Mol. Cells* **2016**, *39*, 103–110. [PubMed]

57. Li, Z.C.; Han, N.; Li, X.; Li, G.; Liu, Y.Z.; Sun, G.X.; Wang, T.; Chen, G.T.; Li, G.F. Decreased expression of microRNA-130a correlates with TNF-α in the development of osteoarthritis. *Int. J. Clin. Exp. Pathol.* **2015**, *8*, 2555–2564. [PubMed]

58. Li, X.; He, P.; Li, Z.; Wang, H.; Liu, M.; Xiao, Y.; Xu, D.; Kang, Y.; Wang, H. Interleukin-1ß-mediated suppression of microRNA-27a-3p activity in human cartilage via MAPK and NF-κB pathways: A potential mechanism of osteoarthritis pathogenesis. *Mol. Med. Rep.* **2018**. [CrossRef]

59. Kang, L.; Yang, C.; Song, Y.; Liu, W.; Wang, K.; Li, S.; Zhang, Y. MicroRNA-23a-3p promotes the development of osteoarthritis by directly targeting SMAD3 in chondrocytes. *Biochem. Biophys. Res. Commun.* **2016**, *478*, 467–473. [CrossRef] [PubMed]

60. Cui, X.; Wang, S.; Cai, H.; Lin, Y.; Zheng, X.; Zhang, B.; Xia, C. Overexpression of microRNA-634 suppresses survival and matrix synthesis of human osteoarthritis chondrocytes by targeting PIK3R1. *Sci. Rep.* **2016**, *6*, 23117. [CrossRef] [PubMed]

61. Park, K.W.; Lee, K.M.; Yoon, D.S.; Park, K.H.; Choi, W.J.; Lee, J.W.; Kim, S.H. Inhibition of microRNA-449a prevent IL-1ß-induced cartilage destruction via SIRT1. *Osteoarthr. Cartil.* **2016**, *24*, 2153–2161. [CrossRef] [PubMed]

62. Makki, M.S.; Haseeb, A.; Haqqi, T.M. MicroRNA-9 promotion of interleukin-6 expression by inhibiting monocyte chemoattractant protein-induced protein 1 expression in interleukin-1ß-stimulated human chondrocytes. *Arthritis Rheumatol.* **2015**, *67*, 2117–2128. [CrossRef] [PubMed]

63. Le, T.; Swingler, T.E.; Crowe, N.; Vincent, T.L.; Barter, M.J.; Donell, S.T.; Delany, A.M.; Dalmay, T.; Young, D.A.; Clark, I.M. The microRNA-29 family in cartilage homeostasis and osteoarthritis. *J. Mol. Med.* **2016**, *94*, 583–596. [CrossRef] [PubMed]

64. Mao, G.; Zhang, Z.; Huang, Z.; Chen, W.; Huang, G.; Meng, F.; Zhang, Z.; Kang, Y. MicroRNA-92a-3p regulates the expression of cartilage-specific genes by directly targeting histone deacetylase 2 in chondrogenesis and degradation. *Osteoarthr. Cartil.* **2017**, *25*, 521–532. [CrossRef] [PubMed]

65. Chen, W.; Sheng, P.; Huang, Z.; Meng, F.; Kang, Y.; Huang, G.; Zhang, Z.; Liao, W.; Zhang, Z. MicroRNA-381 regulates chondrocyte hypertrophy by inhibiting histone deacetylase 4 expression. *Int. J. Mol. Sci.* **2016**, *17*, 1377. [CrossRef] [PubMed]

66. Song, J.; Kim, D.; Chun, C.H.; Jin, E.J. miR-370 and miR-373 regulate the pathogenesis of osteoarthritis by modulating one-carbon metabolism via SHMT-2 and MECP-2, respectively. *Aging Cell* **2015**, *14*, 826–837. [CrossRef] [PubMed]

67. Zhang, Y.; Jia, J.; Yang, S.; Liu, X.; Ye, S.; Tian, H. MicroRNA-21 controls the development of osteoarthritis by targeting GDF5 in chondrocytes. *Exp. Mol. Med.* **2014**, *46*, e79. [CrossRef] [PubMed]

68. Moulin, D.; Salone, V.; Koufany, M.; Clément, T.; Behm-Ansmant, I.; Branlant, C.; Charpentier, B.; Jouzeau, J.Y. MicroRNA-29b contributes to collagen imbalance in human osteoarthritic and dedifferentiated articular chondrocytes. *Biomed. Res. Int.* **2017**, *2017*, 9792512. [CrossRef] [PubMed]

69. Zhang, X.; Wang, C.; Zhao, J.; Xu, J.; Geng, Y.; Dai, L.; Huang, Y.; Fu, S.C.; Dai, K.; Zhang, X. miR-146a facilitates osteoarthritis by regulating cartilage homeostasis via targeting Camk2d and Ppp3r2. *Cell Death Dis.* **2017**, *8*, e2734. [CrossRef] [PubMed]

70. Budd, E.; de Andrées, M.C.; Sanchez-Eisner, T.; Oreffo, R.O.C. MiR-146b is down-regulated during the chondrogenic differentiation of human bone marrow derived skeletal stem cells and up-regulated in osteoarthritis. *Sci. Rep.* **2017**, *7*, 46704. [CrossRef] [PubMed]

71. D'Adamo, S.; Alvarez-Garcia, O.; Muramastu, Y.; Flamigni, F.; Lotz, M.K. MicroRNA-155 suppresses autophagy in chondrocytes by modulating expression of autophagy proteins. *Osteoarthr. Cartil.* **2016**, *24*, 1082–1091. [CrossRef] [PubMed]

72. Chen, Z.; Jin, T.; Lu, Y. AntimiR-30b inhibits TNF-α mediated apoptosis and attenuated cartilage degradation through enhancing autophagy. *Cell Physiol. Biochem.* **2016**, *40*, 883–894. [CrossRef] [PubMed]

73. Song, B.; Song, H.; Wang, W.; Wang, H.; Peng, H.; Cui, J.; Wang, B.; Huang, H.; Wang, W.; Wang, L. Beclin overexpression inhibits chondrocyte apoptosis and downregulates extracellular matrix metabolism in osteoarthritis. *Mol. Med. Rep.* **2017**, *16*, 3958–3964. [CrossRef] [PubMed]

74. Yamasaki, K.; Nakasa, T.; Miyaki, S.; Ishikawa, M.; Deie, M.; Adachi, N.; Yasunaga, Y.; Asahara, H.; Ochi, M. Expression of MicroRNA-146a in osteoarthritic cartilage. *Arthritis Rheum.* **2009**, *60*, 1035–1041. [CrossRef] [PubMed]

75. Jin, L.; Zhao, J.; Jing, W.; Yan, S.; Wang, X.; Xiao, C.; Ma, B. Role of miR-146a in human chondrocyte apoptosis in response to mechanical pressure injury in vitro. *Int. J. Mol. Med.* **2014**, *34*, 451–463. [CrossRef] [PubMed]

76. Yan, S.; Wang, M.; Zhao, J.; Zhang, H.; Zhou, C.; Jin, L.; Zhang, Y.; Qiu, X.; Ma, B.; Fan, Q. MicroRNA-34a affects chondrocyte apoptosis and proliferation by targeting the SIRT1/p53 signaling pathway during the pathogenesis of osteoarthritis. *Int. J. Mol. Med.* **2016**, *38*, 201–209. [CrossRef] [PubMed]

77. Wu, X.F.; Zhou, Z.H.; Zou, J. MicroRNA-181 inhibits proliferation and promotes apoptosis of chondrocytes in osteoarthritis by targeting PTEN. *Biochem. Cell Biol.* **2017**, *95*, 437–444. [CrossRef] [PubMed]

78. Malemud, C.J. PI3K/Akt/PTEN/mTOR signaling: A fruitful target for inducing cell death in rheumatoid arthritis? *Future Med. Chem.* **2015**, *7*, 1137–1147. [CrossRef] [PubMed]

79. Song, J.; Ahn, C.; Chun, C.H.; Jin, E.J. A long non-coding RNA, GAS5, plays a crucial role in the regulation of miR-21 during osteoarthritis. *J. Orthop. Res.* **2014**, *32*, 1628–1635. [CrossRef] [PubMed]

80. Song, J.; Jin, E.H.; Kim, D.; Kim, K.Y.; Chun, C.H.; Jin, E.J. MicroRNA-222 regulates MMP-13 via targeting HDAC-4 during osteoarthritis pathogenesis. *BBA Clin.* **2014**, *3*, 79–89. [CrossRef] [PubMed]

81. Philpot, D.; Guérit, D.; Platano, D.; Chuchana, P.; Olivotto, E.; Espinoza, F.; Dorandeu, A.; Pers, Y.M.; Piette, J.; Borzi, R.M.; et al. P161NK4a and its regulator miR-24 link senescence and chondrocyte terminal differentiation-associated matrix remodeling in osteoarthritis. *Arthritis Res. Ther.* **2014**, *16*, R58. [CrossRef] [PubMed]

82. Meng, F.; Zhang, Z.; Chen, W.; Huang, G.; He, A.; Hou, C.; Long, Y.; Zang, Z.; Zhang, Z.; Liao, W. MicroRNA-320 regulates matrix metalloproteinase-13 expression in chondrogenesis and interleukin-1ß-induced chondrocyte responses. *Osteoarthr. Cartil.* **2016**, *24*, 932–941. [CrossRef] [PubMed]

83. Jin, Y.; Chen, X.; Gao, Z.Y.; Liu, K.; Hou, Y.; Zheng, J. The role of miR-320a and IL-1ß in human chondrocyte degradation. *Bone Jt. Res.* **2017**, *6*, 196–203. [CrossRef] [PubMed]

84. Ji, Q.; Xu, X.; Xu, Y.; Fan, Z.; Kang, L.; Li, L.; Liang, Y.; Guo, J.; Hong, T.; Li, Z.; et al. miR-105/Runx2 axis mediates FGF2-induced ADAMTS5 expression in osteoarthritis. *J. Mol. Med.* **2016**, *94*, 681–694. [CrossRef] [PubMed]

85. Mao, G.; Wu, P.; Zhang, Z.; Zhang, Z.; Liao, W.; Li, Y.; Kang, Y. MicroRNA-92a-3p regulates aggrecanase—1 and aggrecanase-2 expression in chondrogenesis and IL-1ß-induced catabolism in human articular chondrocytes. *Cell Physiol. Biochem.* **2017**, *44*, 38–52. [CrossRef] [PubMed]

86. Ji, Q.; Xu, X.; Zhang, Q.; Kang, L.; Xu, Y.; Zhang, K.; Li, L.; Liang, Y.; Hong, T.; Ye, Q.; et al. The IL-1ß/AP-1/miR-30a/ADAMTS-5 axis regulates cartilage matrix degradation in human osteoarthritis. *J. Mol. Med. (Berl.)* **2016**, *94*, 771–785. [CrossRef] [PubMed]

87. Pelletier, J.-P.; Fernandes, J.C.; Jovanovic, D.V.; Reboul, P.; Martel-Pelletier, J. Chondrocyte death in experimental osteoarthritis is mediated by MEK1/2 and p38 pathways: Role of cyclooxygenase-2 and inducible nitric oxide synthase. *J. Rheumatol.* **2001**, *28*, 2509–2519. [PubMed]

88. Malemud, C.J.; Islam, N.; Haqqi, T.M. Pathophysiological mechanisms in osteoarthritis lead to novel therapeutic strategies. *Cells Tissues Organs* **2003**, *174*, 34–48. [CrossRef] [PubMed]

89. Serrano, R.L.; Chen, L.Y.; Lotz, M.K.; Liu-Bryan, R.; Terkeltaub, R. Impaired proteasomal function in human osteoarthritic chondrocytes can contribute to decreased levels of SOX9 and aggrecan. *Arthritis Rheumatol.* **2018**, *70*, 1030–1041. [CrossRef] [PubMed]

cells

MDPI

Article

Association of NF-κB and AP-1 with MMP-9 Overexpression in 2-Chloroethanol Exposed Rat Astrocytes

Tong Wang [1], Xiaoxia Jin [1], Yingjun Liao [2], Qi Sun [1], Chaohong Luo [1], Gaoyang Wang [1], Fenghong Zhao [1] and Yaping Jin [1,*]

[1]   Department of Occupational and Environmental Health, School of Public Health, China Medical University, No. 77 Puhe Road, Shenyang North New Area, Shenyang 110122, Liaoning, China; wang1234.sm@163.com (T.W.); xxjincmu@163.com (X.J.); sunqi@cmu.edu.cn (Q.S.); 13940522351@163.com (C.L.); gywang@cmu.edu.cn (G.W.); fhzhao@cmu.edu.cn (F.Z.)
[2]   Department of Physiology, China Medical University, Shenyang 110122, Liaoning, China; yjliao@cmu.edu.cn
*   Correspondence: ypjin@cmu.edu.cn; Tel.: +86-024-3193-9406

Received: 16 July 2018; Accepted: 3 August 2018; Published: 7 August 2018

Abstract: Subacute poisoning of 1,2-dichloroethane (1,2-DCE) has become a serious occupational problem in China, and brain edema is its main pathological consequence, but little is known about the underlying mechanisms. As the metabolite of 1,2-DCE, 2-chloroethanol (2-CE) is more reactive, and might play an important role in the toxic effects of 1,2-DCE. In our previous studies, we found that matrix metalloproteinases-9 (MMP-9) expression was enhanced in mouse brains upon treatment with 1,2-DCE, and in rat astrocytes exposed to 2-CE. In the present study, we analyzed the association of nuclear factor kappa B (NF-κB) and activator protein-1 (AP-1) with MMP-9 overexpression in astrocytes treated with 2-CE. MMP-9, p65, c-Jun, and c-Fos were significantly upregulated by 2-CE treatment, which also enhanced phosphorylation of c-Jun, c-Fos and inhibitor of κBα (IκBα), and nuclear translocation of p65. Furthermore, inhibition of IκBα phosphorylation and AP-1 activity with the specific inhibitors could attenuate MMP-9 overexpression in the cells. On the other hand, inhibition of p38 mitogen-activated protein kinase (p38 MAPK) signaling pathway suppressed the activation of both NF-κB and AP-1 in 2-CE-treated astrocytes. In conclusion, MMP-9 overexpression induced by 2-CE in astrocytes could be mediated at least in part through the p38 signaling pathway via activation of both NF-κB and AP-1. This study might provide novel clues for clarifying the mechanisms underlying 1,2-DCE associated cerebral edema.

Keywords: 1,2-Dichloroethane poisoning; 2-Chloroethanol; matrix metalloproteinases-9; p38 MAPK signal pathway; nuclear factor-κB; activator protein-1

1. Introduction

The synthetic halohydrocarbon 1,2-dichloroethane (1,2-DCE) is mainly used as the monomer in the manufacture of polyvinyl chloride, and as an industrial solvent and glue thinner. It presents a significant occupational hazard to factory workers as exposure to high concentrations of 1,2-DCE vapors is lethal. In the last thirty years, a number of cases of subacute poisoning of 1,2-DCE have been reported in China [1]. Brain edema is the main pathological consequence of 1,2-DCE poisoning [2,3], but little is known regarding the underlying mechanisms.

Previous studies have shown that 2-chloroethanol (2-CE), a metabolite of 1,2-DCE generated in vivo via microsomal CYP2E1, is more reactive than its parent molecule, and therefore might play an important role in the toxic effects of 1,2-DCE [4–6]. Our studies showed that matrix metalloproteinases-9 (MMP-9) was transcriptionally upregulated in 2-CE-treated astrocytes in vitro [7], as well as in the early phase of brain edema induced by subacute 1,2-DCE poisoning in mice [8]. MMP-9 is a zinc-dependent endopeptidase that degrades components of the extracellular matrix (ECM) in both physiological and pathological processes [9]. Although MMP-9 is normally present at low levels in astrocytes, it can be dramatically enhanced by ischemic and hemorrhagic stroke, and is involved in subsequent brain injury and vasogenic brain edema [10,11]. Therefore, it is reasonable to assume that overexpression of MMP-9 plays an important mechanistic role in cerebral edema induced by 1,2-DCE. Accordingly, it is necessary to elucidate the pathways regulating MMP-9 overexpression in 2-CE-treated astrocytes. In our previous studies, the mitogen-activated protein kinase (MAPK) signaling pathways, including the extracellular signal-regulated kinase 1/2 (ERK 1/2), c-Jun amino terminal kinase (JNK), and p38 MAPK pathways, were shown to be involved in the upregulation of MMP-2 and MMP-9 in 2-CE-treated astrocytes [7,12]. In this study, we further explored the involvement of transcription factors in MMP-9 overexpression in 2-CE-treated astrocytes.

Many reports have in fact indicated important roles of the transcription factors in regulating MMP-9 expression. The promoter region of the MMP-9 gene harbors binding sites for both nuclear factor-κB (NF-κB) and activator protein-1 (AP-1), and activation of both transcription factors is essential for MMP-9 expression [10,13–16]. However, whether they are also activated in 2-CE-treated astrocytes and drive MMP-9 overexpression in these cells is poorly understood. On the other hand, it is well-known that p38 MAPK plays an important role in the inflammation response, the main mechanism underlying brain edema. Therefore, to further investigate the regulatory pathways mediating MMP-9 overexpression induced by 2-CE in rat astrocytes, we treated the cells with specific inhibitors of NF-κB, AP-1 and p38 before 2-CE exposure. Our study might provide novel clues for clarifying the mechanisms underlying 1,2-DCE associated cerebral edema.

2. Materials and Methods

2.1. Reagents

The 2-Chloroethanol was purchased from Sinopharm Chemical Reagent Co., Ltd. (Ningbo, China). Reagents for the primary culture of astrocytes were purchased from Biological Industries (Beit-Haemek, Israel). The quantitative real-time (RT)-PCR assay kit was purchased from Takara, Japan. The enhanced chemiluminescence (ECL) plus kit, bicinchoninic acid (BCA) protein assay kit and NE-PER™ nuclear and cytoplasmic extraction reagents were obtained from Thermo Fisher Scientific (Waltham, MA, USA). SB202190, pyrrolidine dithiocarbamate (PDTC), and SR11302 were purchased from Selleck (Houston, TX, USA) and APExBIO (Houston, TX, USA). Primary antibodies against MMP-9, p65, IκBα, c-Jun, c-Fos, p-c-Jun, p-c-Fos, p-IκBα, and LaminB were products of Abcam (Cambridge, UK) and Cell Signaling Technology (Beverly, MA, USA). Antibodies against glial fibrillary acid protein (GFAP), glyceraldehyde 3-phosphate dehydrogenase (GAPDH), and β-actin were obtained from Millipore (Billerica, MA, USA), Proteintech (Wuhan, China), and ABclonal (Wuhan, China), respectively. The secondary antibodies conjugated with Alexa Fluor 488 or tetramethylrhodamine (TRITC), RIPA Lysis Buffer, and DAPI were obtained from Beyotime Biotechnology (Shanghai, China).

2.2. Astrocyte Enrichment and Culture

All experiments were approved by the Animal Care and Use Committee at China Medical University, which complies with the recommendations of the Chinese National Guidelines for the Protection of Laboratory Animals. The project identification code was IACUC: NO.16101. The cerebral cortexes of neonatal rats were isolated and enzymatically digested [7,12]. The dissociated brain cells were then collected and plated in a culture dish pre-coated with poly-L-lysine. The cells were cultured in a humidified incubator at 37 °C under 5% CO_2 until they formed a confluent layer. The oligodendrocytes and microglia were removed by shaking the culture dishes at 250 rpm for 15 h on a horizontal shaker. The remaining cells were resuspended and re-plated for further enrichment of the astrocytes, which were identified by positive immunostaining for the glial fibrillary acid protein (GFAP). The cultures with more than 95% astrocytes were used for subsequent experiments.

2.3. In Vitro Treatments

The 1 M stock solution of 2-CE was prepared with redistilled water, and then diluted to the final concentrations by Dulbecco's Modified Eagle Medium (DMEM) containing 5% fetal bovine serum (FBS) before use. The primary astrocytes were treated with 30 mM 2-CE for 1, 2, 4, 8, 12, 24, and 48 h, and with 0, 7.5, 15, and 30 mM 2-CE for 12 h to determine the effect of different 2-CE concentrations and treatment durations on MMP-9 expression. To test the effects of NF-κB, AP-1, and p38 MAPK inhibition, astrocytes were pretreated with pyrrolidine dithiocarbamate (PDTC; 10, 25 μM for 2 h), SR11302 (5, 10 μM for 1 h), and SB202190 (1, 10, 30 μM for 1 h), respectively, before a 12 h treatment with 30 mM 2-CE. In addition, untreated controls, inhibitor controls, and solvent controls were also included. The different inhibitors were dissolved in water or dimethyl sulfoxide (DMSO) for preparing the stock solution, and then diluted by the culture media to the final dose. Solvent control cells were treated with 0.1% DMSO, and inhibitor controls with only PDTC (25 μM), SR11302 (10 μM), or SB202190 (30 μM).

2.4. Immunofluorescence

The cells were fixed and permeabilized as described [7], and then blocked with goat serum at room temperature. Thereafter, the cells were incubated overnight with rabbit anti-rat p65 and mouse anti-rat GFAP antibodies at 4 °C, followed by a 30 min incubation with goat anti-rabbit-Alexa Fluor 488 and goat anti-mouse TRITC secondary antibodies at 37 °C. The stained cells were observed under a fluorescence microscope (Olympus IX71, Tokyo, Japan), and imaged using a digital camera system (Olympus SC35, Tokyo, Japan). Negative controls lacking primary antibodies were also included.

2.5. Western Blotting

Total proteins were extracted with RIPA buffer, and the nuclear and cytoplasmic protein fractions were extracted with the NE-PER™ nuclear and cytoplasmic extraction reagents, respectively. Protein concentrations of the samples were determined by BCA protein assay kit. Equal amounts of sample proteins were separated by SDS-PAGE, and then transferred to a polyvinylidene difluoride (PVDF) membrane (Millipore, Burlington, MA, USA). The blots were incubated overnight with primary antibodies against MMP-9, c-Jun, c-Fos, IκBα, p65, p-c-Jun, p-c-Fos, p-IκBα, and GAPDH (internal control) at 4 °C. On the following day, the membranes were incubated with the secondary antibodies at room temperature for 1 h. The bands were imaged using Azure c300 Chemiluminescent Western Blot Imaging System (Azure Biosystems, Dublin, CA, USA) with the ECL Western blot chemiluminescent detection reagents. The intensity of the bands were analyzed semi-quantitatively via densitometry using the Gel-Pro analyzer v4.0 software (Meyer Instruments, Houston, TX, USA), and the relative protein expression levels were normalized to GAPDH.

2.6. Quantitative Real-Time (RT)-PCR

Total RNA was extracted using Trizol Reagent (Takara, Japan) and cDNA was synthesized by reverse transcription using PrimeScript RT reagent kit (Takara, Japan). MMP-9, p65, IκBα, c-Jun, c-Fos, and GAPDH (constitutive gene) fragments were then amplified using the specific primer pairs detailed in Table 1. The reaction was carried out for 40 cycles at 95 °C for 5 s and 60 °C for 34 s on a QuantStudio 6 Flex real-time PCR System (Life Technologies, Carlsbad, CA, USA) using the SYBR Premix Ex Taq II (Takara, Japan). The relative mRNA levels were analyzed using the comparative Ct method and expressed as $2^{-\Delta\Delta Ct}$ formula. GAPDH mRNA was used as an internal control for normalization of gene expression.

Table 1. Primer sequences for the target genes. MMP-9—matrix metalloproteinases-9; IκBα—inhibitor of κBα.

Gene		Primer Sequence (5′-3′)
MMP-9	Forward	5′-ATCCGCAGTCCAAGAAGATT-3′
	Reverse	5′-GCCAGAGAACTCGTTATCCA-3′
p38	Forward	5′-CCGAGCGATACCAGAACCT-3′
	Reverse	5′-AACACATCCAACAGACCAATCA-3′
p65	Forward	5′-TTAGCCATCATCCACCTTC-3′
	Reverse	5′-AGTCCTCCACCACATCTT-3′
IκBα	Forward	5′-GAGGATTACGAGCAGATGG-3′
	Reverse	5′-ATGGTCAGTGTCTTCTCTTC-3′
c-Jun	Forward	5′-ACGACCTTCTACGACGAT-3′
	Reverse	5′-CATTGCTGGACTGGATGAT-3′
c-Fos	Forward	5′-TCCGAAGGGAAAGGAATAAG-3′
	Reverse	5′-AGTCAAGTCCAGGGAGGTC-3′
GAPDH	Forward	5′-GCAAGAGAGAGGCCCTCAG-3′
	Reverse	5′-TGTGAGGGAGATGCTCAGTG-3′

2.7. ELISA

Culture supernatants were centrifuged at 10,000 rpm for 5 min at 4 °C. Secreted MMP-9 levels in the supernatants were measured using the rat total MMP-9 ELISA Kit (Wuhan Boster Biological Technology Co., Ltd., Wuhan, China) according to the manufacturer's instructions.

2.8. Statistical Analysis

The results are represented as the mean ± standard deviation (SD) of at least four independent experiments. Statistical analyses were carried out using SPSS for Windows, version 20.0 (SPSS, Armonk, NY, USA). Data were evaluated by one-way analysis of variance (ANOVA) followed by the Student–Newman–Keuls (SNK) test. A p-value of less than 0.05 was considered significant.

3. Results

3.1. Upregulated MMP-9 Expression in 2-CE Treated Rat Astrocytes along with the Exposure Duration and Concentrations

Although our previous study indicated that MMP-9 expression could be augmented in astrocytes upon 2-CE treatment, the transient expression profile of MMP-9 was unclear. Therefore, in this study, we tracked MMP-9 expression in the cells treated with 30 mM 2-CE for 1 h to 48 h.

The protein levels of MMP-9 increased in the astrocytes within 8 h of 2-CE exposure, and remained significantly above the control levels at 24 h (Figure 1A). Moreover, those in the culture media also increased significantly by 8 h, peaked at 24 h, and remained significantly above the control levels even after 48 h (Figure 1B). Therefore, in subsequent in vitro experiments, the cells were exposed to 2-CE for 12 h. In addition, the MMP-9 protein levels in both cells and culture media were significantly higher in the 15 and 30 mM groups compared with the control group or the 7.5 mM group, and were significantly higher in the culture media of the 30 mM group compared with the 15 mM group (Figure 1C,D). The MMP-9 mRNA levels also increased significantly upon 2-CE treatment in a dose dependent manner (Figure 1E).

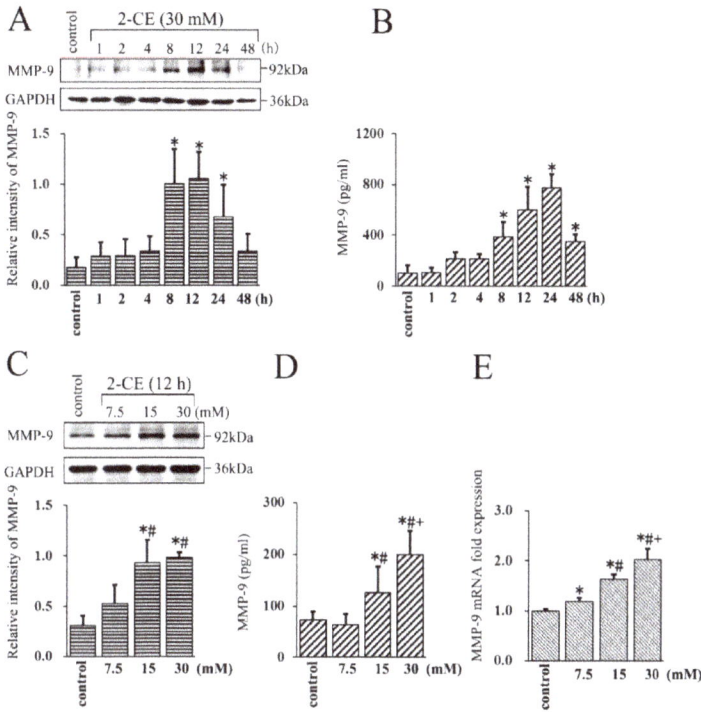

Figure 1. Effects of 2-chloroethanol (2-CE) on matrix metalloproteinases-9 (MMP-9) expression in astrocytes. (**A,B**) Astrocytes were treated with 30 mM 2-CE for the indicated times; (**C–E**) Astrocytes were treated with 0, 7.5, 15, and 30 mM 2-CE for 12 h. (**A,C**) MMP-9 protein levels in astrocytes measured by Western blotting; (**B,D**) MMP-9 protein levels in the culture media measured by ELISA; (**E**) MMP-9 mRNA levels in astrocytes analyzed by real-time (RT)-PCR. ($n = 4$; mean $\pm$ SD; * $p < 0.05$ vs. control group, # $p < 0.05$ vs. 7.5 mM of 2-CE, + $p < 0.05$ vs. 15 mM of 2-CE; one-way analysis of variance (ANOVA) followed by Student–Newman–Keuls (SNK) tests.).

3.2. Increased Expression and Nuclear Translocation of p65 in 2-CE Treated Rat Astrocytes along with the Exposure Duration and Concentrations

The protein levels of p65 increased in the whole cell lysate within 8 h of 2-CE exposure, and remained significantly above the control levels at 12 h (Figure 2A). Moreover, those were significantly higher in the 15 and 30 mM groups compared with the control group or 7.5 mM group, and were also significantly higher in the 30 mM group compared with the 15 mM group (Figure 2B). In addition, the p65 mRNA levels also increased with 2-CE treatment (Figure 2C). The nuclear p65 protein levels in the 15 and 30 mM 2-CE treated cells, and the nuclear translocation ratios of p65 in cells treated with every 2-CE dose, were significantly higher than in the untreated control cells (Figure 2D–F). However, 2-CE treatment did not have an obvious effect on cytosolic p65 protein levels.

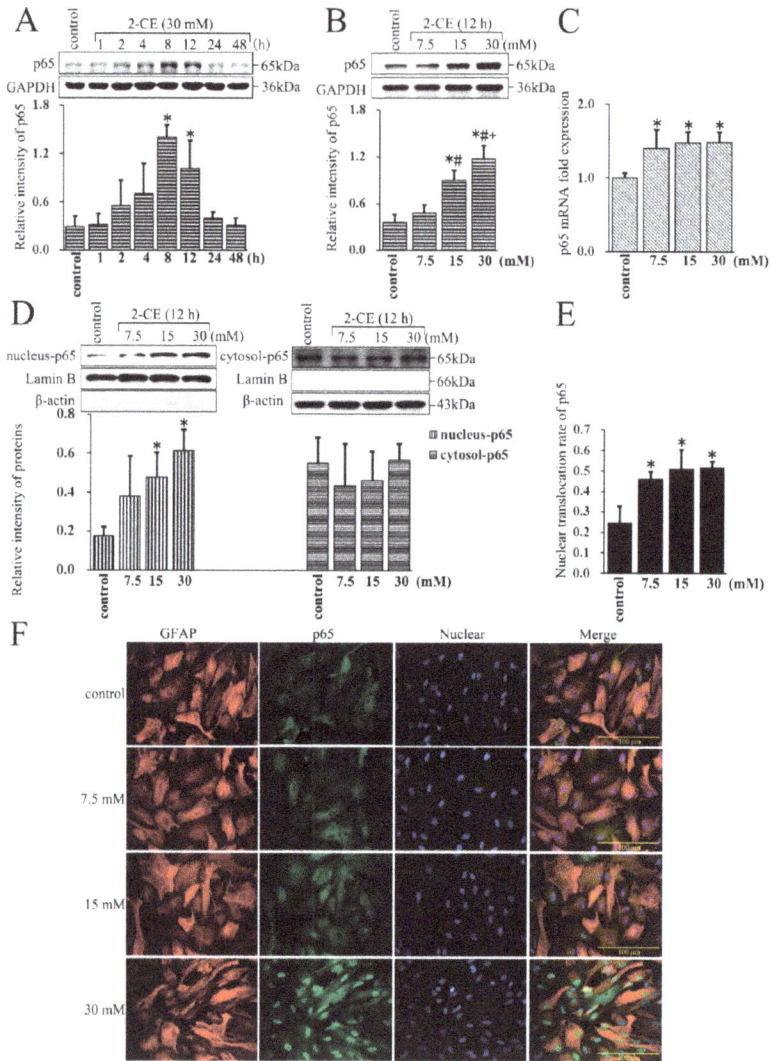

Figure 2. The nuclear translocation of p65 affected by 2-CE in astrocytes. (**A**) Astrocytes were treated with 30 mM of 2-CE for the indicated times, and (**B–F**) with 0, 7.5, 15, and 30 mM 2-CE for 12 h; (**A,B**) The levels of p65 in whole cell lysate measured by Western blotting; (**C**) The p65 mRNA levels analyzed by RT-PCR; (**D**) The cytosolic and nuclear p65 fractions measured by Western blotting, with β-actin and Lamin B as the respective internal standards; (**E**) The nuclear translocation rate of p65 was expressed as (relative intensity of p65 in nuclear fraction/[relative intensity of p65 in nuclear fraction + relative intensity of p65 in cytoplasmic fraction]); (**F**) Representative immunofluorescence images of p65 and glial fibrillary acid protein (GFAP) in primary culture astrocytes (400×). ($n = 4$; mean $\pm$ SD; * $p < 0.05$ vs. control group, # $p < 0.05$ vs. 7.5 mM of 2-CE, + $p < 0.05$ vs. 15 mM of 2-CE; one-way ANOVA followed by SNK tests.).

3.3. Increased Expression and Phosphorylation of IκBα in 2-CE Treated Rat Astrocytes along with the Exposure Concentrations

The IκBα protein levels decreased significantly in 2-CE treated astrocytes, whereas their IκBα mRNA levels increased significantly. Furthermore, the level of p-IκBα in the 30 mM group was significantly elevated compared with the control group and other exposure groups (Figure 3A,B). This clearly indicated that 2-CE treatment could enhance phosphorylation and degradation of IκBα, and activation of NF-κB in the astrocytes. In addition, our results also demonstrated that the transcriptional activation of IκBα was induced in 2-CE-treated astrocytes, which was probably due to IκBα degradation and NF-κB activation.

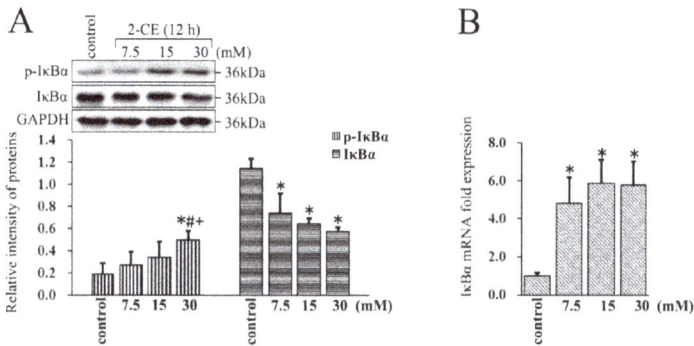

Figure 3. The phosphorylation of IκBα affected by 2-CE in astrocytes. (**A**,**B**) Astrocytes were treated with 0, 7.5, 15, and 30 mM 2-CE for 12 h; (**A**) The levels of p-inhibitor of κBα (IκBα) and IκBα in astrocytes measured by Western blotting; (**B**) The IκBα mRNA levels analyzed by RT-PCR. (n = 4; mean ± SD; * $p < 0.05$ vs. control group, # $p < 0.05$ vs. 7.5 mM of 2-CE, + $p < 0.05$ vs. 15 mM of 2-CE; one-way ANOVA followed by SNK tests.).

3.4. Upregulated Expression and Phosphorylation of c-Jun and c-Fos in 2-CE Treated Rat Astrocytes along with the Exposure Duration and Concentrations

The c-Jun protein levels increased significantly in 2-CE treated astrocytes as early as 2 h, peaked at 8 h, and remained significantly above the control levels at 12 h. The levels of p-c-Jun increased remarkably by 4 h, and remained significantly above the control levels at 12 h (Figure 4A). Furthermore, the levels of both c-Jun and p-c-Jun in 2-CE-treated cells increased significantly in a dose dependent manner (Figure 4B). In addition, the c-Jun mRNA levels in these cells also increased significantly (Figure 4C). Similarly, the levels of both c-Fos and p-c-Fos increased rapidly as early as 1 h, and remained significantly high at 2 h (Figure 4D). Therefore, to explore the changes in c-Fos expression due to different 2-CE concentrations, the astrocytes were treated with 2-CE for 1 h. Furthermore, the levels of c-Fos and p-c-Fos increased after treatment with 2-CE in a dose dependent manner (Figure 4E). Likewise, the mRNA levels of c-Fos in 2-CE-treated cells increased significantly (Figure 4F).

We therefore hypothesized that 2-CE induced overexpression and phosphorylation of c-Jun and c-Fos would activate AP-1 in the 2-CE-treated astrocytes. However, both c-Jun and c-Fos were overexpressed prior to their phosphorylation. In addition, as an immediate early gene, activation of c-Fos occurred earlier and was sustained for a very short time compared with c-Jun.

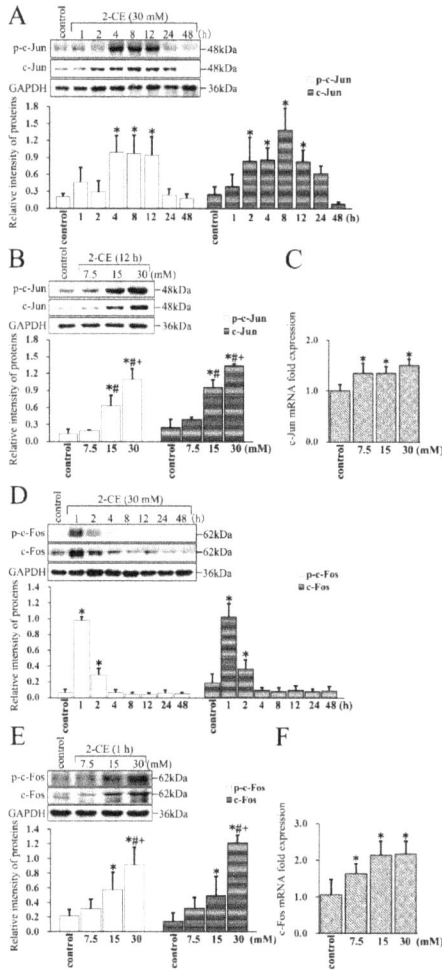

Figure 4. The expression and phosphorylation of c-Jun and c-Fos in astrocytes induced by 2-CE. (**A,D**) Astrocytes were treated with 30 mM 2-CE for the indicated times, and (**B,C,E,F**) with 0, 7.5, 15, and 30 mM 2-CE for 12 h; (**A,B,D,E**) Levels of p-c-Jun, p-c-Fos, c-Jun, and c-Fos in astrocytes measured by Western blotting; (**C,F**) The c-Jun and c-Fos mRNA levels analyzed by RT-PCR. ($n = 4$; mean $\pm$ SD; * $p < 0.05$ vs. control group, # $p < 0.05$ vs. 7.5 mM of 2-CE, + $p < 0.05$ vs. 15 mM of 2-CE; one-way ANOVA followed by SNK tests.).

3.5. MMP-9 Overexpression Mediated by NF-κB and AP-1 in 2-CE-Treated Astrocytes

To determine the role of NF-κB or AP-1 activation in MMP-9 overexpression mediated by 2-CE, the cells were pre-treated for 2 h with PDTC or 1 h with SR11302, specific inhibitors of IκBα phosphorylation and AP-1 activity, respectively, before a 12 h treatment with 30 mM 2-CE. Pre-treatment of astrocytes with 25 μM PDTC significantly lowered the levels of p-IκBα, while both 10 and 25 μM PDTC downregulated IκBα mRNA levels induced by 2-CE exposure. In contrast, both 10 and 25 μM PDTC restored the IκBα protein levels reduced by 2-CE treatment. As a result of the changes in p-IκBα and IκBα levels, MMP-9 expression was also significantly lower in cells pre-treated with PDTC compared with those treated only with 2-CE (Figure 5A,B).

Similarly, pre-treatment with SR11302 also downregulated MMP-9 expression in a dose dependent manner (Figure 5C,D).

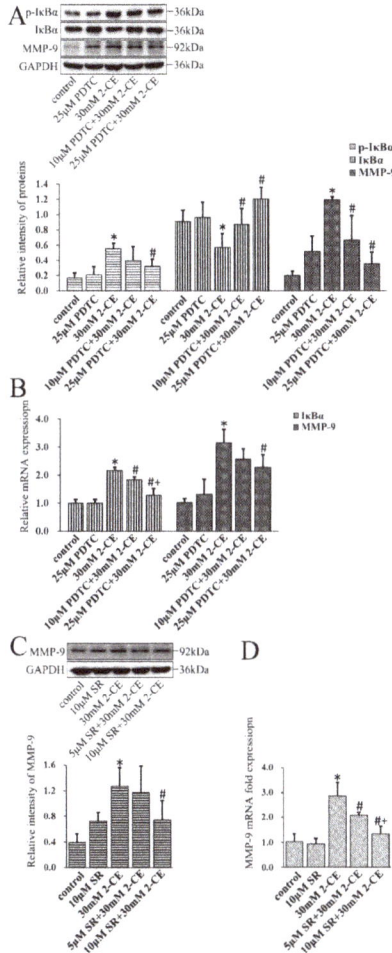

Figure 5. Role of nuclear factor kappa B (NF-κB) and activator protein-1 (AP-1) in 2-CE induced MMP-9 overexpression in astrocytes. (**A,B**) Astrocytes were pre-treated with 10 and 25 μM of pyrrolidine dithiocarbamate (PDTC) for 2 h and then treated with 30 mM 2-CE for 12 h; (**C,D**) Astrocytes were pre-treated with 5 and 10 μM SR for 1 h and then treated with 30 mM 2-CE for 12 h; (**A,C**) The levels of p-IκBα, IκBα, and MMP-9 in astrocytes measured by Western blotting; (**B,D**) The IκBα and MMP-9 mRNA levels analyzed by RT-PCR. ($n = 4$; mean ± SD; * $p < 0.05$ vs. control group, # $p < 0.05$ vs. 30 mM of 2-CE, + $p < 0.05$ vs. 10 μM of PDTC or 5 μM of SR; one-way ANOVA followed by SNK tests.).

3.6. Roles of p38 MAPK Signaling Pathway in NF-κB and AP-1 Activation in 2-CE Treated Astrocytes

To determine the association of the p38 MAPK signaling pathway with activating NF-κB and AP-1 in 2-CE treated astrocytes, cells were pre-treated for 1 h with SB202190, a specific inhibitor of p38 MAPK, before a 12 h treatment with 30 mM 2-CE. Inhibition of p38 MAPK significantly decreased the p65 protein and mRNA levels compared with the cells treated with 2-CE alone (Figure 6A,B). In addition, the level of nuclear p65 protein and its nuclear translocation ratio were decreased significantly upon pretreatment with 1, 10, and 30 μM SB202190, while the cytoplasmic p65 protein levels were significantly increased (Figure 6C–E). SB202190 also significantly decreased the levels of p-IκBα and IκBα mRNA, and increased IκBα protein levels. Surprisingly, treatment with SB202190 alone increased the phosphorylation and transcriptional expression of IκBα in astrocytes (Figure 7A,B).

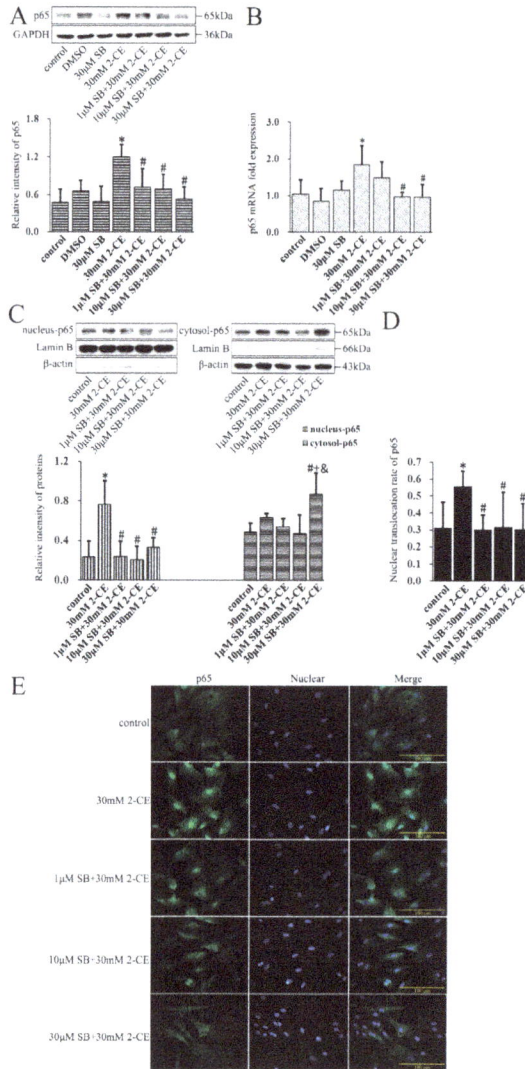

Figure 6. Role of p38 mitogen-activated protein kinase (p38 MAPK) signaling pathway in 2-CE induced nuclear translocation of p65 in astrocytes. Astrocytes were pre-treated with 1, 10, and 30 μM SB for 1 h and then treated with 30 mM of 2-CE for 12 h. (**A**) The levels of p65 in astrocytes measured by Western blotting; (**B**) The p65 mRNA levels analyzed by RT-PCR; (**C**) The cytosolic and nuclear fraction of p65 measured by Western blotting; (**D**) The nuclear translocation rate of p65; (**E**) Immunofluorescence staining for p65 (400×). ($n = 4$; mean ± SD; * $p < 0.05$ vs. control group, [#] $p < 0.05$ vs. 30 mM of 2-CE, [+] $p < 0.05$ vs. 1 μM of SB, [&] $p < 0.05$ vs. 10 μM of SB; one-way ANOVA followed by SNK tests.).

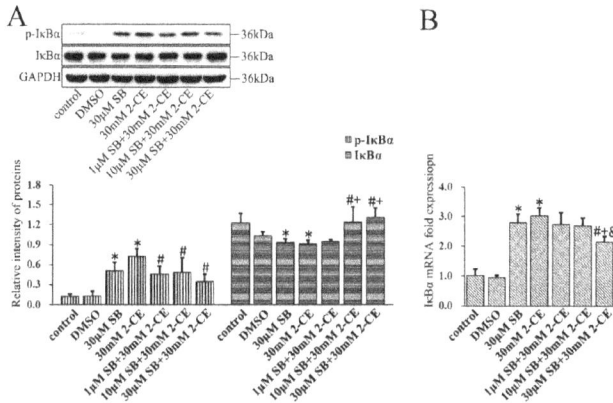

Figure 7. Role of p38 MAPK signaling pathway in 2-CE induced phosphorylation of IκBα in astrocytes. Astrocytes were pre-treated with 1, 10, and 30 μM SB for 1 h and then treated with 30 mM of 2-CE for 12 h. (**A**) The levels of p-IκBα and IκBα in astrocytes measured by Western blotting; (**B**) The IκBα mRNA levels were analyzed by RT-PCR. ($n = 4$; mean ± SD; * $p < 0.05$ vs. control group, # $p < 0.05$ vs. 30 mM of 2-CE, + $p < 0.05$ vs. 1 μM of SB, & $p < 0.05$ vs. 10 μM of SB; one-way ANOVA followed by SNK tests.).

Finally, SB202190 mediated inhibition of p38 MAPK significantly decreased the levels of p-c-Jun and protein and mRNA levels of c-Jun in 2-CE-treated astrocytes (Figure 8). However, because c-Fos expression and phosphorylation occurred much earlier than the phosphorylation of p38 (data as shown in Figure S1), it is possible that activation of c-Fos might not be mediated by the p38 MAPK signaling pathway.

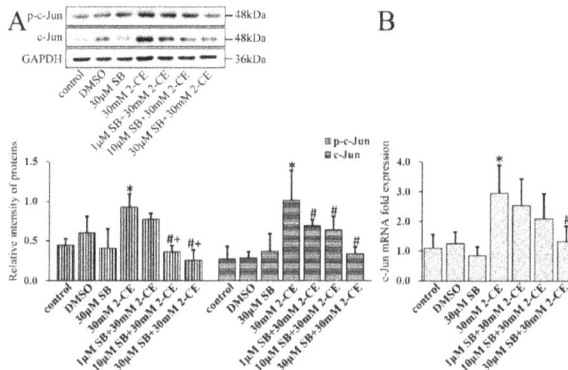

Figure 8. Role of p38 MAPK signal pathway in 2-CE induced activation of AP-1 in astrocytes. Astrocytes were pre-treated with 1, 10, and 30 μM SB for 1 h and then treated with 30 mM 2-CE for 12 h. (**A**) The levels of p-c-Jun and c-Jun in astrocytes measured by Western blotting; (**B**) The c-Jun mRNA levels analyzed by RT-PCR. ($n = 4$; mean ± SD; * $p < 0.05$ vs. control group, # $p < 0.05$ vs. 30 mM of 2-CE, + $p < 0.05$ vs. 1 μM of SB; one-way ANOVA followed by SNK tests.).

4. Discussion

Accumulated evidence has demonstrated that astrocytes are necessary for neuronal survival and function by maintenance of blood brain barrier (BBB) integrity and extracellular homeostasis.

Astrocytes can be activated in pathophysiological conditions, and secrete a variety of proinflammatory cytokines, such as MMP-9, which then degrades the proteins in the tight junctions of BBB, thus leading to breakdown in BBB integrity and brain edema formation.

We previously investigated the association of MMP-9 overexpression with MAPK signaling pathways in 2-CE-treated rat astrocytes [7], but the transient expression of MMP-9 with the exposure duration was unclear. This point is significant because MMP-9 expression is inducible by this stimuli. To the best of our knowledge, this is the first report to show the transient expression profile of MMP-9 in either 2-CE-treated astrocytes or culture media. MMP-9 protein levels increased markedly in both astrocytes and their media within 8 h of 2-CE exposure, and remained high in the cells until 24 h, whereas high levels were detected in the culture media for 48 h, indicating that the overexpressed MMP-9 in astrocytes is simultaneously secreted. Furthermore, MMP-9 levels peaked at 12 h and 24 h in the astrocytes and culture media, respectively. As a result of a lack of MMP-9 specific proteases in the culture media, it could not be eliminated rapidly, and thus accumulated in the culture media. Therefore, MMP-9 was retained longer in the media, and peaked later than that in the astrocytes.

Recent studies have demonstrated that activation of MAPK signaling pathways can contribute to upregulation of MMP-9 expression in the astrocytes [17,18]. As overexpression of MMP-9 can result in breakdown of BBB in traumatic, hemorrhagic, and ischemic brain injury, studying the regulatory pathways is necessary for uncovering the mechanisms of 1,2-DCE induced brain edema. Although we also showed a positive association of MAPK signaling pathways with MMP-9 overexpression in 2-CE-treated astrocytes previously, we did not focus on the transcription factors involved. This is also the first report to show activation of NF-κB and AP-1 in 2-CE-treated astrocytes concomitant with MMP-9 overexpression. Both NF-κB and AP-1 are the key transcriptional regulators in pathways of inflammation, differentiation, proliferation, and apoptosis. Deregulation of NF-κB and AP-1 may result in the overexpression of downstream genes, including cytokines, chemokines, and effector proteins, and lead to cellular damage [19].

The NF-κB consists of five subunits in mammalian cells: c-Rel, RelA (p65), RelB, NF-κB1 (p50), and NF-κB2 (p52) [20]. The major form of NF-κB in cells is a heterodimer consisting of the DNA binding subunit p50 and the trans-activator p65 [21]. In quiescent cells, NF-κB is restricted to the cytoplasm by its binding with IκBα [22]. Upon stimulation by different upstream signals, IκBα is phosphorylated and degraded through an ubiquitin-dependent process, which releases NF-κB and allows its nuclear translocation to activate gene transcription [23,24]. Accordingly, as we found IκBα phosphorylation was enhanced in 2-CE-treated astrocytes, it could be speculated that decreased levels of IκBα protein and increased nuclear translocation of p65 might be the consequence of IκBα phosphorylation. In addition, p65 total protein and mRNA expression were also increased after 2-CE treatment. Furthermore, in response to suppressed IκBα phosphorylation, the overexpression of MMP-9 induced by 2-CE was ameliorated, indicating an involvement of NF-κB activation in MMP-9 overexpression mediated by 2-CE in astrocytes.

AP-1 consists of a set of structurally and functionally related members of the Jun protein family (c-Jun, JunB, and JunD) and Fos protein family (c-Fos, FosB, Fra-1, and Fra-2) [25]. However, c-Fos and c-Jun are the most common subunits of AP-1, which may be assembled as homo- or hetero-dimers of AP-1 [26,27]. Regulation of AP-1 subunits can be achieved via gene transcription, mRNA stability, and post-translational processing. The most important post-translational control is the phosphorylation induced by MAPK signaling pathways and cellular stress [28]. While c-Fos was rapidly and transiently upregulated and phosphorylated in 2-CE-treated astrocytes, both upregulation and phosphorylation of c-Jun occurred much later than c-Fos. In addition, as with NF-κB inhibition, blocking AP-1 also ameliorated MMP-9 overexpression in astrocytes treated by 2-CE, suggesting that activation of AP-1 was also involved in eliciting MMP-9 in 2-CE-treated astrocytes.

The MAPK signal pathways are involved in a variety of fundamental cellular processes, among which the p38 MAPK pathway is involved in response to a wide range of extracellular stimuli, and plays a key role in the regulation of pro-inflammatory networks and biosynthesis of TNF-α, IL-1β, and MMP-9. Previous studies have shown that overexpression of MMP-9 elicited by inflammatory response in astrocytes could be regulated by the p38 MAPK signaling pathway [29]. Other studies indicated that MMP-9 expression could be modulated via activation of p38 MAPK signaling pathway in various cell types [30,31]. Therefore, we also explored the role of this pathway in the activation of NF-κB and AP-1 in 2-CE-treated astrocytes. Inhibition of p38 MAPK markedly reduced p65 overexpression and IκBα phosphorylation in 2-CE-treated astrocytes. Furthermore, the increased nuclear translocation of p65 and decreased IκBα protein levels in 2-CE-treated astrocytes were also ameliorated because of the suppressed phosphorylation of IκBα. Therefore, our results suggested that the p38 MAPK signaling pathway is involved in p65 expression and IκBα phosphorylation induced by 2-CE in astrocytes. In addition, p38 MAPK inhibition also markedly reduced the enhanced c-Jun expression and phosphorylation in 2-CE-treated astrocytes. As c-Fos upregulation and phosphorylation occurred ahead of p38 MAPK phosphorylation (levels of p-p38 in 2-CE-treated astrocytes increased as early as 2 h, and remained significantly high till 24 h, data as shown in Figure S1), we hypothesized that p38 MAPK signaling pathway was not involved in the regulation of c-Fos. Although the mechanisms by which c-Fos was upregulated and phosphorylated in 2-CE-treated astrocytes are still unclear, the ERK1/2 signaling pathway suggested by the recent studies might be involved in the immediately regulation of c-Fos [13]. The proposal schematic diagram was shown in Figure 9.

Taken together, our results suggested that MMP-9 expression was upregulated through p38 MAPK mediated activation of NF-κB and AP-1 in 2-CE-treated astrocytes. Our findings might provide novel clues for clarifying the mechanisms underlying 1,2-DCE associated cerebral edema.

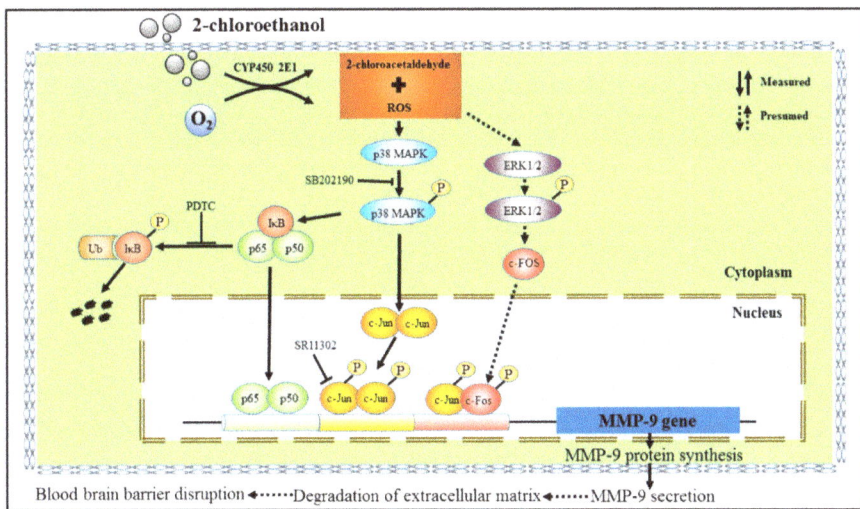

Figure 9. Schematic diagram.

Supplementary Materials: The following are available online at http://www.mdpi.com/2073-4409/7/8/96/s1, Figure S1: Alteration in expression and phosphorylation of p38 MAPK in 2-CE exposed rat astrocytes along with the exposure time and 2-CE concentrations.

Author Contributions: Conceptualization, Y.L. and Y.J.; Data curation, X.J. and Y.L.; Formal analysis, T.W.; Funding acquisition, Y.J.; Investigation, T.W., X.J., and C.L.; Methodology, T.W., X.J., Q.S., and G.W.; Project administration, Q.S., Y.L., F.Z., and Y.J.; Software, T.W. and X.J.; Supervision, Y.L., G.W., F.Z., and Y.J.; Validation, T.W.; Visualization, T.W.; Writing—Original draft, T.W.; Writing—Review & editing, Y.J.

Funding: This research was supported by the grants of the National Natural Science Foundation of China [NSFC 81573105, 81172644], Program for Liaoning Excellent Talents in University [LR2013038], and Program for Liaoning Innovative Research Team in University [LT2015028].

Acknowledgments: We acknowledge financial support from the National Natural Science Foundation of China [NSFC], Program for Liaoning Excellent Talents in University as well as Program for Liaoning Innovative Research Team in University.

Conflicts of Interest: The authors declare no conflicts of interest.

Abbreviations

1,2-DCE	1,2-dichloroethane
2-CE	2-chloroethanol
NF-κB	nuclear factor kappa B
IκBα	inhibitor of κBα
AP-1	activator protein-1
MMP-9	matrix metalloproteinase-9
MAPK	mitogen-activated protein kinase
PDTC	pyrrolidine dithiocarbamate
SR	SR11302
SB	SB202190

References

1. Liu, J.R.; Fang, S.; Ding, M.P.; Chen, Z.C.; Zhou, J.J.; Sun, F. Toxic encephalopathy caused by occupational exposure to 1,2-Dichloroethane. *J. Neurol. Sci.* **2010**, *292*, 111–113. [CrossRef] [PubMed]
2. Zhang, Q.; Niu, Q.; Li, L.Y.; Yang, L.; Guo, X.L.; Huang, J.X. Establishment of a poisoned animal model of toxic encephalopathy induced by 1,2-dichloroethane. *Int. J. Immunopathol. Pharmacol.* **2011**, *24*, 79S–83S. [PubMed]
3. Chen, S.; Zhang, Z.; Lin, H.; Chen, Z.; Wang, Z.; Wang, W. 1,2-Dichloroethane-induced toxic encephalopathy: a case series with morphological investigations. *J. Neurol. Sci.* **2015**, *351*, 36–40. [CrossRef] [PubMed]
4. Guengerich, F.P.; Crawford, W.M., Jr.; Domoradzki, J.Y.; Macdonald, T.L.; Watanabe, P.G. In vitro activation of 1,2-dichloroethane bymicrosomal and cytosolic enzymes. *Toxicol. Appl. Pharmacol.* **1980**, *55*, 303–317. [CrossRef]
5. Igwe, O.J.; Que Hee, S.S.; Wagner, W.D. Inhalation pharmacokinetics of 1,2-dichloroethane after different dietary pretreatments of male Sprague-Dawley rats. *Arch. Toxicol.* **1986**, *59*, 127–134. [CrossRef] [PubMed]
6. Raucy, J.L.; Kraner, J.C.; Lasker, J.M. Bioactivation of halogenated hydrocarbons by cytochrome P4502E1. *Crit. Rev. Toxicol.* **1993**, *23*, 1–20. [CrossRef] [PubMed]
7. Wang, T.; Liao, Y.; Sun, Q.; Tang, H.; Wang, G.; Zhao, F.; Jin, Y. Upregulation of Matrix Metalloproteinase-9 in Primary Cultured Rat Astrocytes Induced by 2-Chloroethanol Via MAPK Signal Pathways. *Front. Cell. Neurosci.* **2017**, *11*, 218. [CrossRef] [PubMed]
8. Wang, G.; Yuan, Y.; Zhang, J.; Gao, L.; Tan, X.; Yang, G.; Lv, X.; Jin, Y. Roles of aquaporins and matrix metalloproteinases in mouse brain edema formation induced by subacute exposure to 1,2-dichloroethane. *Neurotoxicol. Teratol.* **2014**, *44*, 105–112. [CrossRef] [PubMed]
9. Haorah, J.; Ramirez, S.H.; Schall, K.; Smith, D.; Pandya, R.; Persidsky, Y. Oxidative stress activates protein tyrosinkinase and matrix metalloproteinases leading to blood-brain barrier dysfunction. *J. Neurochem.* **2007**, *101*, 566–576. [CrossRef] [PubMed]
10. Rosenberg, G.A. Matrix metalloproteinases in neuroinflammation. *Glia* **2002**, *39*, 279–291. [CrossRef] [PubMed]
11. Rosell, A.; Ortega-Aznar, A.; Alvarez-Sabín, J.; Fernández-Cadenas, I.; Ribó, M.; Molina, C.A.; Lo, E.H.; Montaner, J. Increased brain expression of matrix metalloproteinase-9 after ischemic and hemorrhagic human stroke. *Stroke* **2006**, *37*, 1399–1406. [CrossRef] [PubMed]
12. Sun, Q.; Liao, Y.; Wang, T.; Tang, H.; Wang, G.; Zhao, F.; Jin, Y. 2-Chloroethanol induced upregulation of matrix metalloproteinase-2 in primary cultured rat astrocytes via MAPK signal pathways. *Front. Neurosci.* **2017**, *10*, 593. [CrossRef] [PubMed]
13. Wang, H.H.; Hsieh, H.L.; Wu, C.Y.; Sun, C.C.; Yang, C.M. Oxidized low-density lipoprotein induces matrix metalloproteinase-9 expression via a p42/p44 and JNK-dependent AP-1 pathway in brain astrocytes. *Glia* **2009**, *57*, 24–38. [CrossRef] [PubMed]

14. Speidl, W.S.; Kastl, S.P.; Hutter, R.; Katsaros, K.M.; Kaun, C.; Bauriedel, G.; Maurer, G.; Huber, K.; Badimon, J.J.; Wojta, J. The complement component C5a is present in human coronary lesions in vivo and induces the expression of MMP-1 and MMP-9 in human macrophages in vitro. *FASEB J.* **2011**, *25*, 35–44. [CrossRef] [PubMed]

15. Hwang, Y.P.; Yun, H.J.; Choi, J.H.; Han, E.H.; Kim, H.G.; Song, G.Y.; Kwon, K.I.; Jeong, T.C.; Jeong, H.G. Suppression of EGF-induced tumor cell migration and matrix metalloproteinase-9 expression by capsaicin via the inhibition of EGFR-mediated FAK/Akt, PKC/Raf/ERK, p38 MAPK, and AP-1. *Mol. Nutr. Food Res.* **2011**, *55*, 594–605. [CrossRef] [PubMed]

16. Park, J.; Kwak, C.H.; Ha, S.H.; Kwon, K.M.; Abekura, F.; Cho, S.H.; Chang, Y.C.; Lee, Y.C.; Ha, K.T.; Chung, T.W.; et al. Ganglioside GM3 suppresses lipopolysaccharide-induced inflammatory responses in rAW 264.7 macrophage cells through NF-κB, AP-1, and MAPKs signaling. *J. Cell. Biochem.* **2018**, *119*, 1173–1182. [CrossRef] [PubMed]

17. Hsieh, C.C.; Papaconstantinou, J. Akt/PKB and p38 MAPK signaling, translational initiation and longevity in Snell dwarf mouse livers. *Mech. Ageing Dev.* **2004**, *125*, 785–798. [CrossRef] [PubMed]

18. Wu, C.Y.; Hsieh, H.L.; Jou, M.J.; Yang, C.M. Involvement of p42/p44 MAPK, p38 MAPK, JNK and nuclear factor-kappa B in interleukin-1beta-induced matrix metalloproteinase-9 expression in rat brain astrocytes. *J. Neurochem.* **2004**, *90*, 1477–1488. [CrossRef] [PubMed]

19. Herbein, G.; Varin, A.; Fulop, T. NF-κB, AP-1, Zinc-deficiency and aging. *Biogerontology* **2006**, *7*, 409–419. [CrossRef] [PubMed]

20. Gosselin, K.; Abbadie, C. Involvement of Rel/NF-kappa B transcription factors in senescence. *Exp. Gerontol.* **2003**, *38*, 1271–1283. [CrossRef] [PubMed]

21. Liu, X.; Kumar, A. Differential signaling mechanism for HIV-1 Nef-mediated production of IL-6 and IL-8 in human astrocytes. *Sci. Rep.* **2015**, *5*, 9867. [CrossRef] [PubMed]

22. Shih, R.H.; Wang, C.Y.; Yang, C.M. NF-kappaB Signaling Pathways in Neurological Inflammation: A Mini Review. *Front. Mol. Neurosci.* **2015**, *8*, 77. [CrossRef] [PubMed]

23. Carter, A.B.; Knudtson, K.L.; Monick, M.M.; Hunninghake, G.W. The p38 mitogen-activated protein kinase is required for NF-kappa B-dependent gene expression. The role of TATA-binding protein (TBP). *J. Biol. Chem.* **1999**, *274*, 30858–30863. [CrossRef] [PubMed]

24. Huang, C.W.; Feng, W.; Peh, M.T.; Peh, K.; Dymock, B.W.; Moore, P.K. A novel slow-releasing hydrogen sulfide donor, FW1256, exerts anti-inflammatory effects in mouse macrophages and in vivo. *Pharmacol. Res.* **2016**, *113*, 533–546. [CrossRef] [PubMed]

25. Ding, J.; Huang, Y.; Ning, B.; Gong, W.; Li, J.; Wang, H.; Chen, C.Y.; Huang, C. TNF-alpha induction by nickel compounds is specific through ERKs/AP-1-dependent pathway in human bronchial epithelial cells. *Curr. Cancer Drug Targets* **2009**, *9*, 81–90. [CrossRef] [PubMed]

26. Karin, M.; Liu, Z.; Zandi, E. AP-1 function and regulation. *Curr. Opin. Cell Biol.* **1997**, *9*, 240–246. [CrossRef]

27. Shaulian, E.; Karin, M. AP-1 as a regulator of cell life and death. *Nat. Cell Biol.* **2002**, *4*, E131. [CrossRef] [PubMed]

28. Bogoyevitch, M.A.; Kobe, B. Uses for JNK: The many and varied substrates of the c-Jun N-terminal kinases. *Microbiol. Mol. Biol. Rev.* **2006**, *70*, 1061–1095. [CrossRef] [PubMed]

29. Miraglia, M.C.; Scian, R.; Samartino, C.G.; Barrionuevo, P.; Rodriguez, A.M.; Ibañez, A.E.; Coria, L.M.; Velásquez, L.N.; Baldi, P.C.; Cassataro, J.; et al. Brucella abortus induces TNF-α-dependent astroglial MMP-9 secretion through mitogen-activated protein kinases. *J. Neuroinflamm.* **2013**, *10*, 819. [CrossRef] [PubMed]

30. Lappas, M.; Riley, C.; Lim, R.; Barker, G.; Rice, G.E.; Menon, R.; Permezel, M. MAPK and AP-1 proteins are increased in term pre-labour fetal membranes overlying the cervix: Regulation of enzymes involved in the degradation of fetal membranes. *Placenta* **2011**, *32*, 1016–1025. [CrossRef] [PubMed]

31. Lian, S.; Xia, Y.; Khoi, P.N.; Ung, T.T.; Yoon, H.J.; Kim, N.H.; Kim, K.K.; Jung, Y.D. Cadmium induces matrix metalloproteinase-9 expression via ROS-dependent EGFR, NF-κB, and AP-1 pathways in human endothelial cells. *Toxicology* **2015**, *338*, 104–116. [CrossRef] [PubMed]

cells

MDPI

Review

Agonist-Biased Signaling via Matrix Metalloproteinase-9 Promotes Extracellular Matrix Remodeling

Bessi Qorri [1], Regina-Veronicka Kalaydina [1], Aleksandra Velickovic [1], Yekaterina Kaplya [1], Alexandria Decarlo [2] and Myron R. Szewczuk [1,*]

[1]   Department of Biomedical and Molecular Sciences, Queen's University, Kingston, ON K7L 3N6, Canada; bessi.qorri@queensu.ca (B.Q.); nicka.kalaydina@queensu.ca (R.-V.K.); av36@queensu.ca (A.V.); 13yk5@queensu.ca (Y.K.)

[2]   Department of Biology, Biosciences Complex, Queen's University, Kingston, ON K7L 3N6, Canada; 14ald4@queensu.ca

*   Correspondence: szewczuk@queensu.ca; Tel.: +1-613-770-6403; Fax: +1-613-533-6796

Received: 30 June 2018; Accepted: 23 August 2018; Published: 26 August 2018

Abstract: The extracellular matrix (ECM) is a highly dynamic noncellular structure that is crucial for maintaining tissue architecture and homeostasis. The dynamic nature of the ECM undergoes constant remodeling in response to stressors, tissue needs, and biochemical signals that are mediated primarily by matrix metalloproteinases (MMPs), which work to degrade and build up the ECM. Research on MMP-9 has demonstrated that this proteinase exists on the cell surface of many cell types in complex with G protein-coupled receptors (GPCRs), and receptor tyrosine kinases (RTKs) or Toll-like receptors (TLRs). Through a novel yet ubiquitous signaling platform, MMP-9 is found to play a crucial role not only in the direct remodeling of the ECM but also in the transactivation of associated receptors to mediate and recruit additional remodeling proteins. Here, we summarize the role of MMP-9 as it exists in a tripartite complex on the cell surface and discuss how its association with each of the TrkA receptor, Toll-like receptors, epidermal growth factor receptor, and the insulin receptor contributes to various aspects of ECM remodeling.

Keywords: GPCR bias agonism; MMP-crosstalk; extracellular matrix; biased signaling; functional selectivity; insulin receptor; EGFR; Toll-like receptor; GPCR

---

1. Introduction

The extracellular matrix (ECM) is the dynamic non-cellular structure present in all tissues and organs of the body that provides the essential physical scaffolding for cells and initiates necessary biochemical signaling required for the maintenance of tissue homeostasis [1]. Fundamentally, the ECM is composed of over 300 proteins that are collectively referred to as the core matrisome, consisting of collagen, elastin, fibronectin, proteoglycans, glycosaminoglycans, and glycoproteins [2]. However, each tissue has a unique ECM composition with specific structures that generate continuous remodeling and reciprocal signaling between them and the ECM. Collectively, these interactions have been implicated in the regulation of several vital processes, including cell survival, growth, migration, and differentiation, all of which are necessary to maintain tissue homeostasis [3].

The development and remodeling of the heterogeneous ECM rely on all cell types, including epithelial, fibroblasts, immune cells, and endothelial cells, to synthesize and secrete matrix macromolecules [3]. In addition to these cellular secretions, there are ECM-associated proteins such as growth factors, cytokines, mucins, and ECM-modifying enzymes, all of which collectively contribute to the dynamic ECM network [4]. The ECM provides a structural network for the body through binding of the ECM components to each other and other cells through receptors such as integrins, which allow for signal transduction from the ECM to cells to regulate functions that are vital to the maintenance of homeostasis [4,5]. More recently, adhesion to ECM components has been implicated in cell migration through the ECM, underlying countless critical physiological processes, including morphogenesis and wound healing, and in a more deregulated state, malignant cell invasion, and metastasis [6].

The nature of the ECM composition and structure contributing to the status of ECM remodeling has been associated with the development of several pathological conditions [3,4]. Abnormally high ECM deposition has been associated with fibrosis and cancer, whereas excessive ECM degradation has been linked to the development of osteoarthritis [7]. The intricate balance between ECM deposition and degradation in the process of ECM remodeling is mediated by enzymes degrading the matrix, such as matrix metalloproteinases (MMPs), a disintegrin and metalloproteinases (ADAMs), ADAMs with thrombospondin motifs (ADAMTSs), plasminogen activators, and heparinases during both normal and pathological conditions [4]. Of these mediators, MMPs have been studied as the primary regulators of ECM composition; however, more recently, MMPs play additional roles in controlling the interactions between cells and responses to the environment [8]. This review will focus on the important role played by MMP-9 in the process of ECM remodeling, on maintaining the architecture of various tissues, and in regulating homeostatic functions. We aim to highlight the effects of MMP-9 on ECM molecules and discuss how these effects translate to disruptions and changes in cell-matrix and cell-cell interactions in both standard and pathological conditions.

2. MMPs

MMPs are the primary enzymes in the cleaving components of the ECM and have been long associated with playing a central role in tissue remodeling [8]. Under normal conditions, MMP activity is negligible; however, during repair or remodeling processes such as wound healing, inflammation, or diseased tissue, MMP activity is significantly increased [9]. There are currently 23 mammalian MMPs identified that are classified into six groups: collagenases, gelatinases, stromelysins, matrilysins, membrane-type MMPs (MT-MMPs), and others by substrate specificity, sequence similarity, and domain organization [10]. All mammalian MMPs share structural similarities, characterized by a conserved catalytic metalloproteinase domain structure and an autoinhibitory prodomain that consists of the cysteine-switch motif PRCGXPD that coordinates the active-site zinc-binding motifs to prevent the functioning of the catalytic domain [10,11]. Most MMPs are secreted as zymogens and are subsequently activated in the extracellular space via proteolytic cleavage by Ser proteases or other MMPs to remove the prodomain and make the active site available for catalysis [7].

Two gelatinases within the MMP family, gelatinase A and B (MMP-2 and -9, respectively), contain three repeats of a fibronectin type II motif in the metalloproteinase domain, which mediate binding to collagens [10,11]. These gelatinases are capable of digesting some ECM molecules, including type IV, V, and XI collagens, laminin, and aggrecan core protein [12]. MMP-2 is known to be involved in physiological collagen turnover, whereas MMP-9 has been associated with degradation of the ECM and initiating and promoting vessel formation [13]. Due to the additional roles of MMP-9 proteolysis, particularly as it relates to regulation of tissue architecture and vascular remodeling, this review will focus on the essential roles of MMP-9 in ECM remodeling throughout the body, and the comparable mechanisms at work that rely on its activity.

MMPs are found mainly as membrane-bound or soluble secreted inactive proenzymes. The soluble forms become active in the extracellular matrix [14]. A variety of cells produce MMPs, including epithelial cells, fibroblasts, inflammatory cells, and endothelial cells. Their activities are regulated by the family of tissue inhibitors of metalloproteinase (TIMPs) [15]. Under normal and pathological conditions, MMP-1, -2, -3, -7, and -9 forms are upregulated in endothelial cells.

The MMPs that are linked to the cell transmembrane and glycosylphosphatidylinositol (GPI)-domains are not restricted only to the plasma membrane, since the secreted forms, such as

MMP-1, -2, -7, -9, -13 and -19, can also bind to the plasma membrane [16,17]. Interestingly, Fridman et al. [16] reviewed the evidence for MMP-9 docking on the cell surface, which involves a distinct array of surface proteins regulating the localization, inhibition, and internalization of the enzyme. These unique structural and functional characteristics of MMP-9 associated with cell-surface proteins provide novel conceptual challenges to its cellular functions and ECM remodeling. For example, the plasma membrane-bound MMPs can hydrolyze cell-associated substrates, activate and concentrate their activity in discrete areas, maintain their activity from inhibition, and silence the activity. To this end, MMP-9 has a high affinity for type IV collagen α2 chain [18], the intercellular adhesion molecule-1 (ICAM) [19], the beta 1-integrin in focal contacts [20], the hyaluronan receptor CD44 [21], and the scavenger receptor LRP (low-density lipoprotein receptor-related protein) [22]. From early studies, membrane-bound MMP9 represented only a very small fraction of the enzyme, while the secreted form is more biologically relevant in tumor development [23].

Membrane-bound MMPs can be activated or inhibited in caveolae, cholesterol-rich plasma-membrane invaginations, where MMP activators such as MMP14 can concentrate MMP2 partition in caveolae domains or rafts [24]. MMP inhibitors like GPI-anchored RECK (reversion inducing cysteine-rich protein with Kazal motifs) are also found in these rafts [25]. Thus, MMPs have multiple functional roles having substrates other than components of the ECM, and can function before invasion in the development of cancer. For example, membrane-type 1 MMP (MT1-MMP) localization at the front of migrating cells has been shown to focus its proteolytic activity in specific cellular areas [26]. MT1-MMP localizes at the front of migrating cells and degrades the extracellular matrix barrier during cancer invasion, but how the polarized distribution of MT1-MMP at the migration front is regulated was unknown until now. Mori et al. [26] have shown that MT1-MMP forms a complex with CD44H via the hemopexin-like (PEX) domain. The cytoplasmic tail of CD44H, which tethers with the actin cytoskeleton, localizes at the lamellipodia. Thus, CD44H is a linker connecting MT1-MMP to the actin cytoskeleton to direct MT1-MMP to the migration front.

Our recent studies have demonstrated another essential location of MMP-9, which exists in a complex with mammalian membrane-associated sialidase neuraminidase-1 (Neu-1) and receptor tyrosine kinases [27].

2.1. Matrix Metalloproteinase-9 Crosstalk with Neuraminidase-1 on Cell-Surface Receptors

We have reported that associated G-protein-coupled receptor (GPCR) signaling potentiates MMP-9-Neu-1 crosstalk, which forms a complex with various receptor tyrosine kinases (RTKs), including the nerve growth factor TrkA receptor [28], the epidermal growth factor receptor (EGFRs) [29], and the insulin receptor (IR) [30], as well as Toll-like receptors-2 and -4 (TLR-2, -4) [31]. This same receptor-signaling complex has been observed across these receptors, suggesting that MMP-9 activity is ubiquitous among these different receptors.

Briefly, ligand binding to its specific receptor (RTK or TLR) induces GPCR-signaling processes via the $G\alpha_i$ subunit and MMP-9 activation to induce Neu-1 activity, which collectively forms a complex with the RTK/TLR on the cell surface. Activated MMP-9 removes elastin-binding protein (EBP) to induce Neu-1 activity. Interestingly, it has been shown that during monocyte differentiation into macrophages, Neu-1 tends to relocate from lysosomes to the cell surface, but other sialidases' (Neu-2, Neu-3, and Neu-4) expression does not change [32]. In addition to Neu-1 on the cell surface, cathepsin A, a lysosomal carboxypeptidase and EBP form a complex on the cell surface [33]. The association of Neu-1 with the multienzymatic complex containing β-galactosidase and cathepsin A [34] and EBP with the ectodomain of TLRs and RTKs is thought to be due to a unique orientation of Neu-1 on the cell surface. Neu-1 was shown to be associated tightly with a subunit of cathepsin A, with the complex influencing sialic acid levels on the cell surface of activated cells [35].

Activated Neu-1 hydrolyzes α-2,3-sialic acid residues on glycosylated receptors to remove steric hindrance and facilitate receptor dimerization and activation, as seen in Figure 1 [36]. This process sets the stage for downstream signaling and its involvement in multiple pathological conditions, such as tumorigenesis, inflammation, insulin resistance, and synaptic plasticity [28–31]. Figure 1 depicts for the first time the ability of transcriptional factor Snail in mediating ovarian tumor neovascularization [37]. Abdulkhalek et al. [37] provided supporting evidence to show that silencing Snail in ovarian carcinoma cells resulted in the absence of massive tumor vascularization associated with heterotopic xenografts in immunodeficient mice with a concomitant no tumor growth and metastatic burden in the lungs. Snail and MMP-9 expressions in invasive tumors like ovarian cancers are closely associated since they have similar invasive processes involving extracellular matrix remodeling [38].

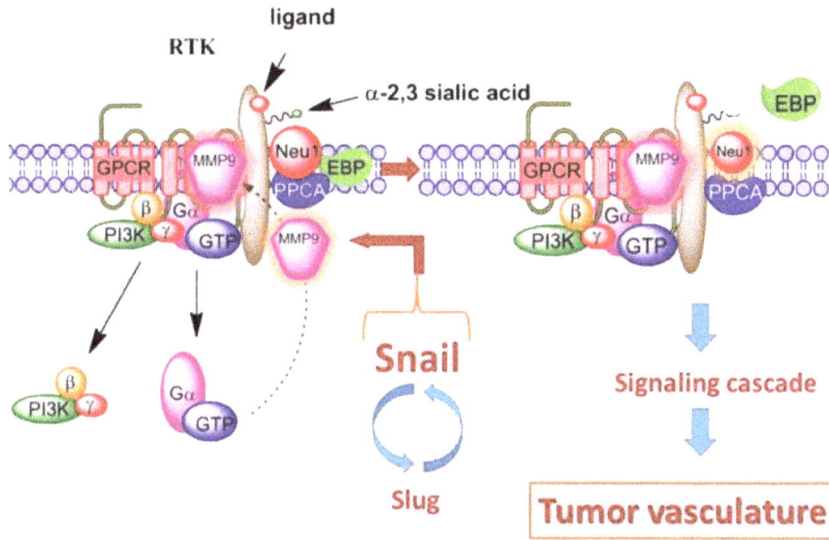

Figure 1. Snail and matrix metalloproteinase-9 (MMP-9) signaling axis in facilitating a neuraminidase-1 (Neu1) and MMP-9 crosstalk in regulating receptor tyrosine kinases (RTKs) to promote tumor neovascularization. Notes: For ovarian cancers, Snail and MMP-9 expressions are closely connected to similar invasive tumor processes. Snail induces MMP-9 secretion through oncogenic H-Ras (RasV12) and other multiple signaling pathways. Snail also leads to the transcriptional upregulation of MMP-9. This Snail-MMP-9 signaling axis is the connecting link in promoting the modification of growth-factor-receptor glycosylation involving the subsequent receptor-signaling platform of Neu1-MMP-9 crosstalk in complex with RTKs within their ectodomains. Activated MMP-9 removes the elastin-binding protein (EBP) within the molecular multienzymatic complex consisting of β-galactosidase/Neu1 and protective protein cathepsin A (PPCA). Activated Neu1 hydrolyzes α-2,3-sialic acid residues of RTKs to remove steric hindrance leading to receptor association and activation. Here, the stage for Snail's role in tumor neovascularization is established. Abbreviations: GPCR, G-protein coupled receptor; Pi3K, phosphatidylinositol 3-kinase; GTP, guanine triphosphate. Citation: © 2014 Abdulkhalek et al. [37]. Licensee and published by Springer. Under the Creative Commons Attribution License, this is an Open Access article (http://creativecommons.org/licenses/by/4.0), which permits unrestricted use, distribution, and reproduction in any medium provided the original work be appropriately credited.

Moreover, in cooperation with oncogenic RasV12 and other signaling pathways, Snail induces MMP-9 secretion through the upregulation of MMP-9 transcription [39]. Collectively, it is proposed that the Neu1-MMP-9 crosstalk may in fact be the invisible link connecting the Snail–MMP-9 signaling axis through the modification of the growth-factor-receptor glycosylation [37]. The progressive growth of a tumor to a larger size requires them to induce revascularization to ensure constant nutrient supply.

Cancer cells induce ECM remodeling and local ECM reorganization to acquire additional tumor space, thereby promoting tumor growth.

Indeed, extracellular matrix remodeling and cellular changes in adhesion molecules are necessary for a cancer cell to become motile. Rearrangement of the actin cytoskeleton promotes cell motility and plasticity together with downregulation of adhesion molecules, which facilitate the binding to the ECM. Following ECM binding, integrins can activate MMP-9 synthesis and regulate its expression [40].

Recognition of surface integrins followed by the extracellular matrix, collectively termed integrin-guided proteolysis, is an essential mechanism in cell invasion and cancer metastasis [41]. The α3β1 integrin is coexpressed with MMP-9 in epithelial-cell carcinomas as well as in epithelial wound healing, with α3β1 signaling required for the sustained production of MMP-9. The sustained MMP-9 expression is responsible for the invasive nature of tumor cells [42]. The bidirectional signaling of integrins in mediating cell-to-ECM interactions involves an arginine-glycine-aspartic acid (RGD) binding motif that has led to our understanding of tumor growth using the ECM remodeling. The RGD motif is characteristic of fibronectin, a significant component of the ECM [43]. To this end, we have engineered a peptide consisting of cyclic Arg-Gly-Asp-D-Phe-Lys (cycloRGDfK) conjugated with triphenylphosphonium cation (TPP), referred to as cyclo-RGDfK(TPP), which we have previously reported to form 3D multicellular spheroids [44]. Briefly, cyclo-RGDfK(TPP) peptide interacts with α5β1 integrins on the cell surface. This interaction stimulates expression of E-cadherin, an adhesion molecule which has been shown to facilitate the formation of compact, tight spheroids (Figure 2). The report also proposed that these cyclo-RGDfK(TPP) peptides mimic the natural ECM protein's ability to induce cell aggregation via α5β1 integrin. Interestingly, the relative levels of specific sialoglycan structures on the cell surface correlated significantly with the ability of the cancer cells to form avascular multicellular tumor spheroids as well with xenograft tumors.

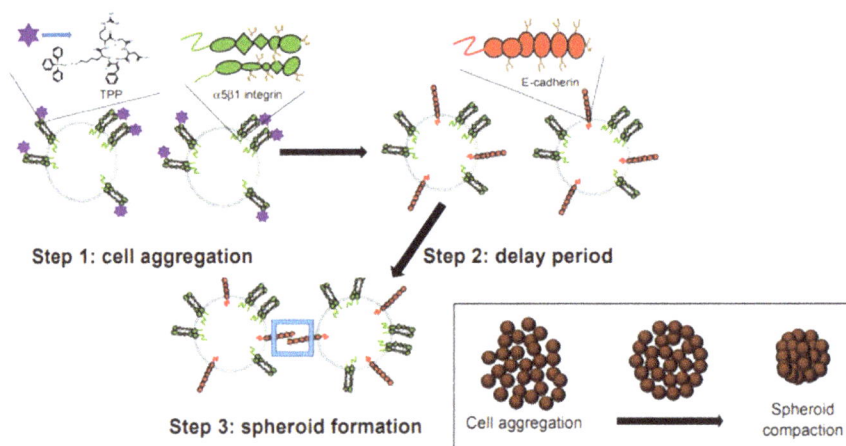

Figure 2. Formation steps of MCTS by the cyclo-RGDfK(TPP) peptide-based biochemical method. Step 1: formation of loose cell aggregates via α5β1 integrin—cyclo-RGDfK(TPP) peptide binding; step 2: a delay period for E-cadherin expression and accumulation; step 3: formation of compact MCTS through E-cadherin—E-cadherin interactions. Abbreviations: MCTS, multicellular tumor spheroid; cyclo-RGDfK(TPP), cyclic Arg-Gly-Asp-D-Phe-Lys peptide modified with 4-carboxybutyl-triphenylphosphonium bromide. Citation: © 2017 Haq et al. [45]. Licensed and published by Dove Medical Press Limited. Noncommercial reproduction of the work is permitted without any further permission from Dove Medical Press Limited, provided the work be cited appropriately.

2.2. Integrins and Matrix Metalloproteinase-9

Integrins are transmembrane receptors that facilitate cell–ECM adhesion. Activation of integrins by their respective ligands induces signal cascades that mediate the reorganization of the intracellular cytoskeleton during the cell cycle, providing movement of new receptors to the cell membrane. Integrins are heterodimeric transmembrane adhesion receptors composed of α and β subunits that mediate bidirectional cell-to-cell and cell-to-ECM interactions [46]. Interestingly, integrins can interact with MMPs to induce multiple signaling pathways that modulate cell proliferation, differentiation, and migration [47]. For example, MMP-9 has been found to play a crucial role in long-term potentiation (LTP) of the nervous system, synaptic plasticity, and maintenance of neuronal dendritic spines [48–51]. Multiple mechanisms have been proposed to explain MMP-9-induced ECM remodeling within the nervous system with a particular focus on integrin receptors [49,52].

When the classical integrin-binding motif arginine-glycine-aspartic acid (RGD) binds to the β1 subunit of integrin receptors, increased surface diffusion of the N-methyl-D-aspartate receptor (also known as the NMDA receptor or NMDAR) occurs [52]. The NMDA receptor controls synaptic plasticity and memory function. It is a specific type of ionotropic glutamate receptor because NMDA selectively binds to it, and not to any other glutamate receptors. NDMA receptors also coexpress with integrins, implicating them in synaptic plasticity [53]. RGD-containing ligands binding to integrins require MMP-9 activity to expose the RGD motif, also known as the integrin-activating epitope [51]. It is noteworthy that the tropomyosin receptor kinase A (TrkA) receptor and α9β1 integrins are coexpressed on the same cells [54]. Thus, MMP-9 may be involved in regulating neuronal growth and maintenance through the TrkA- and β1 integrin-signaling pathways.

2.3. Matrix Metalloproteinase-9 and the TrkA Receptor

The TrkA receptor is a high-affinity neurotrophin RTK associated with nerve growth factor (NGF) binding. It maintains neuronal function, including development, survival, maturation, proliferation and synaptic plasticity of neuronal cells [55–60]. The TrkA receptor is homologous to other Trk receptors, consisting of a transmembrane domain, an intracellular tyrosine kinase domain, and a five-subunit extracellular domain. The extracellular domain includes a leucine-rich region (LRR) that is flanked by two cysteine-rich subdomains with two immunoglobulins (Ig)-like subdomains, Ig-C1 and Ig-C2 [61]. NGF binding to TrkA at the Ig-C2 subdomain results in receptor homodimerization and autophosphorylation, which subsequently activate the receptor [57]. The mechanism(s) of this neuronal TrkA homodimerization and autophosphorylation signaling by NGF was unknown until now. We reported that NGF binding to its receptor induces membrane-associated sialidase activity that hydrolyzes α-2-3-sialyl residues of TrkA receptors [62]. This desialylation process is the initial step for receptor dimerization, internalization, and subsequent activation of TrkA-expressing neuronal cells as well as primary cortical neurons [62]. The subsequent development of neurite outgrowth in neuronal cells and their survival responses against cell death caused by oxidative stress, hypoxia-induced neurite retraction, and serum/glucose deprivation, is regulated by this modification of receptor glycosylation process [63]. Collectively, a prerequisite desialylation of Trk receptors by the membrane sialidase enables the removal of steric hindrance to receptor association. Jayanth et al. [28] have identified Neu1 as the membrane sialidase involved in the mechanism initiated by NGF binding to TrkA. The NGF binding to its receptor potentiates an unprecedented GPCR signaling via membrane Gαi subunit proteins and MMP-9 activation to induce the activation of Neu1 sialidase in live primary neurons and TrkA- and TrkB-expressing cell lines, as depicted in Figure 1. Central to this activation process is that the Neu1/MMP-9 complex is bound to TrkA on the cell surface of naïve primary neurons and Trk-expressing cells. These findings support the concept of MMP-9 and Neu1 crosstalk playing a crucial role in regulating NGF-induced TrkA activation. This receptor-desialylation process mediated by Neu1 activation is depicted in Figure 1. Here, these receptors include TrkA [28,62], IR [30,64,65], insulin growth factor receptor-1 (IGF-R1) [65], TLRs [27,31,66,67], EGFR [29], and others [68,69].

Pshezhetsky and Hinek [69] reported a new dimension for cellular signaling and molecular targeting, which involves the desialylation of cell-surface receptors.

2.4. G protein-Coupled Receptors Biased Agonism to Activate TrkA

It is well known that the association of GPCR and RTK signaling including Trk and insulin receptors upon ligand binding is eloquently reviewed by Pyne and colleagues [70–72], Abdulkhalek et al. [36], and Haxho et al. [73]. Onfroy et al. [74] proposed a mechanism dictating biased agonism involving G protein stoichiometry through distinct partitioning of receptor-G protein integration. Here, the expression levels of Gα proteins influence the biased profiling of β-agonists and antagonists by affecting different membrane distribution of receptor-G protein populations, in that they determine both their activity and efficacy. The level of Gα expression in the naïve state influences the partitioning of not only Gα but also the coexpressed receptor in different membrane domains [74]. Indeed, GPCRs can select more than one active state that is called 'biased agonism', 'functional selectivity', or 'ligand-directed signaling' [75,76]. Similarly, an array of allosteric ligands can have different degrees of modulation where they facilitate 'biased modulation' and can vary dramatically in a probe- and pathway-specific manner [75,77,78]. This biased modulation is not due to differences in orthosteric ligand efficacy or stimulus-response coupling.

The phenomenon of GPCR-biased agonism has led to an increased interest in the potential mechanism(s) of transactivation of the TrkA receptor. GPCR pleiotropy has resulted in the concept of biased agonism offering the potential to exploit cell functioning to achieve desired effects through cell signaling while avoiding those that are linked to adverse effects [79]. For example, adenosine and pituitary adenylate cyclase-activating polypeptides (PACAP) have been identified as GPCR ligands that elicit comparable downstream neurotrophic effects in the absence of neurotrophins [80,81]. Adenosine binding to A2A GPCR has been shown to result in the activation of the phosphatidylinositol 3-kinase (PI3K)/Akt pathway, involved in regulating neuronal survival [80,82]. This signaling activity may be due to adenosine-mediated activation of the TrkA receptor, which may have clinical implications in the treatment of neurodegenerative diseases, such as Alzheimer's disease. Recently, it has been demonstrated in animal models that Alzheimer's disease may be associated with a shortage of endogenous NGF supply [4]. Based on this association, various therapies have focused on the use of exogenous NGF to treat and mitigate the symptoms of Alzheimer's. However, there are obstacles associated with the use of neurotrophic molecules, which stem from difficulties in their delivery and pharmacokinetics [83]. The idea of allosteric modulators has been proposed to confer agonist-biased GPCR signaling and selectively modulate specific signaling pathways while having little to no effect on other parallel pathways [84]. Therefore, the exploitation of agonist adenosine through GPCR-biased agonism could instead be used to create a novel therapy that will not only aid in regulating and slowing the progression of Alzheimer's and potentially other neurodegenerative diseases but will also promote ECM remodeling of injured tissue.

2.5. TrkA, MMP-9, and Angiogenesis

In addition to acting as a neurotrophic factor in the nervous system, NGF also plays a role in the induction of angiogenesis [85–87]. Angiogenesis is the formation of new vascular networks from pre-existing ones throughout embryonal development and in several physiological and pathological conditions, such as in wound healing, inflammation, and cancer and metastasis [88]. NGF was shown to promote tissue healing in limb ischemia, the results of which demonstrated a neural-drive mechanism for angiogenesis [89]. NGF-mediated angiogenesis is due to expression of the TrkA receptor on the surface of endothelial cells and vascular smooth muscle cells [90,91]. NGF was found to potentiate reparative angiogenesis in diabetic wounds and stimulating epithelial-cell proliferation [92]. NGF binding to TrkA results in activation of intracellular signaling cascades including the protein kinase ERK pathway, PI3K/Akt pathway, and phospholipase C pathway that promote cell-cycle progression [92,93]. MMP-9 plays an essential role in angiogenesis in which ECM remodeling permits migration of endothelial cells and smooth muscle cells throughout the tissue, and promotes the release of sequestered growth factors, including angiogenic growth factors [88,94]. TrkA activation by NGF binding in smooth muscle cells induces MMP-9 expression, which is required to degrade the ECM surrounding the smooth muscle to promote migration [94]. Vessel remodeling requires ECM alterations that are a result of MMP-9 activity. The RGD site expressed in fibronectin and other ECM proteins that can bind to integrins has been shown to impact epithelial cell adhesion, migration, proliferation, survival, and cell-to-cell interactions that occur during angiogenesis [95].

Recently, Su et al. [96] have reviewed neurotrophins and their receptors involved in regulating tissue formation and healing in skeletal tissues. Here, neurotrophin NT-3 ligand, which specifically binds the TrkC receptor, can be an osteogenic and angiogenic factor. NT3 is known to enhance the expression of the essential osteogenic factor, BMP-2, and the major angiogenic factor, vascular endothelial growth factor (VEGF), to promote bone formation, vascularization, and healing of the injury site [97]. Saran et al. [98] reviewed the connection of vascularization during bone healing and remodeling, and the insight into the current therapeutic strategies adapted to promote angiogenesis. Bone repair and remodeling together are essential for the activation as well as the interaction between angiogenic and osteogenic signaling pathways. Interestingly, angiogenesis precedes the onset of osteogenesis.

2.6. MMP-9 and EGFR Signaling

The EGFR is an RTK receptor that is implicated in pathways that regulate cell survival, proliferation, and differentiation of mammalian cells [99]. The EGFR belongs to the ErbB family of RTKs and is comprised of an extracellular domain-binding ligand, a hydrophobic transmembrane domain, and a cytoplasmic domain that contains tyrosine kinase activity [100]. Once epidermal growth factor (EGF) ligand binds to the extracellular domain of the EGFR, this induces the formation of homo- or heterodimers, which subsequently activate cytoplasmic tyrosine kinase activity [100,101]. Several ligand precursors for the EGFR are produced as membrane-bound proteins that require cleavage to be released [102–104]. MMPs facilitate the release of several EGFR ligands from the cell surface, implicating MMPs with the invasiveness of tumor growth [105,106]. In this way, MMP overcomes ECM sequestering of growth factors, and can locally release factors including EGF, fibroblast growth factor (FGF), and VEGF [4].

2.7. Biased Agonism of G-Protein Coupled Receptors to Activate EGFRs

Ligand-dependent or ligand-independent mechanisms can mediate transactivation of EGFR. Ligand-dependent transactivation requires the cleavage and release of EGF ligands by paracrine or endocrine mechanisms, whereas ligand-independent activation relies on GPCR kinases [107]. Interestingly, activation of EGFR upregulates MMP-9, which, in turn, degrades E-cadherin, a crucial facilitator of cell–cell adhesion and differentiation. This crosstalk has been implicated in many cancer types, particularly in the development of metastasis [108,109]. Transactivation of EGFR following ligand binding to a GPCR has been shown to involve pro-HB-EGF and MMP activity that is rapidly induced following the ligand binding to the GPCR [104]. Gilmour et al. [29] have identified a novel molecular EGFR-signaling paradigm. Here, Neu1 and MMP-9 are found already in complex with naïve EGFRs, and are rapidly activated by EGF stimulation of the receptor. Neu1 specifically hydrolyzes α-2,3-sialyl residues that are distant from ligand binding; this process enables removal of receptor steric hindrance to association, activation, and subsequent signaling pathways. Furthermore, the neuromedin B GPCR is found to be associated with EGF-induced Neu1 activity in live 3T3–hEGFR cell line. This novel Neu1 and MMP-9 crosstalk, together with GPCR neuromedin B, is an essential signaling platform for EGF-induced receptor activation and cellular signaling. Indeed, Moody et al. [110] have also reported that the EGF-receptor transactivation is regulated by neuromedin B GPCR, a mechanism dependent on Src as well as MMP activation. In support of this EGFR-signaling paradigm, Lillehoj et al. [111] have provided evidence to show that Neu1 associates with EGFR as well with the cell-surface-associated mucin-1 (MUC1) in respiratory-airway epithelial cells. EGF stimulation regulates this Neu1–EGFR association, and when tested in vivo, it played a role in airway epithelial repair, wound healing, tumorigenesis, and metastatic potential [111]. Elevated EGFR levels are associated with reduced disease outcomes in several cancer types, including non-small-cell lung carcinoma (NSCLC), ovarian cancer, and head and neck cancer [112]. MMP-9 activity is essential for the release of EGFR ligands in preovulatory ovarian follicles stimulated with pituitary peptide hormone-luteinizing hormone (LH), and in gonadotropin-releasing hormone-stimulated (GnRH) gonadotropic cells [113].

2.8. MMP-9 and IR Signaling

The IR is a transmembrane tyrosine kinase receptor that is activated by insulin and insulin growth factors I and II [114]. The IR is a heterotetrameric protein consisting of two extracellular α subunits joined by disulphide bonds to two transmembranes β subunits. These subunits are generated from a large precursor molecule by proteolytic cleavage [115]. Similarly to other ligand-binding induced activations discussed, insulin binding to its receptor results in a conformational change that induces activation of kinase activity in the IR-β subunits [116]. The resulting downstream effects of IR activation involve glucose homeostasis and broader pleiotropic actions due to IR expression on a wide range of cells, including liver, skeletal muscle, fat, and brain cells [117,118]. Insulin resistance as a result of decreased sensitivity to the ligand and decreased ligand availability presents as a significant problem in many pathological conditions [118]. The precise mechanism(s) involved in insulin resistance is not well understood [119]. However, insulin resistance may be due to increased plasma-free fatty acid levels [120], elastin-derived peptides [121], subclinical chronic inflammation [119], oxidative and nitrative stress, altered gene expression, and mitochondrial dysfunction [119,120].

Recently, Lukong et al. [34] proposed that insulin binding to its receptor rapidly facilitates the interaction of Neu1 sialidase with the insulin receptor, which then hydrolyzes sialic acid residues of IR and, consequently, induces receptor activation. They showed that Neu1-deficient mice exposed to a high-fat diet developed hyperglycemia and insulin resistance twice as fast as the control cohort. Based on these results, endogenous Neu1 sialidase activity plays an important role in the regulation of the insulin receptor. Blaise et al. [121] provided additional evidence to support that Neu1 interacts with IRβ and desialylates the receptor. Alghamdi et al. [30] provided additional evidence for a novel GPCR signaling platform regulating the IRβ subunits. Here, the findings in their report showed data to support a novel Neu1 and MMP-9 crosstalk in alliance with neuromedin B GPCR tethered to IRβ subunit on the cell surface, all of which are essential for insulin-induced IR activation and cellular signaling. It is noteworthy that insulin can mediate increases in MMP-9 via IR activation [122], in support of the molecular-signaling platform of Neu1-MMP-9 crosstalk in alliance with neuromedin B GPCR in regulating insulin-induced receptors as depicted in Figure 3. GPCR and IR have also been shown for β-adrenergic receptors tethered to IR in adipocytes [123–126]. Here, insulin binding IR stimulates the phosphorylation of the β-adrenergic receptor on Tyr-350, the process of which facilitates IR tethering to β-adrenergic receptor via growth factor receptor-bound protein 2 (Grb-2). This molecular IR/β-adrenergic receptor/Grb-2 tripartite complex is essential for β-adrenergic agonist amplification of insulin-dependent activation of p42/p44 MAPK.

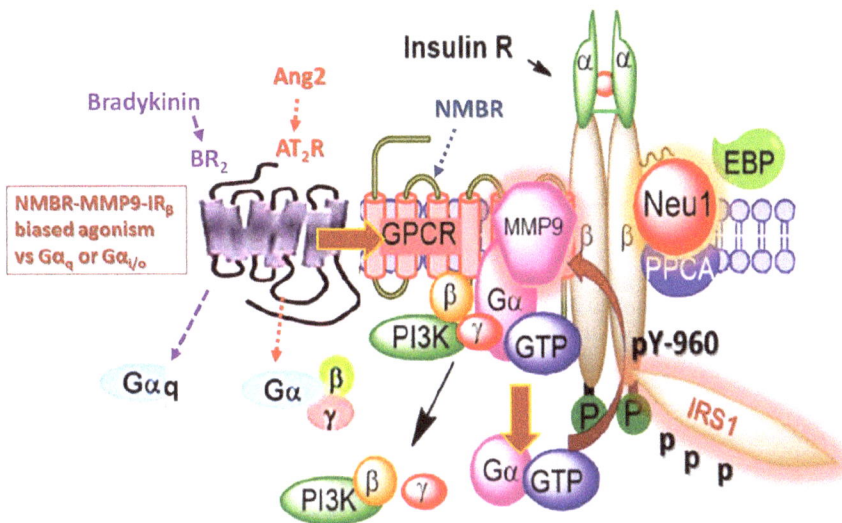

Figure 3. Mechanism of GPCR bias agonism involves insulin receptor activation. Bradykinin and angiotensin II form a complex with the neuromedin B receptor (NMBR), IRβ, and Neu-1. Bradykinin and angiotensin II preferentially lead to insulin-receptor signaling by first forming a complex with NMBR. This heterodimerization leads Gα, β, and γ to activate MMP-9. Upon activation, MMP-9 removes EBP, which in turn activates Neu-1. Crosstalk between these activated components leads to the phosphorylation and subsequent activation of insulin receptor substrate 1 (IRS1), initiating the phosphoinositide 3-kinase-protein kinase B (PI3K–AKT) pathway, in addition to others, without insulin. Citation: Taken from Haxho et al. [127] with permission, Cellular Signalling, Volume 43, Issue 6, March 2018, Pages 71–84. 0898-6568/© 2017 Published by Elsevier Inc., Open access under CC BY-NC-ND license. The article is an Open Access article that permits unrestricted noncommercial use, provided the original work be appropriately cited.

2.9. Biased Agonism of G-Protein-Coupled Receptors to Activate the Insulin Receptor

More recently, Haxho et al. [127] provided strong supporting evidence for the novel concept of biased GPCR agonist-induced IRβ transactivation signaling axis (Figure 3). GPCRs are ubiquitous throughout the body and participate in numerous physiological processes, making them important targets for therapy [128]. To that end, Haxho et al. [127] found that bradykinin (BR2) and angiotensin II receptor type I (AT2R) exist in a multimeric receptor complex with neuromedin B, IRβ, and Neu1 in naïve and stimulated IR-expressing HTC cells. This novel molecular-signaling platform regulates the interaction and signaling mechanism(s) between these players on the cell surface, uncovering the missing link for a biased GPCR agonist-induced IRβ transactivation-signaling axis, mediated by Neu1 sialidase and the modification of insulin receptor glycosylation. The biased GPCR-signaling platform potentiates Neu1 and MMP-9 crosstalk on the cell surface. This signaling platform was deemed essential for the activation of the IRβ tyrosine kinases. Indeed, biased GPCR agonism may, therefore, contribute to the mechanism of ECM remodeling occurring due to upregulation of MMP-9 activity. For instance, MMP activity has been reported to increase during chronic hyperglycemia resulting in excess degradation of tissue matrix [129]. A prominent clinical manifestation of MMP-9-induced ECM remodeling is diabetic retinopathy in type 2 diabetes mellitus (T2DM). This phenomenon of GPCR bias agonism has been observed with the use of various agonists towards the activation of several receptors and presents the potential for novel therapeutics [130,131]. For example, diabetic retinopathy (DR) is a complication of T2DM resulting from persistent hyperglycemia that leads to chronic subclinical inflammation, with long-term effects on the vascular dysfunction of the retinal microvasculature [132]. The ensuing visual loss associated with diabetic retinopathy is attributed to the increased permeability of retinal vessels or due to the proliferation of new retinal vessels [133]. Here, MMP-9 would weaken the blood–retinal barrier, thereby breaking down endothelial tight junction protein cadherin and occluding in the early stages of diabetic retinopathy [134,135]. As such, MMP-9 is considered to be a critical mediator of retinal ischemia-induced angiogenesis and nonperfusion-mediated tissue injury [129]. Preclinical studies have shown that high levels of glucose can result in transcriptional overexpression of MMP-9 [136–138]. Elevations of MMP-9 have been recorded in diabetic individuals, with both T2DM and diabetic retinopathy [139], and correlated with increased severity of T2DM [140]. Recently, Jayashree reported that MMP-9 was shown to be significantly increased in patients with T2DM with diabetic retinopathy compared to individuals with T2DM without retinopathy [129]. The proinflammatory state that exists in patients with diabetic retinopathy was suggested to cause overexpression of MMP-9 [129]. However, in light of recent advances in our understanding of GPCR-biased agonism, perhaps another explanation could be that increased levels of GPCR agonists are contributing to the activation of MMP-9, which could partially account for its role in diabetic retinopathy.

2.10. IR and Cardiac Extracellular Matrix Remodeling

Concerning GPCR-biased agonism, angiotensin I and II agonists previously mentioned have been directly implicated in ECM remodeling [141]. Here, angiotensin I and II are known to be involved in cardiac remodeling by acting through the AT1 and AT2 receptors. Left-ventricular remodeling has been associated with the activated renin–angiotensin–aldosterone system (RAAS), while the inhibition of RAAS has been shown to mitigate ventricular remodeling in a failing heart [142]. Changes in the extracellular collagen matrix of the myocardium are crucial to the remodeling process following acute myocardial infarction [143]. At the cellular level, ECM remodeling has been suggested to be observed due to increased myocyte hypertrophy, fibroblast hyperplasia, and increased collagen deposition [144]. Ducharme et al. [145] showed that targeted deletion of the MMP-9 gene attenuated left-ventricular enlargement after experimental myocardial infarction (MI) in mice with decreased collagen content. More recently, Zheng et al. [146] have shown that renal sympathetic denervation (RSD) may improve postmyocardial infraction through upregulation of TIMP-1 in rats. A higher concentration of MMP-9 and decreased TIMP-1 protein expression in cardiac tissue following MI has been postulated to result in proteolytic imbalance [146]. A possible explanation for RSD improving cardiac remodeling following MI could be attributed to the development of a feed-forward loop whereby the tumor growth factor β1 (TGF-β1) increases MMP-9 and MMP-2; however, both MMP-9 and MMP2 can also cleave latent TGF-β1 and release its active form, which activates the transcription of TIMPs [147]. Given that patients suffering from insulin resistance are at an increased risk of cardiovascular pathology [148], ECM remodeling following MI is of particular importance in the context of diabetes. The open link between GPCR-biased agonism and the IR may further explain the development of severe cardiovascular pathology in insulin-resistant patients, as reviewed by Liauchonak et al. [149]. Furthermore, cardiac ECM remodeling in the context of insulin-resistance may be occurring through biased GPCR agonist-IR crosstalk in part due to the activity of MMP-9.

2.11. Matrix Metalloproteinase-9 and TLRs

TLRs are pattern-recognition receptors (PRRs) that can recognize a broad range of molecular patterns including exogenous pathogen-associated molecular patterns (PAMPs) and endogenous damage-associated molecular patterns (DAMPs) [150]. TLRs are highly conserved type I transmembrane glycoproteins that consist of an extracellular domain, containing leucine-rich repeat (LRR) motifs, and an intracellular Toll/interleukin-1 (IL-1) receptor (TIR) domain [150]. Cell surface TLRs function to recognize extracellular microbes (TLR-1, -2, -4, -5, and -6) or intracellular TLRs localized within the endosomal compartment of the cell that are known to recognize nucleic acids

(TLR-3, -7, -8, and -9) [151]. Collectively, TLRs play critical roles in the immune response, particularly in inflammation and tissue damage.

Ligand binding to TLRs induces receptor oligomerization and, in turn, dimerization of the intracellular domain, which subsequently triggers the activation of downstream-signaling cascades [150]. The TIR domain is required for complex formation with four activating adapter molecules, myeloid differentiating factor-88 (MyD88), MyD88 adapter-like (Mal), TIR domain-containing adapter-inducing IFN-β (TRIF), and TRIF-related adapter molecule (TRAM) [151]. The binding adapters induce downstream activation of transcription factors, including NF-κB and type I interferons (IFN) [151]. MMP-9 has been shown to contribute to TLR activation, by further instigating the stimulation of an inflammatory response, with TLR activation being recognized as playing a role in chronic pain initiated by the immune signaling [152]. Indeed, studies have provided evidence to show that CpG oligodeoxynucleotide (ODN) induces transcriptional TNFα and TNFR-II expressions, which are involved in the expression of MMP9 in the supernatants derived from murine macrophage cell line by a TLR9 and a serine/threonine-specific protein kinase B (Akt)-mediated mechanism [153,154]. This ODN-induced MMP9 expression fits well within our novel Neu-1 and MMP9 crosstalk regulating TLR receptors depicted in Figure 4 and reported by Abdulkalek et al. [27]. That report describes the key players involved in the activation of nucleic acid sensing intracellular TLR-7 and TLR-9 receptors against imiquimod and CpG oligodeoxynucleotide (ODN), respectively [27].

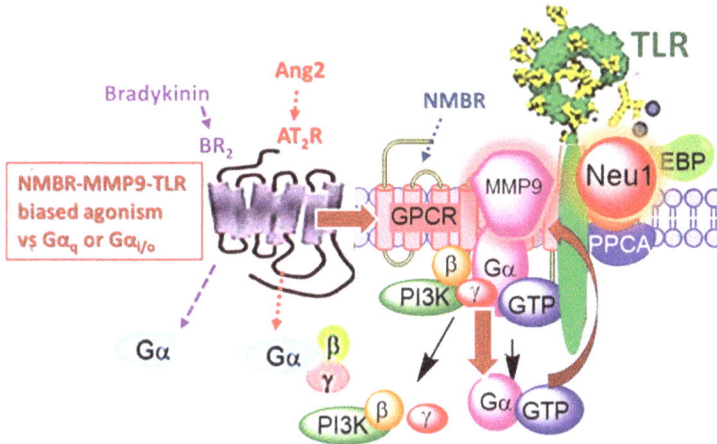

Figure 4. Bradykinin (BR2) and Angiotensin II receptor type I (AT2R) exist in a multimeric receptor complex with NMBR, IRβ, and Neu1 in naïve (unstimulated) and stimulated RAW-blue macrophage cells. Here, a molecular link regulating the interaction and signaling mechanism(s) between these molecules on the cell surface uncover a biased GPCR agonist-induced cell surface and intracellular Toll-like receptor (TLR) transactivation-signaling axis, mediated by Neu1 sialidase and the glycosylation modification of TLRs. The biased GPCR-signaling platform here potentiates Neu1 and MMP-9 crosstalk on the cell surface that is essential for the transactivation of TLRs and subsequent cellular signaling. Notes: TLR ligand, as well as GPCR agonists, can potentiate biased NMBR-TLR signaling and subsequently induce MMP-9 activation and Neu1 sialidase activity. Activated MMP-9 is proposed here to remove the EBP as part of the molecular multienzymatic complex that contains β-galactosidase/Neu1 and PPCA. Activated Neu1 then hydrolyzes α-2,3 sialyl residues of TLR at the ectodomain to remove steric hindrance to facilitate TLR association and subsequent recruitment of MyD88 and downstream signaling. Citation: Adapted from Abdulkhalek et al. [66].

2.12. Biased G-Protein-Coupled Receptors Agonism to Activate TLRs

Conformational changes induced by ligand binding of bombesin-related neuromedin B receptor to TLR-7 and -9 allows for the initiation of GPCR signaling through membrane-bound MMP-9 activation by the Gαi subunits [27]. GPCR-mediated MMP-9 activation, in turn, induces Neu-1 sialidase, creating a three-part complex with TLR-7 and -9, which triggers TLR dimerization and recruitment of MyD88 adapter molecule [27]. A similar crosstalk is seen in the TLR-4 receptor when binding endotoxin lipopolysaccharide (LPS) [31]. Once again, ligand-induced TLR conformational changes trigger GPCR signaling through MMP-9 and Gαi subunits and, in turn, induce Neu-1 sialidase activity [31]. The interplay of these molecules and receptors are essential to creating a unique molecular signaling platform that is required for ligand-induced TLR activation as seen in Figure 4 [31]. These studies further demonstrate the biased agonism or functional selectivity phenomenon, as GPCR appear to favor activation of the unique downstream pathway by different ligands [155].

2.13. TLR Induced MMP-9 Activity in Chronic Inflammation

TLR4 is known to promote a proinflammatory response due to its primary ligand binding, LPS [156]. LPS promotes the release of vasoactive inflammatory mediators from vascular smooth muscle cells, thought to contribute to chronic inflammation. MMP-9 is known to be involved in smooth muscle cell migration, and LPS binding to TLR4 was found to upregulate MMP-9 expression in human aortic smooth muscle cells [157].

Angiotensin II (Ang II) has been shown to play a proinflammatory role in the development of atherosclerosis. Ang II was shown to upregulate TLR-4 expression in cells and upregulate MMP-9 activity. This finding suggests that Ang II stimulates inflammation in vascular smooth muscle cells through regulation of inflammatory factors in an Ang-II-dependent manner [158]. The proinflammatory effects of Ang II have been thoroughly studied, with Ang II promoting immune system activation by upregulating TLR-4. However, a mechanism remains to be elucidated [159,160]. Thus, these proinflammatory effects of Ang II may be the result of bias of the angiotensin II receptor (AT2R) to activate TLR-4 to stimulate tumor necrosis factor-α (TNF-α) preferentially and induce NF-κB activity. Similar results have been obtained with bradykinin, whereby bradykinin upregulates TLR-4 expression and promotes an inflammatory response [161].

Cellular fibronectin is produced in response to tissue injury and contains an extra domain A (EDA). EDA-containing fibronectin produces cellular responses similar to those induced by bacterial LPS. EDA was found to activate TLR-4 and persisted in the absence of LPS antagonists. These findings suggest a mechanism by which EDA-containing fibronectin promotes an inflammatory response [162]. To this end, in the context of airway smooth muscle (ASM) contractility/relaxation, GPCR receptors alone, or together with receptor tyrosine kinases, can contribute to the functionality of ASM involving cellular proliferation, growth, and subsequent secretion of growth factors and inflammatory mediators. These processes will thus influence airway remodeling and the local inflammatory milieu as eloquently reviewed by Prakash [163]. Surrounding the cells in the airway, there is a network of collagenous and noncollagenous ECM protein structures, whereby the density and the composition of these ECM structures can influence the cellular functions such as proliferation, migration, differentiation, and survival. The ECM alone can regulate the formation and release of growth factors and MMPs where they can modify several extracellular proteins involved in ECM remodeling. For example, alterations in ECM asthmatic-airway remodeling can involve enhanced deposition of collagens I, III, and V, fibronectin, tenascin, hyaluronan, versican, and perlecan with a concomitant decrease in other proteins such as collagen IV and elastin [164]. In addition, this altered ECM asthmatic airway could facilitate inflammatory mediators produced by surrounding cells including growth factors. Here, ASM not only responds to inflammatory mediators but is also a source of a wide variety of pro- and anti-inflammatory factors. This ASM immunomodulatory function and property of ASM can be induced by inflammation, infection and microbial products [163].

3. Conclusions

Matrix metalloproteinase-9 plays a crucial role in the remodeling of the extracellular matrix. With the novel discovery of receptor transactivation or GPCR agonist-bias signaling, MMP-9 has been shown to play a more prominent role in ECM remodeling. Through activation of associated GPCRs, such as the angiotensin AT1R and bradykinin receptors, MMP-9 is activated to induce activation of RTKs such as the TrkA receptor, EGFR, and IR as well as Toll-like receptors in the absence of their respective ligands. These discovered roles of MMP-9 provide countless options for novel therapeutic targets in the treatment of pathological conditions including cancer, atherosclerosis, diabetes, inflammation, and wound healing. It is noteworthy that the work highlighted here is by no means all-encompassing of the substantial body of the recent literature.

Author Contributions: Conceptualization, B.Q., R.-V.K., and M.R.S.; software, B.Q., R.-V.K., and M.R.S.; validation, B.Q., R.-V.K., and M.R.S.; formal analysis, B.Q., R.-V.K., and M.R.S.; investigation, B.Q., R.-V.K., A.V., Y.K., A.D., and M.R.S.; resources, M.R.S.; data curation, B.Q., R.-V.K., A.V., Y.K., A.D., and M.R.S.; writing—original draft preparation, B.Q., R.-V.K., A.V., Y.K., A.D., and M.R.S.; writing—review and editing, B.Q., R.-V.K., A.V., Y.K., A.D., and M.R.S.; visualization, supervision, project administration, and funding acquisition, M.R.S.

Funding: This work is acknowledged by grants to M.R. Szewczuk from the Natural Sciences and Engineering Research Council of Canada (NSERC), a private-sector cancer funding from the Josefowitz Family and Encyt Technologies, Inc. B. Qorri is the recipient of the 2017 Queen's Graduate Award (QGA), 2017 Terry Fox Research Institute Transdisciplinary Training Program in Cancer Research and the 2018 Dean's Doctoral Award. R.-V. Kalaydina is the recipient of the QGA.

Conflicts of Interest: The authors declare no conflict of interest.

Abbreviations

ADAM	a disintegrin and metalloproteinase
ADAMTS	ADAMs with thrombospondin motifs
Akt	protein kinase B
Ang II	angiotensin II
ASM	airway smooth muscle
AT2R	angiotensin II receptor type I
BR2	bradykinin
cycloRGDfK	cyclic Arg-Gly-Asp-D-Phe-Lys
DAMP	damage-associated molecular pattern
DR	diabetic retinopathy
EBP	elastin binding protein
ECM	extracellular matrix
EDA	extra domain A
EGF	epidermal growth factor
EGFR	epidermal growth factor receptor
FGF	fibroblast growth factor
GnRH	gonadotropin-releasing hormone
GPCR	G protein-coupled receptor
GPI	glycosylphosphatidylinositol
Grb-2	growth factor receptor-bound protein 2
H-RAS	RasV12
ICAM	intercellular adhesion molecule-1
IFN	type I interferon
Ig	immunoglobulin
IGF-R1	insulin growth factor receptor-1
IL-1	interleukin-1
IR	insulin receptor
IRS1	insulin receptor substrate 1
IRβ	insulin receptor β subunit

LH	leuteinizing hormone
LPS	lipopolysaccharide
LRP	low density lipoprotein receptor-related protein
LRR	leucine-rich region
LTP	long-term potentiation
Mal	MyD88 adaptor-like
MI	myocardia infarction
MMP	matrix metalloproteinase
MMP-9	matrix metalloproteinase-9
MT-MMP	membrane-type MMP
MT1-MMP	membrane-type 1 MMP
MUC1	mucin-1
MyD88	myeloid differentiating factor-88
Neu-1	neuraminidase-1
NGF	nerve growth factor
NMBR	neuromedin B receptor
NSCLC	non-small cell lung cancer
ODN	oligodeoxynucleotide
PACAP	pituitary adenylate cyclase-activating polypeptide
PAMP	pathogen-associated molecular pattern
PEX	hemopexin-like
PI3K	phophatidylinositol 3-kinase
PI3K-AKT	phosphoinositide 3-kinase-protein kinase B
PPCA	protective protein cathepsin A
PRR	pattern-recognition receptor
RAAS	renin-angiotensin-aldosterone system
RECK	reversion inducing cysteine-rich protein with Kazal motifs
RGD	arginine-glycine-aspartic acid
RSD	renal sympathetic denervation
RTK	receptor tyrosine kinase
T2DM	type 2 diabetes mellitus
TGF-β1	tumor growth factor β1
TIMP	tissue inhibitors of metalloproteinase
TIR	Toll/interleukin-1 receptor
TLR	Toll-like receptor
TLR-2	Toll-like receptor-2
TLR-4	Toll-like receptor-4
TNF-α	tumor necrosis factor-α
TPP	triphenylphosphonium cation
TRAM	TRIF-related adaptor molecule
TRIF	TIR domain-containing adaptor-inducing IFN-β
TrkA	tropomyosin receptor kinase A
VEGF	vascular endothelial growth factor

References

1. Hynes, R.O. The extracellular matrix: Not just pretty fibrils. *Science* **2009**, *326*, 1216–1219. [CrossRef] [PubMed]
2. Hynes, R.O.; Naba, A. Overview of the matrisome—An inventory of extracellular matrix constituents and functions. *Cold Spring Harb. Perspect. Biol.* **2012**, *4*, a004903. [CrossRef] [PubMed]
3. Theocharis, A.D.; Skandalis, S.S.; Gialeli, C.; Karamanos, N.K. Extracellular matrix structure. *Adv. Drug Deliv. Rev.* **2016**, *97*, 4–27. [CrossRef] [PubMed]
4. Bonnans, C.; Chou, J.; Werb, Z. Remodelling the extracellular matrix in development and disease. *Nat. Rev. Mol. Cell Biol.* **2014**, *15*, 786–801. [CrossRef] [PubMed]

5. Frantz, C.; Stewart, K.M.; Weaver, V.M. The extracellular matrix at a glance. *J. Cell Sci.* **2010**, *123*, 4195–4200. [CrossRef] [PubMed]

6. Schmidt, S.; Friedl, P. Interstitial cell migration: Integrin-dependent and alternative adhesion mechanisms. *Cell Tissue Res.* **2009**, *339*, 83. [CrossRef] [PubMed]

7. Lu, P.; Takai, K.; Weaver, V.M.; Werb, Z. Extracellular matrix degradation and remodeling in development and disease. *Cold Spring Harb. Perspect. Biol.* **2011**, *3*, a005058. [CrossRef] [PubMed]

8. Ivan, S. Extracellular matrix remodelling: The role of matrix metalloproteinases. *J. Pathol.* **2003**, *200*, 448–464.

9. Nagase, H.; Visse, R.; Murphy, G. Structure and function of matrix metalloproteinases and timps. *Cardiovasc. Res.* **2006**, *69*, 562–573. [CrossRef] [PubMed]

10. Wolfram, B.; Franz-Xaver, G.-R.; Walter, S. Astacins, serralysins, snake venom and matrix metalloproteinases exhibit identical zinc-binding environments (HEXXHXXGXXH and Met-turn) and topologies and should be grouped into a common family, the 'metzincins'. *FEBS Lett.* **1993**, *331*, 134–140.

11. Page-McCaw, A.; Ewald, A.J.; Werb, Z. Matrix metalloproteinases and the regulation of tissue remodelling. *Nat. Rev. Mol. Cell Biol.* **2007**, *8*, 221–233. [CrossRef] [PubMed]

12. Xu, J.; Rodriguez, D.; Petitclerc, E.; Kim, J.J.; Hangai, M.; Yuen, S.M.; Davis, G.E.; Brooks, P.C. Proteolytic exposure of a cryptic site within collagen type IV is required for angiogenesis and tumor growth in vivo. *J. Cell Biol.* **2001**, *154*, 1069–1080. [CrossRef] [PubMed]

13. Galasso, O.; Familiari, F.; De Gori, M.; Gasparini, G. Recent findings on the role of gelatinases (matrix metalloproteinase-2 and-9) in osteoarthritis. *Adv. Orthop.* **2012**, *2012*, 834208. [CrossRef] [PubMed]

14. Neve, A.; Cantatore, F.P.; Maruotti, N.; Corrado, A.; Ribatti, D. Extracellular matrix modulates angiogenesis in physiological and pathological conditions. *BioMed Res. Int.* **2014**, *2014*, 756078. [CrossRef] [PubMed]

15. Gomez, D.E.; Alonso, D.F.; Yoshiji, H.; Thorgeirsson, U.P. Tissue inhibitors of metalloproteinases: Structure, regulation and biological functions. *Eur. J. Cell Biol.* **1997**, *74*, 111–122. [PubMed]

16. Fridman, R.; Toth, M.; Chvyrkova, I.; Meroueh, S.O.; Mobashery, S. Cell surface association of matrix metalloproteinase-9 (gelatinase B). *Cancer Metastasis Rev.* **2003**, *22*, 153–166. [CrossRef] [PubMed]

17. Guo, H.; Li, R.; Zucker, S.; Toole, B.P. EMMPRIN (CD147), an inducer of matrix metalloproteinase synthesis, also binds interstitial collagenase to the tumor cell surface. *Cancer Res.* **2000**, *60*, 888–891. [PubMed]

18. Toth, M.; Sado, Y.; Ninomiya, Y.; Fridman, R. Biosynthesis of α2(IV) and α(IV) chains of collagen IV and interactions with matrix metalloproteinase-9. *J. Cell. Physiol.* **1999**, *180*, 131–139. [CrossRef]

19. Fiore, E.; Fusco, C.; Romero, P.; Stamenkovic, I. Matrix metalloproteinase 9 (MMP-9/gelatinase B) proteolytically cleaves ICAM-1 and participates in tumor cell resistance to natural killer cell-mediated cytotoxicity. *Oncogene* **2002**, *21*, 5213–5223. [CrossRef] [PubMed]

20. Partridge, C.A.; Phillips, P.G.; Niedbala, M.J.; Jeffrey, J.J. Localization and activation of type IV collagenase/gelatinase at endothelial focal contacts. *Am. J. Physiol. Lung Cell. Mol. Physiol.* **1997**, *272*, L813–L822. [CrossRef] [PubMed]

21. Yu, Q.; Stamenkovic, I. Localization of matrix metalloproteinase 9 to the cell surface provides a mechanism for CD44-mediated tumor invasion. *Genes Dev.* **1999**, *13*, 35–48. [CrossRef] [PubMed]

22. Hahn-Dantona, E.; Ruiz, J.F.; Bornstein, P.; Strickland, D.K. The low density lipoprotein receptor-related protein modulates levels of matrix metalloproteinase 9 (MMP-9) by mediating its cellular catabolism. *J. Biol. Chem.* **2001**, *276*, 15498–15503. [CrossRef] [PubMed]

23. Ramos-DeSimone, N.; Hahn-Dantona, E.; Sipley, J.; Nagase, H.; French, D.L.; Quigley, J.P. Activation of matrix metalloproteinase-9 (MMP-9) via a converging plasmin/stromelysin-1 cascade enhances tumor cell invasion. *J. Biol. Chem.* **1999**, *274*, 13066–13076. [CrossRef] [PubMed]

24. Puyraimond, A.; Fridman, R.; Lemesle, M.; Arbeille, B.; Menashi, S. MMP-2 colocalizes with caveolae on the surface of endothelial cells. *Exp. Cell Res.* **2001**, *262*, 28–36. [CrossRef] [PubMed]

25. Egeblad, M.; Werb, Z. New functions for the matrix metalloproteinases in cancer progression. *Nat. Rev. Cancer* **2002**, *2*, 161–174. [CrossRef] [PubMed]

26. Mori, H.; Tomari, T.; Koshikawa, N.; Kajita, M.; Itoh, Y.; Sato, H.; Tojo, H.; Yana, I.; Seiki, M. CD44 directs membrane-type 1 matrix metalloproteinase to lamellipodia by associating with its hemopexin-like domain. *EMBO J.* **2002**, *21*, 3949–3959. [CrossRef] [PubMed]

27. Abdulkhalek, S.; Szewczuk, M.R. NEU1 sialidase and matrix metalloproteinase-9 cross-talk regulates nucleic acid-induced endosomal toll-like receptor-7 and -9 activation, cellular signaling and pro-inflammatory responses. *Cell. Signal.* **2013**, *25*, 2093–2105. [CrossRef] [PubMed]

28. Jayanth, P.; Amith, S.R.; Gee, K.; Szewczuk, M.R. NEU1 sialidase and matrix metalloproteinase-9 cross-talk is essential for neurotrophin activation of Trk receptors and cellular signaling. *Cell. Signal.* **2010**, *22*, 1193–1205. [CrossRef] [PubMed]

29. Gilmour, A.M.; Abdulkhalek, S.; Cheng, T.S.W.; Alghamdi, F.; Jayanth, P.; O'Shea, L.K.; Geen, O.; Arvizu, L.A.; Szewczuk, M.R. A novel epidermal growth factor receptor-signaling platform and its targeted translation in pancreatic cancer. *Cell. Signal.* **2013**, *25*, 2587–2603. [CrossRef] [PubMed]

30. Alghamdi, F.; Guo, M.; Abdulkhalek, S.; Crawford, N.; Amith, S.R.; Szewczuk, M.R. A novel insulin receptor-signaling platform and its link to insulin resistance and type 2 diabetes. *Cell. Signal.* **2014**, *26*, 1355–1368. [CrossRef] [PubMed]

31. Abdulkhalek, S.; Amith, S.R.; Franchuk, S.L.; Jayanth, P.; Guo, M.; Finlay, T.; Gilmour, A.; Guzzo, C.; Gee, K.; Beyaert, R.; et al. NEU1 sialidase and matrix metalloproteinase-9 cross-talk is essential for toll-like receptor activation and cellular signaling. *J. Biol. Chem.* **2011**, *286*, 36532–36549. [CrossRef] [PubMed]

32. Liang, F.; Seyrantepe, V.; Landry, K.; Ahmad, R.; Ahmad, A.; Stamatos, N.M.; Pshezhetsky, A.V. Monocyte differentiation up-regulates the expression of the lysosomal sialidase, NEU1, and triggers its targeting to the plasma membrane via major histocompatibility complex class II-positive compartments. *J. Biol. Chem.* **2006**, *281*, 27526–27538. [CrossRef] [PubMed]

33. Hinek, A.; Pshezhetsky, A.V.; von Itzstein, M.; Starcher, B. Lysosomal sialidase (neuraminidase-1) is targeted to the cell surface in a multiprotein complex that facilitates elastic fiber assembly. *J. Biol. Chem.* **2006**, *281*, 3698–3710. [CrossRef] [PubMed]

34. Lukong, K.E.; Elsliger, M.-A.; Chang, Y.; Richard, C.; Thomas, G.; Carey, W.; Tylki-Szymanska, A.; Czartoryska, B.; Buchholz, T.; Criado, G.R. Characterization of the sialidase molecular defects in sialidosis patients suggests the structural organization of the lysosomal multienzyme complex. *Hum. Mol. Genet.* **2000**, *9*, 1075–1085. [CrossRef] [PubMed]

35. Nan, X.; Carubelli, I.; Stamatos, N.M. Sialidase expression in activated human T lymphocytes influences production of IFN-γ. *J. Leukoc. Biol.* **2007**, *81*, 284–296. [CrossRef] [PubMed]

36. Abdulkhalek, S.; Hrynyk, M.; Szewczuk, M.R. A novel G-protein-coupled receptor-signaling platform and its targeted translation in human disease. *Res. Rep. Biochem.* **2013**, *2013*, 17–30.

37. Abdulkhalek, S.; Geen, O.D.; Brodhagen, L.; Haxho, F.; Alghamdi, F.; Allison, S.; Simmons, D.J.; O'Shea, L.K.; Neufeld, R.J.; Szewczuk, M.R. Transcriptional factor snail controls tumor neovascularization, growth and metastasis in mouse model of human ovarian carcinoma. *Clin. Transl. Med.* **2014**, *3*, 28. [CrossRef] [PubMed]

38. Blanco, M.J.; Moreno-Bueno, G.; Sarrio, D.; Locascio, A.; Cano, A.; Palacios, J.; Nieto, M.A. Correlation of snail expression with histological grade and lymph node status in breast carcinomas. *Oncogene* **2002**, *21*, 3241–3246. [CrossRef] [PubMed]

39. Jorda, M.; Olmeda, D.; Vinyals, A.; Valero, E.; Cubillo, E.; Llorens, A.; Cano, A.; Fabra, A. Upregulation of MMP-9 in MDCK epithelial cell line in response to expression of the Snail transcription factor. *J. Cell Sci.* **2005**, *118*, 3371–3385. [CrossRef] [PubMed]

40. Yue, J.; Zhang, K.; Chen, J. Role of integrins in regulating proteases to mediate extracellular matrix remodeling. *Cancer Microenviron.* **2012**, *5*, 275–283. [CrossRef] [PubMed]

41. Ivaska, J.; Heino, J. Adhesion receptors and cell invasion: Mechanisms of integrin-guided degradation of extracellular matrix. *Cell. Mol. Life Sci.* **2000**, *57*, 16–24. [CrossRef] [PubMed]

42. DiPersio, C.M.; Shao, M.; Di Costanzo, L.; Kreidberg, J.A.; Hynes, R.O. Mouse keratinocytes immortalized with large T antigen acquire alpha3beta1 integrin-dependent secretion of MMP-9/gelatinase B. *J. Cell Sci.* **2000**, *113*, 2909–2921. [PubMed]

43. Pierschbacher, M.D.; Ruoslahti, E. Cell attachment activity of fibronectin can be duplicated by small synthetic fragments of the molecule. *Nature* **1984**, *309*, 30–33. [CrossRef] [PubMed]

44. Sambi, M.; Qorri, B.; Frank, S.S.; Mouhamed, Y.; Kalaydina, R.-V.; Mendonza, N.; Szewczuk, M.R. Novel use of peptides to facilitate the formation of 3D multicellular tumor spheroids. *Curr. Top. Pept. Protein Res.* **2017**, *18*, 25–34.

45. Haq, S.; Samuel, V.; Haxho, F.; Akasov, R.; Leko, M.; Burov, S.V.; Markvicheva, E.; Szewczuk, M.R. Sialylation facilitates self-assembly of 3D multicellular prostaspheres by using cyclo-RGDFK(TPP) peptide. *OncoTargets Ther.* **2017**, *10*, 2427–2447. [CrossRef] [PubMed]

46. Hynes, R.O. Integrins: Bidirectional, allosteric signaling machines. *Cell* **2002**, *110*, 673–687. [CrossRef]

47. Stefanidakis, M.; Koivunen, E. Cell-surface association between matrix metalloproteinases and integrins: Role of the complexes in leukocyte migration and cancer progression. *Blood* **2006**, *108*, 1441–1450. [CrossRef] [PubMed]

48. Stawarski, M.; Stefaniuk, M.; Wlodarczyk, J. Matrix metalloproteinase-9 involvement in the structural plasticity of dendritic spines. *Front. Neuroanat.* **2014**, *8*, 68. [CrossRef] [PubMed]

49. Bijata, M.; Labus, J.; Guseva, D.; Stawarski, M.; Butzlaff, M.; Dzwonek, J.; Schneeberg, J.; Bohm, K.; Michaluk, P.; Rusakov, D.A.; et al. Synaptic remodeling depends on signaling between serotonin receptors and the extracellular matrix. *Cell Rep.* **2017**, *19*, 1767–1782. [CrossRef] [PubMed]

50. Szklarczyk, A.; Lapinska, J.; Rylski, M.; McKay, R.D.; Kaczmarek, L. Matrix metalloproteinase-9 undergoes expression and activation during dendritic remodeling in adult hippocampus. *J. Neurosci.* **2002**, *22*, 920–930. [CrossRef] [PubMed]

51. Wang, X.-b.; Bozdagi, O.; Nikitczuk, J.S.; Zhai, Z.W.; Zhou, Q.; Huntley, G.W. Extracellular proteolysis by matrix metalloproteinase-9 drives dendritic spine enlargement and long-term potentiation coordinately. *Proc. Natl. Acad. Sci. USA* **2008**, *105*, 19520–19525. [CrossRef] [PubMed]

52. Michaluk, P.; Mikasova, L.; Groc, L.; Frischknecht, R.; Choquet, D.; Kaczmarek, L. Matrix metalloproteinase-9 controls nmda receptor surface diffusion through integrin $\beta1$ signaling. *J. Neurosci.* **2009**, *29*, 6007–6012. [CrossRef] [PubMed]

53. Bellone, C.; Nicoll, R.A. Rapid bidirectional switching of synaptic nmda receptors. *Neuron* **2007**, *55*, 779–785. [CrossRef] [PubMed]

54. Staniszewska, I.; Sariyer, I.K.; Lecht, S.; Brown, M.C.; Walsh, E.M.; Tuszynski, G.P.; Safak, M.; Lazarovici, P.; Marcinkiewicz, C. Integrin $\alpha9\beta1$ is a receptor for nerve growth factor and other neurotrophins. *J. Cell Sci.* **2008**, *121*, 504–513. [CrossRef] [PubMed]

55. Bothwell, M. Functional interactions of neurotrophins and neurotrophin receptors. *Ann. Rev. Neurosci.* **1995**, *18*, 223–253. [CrossRef] [PubMed]

56. Snider, W.D. Functions of the neurotrophins during nervous system development: What the knockouts are teaching us. *Cell* **1994**, *77*, 627–638. [CrossRef]

57. Jing, S.; Tapley, P.; Barbacid, M. Nerve growth factor mediates signal transduction through trk homodimer receptors. *Neuron* **1992**, *9*, 1067–1079. [CrossRef]

58. Johnson, D.; Lanahan, A.; Buck, C.R.; Sehgal, A.; Morgan, C.; Mercer, E.; Bothwell, M.; Chao, M. Expression and structure of the human NGF receptor. *Cell* **1986**, *47*, 545–554. [CrossRef]

59. Kaplan, D.R.; Hempstead, B.L.; Martin-Zanca, D.; Chao, M.V.; Parada, L.F. The trk proto-oncogene product: A signal transducing receptor for nerve growth factor. *Science* **1991**, *252*, 554–558. [CrossRef] [PubMed]

60. Klein, R.; Jing, S.Q.; Nanduri, V.; O'Rourke, E.; Barbacid, M. The trk proto-oncogene encodes a receptor for nerve growth factor. *Cell* **1991**, *65*, 189–197. [CrossRef]

61. Schneider, R.; Schweiger, M. A novel modular mosaic of cell adhesion motifs in the extracellular domains of the neurogenic trk and trkB tyrosine kinase receptors. *Oncogene* **1991**, *6*, 1807–1811. [PubMed]

62. Woronowicz, A.; Amith, S.R.; De Vusser, K.; Laroy, W.; Contreras, R.; Basta, S.; Szewczuk, M.R. Dependence of neurotrophic factor activation of trk tyrosine kinase receptors on cellular sialidase. *Glycobiology* **2007**, *17*, 10–24. [CrossRef] [PubMed]

63. Woronowicz, A.; Amith, S.R.; Davis, V.W.; Jayanth, P.; De Vusser, K.; Laroy, W.; Contreras, R.; Meakin, S.O.; Szewczuk, M.R. Trypanosome trans-sialidase mediates neuroprotection against oxidative stress, serum/glucose deprivation, and hypoxia-induced neurite retraction in Trk-expressing PC12 cells. *Glycobiology* **2007**, *17*, 725–734. [CrossRef] [PubMed]

64. Dridi, L.; Seyrantepe, V.; Fougerat, A.; Pan, X.; Bonneil, É.; Thibault, P.; Moreau, A.; Mitchell, G.A.; Heveker, N.; Cairo, C.W.; et al. Positive regulation of insulin signaling by neuraminidase 1. *Diabetes* **2013**, *62*, 2338–2346. [CrossRef] [PubMed]

65. Arabkhari, M.; Bunda, S.; Wang, Y.; Wang, A.; Pshezhetsky, A.V.; Hinek, A. Desialylation of insulin receptors and IGF-1 receptors by neuraminidase-1 controls the net proliferative response of l6 myoblasts to insulin. *Glycobiology* **2010**, *20*, 603–616. [CrossRef] [PubMed]

66. Abdulkhalek, S.; Guo, M.; Amith, S.R.; Jayanth, P.; Szewczuk, M.R. G-protein coupled receptor agonists mediate neu1 sialidase and matrix metalloproteinase-9 cross-talk to induce transactivation of toll-like receptors and cellular signaling. *Cell. Signal.* **2012**, *24*, 2035–2042. [CrossRef] [PubMed]

67. Amith, S.R.; Jayanth, P.; Finlay, T.; Franchuk, S.; Gilmour, A.; Abdulkhalek, S.; Szewczuk, M.R. Detection of Neu1 sialidase activity in regulating toll-like receptor activation. *J. Vis. Exp.* **2010**, *43*, 2142. [CrossRef] [PubMed]

68. Pshezhetsky, A.V.; Ashmarina, L.I. Desialylation of surface receptors as a new dimension in cell signaling. *Biochemistry* **2013**, *78*, 736–745. [CrossRef] [PubMed]

69. Pshezhetsky, A.V.; Hinek, A. Where catabolism meets signalling: Neuraminidase 1 as a modulator of cell receptors. *Glycoconj. J.* **2011**, *28*, 441–452. [CrossRef] [PubMed]

70. Pyne, N.J.; Pyne, S. Receptor tyrosine kinase–G-protein-coupled receptor signalling platforms: Out of the shadow? *Trends Pharmacol. Sci.* **2011**, *32*, 443–450. [CrossRef] [PubMed]

71. Pyne, N.J.; Waters, C.; Moughal, N.A.; Sambi, B.S.; Pyne, S. Receptor tyrosine kinase–GPCR signal complexes. *Biochem. Soc. Trans.* **2003**, *31*, 1220–1225. [CrossRef] [PubMed]

72. Pyne, N.J.; Waters, C.M.; Long, J.S.; Moughal, N.A.; Tigyi, G.; Pyne, S. Receptor tyrosine kinase-G-protein coupled receptor complex signaling in mammalian cells. *Adv. Enzym. Regul.* **2007**, *47*, 271–280. [CrossRef] [PubMed]

73. Haxho, F.; Alghamdi, F.; Neufeld, R.J.; Szewczuk, M.R. Novel insulin receptor signaling platform. *Int. J. Diabetes Clin. Res.* **2014**, *1*, 005. [CrossRef]

74. Onfroy, L.; Galandrin, S.; Pontier, S.M.; Seguelas, M.H.; N'Guyen, D.; Senard, J.M.; Gales, C. G protein stoichiometry dictates biased agonism through distinct receptor-G protein partitioning. *Sci. Rep.* **2017**, *7*, 7885. [CrossRef] [PubMed]

75. Khoury, E.; Clément, S.; Laporte, S.A. Allosteric and biased G protein-coupled receptor signaling regulation: Potentials for new therapeutics. *Front. Endocrinol.* **2014**, *5*, 68. [CrossRef] [PubMed]

76. Luttrell, L.M.; Maudsley, S.; Bohn, L.M. Fulfilling the promise of "biased" G protein–coupled receptor agonism. *Mol. Pharmacol.* **2015**, *88*, 579–588. [CrossRef] [PubMed]

77. Lane, J.R.; May, L.T.; Parton, R.G.; Sexton, P.M.; Christopoulos, A. A kinetic view of GPCR allostery and biased agonism. *Nat. Chem. Biol.* **2017**, *13*, 929–937. [CrossRef] [PubMed]

78. Edelstein, S.J.; Changeux, J.-P. Biased allostery. *Biophys. J.* **2016**, *111*, 902–908. [CrossRef] [PubMed]

79. Sengmany, K.; Singh, J.; Stewart, G.D.; Conn, P.J.; Christopoulos, A.; Gregory, K.J. Biased allosteric agonism and modulation of metabotropic glutamate receptor 5: Implications for optimizing preclinical neuroscience drug discovery. *Neuropharmacology* **2017**, *115*, 60–72. [CrossRef] [PubMed]

80. Lee, F.S.; Chao, M.V. Activation of trk neurotrophin receptors in the absence of neurotrophins. *Proc. Natl. Acad. Sci. USA* **2001**, *98*, 3555–3560. [CrossRef] [PubMed]

81. Rajagopal, R.; Chen, Z.-Y.; Lee, F.S.; Chao, M.V. Transactivation of trk neurotrophin receptors by G-protein-coupled receptor ligands occurs on intracellular membranes. *J. Neurosci.* **2004**, *24*, 6650–6658. [CrossRef] [PubMed]

82. Stephens, R.M.; Loeb, D.M.; Copeland, T.D.; Pawson, T.; Greene, L.A.; Kaplan, D.R. Trk receptors use redundant signal transduction pathways involving SHC and PLC-gamma 1 to mediate NGF responses. *Neuron* **1994**, *12*, 691–705. [CrossRef]

83. Nickols, H.H.; Conn, P.J. Development of allosteric modulators of gpcrs for treatment of cns disorders. *Neurobiol. Dis.* **2014**, *61*, 55–71. [CrossRef] [PubMed]

84. Foster, D.J.; Conn, P.J. Allosteric modulation of gpcrs: New insights and potential utility for treatment of schizophrenia and other cns disorders. *Neuron* **2017**, *94*, 431–446. [CrossRef] [PubMed]

85. Yancopoulos, G.D.; Klagsbrun, M.; Folkman, J. Vasculogenesis, angiogenesis, and growth factors: Ephrins enter the fray at the border. *Cell* **1998**, *93*, 661–664. [CrossRef]

86. Calza, L.; Giardino, L.; Giuliani, A.; Aloe, L.; Levi-Montalcini, R. Nerve growth factor control of neuronal expression of angiogenetic and vasoactive factors. *Proc. Natl. Acad. Sci. USA* **2001**, *98*, 4160–4165. [CrossRef] [PubMed]

87. Verslegers, M.; Lemmens, K.; Van Hove, I.; Moons, L. Matrix metalloproteinase-2 and -9 as promising benefactors in development, plasticity and repair of the nervous system. *Prog. Neurobiol.* **2013**, *105*, 60–78. [CrossRef] [PubMed]

88. Nico, B.; Mangieri, D.; Benagiano, V.; Crivellato, E.; Ribatti, D. Nerve growth factor as an angiogenic factor. *Microvasc. Res.* **2008**, *75*, 135–141. [CrossRef] [PubMed]

89. Emanueli, C.; Salis, M.B.; Pinna, A.; Graiani, G.; Manni, L.; Madeddu, P. Nerve growth factor promotes angiogenesis and arteriogenesis in ischemic hindlimbs. *Circulation* **2002**, *106*, 2257–2262. [CrossRef] [PubMed]

90. Cantarella, G.; Lempereur, L.; Presta, M.; Ribatti, D.; Lombardo, G.; Lazarovici, P.; Zappalà, G.; Pafumi, C.; Bernardini, R. Nerve growth factor–endothelial cell interaction leads to angiogenesis in vitro and in vivo. *FASEB J.* **2002**, *16*, 1307–1309. [CrossRef] [PubMed]

91. Wang, S.; Bray, P.; McCaffrey, T.; March, K.; Hempstead, B.L.; Kraemer, R. P75ntr mediates neurotrophin-induced apoptosis of vascular smooth muscle cells. *Am. J. Pathol.* **2000**, *157*, 1247–1258. [CrossRef]

92. Graiani, G.; Emanueli, C.; Desortes, E.; Van Linthout, S.; Pinna, A.; Figueroa, C.; Manni, L.; Madeddu, P. Nerve growth factor promotes reparative angiogenesis and inhibits endothelial apoptosis in cutaneous wounds of type 1 diabetic mice. *Diabetologia* **2004**, *47*, 1047–1054. [CrossRef] [PubMed]

93. Park, M.-J.; Kwak, H.-J.; Lee, H.-C.; Yoo, D.-H.; Park, I.-C.; Kim, M.-S.; Lee, S.-H.; Rhee, C.H.; Hong, S.-I. Nerve growth factor induces endothelial cell invasion and cord formation by promoting matrix metalloproteinase-2 expression through the phosphatidylinositol 3-kinase/akt signaling pathway and AP-2 transcription factor. *J. Biol. Chem.* **2007**, *282*, 30485–30496. [CrossRef] [PubMed]

94. Kraemer, R.; Hempstead, B.L. Neurotrophins: Novel mediators of angiogenesis. *Front. Biosci.* **2003**, *8*, s1181–s1186. [PubMed]

95. Arroyo, A.G.; Iruela-Arispe, M.L. Extracellular matrix, inflammation, and the angiogenic response. *Cardiovasc. Res.* **2010**, *86*, 226–235. [CrossRef] [PubMed]

96. Su, Y.W.; Zhou, X.F.; Foster, B.K.; Grills, B.L.; Xu, J.; Xian, C.J. Roles of neurotrophins in skeletal tissue formation and healing. *J. Cell. Physiol.* **2018**, *233*, 2133–2145. [CrossRef] [PubMed]

97. Su, Y.W.; Chung, R.; Ruan, C.S.; Chim, S.M.; Kuek, V.; Dwivedi, P.P.; Hassanshahi, M.; Chen, K.M.; Xie, Y.; Chen, L.; et al. Neurotrophin-3 induces BMP-2 and VEGF activities and promotes the bony repair of injured growth plate cartilage and bone in rats. *J. Bone Miner. Res.* **2016**, *31*, 1258–1274. [CrossRef] [PubMed]

98. Saran, U.; Gemini Piperni, S.; Chatterjee, S. Role of angiogenesis in bone repair. *Arch. Biochem. Biophys.* **2014**, *561*, 109–117. [CrossRef] [PubMed]

99. Oda, K.; Matsuoka, Y.; Funahashi, A.; Kitano, H. A comprehensive pathway map of epidermal growth factor receptor signaling. *Mol. Syst. Biol.* **2005**, *1*. [CrossRef] [PubMed]

100. Olayioye, M.A.; Neve, R.M.; Lane, H.A.; Hynes, N.E. The erbb signaling network: Receptor heterodimerization in development and cancer. *EMBO J.* **2000**, *19*, 3159–3167. [CrossRef] [PubMed]

101. Normanno, N.; De Luca, A.; Bianco, C.; Strizzi, L.; Mancino, M.; Maiello, M.R.; Carotenuto, A.; De Feo, G.; Caponigro, F.; Salomon, D.S. Epidermal growth factor receptor (EGFR) signaling in cancer. *Gene* **2006**, *366*, 2–16. [CrossRef] [PubMed]

102. Hiratsuka, T.; Fujita, Y.; Naoki, H.; Aoki, K.; Kamioka, Y.; Matsuda, M. Intercellular propagation of extracellular signal-regulated kinase activation revealed by in vivo imaging of mouse skin. *eLife* **2015**, *4*, e05178. [CrossRef] [PubMed]

103. Gschwind, A.; Prenzel, N.; Ullrich, A. Lysophosphatidic acid-induced squamous cell carcinoma cell proliferation and motility involves epidermal growth factor receptor signal transactivation. *Cancer Res.* **2002**, *62*, 6329–6336. [PubMed]

104. Prenzel, N.; Zwick, E.; Daub, H.; Leserer, M.; Abraham, R.; Wallasch, C.; Ullrich, A. EGF receptor transactivation by G-protein-coupled receptors requires metalloproteinase cleavage of proHB-EGF. *Nature* **1999**, *402*, 884. [CrossRef] [PubMed]

105. Paye, A.; Truong, A.; Yip, C.; Cimino, J.; Blacher, S.; Munaut, C.; Cataldo, D.; Foidart, J.M.; Maquoi, E.; Collignon, J. EGFR activation and signaling in cancer cells are enhanced by the membrane-bound metalloprotease MT4-MMP. *Cancer Res.* **2014**, *74*, 6758–6770. [CrossRef] [PubMed]

106. Kim, D.; Dai, J.; Park, Y.-h.; Fai, L.Y.; Wang, L.; Pratheeshkumar, P.; Son, Y.-O.; Kondo, K.; Xu, M.; Luo, J. Activation of epidermal growth factor receptor/p38/hypoxia-inducible factor-1α is pivotal for angiogenesis and tumorigenesis of malignantly transformed cells induced by hexavalent chromium. *J. Biol. Chem.* **2016**, *291*, 16271–16281. [CrossRef] [PubMed]

107. Overland, A.C.; Insel, P.A. Heterotrimeric G proteins directly regulate MMP14/membrane type-1 matrix metalloprotease: A novel mechanism for GPCR-EGFR transactivation. *J. Biol. Chem.* **2015**, *290*, 9941–9947. [CrossRef] [PubMed]
108. Dahl, K.D.C.; Symowicz, J.; Ning, Y.; Gutierrez, E.; Fishman, D.A.; Adley, B.P.; Stack, M.S.; Hudson, L.G. Matrix metalloproteinase 9 is a mediator of epidermal growth factor–dependent E-cadherin loss in ovarian carcinoma cells. *Cancer Res.* **2008**, *68*, 4606–4613. [CrossRef] [PubMed]
109. Kessenbrock, K.; Plaks, V.; Werb, Z. Matrix metalloproteinases: Regulators of the tumor microenvironment. *Cell* **2010**, *141*, 52–67. [CrossRef] [PubMed]
110. Moody, T.W.; Berna, M.J.; Mantey, S.; Sancho, V.; Ridnour, L.; Wink, D.A.; Chan, D.; Giaccone, G.; Jensen, R.T. Neuromedin B receptors regulate egf receptor tyrosine phosphorylation in lung cancer cells. *Eur. J. Pharmacol.* **2010**, *637*, 38–45. [CrossRef] [PubMed]
111. Lillehoj, E.P.; Hyun, S.W.; Feng, C.; Zhang, L.; Liu, A.; Guang, W.; Nguyen, C.; Luzina, I.G.; Atamas, S.P.; Passaniti, A.; et al. NEU1 sialidase expressed in human airway epithelia regulates epidermal growth factor receptor (EGFR) and MUC1 protein signaling. *J. Biol. Chem.* **2012**, *287*, 8214–8231. [CrossRef] [PubMed]
112. Hudson, L.G.; Moss, N.M.; Stack, M.S. EGF-receptor regulation of matrix metalloproteinases in epithelial ovarian carcinoma. *Future Oncol.* **2009**, *5*, 323–338. [CrossRef] [PubMed]
113. Cattaneo, F.; Guerra, G.; Parisi, M.; De Marinis, M.; Tafuri, D.; Cinelli, M.; Ammendola, R. Cell-surface receptors transactivation mediated by G protein-coupled receptors. *Int. J. Mol. Sci.* **2014**, *15*, 19700–19728. [CrossRef] [PubMed]
114. Ward, C.W.; Lawrence, M.C. Ligand-induced activation of the insulin receptor: A multi-step process involving structural changes in both the ligand and the receptor. *Bioessays* **2009**, *31*, 422–434. [CrossRef] [PubMed]
115. Siddle, K. Molecular basis of signaling specificity of insulin and igf receptors: Neglected corners and recent advances. *Front. Endocrinol.* **2012**, *3*, 34. [CrossRef] [PubMed]
116. Boucher, J.; Kleinridders, A.; Kahn, C.R. Insulin receptor signaling in normal and insulin-resistant states. *Cold Spring Harb. Perspect. Biol.* **2014**, *6*, a009191. [CrossRef] [PubMed]
117. De Meyts, P. The insulin receptor and its signal transduction network. In *Endotext*; De Groot, L.J., Chrousos, G., Dungan, K., Feingold, K.R., Grossman, A., Hershman, J.M., Koch, C., Korbonits, M., McLachlan, R., New, M., et al., Eds.; MDText.com, Inc.: South Dartmouth, MA, USA, 2000.
118. Tokarz, V.L.; MacDonald, P.E.; Klip, A. The cell biology of systemic insulin function. *J. Cell Biol.* **2018**. [CrossRef] [PubMed]
119. Hirabara, S.M.; Gorjao, R.; Vinolo, M.A.; Rodrigues, A.C.; Nachbar, R.T.; Curi, R. Molecular targets related to inflammation and insulin resistance and potential interventions. *J. Biomed. Biotechnol.* **2012**, *2012*, 379024. [CrossRef] [PubMed]
120. Martins, A.R.; Nachbar, R.T.; Gorjao, R.; Vinolo, M.A.; Festuccia, W.T.; Lambertucci, R.H.; Cury-Boaventura, M.F.; Silveira, L.R.; Curi, R.; Hirabara, S.M. Mechanisms underlying skeletal muscle insulin resistance induced by fatty acids: Importance of the mitochondrial function. *Lipids Health Dis.* **2012**, *11*, 30. [CrossRef] [PubMed]
121. Blaise, S.; Romier, B.; Kawecki, C.; Ghirardi, M.; Rabenoelina, F.; Baud, S.; Duca, L.; Maurice, P.; Heinz, A.; Schmelzer, C.E.; et al. Elastin-derived peptides are new regulators of insulin resistance development in mice. *Diabetes* **2013**, *62*, 3807–3816. [CrossRef] [PubMed]
122. Fischoeder, A.; Meyborg, H.; Stibenz, D.; Fleck, E.; Graf, K.; Stawowy, P. Insulin augments matrix metalloproteinase-9 expression in monocytes. *Cardiovasc. Res.* **2007**, *73*, 841–848. [CrossRef] [PubMed]
123. Karoor, V.; Wang, L.; Wang, H.Y.; Malbon, C.C. Insulin stimulates sequestration of beta-adrenergic receptors and enhanced association of beta-adrenergic receptors with GRB2 via tyrosine 350. *J. Biol. Chem.* **1998**, *273*, 33035–33041. [CrossRef] [PubMed]
124. Karoor, V.; Malbon, C.C. Insulin-like growth factor receptor-1 stimulates phosphorylation of the beta2-adrenergic receptor in vivo on sites distinct from those phosphorylated in response to insulin. *J. Biol. Chem.* **1996**, *271*, 29347–29352. [CrossRef] [PubMed]
125. Baltensperger, K.; Karoor, V.; Paul, H.; Ruoho, A.; Czech, M.P.; Malbon, C.C. The beta-adrenergic receptor is a substrate for the insulin receptor tyrosine kinase. *J. Biol. Chem.* **1996**, *271*, 1061–1064. [CrossRef] [PubMed]

126. Karoor, V.; Baltensperger, K.; Paul, H.; Czech, M.P.; Malbon, C.C. Phosphorylation of tyrosyl residues 350/354 of the beta-adrenergic receptor is obligatory for counterregulatory effects of insulin. *J. Biol. Chem.* **1995**, *270*, 25305–25308. [CrossRef] [PubMed]

127. Haxho, F.; Haq, S.; Szewczuk, M.R. Biased G protein-coupled receptor agonism mediates NEU1 sialidase and matrix metalloproteinase-9 crosstalk to induce transactivation of insulin receptor signaling. *Cell. Signal.* **2018**, *43*, 71–84. [CrossRef] [PubMed]

128. Kroeze, W.K.; Sheffler, D.J.; Roth, B.L. G-protein-coupled receptors at a glance. *J. Cell Sci.* **2003**, *116*, 4867–4869. [CrossRef] [PubMed]

129. Jayashree, K.; Yasir, M.; Senthilkumar, G.P.; Ramesh Babu, K.; Mehalingam, V.; Mohanraj, P.S. Circulating matrix modulators (MMP-9 and TIMP-1) and their association with severity of diabetic retinopathy. *Diabetes Metab. Syndr. Clin. Res. Rev.* **2018**. [CrossRef] [PubMed]

130. Gentry, P.R.; Sexton, P.M.; Christopoulo, A. Novel allosteric modulators of G protein-coupled receptors. *J. Biol. Chem.* **2015**, *290*, 19478–19488. [CrossRef] [PubMed]

131. Pupo, A.S.; Duarte, D.A.; Lima, V.; Teixeira, L.B.; Parreiras-e-Silva, L.T.; Costa-Neto, C.M. Recent updates on GPCR biased agonism. *Pharmacol. Res.* **2016**, *112*, 49–57. [CrossRef] [PubMed]

132. Chawla, A.; Chawla, R.; Jaggi, S. Microvasular and macrovascular complications in diabetes mellitus: Distinct or continuum? *Indian J. Endocrinol. Metabol.* **2016**, *20*, 546–551. [CrossRef] [PubMed]

133. Sayin, N.; Kara, N.; Pekel, G. Ocular complications of diabetes mellitus. *World J. Diabetes* **2015**, *6*, 92–108. [CrossRef] [PubMed]

134. Navaratna, D.; McGuire, P.G.; Menicucci, G.; Das, A. Proteolytic degradation of VE-cadherin alters the blood-retinal barrier in diabetes. *Diabetes* **2007**, *56*, 2380–2387. [CrossRef] [PubMed]

135. Giebel, S.J.; Menicucci, G.; McGuire, P.G.; Das, A. Matrix metalloproteinases in early diabetic retinopathy and their role in alteration of the blood-retinal barrier. *Lab. Investig.* **2005**, *85*, 597–607. [CrossRef] [PubMed]

136. Kowluru, R.A. Role of matrix metalloproteinase-9 in the development of diabetic retinopathy and its regulation by H-Ras. *Investig. Ophthalmol. Vis. Sci.* **2010**, *51*, 4320–4326. [CrossRef] [PubMed]

137. Kowluru, R.A.; Kowluru, A.; Chakrabarti, S.; Khan, Z. Potential contributory role of H-Ras, a small G-protein, in the development of retinopathy in diabetic rats. *Diabetes* **2004**, *53*, 775–783. [CrossRef] [PubMed]

138. Kowluru, R.A.; Mohammad, G.; dos Santos, J.M.; Zhong, Q. Abrogation of mmp-9 gene protects against the development of retinopathy in diabetic mice by preventing mitochondrial damage. *Diabetes* **2011**, *60*, 3023–3033. [CrossRef] [PubMed]

139. Derosa, G.; D'Angelo, A.; Tinelli, C.; Devangelio, E.; Consoli, A.; Miccoli, R.; Penno, G.; Del Prato, S.; Paniga, S.; Cicero, A.F. Evaluation of metalloproteinase 2 and 9 levels and their inhibitors in diabetic and healthy subjects. *Diabetes Metab.* **2007**, *33*, 129–134. [CrossRef] [PubMed]

140. Jacqueminet, S.; Ben Abdesselam, O.; Chapman, M.J.; Nicolay, N.; Foglietti, M.J.; Grimaldi, A.; Beaudeux, J.L. Elevated circulating levels of matrix metalloproteinase-9 in type 1 diabetic patients with and without retinopathy. *Clin. Chim. Acta* **2006**, *367*, 103–107. [CrossRef] [PubMed]

141. Busche, S.; Gallinat, S.; Bohle, R.M.; Reinecke, A.; Seebeck, J.; Franke, F.; Fink, L.; Zhu, M.; Sumners, C.; Unger, T. Expression of angiotensin AT$_1$ and AT$_2$ receptors in adult rat cardiomyocytes after myocardial infarction: A single-cell reverse transcriptase-polymerase chain reaction study. *Am. J. Pathol.* **2000**, *157*, 605–611. [CrossRef]

142. Stauss, H.M.; Zhu, Y.C.; Redlich, T.; Adamiak, D.; Mott, A.; Kregel, K.C.; Unger, T. Angiotensin-converting enzyme inhibition in infarct-induced heart failure in rats: Bradykinin versus angiotensin II. *J. Cardiovasc. Risk* **1994**, *1*, 255–262. [CrossRef] [PubMed]

143. Kelly, D.; Cockerill, G.; Ng, L.L.; Thompson, M.; Khan, S.; Samani, N.J.; Squire, I.B. Plasma matrix metalloproteinase-9 and left ventricular remodelling after acute myocardial infarction in man: A prospective cohort study. *Eur. Heart J.* **2007**, *28*, 711–718. [CrossRef] [PubMed]

144. Unger, T.; Li, J. The role of the renin-angiotensin-aldosterone system in heart failure. *J. Renin Angiotensin Aldosterone Syst.* **2004**, *5* (Suppl. 1), S7–S10. [CrossRef] [PubMed]

145. Ducharme, A.; Frantz, S.; Aikawa, M.; Rabkin, E.; Lindsey, M.; Rohde, L.E.; Schoen, F.J.; Kelly, R.A.; Werb, Z.; Libby, P.; et al. Targeted deletion of matrix metalloproteinase-9 attenuates left ventricular enlargement and collagen accumulation after experimental myocardial infarction. *J. Clin. Investig.* **2000**, *106*, 55–62. [CrossRef] [PubMed]

146. Zheng, X.-X.; Li, X.-Y.; Lyu, Y.-N.; He, Y.-Y.; Wan, W.-G.; Zhu, H.-L.; Jiang, X.-J. Possible mechanism by which renal sympathetic denervation improves left ventricular remodelling after myocardial infarction. *Exp. Physiol.* **2015**, *101*, 260–271. [CrossRef] [PubMed]

147. Visse, R.; Nagase, H. Matrix metalloproteinases and tissue inhibitors of metalloproteinases: Structure, function, and biochemistry. *Circ. Res.* **2003**, *92*, 827–839. [CrossRef] [PubMed]

148. Mehta, P.K.; Griendling, K.K. Angiotensin II cell signaling: Physiological and pathological effects in the cardiovascular system. *Am. J. Physiol. Cell Physiol.* **2007**, *292*, C82–C97. [CrossRef] [PubMed]

149. Liauchonak, I.; Dawoud, F.; Riat, Y.; Qorri, B.; Sambi, M.; Jain, J.; Kalaydina, R.V.; Mendonza, N.; Bajwa, K.; Szewczuk, M.R. The biased G-protein-coupled receptor agonism bridges the gap between the insulin receptor and the metabolic syndrome. *Int. J. Mol. Sci.* **2018**, *19*, 575. [CrossRef] [PubMed]

150. Jiménez-Dalmaroni, M.J.; Gerswhin, M.E.; Adamopoulos, I.E. The critical role of toll-like receptors—From microbial recognition to autoimmunity: A comprehensive review. *Autoimmun. Rev.* **2016**, *15*, 1–8. [CrossRef] [PubMed]

151. Lavelle, E.C.; Murphy, C.; O'Neill, L.A.J.; Creagh, E.M. The role of TLRs, NLRs, and RLRs in mucosal innate immunity and homeostasis. *Mucosal Immunol.* **2009**, *3*, 17–28. [CrossRef] [PubMed]

152. Nicotra, L.; Loram, L.C.; Watkins, L.R.; Hutchinson, M.R. Toll-like receptors in chronic pain. *Exp. Neurol.* **2012**, *234*, 316–329. [CrossRef] [PubMed]

153. Eun-Jung, L.; Sun-Hye, L.; Jin-Gu, L.; Byung-Ro, C.; Yoe-Sik, B.; Jae-Ryong, K.; Chu-Hee, L.; Suk-Hwan, B. Activation of toll-like receptor-9 induces matrix metalloproteinase-9 expression through AKT and tumor necrosis factor-α signaling. *FEBS Lett.* **2006**, *580*, 4533–4538.

154. Lim, E.-J.; Lee, S.-H.; Lee, J.-G.; Kim, J.-R.; Yun, S.-S.; Baek, S.-H.; Lee, C. Toll-like receptor 9 dependent activation of MAPK and NF-kB is required for the CpG ODN-induced matrix metalloproteinase-9 expression. *Exp. Mol. Med.* **2007**, *39*, 239–245. [CrossRef] [PubMed]

155. Hodavance, S.Y.; Gareri, C.; Torok, R.D.; Rockman, H.A. G protein-coupled receptor biased agonism. *J. Cardiovasc. Pharmacol.* **2016**, *67*, 193–202. [CrossRef] [PubMed]

156. Medzhitov, R.; Janeway, C.A. Innate immunity: The virtues of a nonclonal system of recognition. *Cell* **1997**, *91*, 295–298. [CrossRef]

157. Li, H.; Xu, H.; Liu, S. Toll-like receptors 4 induces expression of matrix metalloproteinase-9 in human aortic smooth muscle cells. *Mol. Biol. Rep.* **2011**, *38*, 1419–1423. [CrossRef] [PubMed]

158. Ji, Y.; Liu, J.; Wang, Z.; Liu, N. Angiotensin ii induces inflammatory response partly via toll-like receptor 4-dependent signaling pathway in vascular smooth muscle cells. *Cell. Physiol. Biochem.* **2009**, *23*, 265–276. [CrossRef] [PubMed]

159. Wolf, G.; Bohlender, J.; Bondeva, T.; Roger, T.; Thaiss, F.; Wenzel, U.O. Angiotensin II upregulates toll-like receptor 4 on mesangial cells. *J. Am. Soc. Nephrol.* **2006**, *17*, 1585–1593. [CrossRef] [PubMed]

160. Benigni, A.; Cassis, P.; Remuzzi, G. Angiotensin II revisited: New roles in inflammation, immunology and aging. *EMBO Mol. Med.* **2010**, *2*, 247–257. [CrossRef] [PubMed]

161. Gutiérrez-Venegas, G.; Arreguín-Cano, J.A.; Hernández-Bermúdez, C. Bradykinin promotes toll like receptor-4 expression in human gingival fibroblasts. *Int. Immunopharmacol.* **2012**, *14*, 538–545. [CrossRef] [PubMed]

162. Okamura, Y.; Watari, M.; Jerud, E.S.; Young, D.W.; Ishizaka, S.T.; Rose, J.; Chow, J.C.; Strauss, J.F. The extra domain a of fibronectin activates toll-like receptor 4. *J. Biol. Chem.* **2001**, *276*, 10229–10233. [CrossRef] [PubMed]

163. Prakash, Y.S. Airway smooth muscle in airway reactivity and remodeling: What have we learned? *Am. J. Physiol. Lung Cell. Mol. Physiol.* **2013**, *305*, L912–L933. [CrossRef] [PubMed]

164. Johnson, P.R.A.; Burgess, J.K.; Underwood, P.A.; Au, W.; Poniris, M.H.; Tamm, M.; Ge, Q.; Roth, M.; Black, J.L. Extracellular matrix proteins modulate asthmatic airway smooth muscle cell proliferation via an autocrine mechanism. *J. Allergy Clin. Immunol.* **2004**, *113*, 690–696. [CrossRef] [PubMed]

![cells logo] *cells*

MDPI

Article

Injured Achilles Tendons Treated with Adipose-Derived Stem Cells Transplantation and GDF-5

Andrea Aparecida de Aro [1,2,*], Giane Daniela Carneiro [1], Luis Felipe R. Teodoro [1], Fernanda Cristina da Veiga [3], Danilo Lopes Ferrucci [1], Gustavo Ferreira Simões [1], Priscyla Waleska Simões [4], Lúcia Elvira Alvares [3], Alexandre Leite R. de Oliveira [1], Cristina Pontes Vicente [1], Caio Perez Gomes [5], João Bosco Pesquero [5], Marcelo Augusto M. Esquisatto [2], Benedicto de Campos Vidal [1] and Edson Rosa Pimentel [1]

[1]   Department of Structural and Functional Biology, Institute of Biology, State University of Campinas–UNICAMP, Charles Darwin, s/n, CP 6109, 13083-970 Campinas, SP, Brazil; gianedc@gmail.com (G.D.C.); teo.luisfelipe@gmail.com (L.F.R.T.); daniloferrucci@yahoo.com.br (D.L.F.); gfsimoes2@gmail.com (G.F.S.); alroliv@unicamp.br (A.L.R.d.O.); crpvicente@gmail.com (C.P.V.); camposvi@unicamp.br (B.d.C.V.); pimentel@unicamp.br (E.R.P.)

[2]   Biomedical Sciences Graduate Program, Herminio Ometto University Center–UNIARARAS, 13607-339 Araras, SP, Brazil; marcelosquisatto@uniararas.br

[3]   Department of Biochemistry and Tissue Biology, Institute of Biology, State University of Campinas–UNICAMP, Charles Darwin, s/n, CP 6109, 13083-970 Campinas, SP, Brazil; fernandaveiga6@gmail.com (F.C.d.V.); lealvare@unicamp.br (L.E.A.)

[4]   Engineering, Modeling and Applied Social Sciences Center (CECS), Biomedical Engineering Graduate Program (PPGEBM), Universidade Federal do ABC (UFABC), Alameda da Universidade s/n, 09606-045 São Bernardo do Campo, SP, Brazil; pritsimoes@gmail.com

[5]   Department of Biophysics, Federal University of Sao Paulo–Unifesp, Pedro de Toledo, 699, 04039-032 Sao Paulo, SP, Brazil; caiopgomes@hotmail.com (C.P.G.); jbpesquero@unifesp.br (J.B.P.)

*   Correspondence: andreaaro80@gmail.com; Tel.: +55-19-3543-1423

Received: 19 July 2018; Accepted: 23 August 2018; Published: 31 August 2018

Abstract: Tendon injuries represent a clinical challenge in regenerative medicine because their natural repair process is complex and inefficient. The high incidence of tendon injuries is frequently associated with sports practice, aging, tendinopathies, hypertension, diabetes mellitus, and the use of corticosteroids. The growing interest of scientists in using adipose-derived mesenchymal stem cells (ADMSC) in repair processes seems to be mostly due to their paracrine and immunomodulatory effects in stimulating specific cellular events. ADMSC activity can be influenced by GDF-5, which has been successfully used to drive tenogenic differentiation of ADMSC in vitro. Thus, we hypothesized that the application of ADMSC in isolation or in association with GDF-5 could improve Achilles tendon repair through the regulation of important remodeling genes expression. Lewis rats had tendons distributed in four groups: Transected (T), transected and treated with ADMSC (ASC) or GDF-5 (GDF5), or with both (ASC+GDF5). In the characterization of cells before application, ADMSC expressed the positive surface markers, CD90 (90%) and CD105 (95%), and the negative marker, CD45 (7%). ADMSC were also differentiated in chondrocytes, osteoblast, and adipocytes. On the 14th day after the tendon injury, GFP-ADMSC were observed in the transected region of tendons in the ASC and ASC+GDF5 groups, and exhibited and/or stimulated a similar genes expression profile when compared to the in vitro assay. ADMSC up-regulated *Lox*, *Dcn*, and *Tgfb1* genes expression in comparison to T and ASC+GDF5 groups, which contributed to a lower proteoglycans arrangement, and to a higher collagen fiber organization and tendon biomechanics in the ASC group. The application of ADMSC in association with GDF-5 down-regulated *Dcn*, *Gdf5*, *Lox*, *Tgfb1*, *Mmp2*, and *Timp2* genes expression, which contributed to a lower hydroxyproline concentration, lower collagen fiber organization, and to an improvement of the rats' gait 24 h after the injury.

In conclusion, although the literature describes the benefic effect of GDF-5 for the tendon healing process, our results show that its application, isolated or associated with ADMSC, cannot improve the repair process of partial transected tendons, indicating the higher effectiveness of the application of ADMSC in injured Achilles tendons. Our results show that the application of ADMSC in injured Achilles tendons was more effective in relation to its association with GDF-5.

Keywords: repair; extracellular matrix; collagen; gait; biomechanics; gene expression

---

1. Introduction

Tendon injuries are very common and their natural repair process is extremely slow, complex, and inefficient due to their intrinsic hypocellularity and hypovascularity, representing a clinical challenge to orthopedists mainly because these injuries often respond poorly to treatment [1]. As reviewed by Zabrzyński et al. [2], the occurrence of tendon injuries is associated with sports, aging, tendinophaties, hypothyroidism, hypertension, diabetes mellitus, arthropathies, corticosteroids, vitamin C deficiency, and others. Collagen fibers are the main component of the abundant extracellular matrix (ECM) of tendons, and with the proteoglycans (PG), the collagen fibers form a highly organized supramolecular structure, able to attend to the biomechanical demands on the tissue [3,4]. Particularly, small leucine-rich proteoglycans (SLRPs), such as decorin, fibromodulim, and biglycan, are the most abundant PG in tendons, with decorin representing 80% of the total proteoglycan content of the tissue [5]. SLRPs interact with the collagen fibrils, acting in the fibrillogenesis of collagen, and probably regulating the growth in diameter of these fibrils [6]. Besides the parallel bundles of predominantly type I collagen fibers and PG, non-collagenous glycoproteins, matrix metalloproteinases (MMP), growth factors, and cells also comprise elements of the tendons [5,7–9].

The specific mechanical tendon properties are directly related to the high organization of collagen bundles [10]. After injuries, the structural organization and composition of tendons are not completely restored, with a fibrous scar formation that can cause significant dysfunction and joint movement inability [11,12]. After the repair process, tendons become biomechanically weakened, making it more prone to re-rupture [9,13]. Clearly, new treatments are needed with the objective of improving tendon repair, considering that these injuries reach the population that is economically active, and that the use of cell therapy using mesenchymal stem cells (MSC) could be a good strategy.

Adult tissues are attractive MSC sources, which are characterized as undifferentiated cells, with mesodermal differentiation potential and self-renew and high proliferative capacities [14]. Adipose tissue is an alternative source of MSC that can be obtained by a less invasive method under local anesthesia, with little associated patient discomfort and in larger quantities as compared with bone marrow [15]. The effectiveness of adipose-derived mesenchymal stem cells (ADMSC) in regenerative medicine seems to be due to their paracrine effects that stimulate specific cellular events, like the growth factors and cytokines delivery [16], as well as their immunomodulatory effects to promote tendon regeneration [17]. Besides recruiting progenitor cells from some tissues, the ADMSC could also differentiate specific cells during tissue repair [18–20]. However, the literature is controversial about the effects of ADMSC transplantation during tendon repair. In some animal models, there was greater structural organization, biomechanical properties, density of collagen fibers, collagen types I and III genes expression, and growth factors synthesis as fibroblast growth factor (FGF), vascular endothelial growth factor (VEGF), and transforming growth factor (TGF-β) after ADMSC application in relation to untreated tendons [7,21–23]. Conversely, no differences were observed after this cellular therapy in the biomechanical parameters or collagen production [7,17,23].

It is widely accepted that the control of stem cell activity is influenced by several different environmental factors [24], including growth factors [1]. To drive tenogenic ADMSC in in vitro differentiation, insulin like growth factor (IGF)-1 or TGF-β in a co-culture with primary tenocytes and growth differentiation factor-5 (GDF-5) have been used successfully [19,20]. GDF-5 belongs to the TGF-β superfamily and it is known as cartilage-derived morphogenetic protein-1 (CDMP-1) and bone-derived morphogenetic protein-14 (BMP-14). Storm et al. [25] documented two other members of this superfamily, such as GDF-6 (BMP-13) and GDF-7 (BMP-12). GDF family members are also called bone-derived morphogenetic proteins (BMP) because they are found predominantly during the development of endochondral bones [26] and joint formation [27]. Wolfman et al. [28] demonstrated that the three members of BMP induce the formation of connective tissue rich in collagen I, with an important role in the process of tendon healing [29,30]. From this study, the BMP ceased to be considered as only chondrogenic and osteogenic factors, as suggested by the name, and were considered tendinogenic factors. Forslund and colleagues [30], who showed increased Achilles tendon resistance after rupture, reported the initial success of BMP-12, -13, and -14 in tendon healing. In a study of Chhabra et al. [31], mice deficient in the *GDF-5* gene had a poor healing process, with lesser structural organization and decreased biomechanical properties of tendons, evidencing the importance of this growth factor during tendon repair processes. Currently, cell therapy using the ADMSC associated with the exogenous application of growth factors represents a great potential in the process of tendon repair. Despite promising studies in animals, no treatment associated with the application of ADMSC in tendon injuries has been used in clinics due to the lack of knowledge on molecular aspects involving those therapies.

The objective of the present study was to test the hypothesis that the application of ADMSC in isolation or associated with GDF-5 could improve Achilles tendon repair. The use of GDF-5 was based on the literature that demonstrates its importance during tendon healing and the role of GDF-5 in modulating ADMSC tenogenic differentiation in vitro. Thus, the down- or up-regulation of remodeling genes expression in response to ADMSC and GDF-5 application were analyzed, and the involvement of those genes in the restoration of the structural, biomechanical, and functional properties of Achilles tendons after partial transection.

2. Materials and Methods

2.1. In Vitro Experiments

2.1.1. Isolation of ADMSC and Cell Culture

The procedure was done according to Yang et al. [32] with some modifications. Adipose tissue was obtained from the inguinal region of 10 male Lewis rats between 90–120 days. Adipose tissue was cut and washed in Dulbecco's modified phosphate buffered saline solution (DMPBS Flush without calcium and magnesium) containing 2% streptomycin/penicillin to remove contaminating blood cells. Then, 0.2% collagenase (Sigma-Aldrich® Inc., Saint Louis, MO, USA) was added to degradation of the ECM and the solution was maintained at 37 °C under gentle stirring for 1 h to separate the stromal cells from primary adipocytes. Dissociated tissue was filtered using cell strainers (40 μm) and the inactivation of collagenase was then done by the addition of an equal volume of Dulbecco's modified Eagle's medium (DMEM) supplemented with 15% fetal bovine serum (FBS), followed by centrifugation at 1800 rpm for 10 min. The suspending portion containing lipid droplets was discarded and the pellet was resuspended in DMEM (containing 50 mg/L penicillin and 50 mg/L streptomycin) with 15% FBS, and transferred to 25 cm² flasks for 48 hours. After confluence, cells were transferred to 75 cm² flasks (1st passage). The medium was replaced after 48 h and then every 3 days. Cultures were maintained at 37 °C with 5% CO_2 until the 5th passage (5P), always at up to 80% confluency.

2.1.2. Flow Cytometry

ADMSC at 5P ($n = 4$) were trypsinized and centrifuged at 1800 rpm for 10 min, and counted using the Neubauer chamber. 1×10^6 ADMSC were resuspended in 200 uL of DMPBS Flush with 2% BSA (bovine serum albumin). For the immunophenotypic panel, the following antibodies were used:

CD90-APC, CD105-PE, and CD45-APC double conjugated (eBioscience® Inc., San Diego, CA, USA) were diluted 1:200 and incubated for 40 min at room temperature. Subsequently, ADMSC were washed twice with 500 μL of DMPBS Flush and centrifuged at 2000 rpm for 7 min. The ADMSC were resuspended in DMPBS Flush with 2% BSA, followed by the flow cytometry analysis.

2.1.3. In Vitro Differentiation Potential of ADMSC

Using different media for each type of differentiation, ADMSC (5P) were cultured (2×10^4 cells) according to Yang et al. [32] with some modifications. Osteogenic differentiation ($n = 3$): DMEM supplemented with 10% FBS, 0.1 μM dexamethasone, 200 μM ascorbic acid, and 10 mM β-glycerol phosphate. Adipogenic differentiation ($n = 3$): DMEM supplemented with 10% FBS, 1 μmol/L dexamethasone, 50 μmol/L indomethacin, 0.5 mM 3-isobutyl-1-methyl-xanthine, and 10 μM insulin. Condrogenic differentiation ($n = 3$): DMEM supplemented with 10% FBS, acid free 15 mM HEPES, 6.25 μg/mL insulin, 10 ng/mL TGF-β1, and 50 nM AsAP. Cultures were maintained at 37 °C with 5% CO_2, and the complete mediums were replaced twice a week. At the end of four weeks, cells were fixed with 4% paraformaldehyde for 20 min and stained with 2% Alizarin Red S (pH 4.1) during 5 min for calcium detection, with 0.025% Toluidine blue in McIlvaine buffer (0.03 M citric acid, 0.04 M sodium phosphate dibasic, pH 4.0) during 10 min for proteoglycans detection, and with 1% Sudan IV during 5 min to show lipid droplets. Samples were imaged on the Axiovert S100 (ZEISS) inverted microscope.

2.1.4. Cell Viability

Flasks of 75 cm^2 with ADMSC in 5P ($n = 3$) were trypsinized and centrifuged at 1800 rpm for 10 min. The pellet obtained from each flasks was resuspended in 1 mL of DMPBS Flush. Then, an aliquot of 10 μL of each culture was stained with 0.4% Trypan Blue, and the ADMSC were placed for analysis in the Countess II FL (Life Technologies® Inc., Carlsbad, CA, USA) equipment for cell concentration and viability measurements.

2.1.5. Contrast by Differential Interference (DIC)

For the birefringence analyses of nuclei, ADMSC (5P) in culture ($n = 3$) were stained with 0.025% toluidine blue solution in 0.1 M McIlvaine phosphate buffer at pH 4.3 for 30 min, followed by analyses in the Olympus BX-51 (Olympus America, Center Vallery, PA, USA) polarizing microscope equipped with a Q-color 5 camera (Olympus America), and using the Image Pro-plus v.6.3 software for Windows™ (Media Cybernetics, Silver Spring, MD, USA) [4,33,34]. With the microscope and the software, it is possible to carry out analysis of DIC and the anisotropic properties, both individually and in combination [34]. DIC observations were performed after addition of the two condenser's Wollaston prisms in the light path. With DIC-PLM, the optical path differences in samples can be detected through the sliding of another Wollaston prism that is positioned under the analyzer. The resulting colors in the sample are compared with the colors in a Michel-Lévy chart.

2.1.6. Real Time-PCR Array for Analysis of ADMSC Gene Expression

For the total RNA extraction of the ADMSC (5P) using Trizol® reagent, Invitrogen, Carlsbad, CA, USA, three different cultures were analyzed. A spectrophotometer (NanoDrop® ND-1000, Thermo Fisher Scientific®, Waltham, MA, USA) was used to quantify the RNA in each sample by determining the absorbance ratio at 260 and 280 nm. 0.5 μg from the total extracted RNA of each sample was used for the synthesis of cDNA using the RT^2 First Strand Kit (QIAGEN®, Hilden, Germany) and thermocycler Mastercycler Pro (Eppendorf®, Hamburg, Germany), also following the manufacturer's instructions. The cDNA was frozen at −20 °C until tested. The RT-PCR array reaction was performed using the RT^2 Profiler PCR Arrays (A format) kit in combination with the RT^2 SYBR Green Mastermixes (QIAGEN®, Hilden, Germany) on the thermocycler apparatus 7300 (ABI Applied Biosystems®, Foster City, CA, USA), following the manufacturer's instructions. For each culture sample, three types of reaction controls were used:

1. Positive PCR control; 2. Reverse transcriptase control; and 3. Control for contamination of rat genomic DNA. The *Glyceraldehyde-3-phosphate dehydrogenase* (*Gapdh*, NM_017008) was used as an endogenous control for each sample. The following genes were analyzed (QIAGEN®): *Scleraxis* (*Scx*, NM_001130508); *Tenomodulin* (*Tnmd*, NM_022290); *Tumor necrosis factor* (*TNF superfamily, member 2*) (*Tnf*, NM_012675); *Interleukin 1 beta* (*Il1b*, NM_031512); *Transforming growth factor, beta 1* (*Tgfb1*, NM_021578); *Matrix metallopeptidase 2* (*Mmp2*, NM_031054); *Matrix metallopeptidase 9* (*Mmp9*, NM_031055); *Matrix metallopeptidase 8* (*Mmp8*, NM_022221); *TIMP metallopeptidase inhibitor 2* (*Timp2*, NM_021989); *Decorin* (*Dcn*, NM_024129); *Lysyl oxidase* (*Lox*, NM_017061); and *Growth differentiation factor 5* (*Gdf5*, XM_001066344). Reactions were made in a single cDNA pipetting for each gene, including the endogenous control. ΔCT values were obtained by the difference between the CT values of the target genes and the *Gapdh* gene. The $2^{-\Delta CT}$ method [35] was used to calculate the gene expression for each target gene.

2.2. In Vivo Experiments

2.2.1. Experimental Groups

A total of 110 male Lewis rats (120-day-old), with free access to food and water, were divided into 5 experimental groups (22 animals for each group): Normal (N): Rats with tendons without transection; Transected (T): Rats with partially transected tendons and treated with topical application of DMPBS Flush in the transected region; Mesenchymal stem cells derived from adipose tissue (ASC): Rats with transected tendons with subsequent transplant of ADMSC (3.7×10^5 cells) in the transected region; GDF-5 (GDF5): Rats with transected tendons and treated with topical application of DMPBS Flush in the transected region + GDF5 application (500 ng) 24 h after partial transection; and with ADMSC and GDF-5 (ASC+GDF5): Rats with transected tendons and treated with subsequent transplant of ADMSC (3.7×10^5 cells) in the transected region + GDF5 application (500 ng) 24 h after partial transection. Animals were euthanized on the 14th day after transection by an overdose of anesthetic (Ketamine and Xylazine). Animals of group N were euthanized at 134 days and the tendons without transection were collected for analysis.

2.2.2. Partial Transection of the Calcaneal Tendon and Application of ADMSC and GDF-5

The animals were anesthetized with intraperitoneal injection of Ketamine (90 mg/Kg) and Xylazine (12 mg/Kg), and the right lower paws submitted to antisepsis and trichotomy. For the exposure of the calcaneus tendon, a longitudinal incision was made in the animal's skin, followed by a transverse partial transection performed in the proximal tendon region located at a distance of 4 mm from its insertion in the calcaneus bone [36,37]. Approximately 3.7×10^5 ADMSC (5P) were resuspended in 15 μL of DMPBS Flush and transplanted in the transected region of tendons in the ASC group using a pipette. 15 μL of DMPBS Flush was applied in tendons of the T group, and the GDF5 group tendons received an application of 15 μL of DMPBS Flush + 500 ng of GDF-5 24 h after the tendon transection. All applications were made in the region of the tendon where the partial transection was performed. Then, the skin was sutured with nylon thread (Shalon 5-0) and a needle (1.5 cm). All surgical and experimental protocols were approved by the Institutional Committee for Ethics in Animal Research of the State University of Campinas-UNICAMP-Brazil (Protocol n° 2905-1).

2.2.3. Preparation of Sections in Freezing

Tendons were placed in Tissue-Tek®, frozen, and cut in cryostat (serial longitudinal cuts of 7 μm thickness). The sections were fixed using a 4% formaldehyde solution in Millonig buffer (0.13 M sodium phosphate and 0.1 M sodium hydroxide, pH 7.4) for 20 min, followed by birefringence and linear dichroism analysis.

2.2.4. DAPI Staining

Immediately after fixation, the sections (n = 4) were incubated with DAPI (4′,6-Diamidino-2-phenylindole dihydrochloride) (0.1 mg/mL in methanol) for 5 min at 37 °C. The sections were analyzed by fluorescence microscope (Olympus BX60) and the images captured by the QCapture 4.0 program.

2.2.5. Polarization Microscopy: Birefringence Measurements

After fixation, image analyses of the tendons (n = 4) were evaluated to detect differences in morphology based on the aggregation and organization of the collagen bundles, which reflect the variation of birefringence intensity. Birefringence properties were studied using an Olympus BX53 polarizing microscope and an image analyzer (Life Science Imaging Software, Version 510_UMA_cellSens16_Han_en_00). Because the birefringence appears visually as brilliance, this phenomenon was measured with an image analyzer and expressed as gray average (GA) values in pixels (8 bits = 1 pixel). The larger tendon axis was positioned at 45° to the crossed analyzer and polarizer. As collagen bundles exhibit two types of birefringence, intrinsic birefringence (Bi) and form or textural birefringence (Bf) [38], the total birefringence (the sum of Bi and Bf) was used in this study. Measurements of the tendons in each experimental group were made after immersing the sections in water, a condition in which total birefringence is highly detectable [4,33,34,38]. The number of measurements of GA was represented as the median and they were chosen at random in 16 sections from four tendons of each group.

2.2.6. Linear Dichroism Measurements

Linear dichroism measurements (n = 4) were obtained from toluidine blue-stained sections. Linear dichroism measurements have shown that GAG chains present in collagen fiber PGs are linearly distributed and predominantly parallel to the longest fiber axis [3,39,40]. In this case, linear dichroism is an extrinsic phenomenon, resulting from the arrangement of toluidine blue molecules that are electrostatically bound to the anionic link sites of the oriented substrate. The dichroic ratio ($DR=d/d_\perp$) was determined by the toluidine blue absorbance in the parallel (d) and perpendicular ($d_\perp$) positions of the tendon's longest axis, with regard to the polarized light plane (PLP) [39,40]. Linear dichroism was measured using an Olympus BX53 polarizing microscope (Objective: Olympus UPlanFL N 40×; Camera: Olympus Q-color 5; Polarizer: Olympus U-POT) and an image analyzer (Life Science Imaging Software, Version 510_UMA_-cellSens16_Han_en_00). The number of measurements (~100) of GA was represented as the median and they were chosen at random in 16 sections from four tendons of each group.

2.2.7. Real Time-PCR Array

The collected tendons (n = 4) were placed in stabilizing solution (RNA-later, QIAGEN® Hilden, Germany) and maintained at −20 °C. For the total RNA extraction, the tendons were sprayed using liquid nitrogen and then homogenized in a tube containing 5 stainless steel balls (2.3 mm diameter, Biospec) by being shaken in a TissueLyser LT instrument (QIAGEN® Hilden, Germany), with 2 repetitions (60 s) intercaleted with ice cooling (2 min) between each shaking step (5). Total RNA was isolated from each sample using the RNeasy® Fibrous Tissue Mini Kit (QIAGEN® Hilden, Germany), following the manufacturer's instructions. A spectrophotometer (NanoDrop® ND-1000, Thermo Fisher Scientific®, Waltham, Massachusetts, USA) was used to quantify the RNA in each sample by determining the absorbance ratio at 260 and 280 nm. 0.5 µg from the total extracted RNA of each sample was used for the synthesis of cDNA using the RT^2 First Strand Kit (QIAGEN® Hilden, Germany) and thermocycler Mastercycler Pro (Eppendorf® Hamburg, Germany), also following the manufacturer's instructions. The cDNA was frozen at −20 °C until tested. The RT-PCR array reaction was performed using the RT^2 Profiler PCR Arrays (A format) kit in combination with the RT^2 SYBR

Green Mastermixes (QIAGEN® Hilden, Germany) on the thermocycler apparatus 7300 (ABI Applied Biosystems®, Foster City, CA, USA), following the manufacturer's instructions. For each animal sample, three types of reaction controls were used: 1. Positive PCR control; 2. Reverse transcriptase control; and 3. Control for contamination of rat genomic DNA. The *Glyceraldehyde-3-phosphate dehydrogenase* (*Gapdh*, NM_017008) was used as an endogenous control for each sample. The following genes were analyzed (QIAGEN® Hilden, Germany): *Scleraxis* (*Scx*, NM_001130508); *Tenomodulin* (*Tnmd*, NM_022290); *Tumor necrosis factor* (*TNF superfamily, member 2*) (*Tnf*, NM_012675); *Interleukin 1 beta* (*Il1b*, NM_031512); *Transforming growth factor, beta 1* (*Tgfb1*, NM_021578); *Matrix metallopeptidase 2* (*Mmp2*, NM_031054); *Matrix metallopeptidase 9* (*Mmp9*, NM_031055); *Matrix metallopeptidase 8* (*Mmp8*, NM_022221); *TIMP metallopeptidase inhibitor 2* (*Timp2*, NM_021989); *Decorin* (*Dcn*, NM_024129); *Lysyl oxidase* (*Lox*, NM_017061); and *Growth differentiation factor 5* (*Gdf5*, XM_001066344). Reactions were made in a single cDNA pipette for each gene, including the endogenous control. ΔCT values were obtained by the difference between the CT values of the target genes and the *Gapdh* gene. These values were normalized by subtracting the ΔCT value of the calibrator sample (T group) to obtain $\Delta\Delta$CT values. A $2^{-\Delta\Delta CT}$ method was used to calculate the relative expression level (fold change) for each target gene. Results were represented as the relative gene expression in comparison to the calibrator sample that is equal to 1.

2.2.8. Dosage of Hydroxyproline

Hydroxyproline was used as an indicator of the amount of total collagen in the tendons of the different groups (*n* = 5) used previously in the biomechanical assay. The tendons were cut and immersed in acetone for 48 h, followed by a solution containing chloroform:ethanol (2:1) also for 48 h. After dehydration, the samples were placed for drying in the oven at 37 °C. Samples were weighed and hydrolyzed in 6N HCl (1 mL/10 mg of tissue) for 4 h at 130 °C according to Stegemann and Stalder [41] with some modifications, and neutralized with 6N NaOH. The absorbance of the samples was measured at 550 nm using a microplate reader (Expert Plus, Asys®, Holliston, MA, USA).

2.2.9. Evaluation of the Max Contact Intensity of the Rat Paw after Partial Transection

The CatWalk system (Noldus Inc., Wageningen, The Netherlands) was used to analyze the gait recovery of the animals (*n* = 5). In this protocol, the rats crossed a walkway (100 cm length, 5 cm width, and 0.6 cm thickness) with a glass floor illuminated from the long edge in a dark room. Data acquisition was performed with a high-speed camera (Pulnix TM-765E CCD), and the paw prints were automatically classified by the software. The paw prints were obtained during the 3 days before the partial transection of the tendons to assess the normal standard gait of the animals, and they were collected again after the lesions. Post-operative data were assessed on the 1st, 3rd, 5th, 7th, 9th, 11th, and 13th days following surgical lesion. The parameters used herein were "Max Contact Intensity", corresponding to the pressure exerted by the paw on the glass floor during gait. The intensity of magnification can vary from 0 to 255 pixels.

2.2.10. Biomechanical Parameters

Tendons from experimental groups (*n* = 5) were collected and stored at −20 °C until tested. Before the biomechanical test, the tendons were thawed and measured with pachymeter, considering their length, width, and thickness. For the biomechanical assay, the tendons were maintained in PBS to prevent their fibers from drying out. Then, the tendons were fixed to metal claws by the myotendinous junction and by the osteotendinous junction for correct alignment in the equipment (Texturometer, MTS model TESTSTAR II). In each biomechanical assay, tendons were subjected to a gradual increase of load at a displacement velocity of 1 mm/s by using a load 0.05 N until the tendon ruptured. Biomechanical parameters were analyzed according to Biancalana et al. [42] and Tomiosso et al. [9], such as maximum force (N) and maximum displacement (mm), which were used to calculate the maximum stress (Mpa) and maximum strain (L) of tendons from the experimental groups. The cross-sectional area of the calcaneus tendon was calculated by assuming an elliptical approximation (A = πWd/4) using measurements of width (W) and thickness (d) values from the same Sparrow et al. [43]. The maximum stress value (MPa) was estimated by the ratio between the maximum load (N) and the cross-sectional area (mm^2). The maximum deformation (L) was calculated through (L = L_f − L_i/L_i), where (L_f) is the value of the final length before rupture, and (L_i) is the initial tendon length value. Stress-strain curves were constructed using the mean of each mechanical property obtained for each group.

2.2.11. Statistical Analysis

All results were presented as the mean and standard deviation for the values with a normal distribution (or interquartile range and median for the values that did not adhere to the Gaussian distribution). For the data with normal distribution, the analysis of variance (ANOVA) was used, followed by the Tukey post-hoc test for intra-group analysis (in the case of statistical significance), or the Student's *t* Test preceded by the Levene Test. For data that did not adhere to the Gaussian distribution, the non-parametric test of the Kruskal-Wallis test followed by the post-hoc test of Dunn for intra-group analysis (in the case of statistical significance) was used or the U Test of Mann-Whitney. Statistical analysis was performed in the software Statistical Package for Social Sciences (SPSS) version 22.0 and for all the aforementioned tests the significance level α = 0.05 and power of the test of 95% were considered.

3. Results

3.1. In Vitro

3.1.1. In Vitro Differentiation Potential of ADMSC and Flow Cytometry

In vitro ADMSC staining with Toluidine Blue, Alizarin Red S, and Sudan IV showed differentiation of ADMSC at the 5P in chondrocytes, osteoblasts, and adipocytes, respectively (Figure 1A–C). Flow cytometric analysis (Figure 1D,E) showed the presence of CD90 (90%) and CD105 (95%) positive surface markers, and the negative marker, CD45 (75%).

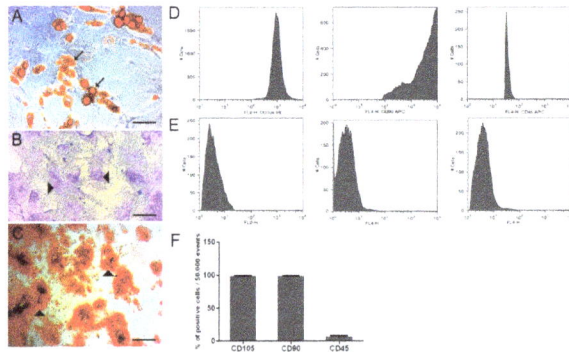

Figure 1. In vitro adipogenic (**A**), condrogenic (**B**), and osteogenic (**C**) differentiation of adipose-derived mesenchymal stem cells (ADMSC) in 5P: Observe intracellular lipid droplets stained with Sudan IV (→), proteoglycans stained with toluidine blue (▶), and extracellular calcium stained with alizarin red S (▶). Bars = A and C: 200 μm; B: 100 μm. (**D**) Histograms demonstrate the x-axis fluorescence scale considered positive when the cell peak is above 10^1 (CD45) or 10^2 (CD105 and CD90). (**E**) Control for −PE and −APC (with very low fluorescence), corresponding to non-marked cells. (**F**) Flow cytometry of ADMSC for CD105, CD90, and CD45 surface markers.

3.1.2. Fluorescence, Birefringence, and Contrast by Differential Interference (DIC)

Analysis using fluorescence microscopy showed ADMSC-GFP on 5P with fusiform fibroblast-like morphology (Figure 2A). When DIC was used, nucleoli were observed (Figure 2B). On AT (pH 4.3) staining, most of the nuclei were stained in blue, with the presence of granules of various sizes in some regions being highly stained in blue (Figure 2C). Under polarizing microscopy, the granules exhibited abnormal interference colors due to differences in the high packing of DNA (Figure 2D).

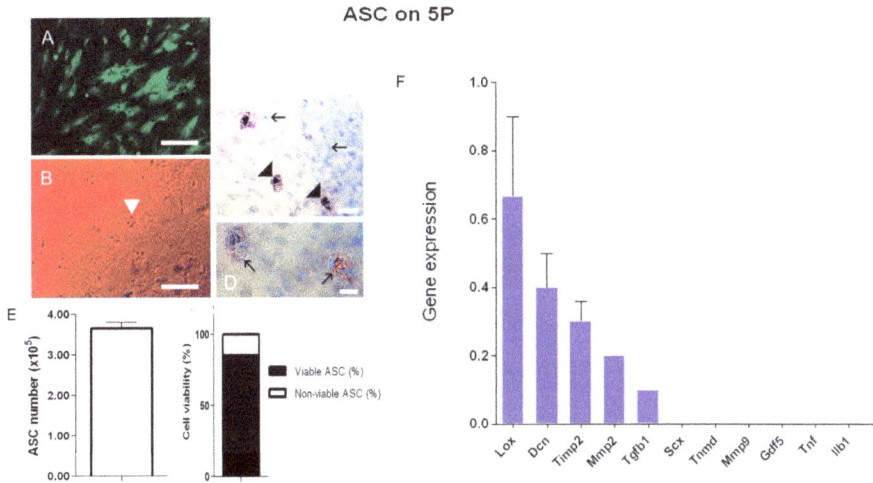

Figure 2. Morphology of ADMSC on 5P: (**A**) ADMSC-GFP: Fibroblast-like morphology, with a fusiform shape. (**B**) Contrast by Differential Interference (DIC): The red to blue band showing the interference effect of the nucleus due to higher concentrations of material. Nucleoli presence seen in the initial of the blue band (▶). (**C**) ADMSC stained with AT (pH 4.3): Note the presence of nuclei stained in blue (→), and the presence of granules of various sizes highly stained in blue in some regions (▶) due to the disponibility of phosphate groups. (**D**) Polarization microscopy: In the detailed image, note the abnormal interference colors due to abnormal dispersion of birefringence because of the differences in the high packing of DNA. Bars = 50 μm (**A**), 70 μm (**B,C**), and 35 μm (**D**). (**E**) Number of ADMSC used for application in tendons, presenting about 80% viability after tripsinization. (**F**) RT-PCR array of ADMSC on 5P in vitro showing the expression profile of the genes *Lox*, *Dcn*, *Timp2*, *Mmp2*, and *Tgfb1*. No expression was observed for *Scx*, *Tnmd*, *Mmp9*, *Gdf5*, *Tnf*, and *Ilb1* genes.

3.1.3. Cell Viability

Figure 1E showed a mean of 3.7×10^5 ADMSC used for tendons' application of the ADMSC and ASC+GDF5 groups, with a mean of 80% of viable ADMSC (Figure 2E).

3.1.4. Real-Time PCR Array

In the RT-PCR array analysis of ADMSC on 5P, the expression profile of the genes, *Lox*, *Dcn*, *Timp2*, *Mmp2*, and *Tgfb1* was observed. No expression was observed for the *Scx*, *Tnmd*, *Mmp9*, *Gdf5*, *Tnf*, and *Ilb1* genes. The expression of the *Mmp8* gene was not represented because only one culture expressed it (Figure 2F).

3.2. In Vivo

3.2.1. Immunofluorescence

In the present study, significant alterations in the ECM of the Achilles tendon were observed after 14 days since the partial transection, both from the application of the ADMSC isolated in the injured region and from the application of the ADMSC associated with GDF-5. The cell migration assay demonstrated the presence of ADMSC-GFP in tendon sections of both groups, ASC and ASC+GDF5, on the 3rd and 14th days after injury (Figure 3).

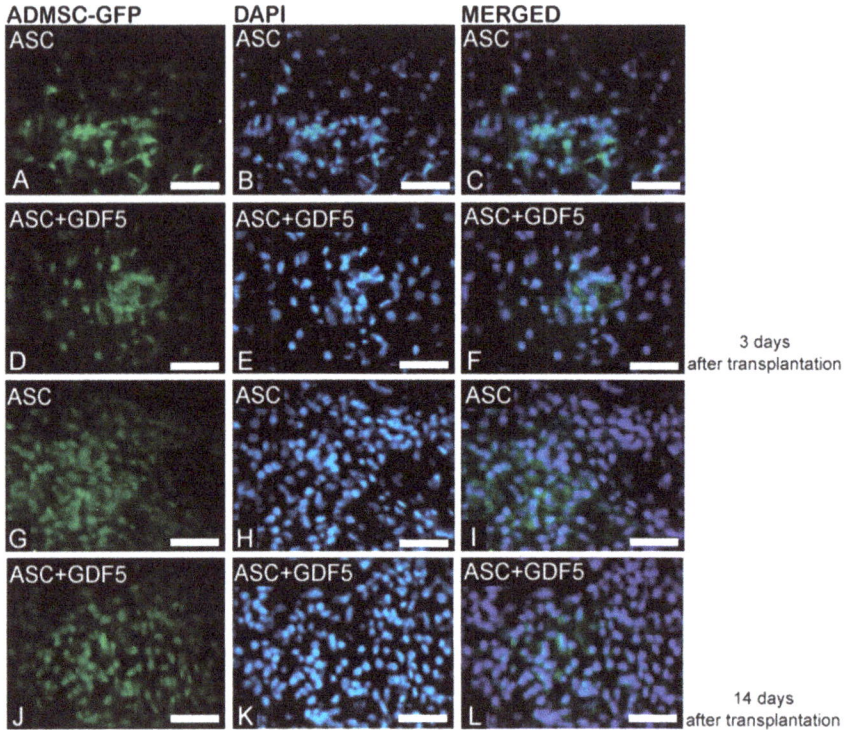

Figure 3. ADMSC-GFP Migration to TR on the 3rd and 14th days after injury: Observe the presence of ADMSC-GFP (**A,D,G,J**) in the ASC and ASC+GDF5 groups. Visible higher numbers of cells can be observed on the 14th day in both groups. DAPI: Nuclei marking (**B,E,H,K**). Merged images of ADMSC-GFP with nuclei marked with DAPI (**C,F,I,L**). Bar = 50 μm.

3.2.2. Real-Time PCR Array

The 12 genes expression analysis, obtained through the RT-PCR array (Figure 4), showed that the ADMSC application up-regulated the expression of the *Lox*, *Dcn*, and *Tgfb1* genes compared to the T group, and of the *Mmp2*, *Timp2*, and *Gdf5* genes in relation to the ASC+GDF5 group.

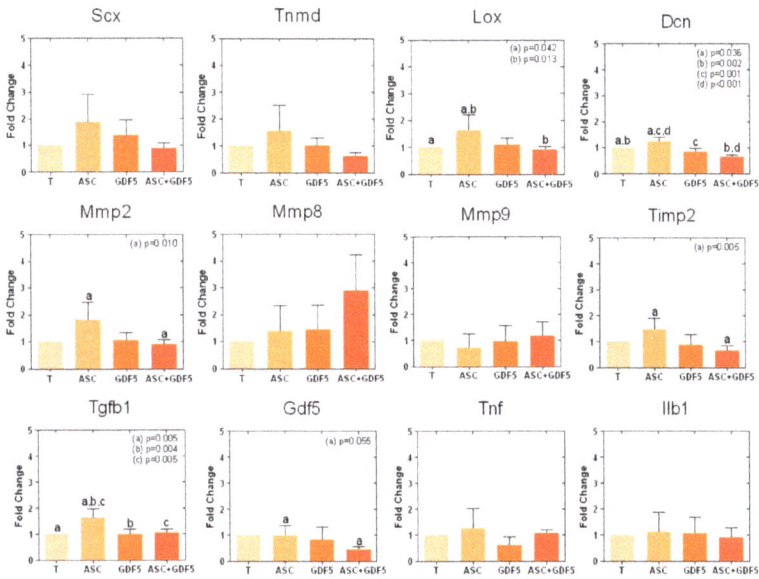

Figure 4. RT-PCR array for expression analysis of 12 genes in transected tendons: 50% of the genes analyzed were altered. The ADMSC increased expression of *Lox*, *Dcn*, and *Tgfb1* when compared to the other groups. Compared only to the ASC+GDF5 group, ADMSC increased the expression of *Mmp2*, *Timp2*, and *Gdf5*. The same letter between the groups corresponds to a significant difference between them.

3.2.3. Dosage of Hydroxyproline

Regarding the hydroxyproline dosage (Figure 5), which infers the concentration of total tissue collagen (mg/g of tissue), a lower value was demonstrated in the ASC+GDF5 group in relation to all other groups.

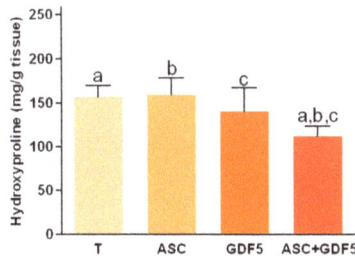

Figure 5. Tendons hydroxyproline concentration: Observe the lower value in the ASC+GDF5 group in relation to the other groups. The same letter among groups corresponds to a significant difference between them ($p \leq 0.05$).

3.2.4. Birefringence Measurements

In the birefringence measurements obtained through polarization microscopy (Figure 6), differences were observed in the collagen fiber organization of the tendons of the different groups. The tendons analyzed areas were characterized as follows: Transection region (TR), where partial transection of the collagen bundles was performed; and proximal and distal transition region

(T1), which are located in the adjacency of the TR. Considering the TR, the T group presented higher birefringence values (gray average values in pixels) in relation to the other transected groups. The ASC group, in relation to the groups, GDF5 and ASC+GDF5, had a higher value of birefringence. Regarding the organization of the collagen bundles in T1, the ASC group presented greater birefringence in relation to the other groups, and GDF5 presented a lower value also in relation to the T group. Regarding the organization pattern of the crimp, no significant difference was observed between groups.

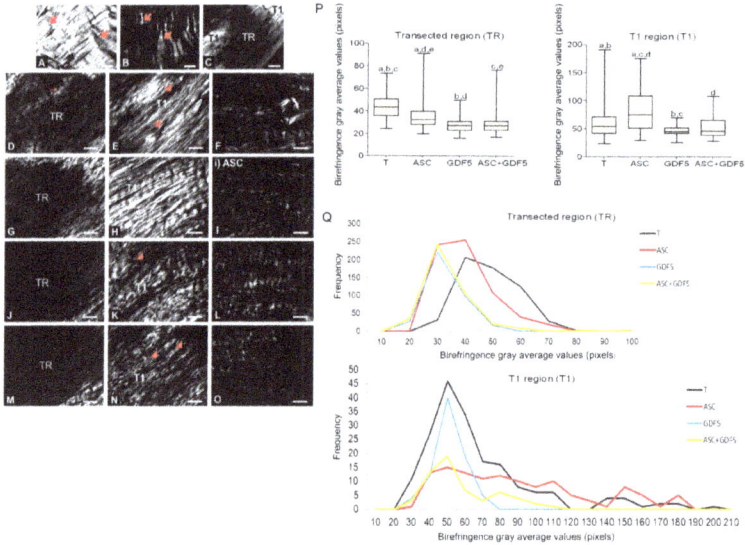

Figure 6. Images of tendons using polarization microscopy. (**A**) group N: Birefringence of the collagen fibers in the proximal region of the calanear tendon. The variation in gray levels is due to the crimp and the degree of aggregation of the collagen fibers (☪); observe the crimp (**B**) by positioning the largest axis of the tendon parallel to one of the polarizers: The same region observed in (**A**). (**C**) Panoramic image of the transected tendon for identification of the transection region (TR) and the proximal and distal transition region (T1). Groups T (**D–F**), ASC (**G–I**), GDF5 (**J–L**), and ASC+GDF5 (**M–O**). Observe the complete disorganization of collagen fibers in TR. The TR from different groups (**D,G,J,M**): Observe freshly formed collagen fibrils and an overlapping (↘) of this region with the thicker fibers present in T1. T1 (**E,H,K,N**): Collagen fibers with a greater organization in relation to TR, however, with fragmentation presence (◢) mainly in groups T (**E**), GDF5 (**K**), and ASC+GDF5 (**N**). *Crimp* (**F,I,L,O**) from the collagen fibers observed on T1: Observe similar undulation patterns of the collagen fibers between the groups, represented by light and dark regions. The largest axis of the tendon was positioned at 45° in relation to the crossed polarizers as parallel to one of the polarizers (**B,F,I,L,O**). (**P**) TR birefringence measurements in T1: Same letter represents significant differences between groups ($p \leq 0.05$). (**Q**) Histogram of the frequency and birefringence values showing differences in the distribution of values in the different groups. Bars = 100 μm and 200 μm (a).

3.2.5. Linear Dichroism Measurements

The dichroic ratio (DR) calculated from the linear dichroism measurements performed in sections stained with toluidine blue (Figure 7), through the use of polarization microscopy, showed differences in the organization of the PG present in the TR of the tendons. The T group presented higher DR when compared to the ASC and ASC+GDF5 groups, followed by the GDF5 group, with a higher value in relation to the ASC group.

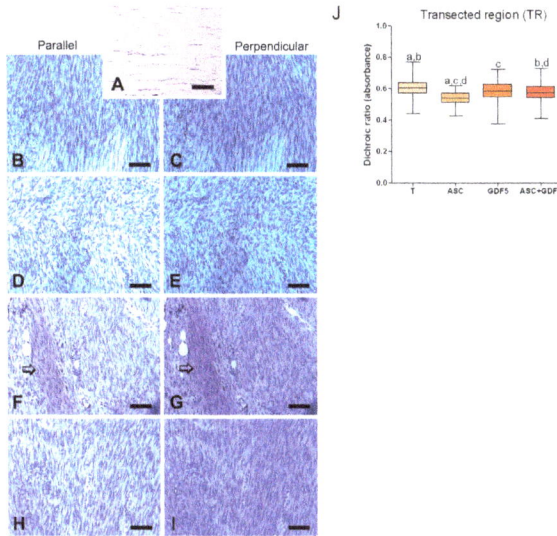

Figure 7. Linear dichroism (DL) of AT-stained tendon sections and analyzed under polarization microscopy. The largest axis of the tendon was placed in the parallel (**B,D,F,H**) and perpendicular (**C,E,G,I**) position relative to the polarized light plane. Observed DL is typically more intense in the cuts in the perpendicular position. Group N (**A**) is observed under common light microscopy due to the low amount of PG. (**J**) Dichroic index (ID) calculated through linear dichroism measurements (absorbance) performed on tendons TR: Greatest value observed in group T. Same letter between groups corresponds to a significant difference between them ($p \leq 0.05$). Bar = 100 μm.

3.2.6. CatWalk System

In the functional analysis obtained through CatWalk (Figure 8), differences were observed in the rats' gait in the different groups. Considering the maximum contact intensity of the paw (pixels) during gait, the ASC+GDF5 group presented a higher value in relation to the T and GDF5 groups 24 h after the transection. Between the 5th and 13th days after transection, the ASC group presented higher values in relation to the GDF-5 treated groups, although without a significant difference in relation to the T group.

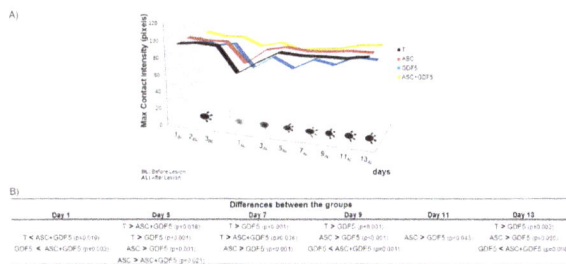

Figure 8. (**A**) Maximum contact intensity of the paw of the animals during walking, obtained through the CatWalk system. The measurements were taken three days before the injury to obtain the normal walk pattern, and on the 1st, 3rd, 5th, 7th, 9th, 11th, and 13th days after the tendon transection. Observe the marked decrease in contact pressure of the animals' paw on the day after surgery, with a higher value of the ASC+GDF5 group compared to the T group. Except for the GDF5 group, observe the complete recovery of the normal walk pattern of the animals in the other groups on the 13th day. (**B**) Comparisons between groups with significant differences observed between the 1st and 13th days.

3.2.7. Biomechanical Parameters

The biomechanical analyses of the Achilles tendon (Figure 9A) showed significant differences between the groups considering some parameters. Regarding the maximum rupture load, the ASC group presented a higher value in relation to the GDF5 and ASC+GDF5 groups. Considering the displacement and strain parameters, the groups treated with GDF-5 presented higher values in relation to the T and ASC groups. No difference was observed in the cross-sectional area between the groups. In the stress-strain curve (Figure 9B), tendons treated with ADMSC presented lower deformation at higher stress in comparison to the other groups.

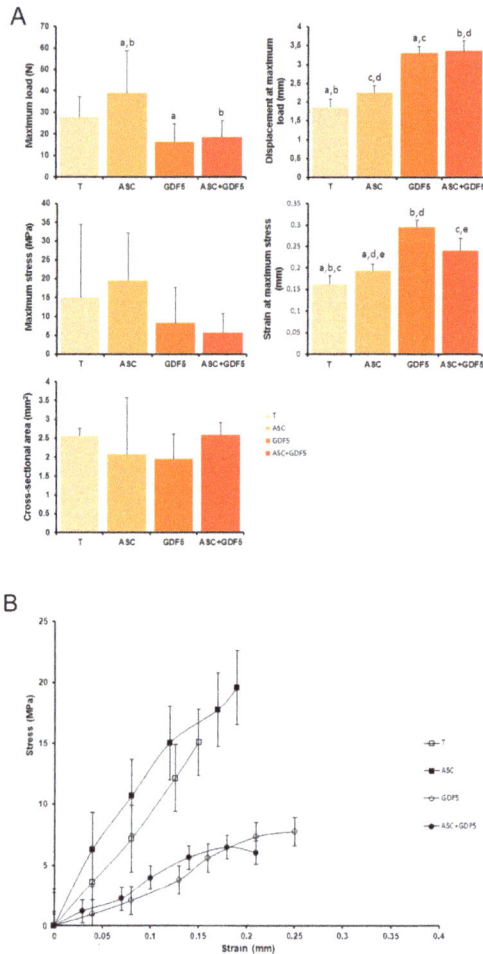

Figure 9. (**A**) Biomechanical properties of the tendons: Significant differences can be observed between the groups for the parameters of maximum load, displacement, and strain. The same letter between the groups corresponds to a significant difference between them ($p \leq 0.001$). (**B**) Stress-strain curve: Tendons treated with ADMSC presented lower deformation at higher stress in comparison to the other groups. Standard deviations are represented by vertical bars.

4. Discussion

Several current therapeutic techniques have been applied with the objective of improving tendon healing with partially satisfactory results considering its functional repair. The healed tendon is often characterized by functional impairment and a risk of re-rupture, mainly at the site of injury or near the injury region [13]. Recently, ADMSC have been proposed as a new treatment to improve this repair process due to its multipotency, cultivation facility, high yield, and, because they originate from adult donors, lack of ethical issues. In the present study, in the characterization of the cells before application, ADMSC expressed the positive surface markers, CD90 and CD105, and low expression of the negative marker, CD45. ADMSC also differentiated in chondrocytes, osteoblasts, and adipocytes, showing multilineage differentiation. Nucleoli were also observed in the ADMSC, indicating their synthesis activity, as an interesting profile of genes expression was exhibited by the ADMSC in vitro due to the expression of *Lox, Dcn, Mmp2, Timp2,* and *Tgfb1* genes, and no expression of the genes, *Scx, Tnmd, Mmp9, Gdf5, Tnf,* and *Ilb1*.

The transplanted ADMSC-GFP migrated to the transected tendon region in response to the specific microenvironment of the injury, characterized by the intense initial inflammatory process [44,45], and remained until the 14th day of the repair process. Thus, the alterations found in the groups after ADMSC transplantation are directly associated with the incorporation of these cells into the tendon. It is important to mention that at the time of tendons application, a mean of 80% of viable ADMSC were transplanted. According to Gimble et al. [18], the use of ADMSC is promising in several repair processes because of their potential for differentiation in different cell lines, as well as the secretion of growth factors and other signaling molecules.

In the present study, ADMSC increased the organization of collagen fibers in the injury adjacent region, which was reflected in the biomechanical properties of the tendons. Compared to the control transected tendon, tendons treated with ADMSC were apparently more resistance to traction, with lower deformation at higher stress. The organization and mechanical properties of the tendon are determined mainly by the orientation of fibrils and collagen fibers, fiber diameter, number of intra- and inter-molecular crosslinks that guarantee high tensile strength, total collagen content, the amount of PG, and the presence of other non-collagen proteins, such as cytokines, growth factors, and structural proteins [8,9,46–48]. Some studies have shown the effectiveness of ADMSC in some models of tendon injury. Results from Del Bue et al. [49] showed the benefits of ADMSC application, in association with a platelet concentrate, in the functional recovery of horses' tendons. In a model of rabbits' calcaneal tendon injuries, the application of ADMSC associated with a gel composed of a plasma rich in platelets increased the biomechanical resistance of the tendons in the 4th week of the healing process and increased the amount of collagen type I, VEGF, and FGF [7]. Uysal and Mizuno [50] also showed the effectiveness of the application of ADMSC in isolation at the site of the tendon injury, in which there was an increase in tendon resistance to tension, increase of angiogenic growth factors, and differentiation of stem cells into tenocytes and endothelial cells.

The application of ADMSC in association to GDF-5 led to a lower concentration of total collagen and a low degree of organization of ECM tendon, evidenced by the low birefringence of the collagen fibers when compared to the group that received only the ADMSC. Our results indicate that the application of GDF-5 alone and when associated with ADMSC both reduce tissue organization, resulting in inferior biomechanical properties when compared to control transected tendons, as well as in relation to the tendons transected and the ADMSC treated ones. However, several studies have demonstrated the modulatory action of GDF-5 on the tendon healing process due to its effect on cell migration and adhesion, differentiation, proliferation, and angiogenesis [28,29,31,51].

The beneficial role of GDF-5 was also demonstrated in the study by Mikic et al. [52] using deficient GDF-5 mutant mice, in which the tendons of these animals showed 40% less collagen compared to control tendons, as well as biomechanically less resistant tendons. A study by Chhabra et al. [31] reported that deficient GDF-5 mice presented until the 8th week of the healing process of the calcaneal tendon, a delay of one to two weeks in this process, characterized by a lower collagen concentration and smaller fibrils diameter collagen, lower amount of glycosaminoglycans (GAG), lower cell and blood vessel density, greater amount of fat, and lower biomechanical resistance until the 12th week. Rickert et al. [53] used gene therapy for the endogenous production of GDF-5 by genetically modified cells, which were applied to the injured tendon. Although the tissue quality was slightly greater than the control tendon, the presence of cartilaginous tissue was observed in the 8th week, disappearing in the 12th week of the healing process. Thus, it is possible to conclude that the lower biomechanical properties observed in the GDF-5 treated groups are directly related to the smaller organization of the fibers [48], as well as to the lowest total collagen content.

A lower dichroic ratio was observed in the ADMSC treated groups compared to the only transected group, indicating changes in the GAG arrangement in the collagen fibers present in the tendons of these groups. GAG chains are constituents of PG, which may be associated with collagen fibers through non-covalent interactions. Additionally, in normal tendons, the GAG chains present in PG are predominantly parallel to the major axis of the tendon [3]. A study by Mello and Vidal 40] showed differences in the arrangement of GAG up to 110 days after calcaneal tendon injury in rats. Thus, the lower dichroic ratio observed in the group treated with ADMSC alone suggests that not all GAG chains are neatly parallel to the major axis of the tendon, corroborating to the smaller organization of newly formed collagen fibers in the TR of the same group. The small PG play an important role in the fibrilogenesis of the collagen, regulating the growth in the fiber diameter [6]. It is important to emphasize that both groups treated with GDF-5 presented a higher dichroic ratio in relation to the ASC group, although less organization of the collagen fibers was observed in these groups treated with GDF-5.

Growth factors are small peptides, which generally transmit signals between cells, and are thus molecules that modulate cellular activity. In general, growth factors can regulate cell activity through several mechanisms, such as mitogenic activity, cell differentiation, cell migration, and gene regulation; and also play an important role in cellular chemotaxis and ECM synthesis [54,55]. The application of growth factors with the goal of improving the tendon healing process is still experimental, and has been restrictive to in vitro experiments and animal models [55,56]. The association of GDF-5 to the ADMSC used in this study was mainly based on the in vitro results of Park et al. [19]. An increase in ADMSC proliferation after GDF-5 application, and an increase in the amount of type I collagen, decorin, and aggrecan was seen, as well as markers of tendon cells, such as Scx, Tnmd, and tenascin-C, which have indicated differentiation of ADMSC in tenocytes [19]. However, in vivo, our results demonstrated low effectiveness of both the application of GDF-5 alone and after its application in association with ADMSC.

The literature points to a limitation in the application form and frequency of growth factors during tendon healing due to the relatively short half-life of these factors, which would only allow the short-term modulation of their biological effects [53]. Results from Forslund et al. [30] showed interesting data after a single injection (six h after the tendon injury) of different doses (0.4, 2, and 10 µg) of GDF-5, -6, and -7. On the 8th day after surgery, all factors increased the biomechanical resistance of the rats' calcaneal tendon after the application of the higher doses of 2 and 10 µg, in addition to observing a decrease in the inflammatory process. Yet, after the application of 10 µg of the factors in this same study, bone and cartilaginous formation in the tendons was observed. In the present study, 500 ng of GDF-5 was used based on the data of Park et al. [19] that treated the culture of ADMSC with different concentrations of GDF-5 (0–1000 ng/mL).

The application of GDF-5 in this study occurred 24 hours after tendon transection. Considering the important participation of the GDF-5 in tendon repair, mentioned above in several studies, and as most of our results indicate a greater effectiveness of applying only ADMSC during this phase of the tendon repair process, important variables should be GDF-5: The animal model used, the model of tendon injury, including the extent of the injured area, the origin of the growth factor, and the dosage and form of administration of the same, as well as the period elapsed between injury and the application of the growth factor [57]. According to some authors, the administration of the growth factor after one week of surgery results in an improvement in the quality of the healing process, including improvement in the biomechanical properties of the tendon and the structural organization of its ECM [58–61].

The most striking result in the functional analysis obtained through CatWalk showed a greater intensity of the rats' paws' contact pressure during gait 24 h after transection, possibly due to the decrease of pain after treatment with the ADMSC associated with GDF-5 in comparison to the control group. Treatment based on this association may have contributed to the reduction of acute inflammation and post-surgical edema, as well as decreased pain due to the reduction in the concentration of prostaglandins E2, nitric oxide activity, and reduction of free radicals [62]. These previously mentioned factors, although speculative, corroborate with the data of Forslund et al. [30], which showed that a single injection of GDF-5 during tendon repair decreased the inflammatory process, and with the data of Schneider et al. [17] that described immunomodulatory effects of the ADMSC.

The expression of 12 different genes related to inflammation, the remodeling of ECM, and the differentiation of ADMSC into tenocytes were studied to elucidate some molecular mechanisms involved in the repair of tendons. Our results demonstrated that ADMSC up-regulated the *Dcn* gene expression in comparison to other groups. The *Dcn* is a PG with a well-established structural function as it mediates the lateral fusion of collagen fibrils, contributing to the formation of mature collagen fiber with a larger diameter and, therefore, greater biomechanical resistance [63]. The lower expression of *Dcn* observed in the tendons treated with the ADMSC in association with GDF-5 may be directly related to the lower organization of the collagen fibers, resulting in lower biomechanical properties when compared to the control transected tendons, in relation to the ADMSC treated tendons. The application of ADMSC in association with GDF-5 also down-regulated the *Lox* gene expression compared to the ASC group, corroborating to the lower organization of collagen fibers in the ASC+GDF5 group. Herchenhan et al. [64] have demonstrated the direct role of *Lox* in collagen fibrilogenesis, showing that the inhibition of this enzyme activity harms the formation of the characteristic pattern of fibrils, leading to a decrease in the biomechanical resistance of tendon-like tissues constructed through bioengineering. It is worth mentioning that the *Lox* expression was higher after the treatment with ADMSC in comparison to the control transected tendon. It seems clear that there is a relation of the lower *Gdf5* gene expression in the ASC+GDF5 group with the lower ECM organization and biomechanical parameters observed when compared to the ASC group [28–31].

Regarding the expression of genes related to tissue remodeling, *Mmp2*, *Mmp8*, *Mmp9*, and *Timp2*, higher expression of *Mmp2* and *Timp2* was observed in the ASC group in relation only to the ASC+GDF5 group. The literature describes the increased expression and activity of these MMPs during tendon repair processes [8], since previous studies of our group showed a direct relationship between the greater activity of MMP-2 and greater organization of the ECM of tendons [36,65,66], corroborating with the results from the present study. It is important to emphasize that the larger organization of the collagen fibers in the T1 region was also observed in the group treated with ADMSC only.

The application of ADMSC also up-regulated the *Tgfb1* gene expression, another important gene whose expression modulates important processes during tendon repair. *Tfb1* acts on the regulation of cell proliferation, differentiation, and migration, apoptosis, GAG deposition, and stimulates the production of collagen by fibroblasts [67–70], influencing the tissue healing cascade [71]. *Tgfb1* is also involved in increased expression of tissue inhibitors of MMP (*Timp*) and *Lox* in different cell types, including in vitro fibroblasts [72], corroborating our results that showed a relation between the higher expression of *Tgfb1*, and increased expression of *Lox* and *Timp2* after ADMSC application.

We observed a strong trend towards higher expression of the *Scx* and *Tnmd* genes in tendons after ADMSC application in relation to the control group, supposing differentiation of ADMSC applied and/or endogenous stem cells into tenocytes. ADMSC have proven to modulate the host's "stem cell niche" by stimulating and recruiting endogenous stem cells to the injured site, promoting their differentiation [18]. *Scx* has recently been reported as a marker of progenitor cell populations in tendons, being a transcription factor present in tendons from development to adulthood [73]. *Scx* is a key regulator in the differentiation of tenocytes, whose expression is highly induced through the signaling pathway involving *Tgfb1* [74], which was increased in the ASC group. According to Shukunami et al. [75], *Scx* positively regulates the expression of *Tnmd*, which is a transmembrane protein specifically expressed in dense connective tissues, such as tendons, ligaments, epimysium of skeletal muscle, cornea, and sclera [76–80], indicating the differentiation of ADMSC into tenocytes.

Our data showed the influence of the application of ADMSC isolated and in association with GDF-5 in the expression of 50% of the analyzed genes, *Dcn*, *Gdf5*, *Lox*, *Tgfb1*, *Mmp2*, and *Timp2*. ADMSC in vitro presented, with exception of the *Gdf5* gene, a similar gene expression profile of those five genes in comparison to the *Scx*, *Tnmd*, *Mmp9*, *Tnf*, and *Ilb1* genes. The literature describes that the ADMSC behavior seems to be modulated by the presence of molecules at the site where they are injected, such as cytokines, chemokines, peptides, and, mainly, growth factors [67,68]. Other studies also demonstrate the paracrine and immunomodulatory effects of ADMSC at the site of injury to promote tendon regeneration [16,17].

In conclusion, the application of ADMSC up-regulated the *Dcn*, *Lox*, and *Tgfb1* genes expression, which contributed to a higher collagen fiber organization and tendon biomechanics. The association of ADMSC with GDF-5 down-regulated the *Dcn*, *Gdf5*, *Lox*, *Tgfb1*, *Mmp2*, and *Timp2*, which contributed to an improvement of the rats' gait 24 h after the injury and impaired the organization and biomechanics of tendons. Although the literature describes the benefic effect of GDF-5 for the tendon healing process, our results show that its application, isolated or associated with ADMSC, cannot improve the repair process of partial transected tendons, indicating the higher effectiveness of the application of ADMSC in injured Achilles tendons.

Author Contributions: Methodology, A.A.d.A., G.D.C., L.F.R.T., F.C.d.V., D.L.F., G.F.S., B.d.C.V. and E.R.P.; Formal Analysis, A.A.d.A., G.D.C., F.C.d.V., D.L.F., G.F.S., P.W.S., L.E.A., A.L.R.d.O., C.P.G., J.B.P., B.d.C.V., C.P.V. and E.R.P.; Resources, J.B.P. and M.A.M.E.; Writing-Original Draft Preparation, A.A.d.A., M.A.M.E., B.d.C.V. and E.R.P.; Writing-Review & Editing, A.A.d.A. and E.R.P.; Supervision, E.R.P. and C.P.V.

Funding: This research was funded by FAPESP (Fundação de Amparo à Pesquisa do Estado de São Palulo), grant number 2012/14973-8.

Acknowledgments: The authors thank FAPESP for the financial support and for the fellowship awarded to A. A. de Aro (2012/14973-8).

Conflicts of Interest: The authors declare no conflict of interest.

References

1. Goncalves, A.I.; Rodrigues, M.T.; Lee, S.J.; Atala, A.; Yoo, J.J.; Reis, R.L.; Gomes, M.E. Understanding the role of growth factors in modulating stem cell tenogenesis. *PLoS ONE* **2013**, *8*, e83734. [CrossRef] [PubMed]
2. Zabrzynski, J.; Lapaj, L.; Paczesny, L.; Zabrzynska, A.; Grzanka, D. Tendon–function related structure, simple healing process and mysterious ageing. *Folia Morphol.* **2018**. [CrossRef] [PubMed]
3. Vidal, B.C.; Mello, M.L. Proteoglycan arrangement in tendon collagen bundles. *Cell Mol. Biol.* **1984**, *30*, 195–204. [PubMed]
4. De Campos Vidal, B.; Mello, M.L. Optical anisotropy of collagen fibers of rat calcaneal tendons: An approach to spatially resolved supramolecular organization. *Acta Histochem.* **2010**, *112*, 53–61. [CrossRef] [PubMed]
5. Marqueti, R.C.; Durigan, J.L.Q.; Oliveira, A.J.S.; Mekaro, M.S.; Guzzoni, V.; Aro, A.A.; Pimentel, E.R.; Selistre-de-Araujo, H.S. Effects of aging and resistance training in rat tendon remodeling. *FASEB J.* **2018**, *32*, 353–368. [CrossRef] [PubMed]
6. Douglas, T.; Heinemann, S.; Bierbaum, S.; Scharnweber, D.; Worch, H. Fibrillogenesis of collagen types I, II, and III with small leucine-rich proteoglycans decorin and biglycan. *Biomacromolecules* **2006**, *7*, 2388–2393. [CrossRef] [PubMed]
7. Uysal, C.A.; Tobita, M.; Hyakusoku, H.; Mizuno, H. Adipose-derived stem cells enhance primary tendon repair: Biomechanical and immunohistochemical evaluation. *J. Plast. Reconstr. Aesthet. Surg.* **2012**, *65*, 1712–1719. [CrossRef] [PubMed]

8. Oshiro, W.; Lou, J.; Xing, X.; Tu, Y.; Manske, P.R. Flexor tendon healing in the rat: A histologic and gene expression study. *J. Hand Surg. Am.* **2003**, *28*, 814–823. [CrossRef]
9. Tomiosso, T.C.; Nakagaki, W.R.; Gomes, L.; Hyslop, S.; Pimentel, E.R. Organization of collagen bundles during tendon healing in rats treated with L-NAME. *Cell Tissue Res.* **2009**, *337*, 235–242. [CrossRef] [PubMed]
10. Christiansen, D.L.; Huang, E.K.; Silver, F.H. Assembly of type I collagen: Fusion of fibril subunits and the influence of fibril diameter on mechanical properties. *Matrix Biol.* **2000**, *19*, 409–420. [CrossRef]
11. Almekinders, L.C.; Almekinders, S.V. Outcome in the treatment of chronic overuse sports injuries: A retrospective study. *J. Orthop. Sports Phys. Ther.* **1994**, *19*, 157–161. [CrossRef] [PubMed]
12. Favata, M.; Beredjiklian, P.K.; Zgonis, M.H.; Beason, D.P.; Crombleholme, T.M.; Jawad, A.F.; Soslowsky, L.J. Regenerative properties of fetal sheep tendon are not adversely affected by transplantation into an adult environment. *J. Orthop. Res.* **2006**, *24*, 2124–2132. [CrossRef] [PubMed]
13. Ni, M.; Lui, P.P.; Rui, Y.F.; Lee, Y.W.; Lee, Y.W.; Tan, Q.; Wong, Y.M.; Kong, S.K.; Lau, P.M.; Li, G.; et al. Tendon-derived stem cells (TDSCs) promote tendon repair in a rat patellar tendon window defect model. *J. Orthop. Res.* **2012**, *30*, 613–619. [CrossRef] [PubMed]
14. Pittenger, M.F.; Mackay, A.M.; Beck, S.C.; Jaiswal, R.K.; Douglas, R.; Mosca, J.D.; Moorman, M.A.; Simonetti, D.W.; Craig, S.; Marshak, D.R. Multilineage potential of adult human mesenchymal stem cells. *Science* **1999**, *284*, 143–147. [CrossRef] [PubMed]
15. Kern, S.; Eichler, H.; Stoeve, J.; Kluter, H.; Bieback, K. Comparative analysis of mesenchymal stem cells from bone marrow, umbilical cord blood, or adipose tissue. *Stem Cells* **2006**, *24*, 1294–1301. [CrossRef] [PubMed]
16. Okamoto, N.; Kushida, T.; Oe, K.; Umeda, M.; Ikehara, S.; Iida, H. Treating Achilles tendon rupture in rats with bone-marrow-cell transplantation therapy. *J. Bone Joint Surg. Am.* **2010**, *92*, 2776–2784. [CrossRef] [PubMed]
17. Schneider, M.; Angele, P.; Jarvinen, T.A.H.; Docheva, D. Rescue plan for Achilles: Therapeutics steering the fate and functions of stem cells in tendon wound healing. *Adv. Drug Deliv. Rev.* **2017**, *129*, 352–375. [CrossRef] [PubMed]
18. Gimble, J.M.; Katz, A.J.; Bunnell, B.A. Adipose-derived stem cells for regenerative medicine. *Circ. Res.* **2007**, *100*, 1249–1260. [CrossRef] [PubMed]
19. Park, A.; Hogan, M.V.; Kesturu, G.S.; James, R.; Balian, G.; Chhabra, A.B. Adipose-derived mesenchymal stem cells treated with growth differentiation factor-5 express tendon-specific markers. *Tissue Eng. Part A* **2010**, *16*, 2941–2951. [CrossRef] [PubMed]
20. Schneider, P.R.; Buhrmann, C.; Mobasheri, A.; Matis, U.; Shakibaei, M. Three-dimensional high-density co-culture with primary tenocytes induces tenogenic differentiation in mesenchymal stem cells. *J. Orthop. Res.* **2011**, *29*, 1351–1360. [CrossRef] [PubMed]
21. Pacini, S.; Spinabella, S.; Trombi, L.; Fazzi, R.; Galimberti, S.; Dini, F.; Carlucci, F.; Petrini, M. Suspension of bone marrow-derived undifferentiated mesenchymal stromal cells for repair of superficial digital flexor tendon in race horses. *Tissue Eng.* **2007**, *13*, 2949–2955. [CrossRef] [PubMed]
22. Vieira, M.H.; Oliveira, R.J.; Eca, L.P.; Pereira, I.S.; Hermeto, L.C.; Matuo, R.; Fernandes, W.S.; Silva, R.A.; Antoniolli, A.C. Therapeutic potential of mesenchymal stem cells to treat Achilles tendon injuries. *Genet. Mol. Res.* **2014**, *13*, 10434–10449. [CrossRef] [PubMed]
23. Valencia Mora, M.; Antuna Antuna, S.; Garcia Arranz, M.; Carrascal, M.T.; Barco, R. Application of adipose tissue-derived stem cells in a rat rotator cuff repair model. *Injury* **2014**, *45* (Suppl. 4), S22–S27. [CrossRef]
24. Guilak, F.; Cohen, D.M.; Estes, B.T.; Gimble, J.M.; Liedtke, W.; Chen, C.S. Control of stem cell fate by physical interactions with the extracellular matrix. *Cell Stem Cell* **2009**, *5*, 17–26. [CrossRef] [PubMed]
25. Storm, E.E.; Huynh, T.V.; Copeland, N.G.; Jenkins, N.A.; Kingsley, D.M.; Lee, S.J. Limb alterations in brachypodism mice due to mutations in a new member of the TGF beta-superfamily. *Nature* **1994**, *368*, 639–643. [CrossRef] [PubMed]
26. Chang, S.C.; Hoang, B.; Thomas, J.T.; Vukicevic, S.; Luyten, F.P.; Ryba, N.J.; Kozak, C.A.; Reddi, A.H.; Moos, M., Jr. Cartilage-derived morphogenetic proteins. New members of the transforming growth factor-beta superfamily predominantly expressed in long bones during human embryonic development. *J. Biol. Chem.* **1994**, *269*, 28227–28234. [PubMed]
27. Zhou, S.; Yates, K.E.; Eid, K.; Glowacki, J. Demineralized bone promotes chondrocyte or osteoblast differentiation of human marrow stromal cells cultured in collagen sponges. *Cell Tissue Bank.* **2005**, *6*, 33–44. [CrossRef] [PubMed]

28. Wolfman, N.M.; Hattersley, G.; Cox, K.; Celeste, A.J.; Nelson, R.; Yamaji, N.; Dube, J.L.; DiBlasio-Smith, E.; Nove, J.; Song, J.J. Ectopic induction of tendon and ligament in rats by growth and differentiation factors 5, 6, and 7, members of the TGF-beta gene family. *J. Clin. Investig.* **1997**, *100*, 321–330. [CrossRef] [PubMed]

29. Forslund, C.; Aspenberg, P. Tendon healing stimulated by injected CDMP-2. *Med. Sci. Sports Exerc.* **2001**,·*33*, 685–687. [CrossRef] [PubMed]

30. Forslund, C.; Rueger, D.; Aspenberg, P. A comparative dose-response study of cartilage-derived morphogenetic protein (CDMP)-1, -2 and -3 for tendon healing in rats. *J. Orthop. Res.* **2003**, *21*, 617–621. [CrossRef]

31. Chhabra, A.; Tsou, D.; Clark, R.T.; Gaschen, V.; Hunziker, E.B.; Mikic, B. GDF-5 deficiency in mice delays Achilles tendon healing. *J. Orthop. Res.* **2003**, *21*, 826–835. [CrossRef]

32. Yang, X.F.; He, X.; He, J.; Zhang, L.H.; Su, X.J.; Dong, Z.Y.; Xu, Y.J.; Li, Y.; Li, Y.L. High efficient isolation and systematic identification of human adipose-derived mesenchymal stem cells. *J. Biomed. Sci.* **2011**, *18*, 59. [CrossRef] [PubMed]

33. Vidal, B.C. Form birefringence as applied to biopolymer and inorganic material supraorganization. *Biotech. Histochem.* **2010**, *85*, 365–378. [CrossRef] [PubMed]

34. Vidal, B.C.; Dos Anjos, E.H.; Mello, M.L. Optical anisotropy reveals molecular order in a mouse enthesis. *Cell Tissue Res.* **2015**, *362*, 177–185. [CrossRef] [PubMed]

35. Livak, K.J. Analyzing real-time PCR data by the comparative C(T) method. *Nat. Protoc.* **2008**, *3*, 1101–1108.

36. Aro, A.A.; Freitas, K.M.; Foglio, M.A.; Carvalho, J.E.; Dolder, H.; Gomes, L.; Vidal, B.C.; Pimentel, E.R. Effect of the Arrabidaea chica extract on collagen fiber organization during healing of partially transected tendon. *Life Sci.* **2013**, *92*, 799–807. [CrossRef] [PubMed]

37. Aro, A.A.; Perez, M.O.; Vieira, C.P.; Esquisatto, M.A.; Rodrigues, R.A.; Gomes, L.; Pimentel, E.R. Effect of Calendula officinalis cream on achilles tendon healing. *Anat. Rec.* **2015**, *298*, 428–435. [CrossRef] [PubMed]

38. Vidal, B.d.C. Evaluation of the carbohydrate role in the molecular order of collagen bundles: Microphotometric measurements of textural birefringence. *Cell. Mol. Biol.* **1986**, *32*, 527–535.

39. Feitosa, V.; Vidal, B.C.; Pimentel, E.R. Optical anisotropy of a pig tendon under compression. *J. Anat.* **2002**, *200*, 105–111. [CrossRef] [PubMed]

40. Mello, M.L.; de Campos Vidal, B. Experimental tendon repair: Glycosaminoglycan arrangement in newly synthesized collagen fibers. *Cell. Mol. Biol. (Noisy-le-grand)* **2003**, *49*, 579–585.

41. Stegemann, H.; Stalder, K. Determination of hydroxyproline. *Clin. Chim. Acta* **1967**, *18*, 267–273. [CrossRef]

42. Biancalana, A.; Veloso, L.A.; Gomes, L. Obesity affects collagen fibril diameter and mechanical properties of tendons in Zucker rats. *Connect. Tissue Res.* **2010**, *51*, 171–178. [CrossRef] [PubMed]

43. Sparrow, K.J.; Finucane, S.D.; Owen, J.R.; Wayne, J.S. The effects of low-intensity ultrasound on medial collateral ligament healing in the rabbit model. *Am. J. Sports Med.* **2005**, *33*, 1048–1056. [CrossRef] [PubMed]

44. Yang, X.; Liang, L.; Zong, C.; Lai, F.; Zhu, P.; Liu, Y.; Jiang, J.; Yang, Y.; Gao, L.; Ye, F.; et al. Kupffer cells-dependent inflammation in the injured liver increases recruitment of mesenchymal stem cells in aging mice. *Oncotarget* **2016**, *7*, 1084–1095. [CrossRef] [PubMed]

45. Heissig, B.; Dhahri, D.; Eiamboonsert, S.; Salama, Y.; Shimazu, H.; Munakata, S.; Hattori, K. Role of mesenchymal stem cell-derived fibrinolytic factor in tissue regeneration and cancer progression. *Cell. Mol. Life Sci.* **2015**, *72*, 4759–4770. [CrossRef] [PubMed]

46. Parry, D.A. The molecular and fibrillar structure of collagen and its relationship to the mechanical properties of connective tissue. *Biophys. Chem.* **1988**, *29*, 195–209. [CrossRef]

47. Birk, D.E.; Mayne, R. Localization of collagen types I, III and V during tendon development. Changes in collagen types I and III are correlated with changes in fibril diameter. *Eur. J. Cell Biol.* **1997**, *72*, 352–361. [PubMed]

48. Oryan, A.; Moshiri, A. A long term study on the role of exogenous human recombinant basic fibroblast growth factor on the superficial digital flexor tendon healing in rabbits. *J. Musculoskelet. Neuronal Interact.* **2011**, *11*, 185–195. [PubMed]

49. Del Bue, M.; Ricco, S.; Ramoni, R.; Conti, V.; Gnudi, G.; Grolli, S. Equine adipose-tissue derived mesenchymal stem cells and platelet concentrates: Their association in vitro and in vivo. *Vet. Res. Commun.* **2008**, *32* (Suppl. 1), S51–S55. [CrossRef] [PubMed]

50. Uysal, A.C.; Mizuno, H. Differentiation of adipose-derived stem cells for tendon repair. *Methods Mol. Biol.* **2011**, *702*, 443–451. [PubMed]

51. Bolt, P.; Clerk, A.N.; Luu, H.H.; Kang, Q.; Kummer, J.L.; Deng, Z.L.; Olson, K.; Primus, F.; Montag, A.G.; He, T.C. BMP-14 gene therapy increases tendon tensile strength in a rat model of Achilles tendon injury. *J. Bone Joint Surg. Am.* **2007**, *89*, 1315–1320. [CrossRef] [PubMed]

52. Mikic, B.; Schalet, B.J.; Clark, R.T.; Gaschen, V.; Hunziker, E.B. GDF-5 deficiency in mice alters the ultrastructure, mechanical properties and composition of the Achilles tendon. *J. Orthop. Res.* **2001**, *19*, 365–371. [CrossRef]

53. Rickert, M.; Wang, H.; Wieloch, P.; Lorenz, H.; Steck, E.; Sabo, D.; Richter, W. Adenovirus-mediated gene transfer of growth and differentiation factor-5 into tenocytes and the healing rat Achilles tendon. *Connect. Tissue Res.* **2005**, *46*, 175–183. [CrossRef] [PubMed]

54. Chan, B.P.; Chan, K.M.; Maffulli, N.; Webb, S.; Lee, K.K. Effect of basic fibroblast growth factor. An in vitro study of tendon healing. *Clin. Orthop. Relat. Res.* **1997**, 239–247. [CrossRef]

55. McGeachie, J.; Tennant, M. Growth factors and their implications for clinicians: A brief review. *Aust. Dent. J.* **1997**, *42*, 375–380. [CrossRef] [PubMed]

56. Sharma, P.; Maffulli, N. Tendon injury and tendinopathy: Healing and repair. *J. Bone Joint Surg. Am.* **2005**, *87*, 187–202. [PubMed]

57. Sheng, J.; Zeng, B.; Jiang, P. Effects of exogenous basic fibroblast growth factor on in-sheathed tendon healing and adhesion formation. *Zhongguo xiu fu chong jian wai ke za zhi Chin. J. Reparative Reconstr. Surg.* **2007**, *21*, 733–737.

58. Gospodarowicz, D.; Ferrara, N.; Schweigerer, L.; Neufeld, G. Structural characterization and biological functions of fibroblast growth factor. *Endocr. Rev.* **1987**, *8*, 95–114. [CrossRef] [PubMed]

59. McNeil, P.L.; Muthukrishnan, L.; Warder, E.; D'Amore, P.A. Growth factors are released by mechanically wounded endothelial cells. *J. Cell Biol.* **1989**, *109*, 811–822. [CrossRef] [PubMed]

60. Kannus, P. Structure of the tendon connective tissue. *Scand. J. Med. Sci. Sports* **2000**, *10*, 312–320. [CrossRef] [PubMed]

61. Wurgler-Hauri, C.C.; Dourte, L.M.; Baradet, T.C.; Williams, G.R.; Soslowsky, L.J. Temporal expression of 8 growth factors in tendon-to-bone healing in a rat supraspinatus model. *J. Shoulder Elb. Surg.* **2007**, *16*, S198–S203. [CrossRef] [PubMed]

62. Hu, C.; Ding, Y.; Chen, J.; Liu, D.; Ding, M.; Zhang, Y. Treatment of corneal epithelial wounds in dogs using basic fibroblast growth factor. *Vet. Med. (Praha)* **2009**, *54*, 280–286. [CrossRef]

63. Dunkman, A.A.; Buckley, M.R.; Mienaltowski, M.J.; Adams, S.M.; Thomas, S.J.; Satchell, L.; Kumar, A.; Pathmanathan, L.; Beason, D.P.; Iozzo, R.V.; et al. The tendon injury response is influenced by decorin and biglycan. *Ann. Biomed. Eng.* **2013**, *42*, 619–630. [CrossRef] [PubMed]

64. Herchenhan, A.; Uhlenbrock, F.; Eliasson, P.; Weis, M.; Eyre, D.; Kadler, K.E.; Magnusson, S.P.; Kjaer, M. Lysyl Oxidase Activity Is Required for Ordered Collagen Fibrillogenesis by Tendon Cells. *J. Biol. Chem.* **2015**, *290*, 16440–16450. [CrossRef] [PubMed]

65. Aro, A.A.; Nishan, U.; Perez, M.O.; Rodrigues, R.A.; Foglio, M.A.; Carvalho, J.E.; Gomes, L.; Vidal, B.C.; Pimentel, E.R. Structural and biochemical alterations during the healing process of tendons treated with Aloe vera. *Life Sci.* **2012**, *91*, 885–893. [CrossRef] [PubMed]

66. Lee, C.; An, J.; Kim, J.H.; Kim, E.S.; Kim, S.H.; Cho, Y.K.; Cha, D.H.; Han, M.Y.; Lee, K.H.; Sheen, Y.H. Low levels of tissue inhibitor of metalloproteinase-2 at birth may be associated with subsequent development of bronchopulmonary dysplasia in preterm infants. *Korean J. Pediatr.* **2015**, *58*, 415–420. [CrossRef] [PubMed]

67. Van Buul, G.M.; Villafuertes, E.; Bos, P.K.; Waarsing, J.H.; Kops, N.; Narcisi, R.; Weinans, H.; Verhaar, J.A.; Bernsen, M.R.; van Osch, G.J. Mesenchymal stem cells secrete factors that inhibit inflammatory processes in short-term osteoarthritic synovium and cartilage explant culture. *Osteoarthritis Cartil.* **2012**, *20*, 1186–1196. [CrossRef] [PubMed]

68. Manferdini, C.; Maumus, M.; Gabusi, E.; Piacentini, A.; Filardo, G.; Peyrafitte, J.A.; Jorgensen, C.; Bourin, P.; Fleury-Cappellesso, S.; Facchini, A. Adipose-derived mesenchymal stem cells exert antiinflammatory effects on chondrocytes and synoviocytes from osteoarthritis patients through prostaglandin E2. *Arthritis Rheum.* **2013**, *65*, 1271–1281. [CrossRef] [PubMed]

69. Molloy, T.; Wang, Y.; Murrell, G. The roles of growth factors in tendon and ligament healing. *Sports Med.* **2003**, *33*, 381–394. [CrossRef] [PubMed]

70. Fu, S.C.; Wong, Y.P.; Cheuk, Y.C.; Lee, K.M.; Chan, K.M. TGF-beta1 reverses the effects of matrix anchorage on the gene expression of decorin and procollagen type I in tendon fibroblasts. *Clin. Orthop. Res.* **2005**, 226–232. [CrossRef]

71. Mehta, V.; Mass, D. The use of growth factors on tendon injuries. *J. Hand Ther.* **2005**, *18*, 87–92; quiz 93. [CrossRef] [PubMed]
72. Finnson, K.W.; McLean, S.; Di Guglielmo, G.M.; Philip, A. Dynamics of Transforming Growth Factor Beta Signaling in Wound Healing and Scarring. *Adv. Wound Care* **2013**, *2*, 195–214. [CrossRef] [PubMed]
73. Schweitzer, R.; Chyung, J.H.; Murtaugh, L.C.; Brent, A.E.; Rosen, V.; Olson, E.N.; Lassar, A.; Tabin, C.J. Analysis of the tendon cell fate using Scleraxis, a specific marker for tendons and ligaments. *Development* **2001**, *128*, 3855–3866. [PubMed]
74. Holladay, C.; Abbah, S.A.; O'Dowd, C.; Pandit, A.; Zeugolis, D.I. Preferential tendon stem cell response to growth factor supplementation. *J. Tissue Eng. Regen. Med.* **2016**, *10*, 783–798. [CrossRef] [PubMed]
75. Shukunami, C.; Takimoto, A.; Oro, M.; Hiraki, Y. Scleraxis positively regulates the expression of tenomodulin, a differentiation marker of tenocytes. *Dev. Biol.* **2006**, *298*, 234–247. [CrossRef] [PubMed]
76. Brandau, O.; Meindl, A.; Fassler, R.; Aszodi, A. A novel gene, tendin, is strongly expressed in tendons and ligaments and shows high homology with chondromodulin-I. *Dev. Dyn.* **2001**, *221*, 72–80. [CrossRef] [PubMed]
77. Shukunami, C.; Oshima, Y.; Hiraki, Y. Molecular cloning of tenomodulin, a novel chondromodulin-I related gene. *Biochem. Biophy. Res. Commun.* **2001**, *280*, 1323–1327. [CrossRef] [PubMed]
78. Yamana, K.; Wada, H.; Takahashi, Y.; Sato, H.; Kasahara, Y.; Kiyoki, M. Molecular cloning and characterization of CHM1L, a novel membrane molecule similar to chondromodulin-I. *Biochem. Biophy. Res. Commun.* **2001**, *280*, 1101–1106. [CrossRef] [PubMed]
79. Oshima, Y.; Shukunami, C.; Honda, J.; Nishida, K.; Tashiro, F.; Miyazaki, J. Chondromodulin-I-related angiogenesis inhibitor, in mouse eyes. *Investig. Ophthalmol. Vis. Sci.* **2003**, *44*, 1814–1823. [CrossRef]
80. Pisani, D.F.; Pierson, P.M.; Massoudi, A.; Leclerc, L.; Chopard, A.; Marini, J.F.; Dechesne, C.A. Myodulin is a novel potential angiogenic factor in skeletal muscle. *Exp. Cell Res.* **2004**, *292*, 40–50. [CrossRef] [PubMed]

cells

MDPI

Review

Multifaceted Interweaving Between Extracellular Matrix, Insulin Resistance, and Skeletal Muscle

Khurshid Ahmad [1,†], Eun Ju Lee [1,†], Jun Sung Moon [2], So-Young Park [3] and Inho Choi [1,*]

1   Department of Medical Biotechnology, Yeungnam University, Gyeongsan 38541, Korea; ahmadkhursheed2008@gmail.com (K.A.); gorapadoc0315@ynu.ac.kr (E.J.L.)
2   Department of Internal Medicine, College of Medicine, Yeungnam University, Daegu 42415, Korea; mjs7912@yu.ac.kr
3   Department of Physiology, College of Medicine, Yeungnam University, Daegu 42415, Korea; sypark@med.yu.ac.kr
*   Correspondence: inhochoi@ynu.ac.kr; Tel.: +82-53-810-3024; Fax: +82-53-810-4769
†   These authors contributed equally to this work.

Received: 9 August 2018; Accepted: 17 September 2018; Published: 22 September 2018

Abstract: The skeletal muscle provides movement and support to the skeleton, controls body temperature, and regulates the glucose level within the body. This is the core tissue of insulin-mediated glucose uptake via glucose transporter type 4 (GLUT4). The extracellular matrix (ECM) provides integrity and biochemical signals and plays an important role in myogenesis. In addition, it undergoes remodeling upon injury and/or repair, which is also related to insulin resistance (IR), a major cause of type 2 diabetes (T2DM). Altered signaling of integrin and ECM remodeling in diet-induced obesity is associated with IR. This review highlights the interweaving relationship between the ECM, IR, and skeletal muscle. In addition, the importance of the ECM in muscle integrity as well as cellular functions is explored. IR and skeletal muscle ECM remodeling has been discussed in clinical and nonclinical aspects. Furthermore, this review considers the role of ECM glycation and its effects on skeletal muscle homeostasis, concentrating on advanced glycation end products (AGEs) as an important risk factor for the development of IR. Understanding this complex interplay between the ECM, muscle, and IR may improve knowledge and help develop new ideas for novel therapeutics for several IR-associated myopathies and diabetes.

Keywords: extracellular matrix; insulin resistance; skeletal muscle; advanced glycation end products

1. Introduction

The skeletal muscle constitutes nearly 40% of body mass and is primarily composed of myofiber, multinucleated contractile cells [1,2], and mainly provides mobility, protects and supports the skeleton, and regulates the body temperature and glucose homeostasis within the body [3]. The skeletal muscle is the core metabolic tissue for the insulin-stimulated uptake of glucose, accounting for approximately 85% through glucose transporter type 4 (GLUT4) [4,5]. Therefore, reduced responsiveness of skeletal muscle to insulin, that is, insulin resistance (IR), is a critical aspect of type 2 diabetes mellitus (T2DM) development [6,7].

The skeletal muscle has a diverse population of stem cells known as muscle stem (or satellite) cells (MSCs), which have a remarkable capability of muscle regeneration to the structural and functional integrity of the skeletal muscle [8]. Furthermore, MSCs can be transdifferentiated into adipocytes or osteocytes; hence, they are good models for intramuscular adipogenesis or osteogenesis studies [9].

The extracellular matrix (ECM) is composed of structural glycoproteins like collagens, laminin (s), and fibronectin (FN) bound to proteoglycans (PGs), which all help to maintain skeletal muscle integrity and provide constructional support. Furthermore, the ECM generates biochemical signals for myogenesis regulation [10]. The ECM also works as a growth factor modulator in the process of cell growth and is involved in various cell signaling processes [10,11]. Collagens are found abundantly in the ECM environment, and are essential for the mechanical support of tissues in addition to cell adhesion, wound healing, and differentiation. Types I, III, and IV collagens are expressed strongly in the skeletal muscle, in which types I and III are fibrillary, whereas type IV is expressed mainly in the basement membrane (BM). Furthermore, MSCs are located under the BM (Figure 1) [12]. The ECM, a highly dynamic structure, undergoes remodeling in a number of metabolic tissues because of injury and repair and is allied with diet-induced IR [13]. ECM remodeling and the altered expression of integrin is generally found in disease conditions. Several studies have examined whether ECM remodeling and altered signaling of the integrin receptor in the diet-induced condition is allied with IR [12,14]. In a study, the muscle-specific exclusion of integrin β1 in chow-fed mice resulted in reduced whole-body insulin sensitivity and reduced uptake of insulin-stimulated glucose during hyperinsulinemic-euglycemic clamp experiments (a gold-standard technique to measure insulin sensitivity) [13,15].

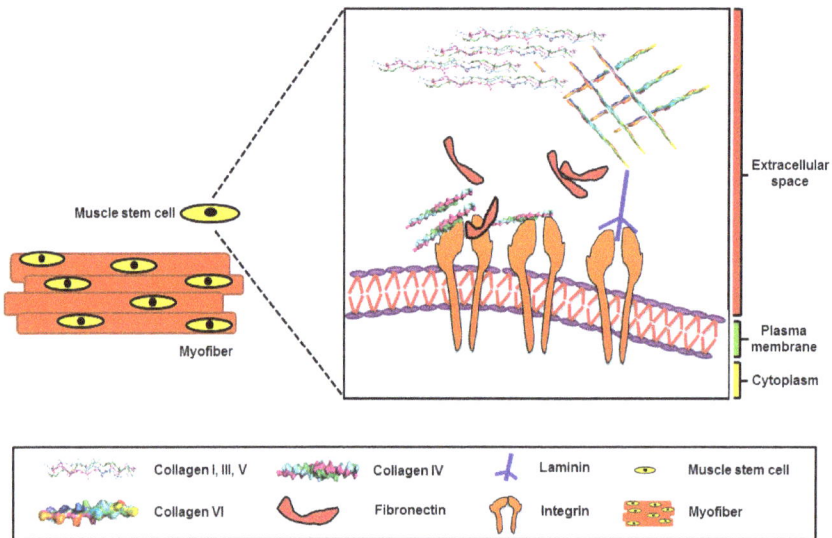

Figure 1. Muscle stem cells and the extracellular matrix (ECM) microenvironment.

This review provides a brief introduction of skeletal muscle development, IR correlations, glycation, and diabetes. In addition, the relationship between ECM remodeling of the skeletal muscle and IR is discussed, and guidelines for the prevention and future directions to combat or minimize the manifestation of the disease conditions are proposed.

2. Extracellular Matrix

The ECM is a complex milieu of diverse structural molecules involved in structural support together with cellular signaling and tissue responses to diseases and injuries [16]. An organization of ECM molecules have their own distinct features and is responsible for the various biological activities [16,17]. A number of muscle-related genetically determined diseases are caused primarily by mutations in the ECM components and their receptors. For example, more than 150 ECM proteins interact with the adhesion site of the integrin receptor [18,19]. Maricela et al. performed a clinical study involving 66 patients with Duchenne/Becker muscular dystrophy (DMD/BMD), hyperinsulinemia, IR, and obesity. They suggested that an alteration in GLUT4 in muscle fibers from DMD/BMD patients might be involved in IR [20].

Different types of collagens are expressed in skeletal muscle with their specific role (Table 1). Collagens can be subdivided broadly into various subfamilies according to the sequence similarities and supramolecular assemblies they form, for example, fibrils, beaded filaments, anchoring fibrils, and networks [21]. Fibrillar collagens (types I, II, III, V, XI, XXIV, and XXVII) generally provide three-dimensional structures for tissues and organs [22]. These networks have mechanical strength and signaling functions by binding to the ECM components and cellular receptors. More than 90% of the collagens were found to be expressed in skeletal muscle and were composed mostly of collagen I, III, and IV [23,24]. Although collagen I and III encompasses fibrillar collagen, collagen IV is the most plentiful structural component of the BM [12,25]. Type I collagen is the most important and ample protein in the vertebrates (including humans), found ubiquitously in connective tissues, and is usually involved in the promotion of membrane proteinase activation, which leads to cellular migration and adhesion [26–28]. The packing and positioning of subfibrillar elements of the collagen structure characterize most of the biologically substantial aspects of the fibrillar collagen structure [26,28]. Seminal studies have shown that changes in the composition of the ECM (generally increased collagen) are a general characteristic of IR human skeletal muscle [12]. FN is a modular protein and an important structural element in the niche of MSCs that plays a vital role in the muscle regeneration process. The loss of FN from the niche affects many pathways and cellular mechanisms involved in MSC aging. In aged skeletal muscle, FN triggers adhesion signaling and functioning of MSCs. Treatment with FN is shown to restore the regenerative capacity of aged muscles [29]. In the previous studies, we explored the role of fibromodulin (FMOD) and matrix gla protein (MGP), the ECM proteins, involved in myoblast differentiation by regulating the interaction of myostatin (MSTN) with its receptor activin receptor type IIB (ACVRIIB). Involvement of FMOD and MGP in the regulation of myogenesis provides a clue for the development of novel therapeutics for the treatment of the different types of muscle diseases because it plays an important role by recruiting more MSCs to the sites of muscle injury [8,30–32].

Table 1. Major types of collagens in skeletal muscle ECM.

Collagen Types	Description	Expression During Diet-Induced IR	Reference
I	Abundantly found in endo-, peri-, and epimysium. Stimulate myogenic differentiation of stem cells.	↑	[16,33]
III	It is more consistently found between endomysium and epimysium.	↑	[34]
IV	Main component of basal lamina. Found to be 4 to 30-fold increase in skeletal muscle ECM mRNA levels	↑	[35]
V	Fibril-forming collagen and found to be increased in skeletal muscle ECM mRNA levels		[35]
VI	Found to be increased in skeletal muscle ECM mRNA levels		[35]
IX	Multiple-epiphyseal-dysplasia-related myopathy is caused due to mutation in collagen IX		[36]
XII	It is the largest member of the fibril-associated collagens with interrupted triple helix (FACIT) family. Important for muscle integrity.		[37]
XIV	A member of FACIT family and involved in muscle metabolism		[38]
XV	Extensively found in the basement membrane and a structural component vital to stabilizing the skeletal muscle		[39]
XVIII	Classified as multiplexins, bind with growth factors and other membranes of basement membrane glycoproteins.		[39]

ECM homeostasis is vital for normal functioning of a cell and for stable communication among cells, and a disruption of this homeostasis may have an adverse effect on the functioning of organ systems and promote many deadly diseases (e.g., fibrotic diseases and cancer) [40]. Furthermore, it has been reported on several occasions that the expression of several ECM proteins is altered during the onset of T2DM, and that these alterations change the ECM networks and consequently cell-to-cell and cell-to-ECM interactions [41]. The ECM components communicate with cells through cell surface receptors, and integrins are the most important. These receptors are composed of heterodimeric (alpha and beta) subunits, which interact with different types of the ECM ligands [13]. Seven alpha (α) subunits (α1, α3, α4, α5, α6, α7, and αv) in association with the β1 subunit are expressed in the skeletal muscle [13,42]. The integrin α7β1 expressed by MSCs is augmented in the myotendinous (MTJ) and neuromuscular junctions in the skeletal muscle.

Muscle Stem Cells and Extracellular Matrix(ECM)

Muscle fibers or myofibers are the functional units of skeletal muscles, and are formed during embryogenesis when myoblasts fuse to form myotubes. MSCs are usually found in the quiescent phase and remain in this form until they are invoked by injury and exercise. Gentle injuries may initiate minimal proliferation, whereas major ones can recruit a larger number of MSCs and induce more proliferation earlier than differentiation [43]. Several factors regulate the activation of MSCs. Among them, some widely explored factors are muscle regulatory factors (MRFs: MYF5, MYOD: Myoblast determination protein 1, myogenin, etc.), hepatocyte growth factor (HGF), and neuronal nitric oxide synthase (NOS) [44]. MSCs are positioned between the sarcolemma (cell membrane) and BM (basal lamina; BL) of the muscle fibers, which are indicated as a well-equipped 'niche'. The balance between the quiescent and activated form of MSCs is sustained mainly by this specific niche [45]. The ability of regeneration of the skeletal muscle is dependent primarily on the interaction between MSCs and their niche. The BL is encompassed by a network of ECM, which is connected directly to MSCs. Type IV collagen and laminin-2 are the main components of BL, and the concentration of these components diverges according to the function of the muscle fiber type. In addition to these two components, collagen, VI, perlecan, nidogen, FN, and other glycoproteins and PGs are the constituents of BL [45,46].

α7 and β1 integrins are typically expressed by MSCs to form a complex in BL and bind with laminin-2, though their expressions are dependent on the functions of activated MSCs [45]. Activated MSCs of mouse express β3 integrin, which probably forms a complex with αv integrin to produce the αv-β3 receptor for proteins having an exposed tripeptide of (Arg-Gly-Asp: RGD) ECM ligands, including FN, collagens, osteopontin, and laminins [47]. Another study showed that the activated MSCs induce confined remodeling of the ECM components and the deposition of laminin (α1 and α5) into the BL. The activation of AMP-activated protein kinase (AMPK), that is, the phosphorylated form, indorses glucose uptake and upsurges insulin sensitivity. The MSCs isolated from the injured muscles of diet-induced obese (DIO) mice determine the reduced AMPK activity and decreased regeneration [48].

3. Insulin Resistance in Skeletal Muscle

IR is defined as a decrease in the metabolic response of the skeletal muscle cell to insulin, which is a protruding feature of obesity and T2DM [49]. Insulin binding to the receptors in the cell membrane activates the signal transduction pathways, insulin receptor substrate (IRS)-1, phosphatidylinositide 3-kinases (PI3K), and AKT (protein kinase B, PKB), which mediates the insulin-stimulated glucose uptake via GLUT4 from the cytoplasm to the plasma membrane. Reduced insulin-induced activation of the signaling pathway and GLUT4 translocation lead to the development of insulin resistance and T2DM [50]. Although the precise mechanisms of IR are unclear, a robust relationship has been found between IR and obesity. Obesity is accompanied by increases in the lipid levels in the plasma and the accumulation of extra lipid, predominantly in the skeletal muscle and liver. The possible mechanisms through which obesity induces IR are increased fatty acid metabolites, oxidative stress, and inflammation, leading to suppression of the insulin signaling pathways. Another possible mechanism involved in the development of insulin resistance in obese subjects is a reduction of the vascular density in the skeletal muscle [51]. In addition, an interaction was reported between ankyrin-1(ANK1) and insulin receptor substrate-1 (IRS1) in skeletal muscle, and IRS1 is a key constituent of insulin signal transduction and arbitrates metabolic and mitogenic responses to insulin. ANK1 has been identified as a candidate gene for T2DM in skeletal muscle [52]. Furthermore, mutations in IRS1 protein are linked with IR, and in one study, the IRS1 gene was detected in T2DM patients that exhibited polymorphisms in over 11 amino acids [53].

Endoplasmic reticulum (ER) stress is associated with the relation between nonesterified fatty acid and IR, and eventually to the progression of T2DM. Direct contact between myotubes and palmitate acid induces ER stress [54]. Panzhinskiy et al. reported protein tyrosine phosphatase 1B (PTP1B), which is found on the ER membrane, acts as a negative regulator of insulin signaling activated by ER stress, and is essential for full activation of ER stress pathways that mediate IR in skeletal muscle [55]. Ijuin et al. demonstrated that skeletal muscle and kidney-enriched inositol polyphosphate phosphatase (SKIP), a key regulator of MSC differentiation, has a specific role in IR progression in skeletal muscle. Increased SKIP expression in the presence of ER stress was found to be significantly higher in the skeletal muscle of high-fat diet (HFD) and db/db mice than in wild-type controls [56].

IR in skeletal muscle is strongly linked with the lipid metabolism [57]. Increased levels of triglycerides (TGs) and fatty acids in the blood circulation and the augmented intracellular accumulation of several lipid intermediates are the hallmarks of this condition. Increased fatty acid uptake or a low rate of oxidation capacity in the presence of IR leads to higher concentrations of lipid intermediates in skeletal muscle cells. Furthermore, numerous studies have shown that an unbalanced diet or an HFD lead to accumulation of TGs and other byproducts of fatty acid oxidation due to incomplete or reduced oxidation of these fatty acids, which eventually causes IR in skeletal muscle [58,59]. It has also been reported that skeletal muscles of individuals with IR and/or T2DM are characterized by decreased oxidative capacity and mitochondrial contents and functions. Actually, oxidative capacity has been reported to better predict insulin sensitivity than intracellular TG or LC-CoA concentration in T2DM patients [60,61].

There are several other imperative factors regulating IR in skeletal muscle. Recently, a study reported that reduced signaling of p38 MAPK/JNK module instead of increased signaling in skeletal muscle endorses IR and metabolic syndrome [62]. Another study based on the Korean population with 14,807 participants (18–65 years of age) suggested that connotation of muscle mass with metabolic syndrome and IR is attenuated by high-fat mass [63].

4. The Extracellular Matrix (ECM) and Insulin Resistance

The ECM in the muscle plays a crucial role in the regulation of glucose homeostasis; a change in the composition of the ECM is a hallmark of IR muscle. Previous studies examined insulin-resistant muscle in diabetic and obese people and reported that collagen deposition was remarkably higher than that in normal people [64,65]. A study on healthy males who gained weight rapidly reported reduced insulin sensitivity and that a number of muscle ECM genes were upregulated. They reported that the reason for the weight gain was not due to local adipose tissues or any systemic inflammation, which eventually indicates the role of muscle ECM in the regulation of glucose homeostasis [13,35]. An imbalanced diet and over nutrition play contributory roles to the changes in the gene expression resulting in long-term obesity obstructions. Several studies have shown that the ECM genes (Type I, III, IV, V, and VI collagen), integrins (ITGBL1, ITGA4, and ITGA5) and matrix metalloproteases (MMP2 and MMP25) are upregulated in response to overfeeding. These findings suggest that ECM remodeling is coupled with the development of diet-induced IR, and is causative to the pathophysiology of T2DM [16,66]. MMPs are crucial for the degradation of the ECM components, and in particular, MMP 9 degrades collagen IV, a major constituent of basement membrane and an important factor in ECM remodeling. Kang et al. reported that the augmented deposition of collagen in diet-induced obese conditions is due in part to the reduced activity of MMP 9, and its genetic deletion increases the deposition of collagen in the muscle and impairs muscle IR in HFD mice [13,67]. It has also been suggested that the activation of growth factors (e.g., TGF-β1) by oxidative stress and inflammation causes ECM remodeling [68,69].

Recently, ECM remodeling in skeletal muscle has been reported in diabetic subjects. The skeletal muscles of diabetic rats showed a reduction in collagen integrity and altered normal triple helical structures [70]. Type I and III collagen levels are also elevated in skeletal muscle in diabetic patients [71] and a link between skeletal muscle ECM remodeling and IR has been proposed [16]. In one study, ECM-related gene expression was increased by a 48 h lipid infusion designed to induce IR [65]. Furthermore, in human subjects, short-term overfeeding induced IR and upregulation of the collagen I/III and MMP2 genes [66]. Although reports on the topic are limited, increasing evidence indicates that a fundamental relationship exists between skeletal muscle, ECM remodeling, and IR. Prior exposure of monkeys to whole-body radiation resulted in the ECM fibrosis and IR in skeletal muscle [71]. Furthermore, in the HFD-induced obese mice model, diet-induced IR was accompanied by increased deposition of hyaluronan/hyaluronic acid (HA) in skeletal muscle ECM, and subsequent long-term hyaluronidase treatment reversed IR by reducing HA levels [72].

The mechanisms of skeletal muscle ECM remodeling by induced IR is still unclear, but several hypotheses have been proposed. The most plausible hypothesis is a scarcity of microvasculature in the fibrotic ECM because a decrease in microvascular density supplies fewer nutrients and hormones to the skeletal muscle [67,71]. A decrease in capillary density has been proposed as one of the causes of IR in obese and older individuals [73–76]. Increased vascular density by angiopoietin-1 in high-fat-fed obese mice prevents the progression of IR in the skeletal muscle [77]. Furthermore, the prevention of muscle IR in the HFD mouse model by overexpressing catalase or by sildenafil (a phosphodiesterase 5a inhibitor) treatment reduced collagen I/III deposition and improved muscle vascularization [12,78,79]. Another hypothesis proposed for IR induction by fibrosis is that some components of remodeled ECM act to induce IR. In one study, integrin α (2) β(1)-null mice fed an HFD did not develop obesity-induced IR [12], and in another, reduction of HA in skeletal muscle ECM reversed HFD-induced IR [72].

Downstream integrin signaling through focal adhesion kinase (FAK) and integrin-linked kinase (ILK) might be a mechanistic connection between the muscle ECM and IR [13]. FAK is a tyrosine kinase with the properties of intracellular signaling, stabilization of cytoskeleton, and focal adhesion turnover, and is regulated by insulin receptors [80]. Bisht et al. reported that FAK is associated with the regulation of insulin action in the muscle because FAK tyrosine phosphorylation is reduced in the muscle from HFD rats [81]. Bisht et al. reported that the knockdown of FAK (in vivo siRNA-mediated) in chow-fed mice resulted in hyperinsulinemia, diminished glucose tolerance, and reduced insulin action [82].

ILK is an intracellular scaffolding protein that interacts with the cytoplasmic domains of b1, b2, and b3integrin [13]. ILK plays a critical role in muscle insulin action. Kang et al. reported that downstream integrin signaling through ILK is hazardous to the pathogenesis of IR. They showed that muscle-specific removal of ILK improves the muscle insulin sensitivity significantly in HFD-IR mice via the augmented phosphorylation of Akt [83].

Skeletal muscles are more susceptible to exercise-induced myofiber injury in the presence of T2DM, and T2DM-mediated changes in skeletal muscle depend on BM structure, and particularly on the activities of enzymes that regulate the synthesis of collagen. In a comparative microarray study of skeletal muscles, several types of collagen (type I, III, IV, V, VI, and XV) were downregulated and PGs (laminin-2, elastin, thrombospondin-1, and decorin), noncollagenous proteins, and connective tissue growth factor (CTGF) were upregulated in streptozotocin-induced diabetic mice compared to normal mice, and these changes eventually affected the basement membrane structure [84].

5. Glycation of Skeletal Muscle ECM

The nonenzymatic binding of a glucose molecule to proteins, lipids, or nucleic acids yields stable 'Amadori products', which undergo additional modifications to form advanced glycation end products (AGEs). AGEs are chemically heterogeneously modified molecules that form through the nonenzymatic glycation of proteins over an individual's lifetime and have been implicated in a number of chronic diseases, such as diabetes [85]. Elevated levels of AGEs have been directly related to degrees of hyperglycemia, which underlies tissue damage and T2DM [86]. The formation of AGEs in diabetic patients is enhanced by high glucose concentrations in blood and is 'impulsive' and rather slow and predominantly affects proteins with comparatively long half-lives with exposed 'lysine' residues. The ECM proteins are usually long-lived and are latent targets of nonenzymatic glycation. Moreover, among the ECM proteins, collagens are highly vulnerable to glycation. Fibrillar collagens have exceptionally long half-lives, for example, type I collagen has half-lives of 1–2 years in bone and 10–15 years in skin, and type II collagen has a half-life exceeding 100 years in cartilage [87]. The glycation of fibrillar collagen is a surface phenomenon that produces cross-linkages, which subsequently modify matrix properties. The glycation of the ECM proteins causes structural alterations and disrupts binding affinities because it modifies the arginine residues of RGD and GFOGER motifs of major ECM components like FN and collagens. Furthermore, the intramolecular cross-links formed may confer proteolytic resistance, eventually leading to basement membrane coagulation [88]. In addition, AGE-mediated alterations in collagens I and IV affect the intensities of their interactions (binding capacity) with other components of the ECM as well as their capability to support cell adhesion. The interactions among the ECM components, such as collagen I and/or IV, FN, and heparin were found to be reduced by AGE-modification (Figure 2) [89–91].

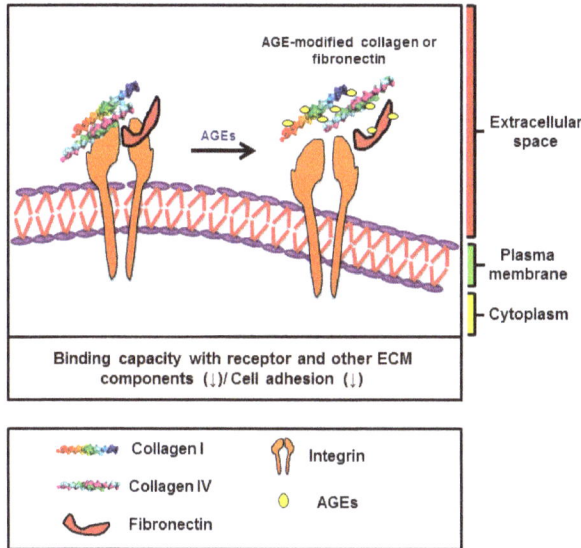

Figure 2. Modifications of ECM components by advanced glycation end products (AGEs) and their effects.

AGEs are an important risk factor for the development of IR because they are discretely correlated with IR in healthy subjects [92]. Though insulin is not a target for AGE-modification as its half-life is short, in vivo and in vitro experiments (cultured in hyperglycemic conditions) have reported glycation sites on insulin [93]. Several AGE inhibitors, both natural and synthetic, have been identified. Our group recently investigated some potent AGE inhibitors, and during the study, we found silver nanoparticles (AgNPs) significantly and concentration-dependently inhibited AGE formation, which suggested they can be considered a candidate for the treatment of diabetes and diabetes-associated problems [85]. We also studied the roles of AGEs in muscle-related myopathies and found AGE production and subsequent receptor for advanced glycation endproducts (RAGE)-AGE binding hinders the myogenesis program. In addition, we found curcumin and gingerol both reduce the effects of AGEs on a muscle development program [94].

6. Insulin Resistance and Skeletal Muscle ECM Remodeling: Clinical Studies

Several clinical studies have shown that skeletal muscle ECM remodeling is closely associated with IR, obesity, and metabolic disorders. Richardson et al. first demonstrated that free fatty acid (FFA) markedly elevated ECM genes and collagen deposition in healthy human skeletal muscle [65]. These changes in the ECM composition are typically found in IR skeletal muscle, which also shows a robust increase in collagen content (types I and III) [64]. Furthermore, it is widely accepted that increased transforming growth factor beta (TGF-β) signaling results in ECM remodeling in IR skeletal muscle [95]. MSTN, a potent antianabolic regulator of muscle mass, has been reported to be upregulated in muscle myotubes and plasma in extremely obese subjects, and to be closely correlated with systemic IR [96]. On the other hand, it was also reported that Smad signaling (Suppressor of Mothers Against Decapentaplegic) was activated in association with a reduction in MYOD transcription, but it was observed that TGF-β1 and MSTN protein levels were not changed significantly [95]. A recent study reported failure of autophagy caused by overweight impaired myogenic differentiation in the elderly, but that MSTN expression did not change significantly [97]. Although studies in human subjects have produced conflicting results regarding ECM-related gene changes, in one study, Smad signaling was suggested to account for ECM remodeling, but neither TGF-β1 nor MSTN alone were found to be implicated in the atrophic effects on skeletal muscle [98].

In a study of overfeeding-induced weight gain (approximately 10%), skeletal remodeling of ECM genes was observed to dramatically increase in association with skeletal muscle inflammation, whereas slight change occurred in adipose tissue. Interestingly, in this study, there was no indication of systemic or local inflammation in adipose tissue despite the presence of IR [35]. Although the increase in body weight was small (as little as 3%) after short-term (4 weeks) overfeeding, the insulin sensitivity was markedly impaired, which was attributed to skeletal muscle ECM remodeling accompanied by increase in the mRNA expression of the ECM-related genes (*COL1α1*, *COL3α1*, *MMP2*). However, no significant changes were found in the expression of *MMP9*, *TIMP1*, *CD68*, and integrin. Thus, this small gain in body weight altered the expressions of genes related to ECM receptor interactions, such as focal adhesion and adherens junctions [66]. These results show skeletal muscle ECM remodeling plays a crucial role in the progression of obesity-induced IR and not in adipose tissue inflammation.

In a population-based study, adults aged over 65 years with elevated circulatory levels of carboxymethyl-lysine (CML) were found to be at a higher risk of impaired muscle quality (as determined by grip strength and gait speed testing), and this relationship remained significant after adjusting for risk factors [99,100]. In healthy middle-aged and older Japanese, AGE accumulation (measured by skin autofluorescence) was significantly correlated with lower muscle mass (skeletal muscle index) [101,102]. However, another small study showed that AGEs were not correlated with muscle mass but with lower limb muscle dysfunction in people with type 1 diabetes [103].

ECM remodeling of the skeletal muscle has attracted attention as a new therapeutic target for obesity and metabolic dysfunction. The skeletal muscle is highly adaptive to exercise, and regular exercise improves peripheral insulin sensitivity. Furthermore, both acute and long-term exercise activated significant amounts of genes in skeletal muscle (approximately 500 and 290, respectively), and ECM-related genes were also increased by the muscle's adaptation mechanism to exercise (5% and 20%, respectively) [104,105]. In particular, serglycin, which is believed to be related to exercise adaptation by blocking serpinE1 (SERPINE1), was among several proteoglycans that were increased significantly by exercise [106]. *MSTN* gene expression was downregulated after acute and long-term exercise in IR skeletal muscles. Interestingly, *MSTN* expression increased in adipose tissues, but not in muscle cells, after 12 weeks of exercise, and this upregulation was found to be positively related to insulin sensitivity markers, indicating a tissue-specific effect [106].

Recently, attempts have been made to ameliorate IR using drugs targeting ECM remodeling. Bimagrumab (BYM338, Novartis) is a human monoclonal antibody that binds ActRII A and B, and thus interferes with their bindings to natural ligands, such as *MSTN*, growth Differentiation Factor11 okno.

(GDF11), and activin, which inhibit muscle growth [107]. A single dose of Bimagrumab showed decreased fat tissue volume by 7.9% and increased thigh muscle volume by around 2.7% in healthy lean subjects after 10 weeks. In IR subjects, Bimagrumab also increased skeletal muscle mass and reduced fat mass without causing body weight changes, and improved insulin sensitivity and metabolic statuses [108]. Although the results of human and animal studies often differ depending on experiment conditions and subject characteristics, ECM remodeling has become a new therapeutic target for metabolic disorders.

7. Concluding Remarks and Future Perspectives

The skeletal muscle is a crucial target for several metabolic syndromes, particularly T2DM, which is caused largely by IR. Insulin signaling is governed by several regulators, which include IRS1, GLUT4, and AKT as leading regulators. ER stress and fatty acid metabolism are the leading factors prompting IR in the skeletal muscle. In tissues, cellular interactions occur in the ECM, a three-dimensional network of polymeric biomolecules. The irregular expression of several ECM components, particularly collagen and MMPs, has been reported in skeletal muscle IR. Alterations in the ECM components perturb insulin signaling (inside-out and outside-in signaling) and alter the effects of insulin. ECM remodeling of skeletal muscle has only been recently proposed to be a mechanism of IR, and more evidence is required to prove the involvement of skeletal muscle ECM remodeling in IR. Future studies are needed to determine the mechanisms responsible for the manifestations of pathologic events in skeletal muscle, IR, and the ECM. An in-depth study of the connection between ECM remodeling of the skeletal muscle, the action mechanism of insulin, and integrin signaling will be a promising innovative line of research to develop novel therapeutics.

Author Contributions: I.C. and K.A. contributed to the systematic review of literature, and the design and writing of the manuscript; K.A., E.J.L., J.S.M., and S.-Y.P. contributed to the systematic review of literature, and the writing of the manuscript; I.C. contributed to the supervision of the manuscript editing, and the revision of major intellectual content.

Funding: This work was supported by the Creative Economy Leading Technology Development Program of the Gyeongsanbuk-Do and Gyeongbuk Science & Technology Promotion Center of Korea (#SF316001A) and by the Next-Generation BioGreen 21 Program (Project No. PJ01324701), Rural Development Administration, Republic of Korea.

Conflicts of Interest: The authors declare no conflict of interest.

Abbreviations

MSCs	muscle stem cells
GLUT4	glucose transporter type 4
IR	insulin resistance
T2DM	type 2 diabetes mellitus
ECM	extracellular matrix
FN	fibronectin
MFRs	muscle regulatory factors
HGF	hepatocyte growth factor
NOS	neuronal nitric oxide synthase
BM	basement membrane
BL	basal lamina
AMPK	AMP-activated protein kinase
GWAS	genome-wide association studies
ANK1	ankyrin-1
IRS	insulin receptor substrate-1
PFKM	phosphofructokinase
ER	endoplasmic reticulum
IRS	insulin receptor substrate

IRE-1	inositol-requiring enzyme 1
TRB3	ER stress upraises tribbles 3
HFD	high-fat diet
PTP1B	protein tyrosine phosphatase 1B
SKIP	inositol polyphosphate phosphatase
TGs	triglycerides
MMP	matrix metalloproteases
FMOD	fibromodulin
MGP	matrix gla protein
MSTN	myostatin
ACVRIIB	activin receptor type II B
CTGF	connective tissue growth factor
MTJ	myotendinous
HA	hyaluronan
TGF	transforming growth factor
AGEs	advanced glycation end products

References

1. Kim, K.M.; Jang, H.C.; Lim, S. Differences among skeletal muscle mass indices derived from height-, weight-, and body mass index-adjusted models in assessing sarcopenia. *Korean J. Intern. Med.* **2016**, *31*, 643. [CrossRef] [PubMed]
2. Yin, H.; Price, F.; Rudnicki, M.A. Satellite cells and the muscle stem cell niche. *Physiol. Rev.* **2013**, *93*, 23–67. [CrossRef] [PubMed]
3. Shadrin, I.; Khodabukus, A.; Bursac, N. Striated muscle function, regeneration, and repair. *Cell. Mol. Life Sci.* **2016**, *73*, 4175–4202. [CrossRef] [PubMed]
4. Richter, E.A.; Hargreaves, M. Exercise, GLUT4, and skeletal muscle glucose uptake. *Physiol. Rev.* **2013**, *93*, 993–1017. [CrossRef] [PubMed]
5. DeFronzo, R.; Jacot, E.; Jequier, E.; Maeder, E.; Wahren, J.; Felber, J. The effect of insulin on the disposal of intravenous glucose: Results from indirect calorimetry and hepatic and femoral venous catheterization. *Diabetes* **1981**, *30*, 1000–1007. [CrossRef] [PubMed]
6. Hansen, D.; de Strijcker, D.; Calders, P. Impact of endurance exercise training in the fasted state on muscle biochemistry and metabolism in healthy subjects: Can these effects be of particular clinical benefit to type 2 diabetes mellitus and insulin-resistant patients? *Sports Med.* **2017**, *47*, 415–428. [CrossRef] [PubMed]
7. Gerich, J.E. Is insulin resistance the principal cause of type 2 diabetes? *Diabetes Obes. Metab.* **1999**, *1*, 257–263. [CrossRef] [PubMed]
8. Lee, E.J.; Jan, A.T.; Baig, M.H.; Ahmad, K.; Malik, A.; Rabbani, G.; Kim, T.; Lee, I.-K.; Lee, Y.H.; Park, S.-Y.; et al. Fibromodulin and regulation of the intricate balance between myoblast differentiation to myocytes or adipocyte-like cells. *FASEB J.* **2017**, *32*, 768–781. [CrossRef] [PubMed]
9. Asakura, A.; Rudnicki, M.A.; Komaki, M. Muscle satellite cells are multipotential stem cells that exhibit myogenic, osteogenic, and adipogenic differentiation. *Differentiation* **2001**, *68*, 245–253. [CrossRef] [PubMed]
10. Frantz, C.; Stewart, K.M.; Weaver, V.M. The extracellular matrix at a glance. *J. Cell Sci.* **2010**, *123*, 4195–4200. [CrossRef] [PubMed]
11. Yue, B. Biology of the extracellular matrix: An overview. *J. Glaucoma* **2014**, *23*, 20–23. [CrossRef] [PubMed]
12. Kang, L.; Ayala, J.E.; Lee-Young, R.S.; Zhang, Z.; James, F.D.; Neufer, P.D.; Pozzi, A.; Zutter, M.M.; Wasserman, D.H. Diet-induced muscle insulin resistance is associated with extracellular matrix remodeling and interaction with integrin alpha2beta1 in mice. *Diabetes* **2011**, *60*, 416–426. [CrossRef] [PubMed]
13. Williams, A.S.; Kang, L.; Wasserman, D.H. The extracellular matrix and insulin resistance. *Trends Endocrinol. Metab.* **2015**, *26*, 357–366. [CrossRef] [PubMed]
14. Zong, H.; Bastie, C.C.; Xu, J.; Fassler, R.; Campbell, K.P.; Kurland, I.J.; Pessin, J.E. Insulin resistance in striated muscle-specific integrin receptor β1-deficient mice. *J. Biol. Chem.* **2009**, *284*, 4679–4688. [CrossRef] [PubMed]
15. Kim, J.K. Hyperinsulinemic-euglycemic clamp to assess insulin sensitivity in vivo. *Methods Mol. Biol.* **2009**, *560*, 221–238. [PubMed]

16. Martinez-Huenchullan, S.; McLennan, S.; Verhoeven, A.; Twigg, S.; Tam, C. The emerging role of skeletal muscle extracellular matrix remodelling in obesity and exercise. *Obes. Rev.* **2017**, *18*, 776–790. [CrossRef] [PubMed]

17. Kaasik, P.; Riso, E.M.; Seene, T. Extracellular matrix and myofibrils during unloading and reloading of skeletal muscle. *Intern. J. Sports Med.* **2011**, *32*, 247–253. [CrossRef] [PubMed]

18. Brown, S.C.; Mueller, U.; Conti, F.J. Integrins in the Development and Pathology of Skeletal Muscle. In *Neuromuscular Disorders*; InTech: London, UK, 2012; Available online: https://www.intechopen.com/books/neuromuscular-disorders/integrins-in-the-development-and-pathology-of-skeletal-muscle (accessed on 13 May 2018).

19. Zaidel-Bar, R.; Itzkovitz, S.; Ma'ayan, A.; Iyengar, R.; Geiger, B. Functional atlas of the integrin adhesome. *Nat. Cell Biol.* **2007**, *9*, 858. [CrossRef] [PubMed]

20. Rodríguez-Cruz, M.; Sanchez, R.; Escobar, R.E.; Cruz-Guzmán, O.d.R.; López-Alarcón, M.; Bernabe;a, M.; Coral-Vázquez, R.; Matute, G.; Velázquez Wong, A.C. Evidence of insulin resistance and other metabolic alterations in boys with duchenne or becker muscular dystrophy. *Inter. J. Endocrinol.* **2015**, *2015*, 867273.

21. Ricard-Blum, S.; Ruggiero, F.; van der Rest, M. The Collagen Superfamily. In *Collagen*; Springer: Berlin/Heidelberg, Germany, 2005; pp. 35–84.

22. Bella, J.; Hulmes, D.J. Fibrillar collagens. In *Fibrous Proteins: Structures and Mechanisms*; Springer: Berlin/Heidelberg, Germany, 2017; pp. 457–490.

23. Gelse, K.; Pöschl, E.; Aigner, T. Collagens—Structure, function, and biosynthesis. *Adv. Drug Deliv. Rev.* **2003**, *55*, 1531–1546. [CrossRef] [PubMed]

24. Huijing, P.A. Muscle as a collagen fiber reinforced composite: A review of force transmission in muscle and whole limb. *J. Biomech.* **1999**, *32*, 329–345. [CrossRef]

25. Ricard-Blum, S.; Ruggiero, F. The collagen superfamily: From the extracellular matrix to the cell membrane. *Pathol. Biol.* **2005**, *53*, 430–442. [CrossRef] [PubMed]

26. Orgel, J.; San Antonio, J.; Antipova, O. Molecular and structural mapping of collagen fibril interactions. *Connect. Tissue Res.* **2011**, *52*, 2–17. [CrossRef] [PubMed]

27. Tam, E.M.; Wu, Y.I.; Butler, G.S.; Stack, M.S.; Overall, C.M. Collagen binding properties of the membrane type-1 matrix metalloproteinase (MT1-MMP) hemopexin c domain the ectodomain of the 44-kDa autocatalytic product of MT1-MMP inhibits cell invasion by disrupting native type I collagen cleavage. *J. Biol. Chem.* **2002**, *277*, 39005–39014. [CrossRef] [PubMed]

28. Sweeney, S.M.; Orgel, J.P.; Fertala, A.; McAuliffe, J.D.; Turner, K.R.; Di Lullo, G.A.; Chen, S.; Antipova, O.; Perumal, S.; Ala-Kokko, L. Candidate cell and matrix interaction domains on the collagen fibril, the predominant protein of vertebrates. *J. Biol. Chem.* **2008**, *283*, 21187–21197. [CrossRef] [PubMed]

29. Lukjanenko, L.; Jung, M.J.; Hegde, N.; Perruisseau-Carrier, C.; Migliavacca, E.; Rozo, M.; Karaz, S.; Jacot, G.; Schmidt, M.; Li, L.; et al. Loss of fibronectin from the aged stem cell niche affects the regenerative capacity of skeletal muscle in mice. *Nat. Med.* **2016**, *22*, 897–905. [CrossRef] [PubMed]

30. Jan, A.T.; Lee, E.J.; Choi, I. Fibromodulin: A regulatory molecule maintaining cellular architecture for normal cellular function. *Int. J. Biochem. Cell. Biol.* **2016**, *80*, 66–70. [CrossRef] [PubMed]

31. Lee, E.J.; Jan, A.T.; Baig, M.H.; Ashraf, J.M.; Nahm, S.S.; Kim, Y.W.; Park, S.Y.; Choi, I. Fibromodulin: A master regulator of myostatin controlling progression of satellite cells through a myogenic program. *FASEB J.* **2016**, *30*, 2708–2719. [CrossRef] [PubMed]

32. Ahmad, S.; Jan, A.T.; Baig, M.H.; Lee, E.J.; Choi, I. Matrix gla protein: An extracellular matrix protein regulates myostatin expression in the muscle developmental program. *Life Sci.* **2017**, *172*, 55–63. [CrossRef] [PubMed]

33. Engler, A.J.; Sen, S.; Sweeney, H.L.; Discher, D.E. Matrix elasticity directs stem cell lineage specification. *Cell* **2006**, *126*, 677–689. [CrossRef] [PubMed]

34. Gillies, A.R.; Lieber, R.L. Structure and function of the skeletal muscle extracellular matrix. *Muscle Nerve* **2011**, *44*, 318–331. [CrossRef] [PubMed]

35. Tam, C.S.; Covington, J.D.; Bajpeyi, S.; Tchoukalova, Y.; Burk, D.; Johannsen, D.L.; Zingaretti, C.M.; Cinti, S.; Ravussin, E. Weight gain reveals dramatic increases in skeletal muscle extracellular matrix remodeling. *J. Clin. Endocrinol. Met.* **2014**, *99*, 1749–1757. [CrossRef] [PubMed]

36. Jackson, G.C.; Marcus-Soekarman, D.; Stolte-Dijkstra, I.; Verrips, A.; Taylor, J.A.; Briggs, M.D. Type ix collagen gene mutations can result in multiple epiphyseal dysplasia that is associated with osteochondritis dissecans and a mild myopathy. *Am. J. Med. Genet. A* **2010**, *152A*, 863–869. [CrossRef] [PubMed]

37. Chiquet, M.; Birk, D.E.; Bonnemann, C.G.; Koch, M. Collagen xii: Protecting bone and muscle integrity by organizing collagen fibrils. *Int. J. Biochem. Cell Biol.* **2014**, *53*, 51–54. [CrossRef] [PubMed]

38. Listrat, A.; Pissavy, A.-L.; Micol, D.; Jurie, C.; Lethias, C.; Pethick, D.; Hocquette, J.-F. Collagens xii and xiv: Two collagen types both associated with bovine muscle and intramuscular lipid metabolism. *Livest. Sci.* **2016**, *187*, 80–86. [CrossRef]

39. Eklund, L.; Piuhola, J.; Komulainen, J.; Sormunen, R.; Ongvarrasopone, C.; Fassler, R.; Muona, A.; Ilves, M.; Ruskoaho, H.; Takala, T.E.; et al. Lack of type XV collagen causes a skeletal myopathy and cardiovascular defects in mice. *Proc. Natl. Acad. Sci. USA* **2001**, *98*, 1194–1199. [CrossRef] [PubMed]

40. Cox, T.R.; Erler, J.T. Remodeling and homeostasis of the extracellular matrix: Implications for fibrotic diseases and cancer. *Dis. Model. Mech.* **2011**, *4*, 165–178. [CrossRef] [PubMed]

41. Law, B.; Fowlkes, V.; Goldsmith, J.G.; Carver, W.; Goldsmith, E.C. Diabetes-induced alterations in the extracellular matrix and their impact on myocardial function. *Microsc. Microanal.* **2012**, *18*, 22–34. [CrossRef] [PubMed]

42. Gullberg, D.; Velling, T.; Lohikangas, L.; Tiger, C.F. Integrins during muscle development and in muscular dystrophies. *Front. Biosci.* **1998**, *3*, D1039–D1050. [CrossRef] [PubMed]

43. Danoviz, M.E.; Yablonka-Reuveni, Z. Skeletal muscle satellite cells: Background and methods for isolation and analysis in a primary culture system. In *Myogenesis*; Springer: Berlin/Heidelberg, Germany, 2012; pp. 21–52.

44. Fu, X.; Wang, H.; Hu, P. Stem cell activation in skeletal muscle regeneration. *Cell. Mol. Life Sci.* **2015**, *72*, 1663–1677. [CrossRef] [PubMed]

45. Gattazzo, F.; Urciuolo, A.; Bonaldo, P. Extracellular matrix: A dynamic microenvironment for stem cell niche. *Biochim. Biophys. Acta.* **2014**, *1840*, 2506–2519. [CrossRef] [PubMed]

46. Thomas, K.; Engler, A.J.; Meyer, G.A. Extracellular matrix regulation in the muscle satellite cell niche. *Connect. Tissue Res.* **2015**, *56*, 1–8. [CrossRef] [PubMed]

47. Liu, H.; Niu, A.; Chen, S.E.; Li, Y.P. Beta3-integrin mediates satellite cell differentiation in regenerating mouse muscle. *FASEB J.* **2011**, *25*, 1914–1921. [CrossRef] [PubMed]

48. Sinha, I.; Sakthivel, D.; Varon, D.E. Systemic regulators of skeletal muscle regeneration in obesity. *Front. Endocrinol.* **2017**, *8*, 29. [CrossRef] [PubMed]

49. Czech, M.P. Insulin action and resistance in obesity and type 2 diabetes. *Nat. Med.* **2017**, *23*, 804–814. [CrossRef] [PubMed]

50. Turcotte, L.P.; Fisher, J.S. Skeletal muscle insulin resistance: Roles of fatty acid metabolism and exercise. *Phys. Ther.* **2008**, *88*, 1279–1296. [CrossRef] [PubMed]

51. Keske, M.A.; Premilovac, D.; Bradley, E.A.; Dwyer, R.M.; Richards, S.M.; Rattigan, S. Muscle microvascular blood flow responses in insulin resistance and ageing. *J. Physiol.* **2016**, *594*, 2223–2231. [CrossRef] [PubMed]

52. Yan, R.; Lai, S.; Yang, Y.; Shi, H.; Cai, Z.; Sorrentino, V.; Du, H.; Chen, H. A novel type 2 diabetes risk allele increases the promoter activity of the muscle-specific small ankyrin 1 gene. *Sci. Rep.* **2016**, *6*, 25105. [CrossRef] [PubMed]

53. Almind, K.; Inoue, G.; Pedersen, O.; Kahn, C.R. A common amino acid polymorphism in insulin receptor substrate-1 causes impaired insulin signaling. Evidence from transfection studies. *J. Clin. Invest.* **1996**, *97*, 2569–2575. [CrossRef] [PubMed]

54. Salvado, L.; Palomer, X.; Barroso, E.; Vazquez-Carrera, M. Targeting endoplasmic reticulum stress in insulin resistance. *Trends Endocrinol. Metab.* **2015**, *26*, 438–448. [CrossRef] [PubMed]

55. Panzhinskiy, E.; Hua, Y.; Culver, B.; Ren, J.; Nair, S. Endoplasmic reticulum stress upregulates protein tyrosine phosphatase 1b and impairs glucose uptake in cultured myotubes. *Diabetologia* **2013**, *56*, 598–607. [CrossRef] [PubMed]

56. Ijuin, T.; Hosooka, T.; Takenawa, T. Phosphatidylinositol 3,4,5-trisphosphate phosphatase skip links endoplasmic reticulum stress in skeletal muscle to insulin resistance. *Mol. Cell. Biol.* **2016**, *36*, 108–118. [PubMed]

57. Bonen, A.; Holloway, G.P.; Tandon, N.N.; Han, X.X.; McFarlan, J.; Glatz, J.F.; Luiken, J.J. Cardiac and skeletal muscle fatty acid transport and transporters and triacylglycerol and fatty acid oxidation in lean and zucker diabetic fatty rats. *Am. J. Physiol. Regul. Integr. Comp. Physiol.* **2009**, *297*, R1202–1212. [CrossRef] [PubMed]

58. Kelley, D.E.; Goodpaster, B.; Wing, R.R.; Simoneau, J.A. Skeletal muscle fatty acid metabolism in association with insulin resistance, obesity, and weight loss. *Am. J. Physiol.* **1999**, *277*, E1130–E1141. [CrossRef] [PubMed]

59. Lopaschuk, G.D. Fatty acid oxidation and its relation with insulin resistance and associated disorders. *Ann. Nutr. Metab.* **2016**, *68* (Suppl. 3), 15–20. [CrossRef]

60. Schrauwen-Hinderling, V.B.; Kooi, M.E.; Hesselink, M.K.; Jeneson, J.A.; Backes, W.H.; van Echteld, C.J.; van Engelshoven, J.M.; Mensink, M.; Schrauwen, P. Impaired in vivo mitochondrial function but similar intramyocellular lipid content in patients with type 2 diabetes mellitus and bmi-matched control subjects. *Diabetologia* **2007**, *50*, 113–120. [CrossRef] [PubMed]

61. Holloway, G.P.; Thrush, A.B.; Heigenhauser, G.J.; Tandon, N.N.; Dyck, D.J.; Bonen, A.; Spriet, L.L. Skeletal muscle mitochondrial FAT/CD36 content and palmitate oxidation are not decreased in obese women. *Am. J. Physiol. Endocrinol. Metab.* **2007**, *292*, E1782–E1789. [CrossRef] [PubMed]

62. Lawan, A.; Min, K.; Zhang, L.; Canfran-Duque, A.; Jurczak, M.J.; Camporez, J.P.G.; Nie, Y.; Gavin, T.P.; Shulman, G.I.; Hernandez-Fernando, C. Skeletal muscle-specific deletion of MKP-1 reveals a p38 MAPK/JNK/Akt signaling node that regulates obesity-induced insulin resistance. *Diabetes* **2018**, *67*, 624–635. [CrossRef] [PubMed]

63. Kim, K.; Park, S.M. Association of muscle mass and fat mass with insulin resistance and the prevalence of metabolic syndrome in korean adults: A cross-sectional study. *Sci. Rep.* **2018**, *8*, 2703. [CrossRef] [PubMed]

64. Berria, R.; Wang, L.; Richardson, D.K.; Finlayson, J.; Belfort, R.; Pratipanawatr, T.; De Filippis, E.A.; Kashyap, S.; Mandarino, L.J. Increased collagen content in insulin-resistant skeletal muscle. *Am. J. Physiol.-Endocrinol. Met.* **2006**, *290*, E560–E565. [CrossRef] [PubMed]

65. Richardson, D.K.; Kashyap, S.; Bajaj, M.; Cusi, K.; Mandarino, S.J.; Finlayson, J.; DeFronzo, R.A.; Jenkinson, C.P.; Mandarino, L.J. Lipid infusion decreases the expression of nuclear encoded mitochondrial genes and increases the expression of extracellular matrix genes in human skeletal muscle. *J. Biol. Chem.* **2005**, *280*, 10290–10297. [CrossRef] [PubMed]

66. Tam, C.S.; Chaudhuri, R.; Hutchison, A.T.; Samocha-Bonet, D.; Heilbronn, L.K. Skeletal muscle extracellular matrix remodeling after short-term overfeeding in healthy humans. *Metabolism* **2017**, *67*, 26–30. [CrossRef] [PubMed]

67. Kang, L.; Mayes, W.H.; James, F.D.; Bracy, D.P.; Wasserman, D.H. Matrix metalloproteinase 9 opposes diet-induced muscle insulin resistance in mice. *Diabetologia* **2014**, *57*, 603–613. [CrossRef] [PubMed]

68. Jha, J.C.; Banal, C.; Chow, B.S.; Cooper, M.E.; Jandeleit-Dahm, K. Diabetes and kidney disease: Role of oxidative stress. *Antioxid. Redox. Signal.* **2016**, *25*, 657–684. [CrossRef] [PubMed]

69. Richter, K.; Konzack, A.; Pihlajaniemi, T.; Heljasvaara, R.; Kietzmann, T. Redox-fibrosis: Impact of TGF β1 on ROS generators, mediators and functional consequences. *Redox. Biol.* **2015**, *6*, 344–352. [CrossRef] [PubMed]

70. Bozkurt, O.; Severcan, M.; Severcan, F. Diabetes induces compositional, structural and functional alterations on rat skeletal soleus muscle revealed by ftir spectroscopy: A comparative study with edl muscle. *Analyst* **2010**, *135*, 3110–3119. [CrossRef] [PubMed]

71. Fanning, K.M.; Pfisterer, B.; Davis, A.T.; Presley, T.D.; Williams, I.M.; Wasserman, D.H.; Cline, J.M.; Kavanagh, K. Changes in microvascular density differentiate metabolic health outcomes in monkeys with prior radiation exposure and subsequent skeletal muscle ECM remodeling. *Am. J. Physiol. Regul. Integr. Comp. Physiol.* **2017**, *313*, R290–R297. [CrossRef] [PubMed]

72. Kang, L.; Lantier, L.; Kennedy, A.; Bonner, J.S.; Mayes, W.H.; Bracy, D.P.; Bookbinder, L.H.; Hasty, A.H.; Thompson, C.B.; Wasserman, D.H. Hyaluronan accumulates with high-fat feeding and contributes to insulin resistance. *Diabetes* **2013**, *62*, 1888–1896. [CrossRef] [PubMed]

73. Groen, B.B.; Hamer, H.M.; Snijders, T.; van Kranenburg, J.; Frijns, D.; Vink, H.; van Loon, L.J. Skeletal muscle capillary density and microvascular function are compromised with aging and type 2 diabetes. *J. Appl. Physiol.* **2014**, *116*, 998–1005. [CrossRef] [PubMed]

74. Wagenmakers, A.J.; Strauss, J.A.; Shepherd, S.O.; Keske, M.A.; Cocks, M. Increased muscle blood supply and transendothelial nutrient and insulin transport induced by food intake and exercise: Effect of obesity and ageing. *J. Physiol.* **2016**, *594*, 2207–2222. [CrossRef] [PubMed]

75. Prior, S.J.; Goldberg, A.P.; Ortmeyer, H.K.; Chin, E.R.; Chen, D.; Blumenthal, J.B.; Ryan, A.S. Increased skeletal muscle capillarization independently enhances insulin sensitivity in older adults after exercise training and detraining. *Diabetes* **2015**, *64*, 3386–3395. [CrossRef] [PubMed]

76. Nascimento, A.R.; Machado, M.; de Jesus, N.; Gomes, F.; Lessa, M.A.; Bonomo, I.T.; Tibirica, E. Structural and functional microvascular alterations in a rat model of metabolic syndrome induced by a high-fat diet. *Obesity* **2013**, *21*, 2046–2054. [CrossRef] [PubMed]

77. Sung, H.K.; Kim, Y.W.; Choi, S.J.; Kim, J.Y.; Jeune, K.H.; Won, K.C.; Kim, J.K.; Koh, G.Y.; Park, S.Y. COMP-angiopoietin-1 enhances skeletal muscle blood flow and insulin sensitivity in mice. *Am. J. Physiol. Endocrinol. Met.* **2009**, *297*, E402–E409. [CrossRef] [PubMed]

78. Anderson, E.J.; Lustig, M.E.; Boyle, K.E.; Woodlief, T.L.; Kane, D.A.; Lin, C.T.; Price, J.W., 3rd; Kang, L.; Rabinovitch, P.S.; Szeto, H.H.; et al. Mitochondrial H_2O_2 emission and cellular redox state link excess fat intake to insulin resistance in both rodents and humans. *J. Clin. Invest.* **2009**, *119*, 573–581. [CrossRef] [PubMed]

79. Ayala, J.E.; Bracy, D.P.; Julien, B.M.; Rottman, J.N.; Fueger, P.T.; Wasserman, D.H. Chronic treatment with sildenafil improves energy balance and insulin action in high fat-fed conscious mice. *Diabetes* **2007**, *56*, 1025–1033. [CrossRef] [PubMed]

80. El Annabi, S.; Gautier, N.; Baron, V. Focal adhesion kinase and src mediate integrin regulation of insulin receptor phosphorylation. *FEBS Lett.* **2001**, *507*, 247–252. [CrossRef]

81. Bisht, B.; Goel, H.; Dey, C. Focal adhesion kinase regulates insulin resistance in skeletal muscle. *Diabetologia* **2007**, *50*, 1058–1069. [CrossRef] [PubMed]

82. Bisht, B.; Srinivasan, K.; Dey, C.S. In vivo inhibition of focal adhesion kinase causes insulin resistance. *J. Physiol.* **2008**, *586*, 3825–3837. [CrossRef] [PubMed]

83. Kang, L.; Mokshagundam, S.; Reuter, B.; Lark, D.S.; Sneddon, C.C.; Hennayake, C.; Williams, A.S.; Bracy, D.P.; James, F.D.; Pozzi, A. Integrin-linked kinase in muscle is necessary for the development of insulin resistance in diet-induced obese mice. *Diabetes* **2016**, *65*, 1590–1600. [CrossRef] [PubMed]

84. Lehti, T.M.; Silvennoinen, M.; Kivela, R.; Kainulainen, H.; Komulainen, J. Effects of streptozotocin-induced diabetes and physical training on gene expression of extracellular matrix proteins in mouse skeletal muscle. *Am. J. Physiol. Endocrinol. Met.* **2006**, *290*, E900–E907. [CrossRef] [PubMed]

85. Ashraf, J.M.; Ansari, M.A.; Khan, H.M.; Alzohairy, M.A.; Choi, I. Green synthesis of silver nanoparticles and characterization of their inhibitory effects on ages formation using biophysical techniques. *Sci. Rep.* **2016**, *6*, 20414. [CrossRef] [PubMed]

86. Singh, V.P.; Bali, A.; Singh, N.; Jaggi, A.S. Advanced glycation end products and diabetic complications. *Korean J. Physiol. Pharmacol.* **2014**, *18*, 1–14. [CrossRef] [PubMed]

87. Gautieri, A.; Passini, F.S.; Silvan, U.; Guizar-Sicairos, M.; Carimati, G.; Volpi, P.; Moretti, M.; Schoenhuber, H.; Redaelli, A.; Berli, M.; et al. Advanced glycation end-products: Mechanics of aged collagen from molecule to tissue. *Matrix. Biol.* **2017**, *59*, 95–108. [CrossRef] [PubMed]

88. Eble, J.A.; de Rezende, F.F. Redox-relevant aspects of the extracellular matrix and its cellular contacts via integrins. *Antioxid. Redox. Signal.* **2014**, *20*, 1977–1993. [CrossRef] [PubMed]

89. Bartling, B.; Desole, M.; Rohrbach, S.; Silber, R.E.; Simm, A. Age-associated changes of extracellular matrix collagen impair lung cancer cell migration. *FASEB J.* **2009**, *23*, 1510–1520. [CrossRef] [PubMed]

90. Tarsio, J.F.; Reger, L.A.; Furcht, L.T. Decreased interaction of fibronectin, type iv collagen, and heparin due to nonenzymic glycation. Implications for diabetes mellitus. *Biochemistry* **1987**, *26*, 1014–1020. [CrossRef] [PubMed]

91. Pastino, A.K.; Greco, T.M.; Mathias, R.A.; Cristea, I.M.; Schwarzbauer, J.E. Stimulatory effects of advanced glycation endproducts (ages) on fibronectin matrix assembly. *Matrix. Biol.* **2017**, *59*, 39–53. [CrossRef] [PubMed]

92. Tan, K.C.; Shiu, S.W.; Wong, Y.; Tam, X. Serum advanced glycation end products (ages) are associated with insulin resistance. *Diabetes Metab. Res. Rev.* **2011**, *27*, 488–492. [CrossRef] [PubMed]

93. Abdel-Wahab, Y.H.; O'Harte, F.P.; Ratcliff, H.; McClenaghan, N.H.; Barnett, C.R.; Flatt, P.R. Glycation of insulin in the islets of langerhans of normal and diabetic animals. *Diabetes* **1996**, *45*, 1489–1496. [CrossRef] [PubMed]

94. Baig, M.H.; Jan, A.T.; Rabbani, G.; Ahmad, K.; Ashraf, J.M.; Kim, T.; Min, H.S.; Lee, Y.H.; Cho, W.-K.; Ma, J.Y. Methylglyoxal and advanced glycation end products: Insight of the regulatory machinery affecting

the myogenic program and of its modulation by natural compounds. *Sci. Rep.* **2017**, *7*, 5916. [CrossRef] [PubMed]

95. Watts, R.; McAinch, A.J.; Dixon, J.B.; O'Brien, P.E.; Cameron-Smith, D. Increased Smad signaling and reduced MRF expression in skeletal muscle from obese subjects. *Obesity* **2013**, *21*, 525–528. [CrossRef] [PubMed]
96. Hittel, D.S.; Berggren, J.R.; Shearer, J.; Boyle, K.; Houmard, J.A. Increased secretion and expression of myostatin in skeletal muscle from extremely obese women. *Diabetes* **2009**, *58*, 30–38. [CrossRef] [PubMed]
97. Potes, Y.; de Luxán-Delgado, B.; Rodriguez-González, S.; Guimarães, M.R.M.; Solano, J.J.; Fernández-Fernández, M.; Bermúdez, M.; Boga, J.A.; Vega-Naredo, I.; Coto-Montes, A. Overweight in elderly people induces impaired autophagy in skeletal muscle. *Free Radic. Biol. Med.* **2017**, *110*, 31–41. [CrossRef] [PubMed]
98. Parker, L.; Caldow, M.K.; Watts, R.; Levinger, P.; Cameron-Smith, D.; Levinger, I. Age and sex differences in human skeletal muscle fibrosis markers and transforming growth factor-β signaling. *Eur. J. Appl. Physiol.* **2017**, *117*, 1463–1472. [CrossRef] [PubMed]
99. Semba, R.D.; Bandinelli, S.; Sun, K.; Guralnik, J.M.; Ferrucci, L. Relationship of an advanced glycation end product, plasma carboxymethyl-lysine, with slow walking speed in older adults: The inchianti study. *Eur. J. Appl. Physiol.* **2010**, *108*, 191. [CrossRef] [PubMed]
100. Dalal, M.; Ferrucci, L.; Sun, K.; Beck, J.; Fried, L.P.; Semba, R.D. Elevated serum advanced glycation end products and poor grip strength in older community-dwelling women. *J. Gerontol. Ser. A: Biomed. Sci. Med. Sci.* **2009**, *64*, 132–137. [CrossRef] [PubMed]
101. Momma, H.; Niu, K.; Kobayashi, Y.; Guan, L.; Sato, M.; Guo, H.; Chujo, M.; Otomo, A.; Yufei, C.; Tadaura, H. Skin advanced glycation end product accumulation and muscle strength among adult men. *Eur. J. Appl. Physiol.* **2011**, *111*, 1545–1552. [CrossRef] [PubMed]
102. Kato, M.; Kubo, A.; Sugioka, Y.; Mitsui, R.; Fukuhara, N.; Nihei, F.; Takeda, Y. Relationship between advanced glycation end-product accumulation and low skeletal muscle mass in japanese men and women. *Geriatr. Gerontol. Int.* **2017**, *17*, 785–790. [CrossRef] [PubMed]
103. Mori, H.; Kuroda, A.; Araki, M.; Suzuki, R.; Taniguchi, S.; Tamaki, M.; Akehi, Y.; Matsuhisa, M. Advanced glycation end-products are a risk for muscle weakness in japanese patients with type 1 diabetes. *J. Diabetes Investig.* **2017**, *8*, 377–382. [CrossRef] [PubMed]
104. Timmons, J.A.; Jansson, E.; Fischer, H.; Gustafsson, T.; Greenhaff, P.L.; Ridden, J.; Rachman, J.; Sundberg, C.J. Modulation of extracellular matrix genes reflects the magnitude of physiological adaptation to aerobic exercise training in humans. *BMC Biol.* **2005**, *3*, 19. [CrossRef] [PubMed]
105. Hjorth, M.; Norheim, F.; Meen, A.J.; Pourteymour, S.; Lee, S.; Holen, T.; Jensen, J.; Birkeland, K.I.; Martinov, V.N.; Langleite, T.M. The effect of acute and long-term physical activity on extracellular matrix and serglycin in human skeletal muscle. *Physiol. Rep.* **2015**, *3*, 12473. [CrossRef] [PubMed]
106. Hjorth, M.; Pourteymour, S.; Görgens, S.; Langleite, T.; Lee, S.; Holen, T.; Gulseth, H.; Birkeland, K.; Jensen, J.; Drevon, C. Myostatin in relation to physical activity and dysglycaemia and its effect on energy metabolism in human skeletal muscle cells. *Acta Physiol.* **2016**, *217*, 45–60. [CrossRef] [PubMed]
107. Lee, S.-J.; McPherron, A.C. Regulation of myostatin activity and muscle growth. *Proc. Nat. Acad. Sci. USA* **2001**, *98*, 9306–9311. [CrossRef] [PubMed]
108. Garito, T.; Roubenoff, R.; Hompesch, M.; Morrow, L.; Gomez, K.; Rooks, D.; Meyers, C.; Buchsbaum, M.; Neelakantham, S.; Swan, T. Bimagrumab improves body composition and insulin sensitivity in insulin-resistant subjects. *Diabetes Obes. Metab.* **2017**, *20*, 94–102. [CrossRef] [PubMed]

cells

MDPI

Article

Tumor–Stroma Cross-Talk in Human Pancreatic Ductal Adenocarcinoma: A Focus on the Effect of the Extracellular Matrix on Tumor Cell Phenotype and Invasive Potential

Patrizia Procacci [1,†], Claudia Moscheni [2,†], Patrizia Sartori [1], Michele Sommariva [1] and Nicoletta Gagliano [1,*]

[1] Department of Biomedical Sciences for Health, Università degli Studi di Milano, via L. Mangiagalli 31, 20133 Milan, Italy; patrizia.procacci@unimi.it (P.P.); patrizia.sartori@unimi.it (P.S.); michele.sommariva@unimi.it (M.S.)

[2] Department of Biomedical and Clinical Sciences "L. Sacco", Università degli Studi di Milano, via G.B. Grassi 74, 20157 Milan, Italy; claudia.moscheni@unimi.it

*   Correspondence: nicoletta.gagliano@unimi.it; Tel.: +39 02 50315374; Fax: +39 02 50315387

†   These authors contributed equally to this work.

Received: 7 September 2018; Accepted: 3 October 2018; Published: 5 October 2018

Abstract: The extracellular matrix (ECM) in the tumor microenvironment modulates the cancer cell phenotype, especially in pancreatic ductal adenocarcinoma (PDAC), a tumor characterized by an intense desmoplastic reaction. Because the epithelial-to-mesenchymal transition (EMT), a process that provides cancer cells with a metastatic phenotype, plays an important role in PDAC progression, the authors aimed to explore in vitro the interactions between human PDAC cells and ECM components of the PDAC microenvironment, focusing on the expression of EMT markers and matrix metalloproteinases (MMPs) that are able to digest the basement membrane during tumor invasion. EMT markers and the invasive potential of HPAF-II, HPAC, and PL45 cells grown on different ECM substrates (fibronectin, laminin, and collagen) were analyzed. While N-cadherin, αSMA, and type I collagen were not significantly affected by ECM components, the E-cadherin/β-catenin complex was highly expressed in all the experimental conditions, and E-cadherin was upregulated by collagen in PL45 cells. Cell migration was unaffected by fibronectin and delayed by laminin. In contrast, collagen significantly stimulated cell migration and the secretion of MMPs. This study's results showed that ECM components impacted cell migration and invasive potential differently. Collagen exerted a more evident effect, providing new insights into the understanding of the intricate interplay between ECM molecules and cancer cells, in order to find novel therapeutic targets for PDAC treatment.

Keywords: epithelial-to-mesenchymal transition; E-cadherin; MMPs; cell migration; extracellular matrix remodeling

1. Introduction

Pancreatic ductal adenocarcinoma (PDAC) is one of the most aggressive carcinomas, characterized by a dismal prognosis due to the high incidence of recurrence and metastases dissemination. PDAC is the fourth leading cause of cancer-related mortality in the Western world, with an estimated incidence of more than 40,000 cases per year in the United States. Due to the high recurrence and malignancy, the overall five-year survival rate for all stages of the disease is <7% [1–3].

During PDAC progression, the pancreatic tumor microenvironment (TME) containing extracellular matrix (ECM) components, growth factors, and soluble mediators, and different non-parenchymal stromal cells including fibroblasts, inflammatory, and pancreatic stellate cells, undergoes evident qualitative and quantitative modifications. In addition, it plays a key role as a modulator of cancer cell phenotype, behavior, and chemoresistance [4,5].

PDAC is characterized by an intense "desmoplastic reaction", defined as the host fibrotic response to the invasive carcinoma, consisting of the abnormal accumulation of ECM components, mostly collagen fibers [6]. The desmoplastic reaction represents the histological hallmark of PDAC, often accounting for 50–80% of the tumor volume [4,7]. Desmoplasia allows for a complex and dynamic interplay among invading tumor cells, normal host epithelial cells, fibroblasts, ECM components, and released cytokines, growth, and angiogenic factors [8]. ECM acts as a physical scaffold, facilitating interactions between different cell types, providing survival and differentiation signals, and affecting resistance to anticancer drugs. Therefore, ECM represents an important mediator of cancer cell behavior, influencing tumor cell proliferation and migration [9], as well as tissue homeostasis.

Key molecules occurring in the pancreatic stroma and the desmoplastic reaction have been identified, such as collagen types I, IV, and V, and fibronectin, laminin, matrix metalloproteinases (MMPs) and their inhibitors, tissue inhibitors of metalloproteinases (TIMPs), and transforming growth factor-β1 (TGF-β1) [10].

Among ECM molecules, type I collagen (COL-I) is the most abundant in pancreatic TME. Previous studies suggested that it was associated with increased integrin-mediated cell–cell adhesion, proliferation, and the migration of PDAC cells [11]. Basement membrane collagen type IV and laminin provide a proper microenvironment for pancreatic cancer cells affecting the cytotoxicity of anticancer drugs, and favoring cancer cell growth [9]. In contrast, the role of type V collagen, a minor component of ECM, is still poorly understood, since it triggers opposite cellular responses depending on the cell type. For instance, in breast cancer type V, collagen promotes breast cancer cell apoptosis [12], and its decrease in lung cancer is associated with increased tumor growth rate, motility, and invasion, as well as increased angiogenesis [13]. Fibronectin is abundant in both chronic pancreatitis and pancreatic cancer, suggesting that this protein may favor the development of pancreatic cancer [14].

MMPs, also considered as markers of epithelial-to-mesenchymal transition (EMT) (especially MMP-2 and MMP-9), render tumor cells able to metastasize in distant organs by breaking down the basement membrane, thus allowing cancer cells to enter the lymphatic or blood vessels [15]. The phenotype of carcinoma cells undergoing EMT is characterized by the loss of epithelial characteristics, especially the down-regulation of E-cadherin, leading to the loss of cell adhesion and polarity. Cytoskeleton reorganization, vimentin, and α-smooth muscle actin (αSMA) expression are also observed, as well as an increased degree of motility and secretion by MMPs [16].

Considering the role of desmoplasia in PDAC, the authors aimed to investigate the effects of single ECM components that are present in the TME on the PDAC cell phenotype, focusing on EMT pathways and MMP expression, in order to better understand the intricate cell–ECM cross-talk.

2. Materials and Methods

2.1. Cell Cultures

Three human pancreatic cancer cell lines (HPAF-II, HPAC, and PL45) from pancreatic ductal adenocarcinoma (PDAC) (American Type Culture Collection, ATCC, Manassas, VA, USA) were studied. PDAC cells were cultured in Dulbecco's Modified Eagle's Medium (DMEM) supplemented with 10% heat-inactivated fetal bovine serum (FBS), 2 mM glutamine, antibiotics (100 U/mL penicillin, 0.1 mg/mL streptomycin), and 0.25 µg/mL amphotericin B. Cell viability was determined by Trypan blue staining.

Cells were cultured in duplicate on Petri dishes coated with fibronectin (FN), laminin (LAM), COL-I (COL) (Cellcoat, Greiner Bio-One, Cassina de Pecchi-Milan, Italy), or without coating (NC), in order to characterize the specific effect of these proteins on PDAC cell phenotype.

2.2. Immunofluorescence

The expression of key EMT markers and their localization were assessed by immunofluorescence in PDAC cells grown on 12 mm-diameter rounded coverslips coated with FN, LAM, COL-I (Neuvitro Corporation, Vancouver, WA, USA), or uncoated. When they were at the desired confluence, cells were washed in phosphate-buffered saline (PBS), fixed in 4% paraformaldehyde in PBS-containing 2% sucrose for 10 min at room temperature, post-fixed in 70% ethanol, and stored at $-20\,^{\circ}$C until use. Cells were incubated with the primary antibodies anti-E-cadherin (1:2500, Becton Dickinson, Milan, Italy), anti β-catenin (1:500, Novocastra, Newcastle upon Tyne, UK), anti-N-cadherin (1:200, Santa Cruz Biotechnology, Heidelberg, Germany), anti-COL-I (1:2000, Sigma-Aldrich, Milan, Italy), anti-vimentin (1:200, Novocastra), and anti-αSMA (1:400, Sigma-Aldrich). Secondary antibodies conjugated with Alexa 488 (1:500, Molecular Probes, Invitrogen, Waltham, MA, USA) were applied for 1 h at room temperature in PBS. Negative controls were incubated omitting the primary antibody. Finally, after incubation with 4′,6-Diamidine-2′-phenylindole dihydrochloride (DAPI) (1:100,000, Sigma-Aldrich), the coverslips were mounted on glass slides using Mowiol.

2.3. Western Blot

Cell lysates were prepared in Tris-HCl 50 mM pH 7.6, 150 mM NaCl, 1% Triton X-100, 5 mM ethylenediaminetetraacetic acid (EDTA), 1% sodium dodecyl sulphate (SDS), proteases inhibitors, and 1 mM sodium orthovanadate. Lysates were incubated on ice for 30 min and centrifuged at 14,000 g for 10 min at 4 $^{\circ}$C to remove cell debris. Cell lysates (40 μg of total proteins) were diluted in SDS sample buffer, loaded on 10% SDS polyacrylamide gel, separated under reducing and denaturing conditions at 80 V according to Laemmli, and transferred at 90 V for 90 min to a nitrocellulose membrane in 0.025 M Tris, 192 mM glycine, and 20% methanol, pH 8.3. For E-cadherin evaluation, membranes were incubated for 1 h at room temperature with monoclonal antibodies to E-cadherin (1:2500, Becton Dickinson) and, after washing, in horseradish peroxidase (HRP)-conjugated rabbit anti-mouse serum (1:40,000 dilution, Sigma-Aldrich). To confirm equal loading, membranes were reprobed by monoclonal antibody to α-tubulin (1:2000 dilution, Sigma-Aldrich). Immunoreactive bands were revealed using the Amplified Opti-4CN (Bio Rad, Hercules, CA, USA).

2.4. SDS-Zymography

Serum-free culture media were mixed 3:1 with sample buffer (containing 10% SDS). Samples (5 μg of total protein per sample) were run under non-reducing conditions without heat denaturation on 10% polyacrylamide gel (SDS-PAGE) co-polymerized with 1 mg/mL of type I gelatin. The gels were run at 4 $^{\circ}$C. After SDS-PAGE, the gels were washed twice in 2.5% Triton X-100 for 30 min each, and incubated overnight in a substrate buffer at 37 $^{\circ}$C (Tris-HCl 50 mM, CaCl$_2$ 5 mM, 0.02% NaN$_3$, pH 7.5). MMP gelatinolytic activity, detected after staining the gels with Coomassie brilliant blue R250 as clear bands on a blue background, was quantified by densitometric scanning (UVBand, Eppendorf, Milan, Italy).

2.5. Wound Healing Assay

The cell migration of PDAC cells was analyzed by wound healing assay [17] in confluent cells using 2-well silicone culture-inserts (Ibidi, Martinsried, Germany) in Petri dishes coated with FN, LAM, COL-I, or uncoated (NC). After removal of the insert, migration of cells was assessed by measuring the closure of the wound at different time points. Petri dishes were incubated at 37 $^{\circ}$C and observed under an inverted microscope at different time points. Digital images were captured by a digital camera after 0 and 27 h, and the size of the "scratch" was measured to obtain the migration potential.

2.6. Statistical Analysis

Statistical analysis was performed using GraphPad Prism software (GraphPad Software Inc., version 6.0, La Jolla, CA, USA). Data were obtained from two replicate experiments for each cell line in each experimental condition cultured in duplicate and were expressed as mean ± standard deviation (SD). Comparison of groups was calculated using one-way ANOVA. Differences associated with *p*-values lower than 5% were to be considered significant.

3. Results

3.1. Cell Morphology is not Influenced by ECM Components

HPAF-II, HPAC, and PL45 cell lines, normally characterized by an epithelial phenotype, retained their morphology after culture on FN, LAM, COL-I, or without coating (NC) as demonstrated by the observation at the inverted microscope (see Figure 1), suggesting that these three molecules do not alter cell conformation.

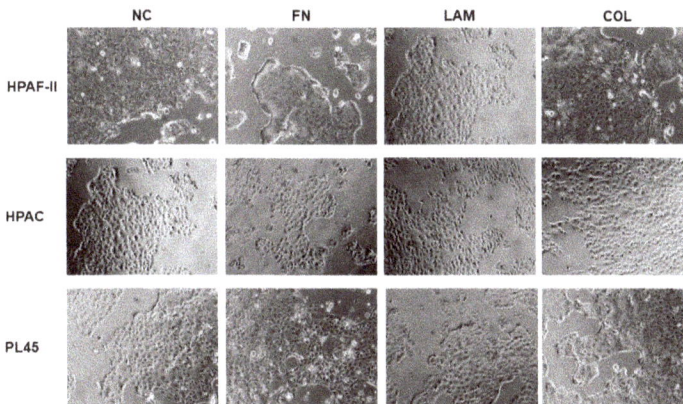

Figure 1. Cell morphology. Photomicrographs at the inverted microscope showing the epithelial morphology of HPAF-II, HPAC, and PL45 cells grown on fibronectin (FN), laminin (LAM), COL-I (COL), or without coating (NC) (original magnification 10×).

3.2. EMT Markers Are Differently Expressed in Cells Grown on Different ECM Components

Immunofluorescence analysis revealed that E-cadherin is strongly expressed at cell boundaries in HPAC cells cultured on FN, LAM, COL-I, or without coating (NC), suggesting the presence of functional adherens junctions (see Figure 2). A similar pattern was observed for β-catenin immunoreactivity, detectable at the plasma membrane in all experimental conditions. These results indicated that the E-cadherin/β-catenin complex was not modified by the considered ECM components (see Figure 2). The same immunoreactivity was observed in HPAF-II and PL45 cells (data not shown).

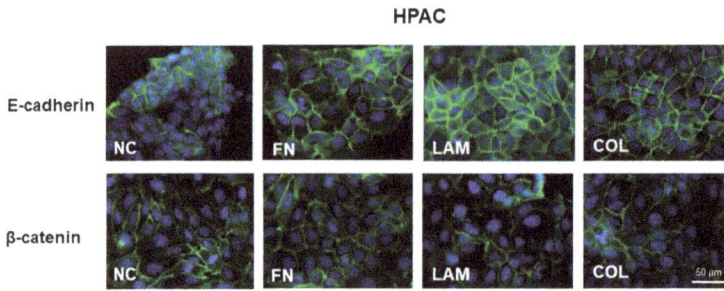

Figure 2. Immunofluorescence analysis of epithelial markers. Photomicrographs showing the expression of the epithelial markers E-cadherin and β-catenin in HPAC cells cultured on FN, LAM, COL, or without coating (NC) (original magnification 60×).

Since immunofluorescence does not allow a precise quantification of protein expression, E-cadherin protein levels were quantified by Western blot. According to immunofluorescence analysis, E-cadherin expression was not significantly modulated in HPAF-II and HPAC cells cultured on FN, LAM, COL-I, or without coating (NC). In contrast, it was strongly induced in PL45 cells that were grown on COL compared to NC controls ($p < 0.01$ vs. NC, $p < 0.05$ vs. FN, LAM) (see Figure 3).

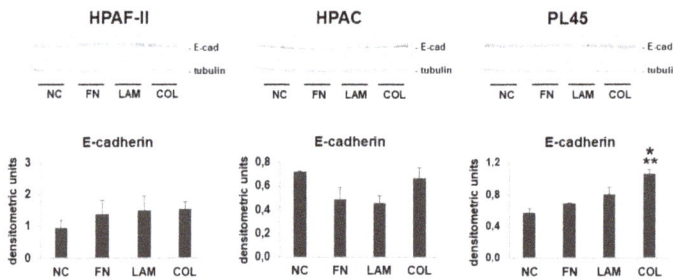

Figure 3. E-cadherin protein levels. Representative Western blot analysis and bar graphs showing E-cadherin expression in whole cell lysates of HPAF-II, HPAC, and PL45 cells cultured on FN, LAM, COL, or without coating (NC). Data are means ± SD. * $p < 0.01$ vs. NC; ** $p < 0.05$ vs. FN, LAM.

The analysis of mesenchymal markers in HPAC cells showed that N-cadherin was expressed at very low levels in HPAC cells, although this seemed to be slightly more evident in the cytoplasm of some cells that were cultured on FN and COL-I (see Figure 4). Vimentin was undetectable under all experimental conditions, whereas αSMA was highly expressed in all PDAC cells under the different experimental conditions (see Figure 4). COL-I immunoreactivity was detectable in cells cultured on all the substrates, but immunoreactivity seemed more evident in cells that were grown on LAM and COL (see Figure 4).

HPAC

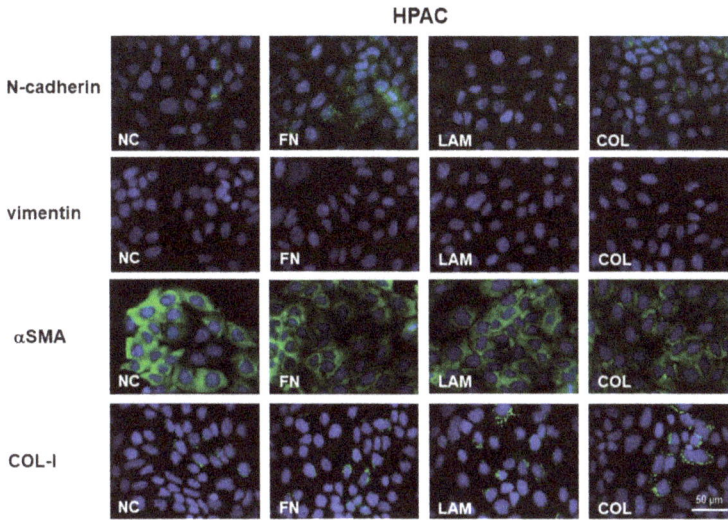

Figure 4. Immunofluorescence analysis of mesenchymal markers. Photomicrographs showing the expression of mesenchymal markers N-cadherin, vimentin, αSMA, and COL-I in HPAC cells cultured on FN, LAM, COL, or without coating (NC) (original magnification 60×).

3.3. Cell Migration is Differently Stimulated by ECM Components

The data of cell migration assessed by a wound healing assay at the considered time points are shown in Figure 5. Although migration appeared slightly reduced in HPAF-II cells, a similar trend was observed in HPAF-II and HPAC cells cultured on FN (38%, 73%, 89%, and 100% closure for HPAF-II after, respectively, 4 h, 6 h, 8 h, and 27 h; 88%, 100%, 100%, and 100% closure after for HPAC, respectively, 4 h, 6 h, 8 h, and 27 h). A different pattern was observed for LAM. In fact, migration was very slow in both HPAF-II and HPAC cells, compared not only to NC, but also to FN and COL. In contrast, PL45 exhibited a slower migration, compared to HPAF-II and HPAC cells, and in these cells, migration was similar at all the considered time points for cells grown on FN, as well as on LAM. When analyzing the effect of COL, increased migration was evident, especially in HPAF-II and HPAC cells (79%, 99%, 100%, and 100% closure after, respectively, 4 h, 6 h, 8 h, and 27 h for HPAF-II; 80%, 100%, 100%, and 100% closure after, respectively, 4 h, 6 h, 8 h, and 27 h for HPAC cells). For PL45, although at a lower extent, migration was strongly induced by COL, compared to NC, FN, and LAM (45%, 56%, 60%, and 100% closure after, respectively, 4 h, 6 h, 8 h, and 27 h for PL45 cells).

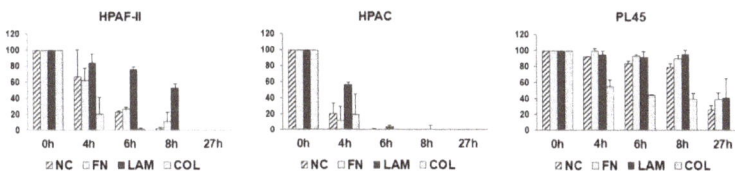

Figure 5. Cell migration. Bar graphs showing the closure of the wound at the indicated time points in HPAF-II, HPAC, and PL45 cells cultured on FN, LAM, COL, or without coating (NC). Bars represent the open wound, and data are expressed as the % vs. time point 0 h. Data are mean ± SD.

3.4. MMP Levels and Activity Are Affected by ECM Components

The invasive potential of PDAC cells was assayed by SDS-zymography. The results show that HPAF-II and PL45 cells expressed mostly MMP-2, whereas HPAC cells were characterized by a high expression of MMP-9. Both MMP-2 and MMP-9 were similarly expressed in HPAF-II cells cultured on FN, LAM, or without coating (NC), but they were significantly induced by COL-I ($p < 0.001$ for proMMP-2 vs. NC, FN, LAM) (see Figure 6a). A similar pattern was observed in HPAC for MMP-2, whereas MMP-9 was significantly induced when HPAC cells were cultured on FN ($p < 0.05$ vs. NC, LAM) (see Figure 6b). In PL45, MMP-2 activity was significantly reduced by LAM ($p < 0.05$ vs. NC), as well as by COL ($p < 0.05$ vs. NC). However, COL-I induced the expression of the active MMP-2 and a strong increase of MMP-9 (see Figure 6c). Collectively, MMP secretion resulted in a strong induction in all three cell lines cultured on COL-I. Moreover, in cells grown on COL-I, the active forms of MMPs were detectable.

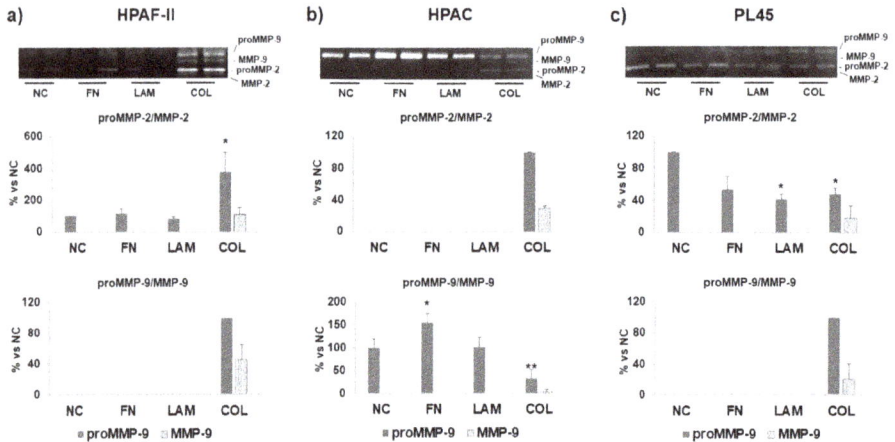

Figure 6. Matrix metalloprotenases-2 (MMP-2) and -9 (MMP-9) activity. Representative gelatin zymograms and bar graphs showing MMP-2 and MMP-9 activity in supernatants from (**a**) HPAF-II, (**b**) HPAC, and (**c**) PL45 cells cultured on FN, LAM, COL, or without coating (NC). Matrix metalloproteinases (MMPs) activity was induced, especially by COL. Data are expressed as % vs. NC and are means ± SD. (**a**) * $p < 0.001$ vs. NC, FN, LAM; (**b**) * $p < 0.05$ vs. NC, LAM; ** $p < 0.01$ vs. NC, FN, LAM; (**c**) * $p < 0.05$ vs. NC.

4. Discussion

PDAC is characterized by an intense desmoplastic reaction, a fibrotic lesion determined by an abnormal accumulation of ECM components, especially COL-I [4]. Interestingly, it has been demonstrated that primary tumors and metastatic lesions exhibited similar levels of desmoplasia, including high levels of some ECM components such as COL-I, COL-III, and COL-IV [18], and the expression of markers of desmoplasia has also been detected in metastatic sites [19]. Therefore, metastatic lesions are also fibrotic, like primary tumors, thus suggesting a key role for ECM components in the desmoplastic reaction of PDAC.

ECM is a dynamic structure, acting also as a physical scaffold, because the interaction between cells and the ECM through integrins or other cell surface receptors triggers intracellular signaling pathways that can influence cell survival, differentiation, chemoresistance, angiogenesis proliferation, migration, and invasion, which are all processes that contribute to cancer progression [20,21]. COL-I is the most abundant component of ECM in the tumor stroma, and it is highly expressed in metastatic tumors [22,23]. Previous studies investigated the effect of COL-I on pancreatic carcinoma cells, showing that high levels of COL-I significantly correlated with a reduced overall survival of PDAC patients [18], and influenced E-cadherin expression [24,25].

In this study, the authors characterized the effect of single ECM components on the PDAC cell phenotype, focusing their attention especially on E-cadherin and MMP expression. E-cadherin down-regulation is considered to be a key event during EMT, as demonstrated in vivo and in different cancer cell lines, including lung, breast, colorectal, and ovarian cancer [26–28]. The so-called "cadherin switch", a reduced expression of E-cadherin paralleled with an increased expression of N-cadherin, was described in PDAC [29,30]. Although E-cadherin down-regulation is considered an early essential event in EMT, experimental evidence demonstrates that six out of seven PDAC commercial cell lines maintain E-cadherin expression on the cell membrane [31], supporting the importance of studies investigating the role of E-cadherin, and more generally of EMT markers, in PDAC progression.

In this study, the authors showed that E-cadherin was strongly expressed in PDAC cells cultured on the different substrates and was significantly upregulated in PL45 grown on COL-I, suggesting that this ECM component elicited an important effect on tumor cells. According to their previous study [32], the maintenance of E-cadherin at cell–cell boundaries in PDAC cells showed that adherens junctions, and therefore cell adhesion, were preserved. This phenotypic characteristic of PDAC cells was needed to ensure tissue integrity during collective cell migration [33], and this study's results suggested that COL-I in the tumor stroma could contribute to PDAC malignant behavior.

ECM in the tumor stroma is a dynamic structure and stromal ECM undergoes a finely regulated dynamic turnover mediated by MMP enzymatic activity [15]. ECM degradation allows the migration of invasive cells into the surrounding tissue and vasculature [34]. The authors investigated the effect of ECM components on MMP-2 and MMP-9, because they are the key effectors that are involved in the degradation of the basement membrane, an important step during tumor invasion. Moreover, the roles of MMP-2 and MMP-9 in metastasis in pancreatic cancer are also demonstrated by a positive correlation between the expression of these two MMPs and the microvessel density, suggesting their involvement in the angiogenic processes [35]. In pancreatic cancer, MMP-2 is produced and secreted by both tumor and stromal cells [36], and a strong correlation between MMP-2 expression and the invasive potential of pancreatic cell lines has been observed [37]. MMP-9 expression is associated with lymph node invasion and the occurrence of distant metastases. Moreover, a correlation between MMP-9 expression and a worse prognosis in PDAC patients was found [38]. Previous studies indicated that MMP-9 production was influenced by the TME [39], suggesting that the stroma may act as a key facilitator of tumor invasion.

This study's results showed that PDAC cells had a different MMP secretion profile and activity, because HPAF-II and PL45 cells expressed mostly MMP-2, whereas HPAC cells expressed mostly MMP-9. Independently of the MMP expressed, the authors observed a different effect of ECM components on the PDAC cells considered. COL-I strongly induced MMP-2 and MMP-9 activity in HPAF-II cells, MMP-2 in HPAC, and MMP-9 in PL45, pointing to a pivotal influence of COL-I on the invasive potential of PDAC cells. These data are consistent with the wound healing assay results, showing an increased migration of cells cultured on COL-I. Because migration and invasive potential are both necessary to allow tumor invasion, and are both strongly stimulated by COL-I, the authors can hypothesize that COL-I is the component of the ECM that plays a pivotal role in influencing tumor cell behavior. MMP levels were differently affected by FN and LAM, depending on cell type, leading to the hypothesis that the effect of ECM components is dependent on tumor cell phenotype, and therefore, on the differentiation grade. However, a relationship between differentiation grade, cell migration, and invasion potential has not yet been clearly defined [40].

This study's findings, summarized in Figure 7, point to COL-I as a key influencer of PDAC cell phenotype and behavior. In PDAC stroma, COL-I is secreted by pancreatic stellate cells and myofibroblasts, acting as a crucial player in the development and maintenance of desmoplasia [4]. In this study, the authors showed that ECM components stimulated COL-I expression in PDAC cells, which very likely contributed to the secretion and deposition of COL-I in the tumor microenvironment.

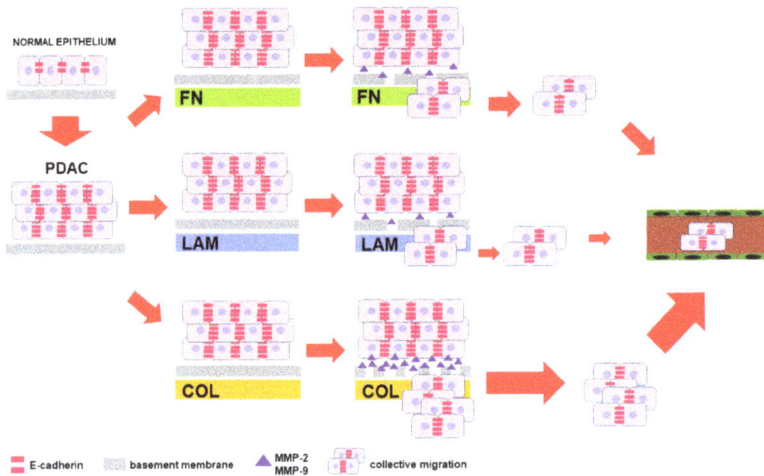

Figure 7. Extracellular matrix (ECM) components differentially affect pancreatic ductal adenocarcinoma (PDAC) cell migration and invasive properties. Diagram summarizing the hypothesis based on the most striking characteristics of PDAC cells grown on different ECM components used as a substrate. High-level expression of E-cadherin at cell boundaries is preserved and allows collective cell migration and invasion, favored by COL occurring in the pancreatic tumor microenvironment (TME), which stimulates both migration and secretion of MMPs.

5. Conclusions

Collectively, the authors' results are consistent with the pivotal role of desmoplasia and ECM components, occurring in the PDAC, in influencing tumor cell behavior, especially migration and invasive potential. This is achieved possibly by triggering signaling pathways and affecting anti-tumor drug penetration through the tumor. Although targeted therapies directed against stroma components have appeared to be an appealing therapeutic approach for the treatment of PDAC [41], recent studies have demonstrated that the depletion of the stroma [42] or of activated myofibroblasts [43] elicited a paradoxical result, rendering the tumor more aggressive. Moreover, it was suggested that the final tumor-promoting or tumor-suppressive effect of stroma components depended on the differentiation grade of cancer cells [42], further demonstrating the complexity of the microenvironment and of the bidirectional and mutual tumor–cell cross-talk. More recently, it was suggested that pharmacological stromal "normalization", used in order to achieve the homeostatic restoration of desmoplastic stroma, may represent a novel and promising therapeutic approach for PDAC [44]. In this complex context, the authors' data contribute to the understanding of the intricate cellular interactions between stromal ECM components and cancer cells, and stimulate further research on the ECM components in PDAC progression, in order to find more effective therapeutic tools for PDAC treatment.

Author Contributions: Conceptualization, N.G.; Methodology, N.G.; Investigation, N.G., P.P., C.M., P.S., and M.S.; Data Curation, N.G.; Writing—Original Draft Preparation, N.G.; Writing—Review and Editing, N.G., P.P., C.M., P.S., and M.S.

Funding: This research was funded by the University of Milan (Project B grant).

Acknowledgments: We thank Cinzia Lozio (Department of Biomedical Sciences for Health) for her administrative support.

Conflicts of Interest: The authors declare no conflict of interest.

References

1. Siegel, R.; Naishadham, D.; Jemal, A. Cancer statistics. *CA Cancer J. Clin.* **2013**, *63*, 11–30. [CrossRef] [PubMed]
2. Li, D.; Xie, K.; Wolff, R.; Abbruzzese, J.L. Pancreatic cancer. *Lancet* **2004**, *363*, 1049–1057. [CrossRef]
3. Ghaneh, P.; Costello, E.; Neoptolemos, J.P. Biology and management of pancreatic cancer. *Gut* **2007**, *56*, 1134–1152. [CrossRef] [PubMed]
4. Chu, G.C.; Kimmelman, A.C.; Hezel, A.F.; DePinho, R.A. Stromal biology of pancreatic cancer. *J. Cell. Biochem.* **2007**, *101*, 887–907. [CrossRef] [PubMed]
5. Bissell, M.J.; Radisky, D. Putting tumours in context. *Nat. Rev. Cancer* **2001**, *1*, 46–54. [CrossRef] [PubMed]
6. Nielsen, M.F.B.; Mortensen, M.B.; Detlefsen, S. Key players in pancreatic cancer-stroma interaction: Cancer-associated fibroblasts, endothelial and inflammatory cells. *World J. Gastroenterol.* **2016**, *22*, 2678–2700. [CrossRef] [PubMed]
7. Erkan, M.; Hausmann, S.; Michalski, C.W.; Fingerle, A.A.; Dobritz, M.; Kleeff, J.; Friess, H. The role of stroma in pancreatic cancer: Diagnostic and therapeutic implications. *Nat. Rev. Gastroenterol. Hepatol.* **2012**, *9*, 454–467. [CrossRef] [PubMed]
8. Ren, B.; Cui, M.; Yang, G.; Wang, H.; Feng, M.; You, L.; Zhao, Y. Tumor microenvironment participates in metastasis of pancreatic cancer. *Mol. Cancer* **2018**, *17*, 108. [CrossRef] [PubMed]
9. Miyamoto, H.; Murakami, T.; Tsuchida, K.; Sugino, H.; Miyake, H.; Tashiro, S. Tumor-stroma interaction of human pancreatic cancer: Acquired resistance to anticancer drugs and proliferation regulation is dependent on extracellular matrix proteins. *Pancreas* **2004**, *28*, 38–44. [CrossRef] [PubMed]
10. Park, C.C.; Bissell, M.J.; Barcellos-Hoff, M.H. The influence of the microenvironment on the malignant phenotype. *Mol. Med. Today* **2000**, *6*, 324–329. [CrossRef]
11. Grzesiak, J.J.; Bouvet, M. The alpha2beta1 integrin mediates the malignant phenotype on type I collagen in pancreatic cancer cell lines. *Br. J. Cancer* **2006**, *94*, 1311–1319. [CrossRef] [PubMed]
12. Luparello, C.; Sirchia, R. Type V collagen regulates the expression of apoptotic and stress response genes by breast cancer cells. *J. Cell. Physiol.* **2005**, *202*, 411–421. [CrossRef] [PubMed]
13. Souza, P.; Rizzardi, F.; Noleto, G.; Atanazio, M.; Bianchi, O.; Parra, E.R.; Teodoro, W.R.; Carrasco, S.; Velosa, A.P.; Fernezlian, S.; et al. Refractory remodeling of the microenvironment by abnormal type V collagen, apoptosis, and immune response in non-small cell lung cancer. *Hum. Pathol.* **2010**, *41*, 239–248. [CrossRef] [PubMed]
14. Binkley, C.E.; Zhang, L.; Greenson, J.K.; Giordano, T.J.; Kuick, R.; Misek, D.; Hanash, S.; Logsdon, C.D.; Simeone, D.M. The molecular basis of pancreatic fibrosis: Common stromal gene expression in chronic pancreatitis and pancreatic adenocarcinoma. *Pancreas* **2004**, *29*, 254–263. [CrossRef] [PubMed]
15. Knapinska, A.M.; Estrada, C.A.; Fields, G.B. The Roles of Matrix Metalloproteinases in Pancreatic Cancer. *Prog. Mol. Biol. Transl. Sci.* **2017**, *148*, 339–354. [CrossRef] [PubMed]
16. Thiery, J.P. Epithelial-mesenchymal transitions in tumour progression. *Nat. Rev. Cancer* **2002**, *2*, 442–454. [CrossRef] [PubMed]
17. Liang, C.C.; Park, A.Y.; Guan, J.L. In vitro scratch assay: A convenient and inexpensive method for analysis of cell migration in vitro. *Nat. Protoc.* **2007**, *2*, 329–333. [CrossRef] [PubMed]
18. Whatcott, C.J.; Diep, C.H.; Jiang, P.; Watanabe, A.; LoBello, J.; Sima, C.; Hostetter, G.; Shepard, H.M.; Von Hoff, D.D.; Han, H. Desmoplasia in primary tumors and metastatic lesions of pancreatic cancer. *Clin. Cancer Res.* **2015**, *21*, 3561–3568. [CrossRef] [PubMed]
19. Duda, D.G.; Duyverman, A.M.; Kohno, M.; Snuderl, M.; Steller, E.J.; Fukumura, D.; Jain, R.K. Malignant cells facilitate lung metastasis by bringing their own soil. *Proc. Natl. Acad. Sci. USA* **2010**, *107*, 21677–21682. [CrossRef] [PubMed]
20. Erler, J.T.; Weaver, V.M. Three-dimensional context regulation of metastasis. *Clin. Exp. Metast.* **2009**, *26*, 35–49. [CrossRef] [PubMed]
21. Liotta, L.A.; Kohn, E.C. The microenvironment of the tumour-host interface. *Nature* **2001**, *411*, 375–379. [CrossRef] [PubMed]
22. Van Kempen, L.C.; Ruiter, D.J.; van Muijen, G.N.; Coussens, L.M. The tumor microenvironment: A critical determinant of neoplastic evolution. *Eur. J. Cell Biol.* **2003**, *82*, 539–548. [CrossRef] [PubMed]

23. Linder, S.; Castanos-Velez, E.; von Rosen, A.; Biberfeld, P. Immunohistochemical expression of extracellular matrix proteins and adhesion molecules in pancreatic carcinoma. *Hepatogastroenterology* **2001**, *48*, 1321–1327. [PubMed]

24. Cheng, J.C.; Leung, P.C. Type I collagen down-regulates E-cadherin expression by increasing PI3KCA in cancer cells. *Cancer Lett.* **2011**, *304*, 107–116. [CrossRef] [PubMed]

25. Koenig, A.; Mueller, C.; Hasel, C.; Adler, G.; Menke, A. Collagen type I induces disruption of E-cadherin-mediated cell-cell contacts and promotes proliferation of pancreatic carcinoma cells. *Cancer Res.* **2006**, *66*, 4662–4671. [CrossRef] [PubMed]

26. Kalluri, R.; Weinberg, R.A. The basics of epithelial-mesenchymal transition. *J. Clin. Invest.* **2009**, *119*, 1420–1428. [CrossRef] [PubMed]

27. Hotz, B.; Arndt, M.; Dullat, S.; Bahrgava, S.; Buhr, H.J.; Hotz, H.G. Epithelial to Mesenchymal Transition: Expression of the regulators snail, slug, and twist in pancreatic cancer. *Clin. Cancer Res.* **2007**, *13*, 4769–4776. [CrossRef] [PubMed]

28. Celesti, G.; Di Caro, G.; Bianchi, P.; Grizzi, F.; Basso, G.; Marchesi, F.; Doni, A.; Marra, G.; Roncalli, M.; Mantovani, A.; et al. Presence of Twist1-positive neoplastic cells in the stroma of chromosome-unstable colorectal tumors. *Gastroenterology* **2013**, *145*, 647–657. [CrossRef] [PubMed]

29. Nakajiama, S.; Doi, R.; Toyoda, E.; Tsuji, S.; Wada, M.; Koizumi, M.; Tulachan, S.S.; Ito, D.; Kami, K.; Mori, T.; et al. N-cadherin expression and epithelial-mesenchymal transition in pancreatic carcinoma. *Clin. Cancer Res.* **2004**, *10*, 4125–4133. [CrossRef] [PubMed]

30. Joo, Y.E.; Rew, J.S.; Park, C.S.; Kim, S.J. Expression of E-cadherin, alpha and beta-catenins in patients with pancreatic adenocarcinoma. *Pancreatology* **2002**, *2*, 129–137. [CrossRef] [PubMed]

31. Cates, J.M.; Byrd, R.H.; Fohn, L.E.; Tatsas, A.D.; Washington, M.K.; Black, C.C. Epithelial-mesenchymal transition markers in pancreatic ductal adenocarcinoma. *Pancreas* **2009**, *38*, E1–E6. [CrossRef] [PubMed]

32. Gagliano, N.; Celesti, G.; Tacchini, L.; Pluchino, S.; Sforza, C.; Rasile, M.; Valerio, V.; Laghi, L.; Conte, V.; Procacci, P. Epithelial-to-mesenchymal transition in pancreatic ductal adenocarcinoma: Characterization in a 3D-cell culture model. *World J. Gastroenterol.* **2016**, *22*, 4466–4483. [CrossRef] [PubMed]

33. Yilmaz, M.; Christofori, G. Mechanisms of motility in metastasizing cells. *Mol. Cancer Res.* **2010**, *8*, 629–642. [CrossRef] [PubMed]

34. Collins, C.; Nelson, W.J. Running with neighbors: Coordinating cell migration and cell-cell adhesion. *Curr. Opin. Cell Biol.* **2015**, *36*, 62–70. [CrossRef] [PubMed]

35. Xiang, T.; Xia, X.; Yan, W. Expression of matrix metalloproteinases-2/-9 is associated with microvessel density in pancreatic cancer. *Am. J. Ther.* **2017**, *24*, e431–e434. [CrossRef] [PubMed]

36. Määttä, M.; Soini, Y.; Liakka, A.; Autio-Harmainen, H. Differential expression of matrix metalloproteinase (MMP)-2, MMP-9, and membrane type 1-MMP in hepatocellular and pancreatic adenocarcinoma: Implications for tumor progression and clinical prognosis. *Clin. Cancer Res.* **2000**, *6*, 2726–2734. [PubMed]

37. Ellenrieder, V.; Alber, B.; Lacher, U.; Hendler, S.F.; Menke, A.; Boeck, W.; Wagner, M.; Wilda, M.; Friess, H.; Büchler, M.; et al. Role of MT-MMPs and MMP-2 in pancreatic cancer progression. *Int. J. Cancer* **2000**, *85*, 14–20. [CrossRef]

38. Pryczynicz, A.; Guzinska-Ustymowicz, K.; Dymicka-Piekarska, V.; Czyzewska, J.; Kemona, A. Expression of matrix metalloproteinase in pancreatic ductal carcinoma is associated with tumor metastasis formation. *Folia Histochem. Cytobiol.* **2007**, *45*, 37–40. [PubMed]

39. Qian, X.; Rothman, V.L.; Nicosia, R.F.; Tuszynski, G.P. Expression of thrombospondin-1 in human pancreatic adenocarcinomas: Role in matrix metalloproteinase-9 production. *Pathol. Oncol. Res.* **2001**, *7*, 251–259. [CrossRef] [PubMed]

40. Deer, E.L.; Gonzalez-Hernandez, J.; Jill, D.; Coursen, J.D.; Shea, J.E.; Ngatia, J.; Scaife, C.L.; Firpo, M.A.; Mulvihill, S.J. Phenotype and genotype of pancreatic cancer cell lines. *Pancreas* **2010**, *39*, 425–435. [CrossRef] [PubMed]

41. Stromnes, I.M.; DelGiorno, K.E.; Greenberg, P.D.; Hingorani, S.R. Stromal reengineering to treat pancreas cancer. *Carcinogenesis* **2014**, *35*, 1451–1460. [CrossRef] [PubMed]

42. Rhim, A.D.; Oberstein, P.E.; Thomas, D.H.; Mirek, E.T.; Palermo, C.F.; Sastra, S.A.; Dekleva, E.N.; Saunders, T.; Becerra, C.P.; Tattersall, I.W.; et al. Stromal elements act to restrain, rather than support, pancreatic ductal adenocarcinoma. *Cancer Cell* **2014**, *25*, 735–747. [CrossRef] [PubMed]

43. Özdemir, B.C.; Pentcheva-Hoang, T.; Carstens, J.L.; Zheng, X.; Wu, C.C.; Simpson, T.R.; Laklai, H.; Sugimoto, H.; Kahlert, C.; Novitskiy, S.V.; et al. Depletion of carcinoma-associated fibroblasts and fibrosis induces immunosuppression and accelerates pancreas cancer with reduced survival. *Cancer Cell* **2014**, *25*, 719–734. [CrossRef] [PubMed]
44. Froeling, F.E.; Kocher, H.M. Homeostatic restoration of desmoplastic stroma rather than its ablation slows pancreatic cancer progression. *Gastroenterology* **2015**, *148*, 849–850. [CrossRef] [PubMed]

cells MDPI

Article

Matrix Metalloproteinase-1 and Acid Phosphatase in the Degradation of the Lamina Propria of Eruptive Pathway of Rat Molars

José Paulo de Pizzol Júnior [1], Estela Sasso-Cerri [2] and Paulo Sérgio Cerri [2,*]

[1] Department of Morphology and Genetics, Federal University of São Paulo (UNIFESP), 04021-001 São Paulo, SP, Brazil; jpaulopizzol@gmail.com
[2] Laboratory of Histology and Embryology-Araraquara, School of Dentistry, São Paulo State University (UNESP), 1680 Centro, CEP 14801–903 Araraquara, SP, Brazil; estela.sasso@unesp.br
* Correspondence: paulo.cerri@unesp.br; Tel.: +55-16-33016497; Fax: +55-16-33016433

Received: 12 September 2018; Accepted: 8 November 2018; Published: 10 November 2018

Abstract: The comprehension of dental pathogenesis and disorders derived from eruption failure requires a deep understanding of the molecular mechanisms underlying normal tooth eruption. As intense remodelling is needed during tooth eruption, we hypothesize that matrix metalloproteinase-1 (MMP-1) and acid phosphatase (ACP) play a role in the eruptive pathway degradation. We evaluated MMP-1-immunoexpression and the collagen content in the lamina propria at different eruptive phases. Immunohistochemistry and ultrastructural cytochemistry for detection of ACP were also performed. In the maxillary sections containing first molars of 9-, 11-, 13-, and 16-day-old rats, the birefringent collagen of eruptive pathway was quantified. MMP-1 and ACP-2 immunohistochemical reactions were performed and the number of MMP-1-immunolabelled cells was computed. Data were analyzed by one-way ANOVA and Tukey post-test ($p \leq 0.05$). ACP cytochemistry was evaluated in specimens incubated in sodium β-glycerophosphate. In the eruptive pathway of 13- and 16-day-old rats, the number of MMP-1-immunolabelled cells increased concomitantly to reduction of collagen in the lamina propria. Enhanced ACP-2-immunolabelling was observed in the lamina propria of 13- and 16-day-old rats. Fibroblasts and macrophages showed lysosomes and vacuoles containing fragmented material reactive to ACP. MMP-1 degrades extracellular matrix, including collagen fibers, being responsible for the reduction in the collagen content during tooth eruption. The enhanced ACP activity at the mucosal penetration stage indicates that this enzyme plays a role in the degradation of remnant material, which is engulfed by macrophages and fibroblasts of the eruptive pathway. Therefore, enzymatic failure in the eruptive pathway may disturbs tooth eruption.

Keywords: eruptive pathway; matrix metalloproteinase-1; acid phosphatase; ultrastructure; lamina propria; birefringent collagen

---

1. Introduction

In the orthodontic clinic, the adequate diagnosis of eruption disorders is essential for the correct management of the orthodontic problem [1,2]. The eruption failure cannot be classified based only on clinical characters, but the mechanism by which this pathological condition occurs, including possible physical and/or biological defects in the eruption process, should be considered and investigated [3]. Eruption disorders are in general difficult to diagnose given the scarce studies on the eruptive process [4]. Therefore, the better understanding of the pathogenic conditions and disorders derived from eruption failure depends on a complete understanding of the specific molecular mechanisms underlying normal eruption [3,5].

Tooth eruption is the phenomenon of movement of the tooth germ from it intraosseous position into occlusion position [6,7]. The tooth eruption is divided into five phases: pre-eruptive, intraosseous, mucosal penetration, pre-occlusal and post-occlusal eruption [6,8]. Accentuated structural changes in the eruptive pathway occur during the intraosseous and mucosal penetration stages [9,10]. The intraosseous stage begins with the axial movements of the tooth germ leading to the process of bone resorption, mainly in the occlusal portion of the bone crypt [6,8]. At the intraosseous phase, it has been demonstrated that the cells of the dental follicle release several factors and cytokines such as colony-stimulating factor-1 (CSF-1), monocyte chemotactic protein-1 (MCP-1) and receptor activator of NF kappa B ligand (RANKL) [11–13]. These molecules stimulate the recruitment and differentiation of mononuclear cells into osteoclasts, which promote an intense bone resorption, particularly in the occlusal portion of the bone crypt [14]. Moreover, we demonstrated a significant concomitant increase in the number of mast cells and osteoclasts in the eruptive pathway, indicating that mast cells seem to participate in the recruitment of osteoclasts and, consequently, in the bone resorption [9]. During the mucosal penetration phase, intense structural changes occur in the lamina propria of the eruptive pathway to allow the passage of teeth. Among these changes, an accentuated cell death by apoptosis and reduction in the blood vessel profiles were reported during the eruption of rat molars [10].

Therefore, the tooth eruption leads to a complex series of structural changes that occur in an extremely coordinated way avoiding an inflammatory reaction, which could compromise the integrity of dental tissues. Matrix metalloproteinases (MMPs) constitute a family of zinc-dependent endopeptidase responsible for the degradation of extracellular matrix (ECM) components [15]. MMPs are necessary for the development and maintenance of tissues and/or organs, exerting an important role in the tissue turnover/remodelling [15,16]. Moreover, MMPs are also detected under pathological conditions, cleaving the ECM components and leading to tissue degradation [17–19]. In rat molars, an enhanced MMP-9 immunoexpression was found in the lamina propria during the mucosal penetration stage of the tooth eruption [9]. As MMP-9 (Gelatinase B) cleaves denatured collagen, in particular, type IV collagen of the basal lamina, as well as degrades amorphous components of the ECM [20,21], the high immunoexpression of this metalloproteinase in the lamina propria was associated with tissue degradation and remodelling necessary for the establishment of the eruptive pathway [9]. MMP-1, other member of the MMPs superfamily, is a neutral proteinase that cleave native fibrillar collagens and, therefore, this MMP plays a key role in the degradation of the collagenous matrix [22,23]. MMP-1 can be released by several cells, including fibroblasts, macrophages and endothelial cells [19,24].

Delayed tooth eruption has been demonstrated in membrane type-1 matrix metalloproteinase deficient mice. In normal animals at 18–20 days of development, the first molars are already erupted. However, in the type-1 matrix metalloproteinase deficient mice, the molars show crowns with normal dentin and enamel, but the roots are truncated and the teeth are not erupted [25]. Another study in mice model with hyperplastic dental follicle (HDFs) showed reduced expression of MMPs (MMP-1 and MMP-3), whereas the expression of tissue inhibitor of metalloproteinases (TIMPs) and collagen increased, leading to thickening of fibrous components in HDFs. These results indicate that an appropriate control of the connective tissue remodelling is essential for the normal tooth eruption, and that MMPs, including MMP-1 and MMP-3, play important role during tooth eruption [26].

The degradation and remodelling of the eruptive pathway may also involve the phagocytosis and intracellular digestion processes. After the action of MMPs in the extracellular space, the remnants of the ECM components may be engulfed and digested by phagocytic cells. Although professional phagocytic cells such as macrophages exhibit high acid phosphatase content inside lysosomes [27], it has been shown that, in certain circumstances, ECM components [28,29] and apoptotic bodies can also be engulfed by fibroblasts [10,30]. Furthermore, the acid phosphatase activity inside phagocytic vacuoles of fibroblasts reinforces the concept that these cells can also act as phagocytic cells [28,29]. It has been described at least six types of acid phosphatase (ACP1, ACP2, ACP3, ACP4, ACP5, and ACP6) [31,32]. ACP2 is the lysosomal acid phosphatase, which degrades the intracellular collagen and plays an important role in tissue remodelling [29,33].

Few studies have focused on the importance of collagen remodelling for the tooth eruption, and the understanding of the specific molecular mechanisms underlying normal eruption is essential for the clinical diagnosis of eruptive disturbs. Considering that the lamina propria is a loose connective tissue containing mainly fibroblasts and macrophages intermingled with collagen fibres (predominantly types I and III) and amorphous substance [9,34,35] and the essential function of the MMP-1 in the cleavage of ECM components, we hypothesize that MMP-1 and acid phosphatase could have a role in the degradation of the eruptive pathway. Therefore, here, we evaluated the number of MMP-1-immunolabelled cells in the lamina propria and correlated this parameter with collagen content at the different eruptive phases. We also investigated whether acid phosphatase is involved in the degradation process of the lamina propria during tooth eruption.

2. Materials and Methods

Animal treatment was performed in accordance with Brazilian animal care and national laws on animal use. Our research protocol was authorized by the Ethical Committee for Animal Research of the São Paulo State University, Brazil (CEUA, Dental School-UNESP, Araraquara, protocol number 13/2013 approved on 14 June 2013).

Forty Holtzman postnatal male rats (*Rattus norvegicus albinus*) were divided into four groups (*n* = 10) according to their age: 9-, 11-, 13-, and 16-day-old. At these ages, the teeth are at specific eruptive phases. Therefore, the first molar germs of 9- and 11-day-old rats are at the intraosseous phase, whereas the first molars of 13- and 16-day-old rats are at the mucosal penetration phase [9,10]. The animals were housed in polypropylene cages that were filled with a layer of white pine shavings. During the experiment, one mother plus five male pups were housed per cage in a room with controlled temperature (23 ± 2 °C) and humidity (55 ± 10%). Rats were maintained under a 12:12 light/dark cycle with light onset at 07:00 h. Standardized chow (Guabi Rat Chow, Paulinia, SP, Brazil) and water were provided ad libitum.

The rats were killed by an overdose of ketamine hydrochloride and xylazine hydrochloride, decapitated and the upper maxillae were removed. The fragments of the maxilla from 5 rats per group were fixed and processed for light microscopy and molecular analysis (Western blot), whereas the fragments of maxilla from the other 5 rats were fixed and processed for transmission electron microscopy.

Using a stereoscopic microscope (Wild M7; Wild Heerbrugg, Heerbrugg, Switzerland), fragments of maxilla containing the first right molars were removed and placed in the fixative solution for light microscopy while the left maxillae were used for molecular analysis.

2.1. Light Microscopy

The fragments of maxilla containing the first right molar germs were fixed for 48 h at room temperature in 4% formaldehyde (freshly prepared from paraformaldehyde) buffered at pH 7.2 with 0.1 M sodium phosphate. After decalcification for 10 days in a 7% solution of EDTA (ethylenediaminetetraacetic acid) containing 0.5% formaldehyde (4), buffered at pH 7.2 with 0.1 M sodium phosphate-saline (PBS), the fragments of maxilla were dehydrated in graded concentrations of ethanol and embedded in paraffin. From each maxilla, fifty sagittal 6 μm-thick sections were collected onto slides. Five non-serial sections were stained with Carazzi's haematoxylin and eosin (HE) whereas three non-serial sections were subjected to picrosirius-red method to estimate the collagen content. Other sections were adhered to silanized slides and submitted to immunohistochemistry for detection of MMP-1 and acid phosphatase (ACP-2).

2.2. Collagen Content Measurement in the Eruptive Pathway

For the collagen content estimation in the lamina propria of the eruptive pathway, the sections were submitted to the picrosirius-red method and analyzed under polarized light [36].

In each rat, three non-serial picrosirius-red-stained sections containing the lamina propria of the first molar were captured under polarized light at ×695 magnification; the smallest distance between the sections was 100 μm. In each section, three standardized fields were captured using an Olympus camera (DP-71, Olympus, Tokyo, Japan) attached to a light microscope (BX51, Olympus, Tokyo, Japan) at ×695 magnification, totalling a standardized area of 0.27 mm^2 per section.

The birefringent collagen frequency was estimated using the following hue definition: red/orange 2–38 and 230–256, yellow 39–51 and green 52–128 [37]. The collagen content was calculated as a percentage of the area of each image (expressed in pixels) using ImageJ$^®$ (NIH) as previously described [38]. Images were loaded and the hues were isolated using the hue histogram filter available in "Threshold Colour". A black/white picture was created in which black pixels represented the hue in analysis (red/orange e.g.,) and the white pixels were the remaining hues. Total of black pixels was obtained and the percentage of birefringent collagen was calculated in the total area [39].

2.3. Immunohistochemical Detection of MMP-1

To unmask the antigenic sites, deparaffinized sections were immersed in 10 mM sodium citrate buffer pH 6.0, and placed into a microwave oven at 90–94 °C for 30 min. After a cooling-off period, the endogenous peroxidase was blocked with 3% hydrogen peroxide for 20 min. The slides were washed in 0.1 M PBS pH 7.2 and incubated for 30 min with 2% bovine serum albumin (BSA, Sigma-Aldrich Chemie, Munich, Germany). The sections were incubated overnight in a humidified chamber at 4 °C with mouse anti-MMP-1 primary antibody (MAB901; R&D System, Minneapolis, MN, USA), diluted 1:400. Sections were washed with PBS and the immunoreaction was detected by the Labelled Streptavidin-Biotin system (LSAB-plus Kit; DAKO Corporation, Carpinteria, CA, USA). The sections were incubated for 30 min at room temperature with a multi-link solution containing biotinylated anti-mouse/rabbit/goat secondary antibodies. After washing with PBS, the sections were incubated with the streptavidin-peroxidase complex for 30 min at room temperature. The sections were washed with PBS and the immunoreaction was revealed by 3,3'-diaminobenzidine (DAB-BiocareMedical, Concord, CA, USA); the sections were counterstained with Carazzi's haematoxylin. For negative controls, the sections were incubated in non-immune serum (Sigma-Aldrich Chemie, Munich, Germany) instead of primary antibody.

2.4. Numerical Density of MMP-1-Immunolabeled Cells

Two non-serial sections exhibiting the lamina propria of the eruptive pathway of the first molar from each animal were used. The shortest distance between the sections was 100 μm. In each section, three standardized fields (0.09 mm^2 each field) were captured using an Olympus camera (DP-71, Olympus, Tokyo, Japan) attached to a light microscope (BX51, Olympus, Tokyo, Japan) at ×695 magnification. The number of immunolabelled cells (brown-yellow colour) was computed by one blinded and calibrated examiner using an image analysis system (Image Pro-Express 6.0, Olympus, Silver Spring, MD, USA). The number of MMP-1-positive cells/mm^2 of the lamina propria per animal was calculated [10,19].

2.5. Immunohistochemical Detection of Acid Phosphatase (ACP-2)

For antigen retrieval, deparaffinized sections were immersed in distilled water containing 0.1% CaCl$_2$ and 0.5% trypsin for 1 h and 30 min at 37 °C. After the inactivation of endogenous peroxidase in 3% hydrogen peroxide for 20 min, the sections were washed in 0.1 M PBS pH 7.2 and incubated for 30 min. with 2% BSA (Sigma-Aldrich, St. Louis, MO, USA). Subsequently, the sections were incubated overnight in a humidified chamber at 4 °C with mouse anti-acid phosphatase primary antibody (sc-100344; ACP-2-Santa Cruz Biotechnology, Inc$^®$, Santa Cruz, CA, USA) diluted 1:100. Sections were washed with PBS and the immunoreaction was amplified using the Labelled Streptavidin-Biotin system (LSAB-plus Kit; DAKO Corporation, Carpinteria, CA, USA), as described above. Peroxidase activity was revealed by 3,3'-diaminobenzidine (DAB-BiocareMedical, Concord, CA, USA) and the sections were counterstained with Carazzi's haematoxylin. As negative controls, the sections were incubated in non-immune serum (Sigma-Aldrich Chemie, Munich, Germany) instead of primary antibody.

2.6. Protein Extraction and Western Blot for MMP-1 and ACP-2

For molecular analysis, small fragments of the eruptive pathway were obtained from the 16-day-old rats. Using a scalpel (Fibra Cirúrgica, Joinvile, Brazil), the oral mucosa overlaying the upper first molars was carefully removed with the help of a stereoscopic microscope (Wild M7; Wild Heerbrugg, Switzerland) at ×60. The small fragments of oral mucosa of the eruptive pathway were frozen at −80 °C for Western blot.

Frozen fragments of oral mucosa of the eruptive pathway were homogenized directly into lysis buffer (50 mM Tris pH 8.0, 150 mM NaCl, 1 mM EDTA, 10% glycerol, 1% Triton X-100, 1 mM phenylmethylsulfonyl fluoride (PMSF)), containing 5 ng/mL of each of the following protease inhibitors: Pepstatin, Leupeptin, Aprotinin, Antipain, and Chymostatin (Sigma-Aldrich®; P8340). The crude extracts were clarified by centrifugation at 10,000 rpm for 20 min and 4 °C, and the supernatant was collected. Protein concentration was determined using Bradford assay (Sigma-Aldrich®; B6916), and Western blot for MMP-1 and ACP-2 detection was performed. The same amount of proteins (20 μg) was mixed with an equivalent volume of Laemmli buffer and heated at 95 °C for 5 min. before electrophoresis. The samples were separated in sodium dodecyl sulfate polyacrylamide gels (SDS-PAGE; 12% polyacrylamide), and then transferred to a nitrocellulose membrane (GE Healthcare®, Chicago, IL, USA). The membrane was treated with blocking solution containing 5% non-fat milk, and then incubated overnight at 4 °C with mouse anti-MMP-1 primary antibody (MAB901; 1:400; R&D System, Minneapolis, MN, USA) or with mouse anti-acid phosphatase primary antibody (sc-100344; 1:100; ACP2, Santa Cruz Biotechnology, Inc®, Santa Cruz, CA, USA) diluted in PBS containing 0.05% Tween 20 (PBS/T). After washing, the membranes were incubated for 1 h with anti-rabbit (Sigma-Aldrich, St. Louis, MO, USA) or anti-mouse (Sigma-Aldrich, USA, A9044) peroxidase antibodies diluted with PBS/T solution (1:1250). The reactions were detected using an enhanced chemiluminescence system (ECL) and the bands were visualized using a digital documentation system (GelDoc XR, Bio-Rad Laboratories, Hercules, CA, USA). As loading control, the membranes were probed with rabbit anti-actin antibody (1:8000; Sigma-Aldrich, Sigma-Aldrich, St. Louis, MO, USA). The assays were performed in triplicate for each protein.

2.7. Transmission Electron Microscopy

For ultrastructural analysis, five right maxilla per group were used. The fragments of maxilla containing the lamina propria overlaying the first molar germs were fixed for 16 h in a solution of 4% glutaraldehyde and 4% formaldehyde buffered at pH 7.2 with 0.1 M sodium cacodylate [40]. After decalcification for 10 days in a solution of 7% EDTA buffered at pH 7.2 in 0.1 M sodium cacodylate, the specimens were postfixed in sodium cacodylate-buffered 1% osmium tetroxide at pH 7.2 for 1.5 h. Subsequently, the specimens were washed in distilled water and immersed in 2% aqueous uranyl acetate for 2 h. After washing in distilled water, the specimens were dehydrated in graded concentrations of ethanol, treated with propylene oxide and then embedded in Araldite. Semithin sections stained with aqueous solution of 1% toluidine blue and 1% sodium borate were examined under light microscope, and suitable regions were carefully selected for trimming of the blocks. Ultrathin sections were collected onto grids, stained in alcoholic 2% uranyl acetate and in lead citrate solution and examined under a transmission electron microscope (Tecnai G2 Spirit, FEI Company, Hillsboro, OR, USA).

2.8. Ultrastructural Localization of Acid Phosphatase Activity

For ultrastructural localization of acid phosphatase, five left maxilla per group were used. The oral mucosa overlaying the upper first molar germs was removed with the help of a stereoscopic microscope (Wild M7; Wild Heerbrugg, Switzerland) at ×60 magnification. Small fragments of the oral mucosa containing the lamina propria of the eruptive pathway were fixed for 6 h in a solution of 2% formaldehyde and 0.5% glutaraldehyde buffered at pH 7.2 with 0.1 M sodium cacodylate.

After fixation, the specimens were washed in sodium cacodylate buffer and incubated in a medium prepared by dissolving 20 mM lead nitrate (Merck, Darmstadt-DE, Germany) in 0.1 M sodium acetate buffer (pH 5.0) followed by addition of 10 mM sodium β-glycerophosphate (Merck, Darmstadt-DE), as described by Barka [41]. After addition of magnesium chloride, the solution was filtered and the specimens were incubated for 2 h at 37 °C. As control of specificity, some specimens were incubated in substrate-free medium (without β-glycerophosphate). Subsequently, the specimens were fixed for 2 h in 3% glutaraldehyde buffered at pH 7.2 with 0.1 M sodium cacodylate at 4 °C. Then, the specimens were washed and immersed in 1% osmium tetroxide for 1 h. After washing, the specimens were dehydrated in graded concentrations of ethanol, treated with propylene oxide and then embedded in Araldite. Ultrathin sections were examined with Tecnai electron microscope.

2.9. Statistical Analysis

The statistical analyses were performed using Statistical Sigma Stat 3.2 software (Jandel Scientific, Sausalito, CA, USA). The differences in the numerical density of MMP-1-immunolabelled cells and birefringent collagen content were statistically analyzed among all groups (9-, 11-, 13- and 16-day-old rats). One-way ANOVA was used for the analysis of variance followed by Tukey post-hoc test. The significance level was set at $p \leq 0.05$.

The correlation between numerical density of MMP-1-immunolabelled cells and collagen content was evaluated by Pearson product-moment coefficient. The significance level considered was $p \leq 0.05$.

3. Results

3.1. Morphological Findings and Content of Birefringent Collagen

The analysis of sagittal sections of maxillae revealed changes in the arrangement and in the content of collagen of the lamina propria of the eruptive pathway in the rats at different ages (Figures 1 and 2). In the 9- and 11-day-old rats, evident bundles of collagen fibres were distributed throughout lamina propria (Figure 1A,B) while scarce collagen fibres were present in the thin lamina propria of 13- and 16-day-old rats (Figure 1C,D). Under polarized light, the sections subjected to the picrosirius-red staining revealed a different pattern of birefringent colours in the lamina propria in accordance with the eruptive phase of first molars (Figure 2). In the 9- and 11-day-old rats, the lamina propria showed an evident continuous layer of birefringent bundles of collagen with the hue varying from red/orange to yellow colour (Figure 2A,B; insets). Otherwise, few and thin birefringent collagen content was observed in the eruptive pathway of 13- and 16-day-old rats (Figure 2C,D). At these time points, the thin bundles of collagen material exhibited predominantly yellow and green colours (Figure 2C,D, insets). Quantitative analysis revealed a significant decrease in the birefringent collagen content in the lamina propria of 16-day-old rats in comparison to 9- ($p = 0.007$) and 11- ($p = 0.012$) day-old rats (Figure 3).

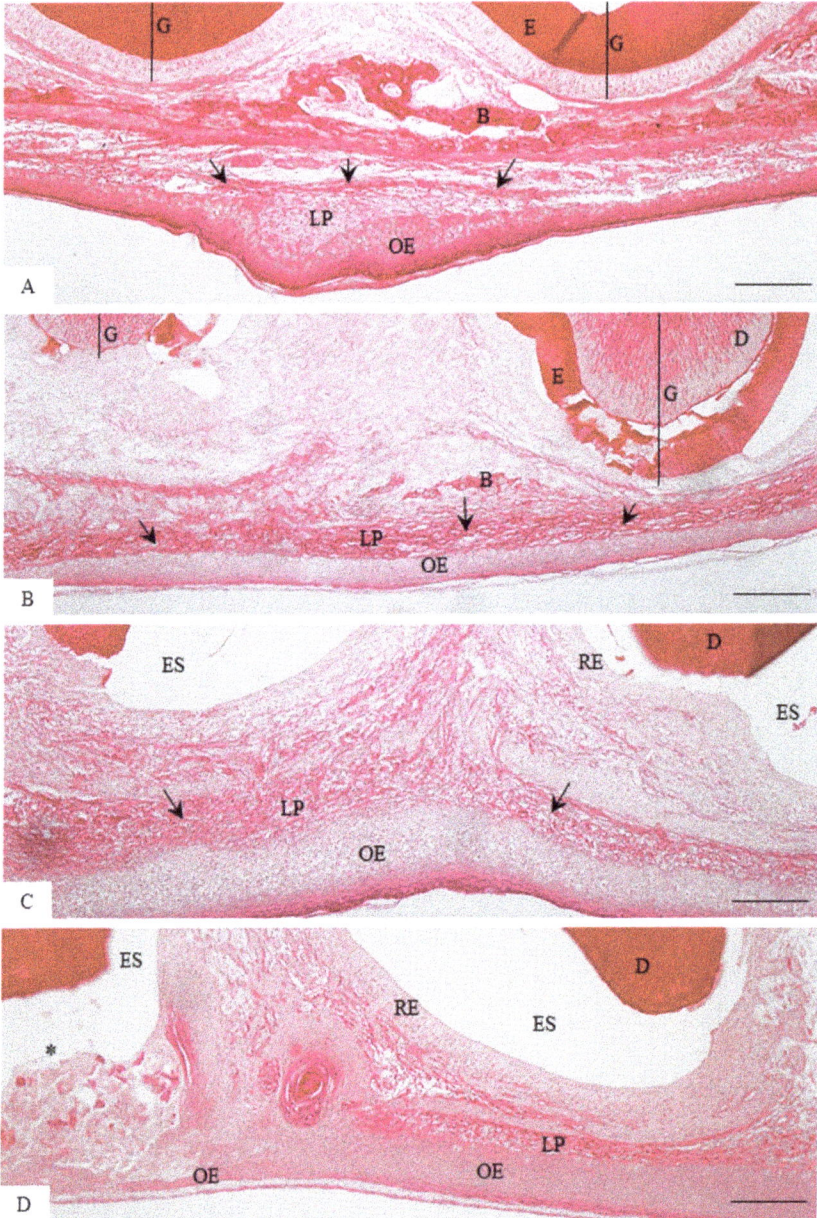

Figure 1. Light micrographs of sagittal sections of maxilla showing portions of the eruptive pathway of the first molar of 9- (**A**), 11- (**B**), 13- (**C**), and 16-day-old (**D**) rats. In **A** and **B**, the lamina propria (LP) contains bundles of collagen fibres (arrows). Note that in the 9-day-old rat (**A**), a continuous layer of bone trabeculae (**B**) is observed between the tooth germ (G) and oral epithelium (OE). In **B**, a thin bone trabecula (**B**) is present in the area between the molar cusps. In **C** and **D**, the delicate collagen fibres (arrows) are irregularly distributed in the thin lamina propria (LP). In **D**, a cusp tip (asterisk) is passing through the oral epithelium (OE). D, dentine; E, enamel matrix; ES, enamel space; RE, reduced enamel epithelium. Stained with picrosirius. Bars: 50 μm.

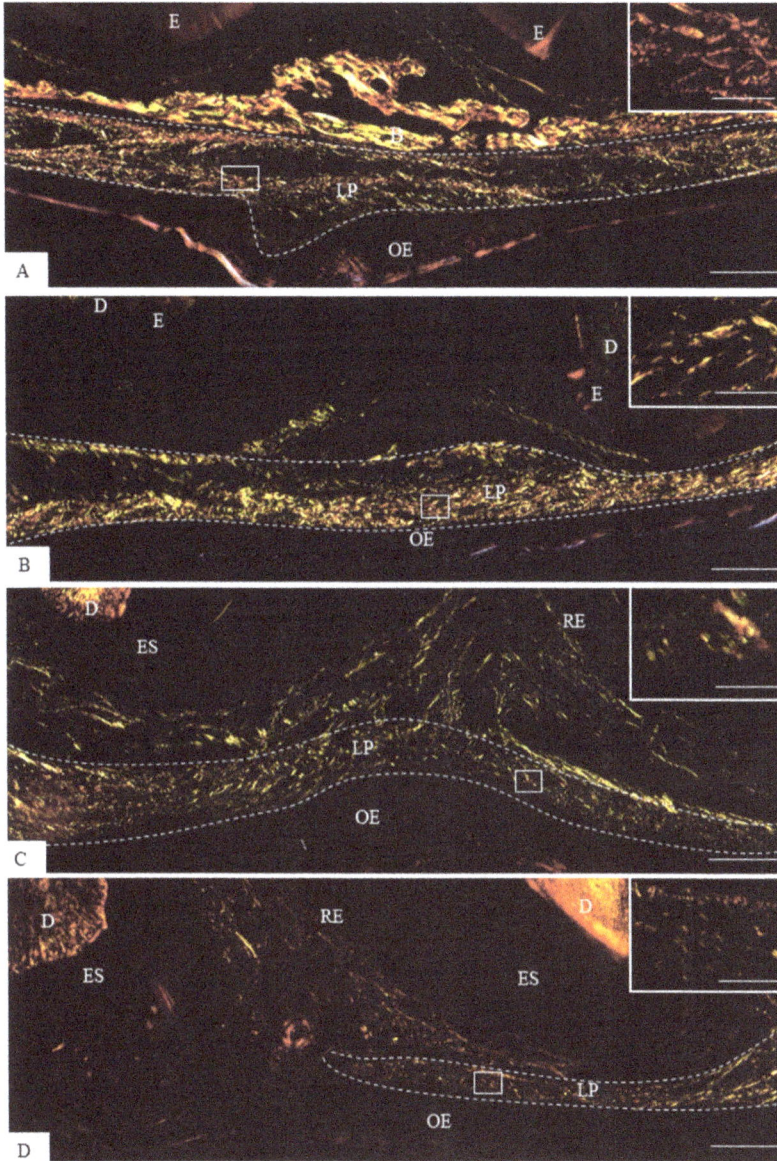

Figure 2. Light micrographs of sagittal sections of maxilla showing portions of the eruptive pathway of the first molar of 9- (**A**), 11- (**B**), 13- (**C**), and 16-day-old (**D**) rats, subjected to the picrosirius-red and analyzed under polarized light. The hatched lines delimit the lamina propria (LP) of the eruptive pathway. Irregularly arranged birefringent collagen fibres are distributed throughout the lamina propria (LP) of 9- (**A**) and 11-day-old (**B**) rats. The lamina propria (LP) contains mainly bundles of birefringent collagen exhibiting red (**A**, inset) and yellow (**B**, inset) colours. **C** and **D**—few birefringent collagen fibres exhibiting the hue varying from yellow to green are present in the lamina propria (LP) of 13- and 16-day-old rats. The insets, outlined areas in **C** and **D**, show thin bundles of birefringent collagen fibres. D, dentine; E, enamel matrix; B, bone trabeculae; OE, oral epithelium; RE, reduced enamel epithelium. Bars: 50 μm and 5 μm (insets).

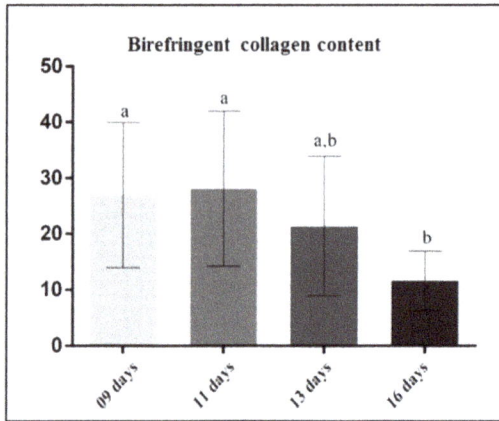

Figure 3. Birefringent collagen content (percentage) in the lamina propria of the eruptive pathway from 9-, 11-, 13-, and 16-day-old rats. Significant reduction in the collagen content is observed in 16-day-old rats in comparison to 9- and 11-day-old rats. Significant difference is not observed between 13- and 16-day-old rats. One way ANOVA and the Tukey post-hoc test ($p \leq 0.05$). Statistically significant difference among groups is indicated by different letters.

Ultrastructural analysis of the lamina propria at the initial phases of tooth eruption exhibited elongated fibroblasts with well-developed rough endoplasmic reticulum and Golgi complex in their large cytoplasm; several bundles of collagen fibrils in the extracellular matrix were observed (Figure 4A). At the advanced stage of tooth eruption (13- and 16-day-old rats), the extracellular matrix showed a granular material and scarce collagen fibrils (Figure 4B). Moreover, fibroblasts exhibiting collagen fibrils in their cytoplasm were also observed (Figure 4C).

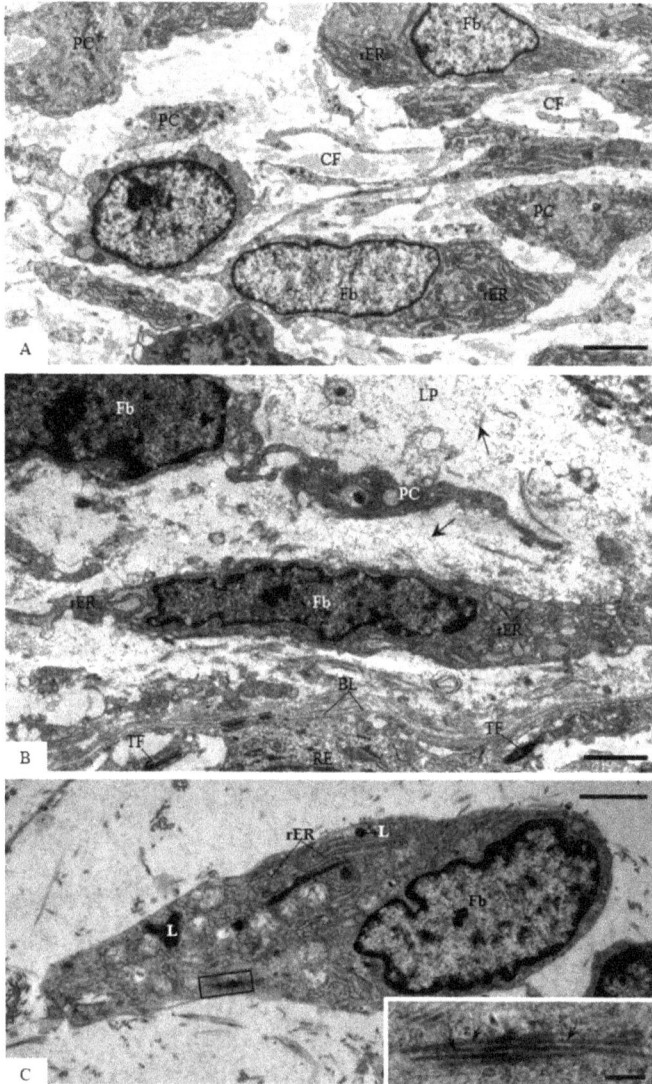

Figure 4. Electron micrographs of portions of lamina propria of the eruptive pathway of first molars of 9- (**A**) and 16- (**B** and **C**) day-old rats. In **A**—fibroblasts (Fb) exhibiting several rough endoplasmic reticulum profiles (rER) in their large cytoplasm are surrounded by numerous collagen fibrils (CF). PC, cell portions. Bar: 5 μm. In **B**—Fibroblasts (Fb) containing few rough endoplasmic reticulum profiles (rER) in the scarce cytoplasm are surrounded by granular material (arrows) in the extracellular matrix. PC, cell portion; RE portion of the reduced enamel epithelium; TF, bundles of tonofilaments; BL, basal lamina. Bar: 5 μm. **C**—a polarized fibroblast (Fb) with irregular nucleus exhibits collagen fibrils (outlined area) apparently internalized in the large cytoplasm. The inset of the outlined area shows the profiles of banded collagen fibrils (arrows). rER, rough endoplasmic reticulum profiles; L, lysosomes. Bars: 3 μm and 0.4 μm (inset).

3.2. MMP-1 Immunoexpression in the Lamina Propria

In all the groups, strong immunolabelling in the cytoplasm (brown-yellow colour) of fibroblasts was observed in the lamina propria (Figure 5). However, an enhanced immunoexpression was evident in the lamina propria of 13- and 16-day-old rats (Figure 5C,D). MMP-1-immunoreaction was also detected in the macrophages and endothelial cells (Figure 5A,B). Moreover, conspicuous immunolabelling was seen in the osteoblasts next to the bone trabeculae located in the eruptive pathway of 9-day-old rats (Figure 5A). In the maxilla sections incubated in the non-immune serum (negative control), no MMP-1-immunolabelled cell was observed (data not shown). According to Figure 6, no significant difference was found in the number of MMP-1-immunolabelled cells/mm^2 between 9- and 11-day-old rats ($p = 0.998$) as well as between 13- and 16-day-old rats ($p = 0.56$). On the other hand, a significant increase ($p < 0.001$) in the immunolabelled cells was observed in the 13- and 16-day-old rats in comparison with the groups of 9- and 11-day-old rats.

A significant correlation ($p \leq 0.01$) was observed between immunoexpression for MMP-1- and collagen content. Moreover, this analysis revealed that these parameters were inversely proportional ($r = -0.614$).

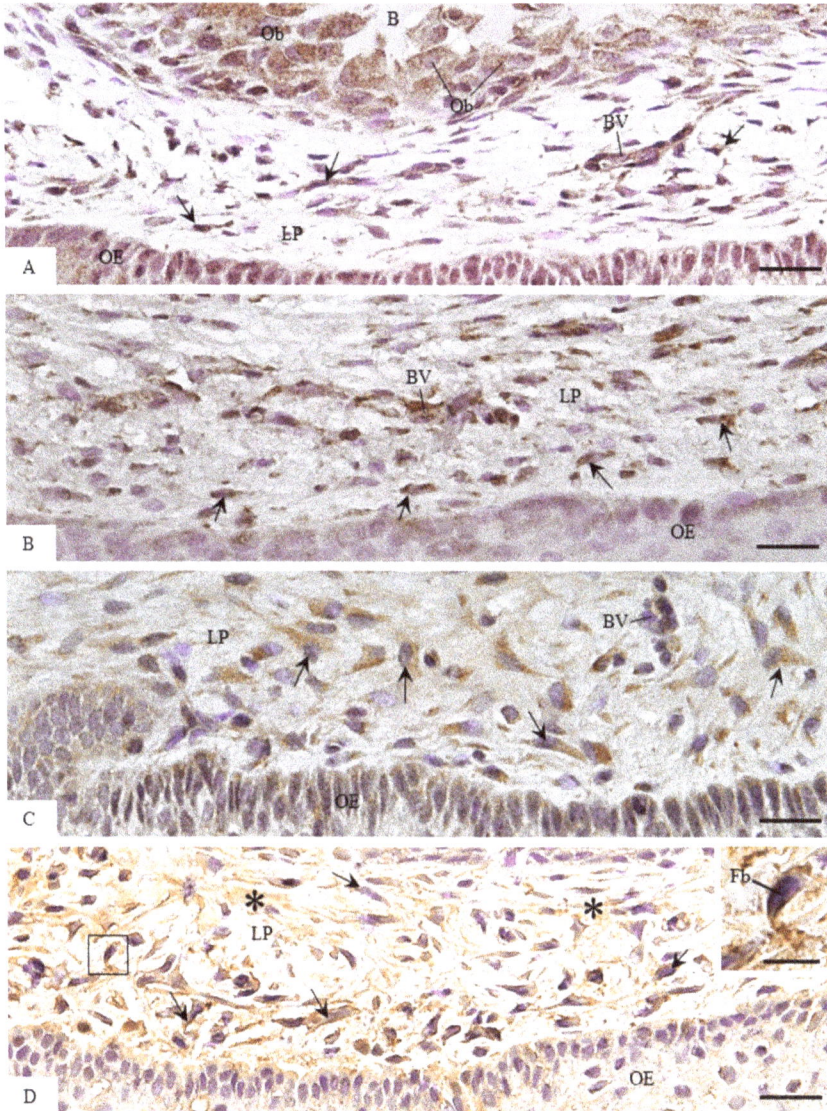

Figure 5. Light micrographs of portions of oral mucosa of the eruptive pathway of first molars of 9- (**A**), 11- (**B**), 13- (**C**), and 16-day-old (**D**) rats. The sections were subjected to the immunohistochemistry for detection of MMP-1 (brown-yellow colour) and counterstained with haematoxylin. Immunolabelled cells (arrows) are observed in the lamina propria (LP) at different stages of tooth eruption. High magnification (inset), outlined area of **D**, shows immunostained fibroblast cytoplasm (Fb). Note an enhanced immunolabelling in the lamina propria (LP) of 13- and 16-day-old rats (**C** and **D**); immunolabelling is observed in the extracellular matrix (asterisks). Ob, osteoblasts; B, bone matrix; OE, oral epithelium; BV, blood vessel. Bars: 20 μm and 6 μm (inset).

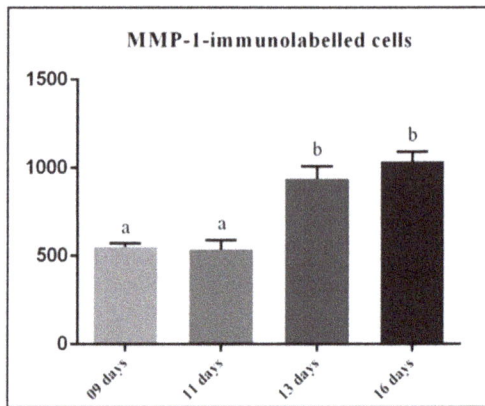

Figure 6. The number of MMP-1-immunolabelled cells per mm^2 of lamina propria of the eruptive pathway from 9-, 11-, 13-, and 16-day-old rats. Significant increase of immunolabelled cells is observed in 13- and 16-day-old rats in comparison to 9- and 11-day-old rats. One way ANOVA and the Tukey post-hoc test ($p \leq 0.05$). Statistically significant difference among groups is indicated by different letters.

3.3. ACP-2 Immunoexpression in the Lamina Propria

Sections subjected to immunohistochemistry for ACP-2 detection revealed a distinct immunolabelling pattern according to the stage of tooth eruption (Figure 7). In the 9- and 11-day-old rats, scarce or no immunolabelled cells were observed in the lamina propria (Figure 7A,B). In the 9-day-old rats, immunolabelling in the osteoclast cytoplasm was often observed (Figure 7A). Occasionally, immunoreaction was also found in the fibroblasts of the eruptive pathway, located next to the dental follicle (Figure 7B). On the other hand, conspicuous immunolabelling was often observed in the cytoplasm of fibroblasts and macrophages in the groups of 13- and 16-day-old rats (Figure 7C,D). In addition, immunostaining was observed in the endothelial cells of the blood vessels in the eruptive pathway (Figure 7C,D). In the maxilla sections used as negative controls, immunolabelled cells were not found (data not shown).

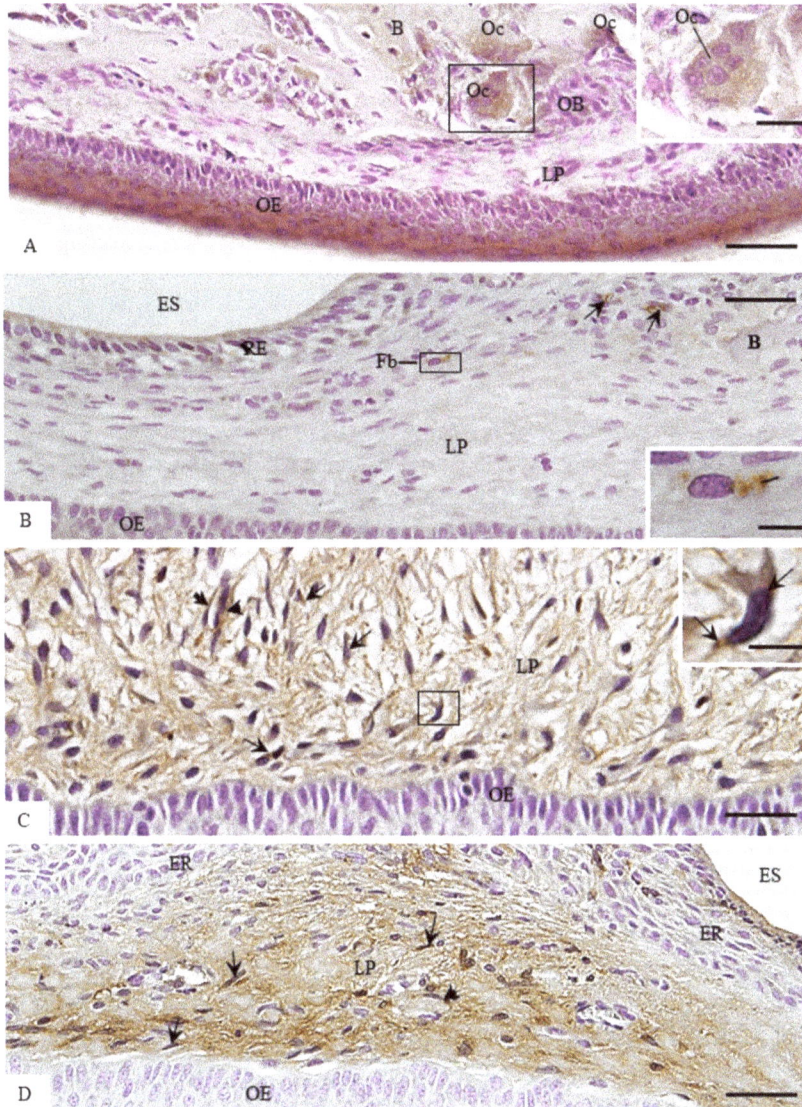

Figure 7. Light micrographs of portions of oral mucosa of the eruptive pathway of first molars subjected to the immunohistochemistry for detection of ACP-2 (brown-yellow colour) and counterstained with haematoxylin. In **A** (9-day-old rat) and **B** (11-day-old rat) scarce immunolabelled cells (arrows) are observed in the eruptive pathway. In **A**, ACP-2 immunolabelled multinucleated osteoclasts (Oc) are present in the bone (**B**) surface. The inset, outlined area in **A**, shows immunolabelled osteoclast (Oc) cytoplasm. LP, lamina propria. Bars: 40 μm and 15 μm (inset). In **B**, an immunolabelled fibroblast (Fb) is observed between the lamina propria (LP) and reduced enamel epithelium (RE). Immunolabelled cells (arrows) are observed next to the bone (B) surface. The inset, outlined area, shows conspicuous immunostaining in the fibroblast cytoplasm (Fb). OE, oral epithelium; ES, enamel space. Bars: 40 μm and 6 μm (inset). In **C** (13-day-old rat) and **D** (16-day-old rat), several immunolabelled fibroblasts (arrows) are seen in the lamina propria (LP). In the inset, outlined area of **C**, strong immunostaining is observed in the cytoplasm of a fibroblast (arrows). ACP-2-positive immunolabelling is observed in the endothelial cells (arrowheads). ES, enamel space; ER, reduced enamel epithelium; OE, oral epithelium. Bars: 40 μm (**C** and **D**) and 6 μm (inset).

3.4. Detection of MMP-1 and ACP-2 by Western Blot

As shown in Figure 8, Western blot analyses of protein extracts from oral mucosa of the eruptive pathway demonstrated evident MMP-1 and ACP2 immunoreactive bands at ~54 KDa and ~55 KDa, respectively, confirming the specificity of the antibodies to rat tissues. A strong band at 42 KDa, corresponding to actin, was also observed (Figure 8).

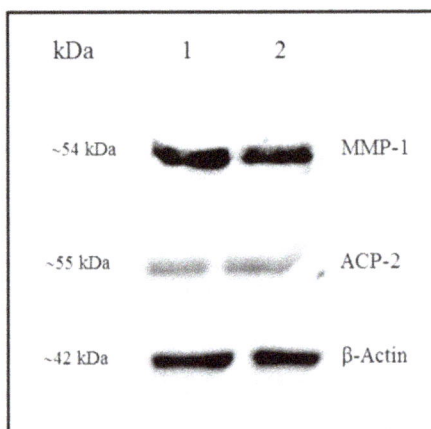

Figure 8. Western blot of MMP-1 and ACP-2 from extracts of oral mucosa of the eruptive pathway obtained from two 16-day-old rats. Bands at 54 kDa and 55 KDa levels, corresponding to MMP-1 and ACP-2 molecular weights, respectively, are observed. Actin bands (~42 KDa) are also seen.

3.5. Ultrastructural Localization of Acid Phosphatase Activity

The ultrathin sections of the specimens incubated in the medium containing β-glycerophosphate revealed electron-opaque deposits of reaction product, i.e., acid phosphatase activity (Figure 9). A distinct pattern of acid phosphatase reaction was observed in the cells of the lamina propria according to the eruptive phase. Few cells exhibiting deposits of reaction product in the lysosomes were observed in the lamina propria during the intraosseous eruptive phase (Figure 9A) whereas conspicuous acid phosphatase activity was seen in the cells during the mucosal penetration phase of tooth eruption (Figure 9B–E). In some irregular-shaped cells—macrophage-like cells, phosphatase acid activity was also seen in the large vacuoles containing partially degraded material (Figure 9C,D). No reaction product was seen in the specimens incubated without β-glycerophosphate (Figure 9F).

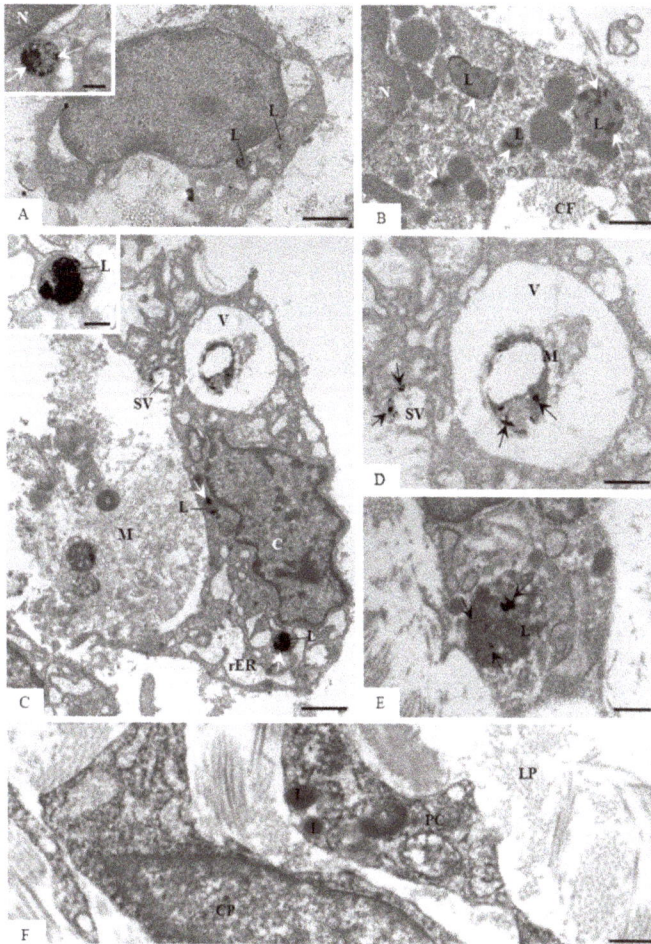

Figure 9. Electron micrographs of portions of lamina propria of the eruptive pathway of first molars of 9- (**A**), 13- (**B** and **F**), 16- (**C–E**) day-old rats. **A–E**—portions of the eruptive pathway of first molars incubated for acid phosphatase reaction. **A**—a round-shaped cell exhibits electron-opaque deposits in the lysosomes (L). In high magnification (inset), conspicuous electron-opaque deposits (arrows)—reaction product of the acid phosphatase activity—is irregularly distributed throughout the lysosome. N, nucleus. Bars: 5 μm and 0.2 μm (inset). In **B**, several lysosomes (L) exhibiting electron-opaque deposits (arrows) are observed in the cytoplasm of a macrophage. N, nucleus; CF, collagen fibrils. Bar: 1.5 μm. **C**—an irregular cell (C) with large vacuole (V) is surrounding partially a heterogeneous material (M). Electron-opaque deposits (arrows) are observed in the lysosomes (L) and in the periphery of a small vacuole (SV). The inset shows conspicuous electron-opaque deposits almost entirely filling a lysosome (L). rER, rough endoplasmic reticulum. Bars: 4 μm and 0.3 μm (inset). **D**—high magnification of the large vacuole (V) observed in the superior portion of the Figure 9C. Fine products of the reaction to acid phosphatase (arrows) are observed on the partially digested material (M). Granular electron-opaque deposits (arrows) are also seen in the periphery of a small vacuole (SV). Bar: 1 μm. Figure 9**E**—a large lysosome (L) exhibiting acid phosphatase-positive deposits (arrows) intermingled with heterogeneous material is present in the fibroblast. Bar: 0.5 μm. **F**—ultrathin section of a portion of the eruptive pathway incubated in substrate-free medium (negative control). No reaction product is observed in the lysosome (L) of cellular portions (CP) in the lamina propria (LP). Bar: 1.5 μm.

4. Discussion

The significant increase in the MMP-1 immunoexpression in parallel to accentuated reduction in the birefringent collagen content during the mucosal penetration phase of tooth eruption indicates that MMP-1 is involved in the degradation of the extracellular matrix (ECM) components of the lamina propria of the eruptive pathway. Our findings also indicate that ECM components degraded by MMP-1 are engulfed by macrophages and fibroblasts of the lamina propria and digested by acid phosphatase activity within these cells, as summarized in Figure 10.

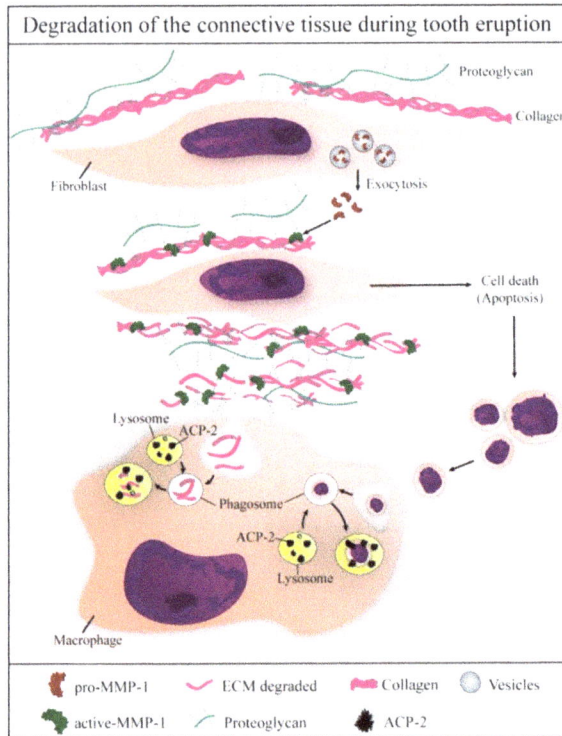

Figure 10. Schematic figure showing some mechanisms involved in the degradation of ECM of eruptive pathway during the normal tooth eruption. Fibroblasts produce and release pro-MMP-1, which is activated in the extracellular microenvironment and becomes active-MMP-1. This enzyme degrades collagen fibrils surrounding fibroblasts and other cells, which undergo apoptosis. Either collagen fragments or apoptotic bodies are engulfed by neighbouring macrophages. Phagosomes containing collagen and/or apoptotic bodies are fused with lysosomes containing lysosomal enzymes, such as ACP-2, which allows the intracellular digestion of these eruptive pathway components.

MMP-1 has an important participation during development as well as in the maintenance, repair and regeneration of several tissues and organs. During growth of fly Mmp-1 mutants, tracheal tubes cannot expand and dilate properly to follow the larval growth and, consequently, they break from the stretching tension. This disturbance in tracheal tube growth is caused by failure of degradation of attachment components between cells and ECM due to lack of MMP-1 [42]. MMP-1 expression by keratinocytes at onset of healing seems to be essential to migration and orientation of these cells during reepithelialization [43]. When MMP-1 activity is blocked by TIMP-1, the apoptosis is inhibited indicating that this metalloproteinase also exerts a control under the cell survival [44]. On the other hand, overexpression of MMP-1 leads to tumour progression and metastasis [45]. Elevated MMP-1 expression has been associated with several tumours such as peritoneal metastasis in gastric cancer [46], colorectal cancer [47], and cutaneous melanoma cancer [47,48].

In the present study, the accentuated MMP-1 immunoexpression in association with the significant reduction in the birefringent collagen content indicates that this enzyme exerts an intense activity in the eruptive pathway mainly at the mucosal penetration stage. Moreover, the reduction in the birefringent content was accompanied by changes in the colour pattern exhibited by the birefringent collagen in the lamina propria. At the intraosseous eruptive phase (9- and 11-day-old rats), the thick bundles of collagen fibres exhibited predominantly red/orange and yellow colours whereas in the mucosal penetration phase (13- and 16-day-old rats), the yellow and green birefringence of thin collagen fibres were often seen. Although the birefringence colour is not useful to identify the molecular collagen nature [49,50], it has been suggested that birefringence colour reflects the collagen fibre diameter. Thus, the birefringence colours from red to orange, to yellow, to green correspond to decreasing fibre diameter [50,51]. Here, few thin collagen bundles exhibiting yellow or green birefringence were often observed at the advanced stage of tooth eruption, indicating that the thick bundles of fibres with red/orange or yellow birefringence present in the intraosseous stage were at least in part degraded during the establishment of the eruptive pathway. In fact, a significant increase in the immunoexpression of MMP-1 was detected in the eruptive pathway of 13- and 16-day-old rats. The specificity of the MMP-1 antibody to rat tissues was confirmed by Western blot, which revealed bands at ~54 KDa, corresponding to MMP-1 [15].

Our findings showed immunolabelling in the cytoplasm of fibroblasts and macrophages of the lamina propria at the different eruptive phases. It is known that MMP-1 is responsible for degradation of collagen type I, II, III, V, and XI [52]; this MMP is produced and released by different cell types, including fibroblasts, macrophages, and neutrophils [19,53,54]. Usually the MMPs are synthesized as inactive proenzymes and are activated by several factors such as FGF (fibroblast growing factor), TGF (transforming growing factor) and IL-1 (interleukin-1) in the extracellular matrix [55–57]. Moreover, tissue inhibitors of metalloproteinases (TIMPs) regulate the proteolytic activity of the MMPs, and the balance between MMP/TIMPs is responsible for extracellular matrix turnover [26,58]. Evidence indicates that MMPs have pivotal role in the tissue degradation of the eruptive pathway [9,26] and decreased expression of these enzymes may impair the tooth eruption and induces, for example, non-syndromic hyperplastic dental follicle [26]. Failure of tooth eruption and odontoma-like structures formation in the op/op (osteopetrotic) mouse were associated with intense disordered ECM remodelling [59], indicating that MMP/TIMP unbalance is responsible for tooth eruption disturbance and may lead to formation of tumour lesions.

During tooth eruption, an accentuated expression of IL-1 [56] and TNF-α [57] has been demonstrated in the dental follicle. Among several functions, these cytokines released by cells from the dental follicle stimulate the osteoclast formation and subsequent bone resorption during tooth eruption [11]. It is possible that these cytokines also promote the cleavage of the pro-MMP-1, activating this metalloproteinase in the eruptive pathway. Differences in the pattern of MMP-9 immunoexpression have also been reported in the eruptive pathway of rat molars; an enhanced immunoexpression has been seen at the mucosal penetration phase, particularly in the 16-day-old rats [9]. MMP-9 is responsible for degradation of amorphous components of the extracellular matrix and denatured type I collagen. Thus, the accentuated immunoexpression at the advanced stage of tooth eruption indicates that MMP-9 acts in the degradation of the lamina propria [9].

Our findings showed a significant increase in the MMP-1-immunolabelled cells over time, indicating that this enzyme may be responsible for collagen breakdown of the lamina propria (Figure 10), leading to reduction of the collagen content as well as changes in the collagen thickness at the mucosal penetration phase of the tooth eruption. This idea is supported by the significant correlation between MMP-1 immunoexpression and collagen content. In fact, the collagen content of the lamina propria reduced 57% from 9 days to 16 days, and this reduction was accompanied by an increase in the number of MMP-1-immunolabelled cells (89%) in the lamina propria. At advanced eruptive stage, the ECM showed an evident immunoreactivity for MMP-1, and the ultrastructural analysis revealed scarce collagen fibrils and granular material surrounding the cells in the lamina propria confirming that the extracellular matrix components are undergoing degradation. It is important to emphasize that at 16 days the first rat molars are passing through the oral mucosa and, therefore, structural changes are necessary in the lamina propria to allow the tooth eruption. Among these structural changes in the lamina propria, we have demonstrated an accentuated apoptosis index and reduction in the blood vessel profiles during mucosal penetration phase of the tooth eruption [10]. In the present study, the increase in the number of MMP-1-immunolabelled cells at the mucosal penetration stage, confirms the role of MMP-1 in the degradation of the eruptive pathway. Thus, the MMPs, including the MMP-1, exert a crucial role in the degradation of the eruptive pathway and the deficiency of these enzymes may impair tooth eruption. This hypothesis is reinforced by the fact that a marked decrease in the expression of MMP-1 and MMP-3 was detected in the thick fibrous connective tissue of hyperplastic dental follicle [26].

Our findings revealed immunolabelling in the endothelial cells suggesting a possible participation of MMP-1 in the microvasculature remodelling. In addition to the well-established role of MMP-1 in the ECM components breakdown, it has been reported that MMPs can also regulate the vascular proliferation [60–62]. Thus, it is conceivable to suggest that MMP-1 has also a role in the rearrangement of the vascular plexus since the involution of blood vessel in the eruptive pathway may be concomitant with vascular proliferation necessary for gingiva development during tooth eruption [10]. It is important to emphasize that a marked increase of blood vessels has been reported in opercular lesions in non-erupted human permanent molars [5]. Thus, there is a consensus regarding the fact that any disturbance in the breakdown of the oral mucosa of the eruptive pathway may delay the tooth eruption [63].

Regarding the ACP-2 immunohistochemical detection, scarce immunolabelled cells were observed in the lamina propria at the intraosseous eruptive phase, whereas an evident immunostaining was detected in the osteoclasts adjacent to the bone surface overlaying the developing tooth germ of 9- and 11-day-old rats. It is known that osteoclasts express high levels of acid phosphatases, including the tartrate-resistant acid phosphatase (TRAP), also named ACP5 [64–66], and the lysosomal acid phosphatase, ACP-2. ACP-2 is an important enzyme for lysosomal function, responsible for the hydrolysis of orthophosphoric monoesters to alcohol and phosphate [33]. Here, the monoclonal antibody raised against recombinant ACP-2 of human origin was reactive to rat tissues, as demonstrated by Western blot analysis, which revealed bands at ~55 KDa [33].

A conspicuous immunostaining for ACP-2 in the fibroblasts and macrophages observed in the 13- and 16-day-old rats indicates that these cells act in the degradation process of the lamina propria during the mucosal penetration phase of tooth eruption. Moreover, ACP-2 immunolabelling was also observed in the endothelial cells as also described in other tissues [67,68]. Here, the presence of ACP-2 in the endothelial cells, particularly at the advanced stages of tooth eruption, indicates that this enzyme is involved in the vascular remodelling, which is necessary during tooth eruption. The apoptosis of vascular cells together with the significant reduction in the blood vessel profiles in the lamina propria indicates that a rearrangement of the microvascular plexus occurs in the eruptive pathway at the mucosal penetration phase [10]. It is known that apoptotic bodies can be recognized and engulfed by neighbouring cells [10,30,40,63,69]. Therefore, it is possible that the ACP-2 immunoexpression in the endothelial cells may reflect an increase of phagocytosis and intracellular digestion of apoptotic vascular cells.

In the present study, the immunolocalization of ACP-2 was consistent with the ultrastructural localization of the acid phosphatase activity. The ultrathin sections from specimens incubated with β-glycerophosphate showed electron-opaque deposits irregularly distributed in the lysosomes of fibroblasts and macrophages, indicating ultrastructural features of reaction product of acid phosphatase [28,41,70,71]. The acid phosphatase activity is found in different cell types and has been associated with intracellular digestion, including the degradation of material internalized by cells [28,33,71,72]. The high activity of this enzyme, also named Lysosomal Acid Phosphatase (LAP), has been detected under certain circumstances, such as formation of tissues and organs including rat nervous tissue [73] and amphibian vitelline vesicle [74], as well as in the tissue remodelling, such as in rat endometrium [29] and periodontium [28], reinforcing the concept that the acid phosphatase plays an important role in the tissue degradation/remodelling. Moreover, the ACP-2 knockout promotes skeletal abnormalities during rat development [75], and mutations in ACP-2 gene are associated with cerebellar malformations, suggesting a critical role of ACP-2 in tissue development [33]. Here, the positive reaction to acid phosphatase was present in the fibroblasts and macrophages, mainly in the 13- and 16-day-old rats, i.e., at mucosal penetration stage of the tooth eruption. Usually, these cells were surrounded by extracellular matrix exhibiting a granular and flocculent material, indicating the occurrence of degradation of its components. The presence of material inside macrophage and fibroblast vacuoles exhibiting acid phosphatase activity indicates that these cells may be digesting the remnants of extracellular matrix. Therefore, the ECM components are degraded by different MMPs, including MMP-1, and the remnants are engulfed and digested by macrophages and fibroblasts. Once inside a vacuole, the remnants are degraded by lysosomal enzymes such as acid phosphatase (Figure 10). Moreover, we cannot exclude the possibility that acid phosphatase reactivity inside large vacuoles may also be involved in the intracellular digestion of apoptotic bodies, as described in other tissues and organs [76,77]. It is known that changes in the ECM affect cell survival and proliferation since the interactions between ECM components and cell surface molecules regulate cell behaviour [42,78]. ECM fragments derived from cleavage by MMPs induce apoptosis of mammary epithelial cells [79], indicating that changes in the ECM microenvironment may promote apoptosis. Here, an enhanced immunoexpression of MMP-1 and ACP-2 was detected in the eruptive pathway at the mucosal penetration phase (13- and 16-day-old rats), suggesting a coordinate interaction between theses enzymes since MMP-1 cleaves ECM substrates, and ACP-2 is responsible for intracellular digestion (Figure 10). In fact, the number of apoptotic cells in the lamina propria increases significantly at the mucosal penetration phase of eruption when macrophages and fibroblasts engulfing apoptotic bodies were found [10], and acid phosphatase activity was observed inside large vacuoles containing digesting remnants material. Therefore, the establishment of the eruptive pathway for the passage of erupting teeth involves a coordinated cascade of cellular and molecular events, culminating in a rapid and programmed degradation of cellular and ECM components of the lamina propria.

In conclusion, the collagen content of the lamina propria reduces significantly during tooth eruption due to degradation of the extracellular matrix components by the MMP-1. The enhanced acid phosphatase activity in the fibroblasts and macrophages of the lamina propria during the mucosal penetration stage points to an important role of this enzyme in the intracellular digestion of extracellular remnants, being involved in the establishment of the eruptive pathway. Although the present study was performed in rodent model, it is well stated that the tooth eruption in rat molars is similar to human, including the stages of eruption. Thus, despite the limitations of extrapolating the results obtained in rodents to humans, our findings indicate that any disturb in MMP-1 and/or ACP-2 expression may delay or impair tooth eruption. Further studies, including the evaluation of the control and participation of MMPs and ACPs during tooth eruption are necessary for the better understanding of the causes of delayed tooth eruption, mainly when this disturb is not of physical origin.

Author Contributions: P.S.C. coordinated the study. J.P.d.P.J. performed the histological processing, morphometric analyses, immunohistochemistry, WB analysis and statistical analyses. J.P.d.P.J. and P.S.C. prepared the specimens for analysis under TEM. P.S.C. and E.S.-C. analyzed and interpreted the ultrastructural images. All authors participate in the writing and editing of the manuscript. All authors read and approved the final manuscript.

Funding: This research was supported by public funding from São Paulo Research Foundation (FAPESP: 2011/23064-9; 2016/09264-9), National Council for Scientific and Technological Development (CNPq) and CAPES (Finance code 001), Brazil.

Acknowledgments: The authors thank Luis Antônio Potenza and Pedro Sérgio Simões for the microtomy of paraffin blocks. Authors also thank Fabiane de Santi for the Western blot assistance and Carlos Rossa and Morgana Rodrigues Guimarães Stabili for the use of the Digital Documentation System (Western Blot analysis).

Conflicts of Interest: The authors declared no potential conflicts of interest with respect to the research, authorship, and/or publication of this article.

References

1. Kurol, J. Impacted and ankylosed teeth: Why, when, and how to intervene. *Am. J. Orthodont. Dent. Orthop.* **2006**, *129*, S86–S90. [CrossRef] [PubMed]
2. Loriato, L.B.; Machado, A.W.; Souki, B.Q.; Pereira, T.J. Late diagnosis of dentoalveolar ankylosis: Impact on effectiveness and efficiency of orthodontic treatment. *Am. J. Orthodont. Dent. Orthop.* **2009**, *135*, 799–808. [CrossRef] [PubMed]
3. Frazier-Bowers, S.A.; Puranik, C.P.; Mahaney, M.C. The etiology of eruption disorders—Further evidence of a 'genetic paradigm'. *Semin. Orthodont.* **2010**, *16*, 180–185. [CrossRef] [PubMed]
4. Kreiborg, S.; Jensen, B.L. Tooth formation and eruption—Lessons learnt from cleidocranial dysplasia. *Eur. J. Oral Sci.* **2018**, *126*, 72–80. [CrossRef] [PubMed]
5. Verma, S.; Arul, A.S.; Arul, A.S.; Chitra, S. Erupted complex odontoma of the posterior maxilla: A rarity. *J. Nat. Sci. Biol. Med.* **2015**, *6*, S167–S169. [PubMed]
6. Marks, S.C., Jr.; Schroeder, H.E. Tooth eruption: Theories and facts. *Anat. Rec.* **1996**, *245*, 374–393. [CrossRef]
7. Kjær, I. Mechanism of human tooth eruption: Review article including a new theory for future studies on the eruption process. *Scientifica* **2014**, *2014*, 341905. [CrossRef] [PubMed]
8. Wang, X.P. Tooth eruption without roots. *J. Dent. Res.* **2013**, *92*, 212–214. [CrossRef] [PubMed]
9. Cerri, P.S.; Pereira-Júnior, J.A.; Biselli, N.B.; Sasso-Cerri, E. Mast cells and MMP-9 in the lamina propria during eruption of rat molars: Quantitative and immunohistochemical evaluation. *J. Anat.* **2010**, *217*, 116–125. [CrossRef] [PubMed]
10. De Pizzol Júnior, J.P.; Sasso-Cerri, E.; Cerri, P.S. Apoptosis and reduced microvascular density of the lamina propria during tooth eruption in rats. *J. Anat.* **2015**, *227*, 487–496. [CrossRef] [PubMed]
11. Wise, G.E. Cellular and molecular basis of tooth eruption. *Orthodont. Craniofac. Res.* **2009**, *12*, 67–73. [CrossRef] [PubMed]
12. Liu, D.; Yao, S.; Wise, G.E. MyD88 expression in the rat dental follicle: Implications for osteoclastogenesis and tooth eruption. *Eur. J. Oral Sci.* **2010**, *118*, 333–341. [CrossRef] [PubMed]
13. Liu, D.; Yao, S.; Wise, G.E. Regulation of SFRP-1 expression in the rat dental follicle. *Connect. Tissue Res.* **2012**, *53*, 366–372. [CrossRef] [PubMed]
14. Castaneda, B.; Simon, Y.; Jacques, J.; Hess, E.; Choi, Y.W.; Blin-Wakkach, C.; Mueller, C.; Berdal, A.; Lézot, F. Bone resorption control of tooth eruption and root morphogenesis: Involvement of the receptor activator of NF-κB (RANK). *J. Cell. Physiol.* **2011**, *226*, 74–85. [CrossRef] [PubMed]
15. Klein, T.; Bischoff, R. Physiology and pathophysiology of matrix metalloproteases. *Amino Acids* **2011**, *41*, 271–290. [CrossRef] [PubMed]
16. Rohani, M.G.; Parks, W.C. Matrix remodeling by MMPs during wound repair. *Matrix Biol.* **2015**, *44–46*, 113–121. [CrossRef] [PubMed]
17. Pittayapruek, P.; Meephansan, J.; Prapapan, O.; Komine, M.; Ohtsuki, M. Role of matrix metalloproteinases in photoaging and photocarcinogenesis. *Int. J. Mol. Sci.* **2016**, *17*, 868. [CrossRef] [PubMed]
18. Golestani, R.; Razavian, M.; Ye, Y.; Zhang, J.; Jung, J.J.; Toczek, J.; Gona, K.; Kim, H.Y.; Elias, J.A.; Lee, C.G.; et al. Matrix metalloproteinase-targeted imaging of lung inflammation and remodeling. *J. Nucl. Med.* **2017**, *58*, 138–143. [CrossRef] [PubMed]
19. De Oliveira, P.A.; de Pizzol-Júnior, J.P.; Longhini, R.; Sasso-Cerri, E.; Cerri, P.S. Cimetidine reduces interleukin-6, matrix metalloproteinases-1 and -9 immunoexpression in the gingival mucosa of rat Molars with induced periodontal disease. *J. Periodontol.* **2017**, *88*, 100–111. [CrossRef] [PubMed]
20. Aalinkeel, R.; Nair, B.B.; Reynolds, J.L.; Sykes, D.E.; Mahajan, S.D.; Chadha, K.C.; Schwartz, S.A. Overexpression of MMP-9 contributes to invasiveness of prostate cancer cell line LNCaP. *Immunol. Invest.* **2011**, *40*, 447–464. [CrossRef] [PubMed]
21. Łukaszewicz-Zając, M.; Mroczko, B.; Słowik, A. Matrix metalloproteinases (MMPs) and their tissue inhibitors (TIMPs) in amyotrophic lateral sclerosis (ALS). *J. Neural Transm.* **2014**, *121*, 1387–1397. [CrossRef] [PubMed]

22. Vincenti, M.P.; White, L.A.; Schroen, D.J.; Benbow, U.; Brinckerhoff, C.E. Regulating expression of the gene for matrix metalloproteinase-1 (collagenase): Mechanisms that control enzyme activity, transcription, and mRNA stability. *Crit. Rev. Eukaryot. Gene Expr.* **1996**, *6*, 391–411. [CrossRef] [PubMed]

23. Visse, R.; Nagase, H. Matrix metalloproteinases and tissue inhibitors of metalloproteinases: Structure, function, and biochemistry. *Circ. Res.* **2003**, *92*, 827–839. [CrossRef] [PubMed]

24. Brinckerhoff, C.E.; Rutter, J.L.; Benbow, U. Interstitial collagenases as markers of tumor progression. *Clin. Cancer Res.* **2000**, *6*, 4823–4830. [PubMed]

25. Bartlett, J.D.; Zhou, Z.; Skobe, Z.; Dobeck, J.M.; Tryggvason, K. Delayed tooth eruption in membrane type-1 matrix metalloproteinase deficient mice. *Connect. Tissue Res.* **2003**, *44*, 300–304. [CrossRef] [PubMed]

26. Kim, S.G.; Kim, M.H.; Chae, C.H.; Jung, Y.K.; Choi, J.Y. Downregulation of matrix metalloproteinases in hyperplastic dental follicles results in abnormal tooth eruption. *BMB Rep.* **2008**, *41*, 322–327. [CrossRef] [PubMed]

27. Bull, H.; Murray, P.G.; Thomas, D.; Fraser, A.M.; Nelson, P.N. Acid phosphatases. *Mol. Pathol.* **2002**, *55*, 65–72. [CrossRef] [PubMed]

28. Deporter, D.A.; Ten Cate, A.R. Fine structural localization of acid and alkaline phosphatase in collagen-containing vesicles of fibroblasts. *J. Anat.* **1973**, *114*, 457–461. [PubMed]

29. Katz, S.G. Extracellular breakdown of collagen by mice decidual cells. A cytochemical and ultrastructural study. *Biocell* **2005**, *29*, 261–270. [PubMed]

30. Cerri, P.S.; Freymüller, E.; Katchburian, E. Apoptosis in the early developing periodontium of rat molars. *Anat. Rec.* **2000**, *258*, 136–144. [CrossRef]

31. Moss, D.W.; Raymond, F.D.; Wile, D.B. Clinical and biological aspects of acid phosphatase. *Crit. Rev. Clin. Lab. Sci.* **1995**, *32*, 431–467. [CrossRef] [PubMed]

32. Suter, A.; Everts, V.; Boyde, A.; Jones, S.J.; Lüllmann-Rauch, R.; Hartmann, D.; Hayman, A.R.; Cox, T.M.; Evans, M.J.; Meister, T.; et al. Overlapping functions of lysosomal acid phosphatase (LAP) and tartrate-resistant acid phosphatase (Acp5) revealed by doubly deficient mice. *Development* **2001**, *128*, 4899–4910. [PubMed]

33. Bailey, K.; Balaei, M.R.; Mannan, A.; Del Bigio, M.R.; Marzban, H. Purkinje cell compartmentation in the cerebellum of the lysosomal Acid phosphatase 2 mutant mouse (nax-naked-ataxia mutant mouse). *PLoS ONE* **2014**, *9*, e94327. [CrossRef] [PubMed]

34. Squier, C.A.; Kremer, M.J. Biology of oral mucosa and esophagus. *J. Natl. Cancer Inst. Monogr.* **2001**, 7–15. [CrossRef]

35. Marynka-Kalmani, K.; Treves, S.; Yafee, M.; Rachima, H.; Gafni, Y.; Cohen, M.A.; Pitaru, S. The lamina propria of adult human oral mucosa harbors a novel stem cell population. *Stem Cells.* **2010**, *28*, 984–995. [CrossRef] [PubMed]

36. Junqueira, L.C.; Bignolas, G.; Brentani, R.R. Picrosirius staining plus polarization microscopy, a specific method for collagen detection in tissue sections. *Histochem. J.* **1979**, *11*, 447–455. [CrossRef] [PubMed]

37. Manni, M.L.; Czajka, C.A.; Oury, T.D.; Gilbert, T.W. Extracellular matrix powder protects against bleomycin-induced pulmonary fibrosis. *Tissue Eng. Part A* **2011**, *17*, 2795–2804. [CrossRef] [PubMed]

38. Rich, L.; Whittaker, P. Collagen and picrosirius red staining: A polarized light assessment of fibrillar hue and spatial distribution. *Braz. J. Morphol. Sci.* **2005**, *22*, 97–104.

39. Koshimizu, J.Y.; Beltrame, F.L.; de Pizzol, J.P., Jr.; Cerri, P.S.; Caneguim, B.H.; Sasso-Cerri, E. NF-kB overexpression and decreased immunoexpression of AR in the muscular layer is related to structural damages and apoptosis in cimetidine-treated rat vas deferens. *Reprod. Biol. Endocrinol.* **2013**, *11*. [CrossRef] [PubMed]

40. Cerri, P.S. Osteoblasts engulf apoptotic bodies during alveolar bone formation in the rat maxilla. *Anat. Rec. A Discov. Mol. Cell. Evol. Biol.* **2005**, *286*, 833–840. [CrossRef] [PubMed]

41. Barka, T. Electron histochemical localization of acid phosphatase activity in the small intestine of mouse. *J. Histochem. Cytochem.* **1964**, *12*, 229–238. [CrossRef] [PubMed]

42. Page-McCaw, A.; Ewald, A.J.; Werb, Z. Matrix metalloproteinases and the regulation of tissue remodelling. *Nat. Rev. Mol. Cell Biol.* **2007**, *8*, 221–233. [CrossRef] [PubMed]

43. Pilcher, B.K.; Dumin, J.A.; Sudbeck, B.D.; Krane, S.M.; Welgus, H.G.; Parks, W.C. The activity of collagenase-1 is required for keratinocyte migration on a type I collagen matrix. *J. Cell Biol.* **1997**, *137*, 1445–1457. [CrossRef] [PubMed]

44. Limb, G.A.; Matter, K.; Murphy, G.; Cambrey, A.D.; Bishop, P.N.; Morris, G.E.; Khaw, P.T. Matrix metalloproteinase-1 associates with intracellular organelles and confers resistance to lamin A/C degradation during apoptosis. *Am. J. Pathol.* **2005**, *166*, 1555–1563. [CrossRef]

45. Bachmeier, B.E.; Nerlich, A.G.; Lichtinghagen, R.; Sommerhoff, C.P. Matrix metalloproteinases (MMPs) in breast cancer cell lines of different tumorigenicity. *Anticancer Res.* **2001**, *21*, 3821–3828. [PubMed]

46. Inoue, T.; Yashiro, M.; Nishimura, S.; Maeda, K.; Sawada, T.; Ogawa, Y.; Sowa, M.; Chung, K.H. Matrix metalloproteinase-1 expression is a prognostic factor for patients with advanced gastric cancer. *Int. J. Mol. Med.* **1999**, *4*, 73–77. [CrossRef] [PubMed]

47. Ghilardi, G.; Biondi, M.L.; Mangoni, J.; Leviti, S.; DeMonti, M.; Guagnellini, E.; Scorza, R. Matrix metalloproteinase-1 promoter polymorphism 1G/2G is correlated with colorectal cancer invasiveness. *Clin. Cancer Res.* **2001**, *7*, 2344–2346. [PubMed]

48. Nikkola, J.; Vihinen, P.; Vlaykova, T.; Hahka-Kemppinen, M.; Kähäri, V.M.; Pyrhönen, S. High expression levels of collagenase-1 and stromelysin-1 correlate with shorter disease-free survival in human metastatic melanoma. *Int. J. Cancer* **2002**, *97*, 432–438. [CrossRef] [PubMed]

49. Piérard, G.E. Sirius red polarization method is useful to visualize the organization of connective tissues but not the molecular composition of their fibrous polymers. *Matrix* **1989**, *9*, 68–71. [CrossRef]

50. Junqueira, L.C.; Montes, G.S.; Sanchez, E.M. The influence of tissue section thickness on the study of collagen by the picrosirius-polarization method. *Histochemistry* **1982**, *74*, 153–156. [CrossRef] [PubMed]

51. Pérez-Tamayo, R.; Montfort, I. The susceptibility of hepatic collagen to homologous collagenase in human and experimental cirrhosis of the liver. *Am. J. Pathol.* **1980**, *100*, 427–442. [PubMed]

52. Amălinei, C.; Căruntu, I.D.; Giuşcă, S.E.; Bălan, R.A. Matrix metalloproteinases involvement in pathologic conditions. *Rom. J. Morphol. Embryol.* **2010**, *51*, 215–228. [PubMed]

53. Butoi, E.; Gan, A.M.; Tucureanu, M.M.; Stan, D.; Macarie, R.D.; Constantinescu, C.; Calin, M.; Simionescu, M.; Manduteanu, I. Cross-talk between macrophages and smooth muscle cells impairs collagen and metalloprotease synthesis and promotes angiogenesis. *Biochim. Biophys. Acta* **2016**, *1863*, 1568–1578. [CrossRef] [PubMed]

54. Du, G.L.; Chen, W.Y.; Li, X.N.; He, R.; Feng, P.F. Induction of MMP-1 and -3 by cyclical mechanical stretch is mediated by IL-6 in cultured fibroblasts of keratoconus. *Mol. Med. Rep.* **2017**, *15*, 3885–3892. [CrossRef] [PubMed]

55. Nagase, H.; Woessner, J.F., Jr. Matrix metalloproteinases. *J. Biol. Chem.* **1999**, *30*, 21491–21494. [CrossRef]

56. Huang, H.; Wise, G.E. Delay of tooth eruption in null mice devoid of the type I IL-1R gene. *Eur. J. Oral. Sci.* **2000**, *108*, 297–302. [CrossRef] [PubMed]

57. Yao, S.; Prpic, V.; Pan, F.; Wise, G.E. TNF-alpha upregulates expression of BMP-2 and BMP-3 genes in the rat dental follicle—Implications for tooth eruption. *Connect. Tissue Res.* **2010**, *51*, 59–66. [CrossRef] [PubMed]

58. Sternlicht, M.D.; Werb, Z. How matrix metalloproteinases regulate cell behavior. *Annu. Rev. Cell Dev. Biol.* **2001**, *17*, 463–516. [CrossRef] [PubMed]

59. Ida-Yonemochi, H.; Noda, T.; Shimokawa, H.; Saku, T. Disturbed tooth eruption in osteopetrotic (op/op) mice: Histopathogenesis of tooth malformation and odontomas. *J. Oral Pathol. Med.* **2002**, *31*, 361–373. [CrossRef] [PubMed]

60. Stetler-Stevenson, W.G. Matrix metalloproteinases in angiogenesis: A moving target for therapeutic intervention. *J. Clin. Invest.* **1999**, *103*, 1237–1241. [CrossRef] [PubMed]

61. Chesler, N.C.; Ku, D.N.; Galis, Z.S. Transmural pressure induces matrix-degrading activity in porcine arteries ex vivo. *Am. J. Physiol.* **1999**, *277*, H2002–H2009. [CrossRef] [PubMed]

62. Galis, Z.S.; Johnson, C.; Godin, D.; Magid, R.; Shipley, J.M.; Senior, R.M; Ivan, E. Targeted disruption of the matrix metalloproteinase-9 gene impairs smooth muscle cell migration and geometrical arterial remodeling. *Circ. Res.* **2002**, *91*, 852–859. [CrossRef] [PubMed]

63. Suri, L.; Gagari, E.; Vastardis, H. Delayed tooth eruption: Pathogenesis, diagnosis, and treatment. A literature review. *Am. J. Orthodont. Dent. Orthop.* **2004**, *126*, 432–445. [CrossRef]

64. Kirstein, B.; Chambers, T.J.; Fuller, K. Secretion of tartrate-resistant acid phosphatase by osteoclasts correlates with resorptive behavior. *J. Cell. Biochem.* **2006**, *98*, 1085–1094. [CrossRef] [PubMed]

65. Solberg, L.B.; Brorson, S.H.; Stordalen, G.A.; Baekkevold, E.S.; Andersson, G.; Reinholt, F.P. Increased tartrate-resistant acid phosphatase expression in osteoblasts and osteocytes in experimental osteoporosis in rats. *Calcif. Tissue Int.* **2014**, *94*, 510–521. [CrossRef] [PubMed]

66. Florencio-Silva, R.; Sasso, G.R.; Sasso-Cerri, E.; Simões, M.J.; Cerri, P.S. Biology of bone tissue: Structure, function, and factors that influence bone cells. *Biomed. Res. Int.* **2015**, *2015*. [CrossRef] [PubMed]
67. Connolly, D.T.; Knight, M.B.; Harakas, N.K.; Wittwer, A.J.; Feder, J. Determination of the number of endothelial cells in culture using an acid phosphatase assay. *Anal. Biochem.* **1986**, *152*, 136–140. [CrossRef]
68. Turner, R.R.; Beckstead, J.H.; Warnke, R.A.; Wood, G.S. Endothelial cell phenotypic diversity. In situ demonstration of immunologic and enzymatic heterogeneity that correlates with specific morphologic subtypes. *Am. J. Clin. Pathol.* **1987**, *87*, 569–575. [CrossRef] [PubMed]
69. Kerr, J.F.; Wyllie, A.H.; Currie, A.R. Apoptosis: A basic biological phenomenon with wide-ranging implications in tissue kinetics. *Br. J. Cancer* **1972**, *26*, 239–257. [CrossRef] [PubMed]
70. Katz, S.G. Demonstration of extracellular acid phosphatase activity in the involuting, antimesometrial decidua in fed and acutely fasted mice by combined cytochemistry and electron microscopy. *Anat. Rec.* **1998**, *252*, 1–7. [CrossRef]
71. Faccioli, C.K.; Chedid, R.A.; Mori, R.H.; Amaral, A.C.; Franceschini-Vicentini, I.B.; Vicentini, C.A. Acid and alkaline phosphatase localization in the digestive tract mucosa of the Hemisorubim platyrhynchos. *Acta Histochem.* **2016**, *118*, 722–728. [CrossRef] [PubMed]
72. Ashtari, N.; Jiao, X.; Rahimi-Balaei, M.; Amiri, S.; Mehr, S.E.; Yeganeh, B.; Marzban, H. Lysosomal acid phosphatase biosynthesis and dysfunction: A mini review focused on lysosomal enzyme dysfunction in brain. *Curr. Mol. Med.* **2016**, *16*, 439–446. [CrossRef] [PubMed]
73. Geier, C.; Kreysing, J.; Boettcher, H.; Pohlmann, R.; von Figura, K. Localization of lysosomal acid phosphatase mRNA in mouse tissues. *J. Histochem. Cytochem.* **1992**, *40*, 1275–1282. [CrossRef] [PubMed]
74. Lemanski, L.F.; Aldoroty, R. Role of acid phosphatase in the breakdown of yolk platelets in developing amphibian embryos. *J. Morphol.* **1977**, *153*, 419–425. [CrossRef] [PubMed]
75. Saftig, P.; Hartmann, D.; Lüllmann-Rauch, R.; Wolff, J.; Evers, M.; Köster, A.; Hetman, M.; von Figura, K.; Peters, C. Mice deficient in lysosomal acid phosphatase develop lysosomal storage in the kidney and central nervous system. *J. Biol. Chem.* **1997**, *272*, 18628–18635. [CrossRef] [PubMed]
76. Pipan, N.; Sterle, M. Cytochemical analysis of organelle degradation in phagosomes and apoptotic cells of the mucoid epithelium of mice. *Histochemistry* **1979**, *59*, 225–232. [CrossRef] [PubMed]
77. Ferguson, D.J.; Anderson, T.J. Ultrastructural observations on cell death by apoptosis in the "resting" human breast. *Virchows Arch. A Pathol. Anat. Histopathol.* **1981**, *393*, 193–203. [CrossRef]
78. Vu, T.H.; Werb, Z. Matrix metalloproteinases: Effectors of development and normal physiology. *Genes Dev.* **2000**, *14*, 2123–2133. [CrossRef] [PubMed]
79. Boudreau, N.; Werb, Z.; Bissell, M.J. Suppression of apoptosis by basement membrane requires three-dimensional tissue organization and withdrawal from the cell cycle. *Proc. Natl. Acad. Sci. USA* **1996**, *93*, 3509–3513. [CrossRef] [PubMed]

cells

MDPI

Review

Heparanase: A Multitasking Protein Involved in Extracellular Matrix (ECM) Remodeling and Intracellular Events

Valentina Masola [1,2], Gloria Bellin [1,3], Giovanni Gambaro [2] and Maurizio Onisto [1,*]

[1] Department of Biomedical Sciences, University of Padova, Viale G. Colombo 3, 35121 Padova, Italy; valentina.masola@unipd.it (V.M.); gloria.bellin@gmail.com (G.B.)

[2] Renal Unit, Department of Medicine, University of Verona, Piazzale Stefani 1, 37126 Verona, Italy; giovanni.gambaro@unicatt.it

[3] Maria Cecilia Hospital, GVM Care and Research, Via Corriera 1, 48033 Cotignola (Ravenna), Italy

* Correspondence: maurizio.onisto@unipd.it; Tel.: +39-049-8276093

Received: 6 October 2018; Accepted: 22 November 2018; Published: 28 November 2018

Abstract: Heparanase (HPSE) has been defined as a multitasking protein that exhibits a peculiar enzymatic activity towards HS chains but which simultaneously performs other non-enzymatic functions. Through its enzymatic activity, HPSE catalyzes the cutting of the side chains of heparan sulfate (HS) proteoglycans, thus contributing to the remodeling of the extracellular matrix and of the basal membranes. Furthermore, thanks to this activity, HPSE also promotes the release and diffusion of various HS-linked molecules like growth factors, cytokines and enzymes. In addition to being an enzyme, HPSE has been shown to possess the ability to trigger different signaling pathways by interacting with transmembrane proteins. In normal tissue and in physiological conditions, HPSE exhibits only low levels of expression restricted only to keratinocytes, trophoblast, platelets and mast cells and leukocytes. On the contrary, in pathological conditions, such as in tumor progression and metastasis, inflammation and fibrosis, it is overexpressed. With this brief review, we intend to provide an update on the current knowledge about the different role of HPSE protein exerted by its enzymatic and non-enzymatic activity.

Keywords: heparanase; extracellular matrix (ECM)

1. Introduction

Heparanase is an endoglycosidase that cleaves heparan sulphate (HS) chains and whose activity contributes to degradation and remodeling of extracellular matrix (ECM). This enzyme is mainly involved in cancer progression [1] but recent studies have added multiple functions to its repertoire [2]. Several extensive reviews addressing the specific roles of heparanase such as in the case of inflammation, autophagy, exosome, and fibrosis [3–6] are available. Thus, the aim of the current review is to give a brief overview summarizing and updating the different aspects of heparanase biology. Collectively, the data presented here support the role of heparanase in multiple biological processes and its involvement in several human diseases beyond cancer.

Extracellular Matrix, Heparan Sulfate Proteoglycans and Heparanase

ECM is composed of two main classes of macromolecules: fibrous proteins and polysaccharide chains belonging to the glycosaminoglycan class (GAG). The fibrous proteins include two groups: one with mainly structural functions (collagen and elastin), and the other with mainly adhesive functions (fibronectin, laminins, nidogens and vitronectin). The GAGs are long linear chains of polysaccharides formed by disaccharide units of acetylated hexosamines (N-acetyl-galactosamine or

N-acetyl-glucosamine) and uronic acids (D-glucuronic acid or L-iduronic acid). When they bind to proteins, they give rise to proteoglycans (PGs) which can be rich in sulfate groups with a high negative charge (chondroitin sulfate, dermatan sulfate, heparansulfate and keratansulfate). The high structural heterogeneity of PGs is essentially due to the number of attached GAG chains and to the level of sulfation. The proteoglycans also have a heterogeneous distribution. Keratansulfate proteoglycans, chondroitinsulfate proteoglycans and dermatansulfate proteoglycans are among the main structural components of the extracellular matrix (ECM), especially of connective tissues where thanks to the presence of highly anionic GAGs, they provide hydration and viscosity of the tissues and promote the diffusion of nutrients, metabolites and growth factors [7].

In particular, heparan sulfate proteoglycans (HSPG) are made up of various types of core proteins that covalently link variable HS chains. The HS proteoglycans are classified on the basis of the core protein and include the syndecans and glypicans (membrane-linked), perlecan, agrin and collagen XVIII (ECM components) and serglycin which is the only intracellular PG. Cell surface HSPG can activate receptors present on the same cell or on neighboring cells as in the case of fibroblast growth factor 2 (FGF-2) which bind to syndecan1 and whose release contributes to activate FGF-2 receptor-1. The biological activity of these proteoglycans can be modulated by proteolytic processing that leads to the shedding of syndecans and glypicans from the cell surface (ectodomain shedding).

There are two main types of HSPGs linked to ECM: agrin which is abundant in most basal membranes, mainly in the synaptic region and perlecan with a diffuse distribution and a very complex modular structure. Several pieces of evidence show that HSPG has the function of inhibiting cell invasion by promoting the interaction between cells and cell-ECM and maintaining the structural integrity and self-assembly of the ECM [8,9]. Together with shedding, the removal of specific sulfate groups by endo-sulfatases and the cleavage of HS chains are other post-biosynthetic modifications of HSPGs. The enzyme that is able to cut HS polysaccharide and release diffusible HS fragments is called heparanase.

Heparanase (HPSE) is an endo-β-D-glucuronidase which cleaves HS. Human HPSE gene (HPSE-1) contains 14 exons and 13 introns. It is located on chromosome 4q21.3 and expressed by alternative splicing as two mRNA, both containing the same open reading frame [10]. Interestingly, the HPSE-2 protein also exists, which shares ~40% similarity with HPSE-1, but does not exert the same activity [11]. HPSE cleaves HS chains on only a limited number of sites. Specifically, it cleaves the β (1,4) glycosidic linkage between GlcA and GlcNS, generating 5–10 kDa HS fragments (10–20 sugar units). Since heparin shares a high structural similarity with HS, HPSE is also able to cleave this substrate, thus generating 5–20 kDa fragments [12].

2. Heparanase Structure and Activity

2.1. Heparanase Processing and Structure

The active form of HPSE is a 58 kDa dimer made up of 50 kDa and 8 kDa subunits non-covalently linked. HPSE is synthesized in the endoplasmic reticulum as a precursor of 68 kDa which, in the Golgi, is then processed in proHPSE (65 kDa) by the elimination of the N-terminal signal peptide. Pro-HPSE is secreted in the extracellular space where it interacts with several membrane molecules (low-density lipoprotein-receptor-related protein, mannose 6-phosphate and membrane HSPGs such as syndecans [13]) for being endocytosed and delivered into lysosomes. In lysosome, cathepsin L protease catalyzes the excision of a 6 kDa linker region giving rise to the two subunits that form the mature enzyme. Active HPSE can have many destinations in the cell: it can be secreted, it can be anchored on the surface of exosomes, it can be included in autophagosomes or it can be shuttled into the nucleus [2] (Figure 1).

Recently, human HPSE crystal structure has been solved [14]. It is composed of a (β/α) 8 domain and a β-sandwich domain. A cleft of ~10 Å in the (β/α) 8 domain of the apo-enzyme was recognized, suggesting that the HS-binding site is contained within this part of the enzyme. Moreover, in this site, the residues Glu_{343} and Glu_{225} [14] are present, which have been identified as the catalytic nucleophile and acid-base of heparanase-cleaving activity [15]. The C-terminal domain of the 50 kDa subunit regulates protein secretion, enzymatic and non-enzymatic activity of HPSE [14].

2.2. Heparanase Enzymatic Activity

Consistent with its primary localization in late endosomes and perinuclear lysosomes, the physiological cellular role of active HPSE is to take part in the degradation and turnover of cell surface HSPGs. However, HPSE localization is not restricted to intracellular vesicles. In response to proper stimuli, mature HPSE can be secreted after the activation of protein kinase A (PKA) and kinase C (PKC) [16].

Extracellular active HPSE contributes to HSPG degradation by the cleavage of HS. HPSE-mediated breakdown of HS affects not only the structure of basal membranes and ECM but also the pool of HS-bound ligands which are released into the surrounding environment. In turn, the remodeling of ECM network and the diffusion of cytokines, growth factors and lipoproteins facilitate cell motility, angiogenesis, inflammation, coagulation and, as shown more recently, the stimulation of autophagy and exosome production [3–5,17].

2.3. Heparanase Non-Enzymatic Activities

Several studies demonstrate that HPSE also exhibits non-enzymatic activity even if receptors that could mediate these effects have not yet been identified. The pro-enzyme of 65 kDa induces signaling cascades that enhance phosphorylation of selected proteins such as Akt, ERK, p38 and Src [18]. For example, endothelial cell migration and invasion are enhanced by proHPSE Akt-phosphorylation and the activation of PI3K [19]. In addition, latent HPSE also induces glioma, lymphoma and T-cell adhesion mediated by β1-integrin and correlated with Akt, PyK2 and ERK activation, Akt/PKB phosphorylation turned out to be mediated by lipid-raft resident components [20].

3. Role of Heparanase in Pathological Conditions

3.1. Heparanase and Cancer Motility, Invasion and Metastasis

Heparanase expression is enhanced in a multiplicity of malignancies: for example, ovarian, pancreatic, gastric, renal, head and neck, colon, bladder, brain, prostate, breast and liver carcinomas, Ewing's sarcoma, multiple myeloma and B-lymphomas [21–24]. The role of HPSE in the development of cancers has been widely investigated and several recent reviews have covered that area in great depth [3]. The role of HPSE in cancer is mainly due to its HS degrading activity, facilitating cell invasion and metastasis dissemination. This hypothesis is also supported by several in vivo studies where HPSE inhibitors reduced tumor growth [25,26].

3.2. Heparanase and Angiogenesis

HPSE releases a combination of HS-bound growth factors (i.e., bFGF, VEGF, HB-EGF and KGF) which sustain neovascularization and wound healing. Indeed, it has been proved that HPSE overexpressing transgenic has an enhanced vascularization [27]. On a vicious loop, the high HPSE level produced by cancerous cells facilitates angiogenesis, which in turn sustains tumor growth [27]. Neovascularization is also increased by the non-enzymatic action of HPSE that up-regulates VEGF expression via p38-phosphorylation and Src kinase [28].

3.3. Heparanase and Coagulation

It has been proved that HPSE up-regulates the expression of the blood coagulation initiator-tissue factor (TF) and directly enhances its activity, which leads to increased factor Xa production and subsequent activation of the coagulation system. Moreover, HPSE interacts with the tissue factor pathway inhibitor (TFPI) on the cell surface of endothelial and tumor cells, leading to dissociation of

TFPI and causing increased cell surface coagulation activity. Consequently, the higher level of thrombin activates platelets which release additional HPSE [29]. Since many cancer types are associated with increased TF-associated hypercoagulable states, the high HPSE levels produced by cancer sustaining this event create a vicious cycle promoting cancer metastasis.

3.4. Heparanase and Inflammation

Inflammation occurs as a response of the body to dangerous stimuli, recruiting leucocytes from the bloodstream into the injured site. HS has a central role in the inflammatory response by controlling the release of pro-inflammatory cytokines (IL-2, IL-8, bFGF and TGF-β), by modulating the interaction between leucocytes and vascular endothelium, favoring leucocyte recruitment, rolling process and extravasation [30–32]. As a consequence, HPSE ends up having an essential role in inflammation. Before cloning the HPSE gene, an HS-degrading activity was discovered in neutrophils and activated T-lymphocytes and it was involved in their extravasation and accumulation in target organs [33]. Subsequently, HPSE non-enzymatic activities were reported to facilitate pro-inflammatory cell adhesion and signal transduction [2]. The main sources of HPSE are endothelial and epithelial cells in several inflammatory diseases including delayed-type hypersensitivity, chronic colitis, Crohn's disease, sepsis-associated lung injury and rheumatoid arthritis [34–36]. In colitis, HPSE from epithelial cells promotes monocyte-to-macrophage activation and its over-expression is able to prevent the regression of inflammation, switching macrophage response to chronic inflammation [34]. Moreover, activated macrophages are able to induce HPSE expression in colonic epithelial cells via tumor necrosis factor α (TNFα) stimulation of early growth response 1 factor (Egr1) [34]. The stimulation of TLRs is among the leading candidate pathways for HPSE-dependent macrophage activation for two main reasons: (i) intact extracellular HS inhibits TLR4 signaling and macrophage activation and, so, its removal relieves the inhibition; (ii) soluble HS released upon HPSE activation is able to stimulate TLR4 [37–39]. Recently, it has been proved that HPSE regulates macrophage polarization and the crosstalk between macrophages and proximal tubular epithelial cells after ischemia/reperfusion (I/R) injury [40]. In particular, I/R injury up-regulates HPSE at both tubular and glomerular levels. HPSE then induces tubular cell apoptosis and Damage Associated Molecular Patterns (DAMPs) production. DAMPs, HPSE-released HS-fragments and molecules generated from necrotic cells activate TLRs both on macrophages and tubular cells. Tubular cells in response to direct hypoxic stimuli and TLR activation produce pro-inflammatory cytokines which attract and activate macrophages and the presence of high levels of HPSE facilitates M1 polarization of infiltrated macrophages which worsen parenchymal damage [40].

3.5. Heparanase and Fibrosis

Tissue fibrosis is a deregulated wound-healing process characterized by the progressive accumulation of ECM together with its reduced remodeling. This event is common in different parenchymal organs such as the kidney, liver and lungs: HPSE seems involved in all of them with different mechanisms [41–43]. In the kidney, HPSE is overexpressed in injured tubular epithelial cells and glomerular cells exposed to several stimuli such as high glucose, advanced glycosylation end products and albuminuria [44], I/R injury [45,46] and elevated HPSE expression levels have been demonstrated to regulate epithelial-to-mesenchymal transition (EMT) of tubular cells [41]. Specifically, HPSE is necessary for FGF-2 to activate the PI3K/AKT pathway leading to EMT and for the establishment of the FGF-2 autocrine loop by the down-regulation of syndecan-1 (SDC1) and the up-regulation of metalloprotease-9 (MMP9) and HPSE [47]. Moreover, HPSE is deeply involved in TGF-β-induced EMT in the kidney since it turned out to be essential for TGF-β response to pro-fibrotic stimuli and its lack delayed tubular cell transdifferentiation and impaired TGF-β autocrine loop [48]. In the liver, the role of HPSE in fibrosis was sometimes controversial. For example, one study showed that the level of HPSE inversely correlates with the stage of liver fibrosis, while another one reported no difference in HPSE expression between cirrhotic and normal livers [49–52]. Our recent findings in a mouse model of chronic liver fibrosis suggested the involvement of HPSE in early phases of reaction to liver damage and inflammatory macrophages as an important source of HPSE. HPSE seems to play a key role in the macrophage-mediated activation of hepatic stellate cells (HSCs), thus suggesting that HPSE targeting could be a new therapeutic option in the treatment of liver fibrosis [38]. In the lungs, it has been reported that DAMPs such as HMGB1 released from necrotic/damaged cells lead to macrophage infiltration-sustaining inflammation. Moreover, HMGB1 is able to activate NF-κB, which then up-regulates heparanase expression. HPSE then releases TGF-beta form HS-proteoglycans creating a fibrotic setting [6].

3.6. Heparanase and Autophagy

Since, after secretion, HPSE is up-taken and stored in lysosomes, it has been proved that here it participates in the autophagy process [3,29]. Specifically, HPSE expression correlates with LC3b levels in cells and tissue of HPSE knockout and overexpressing mice [29] and it seems that this is an mTORC1-dependent mechanism [29]. Since autophagy confers an advantage to tumor-cell, by escaping from cell death, targeting synergistically heparanase and autophagy may be an additional strategy in cancer treatment (Figure 1).

3.7. Heparanase and Exosome Production

Heparanase also participates in the secretion of exosomes, which are membrane-bound extracellular vesicles, and is localized to their surface [5]. Specifically, the syndecan-syntenin-ALIX complex regulates the biogenesis of exosomes [53]. Since this process is regulated by heparan-sulphate, it has been proved that HPSE modulated the syndecan-syntenin-ALIX pathway resulting in enhanced endosomal intraluminal budding and biogenesis of exosomes [54]. Subsequently, it has been proved that exosomes are HPSE carriers, have a membrane localization and retain their ECM-degrading activity [55,56]. This additional HPSE source can significantly impact ECM degradation and growth-factor mobilization in neoplastic and inflammatory sets (Figure 1).

3.8. Heparanase Nuclear Activity

Given the nuclear localization of HSPGs, it is not surprising that HPSE can be translocated into the nucleus. Upon lysosome permeabilization and via interaction with the chaperon heat shock protein 90, active HPSE can translocate in the nucleus where it degrades nuclear HS and regulates gene expression [57]. Two different modes of gene expression regulation have been described for HPSE so far: the promotion of HAT activity by the cleavage of nuclear HS and through direct interaction with DNA [58,59]. HPSE regulates the expression of genes associated with glucose metabolism and inflammation in endothelial cells [60], differentiation in pro-myeloblast and tumorigenesis in melanoma cell lines [59]. In addition to mature HPSE, latent proHPSE has also been detected in the nucleus. Moreover, the observation that exogenously added proHPSE can be translocated in the nucleus and converted in the mature enzyme has led to the hypothesis that HPSE processing may also occur in this compartment [61] (Figure 1).

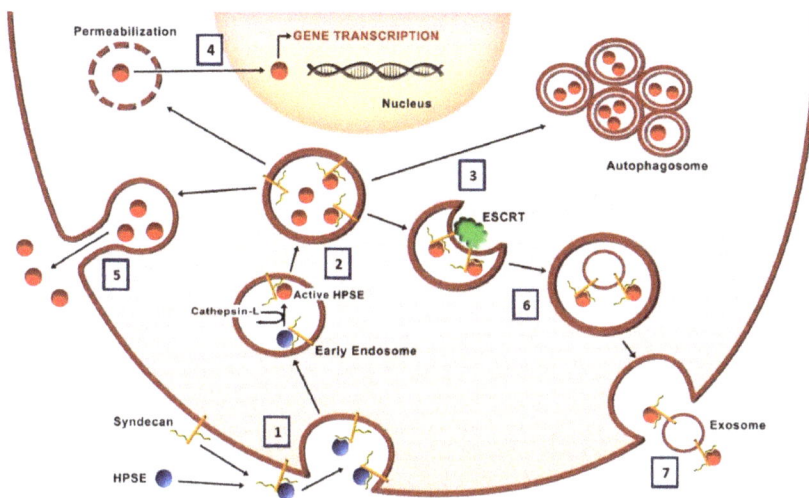

Figure 1. Schematic model of heparanase trafficking. (**1**) The inactive pro-HPSE in the extracellular spaces interacts with HS-proteoglycans such as syndecan-1 and the complex is endocytosed. (**2**) The fusion of endosomes with lysosomes, with the consequent acidification, induces the activation of HPSE exerted by the cleavage by cathepsin-L. (**3**) Here HPSE participates in the formation of autophagosome and thus controls the basal levels of autophagy. (**4**) HPSE can translocate into the nucleus where it can modulate gene transcription or (**5**) it can be secreted in the extracellular space. (**6**) Moreover, HPSE modulates the formation and the release of exosomes and (**7**) active HPSE is also released and anchored to syndecan on exosome surfaces. Collectively, by regulating autophagy and the production of exosomes, HPSE modulate several mechanisms which characterize cancer chemoresistance [62,63].

3.9. Heparanase in Viral Pathogenesis

Several human and non-human viruses utilize HS as an attachment co-receptor to entry into host cells: thus, HPSE, by modulating HS-bioavailability, is involved in viral-disease pathogenesis. It has been proved that HPSE expression and activity are upregulated in response to Herpes Simplex Virus (HSV-1) infection, via NF-kB pathway and, in turn, HPSE facilitates HS shedding from plasma membranes helping the release of surface-bound virions [64]. HPSE-dependent HS degradation similarly facilitates the infection of keratinocytes by Human Papilloma Virus (HPV) [65] and, subsequently, HPV gene E6, by interacting with p53, increases HPSE expression [66]. HPSE is involved in the pathogenesis of several other viral diseases such as Adenovirus, Dengue Virus, Hepatitis C Virus, and some retroviruses [67]. Looking forward, it is important to keep in mind that several cancers are induced by viruses and, thus, the same HPSE inhibitors may represent a useful tool to fight viral infection and associated cancer.

4. Heparanase Inhibition as Pharmacological Strategy

Several classes of HPSE inhibitors were developed in the last two decades ranging from monoclonal antibodies, small-molecules to polysulfated saccharides-molecule inhibitors.

Antibodies against HPSE are an efficient strategy to inhibit its activity. Recently, two monoclonal antibodies were described: one against the KKDC peptide and the other against the full-length heparanase protein. The result was that they were able to neutralize extracellular HPSE and to decrease its intracellular contents [68]. Small-molecule inhibitors are characterized by high variability in molecular weight, relevant functional group and physiochemical properties supporting the idea that HPSE could be inhibited by several mechanisms and several compounds with different structures [4].

However, the only HPSE-inhibitor compounds that have reached the phase of clinical trial belong to the class of polysaccharides. The development of these compounds began by observing heparin capacity to inhibit HPSE activity because of its competition with HS for binding to the enzyme. Currently, four HPSE-inhibitors are being tracked: PI-88, PG545, Roneparstat and M402. PG545 is a fully-sulphated HS mimetic, which is able to inhibit HPSE enzymatic function on HS chain [69,70]. Roneparstat is a semisynthetic heparin-like polymer transformed into a 15–25 kDa glycol-split N-acetyl heparin with reduced anticoagulant properties and a powerful anti-HPSE activity [71]. It has positively completed Phase I study with dexamethasone in patients with advanced multiple myeloma [72]. M402-necuparanid is another glycol-split HS mimetic with low molecular weight (5–8 kDa). It is currently under Phase II trial investigation in patients with pancreatic cancer [4].

5. Conclusions

Initially, HPSE has been identified as an enzyme with glycosidase activity implicated in the invasion of tumor cells. However, over the years, HPSE has been shown to be involved in many other pathological situations. It is now clear that considering its double enzymatic and non-enzymatic function and its intra and extracellular localization, HPSE can be defined as a multifunctional protein whose action is decisive in the establishment and development of numerous diseases. Considering that once the activity of HPSE is inhibited, no other molecule is able to perform a similar function, this enzyme has proved to be more and more eligible as a pharmacological target. HPSE inhibitors are currently being tested in several clinical trials, and some have already shown some antitumor efficacy. It is therefore expected that the next drugs aimed at inhibiting its activity may have therapeutic efficacy not only in the field of oncology but, hopefully, also for other diseases for which HPSE is a determinant etiological factor.

Author Contributions: V.M. and G.B. searched literature; V.M., G.B., G.G. and M.O. discussed and arranged the manuscript.

Funding: This work was supported by a grant from University of Padova (CRDA144519/14).

Conflicts of Interest: The authors declare no conflicts of interest.

References

1. Ramani, V.C.; Purushothaman, A.; Stewart, M.D.; Thompson, C.A.; Vlodavsky, I.; Au, J.L.; Sanderson, R.D. The heparanase/syndecan-1 axis in cancer: Mechanisms and therapies. *FEBS J.* **2013**, *280*, 2294–2306. [CrossRef] [PubMed]
2. Vlodavsky, I.; Singh, P.; Boyango, I.; Gutter-Kapon, L.; Elkin, M.; Sanderson, R.D.; Ilan, N. Heparanase: From basic research to therapeutic applications in cancer and inflammation. *Drug Resist. Updat.* **2016**, *29*, 54–75. [CrossRef] [PubMed]
3. Sanderson, R.D.; Elkin, M.; Rapraeger, A.C.; Ilan, N.; Vlodavsky, I. Heparanase regulation of cancer, autophagy and inflammation: New mechanisms and targets for therapy. *FEBS J.* **2017**, *284*, 42–55. [CrossRef] [PubMed]
4. Meirovitz, A.; Goldberg, R.; Binder, A.; Rubinstein, A.M.; Hermano, E.; Elkin, M. Heparanase in inflammation and inflammation-associated cancer. *FEBS J.* **2013**, *280*, 2307–2319. [CrossRef] [PubMed]
5. Sanderson, R.D.; Bandari, S.K.; Vlodavsky, I. Proteases and glycosidases on the surface of exosomes: Newly discovered mechanisms for extracellular remodeling. *Matrix Biol.* **2017**. [CrossRef] [PubMed]
6. Lv, Q.; Zeng, J.; He, L. The advancements of heparanase in fibrosis. *Int. J. Mol. Epidemiol. Genet.* **2016**, *7*, 137–140. [PubMed]
7. Iozzo, R.V.; Schaefer, L. Proteoglycan form and function: A comprehensive nomenclature of proteoglycans. *Matrix Biol.* **2015**, *42*, 11–55. [CrossRef] [PubMed]
8. Iozzo, R.V. Basement membrane proteoglycans: From cellar to ceiling. *Nat. Rev. Mol. Cell Biol.* **2005**, *6*, 646–656. [CrossRef] [PubMed]
9. Marneros, A.G.; Olsen, B.R. Physiological role of collagen XVIII and endostatin. *FASEB J.* **2005**, *19*, 716–728. [CrossRef] [PubMed]

10. Rivara, S.; Milazzo, F.M.; Giannini, G. Heparanase: A rainbow pharmacological target associated to multiple pathologies including rare diseases. *Future Med. Chem.* **2016**, *8*, 647–680. [CrossRef] [PubMed]

11. Vlodavsky, I.; Gross-Cohen, M.; Weissmann, M.; Ilan, N.; Sanderson, R.D. Opposing Functions of Heparanase-1 and Heparanase-2 in Cancer Progression. *Trends Biochem. Sci.* **2018**, *43*, 18–31. [CrossRef] [PubMed]

12. Levy-Adam, F.; Feld, S.; Cohen-Kaplan, V.; Shteingauz, A.; Gross, M.; Arvatz, G.; Naroditsky, I.; Ilan, N.; Doweck, I.; Vlodavsky, I. Heparanase 2 interacts with heparan sulfate with high affinity and inhibits heparanase activity. *J. Biol. Chem.* **2010**, *285*, 28010–28019. [CrossRef] [PubMed]

13. Gong, F.; Jemth, P.; Escobar Galvis, M.L.; Vlodavsky, I.; Horner, A.; Lindahl, U.; Li, J.P. Processing of macromolecular heparin by heparanase. *J. Biol. Chem.* **2003**, *278*, 35152–35158. [CrossRef] [PubMed]

14. Wu, L.; Viola, C.M.; Brzozowski, A.M.; Davies, G.J. Structural characterization of human heparanase reveals insights into substrate recognition. *Nat. Struct. Mol. Biol.* **2015**, *22*, 1016–1022. [CrossRef] [PubMed]

15. Hulett, M.D.; Hornby, J.R.; Ohms, S.J.; Zuegg, J.; Freeman, C.; Gready, J.E.; Parish, C.R. Identification of active-site residues of the pro-metastatic endoglycosidase heparanase. *Biochemistry* **2000**, *39*, 15659–15667. [CrossRef] [PubMed]

16. Shafat, I.; Vlodavsky, I.; Ilan, N. Characterization of mechanisms involved in secretion of active heparanase. *J. Biol. Chem.* **2006**, *281*, 23804–23811. [CrossRef] [PubMed]

17. Jin, H.; Zhou, S. The functions of heparanase in human diseases. *Mini Rev. Med. Chem.* **2017**, *17*, 541–548. [CrossRef] [PubMed]

18. Fux, L.; Ilan, N.; Sanderson, R.D.; Vlodavsky, I. Heparanase: Busy at the cell surface. *Trends Biochem. Sci.* **2009**, *34*, 511–519. [CrossRef] [PubMed]

19. Gingis-Velitski, S.; Zetser, A.; Flugelman, M.Y.; Vlodavsky, I.; Ilan, N. Heparanase induces endothelial cell migration via protein kinase B/Akt activation. *J. Biol. Chem.* **2004**, *279*, 23536–23541. [CrossRef] [PubMed]

20. Ben-Zaken, O.; Gingis-Velitski, S.; Vlodavsky, I.; Ilan, N. Heparanase induces Akt phosphorylation via a lipid raft receptor. *Biochem. Biophys. Res. Commun.* **2007**, *361*, 829–834. [CrossRef] [PubMed]

21. Ilan, N.; Elkin, M.; Vlodavsky, I. Regulation, function and clinical significance of heparanase in cancer metastasis and angiogenesis. *Int. J. Biochem. Cell. Biol.* **2006**, *38*, 2018–2039. [CrossRef] [PubMed]

22. Vlodavsky, I.; Beckhove, P.; Lerner, I.; Pisano, C.; Meirovitz, A.; Ilan, N.; Elkin, M. Significance of heparanase in cancer and inflammation. *Cancer Microenviron.* **2012**, *5*, 115–132. [CrossRef] [PubMed]

23. Vreys, V.; David, G. Mammalian heparanase: What is the message? *J. Cell. Mol. Med.* **2007**, *11*, 427–452. [CrossRef] [PubMed]

24. Secchi, M.F.; Masola, V.; Zaza, G.; Lupo, A.; Gambaro, G.; Onisto, M. Recent data concerning heparanase: Focus on fibrosis, inflammation and cancer. *Biomol. Concepts* **2015**, *6*, 415–421. [CrossRef] [PubMed]

25. Cassinelli, G.; Favini, E.; Dal Bo, L.; Tortoreto, M.; De Maglie, M.; Dagrada, G.; Pilotti, S.; Zunino, F.; Zaffaroni, N.; Lanzi, C. Antitumor efficacy of the heparan sulfate mimic roneparstat (SST0001) against sarcoma models involves multi-target inhibition of receptor tyrosine kinases. *Oncotarget* **2016**, *7*, 47848–47863. [CrossRef] [PubMed]

26. Ostapoff, K.T.; Awasthi, N.; Cenik, B.K.; Hinz, S.; Dredge, K.; Schwarz, R.E.; Brekken, R.A. PG545, an angiogenesis and heparanase inhibitor, reduces primary tumor growth and metastasis in experimental pancreatic cancer. *Mol. Cancer Ther.* **2013**, *12*, 1190–1201. [CrossRef] [PubMed]

27. Zcharia, E.; Metzger, S.; Chajek-ShaulL, T.; Aingorn, H.; Elkin, M.; Friedmann, Y.; Weinstein, T.; Li, J.P.; Lindahl, U.; Vlodavsky, I. Transgenic expression of mammalian heparanase uncovers physiological functions of heparan sulfate in tissue morphogenesis, vascularization, and feeding behavior. *FASEB J.* **2004**, *18*, 252–263. [CrossRef] [PubMed]

28. Ellis, L.M.; Staley, C.A.; Liu, W.; Fleming, R.Y.; Parikh, N.U.; Bucana, C.D.; Gallick, G.E. Down-regulation of vascular endothelial growth factor in a human colon carcinoma cell line transfected with an antisense expression vector specific for c-src. *J. Biol. Chem.* **1998**, *273*, 1052–1057. [CrossRef] [PubMed]

29. Nadir, Y.; Brenner, B. Heparanase procoagulant activity in cancer progression. *Thromb. Res.* **2016**, *140* (Suppl. 1), S44–S48. [CrossRef]

30. Axelsson, J.; Xu, D.; Kang, B.N.; Nussbacher, J.K.; Handel, T.M.; Ley, K.; Sriramarao, P.; Esko, J.D. Inactivation of heparan sulfate 2-*O*-sulfotransferase accentuates neutrophil infiltration during acute inflammation in mice. *Blood* **2012**, *120*, 1742–1751. [CrossRef] [PubMed]

31. Gotte, M. Syndecans in inflammation. *FASEB J.* **2003**, *17*, 575–591. [CrossRef] [PubMed]

32. Parish, C.R. The role of heparan sulphate in inflammation. *Nat. Rev. Immunol.* **2003**, *6*, 633–643. [CrossRef] [PubMed]

33. Vlodavsky, I.; Eldor, A.; Haimovitz-Friedman, A.; Matzner, Y.; Ishai-Michaeli, R.; Lider, O.; Naparstek, Y.; Cohen, I.R.; Fuks, Z. Expression of heparanase by platelets and circulating cells of the immune system: Possible involvement in diapedesis and extravasation. *Invasion Metastasis* **1992**, *12*, 112–127.

34. Lerner, I.; Hermano, E.; Zcharia, E.; Rodkin, D.; Bulvik, R.; Doviner, V.; Rubinstein, A.M.; Ishai-Michaeli, R.; Atzmon, R.; Sherman, Y.; et al. Heparanase powers a chronic inflammatory circuit that promotes colitis-associated tumorigenesis in mice. *J. Clin. Investig.* **2011**, *121*, 1709–1721. [CrossRef] [PubMed]

35. Li, R.W.; Freeman, C.; Yu, D.; Hindmarsh, E.J.; Tymms, K.E.; Parish, C.R.; Smith, P.N. Dramatic regulation of heparanase activity and angiogenesis gene expression in synovium from patients with rheumatoid arthritis. *Arthr. Rheum.* **2008**, *58*, 1590–1600. [CrossRef] [PubMed]

36. Schmidt, E.P.; Yang, Y.; Janssen, W.J.; Gandjeva, A.; Perez, M.J.; Barthel, L.; Zemans, R.L.; Bowman, J.C.; Koyanagi, D.E.; Yunt, Z.X.; et al. The pulmonary endothelial glycocalyx regulates neutrophil adhesion and lung injury during experimental sepsis. *Nat. Med.* **2012**, *18*, 1217–1223. [CrossRef] [PubMed]

37. Yu, L.; Wang, L.; Chen, S. Endogenous toll-like receptor ligands and their biological significance. *J. Cell. Mol. Med.* **2010**, *14*, 2592–2603. [CrossRef] [PubMed]

38. Johnson, G.B.; Brunn, G.J.; Kodaira, Y.; Platt, J.L. Receptor-mediated monitoring of tissue well-being via detection of soluble heparan sulfate by Toll-like receptor 4. *J. Immunol.* **2002**, *168*, 5233–5239. [CrossRef] [PubMed]

39. Brunn, G.J.; Bungum, M.K.; Johnson, G.B.; Platt, J.L. Conditional signaling by Toll-like receptor 4. *FASEB J.* **2005**, *19*, 872–874. [CrossRef] [PubMed]

40. Masola, V.; Zaza, G.; Bellin, G.; Dall'Olmo, L.; Granata, S.; Vischini, G.; Secchi, M.F.; Lupo, A.; Gambaro, G.; Onisto, M. Heparanase regulates the M1 polarization of renal macrophages and their crosstalk with renal epithelial tubular cells after ischemia/reperfusion injury. *FASEB J.* **2017**, *32*, 742–756. [CrossRef] [PubMed]

41. Masola, V.; Zaza, G.; Onisto, M.; Lupo, A.; Gambaro, G. Impact of heparanase on renal fibrosis. *J. Transl. Med.* **2015**, *13*, 181. [CrossRef] [PubMed]

42. Secchi, M.F.; Crescenzi, M.; Masola, V.; Russo, F.P.; Floreani, A.; Onisto, M. Heparanase and macrophage interplay in the onset of liver fibrosis. *Sci. Rep.* **2017**, *7*, 14956. [CrossRef] [PubMed]

43. He, L.; Sun, F.; Wang, Y.; Zhu, J.; Fang, J.; Zhang, S.; Yu, Q.; Gong, Q.; Ren, B.; Xiang, X.; et al. HMGB1 exacerbates bronchiolitis obliterans syndrome via RAGE/NF-κB/HPSE signaling to enhance latent TGF-β release from ECM. *Am. J. Transl. Res.* **2016**, *8*, 1971–1984. [PubMed]

44. Masola, V.; Gambaro, G.; Tibaldi, E.; Onisto, M.; Abaterusso, C.; Lupo, A. Regulation of heparanase by albumin and advanced glycation end products in proximal tubular cells. *Biochim. Biophys. Acta.* **2011**, *1813*, 1475–1482. [CrossRef] [PubMed]

45. Masola, V.; Zaza, G.; Gambaro, G.; Onisto, M.; Bellin, G.; Vischini, G.; Khamaysi, I.; Hassan, A.; Hamoud, S.; Nativ, O.; et al. Heparanase: A Potential New Factor Involved in the Renal Epithelial Mesenchymal Transition (EMT) Induced by Ischemia/Reperfusion (I/R) Injury. *PLoS ONE* **2016**, *11*, e0160074. [CrossRef] [PubMed]

46. Gil, N.; Goldberg, R.; Neuman, T.; Garsen, M.; Zcharia, E.; Rubinstein, A.M.; van Kuppevelt, T.; Meirovitz, A.; Pisano, C.; Li, J.P.; et al. Heparanase is essential for the development of diabetic nephropathy in mice. *Diabetes* **2012**, *61*, 208–216. [CrossRef] [PubMed]

47. Masola, V.; Gambaro, G.; Tibaldi, E.; Brunati, A.M.; Gastaldello, A.; D'Angelo, A.; Onisto, M.; Lupo, A. Heparanase and syndecan-1 interplay orchestrates fibroblast growth factor-2-induced epithelial-mesenchymal transition in renal tubular cells. *J. Biol. Chem.* **2012**, *287*, 1478–1488. [CrossRef] [PubMed]

48. Masola, V.; Zaza, G.; Secchi, M.F.; Gambaro, G.; Lupo, A.; Onisto, M. Heparanse is a key player in the renal fibrosis by regulating TGF-b expression and activity. *BBA Mol. Cell Res.* **2014**, *1843*, 2122–2128.

49. Ikeguchi, M.; Hirooka, Y.; Kaibara, N. Heparanase gene expression and its correlation with spontaneous apoptosis in hepatocytes of cirrhotic liver and carcinoma. *Eur. J. Cancer* **2003**, *39*, 86–90. [CrossRef]

50. Xiao, Y.; Kleeff, J.; Shi, X.; Büchler, M.W.; Friess, H. Heparanase expression in hepatocellular carcinoma and the cirrhotic liver. *Hepatol. Res.* **2003**, *26*, 192–198. [CrossRef]

51. Goldshmidt, O.; Yeikilis, R.; Mawasi, N.; Paizi, M.; Gan, N.; Ilan, N.; Pappo, O.; Vlodavsky, I.; Spira, G. Heparanase expression during normal liver development and following partial hepatectomy. *J. Pathol.* **2004**, *203*, 594–602. [CrossRef] [PubMed]

52. Ohayon, O.; Mawasi, N.; Pevzner, A.; Tryvitz, A.; Gildor, T.; Pines, M.; Rojkind, M.; Paizi, M.; Spira, G. Halofuginone upregulates the expression of heparanase in thioacetamide-induced liver fibrosis in rats. *Lab Investig.* **2008**, *88*, 627–633. [CrossRef] [PubMed]

53. Shteingauz, A.; Boyango, I.; Naroditsky, I.; Hammond, E.; Gruber, M.; Doweck, I.; Ilan, N.; Vlodavsky, I. Heparanase enhances tumor growth and chemoresistance by promoting autophagy. *Cancer Res.* **2015**, *75*, 3946–3957. [CrossRef] [PubMed]

54. Baietti, M.F.; Zhang, Z.; Mortier, E.; Melchior, A.; Degeest, G.; Geeraerts, A.; Ivarsson, Y.; Depoortere, F.; Coomans, C.; Vermeiren, E.; et al. Syndecan-syntenin-ALIX regulates the biogenesis of exosomes. *Nat. Cell Biol.* **2012**, *14*, 677–685. [CrossRef] [PubMed]

55. Roucourt, B.; Meeussen, S.; Bao, J.; Zimmermann, P.; David, G. Heparanase activates the syndecan-syntenin-ALIX exosome pathway. *Cell Res.* **2015**, *25*, 412–428. [CrossRef] [PubMed]

56. Thompson, C.A.; Purushothaman, A.; Ramani, V.C.; Vlodavsky, I.; Sanderson, R.D. Heparanase regulates secretion, composition, and function of tumor cell-derived exosomes. *J. Biol. Chem.* **2013**, *288*, 10093–10099. [CrossRef] [PubMed]

57. Nobuhisa, T.; Naomoto, Y.; Okawa, T.; Takaoka, M.; Gunduz, M.; Motoki, T.; Nagatsuka, H.; Tsujigiwa, H.; Shirakawa, Y.; Yamatsuji, T.; et al. Translocation of heparanase into nucleus results in cell differentiation. *Cancer Sci.* **2007**, *98*, 535–540. [CrossRef] [PubMed]

58. Purushothaman, A.; Hurst, D.R.; Pisano, C.; Mizumoto, S.; Sugahara, K.; Sanderson, R.D. Heparanase-mediated loss of nuclear syndecan-1 enhances histone acetyltransferase (HAT) activity to promote expression of genes that drive an aggressive tumor phenotype. *J. Biol. Chem.* **2011**, *286*, 30377–30383. [CrossRef] [PubMed]

59. Yang, Y.; Gorzelanny, C.; Bauer, A.T.; Halter, N.; Komljenovic, D.; Bäuerle, T.; Borsig, L.; Roblek, M.; Schneider, S.W. Nuclear heparanase-1 activity suppresses melanoma progression via its DNA-binding affinity. *Oncogene* **2015**, *34*, 5832–5842. [CrossRef] [PubMed]

60. Wang, F.; Wang, Y.; Zhang, D.; Puthanveetil, P.; Johnson, J.D.; Rodrigues, B. Fatty acid-induced nuclear translocation of heparanase uncouples glucose metabolism in endothelial cells. *Arterioscler. Thromb. Vasc. Biol.* **2012**, *32*, 406–414. [CrossRef] [PubMed]

61. Schubert, S.Y.; Ilan, N.; Shushy, M.; Ben-Izhak, O.; Vlodavsky, I.; Goldshmidt, O. Human heparanase nuclear localization and enzymatic activity. *Lab Investig.* **2004**, *8*, 535–544. [CrossRef] [PubMed]

62. Lanzi, C.; Zaffaroni, N.; Cassinelli, G. Targeting Heparan Sulfate Proteoglycans and their Modifying Enzymes to Enhance Anticancer Chemotherapy Efficacy and Overcome Drug Resistance. *Curr. Med. Chem.* **2017**, *24*, 2860–2886. [CrossRef] [PubMed]

63. Bandari, S.K.; Purushothaman, A.; Ramani, V.C.; Brinkley, G.J.; Chandrashekar, D.S.; Varambally, S.; Mobley, J.A.; Zhang, Y.; Brown, E.E.; Vlodavsky, I.; et al. Chemotherapy induces secretion of exosomes loaded with heparanase that degrades extracellular matrix and impacts tumor and host cell behavior. *Matrix Biol.* **2017**, *65*, 104–118. [CrossRef] [PubMed]

64. Agelidis, A.M.; Hadigal, S.R.; Jaishankar, D.; Shukla, D. Viral Activation of Heparanase Drives Pathogenesis of Herpes Simplex Virus-1. *Cell Rep.* **2017**, *20*, 439–450. [CrossRef] [PubMed]

65. Surviladze, Z.; Sterkand, R.T.; Ozbun, M.A. Interaction of human papillomavirus type 16 particles with heparan sulfate and syndecan-1 molecules in the keratinocyte extracellular matrix plays an active role in infection. *J. Gen. Virol.* **2015**, *96*, 2232–2241. [CrossRef] [PubMed]

66. Hirshoren, N.; Bulvik, R.; Neuman, T.; Rubinstein, A.M.; Meirovitz, A.; Elkin, M. Induction of heparanase by HPV E6 oncogene in head and neck squamous cell carcinoma. *J. Cell. Mol. Med.* **2014**, *18*, 181–186. [CrossRef] [PubMed]

67. Thakkar, N.; Yadavalli, T.; Jaishankar, D.; Shukla, D. Emerging Roles of Heparanase in Viral Pathogenesis. *Pathogens* **2017**, *6*, 43. [CrossRef] [PubMed]

68. Weissmann, M.; Arvatz, G.; Horowitz, N.; Feld, S.; Naroditsky, I.; Zhang, Y.; Ng, M.; Hammond, E.; Nevo, E.; Vlodavsky, I.; et al. Heparanase-neutralizing antibodies attenuate lymphoma tumor growth and metastasis. *Proc. Natl. Acad. Sci. USA* **2016**, *113*, 704–709. [CrossRef] [PubMed]

69. Ferro, V.; Liu, L.; Johnstone, K.D.; Wimmer, N.; Karoli, T.; Handley, P.; Rowley, J.; Dredge, K.; Li, C.P.; Hammond, E.; et al. Discovery of PG545: A highly potent and simultaneous inhibitor of angiogenesis, tumor growth, and metastasis. *J. Med. Chem.* **2012**, *55*, 3804–3813. [CrossRef] [PubMed]

70. Winterhoff, B.; Freyer, L.; Hammond, E.; Giri, S.; Mondal, S.; Roy, D.; Teoman, A.; Mullany, S.; Hoffmann, R.; von Bismarck, A.; et al. PG545 enhances anti-cancer activity of chemotherapy in ovarian models and increases surrogate biomarkers such as VEGF in preclinical and clinical plasma samples. *Eur. J. Cancer* **2015**, *51*, 879–892. [CrossRef] [PubMed]
71. Naggi, A.; Casu, B.; Perez, M.; Torri, G.; Cassinelli, G.; Penco, S.; Pisano, C.; Giannini, G.; Ishai-Michaeli, R.; Vlodavsky, I. Modulation of the heparanase-inhibiting activity of heparin through selective desulfation, graded N-acetylation, and glycol splitting. *J. Biol. Chem.* **2005**, *280*, 12103–12113. [CrossRef] [PubMed]
72. Galli, M.; Chatterjee, M.; Grasso, M.; Specchia, G.; Magen, H.; Einsele, H.; Celeghini, I.; Barbieri, P.; Paoletti, D.; Pace, S.; et al. Phase I study of the heparanase inhibitor Roneparstat: An innovative approach for multiple myeloma therapy. *Haematologica* **2018**, *103*, e469–e472. [CrossRef] [PubMed]

cells

MDPI

Article

Effect of a Collagen-Based Compound on Morpho-Functional Properties of Cultured Human Tenocytes

Filippo Randelli [1], Alessandra Menon [2], Alessio Giai Via [1], Manuel Giovanni Mazzoleni [1], Fabio Sciancalepore [2], Marco Brioschi [1] and Nicoletta Gagliano [3,*]

[1] Centro di Chirurgia dell'Anca e Traumatologia, I.R.C.C.S Policlinico San Donato, 20097 San Donato Milanese, Italy; filippo.randelli@fastwebnet.it (F.R.); alessiogiaivia@hotmail.it (A.G.V.); manuelmazzoleni4@gmail.com (M.G.M.); marco.brioschi@unimi.it (M.B.)
[2] Azienda Socio Sanitaria Territoriale Centro Specialistico Ortopedico Traumatologico Gaetano Pini-CTO, 1ˆ Clinica Ortopedica, 20122 Milan, Italy; ale.menon@me.com (A.M.); fabio.sciancalepore@unimi.it (F.S.)
[3] Department of Biomedical Sciences for Health, Università degli Studi di Milano, 20133 Milan, Italy
* Correspondence: nicoletta.gagliano@unimi.it; Tel.: +39-02-50315374; Fax: +39-02-50315387

Received: 28 November 2018; Accepted: 4 December 2018; Published: 6 December 2018

Abstract: Background: Greater Trochanter Pain Syndrome (GTPS) is the main reason for recalcitrant lateral hip pain. Gluteus medius and minimus tendinopathy plays a key role in this setting. An injectable medical compound containing collagen type I (MD-Tissue, Guna) has been produced with the aim to counteract the physiological and pathological degeneration of tendons. In this study we aimed at characterizing the effect of this medical compound on cultured human gluteal tenocytes, focusing on the collagen turnover pathways, in order to understand how this medical compound could influence tendon biology and healing. Methods: Tenocytes were obtained from gluteal tendon fragments collected in eight patients without any gluteal tendon pathology undergoing total hip replacement through an anterior approach. Cell proliferation and migration were investigated by growth curves and wound healing assay, respectively. The expression of genes and proteins involved in collagen turnover were analysed by real-time PCR, Slot blot and SDS-zymography. Results: Our data show that tenocytes cultured on MD-Tissue, compared to controls, have increased proliferation rate and migration potential. MD-Tissue induced collagen type I (COL-I) secretion and mRNA levels of tissue inhibitor of matrix metalloproteinases (MMP)-1 (TIMP-1). Meanwhile, lysyl hydroxylase 2b and matrix metalloproteinases (MMP)-1 and -2, involved, respectively, in collagen maturation and degradation, were not affected. Conclusions: Considered as a whole, our results suggest that MD-Tissue could induce in tenocytes an anabolic phenotype by stimulating tenocyte proliferation and migration and COL-I synthesis, maturation, and secretion, thus favouring tendon repair. In particular, based on its effect on gluteal tenocytes, MD-Tissue could be effective in the discouraging treatment of GTPS. From now a rigorous clinical investigation is desirable to understand the real clinical potentials of this compound.

Keywords: tendinopathy; Greater Trochanter Pain Syndrome; tendon; collagen turnover; matrix metalloproteinases; cytoskeleton; focal adhesion

1. Introduction

Tendinopathy is a chronic, painful condition affecting tendons, characterized by histological modifications such as collagen fibril disorganization, increased proteoglycan and glycosaminoglycan content, and increased non-collagen extracellular matrix components, hypercellularity, and neovascularization [1]. Pain associated with tendinopathy is a very common cause of disability.

Pain around the greater trochanter was in the past attributed to trochanteric bursitis, but different studies ([2,3] and references therein) have suggested that this condition is mainly determined by a tendinopathy of the gluteus medius or minimus tendons, known as Greater Trochanter Pain Syndrome (GTPS). GTPS is a clinical condition characterized by pain and tenderness at or around the greater trochanter, pain with activities such as walking and stair climbing, and lying on the affected side at night [4]. Even if it is a common condition, few effective treatments are available in the clinical practice, and are often empirical [5]. Therefore, the interest in biological and regenerative therapies aiming to improve tendon healing, such as platelet-rich plasma (PRP) or hyaluronic acid [6], has grown during the last decade.

Treatment of tendinopathy, including Greater Trochanter Pain Syndrome and others, remains a big problem for clinicians, because its pathogenesis is still largely misunderstood, and many treatments have no real evidence. The management of GTPS is traditionally first conservative, including rest, ice-packs, non-steroidal anti-inflammatory drugs (NSAIDs), physiotherapy, and local corticosteroids injections. However, recent studies found little evidence to support physical therapy or exercise programs for gluteal tendinopathy [7]. Corticosteroid injections are widely used despite controversies regarding the use of glucocorticoid injections for the treatment of tendinopathy. A recent systematic review of literature showed that they provide short-term pain relief, but no long-term benefits [5,8]. No statistically significant differences have been reported after 1 year compared to oral NSAIDs. Some authors reported about the use of Extracorporeal Shock Wave Therapy (ESWT) with good results [9], but the evidence of most of these studies is low, and no level I studies are published in literature. Therefore, biological therapies aiming to improve tendon healing become very attractive. The only level I study on the PRP showed an improvement of pain and mHHS at 12 weeks after a single injection of PRP, compared to a single injection of corticosteroids [10]. In a recent case series, promising results have been reported with autologous tenocyte injection [11]. However, there is still little evidence to support the use of biological therapies for treatment of GTPS [12].

Tendons play key functions in musculoskeletal system in transferring forces generated by muscle contraction to the skeleton. Their mechanical properties are based on the underlying extracellular matrix (ECM) structure and composition, mainly consisting of type I collagen (COL-I) [13–15]. Tenocytes are tendon specialized fibroblasts interspersed between collagen fibers, responsible for the metabolic activity and structure of tendon. They are involved in collagen turnover pathways and act as mechanosensors playing key roles in modifying gene expression for ECM components in response to mechanical forces acting on tendons [14,16,17].

The aim of this study was to investigate in vitro the effect of MD-Tissue®, an injectable collagen-based medical compound with therapeutic potential (registered as a medical device in different countries) containing swine collagen as main component, on human tenocytes, with particular attention on collagen turnover pathways, in order to understand the molecular mechanisms triggered by this medical compound and, therefore, how it could act in favoring tendon homeostasis and repair. We were particularly interested in clinical situations such as GTPS; for this purpose we analyzed gluteal tenocytes.

The production process of MD-Tissue by the manufacturer allows to obtain a pure product with a standardized molecular weight, having the chemical-physical characteristics to guarantee safety in clinical use. No side effects, allergic reactions, nor drug interactions have been observed. Clinical data reported that treatment with MD-Knee, a collagen-based medical compound similar to MD-Tissue, for up to 6 months was generally well tolerated, and no systemic adverse events or septic complications were observed [18,19].

The characterization in vitro of the molecular mechanisms triggered by MD-Tissue will be pivotal to plan specific clinical trials.

2. Materials and Methods

All subjects gave their informed consent for inclusion before they participated in the study. The study was conducted in accordance with the Declaration of Helsinki, and the protocol was approved by the Local Ethics Committee (San Raffaele Hospital Ethical Committee, Milan, Italy) of the coordinating Institution (IRCCS Policlinico San Donato, Milan, Italy) (63/INT/2017) Inclusion and exclusion criteria are listed in Table 1.

Table 1. Eligibility Criteria.

Inclusion Criteria	Exclusion Criteria
Age ranging 18–70 years.	Patients diagnosed of great trochanter tendinopathy
Indication for total hip arthroplasty.	Patients affected by genetic collagen disorders.
Patients who signed written informed consent for the surgery.	Patients diagnosed of spondyloarthritis with involvement of the affected hip.
Patients able to understand the study conditions and willing to participate for its entire duration.	Patients affected by psoriatic arthritis.
	Drug-addicted Alcohol-addicted Psychiatric disorders Clinical conditions which could compromise the results of the surgical procedure or of the follow-up.
	Informed consent not accepted.
	Pregnant or breastfeeding women.
	Patients affected by diabetes mellitus.
	Patients who had taken fluoroquinolones within 30 days before the surgery.

2.1. Samples

Fragments from human gluteal tendon were obtained from 8 patients (mean age 64.8 ± 7.2 years, 4 males and 4 females) without any gluteal tendon pathology and undergoing total hip replacement through an anterior approach. For each sample we analyzed the mid-substance of the collected tendon, representing the region with the typical structure of the dense regular connective tissue.

2.2. Cell Cultures

Tendon fragments were immediately washed in sterile PBS and plated in T25 flasks, incubated in Dulbecco's Modified Eagle Medium (DMEM) supplemented with 10% heat-inactivated fetal bovine serum (FBS), antibiotics (100 U/mL penicillin, 0.1 mg/mL streptomycin), and ascorbic acid (200 µM) at 37 °C in a humidified atmosphere containing 5% CO_2. When tenocytes grew out from the explant, they were trypsinized (0.025% trypsin-0.02% EDTA) for secondary cultures and plated in T75 flasks. Viability was assessed by the Trypan blue exclusion method. For evaluations, confluent human tenocytes were used between the fourth and fifth passage. For SDS-zymography cells were cultured in serum-free DMEM. Tenocytes derived from each patients and cell supernatants were prepared in duplicate and were analyzed after 24, 48, and 72 h.

2.3. Coating with MD-Tissue

MD-Tissue®, kindly provided by Guna (Milan, Italy), is an injectable medical compound based on swine collagen (100 µg/2 mL ampoules). It contains ascorbic acid, magnesium gluconate, pyridoxin hydrochloride, riboflavin, thiamine hydrochloride, NaCl and water as excipients. Swine collagen is similar to human collagen and due to the high biocompatibility with humans and very low risk of adverse effects, it has been used in various clinical fields. MD-Tissue is an injectable compound but we hypothesized that it could act as a mechanical scaffold to influence tenocyte metabolism. Thus, MD-Tissue was used as a substrate for cell cultures. However, as previously demonstrated [20], some of the collagen used for the coating in cell cultures comes out contributing to the mechanical stimulation of tenocytes.

To obtain a thin coating, MD-Tissue (50 µg/mL) was added to multi-wells or T25 flasks (3 mL per T25 flask and 6-well multi-well plates, ad 500 µL per 24-well multi-well plates). After an incubation of at least 3–4 h at room temperature to ensure that collagen has adhered to the plastic, excess fluid was removed from the coated surface that was dried for at least 2 h under the laminar flux hood. Coated plastic was used immediately or stored at 4 °C.

Cells cultured on uncoated cell culture plastic (NC) were used as untreated controls.

2.4. Growth Curves

Cell growth was assessed by growth curves. Tenocytes were plated in triplicate samples in 6-well multi-well plates at the same cell density (100,000 cells/well). Cell number was determined using a Neubauer chamber after 24, 48, and 72 h in tenocytes in the proliferative phase.

2.5. Immunofluorescence Analysis

For fluorescence microscopy, tenocytes were cultured on uncoated (NC) or MD-Tissue coated 12-mm diameter round coverslips put into 24-well culture plates, as previously described [21]. Briefly, for actin cytoskeleton analysis, cells were incubated with 50 µM rhodamine-phalloidin (Sigma-Aldrich, St. Louis, MO, USA) and for vinculin detection, cells were incubated for 1 h at room temperature with the rabbit polyclonal antibody anti-vinculin (1:600 in PBS, Sigma-Aldrich, St. Louis, MO, USA). The secondary antibody was an anti-rabbit/Alexa488 (1:500, Life Technologies, Carlsbad, CA, USA). Cells were photographed by a digital camera connected to a Nikon Eclipse 80i microscope.

2.6. Real-Time PCR

Gene expression was analyzed by real-time RT-PCR as previously reported in samples run in triplicate [22]. GAPDH was used as endogenous control to normalize the differences in the amount of total RNA in each sample. The primers sequences were the following: Glyceraldehyde 3-phosphate dehydrogenase (GAPDH): sense CCCTTCATTGACCTCAACTACATG, antisense TGGGATTTCCATTGATGACAAGC; Long lysyl hydroxylase 2 (LH2b): sense CCGGAAACATTCCAAATGCTCAG, antisense GCCAGAGGTCATTGTTATAATGGG; Tissue inhibitor of matrix metalloproteinase 1 (TIMP-1): sense GGCTTCTGGCATCCTGTTGTTG, antisense AAGGTGGTCTGGTTGACTTCTGG; vinculin: sense GGAGGTGATTAACCAGCCAAT, antisense AATGATGTCATTGCCCTTGC; Focal adhesion kinase (FAK): sense GTCTGCCTTCGCTTCACG, antisense GAATTTGTAACTGGAAGATGCAAG; Nanog sense CCCCAGCCTTTACTCTTCCTA, antisense CCAGGTTGAATTGTTCCAGGTC. Each sample was analyzed in triplicate in a Bioer LineGene 9600 thermal cycler (Bioer, hangzhou, China) after 40 cycles. The cycle threshold (Ct) was determined and gene expression levels relative to that of GAPDH were calculated. To confirm the reliability of gene expression results, mRNA levels of target genes were normalized on a second housekeeping gene, the 18s ribosomal RNA. The expression pattern obtained was similar (data not shown).

2.7. Slot Blot

Collagen type I and III (COL-I, COL-III), matrix metalloproteinase (MMP)-1 protein levels secreted by tenocytes were assessed in duplicate samples by Slot blot in serum free cell culture medium, as previously detailed [21]. Membranes were incubated for 1 h at room temperature in monoclonal antibody to COL-I (1:1000 in TBST) (Sigma-Aldrich, Milan, Italy), COL-III (1:2000 in TBST) (Sigma-Aldrich, Milan, Italy), MMP-1 (1 µg/mL in TBST) (Millipore, Milan, Italy). Immunoreactive bands, revealed by the Amplified Opti-4CN substrate (Amplified Opti-4CN, Bio Rad, Italy), were scanned densitometrically (UVBand, Eppendorf, Italy).

2.8. SDS-Zymography

SDS-zymography was used to analyze MMP-2 activity of secreted protein in cell culture medium, as previously described [21]. MMP gelatinolytic activity, detected after staining the gels with Coomassie brilliant blue R250 as clear bands on a blue background, were quantified by densitometric scanning (UVBand, Eppendorf, Italy).

2.9. Wound Healing Assay

Cell migration in adult and aging tenocytes was analyzed by wound healing assay in uncoated (NC) and MD-Tissue coated 6-wells multi-well plates [22]. The "scratch" was created in confluent tenocytes using a p 200 pipet tip. After cell debris removal by DMEM washing, multi-well plates were incubated at 37 °C and observed under an inverted microscope at different time points. Digital images were captured by a digital camera after 0 and 24 h, and the size of the "scratch" was measured to obtain the migration potential.

2.10. Statistical Analysis

Statistical analysis was performed using GraphPad Prism v 6.0 software (GraphPad Software Inc., San Diego, CA 92108, USA). Data were obtained from two replicate experiments for each of the patients-derived cell lines cultured in duplicate and were expressed as mean ± standard deviation (SD). Comparison of groups was calculated using independent samples two-tailed t test. Differences associated with p-values lower than 5%, at a confidence level of 95%, were to be considered significant.

3. Results

3.1. Cell Growth

Cell proliferation evaluated by growth curves showed that tenocytes cultured on uncoated (NC) or MD-Tissue coated 6-well multi-well plates had the same proliferation rate after 24 and 48 h. By contrast, cell proliferation significantly increased in tenocytes grown on MD-Tissue after 72 h ($p < 0.05$ vs. 72 h NC) (Figure 1).

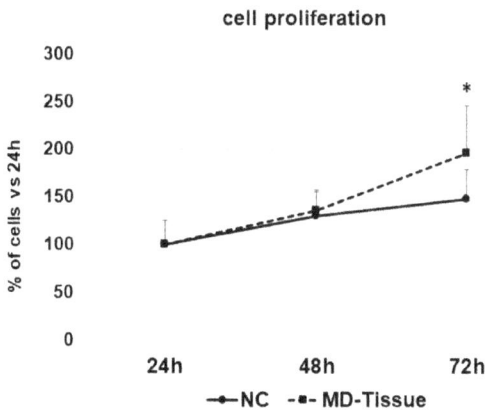

Figure 1. Growth curves of tenocytes grown without (NC) or on MD-Tissue at the indicated time points. Data are expressed as percentages vs. the time point T24 and are mean + SD. * $p < 0.05$ vs. 72 h NC.

3.2. Expression of Genes and Proteins Related to Collagen Turnover

Since it was demonstrated that, using different methods to isolate tenocytes from tendon fragments, the obtained cell population can be a mixed population of terminally differentiated tenocytes and progenitor cells [23], we first characterized our cells determining the expression of Nanog, one of the main mesenchymal stem cell markers [24]. Adipose tissue-derived stem cells (ASCs) (kindly provided by Dr. Anna Brini, University of Milan) were used as positive control. We found that Nanog is almost undetectable in all the considered cell cultures obtained from tendon fragments, while it is highly expressed in ASCs (Figure 2), showing that cells used in this study are mostly differentiated tenocytes.

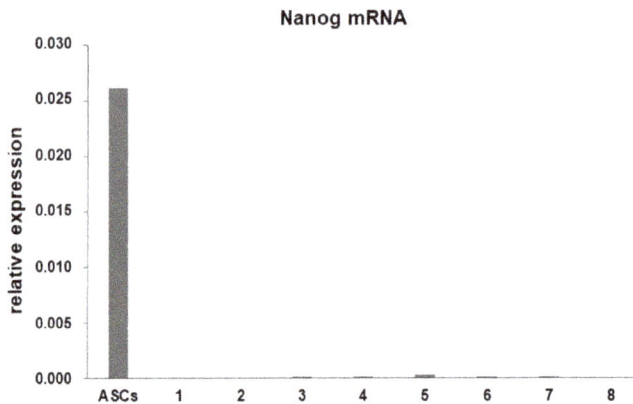

Figure 2. Bar graphs showing mRNA levels for Nanog in adipose tissue-derived stem cells (ASCs) and the eight cell cultures of tenocytes assessed by real-time PCR. Data were normalized on Glyceraldehyde 3-phosphate dehydrogenase (GAPDH) gene expression.

Collagen maturation was assessed by analyzing LH2b gene expression by real time PCR, involved in cross-linking of newly synthetized collagen. LH2b mRNA levels were unaffected by MD-Tissue at all the considered time points (Figure 3). Slot blot analysis revealed that COL-I protein levels secreted in cell culture medium by tenocytes cultured on MD-Tissue were increased (+25%, +19% and +18%, respectively, for COL vs. NC after 24, 48 and 72 h). This increase was statistically significant in cells cultured on COL, compared to NC, at all the considered time points ($p < 0.005$, $p < 0.05$ and $p < 0.05$ for MD-Tissue vs. CT, respectively, at 24, 48 and 72 h) (Figure 4A,B). To demonstrate that collagen expression detected by Slot blot originates from tenocytes and not from the coating, we prepared a Slot blot to analyze COL-I expression in the same volume (100 μL) of tenocyte culture medium, DMEM from a MD-Tissue coated well, DMEM only and TBS buffer. The result shows that an immunoreactive band is detected only for tenocytes, demonstrating that collagen originates only from cells and not from the coating (Figure 3C). By contrast, COL-III protein levels were similarly secreted by NC and MD-Tissue tenocytes (data not shown).

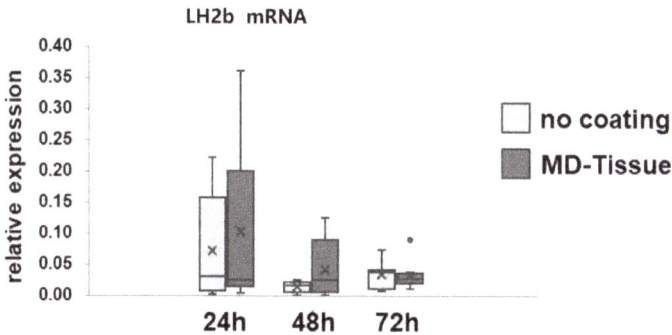

Figure 3. Bar graphs showing mRNA levels for Long lysyl hydroxylase 2 (LH2b) in NC and MD-Tissue tenocytes (COL) assessed by real-time PCR. Data were normalized on GAPDH gene expression and are expressed as mean ± SD for at least two independent experiments for the eight samples run in duplicate. ×: mean; •: outlier sample.

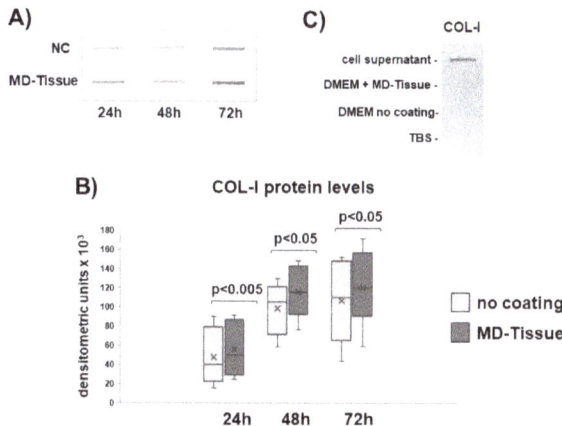

Figure 4. Representative slot blot analysis for collagen type I (COL-I) expression (**A**) expression in cell-culture medium of tenocytes cultured without coating (NC) or on MD-Tissue. Bar graphs displaying COL-I (**B**) protein levels analyzed densitometric scanning of immunoreactive bands in panel A. Data are expressed as mean ± SD for the 8 samples. (**C**) Slot blot analysis for COL-I showing that COL-I expression originates from tenocytes. ×: mean.

Collagen degradation analysis revealed that MMP-1 levels (Figure 5A,B), and MMP-2 activity (Figure 5C,D) were similar in tenocytes cultured without coating (NC) or on MD-Tissue. By contrast, gene expression for TIMP-1, the main inhibitor of MMP-1, was strongly affected by MD-Tissue. In fact, gene expression was significantly up-regulated after 24 h ($p < 0.05$) and tended to a significant up-regulation after 72 h ($p = 0.056$) (Figure 6). Some interindividual differences in MMPs expression and TIMP-1 mRNA levels were observed, showing unchanged, increased or decreased levels. However, a strong correlation ($p = 0.056$) was observed between MMP-1 protein levels and TIMP-1 gene expression for tenocytes cultured on MD-Tissue at 72 h.

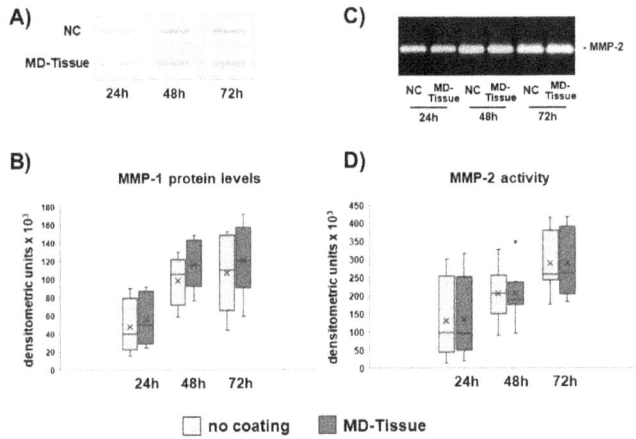

Figure 5. Representative slot blot for matrix metalloproteinase-1 (MMP-1) levels (**A**) and representative SDS-zymography showing MMP-2 activity in serum-free cell supernatants of NC and MD-Tissue tenocytes. Bar graphs showing MMP-1 protein levels (**C**) and MMP-2 activity (**D**) after densitometric analysis of immunoreactive and lytic bands, respectively. Data obtained from the eight samples are expressed as a % of densitometric units vs NC and are $\pm$ SD. NC: no coating. $\times$: mean; $\bullet$: outlier sample.

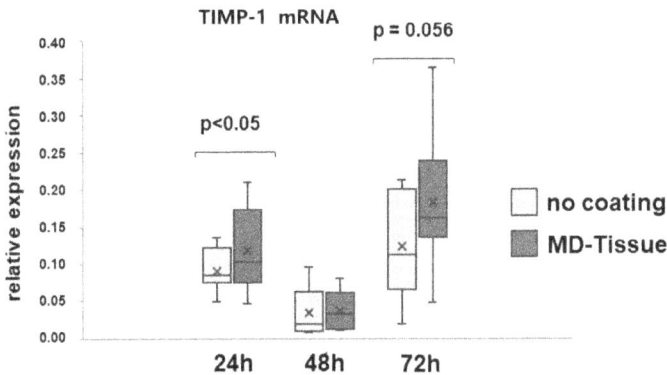

Figure 6. Bar graphs showing tissue inhibitor of matrix metalloproteinase-1 (TIMP-1) gene expression after normalization on GAPDH mRNA levels. Data obtained from the eight samples are expressed as mean $\pm$ SD. $\times$: mean.

3.3. Cytoskeleton Arrangement and Focal Adhesion

Fluorescent microscopy analysis for F-actin (Figure 7A,B) revealed that actin filaments seem unaffected by MD-Tissue. In both experimental conditions, they are long and mostly longitudinally oriented in the cytoplasm.

In order to investigate whether MD-Tissue influences the ability of tenocytes to form focal adhesions needed for cell migration, we analyzed the gene expression for vinculin (VCN) and Focal adhesion kinase (FAK), key proteins involved in the formation of the adhesion plaque. VCN and FAK mRNA levels, although with some interindividual differences, were unaffected by MD-Tissue (Figure 7C,D). Immunofluorescence analysis for VCN shows that the protein is co-localized with actin in correspondence of focal adhesion formation on the substrate, and many focal adhesions were detectable in tenocytes cultured on either NC or MD-Tissue. Very frequently, the region corresponding to the presence of the focal adhesion seem bigger in tenocytes cultured on MD-Tissue compared to NC tenocytes (see arrows in Figure 7B).

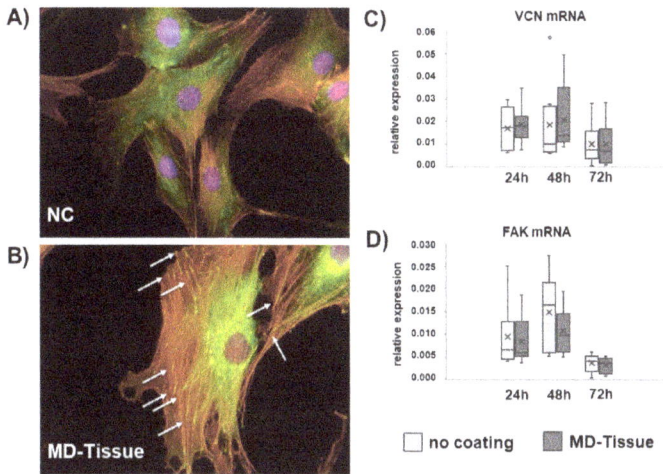

Figure 7. Immunofluorescence analysis for vinculin (green) in control tenocytes (NC) (**A**) and tenocytes cultured on MD-Tissue (**B**). Actin filaments are stained using rhodamine-phalloidin labeling. Nuclei are stained in blue by DAPI. Original magnification: 60×. Bar graphs showing VCN (**C**) and FAK (**D**) gene expression relative to GAPDH mRNA levels in NC and MD-Tissue tenocytes (COL). Data obtained from the eight samples are expressed as mean ± SD. ×: mean; •: outlier sample.

3.4. Wound Healing Assay

Wound healing assay was used to analyze tenocyte migration of tenocytes cultured on MD-Tissue. The comparison of the scratch size revealed that cell migration is significantly increased by MD-Tissue ($p < 0.05$ vs. NC) (Figure 8).

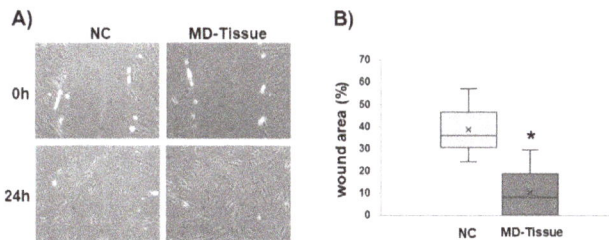

Figure 8. (**A**) Representative micrographs showing wound healing assay in control tenocytes (NC) and tenocytes grown on MD-Tissue at 0 and 24 h after the scratch. Original magnification: 10×. (**B**) Bar graphs showing the area of wound closure, expressed as a % of the area at 0 h, in cultured tenocytes in both experimental conditions 24 h after the scratch. * $p < 0.005$ vs. NC. ×: mean.

4. Discussion

Treatment of tendinopathy, including Greater Trochanter Pain Syndrome and others, remain a big problem for clinicians, because its pathogenesis is still largely misunderstood, and many treatments have no real evidence. We aimed at characterizing in vitro the molecular mechanisms triggered in human tenocytes by MD-Tissue®, and the main findings of this study reveal that this collagen-based injectable medical compound could favor tendon repair by inducing tenocyte proliferation and migration, and stimulating COL-I synthesis, secretion and maturation.

Tendons play key functions in musculoskeletal system serving to transfer forces generated by muscle contraction to the skeleton. Tendon mechanical properties are based on the underlying extracellular matrix (ECM) structure and composition, mainly consisting of type I collagen (COL-I), which constitutes about 60–85% of the dry mass of the tendon and about 95% of the total collagen, and elastin, accounting for 1–2% [13–15]. Collagen and elastin are embedded in a ground substance rich in proteoglycans, glycosaminoglycans, structural glycoproteins, and a wide variety of other small molecules, having water-binding capacity and contributing to the stabilization of the whole tendon structure. The unique structure and composition of tendons provide them the characteristic mechanical stability, and COL-I represents the most important factor for tendon mechanical strength.

Tenocytes are tendon specialized fibroblasts interspersed between collagen fiber bundles and aligned along the long axis of tendon, responsible for the synthesis and degradation of collagen and all components of the ECM. Tenocytes act as mechanosensors playing key roles for the tendon's adaptive response and change their metabolic activities in response to mechanical forces acting on tendons, modifying gene expression for ECM components and, as a consequence, affecting tendon's mechanical properties [14,16,17].

Tendon mechanical properties are based on its structure and composition, allowing tissue mechanical adaptation in response to mechanical forces. Tenocytes are responsible for tendon mechanical adaptation converting mechanical stimuli into biochemical signals that ultimately lead to tendon adaptive physiological or pathological changes. Mechanical loads at physiological levels are usually beneficial to tendons in terms of enhancing their mechanical properties [25–27], and increase collagen synthesis [28]. Therefore, appropriate mechanical loads induce tendon adaptation and have anabolic effects, improving their strength and healing quality after injury.

Collagen is the major component of tendon ECM; its turnover is controlled by tenocytes acting at the level of collagen synthesis, maturation and degradation, thus determining the tendon ability to resist mechanical forces and repair in response to injury [14]. Our data show that COL-I secretion is significantly induced by MD-Tissue® at all considered time points, suggesting that this medical compound is able to stimulate an anabolic phenotype of tenocytes. Newly synthesized collagen undergoes cross-linking, that is an important requirement for collagen maturation providing tendon strength, collagen fibril stabilization and increased tendon tensile strength [29,30]. Furthermore, our data show that this particular medical compound is not able to affect gene expression for LH2b, needed to provide cross-linking during collagen maturation, suggesting that collagen stability is not affected.

Since collagen content is the result of a finely regulated dynamic balance between its synthesis and degradation driven by MMPs, tendon strength is strongly dependent on degradation pathways. COL breakdown is played by MMP-1, able to cleave the intact collagen triple helix, followed by other proteases [31,32]. The role of MMP-1 is boosted by the previously demonstrated inverse correlation between MMP-1 gene/protein expression and the amplitude of tensile mechanical load on tendons, suggesting that low levels of MMP-1 lead to a more stable and less susceptible to damage tendon structure [33]. MMPs activation and activity are regulated by TIMPs [34,35], and TIMP-1 is the main inhibitor of MMP-1. In this study, we show that MMP-1 and MMP-2 were unaffected by MD-Tissue®, but, interestingly, we observed a significant increase in TIMP-1 gene expression induced by the medical compound after 24 h, and a tendency to an up-regulation after 72 h. This result leads to the hypothesis that MD-Tissue is able to stimulate COL secretion by tenocytes, increasing COL content in tendons, and that COL increase is favored by inhibition of its degradation.

Some interindividual differences in fibroblast cell lines were observed, especially in LH2b mRNA, MMPs and TIMP-1 mRNA. In fact, in some tenocytes the expression resulted unchanged, increased or decreased. Interindividual differences were previously described in gingival fibroblasts, suggesting the "cell subpopulation theory" [36] that described that the gingival fibroblast population is composed of different subsets of fibroblasts that respond to external or endogenous stimuli in different way.

Fibroblast heterogeneity was described also recently [37]. According to this theory, we can hypothesize it is valid also for fibroblasts derived from different tissues, such as tenocytes.

ECM homeostasis is influenced by mechanical forces acting on tendons, and tenocytes are key effectors in tendon ECM remodeling and adaptation to the experienced mechanical loading. They are mechanoresponsive cells able to convert mechanical signals into biological events such as gene expression and cell proliferation [38,39]. Tenocytes are able to sense changes in their mechanical environment using a mechanotransduction system based on the actin cytoskeleton, and to respond by modifying their activity [33]. Since actin cytoskeleton arrangement and integrity are the key mechanotransduction apparatus influencing the tensional homeostasis and mechanoresponsiveness needed to maintain ECM balance [40,41], we investigated whether MD-Tissue® affected actin cytoskeleton. Our data show that actin filaments are brightly labelled and longitudinally running in tenocytes cultured on plastic (NC) or on MD-Tissue®, and their arrangement is similar in both experimental conditions, suggesting that this component of the mechanoresponsive apparatus is not modified by the medical compound.

The actin cytoskeleton is also actively involved in cell migration, a dynamic process mediated by a repeated cycle of attachment to the ECM, generation of cytoskeletal forces of propulsion, and subsequent detachment of the cell from the matrix. The attachment of cells to the ECM is mediated by integrins, transmembrane proteins that provide a bridge though which forces can be transmitted between inside and outside of the cells. In fact, the extracellular domain of the integrin binds to ECM components, whereas its cytoplasmic domain links various intracellular proteins interposed between actin filaments and the integrin, forming focal adhesion complexes at the leading edge of the cell [42]. Focal adhesions are not only involved in transmitting the mechanical signals from the ECM into cells to modulate ECM remodeling; they are also necessary to generate the traction required for cell migration [42]. We used a wound healing assay to investigate if tenocyte motility is affected by MD-Tissue® since this test is particularly suitable to investigate the effects of cell–matrix interactions on cell migration, mimicking in vivo cell migration [21], and we found that this medical compound significantly increased cell migration. Interestingly, no difference in actin cytoskeleton between NC and MD-Tissue tenocytes was detected, and also gene expression for FAK and VCN, two main components of the focal adhesion complex, was similar in both experimental conditions, supporting the hypothesis of a similar ability to form focal adhesions. However, interestingly, immunofluorescence analysis of VCN expression revealed that focal adhesions containing VCN seem more evident in tenocytes plated on MD-Tissue, leading to the hypothesis that MD-Tissue® could act as a mechanical scaffold improving tenocyte focal adhesion, allowing a more efficient mechanoresponsiveness and migration ability. This could make tenocytes more efficient in maintaining ECM homeostasis in response to different mechanical load and in generating the traction necessary for cell migration [42]. Since the process of tendon repair and regeneration is reliant on tenocyte migration [43], MD-Tissue® could be effective in favoring tendon healing in tendinopathies, but clinical trials are needed to confirm this suggestion.

A limit of this study is represented by the lacking of studies aimed at characterizing the permanence and rheological and visco-elastic properties of MD-Tissue, however clinical studies are available demonstrating that the collagen-based medical compound is effective in osteoarticular pathologies [18,19].

MD-Tissue could offer some advantages compared to other biological agents since it can be administered singularly or in association with other therapeutic agents and the reduced cost compared to low or high MW HA could allow wider use, resulting in a NSAIDs intake reduction.

5. Conclusions

Considered as a whole, these in vitro findings suggest a mechanism for MD-Tissue likely acting as a mechanical scaffold able to induce an anabolic phenotype in tenocytes, favoring tendon homeostasis and repair. Interestingly, it was demonstrated [20] that the addition of acid-soluble type I collagen to fibroblasts cultured on plastic also affected collagen turnover mechanisms since some of the collagen comes out of solution and precipitates on cell surfaces, still eliciting some effect nonmediated by a cell-substrate interaction. This suggests that little mechanical resistance is sufficient to activate cell receptors, influencing collagen turnover mechanisms and, very likely, other biological activities played by tenocytes. We can hypothesize that MD-Tissue effect could be also mediated by this alternative mechanism. Our results suggest that MD-Tissue® could represent a novel therapeutic approach, and stimulate new clinical trials to investigate the effects of this medical compound to treat difficult tendon pathologies. Based on our in vitro findings obtained on gluteal tenocytes, we will perform in future a clinical trial to prove its utility on GTPS.

Author Contributions: Conceptualization, N.G.; F.R; Methodology, N.G; Investigation, N.G., F.R, A.G.V., A.M, M.G.M., M.B., and F.S; Data Curation, N.G.; Writing—Original Draft Preparation, N.G.; Writing—Review and Editing, N.G., F.R, A.G.V., A.M, M.G.M., M.B., and F.S.

Funding: This study was partially supported by Guna S.p.a. The funder had no role in the design or conduct of the study, in analysis and interpretation of data, or in the preparation of the manuscript.

Acknowledgments: We thank Anna Brini (Department of Biomedical, Surgical and Dental Sciences, University of Milan) for providing ASCs cells, and Lorenzo Castagnoli (Fondazione IRCCS Istituto Nazionale dei Tumori, Milan) for providing primers for Nanog analysis.

Conflicts of Interest: The authors declare that the manuscript is free of conflict of interest. This study was partially supported by Guna S.p.a. The sponsor had no role in the design or conduct of the study, in analysis and interpretation of data, or in preparation of the manuscript.

References

1. Giai Via, A.; Papa, G.; Oliva, F.; Maffulli, N. Tendinopathy. *Curr. Phys. Med. Rehabil. Rep.* **2016**, *4*, 50–55. [CrossRef]
2. Ejnisman, L.; Safran, M.R. Biologics in hip preservation. *Ann. Joint* **2018**, *3*, 50. [CrossRef]
3. Karpinski, M.R.; Piggott, H. Greater trochanteric pain syndrome. A. report of 15 cases. *J. Bone Joint Surg. Br.* **1985**, *67*, 762–763. [CrossRef] [PubMed]
4. Fearon, A.M.; Scarvell, J.M.; Neeman, T.; Cook, J.L.; Cormick, W.; Smith, P.N. Greater trochanteric pain syndrome: Defining the clinical syndrome. *Br. J. Sports Med.* **2013**, *47*, 649–653. [CrossRef]
5. Frizziero, A.; Vittadini, F.; Pignataro, A.; Gasparre, G.; Biz, C.; Ruggieri, P.; Masiero, S. Conservative management of tendinopathies around hip. *Muscles, Ligaments Tendons J.* **2016**, *6*, 281–292. [CrossRef] [PubMed]
6. Reid, D. The management of greater trochanteric pain syndrome: A systematic literature review. *J. Orthop.* **2016**, *13*, 15–28. [CrossRef] [PubMed]
7. Grimaldi, A.; Mellor, R.; Hodges, P.; Bennell, K.; Wajswelner, H.; Vicenzino, B. Gluteal tendinopathy: A review of mechanisms, assessment and management. *Sports Med.* **2015**, *45*, 1107–1119. [CrossRef]
8. Seo, K.H.; Lee, J.Y.; Yoon, K.; Do, J.G.; Park, H.J.; Lee, S.Y.; Park, Y.S.; Lee, Y.T. Long-term outcome of low-energy extracorporeal shockwave therapy on gluteal tendinopathy documented by magnetic resonance imaging. *PLoS ONE* **2018**, *13*, e019746. [CrossRef]
9. Del Buono, A.; Papalia, R.; Khanduja, V.; Denaro, V.; Maffulli, N. Management of the greater trochanteric pain syndrome: A. systematic review. *Br. Med. Bull.* **2012**, *102*, 115–131. [CrossRef]
10. Fitzpatrick, J.; Bulsara, M.K.; O'Donnell, J.; McCrory, P.R.; Zheng, M.H. The effectiveness of platelet-rich plasma injections in gluteal tendinopathy a randomized, double-blind controlled trial comparing a single platelet-rich plasma injection with a single corticosteroid injection. *Am. J. Sports Med.* **2018**, *46*, 933–939. [CrossRef]
11. Thomas, A.; Bucher, T.A.; Ebert, J.R.; Smith, A.; Breidahl, W.; Fallon, M.; Wang, T.; Zheng, M.H.; Janes, G.C. Autologous tenocyte injection for the treatment of chronic recalcitrant gluteal tendinopathy a prospective pilot study. *Orthopaedic J. Sports Med.* **2017**, *5*, 1–10. [CrossRef]
12. Andia, I.; Maffulli, N. How far have biological therapies come in regenerative sports medicine? *Expert Opin. Biol. Ther.* **2018**, *18*, 785–793. [CrossRef]
13. Kannus, P. Structure of the tendon connective tissue. *Scand J. Med. Sci Sports* **2000**, *10*, 312–320. [CrossRef]

14. Kjaer, M. Role of extracellular matrix in adaptation of tendon and skeletal muscle to mechanical loading. *Physiol. Rev.* **2004**, *84*, 649–698. [CrossRef] [PubMed]
15. Riley, G.P.; Harrall, R.L.; Constant, C.R.; Chard, M.D.; Cawston, T.E.; Hazleman, B.L. Glycosaminoglycans of human rotator cuff tendons: Changes with age and in chronic rotator cuff tendinitis. *Ann. Rheum. Dis.* **1994**, *53*, 367–376. [CrossRef] [PubMed]
16. Banes, A.J.; Horesovsky, G.; Larson, C.; Tsuzaki, M.; Judex, S.; Archambault, J.; Zernicke, R.; Herzog, W.; Kelley, S.; Miller, L. Mechanical load stimulates expression of novel genes in vivo and in vitro in avian flexor tendon cells. *Osteoarthritis Cartilage* **1999**, *7*, 141–153. [CrossRef] [PubMed]
17. Benjamin, M.; Ralphs, J.R. The cell and developmental biology of tendons and ligaments. *Int. Rev. Cytol.* **2000**, *196*, 85–130.
18. Martin, L.S.; Massafra, U.; Bizzi, E.; Migliore, A. A double blind randomized active-controlled clinical trial on the intra-articular use of Md-Knee versus sodium hyaluronate in patients with knee osteoarthritis ("Joint"). *BMC Musculoskelet Disord* **2016**, *17*, 94. [CrossRef]
19. Pavelka, K.; Jarosova, H.; Sleglova, O.; Svobodova, R.; Votavova, M.; Milani, L.; Prochazka, Z.; Kotlarova, L.; Kostiuk, P.; Sliva, J.; et al. Chronic Low Back Pain: Current Pharmacotherapeutic Therapies and a New Biological Approach. *Curr. Med. Chem.* **2018**, *25*, 1–8. [CrossRef]
20. Ritty, T.M.; Herzog, J. Tendon cells produce gelatinases in response to type I collagen attachment. *J. Orthop. Res.* **2003**, *21*, 442–450. [CrossRef]
21. Menon, A.; Pettinari, L.; Martinelli, C.; Colombo, G.; Portinaro, N.; Dalle-Donne, I.; d'Agostino, M.C.; Gagliano, N. New insights in extracellular matrix remodeling and collagen turnover related pathways in cultured human tenocytes after ciprofloxacin administration. *Muscles Ligaments Tendons J.* **2013**, *113*, 122–131.
22. Liang, C.C.; Park, A.Y.; Guan, J.L. In vitro scratch assay: A convenient and inexpensive method for analysis of cell migration in vitro. *Nat. Protoc.* **2007**, *2*, 329–333. [CrossRef] [PubMed]
23. Viganò, M.; Perucca Orfei, C.; de Girolamo, L.; Pearson, J.R.; Ragni, E.; De Luca, P.; Colombini, A. Housekeeping Gene Stability in Human Mesenchymal Stem and Tendon Cells Exposed to Tenogenic Factors. *Tissue Eng. Part. C Methods* **2018**, *24*, 360–367. [CrossRef] [PubMed]
24. Calloni, R.; Cordero, E.A.; Henriques, J.A.; Bonatto, D. Reviewing and updating the major molecular markers for stem cells. *Stem Cells Dev.* **2013**, *22*, 1455–1476. [CrossRef] [PubMed]
25. Viidik, A. The effect of training on the tensile strength of isolated rabbit tendons. *Scand J. Plast Reconstr. Surg.* **1967**, *1*, 141–147. [CrossRef] [PubMed]
26. Woo, S.L.; Ritter, M.A.; Amiel, D.; Sanders, T.M.; Gomez, M.A.; Kuei, S.C.; Garfin, S.R.; Akeson, W.H. The biomechanical and biochemical properties of swine tendons–long term effects of exercise on the digital extensors. *Connect. Tissue Res.* **1980**, *7*, 177–183. [CrossRef]
27. Woo, S.L.; Gomez, M.A.; Amiel, D.; Ritter, M.A.; Gelberman, R.H.; Akeson, W.H. The effects of exercise on the biomechanical and biochemical properties of swine digital flexor tendons. *J. Biomech. Eng.* **1981**, *103*, 51–56. [CrossRef]
28. Langberg, H.; Rosendal, L.; Kjaer, M. Training-induced changes in peritendinous type I collagen turnover determined by microdialysis in humans. *J. Physiol.* **2001**, *534*, 297–302. [CrossRef]
29. Silver, F.H.; Christiansen, D.; Snowhill, P.B.; Chen, Y.; Landis, W.J. The role of mineral in the storage of elastic energy in turkey tendons. *Biomacromolecules* **2000**, *1*, 180–185. [CrossRef]
30. Walker, L.C.; Overstreet, M.A.; Yeowell, H.N. Tissue-specific expression and regulation of the alternatively-spliced forms of lysyl hydroxylase 2 (LH2) in human kidney cells and skin fibroblasts. *Matrix. Biol.* **2005**, *23*, 515–523. [CrossRef]
31. Sakai, T.; Gross, J. Some properties of the products of reaction of tadpole collagenase with collagen. *Biochemistry* **1967**, *6*, 518–528. [CrossRef] [PubMed]
32. Woessner, F.J. Matrix metalloproteinases and their inhibitors in connective tissue remodelling. *FASEB J.* **1991**, *5*, 2145–2154. [CrossRef] [PubMed]
33. Arnoczky, S.P.; Tian, T.; Lavagnino, M.; Gardner, K. Ex vivo static tensile loading inhibits MMP-1 expression in rat tail tendon cells through a cytoskeletally based mechanotransduction mechanism. *J. Orthop. Res.* **2004**, *22*, 328–333. [CrossRef]
34. Brew, K.; Dinakarpandian, D.; Nagase, H. Tissue inhibitors of metalloproteinases: Evolution, structure and function. *Biochim. Biophys. Acta* **2001**, *1477*, 267–283. [CrossRef]

35. Murphy, G.; Willenbrock, F.; Crabbe, T.; O'Shea, M.; Ward, R.; Atkinson, S.; O'Connell, J.; Docherty, A. Regulation of matrix metalloproteinase activity. *Ann. N Y Acad. Sci.* **1994**, *732*, 31–41. [CrossRef] [PubMed]
36. Phipps, R.P.; Borrello, M.A.; Blieden, T.M. Fibroblast heterogeneity in the periodontium and other tissues. *J. Periodontal Res.* **1997**, *32*, 159–165. [CrossRef] [PubMed]
37. Archana, A.; Srikanth, V.; Sasireka; Kurien, B.; Ebenezer. Fibroblast Heterogeneity in Periodontium. *Int. J. Dental Sci. Res.* **2014**, *2*, 50–54. [CrossRef]
38. Burridge, K.; Guilluy, C. Focal adhesions, stress fibers and mechanical tension. *Exp. Cell. Res.* **2016**, *343*, 14–20. [CrossRef]
39. Wang, J.H.; Guo, Q.; Li, B. Tendon Biomechanics and Mechanobiology-A Minireview of Basic Concepts and Recent Advancements. *J. Hand Ther.* **2012**, *25*, 133–140. [CrossRef]
40. Docking, S.; Samiric, T.; Scase, E.; Purdam, C.; Cook, J. Relationship between compressive loading and ECM changes in tendons. *Muscle Ligaments Tendons J.* **2013**, *3*, 7–11. [CrossRef]
41. Heinemeier, K.M.; Olesen, J.L.; Haddad, F.; Schjerling, P.; Baldwin, K.M.; Kjaer, M. Effect of unloading followed by reloading on expression of collagen and related growth factors in rat tendon and muscle. *J. Appl. Physiol.* **2009**, *106*, 178–186. [CrossRef] [PubMed]
42. Reed, M.J.; Ferara, N.S.; Vernon, R.B. Impaired migration, integrin function, and actin cytoskeletal organization in dermal fibroblasts from a subset of aged human donors. *Mech. Ageing Dev.* **2001**, *122*, 1203–1220. [CrossRef]
43. Jones, M.E.; Mudera, V.; Brown, R.A.; Cambrey, A.D.; Grobbelaar, A.O.; McGrouther, D.A. The early surface cell response to flexor tendon injury. *J. Hand Surg. Am.* **2003**, *28*, 221–230. [CrossRef] [PubMed]

cells

MDPI

Review

Tendon Remodeling in Response to Resistance Training, Anabolic Androgenic Steroids and Aging

Vinicius Guzzoni [1],*, Heloisa Sobreiro Selistre-de-Araújo [2] and Rita de Cássia Marqueti [3]

[1] Departamento de Biologia Molecular e Celular, Universidade Federal da Paraíba, João Pessoa 58051-970, Paraíba, Brazil

[2] Department of Physiological Sciences, Federal University of São Carlos, São Carlos 13565-205, São Paulo, Brazil; hsaraujo@ufscar.br

[3] Graduate Program of Rehabilitation Science, University of Brasilia, Distrito Federal, Brasília 70840-901, Distrito Federal, Brazil; marqueti@gmail.com

* Correspondence: vinicius.guzzoni@gmail.com or vinicius.guzzoni@cbiotec.ufpb.br; Tel.: +55-016-99706-2846

Received: 16 November 2018; Accepted: 30 November 2018; Published: 7 December 2018

Abstract: Exercise training (ET), anabolic androgenic steroids (AAS), and aging are potential factors that affect tendon homeostasis, particularly extracellular matrix (ECM) remodeling. The goal of this review is to aggregate findings regarding the effects of resistance training (RT), AAS, and aging on tendon homeostasis. Data were gathered from our studies regarding the impact of RT, AAS, and aging on the calcaneal tendon (CT) of rats. We demonstrated a series of detrimental effects of AAS and aging on functional and biomechanical parameters, including the volume density of blood vessel cells, adipose tissue cells, tendon calcification, collagen content, the regulation of the major proteins related to the metabolic/development processes of tendons, and ECM remodeling. Conversely, RT seems to mitigate age-related tendon dysfunction. Our results suggest that AAS combined with high-intensity RT exert harmful effects on ECM remodeling, and also instigate molecular and biomechanical adaptations in the CT. Moreover, we provide further information regarding the harmful effects of AAS on tendons at a transcriptional level, and demonstrate the beneficial effects of RT against the age-induced tendon adaptations of rats. Our studies might contribute in terms of clinical approaches in favor of the benefits of ET against tendinopathy conditions, and provide a warning on the harmful effects of the misuse of AAS on tendon development.

Keywords: tendon; resistance training; anabolic androgenic steroids; aging; extracellular matrix

1. Introduction

The calcaneal tendon (CT), known as the Achilles tendon, is more susceptible to tendinopathy since it is subjected to greater mechanical loading [1–4]. Aging has been shown to induce detrimental effects on tendons [2,5], while exercise training (ET) seems to mitigate these age-induced detrimental effects. In addition, anabolic-androgenic steroids (AAS) have been shown to evoke potential effects on skeletal muscle [6–10], suggesting that tendons might be affected by AAS misuse [11–14]. It has also been demonstrated that aging can cause detrimental effects in tendon composition, however mechanical loading may improve tendon structure and content and diminish the detrimental effects of AAS and aging [15,16]. Thus, the purpose of this narrative review is to provide an in-depth understanding of the major effects of ET, particularly resistance training (RT), AAS and aging on tendon remodeling of rats. Data were gathered from our studies regarding the impact of RT, AAS and aging on CT adaptations. Nevertheless, the evidence of tendon remodeling in response to resistance training, AAS and aging remain inconclusive, further highlighting the importance of this review.

2. Structure and Function of Tendons

Tendons are soft tissues considered as inert and inextensible structures, even though they confer elastic properties that influence the muscle-tendon unit function [17,18]. Tendons constitute important structures of the musculoskeletal system that transmit muscle-generating tensile force to bones, resulting in movement [19–22]. While tendons are attached to muscles through the myotendinous junction, tendons are connected to bone through a fibrocartilaginous tissue called enthesis [23]. The force transmission from muscles to bones is possible because of the complex internal architecture of the tendon [24]. The structure of the tendon is arranged hierarchically by the tropocollagen, collagen fibrils, fibers, and fascicles [25] (Figure 1). A fascicle is a bundle of fibers [25]. Soluble tropocollagen molecules form cross-links to generate insoluble collagen, whose molecules are gradually assembled to arrange collagen fibrils [24]. A group of collagen fibrils, in turn, gives rise to collagen fiber, which corresponds to the basic unit of the tendon [21,24]. The fibrils are oriented within one collagen fiber in a three-dimensional arrangement (longitudinally, horizontally, and transversely), which forms spiral-type plait groups [24,26]. Sheaths of connective tissue surround the collagen fibers (endotendon) and the entire tendon (epitendon) [24]. While the endotendon is a thin reticular network of connective tissue inside the tendon, the epitendon constitutes a dense fibrillar network of collagen [26,27]. The endotendon arrangement enables the fiber groups to glide on each other and drives blood vessels and nerves to a deeper portion of the tendon [28].

Tendon cells (tenoblasts, tenocytes, chondrocytes, synovial cells, and vascular cells) are located between fibril chains and synthesize proteins of the extracellular matrix (ECM), constituted primarily by collagens, large proteoglycans, and small leucine-rich proteoglycans [21]. Type I collagen, elastin, and ECM elements are synthesized by tenocytes, which are fibroblast-like cells situated within collagen fibers and the surrounding endotenon [29,30]. Numerous ECM molecules, including collagens, elastin, proteoglycans, and glycoproteins are involved in the fibrillogenesis of type I collagen [31]. The number and diameter of collagen fibers vary in different tendons [24,32,33]. Collagen fibers and fibrils in tendons present a wavy configuration in the resting state, their shape modifying when the tendon is stretched [27]. A periodic waveform configuration is observed in collagen fibers [34], known as crimp morphology [35–38]. Moreover, collagen fibers are assembled in parallel bundles and aligned along the long axis of the tendon. This arrangement favors the tissue to better respond to mechanical loading [37].

Elastic fibers are sparsely distributed among tendons and account for approximately 1–2% of the dry mass of the tendon [24,39,40]. Elastic fibers are made of fibrillins (fibrillins 1 and 2) and elastin [41], which contribute collectively to the structural integrity and recovery of the wavy configuration of the collagen fibers after stretching [24,42]. As a crucial component in ECM, elastin ensures elastic stretching and recoiling of tissue, cooperating with collagen for tensile resistance [43] and regulating the interactions between cells and the extracellular matrix [44].

The ground substance of tendons consists of proteoglycans (PGs), glycosaminoglycans (GAGs), structural glycoproteins, and water, surrounding the collagen [24]. PGs are core proteins where one or more GAGs are covalently attached [24]. They enable rapid diffusion of water-soluble molecules and migration of cells [24]. The water-binding capacity of proteoglycans and GAGs is essential for stabilization of collagen fibrillogenesis, maintenance of ionic homeostasis, and elasticity of a tendon against shear and compressive forces [24]. Proteoglycans function either as lubricators or organizers of collagen fibril assembly [24,45], retaining water and conferring improved elasticity to tendons [4,24]. Decorin accounts for roughly 80% of the total proteoglycan content in the tissue [46,47]. Laminin, in turn, is found in both the vascular walls and myotendinous junction of tendons [26,48]. Inorganic components correspond to less than 0.2% of the tendon dry mass, including calcium, magnesium, manganese, cadmium, cobalt, copper, zinc, lithium, nickel, fluoride, phosphor, and silicon. These elements play an important role in the growth, development, and metabolism of the tissue. Copper, manganese, and calcium ions seem to assist in the formation of collagen cross-linking and enzymatic reactions related to synthesis of connective tissue molecules [49]. Tendon cells are also involved in energy metabolism, given that tendon cells contain enzymes related to the aerobic Krebs cycle, anaerobic glycolysis, and pentose phosphate shunt [24,50]. In fact, high metabolic activity and intense synthesis of the matrix components has been observed in young tenoblasts [24]. Moreover, physical activity affects human tendons directly through increased metabolic activity [18,51,52] and elevated collagen synthesis [53].

A new type of tendon cells, tendon stem cells (TSPCs) was recently discovered to be present in tendons and ligaments [21]. TSPCs differ from resident tenocytes in shape configuration, proliferation potential, and stem cell specific marker expression [4]. Tendon cell lines express genes of adipogenic, osteogenic, and chondrogenic differentiation pathways, suggesting their capacity to differentiate in vitro [54,55].

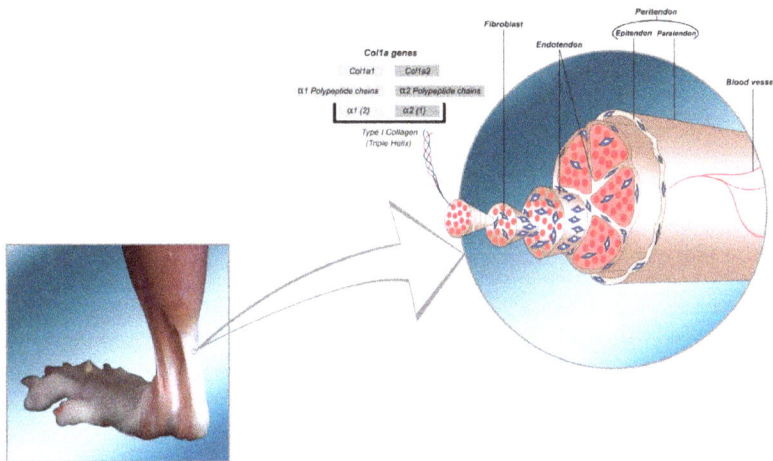

Figure 1. Structure of tendon. The tendon is composed of type I collagen fibers. Type I collagen is the major structural component of the tendon. Col1a1 and Col1a2 code for collagen $\alpha1(I)$ and $\alpha2(I)$ polypeptides, respectively. Type I collagen triple-helical molecules containing two $\alpha1(I)$ and one $\alpha2(I)$ chains assemble into fibrils that combine to form fibers. Tendon fibroblasts reside between collagen fibers. Fibers are surrounded by a connective tissue, the endotendon, which also contains fibroblasts. Fibers combine to form fascicules. Tendons are ensheathed by an outer layer of connective tissue (epitendon), which is surrounded by another layer of connective tissue (paratendon). Together, the epitendon and paratendon external sheaths constitute the peritendon. Adapted from Nourissat et al. 2015 [30] and Lipman et al. 2018 [56].

3. Effect of Training Modalities on Tendon Remodeling

ET is any structured, planned, and repetitive physical activity with the final objective of improving physical fitness [56]. ET is the most common method to apply mechanical loading to tendons [57]. During mechanical loading, ECM supplies tensile strength to the tendon [58]. Tendon cells detect mechanical forces as stimuli that are transduced to biochemical signals, eliciting cellular responses. Mechanical loading results in changes in cytoskeletal components, ECM organization, and gene transcription [30,59] The mechanotransduction mechanism is mediated by growth factors, receptors, intracellular pathways, and transcription factors [30,59]. Mechanotransduction is generally compound into three stages: mechanocoupling (physical load), cell–cell communication and the effector response. The communication throughout a tissue to distribute the loading message and the response at the cellular level to affect the response [60]. Cell responses to the ECM are determined by intrinsic properties that include adhesive affinity, matrix stiffness, fiber alignment, and matrix density [61]. In these terms, integrin-mediated adhesions between cells and ECM are essential for cell function. Cell–matrix adhesions have been described as mechanosensitive since they regulate biochemical signaling. One of the key functions of cell–matrix adhesions is to detect, transmit, and respond to mechanical signals [62]. Accordingly, tendon represents a dynamic, mechanoresponsive tissue [60]. Evidence indicates the existence of a threshold, or set-point at the applied strain magnitude, at which the transduction of the mechanical stimulus seems to impact the tensional homeostasis of the tendons [63]. Furthermore, mechanical forces are involved in type I collagen protein synthesis and ECM components in animal and human tendons [64–67]. In this sense, increased type I collagen formation was observed after acute exercise (2 and 72 h post-training) in the peritendinous tissue of runners [68], suggesting an adaptation to acute physical loading. Accordingly, 4- and 11-weeks post-training, increased turnover of collagen type I was observed in a peritendinous CT region. Interestingly, the authors observed that synthesis and degradation processes elevated after 4 weeks of training, whereas the anabolism was maintained after 11 weeks, generating a net synthesis of type I collagen in the tendon tissue [53]. Collagen synthesis increased in the patellar tendon as a result of a single bout of acute exercise, and this effect was maintained 3 days later [69]. On the other hand, different stress patterns result in different cellular reactions, which depend on the strength of applied stress. For example, repetitive tension applied during one day stimulated proliferation and apoptosis in contrast to extended stress periods [70], which supports the fact of ECM reacts differently depending on the nature and duration of the exercise [71].

Long-term effects of ET were observed on structural and mechanical properties of swine tendons [72], indicating that mechanical forces play a fundamental role in training-induced tendon adaptations. In this sense, striking adaptations in tendons in response to training were documented in a recent systematic review, although the authors highlighted the evident variability between and within studies. Despite the dose-response or time-course of tendon adaptation in response to the first months of training being controversial, larger tendon CSA was associated with long-term (years) training without evidence of differences in material properties [73]. In fact, mechanical loading (and ET) seems to induce both structural and functional adaptations in tendons, including collagen organization, CSA, tendon thickness, elastic energy, force, and stress-strain characteristics (Young's modulus) [63,74–76].

In addition to upregulation of collagen content, treadmill running elevated the expression of mechanical growth factors (MGF) and enhanced the proliferative potential of TSPCs in both the patellar and CT of mice [77]. On the other hand, excessive mechanical loading caused significant differentiation of TSPCs into non-tendon cells [77]. Aberrant mechanical stimulation also favors the production of MMPs, growth factors, and prostaglandins, which can all induce defects in ECM remodeling and, consequently, the induction and progression of tendinosis [78,79]. Mechanical stretching also upregulated MMP gene expression, resulting in elevated interstitial amounts of MMP-2 and MMP-9 in human peritendinous tissue [80]. In fact, pro-MMP-2 levels elevated 3 days after an exercise bout in the peritendinous tissue of young men [81]. On the other hand, Arnoczky and colleagues (2004) observed that ex vivo static tensile loading inhibited the upregulation of MMP-1 expression induced by load deprivation in tendon cells [82]. Moreover, MMP-1 expression in tendon cells can be modulated by different amplitudes and frequencies of cyclic tensile strain [83]. Cyclical load also induced the release of degraded cartilage oligomeric matrix protein (COMP), a non-collagenous ECM protein [58]. In fact, we demonstrated that RT upregulated proteins responsible for ECM organization of tendons, including COL1A1, as well as a series of proteins associated with metabolic/development processes, such as FABPH, GELS, S100A6, TRFE, and serum albumin (ALBU) [84]. Although there is a tendency of upregulation in COMP levels in tendons of young rats submitted to RT, no statistical difference was reported. On the other hand, RT restores the aging-induced downregulation of COMP levels [84].

Further adaptations in tendon tissue have been observed in response to RT, including changes in tissue thickness, strength, resistance to damage, blood flow, and normalization of the fibrillar morphology [75,80,85], although plyometric training did not change force, stiffness, elastic energy, strain, or modulus in CT [76]. RT (or strength training) consists of an ET model that includes concentric and eccentric muscle actions against loads (workload) with the objective of achieving a specific training outcome [86]. RT is the most effective method for developing musculoskeletal strength and has been widely prescribed by the major health organizations [87–89]. In view of divergences concerning studies using experimental models, our research group sought to investigate the effects of RT on the biomechanical properties of aged rat tendons. Moreover, mechanical loading placed on tendon (and muscle) tissue induces collagen expression by the mechanotransduction mechanism, with involvement of TGF-β, CTGF, and IGF-I [90–92]. Together with skeletal muscle responses, we observed that RT significantly elevated the gene expression of key growth factors (TGF-β and CTGF) in gastrocnemius (GAS) and soleus (SOL) muscles of old rats [93].

Whereas tenocytes extracted from flexor and extensor tendons behave similarly when exposed to mechanical strain in vitro, distinct regions of human CT (insertion site, enthesis, and mid-tendon) have been shown to respond differently to mechanical loading. For example, some tendon regions have a high amount of glycosaminoglycans (GAGs), ensuring the ability of the tissue to support strong compressive forces [94].

In our studies on the effects of AAS and aging, we showed that RT improved the biomechanical responses in the CT of rats [95]. RT seems to improve the capacity of tendon tissue to support more stress, although whether different regions of tendons are affected by mechanical loading (and other factors) is still under debate.

In view of tendon repair, the influence of mechanical loading is not the same for different tendon types and tendon regions [96]. The differences in the outcomes probably arise from variation in mechanical loads placed on different tendon areas (for example, mid-tendon versus enthesis) and also on different tendons [30]. In this regard, we recently reported that the CT, SFT, and DFT of trained rats reach values of maximum extension [95]. Furthermore, the CT and SFT of old trained (OT) rats were capable of withstanding more stress, while the DFT showed greater resistance to maximum strain. Maximum extension and strain were observed in the CT of trained groups (either old or young animals) when compared with their sedentary counterparts [95]. Moreover, the CT of young trained rats demonstrated higher capacity to support stress when compared with OT animals, supporting our previous findings regarding the deleterious effects of aging on tendon adaptations. RT also blunted the age-associated low energy absorption in the CT and elevated energy absorption in the SFT.

RT ameliorated the age-induced low energy absorption of the CT and SFT [95]. OT rats presented an increase in elastic modulus (tendon stiffness) of the SFT in relation to young trained animals. RT also promoted greater capacity for producing tendon strength (maximum load) and resisting applied tension (maximum stress) in the SFT of old animals. Moreover, the SFT showed greater capacity to absorb energy to failure and less displacement to maximum load in OT rats when compared with the young trained group. However, RT had no effect on tendon CSA, whereas DFT stiffness was reduced in trained rats. Taken together, we suggest that RT may be considered an effective component that mitigates the age-induced detrimental effects on tendon adaptations [84,95].

In view of transcriptional regulation induced by mechanical loading, we observed that RT induced upregulation of genes linked to ECM homeostasis of tendons, including COL-I, COL-III, CTGF, TGF-β1, IGF-Ia, VEGF, MMP-2, TIMP-1, Bgn, Fmod, tenascin C, and decorin. Whereas RT mitigated the age-associated decrease in IGF-Ia and MMP-2, a dramatic increase in gene expression of CTGF and VEGF were observed in OT animals [97]. In addition, RT promoted higher blood vessel volume density and increased peritendinous sheath cells in young animals, as well as volume density of tendon proper cells being elevated in the proximal region of the CT [97]. Thus, we suggest that RT exerts a protective effect against age-induced tendon adaptations, including the adverse remodeling of ECM proteins and a reduction in the volume density of blood vessels as well.

Proteoglycan content was also elevated in the CT of trained young rats [97], suggesting enhanced elasticity of the CT in response to RT. Moreover, RT prevented age-induced calcification and induced an increase in proteoglycan content in the CT [97]. Furthermore, crimp arrangement of the CT was more intense in response to RT. This was observed quantitatively; however, no significant differences were observed in the birefringence microscopy. Taken together, our studies indicate that RT creates a suitable environment to restore tendons from damage [97].

Considering both collagen synthesis and gene expression analyses, a physiological increase in mechanical loading seems to be beneficial for tendons, whereas a decrease in mechanical loading is detrimental for tendon formation during tissue development [30]. However, RT did not affect the mechanical properties or dimension of the patellar tendon of old individuals [98], even though elevated ECM remodeling has been observed in response to RT [84]. Thus, our results suggest that an RT model might be an effective intervention against aging-induced deleterious effects on the biomechanical and morphological properties of tendons, supporting the use of RT as an important strategy to trigger beneficial adaptations in tendons. Our data are further supported by studies using running training on a treadmill [4,99].

4. Effect of AAS on Tendon Remodeling

A plethora of studies regarding the abuse of AAS have been conducted in subjects (athletes and casual fitness enthusiasts) owing to the potential risk factors for public health [11–14]. In fact, these drugs induce substantial effects on the morphology of skeletal muscle and development of strength [9,100,101]. Furthermore, high doses of AAS have been shown to affect the collagen metabolism [102], which could lead to the supposition that AAS may affect both the mechanical and morphological properties of tendons. These observations led to studies investigating the effects of AAS on tendons, particularly the effects of high doses of AAS on tendon injuries. In the late 1980s and early 1990s, studies investigated the effects of high doses of AAS, concomitant with ET on the structure of tendons [103]. The authors observed collagen dysplasia and reduced volume fraction in tendons of the flexor digitorium longus muscle of mice submitted to ET concomitant with the use of AAS, however, of note, no differences were observed in trained or AAS-treated animals. Changes in the diameter of collagen fibrils were also observed in AAS-treated mice, submitted or not to ET [103,104], although biochemical testing revealed no alterations in fibril diameter, type-III collagen, or fibronectin expressions in another study [105]. In conclusion, the author suggests that AAS-induced collagen abnormalities depend on the duration of AAS treatment and might evoke clinical disorders and tendon rupture [106,107]. Subsequently, morphometric analyses of rat tendons were studied in response to AAS and ET. In combination or not with ET, AAS produced augmented crimp angle and reduced crimp length, which has implications for tendon mechanical properties and functional behavior [108,109]. Tendon elasticity was reduced in rats that received AAS. Finally, the authors suggested that the combination of AAS and ET might predispose tendons to injury and rupture, which has been supported by other studies with humans [110,111]. Biomechanical tests showed that AAS induced a stiffer tendon that presented less elongation, as well as impaired energy absorption, elongation, and toe-limit elongation [105,112]. Interestingly, the effects were reversible following discontinuation of the use of AAS [105]. Publications on the topic of ultrastructural analysis of tendons date from the early 20th century. Considering that tendon ruptures could be associated with tissue fibrosis, investigators have focused on the effects of AAS and ET on collagen metabolism. In this sense, high doses of AAS enhance collagen synthesis in soft connective tissues (muscle, bone, and tendon) [102].

Considering that progress has been made since then on molecular approaches, our research group sought to investigate the effects of AAS and RT, which reflects a mechanical loading condition on tendon homeostasis, particularly events related to ECM remodeling – as will be discussed later in this review. Firstly, the Marqueti study demonstrated that AAS administration (Deca-Durabolin and Durateston) impaired the training-induced increases in MMP activity of CT of rats [113]. It should be noted that a marked elevation in metalloproteinase 2 (MMP-2) activity was observed only in trained rats, suggesting that mechanical loading is a potential stimulus for tendon turnover (synthesis, degradation, and re-synthesis). Skeletal muscle seems to be more sensitive than tendons in response to mechanical loading [90], which is supported by studies demonstrating elevated MMP expression in the muscle [93,114–117]. On the other hand, MMP expression seems to be differently regulated by mechanical loading in distinct regions of tendon (proximal and distal) [16], while, MMPs response might also be dependent on type and frequency of stimuli [118–121]. In this sense, further research is needed to a better understanding of mechanical loading effects on MMP expression in the tendons. Interestingly, a pronounced increase in serum corticosterone levels was observed in trained rats that received high doses of AAS [113]. Considering that AAS might inhibit collagen synthesis as well as which corticosterone treatment suppresses the synthesis of collagen types I and III, the study showed evidence of the detrimental effects of AAS on collagen synthesis by decreasing the MMP-2 proteolytic pathway.

In view of our preliminary results, we were interested in investigating whether MMP-2 activity would be altered in different regions of distinct tendons, since tendon composition and ECM elements vary according to regions (proximal and distal) [122], stage of development, mechanical loading, and aging [15]. Thus, MMP-2 activity was measured in the proximal and distal regions of 3 tendons—CT, superficial (SFT), and deep flexor tendon (DFT)—in response to AAS supplementation and RT [16]. The study showed that the SFT responded more to RT, in association with recruitment of the flexor digitorum superficialis muscle, while DFT was more related to the movement of distal parts of the digits. SFT, in turn, responded more to AAS compared to DFT and CT. However, the disparity in responses among the tendons might be due to the distinct metabolism and loading demands imposed on tendon regions during RT.

Considering our previous findings related to collagen metabolism, we hypothesized that molecular events would affect functional properties of rat tendons. To date, no investigations have been carried out regarding the effects of AAS supplementation and ET. On this basis, Marqueti and colleagues (2011) compared the biomechanical properties in the CT, SFT, and DFT of rats treated with high doses of AAS and submitted to RT [123]. The authors observed that the CT accommodated less energy and resisted tensional load more than the SFT and DFT in the sedentary group. SFT was slightly affected by RT, AAS, or a combination of both. On the other hand, increased elastic modulus of the SFT was observed in trained rats supplemented with AAS in comparison with the sedentary group. The DFT, in turn, supported more stress in response to RT, while AAS alone demonstrated no effect. Furthermore, the DFT showed a reduction in the displacement at maximum load when training and AAS were associated. In other words, the AAS reversed the effect of exercise and induced the DFT to exhibit less deformation. The tendons of AAS-treated rats submitted or not to RT exhibited either a decreased capacity to resist tension (i.e., decreased maximum strain) or accommodate levels of tensile strength (i.e., decreased toe region) together with reduced deformability (i.e., increased elastic modulus). In this sense, elevated stiffness, stress and modulus of patellar tendon were observed in AAS users, suggesting higher risk of tendon injury [124]. Accordingly, the different responses between the tendons are supported by biomechanical analysis in a study using an RT model (jump training), where the authors outlined the movement of rats during training [123]. In fact, AAS supplementation revoked the RT-induced effects on the CT. The study emphasizes that the loss of tendon flexibility might raise the risk of tendon rupture during training in individuals who abusively use AAS.

According to our findings regarding biomechanical properties, AAS supplementation and RT might regulate gene expression of key elements responsible for ECM homeostasis. In this regard, mRNA levels of type I collagen-α1 (COL1A1), type III collagen-α1(COL3A1), tissue inhibitor of metalloproteinase 1 (TIMP-1), tissue inhibitor of metalloproteinase 2 (TIMP-2), MMP-2, (insulin-like grow factor I-Ea) IGF-IEa, (glyceraldehyde-3-Phosphatase Dehydrogenase) GAPDH, (connective tissue growth factor) CTGF, and (transforming growth factor beta 1) TGFβ-1 were evaluated in different regions of the CT, SFT, and DFT of rats [125]. RT did not alter COL1A1, COL3A1, MMP-2, or IGF-IEa mRNA levels in the tendons (except for the distal region of the DFT), which is partially in accordance with other studies using different ET models (voluntary wheel running or squat apparatus) [126], as gene expression of type I collagen (COL-I), type III collagen, (COL-III), and TGFβ-1 expression was upregulated in the CT of trained female rats [90]. On the other hand, ET has been shown to increase collagen turnover, including some degree of collagen synthesis, in human tendons [80]. AAS supplementation, in turn, decreased the RT-induced upregulation of IGF-1 mRNA levels in the intermediate region of the SFT. AAS supplementation reduced the expression of COL1A1 in proximal and distal regions of the CT as well as in the proximal region of the DFT. In the SFT and DFT, AAS combined or not with RT reduced the expression of COL1A1 in both intermediate and distal regions. Similarly, AAS treatment reduced COL3A1 expression in both the distal region of the CT and intermediate region of the DFT. AAS combined or not with RT decreased COL3A1 expression in both intermediate and distal regions of the SFT as well as in the proximal region of the DFT. AAS also reduced MMP-2 expression in the proximal and distal regions of the CT and in the proximal region of the DFT, while AAS combined with RT decreased MMP-2 expression in the distal region of the CT and intermediate region of the DFT. The lower MMP-2 activity observed in different regions of tendons [16], provides further information concerning the harmful effects of AAS on tendons at a transcriptional level [125]. Accordingly, AAS enhanced TIMP-1 mRNA levels in the proximal region of the CT, while AAS combined or not with RT reduced TIMP-1 expression in the intermediate region of the SFT [125]. These findings might support the decreased gene expression and activity of MMP-2 in tendon tissue, as TIMP-1 has been demonstrated to play a pivotal role as an endogenous antagonist of MMPs [127].

As the potential effects of AAS and RT were observed in different tendons (and regions) [125], it is conceivable that the ultrastructure of tendons could be affected. Thus, qualitative analyses were carried out on the CT, SFT, and DFT of rats regarding volume density (Vv%) of the adipose cells, blood vessels (blood vessel lumen, endothelial cells, and perivascular sheath), peritendinous sheath cells, and tendon proper cells (fibroblasts and fibrochondrocyte-like cells) [128]. The major finding of this study relies on the distinct morphological adaptations in response to RT, which was linked to their composition and regional function. For example, different arrangements were found in the intermediate region of the SFT and DFT from the sedentary group, as well as increased cellularity and blood vessels. In addition, the SFT and DFT were similar in respect to material properties, which is consistent with displacement at the maximum load, stress, strain, and elastic modulus [123]. On the other hand, the CT accommodated less energy and resisted the tensional load more promptly than the SFT and DFT in the sedentary group. These findings might suggest that different tendons perform distinct functions in a set of movements by modulating the remodeling of the ECM proteins and adapting differently to new physiological demands [128]. RT altered the tendon proper cells and peritendinous sheath, which affects cells (adipocytes, synovial-like cells, fibroblasts, and fibrochondrocytes) and tendon vascularization (blood vessels) [128]. The Vv% of the tendon proper cells in the SFT and DFT (proximal and distal regions) and cell Vv% of the peritendinous sheath of the SFT (all regions) and DFT (distal region) increased significantly in response to RT. These findings indicate increased metabolism in tendons, which was related to greater blood vessel Vv% and adaptation following tissue remodeling during 7 weeks of training. In contrast, AAS combined with RT seems to mitigate the RT-induced increase in blood vessel Vv% in the CT (proximal region), SFT (intermediated region), and DFT (all regions). This combination (AAS and RT) also induced higher accumulation of adipose cells in the proximal region of the CT. Adipose cells play an important role with respect to the facilitation of movement between fascicles of tendons and dissipation of stress and tension at the attachment sites. Synovial-like cells were observed around the peritendinous sheath of trained rats (SFT, proximal, and intermediate region), AAS-treated rats (CT, distal region), and rats treated with AAS and submitted to RT (intermediate region of SFT, and proximal and intermediate regions of DFT). Synovial cells produce synovial fluid, which reduces tissue friction at the myotendinous interface [25]. The presence of synovial-like cells in the intermediate region of the SFT in trained rats supplemented with AAS suggests the possibility of an injury site. In fact, we have previously reported that AAS and training combined induced stiffer tendons [123], which might lead to the occurrence of tendon injuries.

RT elevated collagen content, as measured by hydroxyproline concentration, in the distal region of the CT, and intermediate and distal regions of the DFT. Curiously, high levels of collagen content were observed in sedentary animals, particularly in regions containing more cells. One feasible explanation for high levels of collagen content might be related to the paw position of rats in their cages [128]. AAS supplementation combined or not with RT reduced collagen content in some tendon regions, which might indicate adverse effects of AAS abuse on collagen metabolism of tendons. Collectively, the negative effects of AAS supplementation observed in this study include a reduction in blood vessels, increased adipose cell Vv%, the presence of synovial-like cells, and a reduction in collagen content [128]. Thus, we demonstrated compelling evidence regarding the effects of abusive AAS supplementation on tendon homeostasis of rats. Taken together, our findings suggest that AAS, particularly when combined with high-intensity RT, exerts harmful effects on tendon tissue in terms of ECM remodeling, and molecular and biomechanical adaptations. In view of these findings, RT might play important role in various aspects of tendon remodeling. On the other hand, a recent systematic review highlights the need for future research on this topic [129].

5. Effect of Aging on Tendon Remodeling

In the early 1980s, a study revealed that aging induced morphological and biochemical alterations in rabbit tendons, including an increase in ECM proteins and collagen concentration and a decrease in water content. Structural alterations in elastic fibers and a decrease in cell numbers of tendons were also observed in response to aging [2]. It should be noted that the drop in cell number and reduced synthetic activity of tendons seem to be attributed to the maturation process rather than the aging phenomenon *per se* [5]. On the other hand, while diameter increases with advancing age, the thickness of fibers is affected in a distinct pattern [2]. It is also becoming evident that age-induced alterations in tendon structure might influence mechanical function [5]. Mechanically, aging appears to be associated with a reduction in the modulus and strength of tendons [5], leading tendons to be more susceptible to injuries [130,131]. In this sense, the low metabolic rate of aged tendons slows the rate of recovery and healing processes after activity and injury, as the metabolism of aged tendons is more dependent on anaerobic energy pathways than aerobic pathways [24].

As the cell-to-matrix ratio gradually decreases with aging, morphological changes occur in the tendon cells. Furthermore, tenoblasts turn into tenocytes (and occasionally vice versa) and become very elongated while the nucleus-to-cytoplasm ratio increases. The cell processes are required to maintain strict contact between the cells and matrix components and then compensate the decreasing number of cells and increasing amount of tendon matrix [24]. Moreover, Gagliano and colleagues (2018) demonstrated the ability of tenocytes to maintain ECM remodeling in aged tendons, supporting the hypothesis that structure and biomechanical properties are preserved with advancing age [132]. He collagen synthesis of tendons reduces drastically with advancing age [24], even though the diameter of collagen fibrils appears to remain unchanged [133]. The relative distribution of collagen fibril sizes may change with aging, while total content of collagen fibrils (volume fraction) remains unaltered [5,133,134]. Furthermore, studies reveal that tendons become more susceptible to weakening [58] and stiffer with advancing age [135,136], although when the effects of aging are separated from those of the maturation process, a decrease in tendon stiffness has been observed [137,138]. In fact, gastrocnemius tendon stiffness was lower in older adults compared to young individuals [139]. The author suggested that more compliant tendons of the elderly allow the muscle fibers to shorten more. In addition, we have observed increases in connective tissue content in the soleus and gastrocnemius of old rats [93], even though no biomechanical analysis was performed. However, our results are in accordance with the Reeves study, suggesting lower stiffness of the gastrocnemius tendon, and more likelihood of injury. Indeed, mechanical properties of tendons are impaired with aging [139].

Substantial alterations in cross sectional area (CSA) and material properties (maximum load, stress, strain, load, elastic modulus, energy at failure, and displacement at maximum load) were observed in the CT, superficial (SFT), and deep (DFT) flexor tendon of older rats, suggesting that aging leads to increased tendon stiffness [95]. Moreover, aging induced low energy absorption in the CT and SFT. Overall, we demonstrated that aging reduced the ability of the CT and SFT to absorb energy, while decreased extension and increased elastic modulus were observed in the DFT of old rats, indicating an important effect of aging in different patterns of tendons.

In experimental models, the normal aging process is characterized by a series of morphological and biochemical alterations in tendons, including decreased collagen turnover [19], increased elastin levels [140], accumulation of partially degraded collagen within the matrix [141], changes in age-related cross-links as a result of glycation reactions [142], and decreased levels of abundant non-collagenous protein, as observed by the lower levels of cartilage oligomeric matrix protein (COMP) [143]. The age of animals used in experimental models varies between species. In this sense, rats ranging in age from 18 to 30 months are usually used in studies addressing this matter [144–146]. Accordingly, we demonstrated that aging caused meaningful reductions in proteins related to ECM organization, including COMP and collagen alpha-2(I) chain (COL1A2) as well as proteins associated with metabolic/development processes, such as carbonic anhydrase (CAH3), fatty acid-binding protein heart (FABPH), (acid-binding protein-4) FABP4, parvalbumin alpha (PRVA), gelsolin (GELS), protein S-100A6 (S100A6), and serotransferrin (TRFE) [84].

Despite recent advances regarding the effects of aging on collagen concentration, compelling evidence has indicated that aging alters collagen cross-linking (both enzymatic and non-enzymatic reactions) rather than collagen concentration in tendons in experimental models on aging [134]. Those non-enzymatic glycation/glycosylation reactions result in production of advanced glycation end products (AGE) in tendon tissue [147]. AGE accumulation is dependent on collagen turnover rates [148] and accumulates to a higher extent in tendon than in skeletal muscle as these tissues show low and high turnover, respectively [149,150]. Glycation increases the distance between collagen molecules within tendon collagen fibrils and, therefore, affects their molecular structure [151]. AGEs also contribute to the loss of water in aged tendons [2], since cross-links lead to dehydration of collagen [152]. The cross-links are important to stabilize the collagen fiber and thereby contribute to the mechanical properties and stiffness of the tendon [142,153]. Collagen concentration was reduced in the patellar tendon of elderly men, even though collagen cross-linking was elevated. This might be considered as a crucial mechanism to maintain the mechanical properties of tendons with advancing age [154]. Aging has been shown to be a potential factor that leads to AGE accumulation [155], resulting in stiffer and more load-resistant tendons [156,157]. Furthermore, recent studies demonstrated a reduction in tendon stiffness with aging [158–160]. While there is no consensus regarding the collagen content in aged tendon, the most remarkable structural alteration is an increase in AGE cross-links, which might lead to an increase in tendon modulus and strength [5]. In vitro observations indicate that AGE favors increased tensile stress and stiffness with aging [161,162].

Effects of aging on tendon mechanical properties (modulus and strength) are still a matter of intense research. For example, stronger and stiffer tendons have been observed with aging [133,163,164], while other studies have shown weaker and more compliant tendons [165–168]. Furthermore, aging seems to affect the viscoelastic properties of tendons [169,170]. Relevant age-dependent differences in ultimate stress, relaxation rate, and percent relaxation were observed in the tail tendon of old rats, where relaxation rate and percent relaxation decreased with age [167]. As mechanical properties of the tendon influence the muscle-tendon unit function [5], it is worth investigating the underlying events of aging-associated tendon development, particularly ECM remodeling. Indeed, dramatic changes in the rate of force development, elastic energy return, and electromechanical delay [17,171,172] might affect balance and mobility in elderly people [160].

As relevant alterations have been observed in terms of functional properties in different aged tendons, we speculate that ECM proteins might be modulated at a molecular level. In fact, we demonstrated that key genes associated with ECM remodeling were downregulated in the tendons of old rats [97]. Our study revealed that COL-I content and GAGS were reduced in the CT of older sedentary (OS) animals, while biglican (Bgn) and fibromodulin (Fmod) content were not altered with aging. Gene expression of COL-I, COL-III, IGF-Ia, MMP-2, TIMP-2, and Bgn were inhibited in aged tendons. MMP-2 activity of tendons was not affected in response to aging. OS animals exhibited higher adipose cell volume density in the proximal compared to distal region of the CT. Aging also reduced blood vessel volume density in both regions (proximal and distal) [97]. Furthermore, OS animals presented substantial calcification in the distal region of the CT [97]. In this regard, reduced levels of Bgn, GAGS, and Fmod seem to regulate ectopic ossification (calcification) in the CT [21,71]. Thus, we speculate that reduction in GAGS content might contribute to CT calcification of old rats [97], therefore, favoring greater CT stiffness in OS animals, which is in accordance with our previous findings [95]. While calcification increases with aging, mechanical loading induced a marked reduction in CT calcification in old mice submitted to ET [99], suggesting that mechanical loading plays a fundamental role in age-related tendon calcification.

In addition to the transcriptional approaches, we carried out proteomic analysis of the CT in old rats. We demonstrated high levels of fatty FABP4 with aging [84]. Thus, the higher adipose cell volume density observed in tendons of old sedentary rats [97] might support, at least in part, our previous findings regarding the FABP4 protein [84]. Considering that FABP4 is a protein associated with obesity and metabolic syndrome [173], we suggest aging-induced accumulation of adipose tissue in tendons, as confirmed by morphologic analysis [97]. In this context, aging has been associated with a lower rate of cell proliferation and a reduction in the number of TSPCs of tendons [5,174–176], while ET can mitigate age-induced deleterious effects on TSPCs proliferation [177], suggesting the importance of ET on tendon adaptations [4,5,77].

6. Conclusions

Our results suggest that AAS, particularly when combined with high-intensity RT, exerts harmful effects on ECM remodeling, and molecular and biomechanical adaptations in rat tendons. Moreover, we provide further information regarding the harmful effects of AAS on tendons at a transcriptional level. We also demonstrated the beneficial effects of RT against the age-induced tendon adaptations of rats. Therefore, our studies might contribute in terms of clinical approaches in favor of the benefits of ET against tendinopathy conditions and provide a warning on the harmful effects of AAS misuse on tendon development. Our results are summarized in Figure 2.

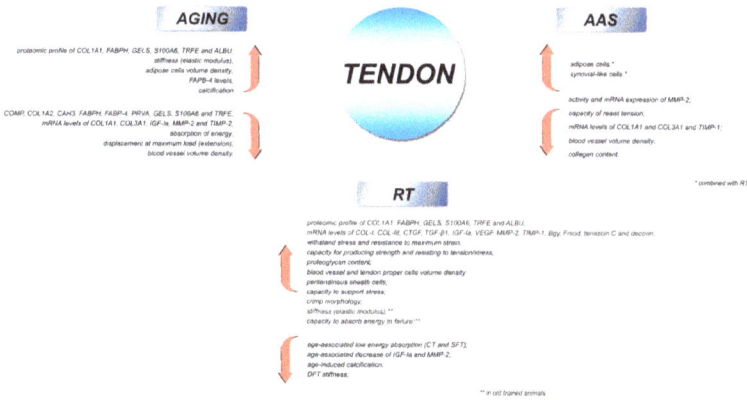

Figure 2. Integrative model based on our findings, indicating the major effects of AAS, RT, and aging on tendons of rats. AAS—anabolic androgenic steroids; RT—resistance training; COL-I—type I collagen; COL-III—type III collagen; COL1A1—type I collagen-α1; COL3A1—type III collagen-α1; TIMP-1—tissue inhibitor of metalloproteinase 1; TIMP-2—tissue inhibitor of metalloproteinase 2; MMP-2—metalloproteinase 2; IGF-IEa—insulin-like growth factor I-Ea; CTGF—connective tissue growth factor; TGFβ-1—transforming growth factor beta 1; VEGF—vascular endothelial growth factor; Bgn—biglycan; Fmod—fibromodulin; COMP—cartilage oligomeric matrix protein; FABPH—acid-binding protein heart; FABP4—acid-binding protein-4; GELS—gelsolin; S100A6—protein S-100A6; TRFE—serotransferrin; ALBU—serum albumin; CAH3—carbonic anhydrase; PRVA—parvalbumin alpha.

Author Contributions: V.G, H.S.S.A. and R.C.M. made substantial contributions to the conception of the Review and approved the submitted version. The authors agree to be personally accountable for the author's own contributions and ensuring that questions related to the accuracy or integrity of any part of the work, even those in which the author was not personally involved, are appropriately investigated, resolved, and documented in the literature.

Funding: This research was funded by Coordenação de Aperfeiçoamento de Pessoal de Nível Superior (CAPES 88887.136367/2017-00), Conselho Nacional de Pesquisa (CNPq 465699/2014-6, 445069/2014-7), Fundação de Amparo à Pesquisa do Estado de São Paulo (FAPESP 2013/00798-2, 2014/50938-8) and Fundação de Apoio à Pesquisa do Distrito Federal (FAPDF 193.000.653/2015).

Acknowledgments: We thank José Pedro Romano Segundo for his assistance in the preparation of the figures.

Conflicts of Interest: The authors declare no conflict of interest.

Abbreviations

AAS	anabolic androgenic steroids
ALBU	serum albumin
Bgn	biglycan
CAH3	carbonic anhydrase
COL1A1	type I collagen-α1
COL3A1	type III collagen-α1
COL-I	type I collagen
COL-III	type III collagen
COMP	cartilage oligomeric matrix protein
CSA	cross-sectional area
CT	calcaneal tendon
CTGF	connective tissue growth factor
DFT	deep flexor tendon
ECM	extracellular matrix
FABP4	acid-binding protein-4
FABPH	acid-binding protein heart
Fmod	fibromodulin
GAGS	glycosaminoglycans
GAPDH	glyceraldehyde-3-Phosphatase Dehydrogenase
GELS	gelsolin
IGF-IEa	insulin-like grow factor I-Ea
MMP-2	metalloproteinase 2
PRVA	parvalbumin alpha
RT	resistance training
S100A6	protein S-100A6
SFT	superficial flexor tendon
TGFβ-1	transforming growth factor beta 1
TIMP-1	tissue inhibitor of metalloproteinase 1
TIMP-2	tissue inhibitor of metalloproteinase 2
TRFE	serotransferrin
TSPC	tendon-derived stem/progenitor cells
VEGF	vascular endothelial growth factor

References

1. Khan, K.M.; Cook, J.L.; Bonar, F.; Harcourt, P.; Astrom, M. Histopathology of common tendinopathies. Update and implications for clinical management. *Sports Med.* **1999**, *27*, 393–408. [CrossRef] [PubMed]
2. Ippolito, E.; Natali, P.G.; Postacchini, F.; Accinni, L.; De Martino, C. Morphological, immunochemical, and biochemical study of rabbit Achilles tendon at various ages. *J. Bone Jt. Surg. Am.* **1980**, *62*, 583–598. [CrossRef] [PubMed]

3. Kader, D.; Mosconi, M.; Benazzo, F.; Maffulli, N. Achilles tendon rupture. In *Tendon Injuries*; Springer: London, UK, 2005.

4. Thampatty, B.P.; Wang, J.H.C. Mechanobiology of young and aging tendons: In vivo studies with treadmill running. *J. Orthop. Res.* **2018**, *36*, 557–565. [CrossRef] [PubMed]

5. Svensson, R.B.; Heinemeier, K.M.; Couppé, C.; Kjaer, M.; Magnusson, S.P. Effect of aging and exercise on the tendon. *J. Appl. Physiol.* **2016**, *121*, 1237–1246. [CrossRef] [PubMed]

6. Pope, H.G.; Kanayama, G.; Athey, A.; Ryan, E.; Hudson, J.I.; Baggish, A. The lifetime prevalence of anabolic-androgenic steroid use and dependence in Americans: Current best estimates. *Am. J. Addict.* **2014**, *23*, 371–377. [CrossRef] [PubMed]

7. Schwingel, P.A.; Cotrim, H.P.; dos Santos, C.R.; dos Santos, A.O.; de Andrade, A.R.C.F.; Carruego, M.V.V.B.; Zoppi, C.C. Recreational anabolic-androgenic steroid use associated with liver injuries among Brazilian young men. *Subst. Use Misuse* **2015**, *50*, 1490–1498. [CrossRef] [PubMed]

8. Copeland, J.; Peters, R.; Dillon, P. Anabolic-androgenic steroid use disorders among a sample of Australian competitive and recreational users. *Drug Alcohol Depend.* **2000**, *60*, 91–96. [CrossRef]

9. Graham, M.R.; Davies, B.; Grace, F.M.; Kicman, A.; Baker, J.S. Anabolic Steroid Use. *Sports Med.* **2008**, *38*, 505–525. [CrossRef] [PubMed]

10. Guzzoni, V.; Cunha, T.S.; das Neves, V.J.; Briet, L.; Costa, R.; Moura, M.J.C.S.; Oliveira, V.; do Carmo Pinho Franco, M.; Novaes, P.D.; Marcondes, F.K. Nandrolone combined with strenuous resistance training reduces vascular nitric oxide bioavailability and impairs endothelium-dependent vasodilation. *Steroids* **2018**, *131*, 7–13. [CrossRef]

11. Haupt, H.A.; Rovere, G.D. Anabolic steroids: A review of the literature. *Am. J. Sports Med.* **1984**, *12*, 469–484. [CrossRef]

12. Yesalis, C.E.; Wright, J.E.; Bahrke, M.S. Epidemiological and policy issues in the measurement of the long-term health effects of anabolic-androgenic steroids. *Sports Med.* **1989**, *8*, 129–138. [CrossRef] [PubMed]

13. Kutscher, E.C.; Lund, B.C.; Perry, P.J. Anabolic steroids: A review for the clinician. *Sports Med.* **2002**, *32*, 285–296. [CrossRef] [PubMed]

14. Kanayama, G.; Hudson, J.I.; Pope, H.G. Illicit anabolic-androgenic steroid use. *Horm. Behav.* **2010**, *58*, 111–121. [CrossRef] [PubMed]

15. Carvalho, H.F.; Felisbino, S.L.; Keene, D.R.; Vogel, K.G. Identification, content, and distribution of type VI collagen in bovine tendons. *Cell Tissue Res.* **2006**, *325*, 315–324. [CrossRef] [PubMed]

16. Marqueti, R.C.; Prestes, J.; Paschoal, M.; Ramos, O.H.; Perez, S.E.; Carvalho, H.F.; Selistre-de-Araujo, H.S. Matrix metallopeptidase 2 activity in tendon regions: Effects of mechanical loading exercise associated to anabolic-androgenic steroids. *Eur. J. Appl. Physiol.* **2008**, *104*, 1087–1093. [CrossRef] [PubMed]

17. Magnusson, S.P.; Narici, M.V.; Maganaris, C.N.; Kjaer, M. Human tendon behaviour and adaptation, in vivo. *J. Physiol.* **2008**, *586*, 71–81. [CrossRef] [PubMed]

18. Hannukainen, J.; Kalliokoski, K.K.; Nuutila, P.; Fujimoto, T.; Kemppainen, J.; Viljanen, T.; Laaksonen, M.S.; Parkkola, R.; Knuuti, J.; Kjær, M. In vivo measurements of glucose uptake in human Achilles tendon during different exercise intensities. *Int. J. Sports Med.* **2005**, *26*, 727–731. [CrossRef]

19. Thorpe, C.T.; Peffers, M.J.; Simpson, D.; Halliwell, E.; Screen, H.R.C.; Clegg, P.D. Anatomical heterogeneity of tendon: Fascicular and interfascicular tendon compartments have distinct proteomic composition. *Sci. Rep.* **2016**, *6*, 20455. [CrossRef]

20. Birch, H.L. Tendon matrix composition and turnover in relation to functional requirements. *Int. J. Exp. Pathol.* **2007**, *88*, 241–248. [CrossRef]

21. Bi, Y.; Ehirchiou, D.; Kilts, T.M.; Inkson, C.A.; Embree, M.C.; Sonoyama, W.; Li, L.; Leet, A.I.; Seo, B.M.; Zhang, L.; et al. Identification of tendon stem/progenitor cells and the role of the extracellular matrix in their niche. *Nat. Med.* **2007**, *13*, 1219–1227. [CrossRef]

22. Provenzano, P.P.; Vanderby, R. Collagen fibril morphology and organization: Implications for force transmission in ligament and tendon. *Matrix Biol.* **2006**, *25*, 71–84. [CrossRef] [PubMed]

23. Zelzer, E.; Blitz, E.; Killian, M.L.; Thomopoulos, S. Tendon-to-bone attachment: From development to maturity. *Birth Defects Res. Part C-Embryo Today Rev.* **2014**, *102*, 101–112. [CrossRef] [PubMed]

24. Kannus, P. Structure of the tendon connective tissue. *Scand. J. Med. Sci. Sports* **2000**, *10*, 312–320. [CrossRef] [PubMed]

25. Benjamin, M.; Kaiser, E.; Milz, S. Structure-function relationships in tendons: A review. *J. Anat.* **2008**, *212*, 211–228. [CrossRef] [PubMed]

26. Jozsa, L.; Kannus, P.; Balint, J.B.; Reffy, A. Three-dimensional ultrastructure of human tendons. *Acta Anat. (Basel)* **1991**, *142*, 306–312. [CrossRef] [PubMed]

27. Rowe, R.W.D. The structure of rat tail tendon fascicles. *Connect. Tissue Res.* **1985**, *14*, 21–30. [CrossRef] [PubMed]

28. Elliott, D. Structure and function of mammalian tendon. *Biol. Rev.* **1965**, *40*, 392–421. [CrossRef] [PubMed]

29. Hess, G.P.; Cappiello, W.L.; Poole, R.M.; Hunter, S.C. Prevention and Treatment of Overuse Tendon Injuries. *Sports Med.* **1989**, *8*, 371–384. [CrossRef] [PubMed]

30. Nourissat, G.; Berenbaum, F.; Duprez, D. Tendon injury: From biology to tendon repair. *Nat. Rev. Rheumatol.* **2015**, *11*, 223–233. [CrossRef] [PubMed]

31. Mienaltowski, M.J.; Birk, D.E. Structure, physiology, and biochemistry of collagens. *Adv. Exp. Med. Biol.* **2014**, *802*, 5–29.

32. Angel, G.; Gheorghe, V. Interferometric evaluation of collagen concentration in tendon fibers. *Connect. Tissue Res.* **1985**, *13*, 323–337. [CrossRef] [PubMed]

33. Moore, M.J.; De Beaux, A. A quantitative ultrastructural study of rat tendon from birth to maturity. *J. Anat.* **1987**, *153*, 163–169. [PubMed]

34. Viidik, A.; Ekholm, R. Light and electron microscopic studies of collagen fibers under strain. *Anat. Embryol. (Berl)* **1968**, *127*, 154–164. [CrossRef]

35. Vidal, C.B. Crimp as part of a helical structure. *C. R. Acad. Sci. III* **1995**, *318*, 173–178.

36. Hansen, K.A.; Weiss, J.A.; Barton, J.K. Recruitment of tendon crimp with applied tensile strain. *J. Biomech. Eng.* **2002**, *124*, 72. [CrossRef]

37. Franchi, M.; Fini, M.; Quaranta, M.; De Pasquale, V.; Raspanti, M.; Giavaresi, G.; Ottani, V.; Ruggeri, A. Crimp morphology in relaxed and stretched rat Achilles tendon. *J. Anat.* **2007**, *210*, 1–7. [CrossRef]

38. Raspanti, M.; Manelli, A.; Franchi, M.; Ruggeri, A. The 3D structure of crimps in the rat Achilles tendon. *Matrix Biol.* **2005**, *24*, 503–507. [CrossRef] [PubMed]

39. Kirkendall, D.T.; Garrett, W.E. Function and biomechanics of tendons. *Scand. J. Med. Sci. Sports* **2007**, *7*, 62–66. [CrossRef]

40. Pang, X.; Wu, J.P.; Allison, G.T.; Xu, J.; Rubenson, J.; Zheng, M.H.; Lloyd, D.G.; Gardiner, B.; Wang, A.; Kirk, T.B. Three-dimensional microstructural network of elastin, collagen, and cells in Achilles tendons. *J. Orthop. Res.* **2017**, *35*, 1203–1214. [CrossRef]

41. Giusti, B.; Pepe, G. Fibrillins in tendon. *Front. Aging Neurosci.* **2016**, *8*, 237. [CrossRef]

42. Mithieux, S.M.; Weiss, A.S. Elastin. In *Advances in Protein Chemistry*; Academic Press: New York, NY, USA, 2005; Volume 70, pp. 437–461.

43. Green, E.M.; Mansfield, J.C.; Bell, J.S.; Winlove, C.P. The structure and micromechanics of elastic tissue. *Interface Focus* **2014**, *4*, 20130058. [CrossRef]

44. Kielty, C.M. Elastic fibres in health and disease. *Expert Rev. Mol. Med.* **2006**, *8*, 1–23. [CrossRef]

45. Yoon, J.H.; Halper, J. Tendon proteoglycans: Biochemistry and function. *J. Musculoskelet. Neuronal Interact.* **2005**, *5*, 22–34.

46. Samiric, T.; Ilic, M.Z.; Handley, C.J. Characterisation of proteoglycans and their catabolic products in tendon and explant cultures of tendon. *Matrix Biol.* **2004**, *23*, 127–140. [CrossRef]

47. Screen, H.R.; Berk, D.E.; Kadler, K.E.; Ramirez, F.; Young, M.F. Tendon functional extracellular matrix. *J. Orthop. Res.* **2015**, *33*, 793–799. [CrossRef]

48. Järvinen, M.; Kannus, P.; Kvist, M.; Isola, J.; Lehto, M.; Jozsa, L. Macromolecular composition of the myotendinous junction. *Exp. Mol. Pathol.* **1991**, *55*, 230–237. [CrossRef]

49. Minor, R.R. Collagen metabolism: A comparison of diseases of collagen and diseases affecting collagen. *Am. J. Pathol.* **1980**, *98*, 225–280. [CrossRef]

50. O'Brien, M. Structure and metabolism of tendons. *Scand. J. Med. Sci. Sports* **1997**, *7*, 55–61. [CrossRef]

51. Bojsen-Møller, J.; Kalliokoski, K.K.; Seppänen, M.; Kjaer, M.; Magnusson, S.P. Low-intensity tensile loading increases intratendinous glucose uptake in the Achilles tendon. *J. Appl. Physiol.* **2006**, *101*, 196–201. [CrossRef]

52. Kalliokoski, K.K.; Langberg, H.; Ryberg, A.K.; Scheede-Bergdahl, C.; Doessing, S.; Kjaer, A.; Boushel, R.; Kjaer, M. The effect of dynamic knee-extension exercise on patellar tendon and quadriceps femoris muscle glucose uptake in humans studied by positron emission tomography. *J. Appl. Physiol.* **2005**, *99*, 1189–1192. [CrossRef]

53. Langberg, H.; Rosendal, L.; Kjær, M. Training-induced changes in peritendinous type I collagen turnover determined by microdialysis in humans. *J. Physiol.* **2001**, *534*, 297–302. [CrossRef]

54. Salingcarnboriboon, R.; Yoshitake, H.; Tsuji, K.; Obinata, M.; Amagasa, T.; Nifuji, A.; Noda, M. Establishment of tendon-derived cell lines exhibiting pluripotent mesenchymal stem cell-like property. *Exp. Cell Res.* **2003**, *287*, 289–300. [CrossRef]

55. De Mos, M.; Koevoet, W.J.L.M.; Jahr, H.; Verstegen, M.M.A.; Heijboer, M.P.; Kops, N.; Van Leeuwen, J.P.T.M.; Weinans, H.; Verhaar, J.A.N.; Van Osch, G.J.V.M. Intrinsic differentiation potential of adolescent human tendon tissue: An in-vitro cell differentiation study. *BMC Musculoskelet. Disord.* **2007**, *8*, 16. [CrossRef]

56. Caspersen, C.J.; Powell, K.E.; Christenson, G.M. Physical activity, exercise, and physical fitness: Definitions and distinctions for health-related research. *Public Health Rep.* **1985**, *100*, 126–131.

57. Lake, S.P.; Ansorge, H.L.; Soslowsky, L.J. Animal models of tendinopathy. *Disabil. Rehabil.* **2008**, *30*, 1530–1541. [CrossRef]

58. Dudhia, J.; Scott, C.M.; Draper, E.R.C.; Heinegård, D.; Pitsillides, A.A.; Smith, R.K. Aging enhances a mechanically-induced reduction in tendon strength by an active process involving matrix metalloproteinase activity. *Aging Cell* **2007**, *6*, 547–556. [CrossRef]

59. Humphrey, J.D.; Dufresne, E.R.; Schwartz, M.A. Mechanotransduction and extracellular matrix homeostasis. *Nat. Rev. Mol. Cell Biol.* **2014**, *15*, 802–812. [CrossRef]

60. Khan, K.M.; Scott, A. Mechanotherapy: How physical therapists' prescription of exercise promotes tissue repair. *Br. J. Sports Med.* **2009**, *43*, 247–252. [CrossRef]

61. Freedman, B.R.; Bade, N.D.; Riggin, C.N.; Zhang, S.; Haines, P.G.; Ong, K.L.; Janmey, P.A. The (dys) functional extracellular matrix. *Biochim. Biophys. Acta-Mol. Cell Res.* **2015**, *1853*, 3153–3164. [CrossRef]

62. Gauthier, N.C.; Roca-Cusachs, P. Mechanosensing at integrin-mediated cell–matrix adhesions: From molecular to integrated mechanisms. *Curr. Opin. Cell Biol.* **2018**, *50*, 20–26. [CrossRef]

63. Arampatzis, A.; Karamanidis, K.; Albracht, K. Adaptational responses of the human Achilles tendon by modulation of the applied cyclic strain magnitude. *J. Exp. Biol.* **2007**, *210*, 2743–2753. [CrossRef]

64. Heinemeier, K.M.; Kjaer, M. In vivo investigation of tendon responses to mechanical loading. *J. Musculoskelet. Neuronal Interact.* **2011**, *11*, 115–123.

65. Wang, J.H.C.; Guo, Q.; Li, B. Tendon biomechanics and mechanobiology-A minireview of basic concepts and recent advancements. *J. Hand Ther.* **2012**, *25*, 133–141. [CrossRef]

66. Shwartz, Y.; Blitz, E.; Zelzer, E. One load to rule them all: Mechanical control of the musculoskeletal system in development and aging. *Differentiation* **2013**, *86*, 104–111. [CrossRef]

67. Heinemeier, K.M.; Olesen, J.L.; Haddad, F.; Schjerling, P.; Baldwin, K.M.; Kjaer, M. Effect of unloading followed by reloading on expression of collagen and related growth factors in rat tendon and muscle. *J. Appl. Physiol.* **2008**, *106*, 178–186. [CrossRef]

68. Langberg, H.; Skovgaard, D.; Petersen, L.J.; Bülow, J.; Kjær, M. Type I collagen synthesis and degradation in peritendinous tissue after exercise determined by microdialysis in humans. *J. Physiol.* **1999**, *521*, 299–306. [CrossRef]

69. Miller, B.F.; Olesen, J.L.; Hansen, M.; Døssing, S.; Crameri, R.M.; Welling, R.J.; Langberg, H.; Flyvbjerg, A.; Kjaer, M.; Babraj, J.A.; et al. Coordinated collagen and muscle protein synthesis in human patella tendon and quadriceps muscle after exercise. *J. Physiol.* **2005**, *567*, 1021–1033. [CrossRef]

70. Barkhausen, T.; van Griensven, M.; Zeichen, J.; Bosch, U. Modulation of cell functions of human tendon fibroblasts by different repetitive cyclic mechanical stress patterns. *Exp. Toxicol. Pathol.* **2003**, *55*, 153–158. [CrossRef]

71. Kilts, T.; Ameye, L.; Syed-Picard, F.; Ono, M.; Berendsen, A.D.; Oldberg, A.; Heegaard, A.M.; Bi, Y.; Young, M.F. Potential roles for the small leucine-rich proteoglycans biglycan and fibromodulin in ectopic ossification of tendon induced by exercise and in modulating rotarod performance. *Scand. J. Med. Sci. Sports* **2009**, *19*, 536–546. [CrossRef]

72. Woo, S.L.; Ritter, M.A.; Amiel, D.; Sanders, T.M.; Gomez, M.A.; Kuei, S.C.; Garfin, S.R.; Akeson, W.H. The biomechanical and biochemical properties of swine tendons-long term effects of exercise on the digital extensors. *Connect. Tissue Res.* **1980**, *7*, 177–183. [CrossRef]

73. Wiesinger, H.P.; Kösters, A.; Müller, E.; Seynnes, O.R. Effects of increased loading on in vivo tendon properties: A systematic review. *Med. Sci. Sports Exerc.* **2015**, *47*, 1885. [CrossRef]

74. Kongsgaard, M.; Reitelseder, S.; Pedersen, T.G.; Holm, L.; Aagaard, P.; Kjaer, M.; Magnusson, S.P. Region specific patellar tendon hypertrophy in humans following resistance training. *Acta Physiol.* **2007**, *191*, 111–121. [CrossRef]

75. Couppe, C.; Kongsgaard, M.; Aagaard, P.; Hansen, P.; Bojsen-Moller, J.; Kjaer, M.; Magnusson, S.P. Habitual loading results in tendon hypertrophy and increased stiffness of the human patellar tendon. *J. Appl. Physiol.* **2008**, *105*, 805–810. [CrossRef]

76. Houghton, L.A.; Dawson, B.T.; Rubenson, J. Effects of plyometric training on Achilles tendon properties and shuttle running during a simulated cricket batting innings. *J. Strength Cond. Res.* **2013**, *27*, 1036–1046. [CrossRef]

77. Zhang, J.; Wang, J.H.-C. The effects of mechanical loading on tendons—An in vivo and in vitro model study. *PLoS ONE* **2013**, *8*, e71740. [CrossRef]

78. Thornton, G.M.; Hart, D.A. The interface of mechanical loading and biological variables as they pertain to the development of tendinosis. *J. Musculoskelet. Neuronal Interact.* **2011**, *11*, 94–105.

79. Davis, M.E.; Gumucio, J.P.; Sugg, K.B.; Bedi, A.; Mendias, C.L. MMP inhibition as a potential method to augment the healing of skeletal muscle and tendon extracellular matrix. *J. Appl. Physiol.* **2013**, *115*, 884–891. [CrossRef]

80. Kjaer, M.; Kjær, M. Role of extracellular matrix in adaptation of tendon and skeletal muscle to mechanical loading. *Physiol. Rev.* **2004**, *84*, 649–698. [CrossRef]

81. Koskinen, S.O.A.; Heinemeier, K.M.; Olesen, J.L.; Langberg, H.; Kjaer, M. Physical exercise can influence local levels of matrix metalloproteinases and their inhibitors in tendon-related connective tissue. *J. Appl. Physiol.* **2004**, *96*, 861–864. [CrossRef]

82. Arnoczky, S.P.; Tian, T.; Lavagnino, M.; Gardner, K. Ex vivo static tensile loading inhibits MMP-1 expression in rat tail tendon cells through a cytoskeletally based mechanotransduction mechanism. *J. Orthop. Res.* **2004**, *22*, 328–333. [CrossRef]

83. Lavagnino, M.; Arnoczky, S.P.; Tian, T.; Vaupel, Z. Effect of amplitude and frequency of cyclic tensile strain on the inhibition of MMP-1 mRNA expression in tendon cells: An in vitro study. *Connect. Tissue Res.* **2003**, *44*, 181–187. [CrossRef]

84. Barin, F.R.; Durigan, J.L.Q.; de S. Oliveira, K.; Migliolo, L.; Almeida, J.A.; Carvalho, M.; Petriz, B.; Selistre-de-Araujo, H.S.; Fontes, W.; Franco, O.L.; et al. Beneficial effects of resistance training on the protein profile of the calcaneal tendon during aging. *Exp. Gerontol.* **2017**, *100*, 54–62. [CrossRef]

85. Kongsgaard, M.; Qvortrup, K.; Larsen, J.; Aagaard, P.; Doessing, S.; Hansen, P.; Kjaer, M.; Magnusson, S.P. Fibril morphology and tendon mechanical properties in patellar tendinopathy. *Am. J. Sports Med.* **2010**, *38*, 749–756. [CrossRef]

86. American College of Sports Medicine. Progression models in resistance training for healthy adults. *Med. Sci. Sports Exerc.* **2009**, *41*, 687–708. [CrossRef]

87. Chodzko-Zajko, W.J.; Proctor, D.N.; Fiatarone Singh, M.A.; Minson, C.T.; Nigg, C.R.; Salem, G.J.; Skinner, J.S. Exercise and physical activity for older adults. *Med. Sci. Sports Exerc.* **2009**, *41*, 1510–1530. [CrossRef]

88. American College of Sports Medicine Position Stand. The recommended quantity and quality of exercise for developing and maintaining cardiorespiratory and muscular fitness, and flexibility in healthy adults. *Med. Sci. Sports Exerc.* **1998**, *30*, 975–991.

89. Fletcher, G.F.; Ades, P.A.; Kligfield, P.; Arena, R.; Balady, G.J.; Bittner, V.A.; Coke, L.A.; Fleg, J.L.; Forman, D.E.; Gerber, T.C.; et al. Exercise standards for testing and training: A scientific statement from the American heart association. *Circulation* **2013**, *128*, 873–934. [CrossRef]

90. Heinemeier, K.M.; Olesen, J.L.; Haddad, F.; Langberg, H.; Kjaer, M.; Baldwin, K.M.; Schjerling, P. Expression of collagen and related growth factors in rat tendon and skeletal muscle in response to specific contraction types. *J. Physiol.* **2007**, *582*, 1303–1316. [CrossRef]

91. Kjær, M.; Langberg, H.; Heinemeier, K.; Bayer, M.L.; Hansen, M.; Holm, L.; Doessing, S.; Kongsgaard, M.; Krogsgaard, M.R.; Magnusson, S.P. From mechanical loading to collagen synthesis, structural changes and function in human tendon. *Scand. J. Med. Sci. Sports* **2009**, *19*, 500–510. [CrossRef]
92. Borst, S.E.; de Hoyos, D.V.; Garzarella, L.; Vincent, K.; Pollock, B.H.; Lowenthal, D.T.; Pollock, M.L. Effects of resistance training on insulin-like growth factor-I and IGF binding proteins. *Med. Sci. Sports Exerc.* **2001**, *33*, 648–653. [CrossRef]
93. Guzzoni, V.; Ribeiro, M.B.T.; Lopes, G.N.; de Cássia Marqueti, R.; de Andrade, R.V.; Selistre-de-Araujo, H.S.; Durigan, J.L.Q. Effect of resistance training on extracellular matrix adaptations in skeletal muscle of older rats. *Front. Physiol.* **2018**, *9*, 374. [CrossRef]
94. Evanko, S.P.; Vogel, K.G. Ultrastructure and proteoglycan composition in the developing fibrocartilaginous region of bovine tendon. *Matrix* **1990**, *10*, 420–436. [CrossRef]
95. De Cassia Marqueti, R.; Almeida, J.A.; Nakagaki, W.R.; Guzzoni, V.; Boghi, F.; Renner, A.; Silva, P.E.; Durigan, J.L.Q.; Selistre-de-Araujo, H.S. Resistance training minimizes the biomechanical effects of aging in three different rat tendons. *J. Biomech.* **2017**, *53*, 29–35. [CrossRef]
96. Killian, M.L.; Cavinatto, L.; Galatz, L.M.; Thomopoulos, S. The role of mechanobiology in tendon healing. *J. Shoulder Elb. Surg.* **2012**, *21*, 228–237. [CrossRef]
97. Marqueti, R.C.; Durigan, J.L.Q.; Oliveira, A.J.S.; Mekaro, M.S.; Guzzoni, V.; Aro, A.A.; Pimentel, E.R.; Selistre-De-Araujo, H.S. Effects of aging and resistance training in rat tendon remodeling. *FASEB J.* **2018**, *32*, 353–368. [CrossRef]
98. Eriksen, C.S.; Henkel, C.; Svensson, R.B.; Agergaard, A.-S.; Couppe, C.; Kjaer, M.; Magnusson, S.P. Lower tendon stiffness in very old compared to old individuals is unaffected by short term resistance training of skeletal muscle. *J. Appl. Physiol.* **2018**, *125*, 205–214. [CrossRef]
99. Wood, L.K.; Brooks, S.V. Ten weeks of treadmill running decreases stiffness and increases collagen turnover in tendons of old mice. *J. Orthop. Res.* **2016**, *34*, 346–353. [CrossRef]
100. Bhasin, S.; Storer, T.W.; Berman, N.; Callegari, C.; Clevenger, B.; Phillips, J.; Bunnell, T.J.; Tricker, R.; Shirazi, A.; Casaburi, R. The effects of supraphysiologic doses of testosterone on muscle size and strength in normal men. *N. Engl. J. Med.* **1996**, *335*, 1–7. [CrossRef]
101. Evans, N.A. Current concepts in anabolic-androgenic steroids. *Am. J. Sports Med.* **2004**, *32*, 534–542. [CrossRef]
102. Parssinen, M.; Karila, T.; Kovanen, V.; Seppälä, T. The effect of supraphysiological doses of anabolic androgenic steroids on collagen metabolism. *Int. J. Sports Med.* **2000**, *21*, 406–411. [CrossRef]
103. Michna, H. Tendon injuries induced by exercise and anabolic steroids in experimental mice. *Int. Orthop.* **1987**, *11*, 157–162. [CrossRef]
104. Michna, H. Organisation of collagen fibrils in tendon: changes induced by an anabolic steroid-I. Functional and ultrastructural studies. *Virchows Arch. B Cell Pathol. Incl. Mol. Pathol.* **1986**, *52*, 75–86. [CrossRef]
105. Inhofe, P.D.; Grana, W.A.; Egle, D.; Min, K.W.; Tomasek, J. The effects of anabolic steroids on rat tendon: An ultrastructural, biomechanical, and biochemical analysis. *Am. J. Sports Med.* **1995**, *23*, 227–232. [CrossRef]
106. Hill, J.A.; Suker, J.R.; Sachs, K.; Brigham, C. The athletic polydrug abuse phenomenon. A case report. *Am. J. Sports Med.* **1983**, *11*, 269–271. [CrossRef]
107. Kramhøft, M.; Solgaard, S. Spontaneous rupture of the extensor pollicis longus tendon after anabolic steroids. *J. Hand Surg. Am.* **1986**, *11*, 87. [CrossRef]
108. Wood, T.O.; Cooke, P.H.; Goodship, A.E. The effect of exercise and anabolic steroids on the mechanical properties and crimp morphology of the rat tendon. *Am. J. Sports Med.* **1988**, *16*, 153–158. [CrossRef]
109. Laseter, J.T.; Russell, J.A. Anabolic steroid-induced tendon pathology: A review of the literature. *Med. Sci. Sports Exerc.* **1991**, *23*, 1–3. [CrossRef]
110. Evans, N.A.; Bowrey, D.J.; Newman, G.R. Ultrastructural analysis of ruptured tendon from anabolic steroid users. *Injury* **1998**, *29*, 769–773. [CrossRef]
111. Kanayama, G.; Deluca, J.; Meehan, W.P.; Hudson, J.I.; Isaacs, S.; Baggish, A.; Weiner, R.; Micheli, L.; Pope, H.G. Ruptured tendons in anabolic-androgenic steroid users. *Am. J. Sports Med.* **2015**, *43*, 2638–2644. [CrossRef]
112. Miles, J.W.; Grana, W.A.; Egle, D.; Min, K.W.; Chitwood, J. The effect of anabolic steroids on the biomechanical and histological properties of rat tendon. *J. Bone Jt. Surg. Am.* **1992**, *74*, 411–422. [CrossRef]

113. Marqueti, R.C.; Parizotto, N.A.; Chriguer, R.S.; Perez, S.E.A.; Selistre-de-Araujo, H.S. Androgenic-anabolic steroids associated with mechanical loading inhibit matrix metallopeptidase activity and affect the remodeling of the Achilles tendon in rats. *Am. J. Sports Med.* **2006**, *34*, 1274–1280. [CrossRef]

114. Urso, M.L.; Pierce, J.R.; Alemany, J.A.; Harman, E.A.; Nindl, B.C. Effects of exercise training on the matrix metalloprotease response to acute exercise. *Eur. J. Appl. Physiol.* **2009**, *106*, 655–663. [CrossRef]

115. De Sousa Neto, I.V.; Durigan, J.L.Q.; Guzzoni, V.; Tibana, R.A.; Prestes, J.; de Araujo, H.S.S.; Marqueti, R.C. Effects of resistance training on matrix metalloproteinase activity in skeletal muscles and blood circulation during aging. *Front. Physiol* **2018**, *9*. [CrossRef]

116. Carmeli, E.; Moas, M.; Lennon, S.; Powers, S.K. High intensity exercise increases expression of matrix metalloproteinases in fast skeletal muscle fibres. *Exp. Physiol.* **2005**, *90*, 613–619. [CrossRef]

117. Rullman, E.; Norrbom, J.; Strömberg, A.; Wågsäter, D.; Rundqvist, H.; Haas, T.; Gustafsson, T. Endurance exercise activates matrix metalloproteinases in human skeletal muscle. *J. Appl. Physiol.* **2009**, *106*, 804–812. [CrossRef]

118. Huisman, E.; Lu, A.; Jamil, S.; Mousavizadeh, R.; McCormack, R.; Roberts, C.; Scott, A. Influence of repetitive mechanical loading on MMP2 activity in tendon fibroblasts. *J. Orthop. Res.* **2016**, *34*, 1991–2000. [CrossRef]

119. Peviani, S.M.; Guzzoni, V.; Pinheiro-Dardis, C.M.; Da Silva, Y.P.; Fioravante, A.C.R.; Sagawa, A.H.; Delfino, G.B.; Durigan, J.L.Q.; Salvini, T.F. Regulation of extracellular matrix elements and sarcomerogenesis in response to different periods of passive stretching in the soleus muscle of rats. *Sci. Rep.* **2018**, *8*, 9010. [CrossRef]

120. Yang, G.; Im, H.J.; Wang, J.H.C. Repetitive mechanical stretching modulates IL-1β induced COX-2, MMP-1 expression, and PGE2 production in human patellar tendon fibroblasts. *Gene* **2005**, *363*, 166–172. [CrossRef]

121. Asundi, K.R.; Rempel, D.M. Cyclic loading inhibits expression of MMP-3 but not MMP-1 in an in vitro rabbit flexor tendon model. *Clin. Biomech.* **2008**, *23*, 117–121. [CrossRef]

122. Waggett, A.D.; Ralphs, J.R.; Kwan, A.P.L.; Woodnutt, D.; Benjamin, M. Characterization of collagens and proteoglycans at the insertion of the human Achilles tendon. *Matrix Biol.* **1998**, *16*, 457–470. [CrossRef]

123. Marqueti, R.C.; Prestes, J.; Wang, C.C.; Ramos, O.H.P.; Perez, S.E.A.; Nakagaki, W.R.; Carvalho, H.F.; Selistre-de-Araujo, H.S. Biomechanical responses of different rat tendons to nandrolone decanoate and load exercise. *Scand. J. Med. Sci. Sports* **2011**, *21*, e91–e99. [CrossRef]

124. Seynnes, O.R.; Kamandulis, S.; Kairaitis, R.; Helland, C.; Campbell, E.-L.; Brazaitis, M.; Skurvydas, A.; Narici, M.V. Effect of androgenic-anabolic steroids and heavy strength training on patellar tendon morphological and mechanical properties. *J. Appl. Physiol.* **2013**, *115*, 84–89. [CrossRef]

125. Marqueti, R.D.C.; Heinemeier, K.M.; Durigan, J.L.Q.; de Andrade Perez, S.E.; Schjerling, P.; Kjaer, M.; Carvalho, H.F.; Selistre-De-Araujo, H.S. Gene expression in distinct regions of rat tendons in response to jump training combined with anabolic androgenic steroid administration. *Eur. J. Appl. Physiol.* **2012**, *112*, 1505–1515. [CrossRef]

126. Legerlotz, K.; Schjerling, P.; Langberg, H.; Brüggemann, G.-P.; Niehoff, A. The effect of running, strength, and vibration strength training on the mechanical, morphological, and biochemical properties of the Achilles tendon in rats. *J. Appl. Physiol.* **2007**, *102*, 564–572. [CrossRef]

127. Minkwitz, S.; Schmock, A.; Kurtoglu, A.; Tsitsilonis, S.; Manegold, S.; Wildemann, B.; Klatte-Schulz, F. Time-dependent alterations of MMPs, TIMPs and tendon structure in human Achilles tendons after acute rupture. *Int. J. Mol. Sci.* **2017**, *18*, 2199. [CrossRef]

128. Marqueti, R.C.; Paulino, M.G.; Fernandes, M.N.; de Oliveira, E.M.; Selistre-de-Araujo, H.S. Tendon structural adaptations to load exercise are inhibited by anabolic androgenic steroids. *Scand. J. Med. Sci. Sports* **2014**, *24*, e39–e51. [CrossRef]

129. Jones, I.A.; Togashi, R.; Hatch, G.F.R.; Weber, A.E.; Vangsness, C.T. Anabolic steroids and tendons: A review of their mechanical, structural, and biologic effects. *J. Orthop. Res.* **2018**. [CrossRef]

130. Clayton, R.A.E.; Court-Brown, C.M. The epidemiology of musculoskeletal tendinous and ligamentous injuries. *Injury* **2008**, *39*, 1338–1344. [CrossRef]

131. Peffers, M.J.; Thorpe, C.T.; Collins, J.A.; Eong, R.; Wei, T.K.J.; Screen, H.R.C.; Clegg, P.D. Proteomic analysis reveals age-related changes in tendon matrix composition, with age- and injury-specific matrix fragmentation. *J. Biol. Chem.* **2014**, *289*, 25867–25878. [CrossRef]

132. Gagliano, N.; Menon, A.; Cabitza, F.; Compagnoni, R.; Randelli, P. Morphological and molecular characterization of human hamstrings shows that tendon features are not influenced by donor age. *Knee Surg. Sports Traumatol. Arthrosc.* **2018**, *26*, 343–352. [CrossRef]

133. Wood, L.K.; Arruda, E.M.; Brooks, S.V. Regional stiffening with aging in tibialis anterior tendons of mice occurs independent of changes in collagen fibril morphology. *J. Appl. Physiol.* **2011**, *111*, 999–1006. [CrossRef]

134. Couppé, C.; Svensson, R.B.; Grosset, J.F.; Kovanen, V.; Nielsen, R.H.; Olsen, M.R.; Larsen, J.O.; Praet, S.F.E.; Skovgaard, D.; Hansen, M.; et al. Life-long endurance running is associated with reduced glycation and mechanical stress in connective tissue. *Age (Omaha)* **2014**, *36*, 9665. [CrossRef]

135. Kubo, K.; Kanehisa, H.; Kawakami, Y.; Fukanaga, T. Growth changes in the elastic properties of human tendon structures. *Int. J. Sports Med.* **2001**, *22*, 138–143. [CrossRef]

136. Shadwick, R.E. Elastic energy storage in tendons: Mechanical differences related to function and age. *J. Appl. Physiol.* **1990**, *68*, 1033–1040. [CrossRef]

137. Noyes, F.R.; Good, E.S. The strength of the anterior cruciate ligament in humans and Rhesus monkeys. *J. Bone Jt. Surg. Am.* **1976**, *58*, 1074–1082. [CrossRef]

138. Nachemson, A.L.; Evans, J.H. Some mechanical properties of the third human lumbar interlaminar ligament (*ligamentum flavum*). *J. Biomech.* **1968**, *1*, 211–220. [CrossRef]

139. Reeves, N.D. Adaptation of the tendon to mechanical usage. *J. Musculoskelet. Neuronal Interact.* **2006**, *6*, 174–180.

140. Vogel, H.G. Species differences of elastic and collagenous tissue-Influence of maturation and age. *Mech. Ageing Dev.* **1991**, *57*, 15–24. [CrossRef]

141. Thorpe, C.T.; Streeter, I.; Pinchbeck, G.L.; Goodship, A.E.; Clegg, P.D.; Birch, H.L. Aspartic acid racemization and collagen degradation markers reveal an accumulation of damage in tendon collagen that is enhanced with aging. *J. Biol. Chem.* **2010**, *285*, 15674–15681. [CrossRef]

142. Avery, N.C.; Bailey, A.J. Enzymic and non-enzymic cross-linking mechanisms in relation to turnover of collagen: Relevance to aging and exercise. *Scand. J. Med. Sci. Sports* **2005**, *15*, 231–240. [CrossRef]

143. Smith, R.K.W.; Birch, H.L.; Goodman, S.; Heinegård, D.; Goodship, A.E. The influence of ageing and exercise on tendon growth and degeneration-Hypotheses for the initiation and prevention of strain-induced tendinopathies. *Comp. Biochem. Physiol. A Mol. Integr. Physiol.* **2002**, *133*, 1039–1050. [CrossRef]

144. Dedkov, E.I.; Kostrominova, T.Y.; Borisov, A.B.; Carlson, B.M. MyoD and myogenin protein expression in skeletal muscles of senile rats. *Cell Tissue Res.* **2003**, *311*, 401–416. [CrossRef]

145. Li, Y.; Zhao, Z.; Cai, J.; Gu, B.; Lv, Y.; Zhao, L. The frequency-dependent aerobic exercise effects of hypothalamic GABAergic expression and cardiovascular functions in aged rats. *Front. Aging Neurosci.* **2017**, *9*, 212. [CrossRef]

146. Yu, T.-Y.; Pang, J.-H.S.; Wu, K.P.-H.; Chen, M.J.-L.; Chen, C.-H.; Tsai, W.-C. Aging is associated with increased activities of matrix metalloproteinase-2 and -9 in tenocytes. *BMC Musculoskelet. Disord.* **2013**, *14*, 2. [CrossRef]

147. Brownlee, M. Advanced protein glycosylation in diabetes and aging. *Annu. Rev. Med.* **1995**, *46*, 223–234. [CrossRef]

148. Bank, R.A.; Tekoppele, J.M.; Oostingh, G.; Hazleman, B.L.; Riley, G.P. Lysylhydroxylation and non-reducible crosslinking of human supraspinatus tendon collagen: Changes with age and in chronic rotator cuff tendinitis. *Ann. Rheum. Dis.* **1999**, *58*, 35–41. [CrossRef]

149. Heinemeier, K.M.; Schjerling, P.; Heinemeier, J.; Magnusson, S.P.; Kjaer, M. Lack of tissue renewal in human adult Achilles tendon is revealed by nuclear bomb 14C. *FASEB J.* **2013**, *27*, 2074–2079. [CrossRef]

150. Haus, J.M.; Carrithers, J.A.; Trappe, S.W.; Trappe, T.A. Collagen, cross-linking, and advanced glycation end products in aging human skeletal muscle. *J. Appl. Physiol.* **2007**, *47306*, 2068–2076. [CrossRef]

151. James, V.J.; Delbridge, L.; McLennan, S.V.; Yue, D.K. Use of X-ray diffraction in study of human diabetic and aging collagen. *Diabetes* **1991**, *40*, 391–394. [CrossRef]

152. Miles, C.A.; Avery, N.C.; Rodin, V.V.; Bailey, A.J. The increase in denaturation temperature following cross-linking of collagen is caused by dehydration of the fibres. *J. Mol. Biol.* **2005**, *346*, 551–556. [CrossRef]

153. Bailey, A.J. Molecular mechanisms of ageing in connective tissues. *Mech. Ageing Dev.* **2001**, *122*, 735–755. [CrossRef]

154. Couppé, C.; Hansen, P.; Kongsgaard, M.; Kovanen, V.; Suetta, C.; Aagaard, P.; Kjaer, M.; Magnusson, S.P. Mechanical properties and collagen cross-linking of the patellar tendon in old and young men. *J. Appl. Physiol.* **2009**, *107*, 880–886. [CrossRef]

155. Dyer, D.G.; Dunn, J.A.; Thorpe, S.R.; Bailie, K.E.; Lyons, T.J.; McCance, D.R.; Baynes, J.W. Accumulation of Maillard reaction products in skin collagen in diabetes and aging. *J. Clin. Investig.* **1993**, *91*, 2463–2469. [CrossRef]

156. Reddy, G.K.; Stehno-Bittel, L.; Enwemeka, C.S. Glycation-induced matrix stability in the rabbit Achilles tendon. *Arch. Biochem. Biophys.* **2002**, *399*, 174–180. [CrossRef]

157. Reddy, G.K. Cross-linking in collagen by nonenzymatic glycation increases the matrix stiffness in rabbit Achilles tendon. *Exp. Diabesity Res.* **2004**, *5*, 143–153. [CrossRef]

158. Karamanidis, K.; Arampatzis, A. Mechanical and morphological properties of human quadriceps femoris and triceps surae muscle-tendon unit in relation to aging and running. *J. Biomech.* **2006**, *39*, 406–417. [CrossRef]

159. Mian, O.S.; Thom, J.M.; Ardigò, L.P.; Minetti, A.E.; Narici, M.V. Gastrocnemius muscle-tendon behaviour during walking in young and older adults. *Acta Physiol.* **2007**, *189*, 57–65. [CrossRef]

160. Onambele, G.L. Calf muscle-tendon properties and postural balance in old age. *J. Appl. Physiol.* **2006**, *100*, 2048–2056. [CrossRef]

161. Bai, P.; Phua, K.; Hardt, T.; Cernadas, M.; Brodsky, B. Glycation alters collagen fibril organization. *Connect. Tissue Res.* **1992**, *28*, 1–12. [CrossRef]

162. Li, Y.; Fessel, G.; Georgiadis, M.; Snedeker, J.G. Advanced glycation end-products diminish tendon collagen fiber sliding. *Matrix Biol.* **2013**, *32*, 169–177. [CrossRef]

163. Nielsen, H.M.; Skalicky, M.; Viidik, A. Influence of physical exercise on aging rats. III. Life-long exercise modifies the aging changes of the mechanical properties of limb muscle tendons. *Mech. Ageing Dev.* **1998**, *100*, 243–260. [CrossRef]

164. Viidik, A.; Nielsen, H.M.; Skalicky, M. Influence of physical exercise on aging rats: II. Life-long exercise delays aging of tail tendon collagen. *Mech. Ageing Dev.* **1996**, *88*, 139–148. [CrossRef]

165. Vogel, H.G. Influence of maturation and age on mechanical and biochemical parameters of connective tissue of various organs in the rat. *Connect. Tissue Res.* **1978**, *6*, 161–166. [CrossRef]

166. Dressler, M.R.; Butler, D.L.; Wenstrup, R.; Awad, H.A.; Smith, F.; Boivin, G.P. A potential mechanism for age-related declines in patellar tendon biomechanics. *J. Orthop. Res.* **2002**, *20*, 1315–1322. [CrossRef]

167. Lacroix, A.S.; Duenwald-Kuehl, S.E.; Brickson, S.; Akins, T.L.; Diffee, G.; Aiken, J.; Vanderby, R.; Lakes, R.S. Effect of age and exercise on the viscoelastic properties of rat tail Tendon. *Ann. Biomed. Eng.* **2013**, *41*, 1120–1128. [CrossRef]

168. Simonsen, E.B.; Klitgaard, H.; Bojsen-Møller, F. The influence of strength training, swim training and ageing on the Achilles tendon and m. Soleus of the rat. *J. Sports Sci.* **1995**, *13*, 291–295. [CrossRef]

169. Hubbard, R.P.; Soutas-Little, R.W. Mechanical properties of human tendon and their age dependence. *J. Biomech. Eng* **1984**, *106*, 144–150. [CrossRef]

170. Flahiff, C.M.; Brooks, A.T.; Hollis, J.M.; Vander Schilden, J.L.; Nicholas, R.W. Biomechanical Analysis of Patellar Tendon Allografts as a Function of Donor Age. *Am. J. Sports Med.* **1995**, *23*, 354–358. [CrossRef]

171. Bojsen-Moller, J. Muscle performance during maximal isometric and dynamic contractions is influenced by the stiffness of the tendinous structures. *J. Appl. Physiol.* **2005**, *99*, 986–994. [CrossRef]

172. Nordez, A.; Gallot, T.; Catheline, S.; Guével, A.; Cornu, C.; Hug, F. Electromechanical delay revisited using very high frame rate ultrasound. *J. Appl. Physiol.* **2009**, *106*, 1970–1975. [CrossRef]

173. Xu, A.; Wang, Y.; Xu, J.Y.; Stejskal, D.; Tam, S.; Zhang, J.; Wat, N.M.S.; Wong, W.K.; Lam, K.S.L. Adipocyte fatty acid-binding protein is a plasma biomarker closely associated with obesity and metabolic syndrome. *Clin. Chem.* **2006**, *52*, 405–413. [CrossRef]

174. Torricelli, P.; Veronesi, F.; Pagani, S.; Maffulli, N.; Masiero, S.; Frizziero, A.; Fini, M. In vitro tenocyte metabolism in aging and oestrogen deficiency. *Age (Omaha)* **2013**, *35*, 2125–2136. [CrossRef]

175. Kohler, J.; Popov, C.; Klotz, B.; Alberton, P.; Prall, W.C.; Haasters, F.; Müller-Deubert, S.; Ebert, R.; Klein-Hitpass, L.; Jakob, F.; et al. Uncovering the cellular and molecular changes in tendon stem/progenitor cells attributed to tendon aging and degeneration. *Aging Cell* **2013**, *12*, 988–999. [CrossRef]

176. Zhou, Z.; Akinbiyi, T.; Xu, L.; Ramcharan, M.; Leong, D.J.; Ros, S.J.; Colvin, A.C.; Schaffler, M.B.; Majeska, R.J.; Flatow, E.L.; et al. Tendon-derived stem/progenitor cell aging: Defective self-renewal and altered fate. *Aging Cell* **2010**, *9*, 911–915. [CrossRef]

177. Zhang, J.; Wang, J.H.C. Moderate exercise mitigates the detrimental effects of aging on tendon stem cells. *PLoS ONE* **2015**, *10*, e0130454. [CrossRef]

![cells logo] *cells*

MDPI

Article

Transected Tendon Treated with a New Fibrin Sealant Alone or Associated with Adipose-Derived Stem Cells

Katleen Frauz [1], Luis Felipe R. Teodoro [1], Giane Daniela Carneiro [1], Fernanda Cristina da Veiga [2],
Danilo Lopes Ferrucci [1], André Luis Bombeiro [1], Priscyla Waleska Simões [3], Lúcia Elvira Alvares [2],
Alexandre Leite R. de Oliveira [1], Cristina Pontes Vicente [1], Rui Seabra Ferreira Jr. [4],
Benedito Barraviera [4], Maria Esméria C. do Amaral [5], Marcelo Augusto M. Esquisatto [5],
Benedicto de Campos Vidal [1], Edson Rosa Pimentel [1] and Andrea Aparecida de Aro [1,5,*]

[1]   Department of Structural and Functional Biology, Institute of Biology, University of Campinas–UNICAMP,
      Charles Darwin, s/n, CP 6109, 13083-970 Campinas, SP, Brazil; kafrauz@hotmail.com (K.F.);
      teo.luisfelipe@gmail.com (L.F.R.T.); gianedc@gmail.com (G.D.C.); daniloferrucci@yahoo.com.br (D.L.F.);
      aobombeiro@gmail.com (A.L.B.); alroliv@unicamp.br (A.L.R.d.O.); crpvicente@gmail.com (C.P.V.);
      camposvi@unicamp.br (B.d.C.V.); pimentel@unicamp.br (E.R.P.)
[2]   Department of Biochemistry and Tissue Biology, Institute of Biology, University of Campinas–UNICAMP,
      Charles Darwin, s/n, CP 6109, 13083-970 Campinas, SP, Brazil; fernandaveiga6@gmail.com (F.C.d.V.);
      lealvare@unicamp.br (L.E.A.)
[3]   Engineering, Modeling and Applied Social Sciences Center (CECS), Biomedical Engineering Graduate
      Program (PPGEBM), Universidade Federal do ABC (UFABC), Alameda da Universidade s/n,
      09606-045 São Bernardo do Campo, SP, Brazil; pritsimoes@gmail.com
[4]   Center for the Study of Venoms and Venomous Animals (CEVAP), São Paulo State University (UNESP –
      Universidade Estadual Paulista), Botucatu, SP, St. José Barbosa de Barros, 1780, Fazenda Experimental
      Lageado, 18610-307 Botucatu, SP, Brazil; rui.ead@gmail.com (R.S.F.J.); bbviera@jvat.org.br (B.B.)
[5]   Biomedical Sciences Graduate Program, Herminio Ometto University Center-UNIARARAS, Av. Dr.
      Maximiliano Baruto, 500, Jd. Universitário, 13607-339 Araras, SP, Brazil; esmeria@fho.edu.br (M.E.C.d.A.);
      marcelosquisatto@fho.edu.br (M.A.M.E.)
*   Correspondence: andreaaro80@gmail.com; Tel.: +55-19-3543-1423

Received: 26 November 2018; Accepted: 7 January 2019; Published: 16 January 2019

Abstract: Tissue engineering and cell-based therapy combine techniques that create biocompatible materials for cell survival, which can improve tendon repair. This study seeks to use a new fibrin sealant (FS) derived from the venom of *Crotalus durissus terrificus*, a biodegradable three-dimensional scaffolding produced from animal components only, associated with adipose-derived stem cells (ASC) for application in tendons injuries, considered a common and serious orthopedic problem. Lewis rats had tendons distributed in five groups: normal (N), transected (T), transected and FS (FS) or ASC (ASC) or with FS and ASC (FS + ASC). The in vivo imaging showed higher quantification of transplanted PKH26-labeled ASC in tendons of FS + ASC compared to ASC on the 14th day after transection. A small number of Iba1 labeled macrophages carrying PKH26 signal, probably due to phagocytosis of dead ASC, were observed in tendons of transected groups. ASC up-regulated the *Tenomodulin* gene expression in the transection region when compared to N, T and FS groups and the expression of *TIMP-2* and *Scleraxis* genes in relation to the N group. FS group presented a greater organization of collagen fibers, followed by FS + ASC and ASC in comparison to N. Tendons from ASC group presented higher hydroxyproline concentration in relation to N and the transected tendons of T, FS and FS + ASC had a higher amount of collagen I and tenomodulin in comparison to N group. Although no marked differences were observed in the other biomechanical parameters, T group had higher value of maximum load compared to the groups ASC and FS + ASC. In conclusion, the FS kept constant the number of transplanted ASC in the transected region until the 14th day after injury. Our data suggest this FS to be a good scaffold for treatment during tendon repair because it was the most effective one regarding tendon organization recovering, followed by the FS treatment associated with ASC and finally by the transplanted ASC on the 21st day. Further investigations in long-term time points of the tendon repair are needed to analyze if the higher tissue organization found with the FS scaffold will improve the biomechanics of the tendons.

Keywords: repair; tenomodulin; collagen; birefringence; scaffold

---

1. Introduction

Tendons are load-bearing structures, which transmit muscle-contraction force to the skeleton so it can maintain posture or produce motion [1]. The abundant tendon extracellular matrix (ECM) is formed by a hierarchical collagen structure composed specially by type I collagen produced by tenocytes, which are associated with numerous non-fibrillar proteins, that are essential to tendons ability of supporting load with stability [2]. The specific mechanical properties of tendons are directly related to the high organization of collagen bundles [3], so as more organized, more resistant to tensile load. As known, any external mechanical load stimulates a mechanotransduction mechanism [4]. It is important for homeostasis and tendon structural integrity as well as for the modulation of synthesis and degradation of ECM components produced by tenocytes [5]. The load transmits different levels and combinations of tensile and compressive forces [6] to tendons and the precise physiologic loads of individual tendons depending on their function, age, sex, location and species [5]. However, abnormal loading can cause tendon injury due to an acute traumatic injury or to degenerative processes [7,8].

Tendon injuries are among the most common orthopedic problems with long-term disability as a frequent consequence due to its prolonged healing time. Further, the repaired tissue presents lower biomechanical resistance, predisposing patients to high rates of recurrence ensuing initial injury [9]. Following an acute rupture, the tendon undergoes a healing process, involving successive steps of inflammation, ECM formation and remodeling [10,11]. Yet, the scar tissue formed during tendon repair is different from the native tendon, presenting one-third of tensile strength observed in native tendons [12].

Several strategies have been studied aiming the tendon repair process to be more effective, attempting to form similar tissue to native tendon. Currently, tissue engineering and cell-based therapy combine techniques to create biocompatible materials for cell survival, which can improve the tissue repair. Considering this field of regenerative medicine based on the cell-based therapy, the investigation of the effects of adipose-derived stem cells (ASC) during wound healing has grown immensely in recent years, due to its high responsiveness to distinct environmental cues and isolation facility [13,14]. Thus, a Gonçalves et al. [15] study confirmed the potential of ASC for tendon regeneration, so proving the role of some growth factors such as EGF (epidermal growth factor), bFGF (basic fibroblast growth factor), PDGF (platelet derived growth factor) and TGF-b1 (beta-1 transforming growth factor) in the tenogenic differentiation of ASC.

The ASC are a population of multipotent cells that can be obtained from subcutaneous adipose tissue through percutaneous or limited open aspiration techniques [16,17]. Under the appropriated conditions, ASC hold the direct differentiation potential towards specific cells lineages as fibroblasts [18,19], osteoblasts, chondroblasts, adipocytes and myoblasts [20–22]. ASC produce important molecules, which play an important role in wound repair as growth factors [23,24], cytokines, matrix metalloproteinases (MMP) [25] and collagen [26,27]. Still, it is extremely small the percentage of ASC that survives after transplantation into a site of tissue injury [28], demonstrating the importance of new scaffolds that can create a good environment for cell functionality. The tenogenic differentiation potential as well as the paracrine secretion of ASC at the injury site during tendon repair have not been extensively described in the literature.

Ideal biodegradable scaffolds for cells should provide them mechanical support, cells adhesion, proliferation and cellular differentiation [29]. According to James et al. [30], scaffolds for tendons should have high porosity, a large surface area and they should also mimic the native tendon ECM architecture, allowing nutrients diffusion and factors secreted by the cells, important for the stimulus of cell proliferation and synthesis of the ECM's components during tissue repair. In the present study, a new fibrin sealant (FS) derived from snake venom from *Crotalus durissus terrificus* was used with a biological three-dimensional scaffolding capacity of maintaining cell survival without interfering in its differentiation and with cell viability rates above 80% [29]. Gasparotto et al. [29] showed an excellent interaction of this FS with the ASC, due to its ability to induce the spontaneous adipogenic, chondrogenic and osteogenic lineages differentiation. This new FS is composed of a fibrinogen-rich cryoprecipitate extracted from the *Bubalus bubalis* buffalo's blood in association with a serine protease (a thrombin-like enzyme) extracted from *Crotalus durissus terrificus* venom [30–33]). According to Ferreira et al. [34], a thrombin-like enzyme, in the presence of calcium, acts upon the fibrinogen molecule transforming it into fibrin monomers forming a stable clot with adhesive, hemostatic and sealant effects [32,33,35].

Fibrin has been used for many years specially because it presents important characteristics like adhesive tissue or sealant to control bleeding, being used for a variety of surgical and repairing processes [29,36,37]. FS has positive effects for bone [38] and cardiac [39] tissue engineering, for peripheral nerve [40] or skin repair [41] among other applications. Still, concerns about the risk transmission of some viral diseases of commercial FS have increased researchers' interest to develop new sealants [34]. Then, the new FS used in the present study has advantages when compared to the commercially available FS products, since it is produced from animal components only, without risk of infectious diseases and lower costs of production [29].

Through the hypothesis of FS being a good scaffold for ASC, as much for tendon graft considering the FS malleability, which is important during limb movement in our model of tendon transection, the goals of this study are: (1) to evaluate the presence of ASC in the FS at the transected region of the tendons until the 21st day after injury; (2) to analyze the cells paracrine secretion through the expression of genes related to tendon remodeling; (3) to measure the organization of the collagen fibers and to quantify the total collagen content; and (4) to test the biomechanical properties of tendons.

2. Materials and Methods

2.1. Isolation of ASC and Ccell Culture

The procedure was done according to Yang et al. [42] with some modifications. Adipose tissue was obtained from the inguinal region of 10 male Lewis rats between 90–120 days. All surgical and experimental protocols were approved (01/12/2015) by the Institutional Committee for Ethics in Animal Research of the State University of Campinas-UNICAMP-Brazil (Protocol n° 3695-1). Adipose tissue was cut and washed in Dulbecco's modified phosphate buffered saline solution (DMPBS Flush without calcium and magnesium) containing 2% streptomycin/penicillin. Then, 0.2% collagenase (Sigma-Aldrich® Inc., Saint Louis, MO, USA) was added to ECM degradation and the solution was maintained at 37 °C under gentle stirring for 1 h to separate the stromal cells from primary adipocytes. Dissociated tissue was filtered using cell strainers (40 μm) and the inactivation of collagenase was then done by the addition of equal volume of Dulbecco's modified Eagle's medium (DMEM) supplemented with 15% fetal bovine serum (FBS), followed by centrifugation at 1800 rpm for 10 min. The suspending portion containing lipid droplets was discarded and the pellet was resuspended in DMEM with 15% FBS and transferred to 25 cm^2 bottle. After confluence, cells were transferred to 75 cm^2 bottle (1st passage) and the cultures were maintained at 37 °C with 5% CO_2 until the 5th passage (5P). For detachment of the adherent cells, it was used 0.25% trypsin-0.02% EDTA and re-plated at a dilution of 1:3.

2.2. Flow Cytometry

ASC at 5P (*n* = 4) were trypsinized and centrifuged at 1800 rpm for 10 min and counted using the Neubauer chamber. 1×10^6 ASCs were resuspended in 200 μL of DMPBS with 2% BSA (bovine serum albumin). For the immunophenotypic panel [29,43], the following antibodies were used: CD90-APC (eBioscience® Inc., San Diego, CA, USA), CD105-PE (BD-Pharmingen™, San Diego, CA, USA) and CD34-FITC double conjugated (eBioscience® Inc., San Diego, CA, USA), were diluted 1:200 and incubated with cells during 1 h at room temperature. Subsequently, ASCs were washed twice with 500 µL of DMPBS and centrifuged at 2000 rpm for 7 min. The ASCs were resuspended in DMPBS with 2% BSA, following for flow cytometry analysis.

2.3. In Vitro Differentiation Potential of ASC

ASC (5P) were cultured (2×10^4 cells) using different media for each type differentiation, according to Yang et al. [42]. Osteogenic differentiation: DMEM supplemented with 10% FBS, 0.1 µM dexamethasone, 200 µM ascorbic acid, 10 mM β-glycerol phosphate. Adipogenic differentiation: DMEM supplemented with 10% FBS, 1 µmol/L dexamethasone, 50 µmol/L indomethacin, 0.5 mM 3-isobutyl-1-methyl-xanthine and 10 µM insulin. Chondrogenic differentiation: DMEM supplemented with 10% FBS, acid free 15 mM HEPES, 6.25 µg/mL insulin, 10 ng/mL TGF-β1 and 50 nM ascorbic acid-2-phophate. Cultures were maintained at 37 °C with 5% CO_2 and the complete mediums were replaced twice a week, for four weeks. At the end of each culture differentiation, the cells were fixed with 4% paraformaldehyde for 20 min and stained with 2% Alizarin Red S (pH 4.1) during 5 min for calcium detection, with 0.025% Toluidine blue in McIlvaine buffer (0.03 M citric acid, 0.04 M sodium phosphate dibasic—pH 4.0) during 10 min for proteoglycans detection and with 1% Sudan IV during 5 min to show lipids droplets. Samples were imaged on the Axiovert S100 (ZEISS) (Carl Zeiss AG, Oberkochen, Germany) inverted microscope.

2.4. Fibrin Sealant (FS) Scaffold

The FS derived from serpent venom (*Crotalus durissus terrificus*) was kindly provided by the Center for the Study of Venoms and Venomous Animals at UNESP (CEVAP at UNESP, São Paulo State University—Brazil) and its components and instructions for use are described in its patents (registration numbers BR1020140114327 and BR1020140114360). This FS, which was developed by the researchers of CEVAP, was produced according to the proposed standardization [29,33,35,44,45]. This new sealant was produced from the thrombin-like enzyme extracted from snake venom and animal fibrinogen. The product was provided in three microtubes that were stored at −20 °C. At the time of use, the components were previously thawed, reconstituted, mixed and applied (9 µL in each transected tendon) to generate a stable clot with a dense fibrin network.

2.5. Confocal Microscope Analysis

Approximately 3.7×10^5 from 1 bottle of 75 cm^2 ASC on 5P ($n = 3$) were trypsinized and centrifuged at 1800 rpm for 10 min. The supernatant was discarded and the pellet containing ASC was fixed with 4% paraformaldehyde. ASC were labeled with phalloidin-FITC at 1.25 µg/mL for 5 min at room temperature to stain the actin cytoskeleton and incubated with DAPI (0.1 mg/mL in methanol) for 5 min at 37 °C to stain the nucleus. Then, the ASC were resuspended in 15 µL of DMPBS + 9 µL of FS and, after clot formation, were analyzed in the National Institute of Science and Technology on Photonics Applied to Cell Biology (INFABIC) at the State University of Campinas (UNICAMP), using a Zeiss LSM 780-NLO confocal on an Axio Observer Z.1 microscope (Carl Zeiss AG, Oberkochen, Germany) using a 10× objective. Images were collected using laser lines 405 nm and 488 nm for excitation DAPI and phalloidin-FITC fluorophores, respectively, with pinholes set to 1 airy unit for each channel, 1024 × 1024 image format.

2.6. Experimental Groups

A total of 110 male Lewis rats (120-day-old) kept at a constant temperature (23 ± 2 °C) and humidity (55%) under a 12/12 h light/dark cycle, with free access to food and water, were divided into 5 experimental groups: Normal (N): rats with tendons without transection; Transected (T): rats with partially transected tendons and treated with topical application of DMPBS in the transected region; Fibrin sealant (FS): rats with transected tendons and treated with FS application in the transected region; Mesenchymal stem cells derived from adipose tissue (ASC): rats with transected tendons with subsequent transplant of ASC (3.7×10^5 cells) in the transected region; And FS with ASC (FS + ASC): rats with transected tendons and treated with application of FS associated with ASC in the transected region. Animals were euthanized on the 21st day after transection by an overdose of anesthetic (ketamine and xylazine). Animals of the N group were euthanized at 141 days and the tendons without transection were collected for analysis.

2.7. Protocol for Partial Transection of the Achilles Tendon and Application of FS and ASC

The animals were anesthetized with intraperitoneal injection of Ketamine (90 mg/Kg) and Xylazine (12 mg/Kg) and the right lower paws submitted to antisepsis and trichotomy. To expose the calcaneus tendon, a longitudinal incision was made in the animal's skin, followed by transverse partial transection performed in the proximal tendon region (predominantly subjected to tension forces) located at a distance of 4 mm from its insertion in the calcaneus bone [11,46–51]. Approximately 3.7×10^5 ASCs in the 5P were resuspended in 15 μL of DMPBS and transplanted in the transected region of tendons in the ASC group, using a pipette. In the FS + ASC group, the ASCs were resuspended in 15 μL of DMPBS + 9 μL of FS and, after clot formation; it was placed on top of the transected region. In each T-group tendon only 15 μL of DMPBS was applied and the SF group tendons received application of 15 μL of DMPBS + 9 μL FS. All applications were made in the region of the tendon where partial transection was performed. Then, the skin was sutured with nylon thread (Shalon 5-0) and needle (1.5 cm). All surgical and experimental protocols were approved (01/12/2105) by the Institutional Committee for Ethics in Animal Research of the UNICAMP—Brazil (Protocol n° 3695-1).

2.8. In Vivo Imaging

The ASC on 5P were trypsinized and centrifuged at 1800 rpm for 10 min. The supernatant was discarded and the pellet containing about 3.7×10^5 cells were labeled with PKH26 (Sigma) as previously described [52]. After labeling with PKH26 the pellet was resuspended in 15 μL of DMPBS and applied to the transected region of the calcaneus tendon of the ASC group. In the FS + ASC group the same procedure was repeated with the addition of 9 μL of FS. In the control rat, sham ASC were applied on the intact tendon and the negative control rat had only its tendon transected without ASC or FS application. In vivo imaging was performed on the 1st, 2nd, 3rd, 7th, 14th and 21st days after injury using the In vivo FX PRO device (BRUKER®, Billerica, Massachusetts, USA) for identification and quantification of the intensity and area of fluorescence of the ASC labeled with PKH26. To this end, the wavelengths of 550 nm and 600 nm were applied for excitation and emission of the fluorophore, respectively, for 1 min. In addition, the animals were submitted to X-ray, aiming at the anatomical location of the marking. During the procedure, the animals of experimental groups ($n = 3$) were kept under anesthesia (3% isoflurane in medicinal air). In order to exclude possible nonspecific markings, negative control was also subjected to imaging. Additionally, the fur was completely removed from the area of interest and the skin was cleaned with 70% ethanol to remove any residues that could interfere with the fluorescence. The following parameters were analyzed: fluorescence area (mm^2) and fluorescence intensity (photons per second per square millimeter, $P/s/mm^2$).

2.9. Real-Time PCR Array

The tendons were collected carefully ($n = 4$), placed in stabilizing solution (RNA-later, QIAGEN®, Hilden, Germany) and maintained at −20 °C. For total RNA extraction, the transected region (TR) of tendons were isolated and sprayed using liquid nitrogen and then homogenized in a tube containing 5 stainless steel balls (2.3 mm diameter, Biospec Products, Inc., Bartlesville, USA) by being shaken in TissueLyser LT instrument (QIAGEN®), with 2 repetitions (60 s) intercalated with ice cooling (2 min) between each shaking step [53]. Total RNA was in isolation from each sample using the RNeasy®

Fibrous Tissue Mini Kit (QIAGEN®), following the manufacturer's instructions. A spectrophotometer (NanoDrop® ND-1000, Thermo Scientific®, Waltham, Massachusetts, USA) was used to quantify RNA in each sample by determining the absorbance ratio at 260 and 280 nm. 0.5 µg from the total extracted RNA of each sample was used for the synthesis of cDNA, using the RT^2 First Strand Kit (QIAGEN®) and thermocycler Mastercycler Pro (Eppendorf®, Hamburg, Germany), also following the manufacturer's instructions. The cDNA was frozen at −20 °C until tested. The RT-PCR array reaction was performed using the RT^2 Profiler PCR Arrays (A format) kit in combination with the RT^2 SYBR Green Mastermixes (QIAGEN®) on the thermocycler apparatus 7300 (ABI Applied Biosystems®, Foster City, CA, USA), following the manufacturer's instructions. For each animal sample, three types of reaction controls were used: 1. Positive PCR control; 2. Reverse transcriptase control; 3. Control for contamination of rat genomic DNA. The *Glyceraldehyde-3-phosphate dehydrogenase* (*Gapdh*, NM_017008) was used as endogenous control for each sample. The following genes were analyzed (QIAGEN®): *Scleraxis* (*Scx*, NM_001130508); *Tenomodulin* (*Tnmd*, NM_022290); *Tumor necrosis factor* (*TNF superfamily, member 2*) (*Tnf*, NM_012675); *Interleukin 1 beta* (*Il1b*, NM_031512); *Transforming growth factor, beta 1* (*Tgfb1*, NM_021578); *Matrix metallopeptidase 2* (*Mmp2*, NM_031054); *Matrix metallopeptidase 9* (*Mmp9*, NM_031055); *TIMP metallopeptidase inhibitor 2* (*Timp2*, NM_021989); *Decorin* (*Dcn*, NM_024129); *Lysyl oxidase* (*Lox*, NM_017061) e *Growth differentiation factor 5* (*Gdf5*, XM_001066344). Reactions were made in a single cDNA pipetting for each gene including endogenous control. ΔCT values were obtained by the difference between the CT values of the target genes and the *Gapdh* gene. These values were normalized by subtracting the ΔCT value of the calibrator sample (N group) to obtain ΔΔCT values. For each target gene, the $2^{-\Delta\Delta CT}$ method was used to calculate the relative expression level (fold change) and the results were represented as the relative gene expression in comparison to the calibrator sample that is equal to 1.

2.10. Dosage of Hydroxyproline

Hydroxyproline was used as an indicator of total collagen amount in tendons of different groups (*n* = 8) used previously in the biomechanical assay. The entire tendons were cut and immersed in acetone for 48h, also followed by a solution containing chloroform: ethanol (2:1) for 48h. After dehydration, the samples were hydrolyzed in 6N HCl (1 mL/10 mg of tissue) for 4h at 130 °C according to Stegemann and Stalder [54], with some modifications. The hydrolysate was neutralized with 6N NaOH, followed by spectrophotometric quantification, according to Jorge et al. [55]. The absorbance of the samples was measured at 550 nm using a microplate reader (Expert Plus, Asys®, Holliston, MA, USA).

2.11. Western Blotting

For protein extraction, 12 entire tendons longitudinal cryosections obtained from 4 different animals of each group were carried out using 50 µL of T-PER™ Tissue Protein Extraction Reagent. The extraction mixture was gently stirred for 30 min at 4 °C, followed by centrifugation at 10,000 rpm for 10 min. The supernatant (total extract) was used for determination of the protein concentration by the biuret method. Aliquots of the supernatant were treated with Laemmli buffer containing 100 mM DTT (Sigma). Samples containing 30 µg of protein were boiled for 5 min and loaded onto 6.5% to 10% SDS-PAGE gels. The gels were run in a Mini-Protean apparatus (Bio-Rad, Hercules, CA, USA) and transferred to PVDF membranes (Bio-Rad). The membranes were washed in basal solution (1 M Trizma base, 5 M NaCl, 0.005% Tween 20 and deionized water) and incubated in blocking solution (basal solution plus 5% Molico skim milk) for 2h. Then, the membranes were incubated with a polyclonal antibody against collagen type I (1:1000; C2456-Sigma), Tenomodulin (1:1000; SAB2108237-Sigma) and Beta-actin used as internal control (1:500; sc-47778-Santa Cruz Biotechnology, California, CA, USA), overnight at 4 °C. Specific protein bands were visualized in the PVDF (Bio-Rad®) membranes incubated with appropriate secondary antibodies at 1:10,000 (Santa Cruz Biotechnology, California, CA, USA) for 2h, followed by exposure to the SuperSignal West Pico Chemiluminescent Substrate kit. Membranes were developed with the Syngene G: BOX documentation system. The band intensities were quantified by optical densitometry, using the free Image J software (National Institutes of Health, Bethesda, MD, USA).

2.12. Preparation of Sections in Freezing

Tendons were placed in Tissue-Tek®, frozen and cut in cryostat (serial longitudinal cuts of 7 μm thickness). The sections were fixed using a 4% formaldehyde solution in Millonig buffer (0.13 M sodium phosphate and 0.1 M sodium hydroxide, 7.4 pH) for 20 min and followed for birefringence and contrast analysis by differential interference and for histology and histomorphometry.

2.13. Birefringence and Contrast by Differential Interference (DIC)

For this procedure the Olympus BX-51 (Olympus America, Center Vallery, PA, USA) equipped with Q-color 5 camera (Olympus America, Center Vallery, PA, USA) was used. For visual evaluation and birefringence measurements, 3 tendons longitudinal cryosections obtained from 5 different animals of each group were carried out and Image a Pro-plus v.6.3 software for Windows™ (Media Cybernetics, Silver Spring, MD, USA) was used. Around 300 measurements were done in tendon sections of each group. With the microscope and the software it is possible to carry out analysis of DIC and anisotropic properties, both individually and in combination [56,57]. To obtain the birefringence of the fibers and collagen bundles without Walston's prism activity, these were removed from the polarized light path by turning the condenser tower to the position of a field of observation free, giving passage only to polarized light. The Nomarski prism was also removed below the analyzer. Under these conditions we simply have a polarizing microscope. In place of the Nomarski prism a Senarmont compensator of 1/4 wavelength was inserted and used for collagen bundles analysis before and after compensation. After studying the fibers morphology and collagen bundles by their birefringence of sections immersed in distilled water, birefringence was measured by captured images analysis. To do this, a standardization of the lighting source and camera sensitivity was used so that the same working conditions were always preserved. In optical terms, 40 × objective and monochromatic light with λ = 546 nm were always used. Longitudinal tendon sections of all groups were examined and measured. As standard to adjust the working conditions, sections of the ASC group were adopted as treatment from which a hypothetically ideal response was expected. The birefringence measurements were in terms of image brightness and expressed in pixels, which allowed a high sampling of measured areas, as well as the detection of the greater variability of birefringence as previously verified [58,59].

2.14. Immunofluorescence

Tendon longitudinal cryosections (*n* = 5) were fixed in acetone (4 °C, 20 min) and washed with PBS (2 × 5 min). For ASC labeling, sections were blocked with PBS/1% BSA (1h, room temperature) and then incubated with anti-rat CD90/CD90.1 (0.5mg/mL, BD-Pharmingen) diluted in PBS/1% BSA (1:200, 2h, room temperature). After washed in PBS (2 × 5 min), sections were incubated with anti-rat FITC (0.5mg/mL, BD-Pharmingen) and PE rat anti-mouse CD105 (0.2mg/mL, BD-Pharmingen), diluted in PBS (both 1:200, 40 min, room temperature). Immediately after washed in PBS (2 × 5 min), the sections were incubated with DAPI (0.1 mg/mL in methanol) for 5 min at 37 °C. The sections were analyzed in a fluorescence microscope (Olympus BX60) and the images were captured by the Q-Capture Pro™ software (QImaging, Surrey, BC, Canada). For macrophages, sections were fixed as above, blocked with PBS/3% BSA (1h, room temperature) and then incubated with anti-Iba1 rabbit IgG (Wako, cat. code: 019-19741) diluted in PBS/1% BSA (1:700, overnight, 4 °C). After washings (3 × 5 min in PBS), sections were incubated with CY2 donkey anti-rabbit IgG (Jackson Immunores., cat. code: 711-225-152) diluted in PBS/1% BSA (1:500, 45min, room temperature). Sections were washed in PBS (3 × 10 min) and then coverslips were mounted using glycerol solution (glycerol and water, 3:1) containing DAPI (1:1000). Slides were photographed in a fluorescence microscope (Leica DM5500B with digital camera Leica DFC345 FX, using Leica Application Suite X software, Leica Microsystems GmbH, Wetzlar, Germany).

For orthogonal sectioning (where indicated) the z stack varied from 20 to 30 layers and when necessary the 3D deconvolution was employed to those projections (total interactions: 10; refractive index: 1.52).

2.15. Histology and Histomorphometry

After fixing the tissue cryosections as described previously, tendons were stained with 0.025% toluidine blue (TB) in McIlvaine buffer (0.03 M citric acid, 0.04 M sodium phosphate dibasic-pH 4.0) [48] for proteoglycans observation and with hematoxylin and eosin (HE) [49] for a panoramic view of the tendon's morphology. The sections on slides were air dried and immersed in xylene, before embedding in entellan (Merck, Rio de Janeiro, Brazil). Tissue sections were analyzed for tendon morphology observation under an Olympus BX53 microscope (Center Vallery, PA, USA).

For the fibroblast count (number of fibroblasts in 10^4 μm^2) of the transected region, longitudinal sections of tendons stained with Harris hematoxylin were used. Three tendons sections from each different animal group (n = 3) were used, in which five samples were taken for fibroblasts count. All images were captured and scanned using the Leica DM2000 Photomicroscope. Measurements were performed on scanned images supported by the SigmaScan Pro 5.0™ program (Systat Software Inc., Chicago, USA).

2.16. Biomechanical Parameters

Tendons from experimental groups (n = 8) were collected and stored at −20°C until tested. Before the biomechanical test, the tendons were thawed and measured with a pachymeter, considering their length, width and thickness. For the biomechanical assay, the tendons were maintained in PBS to prevent their fibers from drying out. Then, the tendons were fixed to metal claws by the myotendinous junction and by the osteotendinous junction, for a correct alignment of the equipment (Texturometer, MTS model TESTSTAR II). In each biomechanical assay tendons were subjected to a gradual increase of load at a displacement velocity of 1 mm/s by using a load 0.05 N, until the tendon ruptured. Biomechanical parameters were analyzed according to Biancalana et al. [60], such as maximum force (N) and maximum displacement (mm), which were used to calculate the maximum stress (Mpa) and maximum strain (L) of tendons from the experimental groups. The cross-sectional area of the calcaneus tendon was calculated by assuming an elliptical approximation ($A = \pi Wd/4$), using measurements of width (W) and thickness (d) and values from the same [61]. The maximum stress value (MPa) was estimated by the ratio between the maximum load (N) and the cross-sectional area (mm^2). The maximum deformation (L) was calculated through ($L = L_f − L_i/L_i$,), where (L_f) is the value of the final length before rupture and (L_i) is the initial tendon length value.

2.17. Statistical Analysis

All results were presented in mean and standard deviation for the values with normal distribution (or interquartile range and median for the values that did not adhere to the Gaussian distribution). For the data with normal distribution, it was used the analysis of variance (ANOVA), followed by the Tukey post-hoc test for intra-group analysis (in the case of statistical significance) or the Student's T Test preceded by Levene's Test (for biomechanical parameters, dosage of hydroxyproline, in vivo imaging, histomorphometry, real-time PCR array and Western blotting). For data that did not adhere to the Gaussian distribution, it was used non-parametric test by the Kruskal-Wallis test followed by the post-hoc test of Dunn for intra-group analysis (in the case of statistical significance) or U Test of Mann-Whitney (for birefringence measurements). Statistical analysis was performed in the software Statistical Package for Social Sciences (SPSS Inc., Chicago, Illinois, USA) version 22.0 and for all the aforementioned tests the significance level α = 0.05 and power of the test of 95% were considered.

3. Results

3.1. In Vitro Adipogenic, Chondrogenic and Osteogenic Differentiation of ASC and Positive ASC Markers

In vitro ASC (5P) staining with Toluidine Blue, Alizarin Red and Sudan IV showed differentiation in chondrocytes, osteoblast and adipocytes, respectively (Figure 1A–D). In flow cytometry, CD90 positive cells (approximately 87%) and CD105 positive cells (approximately 91%) were observed. CD34 cells were not detected (Figure 1E,F).

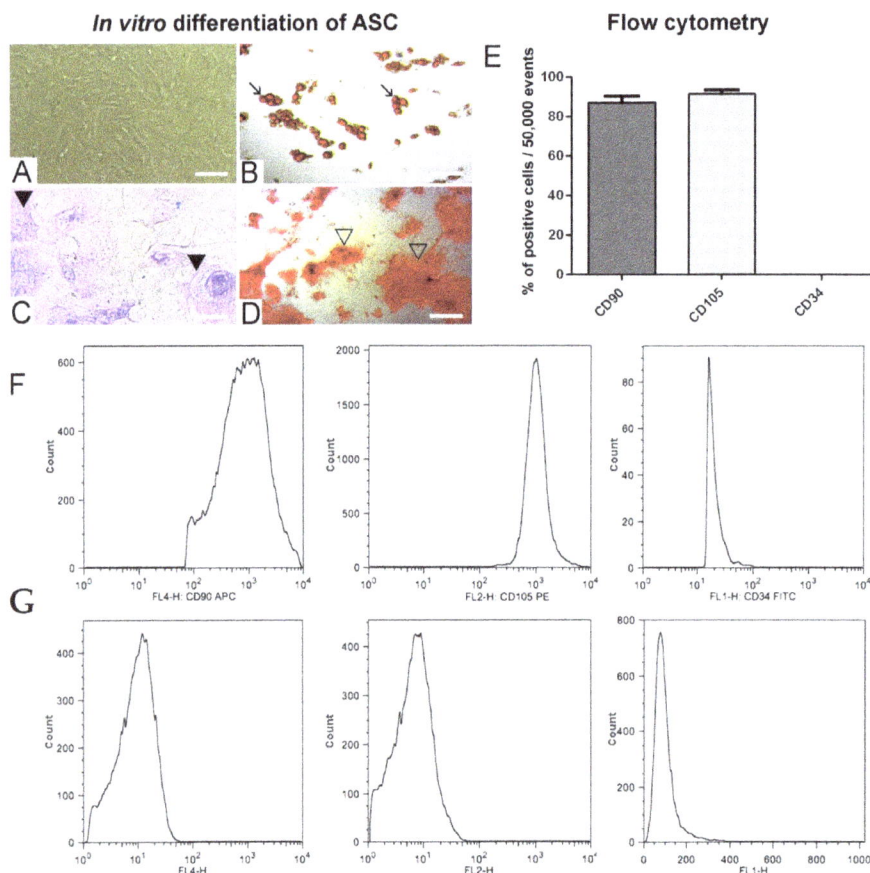

Figure 1. In vitro differentiation potential of ASC (*n* = 4) in 5P (**A**): adipogenic (**B**) and lipid stained with Sudan IV (→); chondrogenic (**C**) and proteoglycans stained with toluidine blue (▶); osteogenic (**D**) and calcium stained with alizarin red (▷). Different cells were stained after 4 weeks of culture. (**E**) Flow cytometry for ASC in 5P (*n* = 4) with positive labeling for CD90 and CD105 and negative labeling for CD34. (**F**) Histograms demonstrate the x-axis fluorescence scale considered positive when the cell peak is above 101 (CD34) or 102 (CD90 and CD105). (**G**) Control for -APC, -PE and -FITC (with very low fluorescence), corresponding to non-marked cells. Bars = A, B, D: 120 μm; C: 40 μm.

3.2. ASC Disposition in the FS and Application on the Transected Tendon

A disposition of approximately 3.7×10^5 ASC in the dense fibrin network formed by the FS was analyzed, showing the homogenous distribution of the cells in the fibrin clot (Figure 2A) before transplantation in the transection tendon region (Figure 2B–D).

ASC distributed in the FS

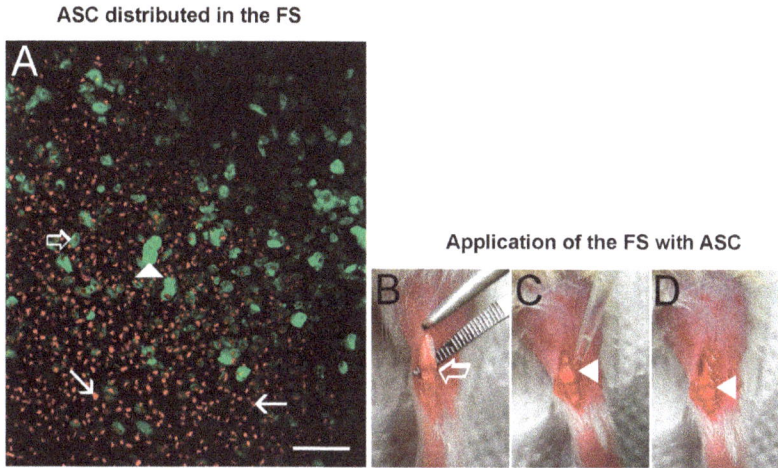

Application of the FS with ASC

Figure 2. (**A**) Confocal microscope image of ASC labeled with phalloidin-FITC (actin cytoskeleton in green) and DAPI (nucleus in red) distributed in the dense fibrin network formed by the FS ($n = 3$). Observe the disposition of 3.7×10^5 cells in the FS before application in the tendon transected region. (▶) superficial cells, (⇕) intermediately positioned cells and cells at the bottom (→). (**B**) Model of tendon injury showing the partial transection (⇕) in the proximal region of the Achilles tendon. (**C**) Application of the FS with ASC using a pipette: note the formation of a clot (▶). (**D**) Representation of the FS with ASC (▶) covering the transected region before the skin suture. Bar = 200 μm.

3.3. In Vivo Imaging for ASC Detection on Tendon

Cell migration was evaluated by in vivo imaging of animals that were injected with ASC labeled with PKH26. PKH26-labeled ASC (group ASC) or with fibrin sealant (group FS + ASC) were observed in lower limb up to 7 days (Figure 3A–D,F–I) or 14 days (Figure 3E,J), respectively, being not observed in any of the groups at the experiment end point (21 days, not shown). A small spot was seen in the sham group but never in the negative control, only on the 1st day (not shown), then its area was discounted from the other groups at the same day. According to our data, labeling intensity peaked at day 1 in the ASC group, decreasing on the 2nd day (1d vs. 2d, $p < 0.05$) with no further variation, while in the FS + ASC group it kept constant through time (Figure 3K). Of importance, comparison of both groups revealed higher fluorescence intensity in ASC treated rats on the 1st day ($p < 0.05$; Figure 3K). Regarding the labeling area, it also peaked on the 1st day in the ASC group, decreasing 93.5% on the 2nd day (1d vs. 2d, $p < 0.05$) after when it oscillated up to 7th day, however, with no significant differences (Figure 3L). No labeling was found on the 14th day (Figure 3E). Variations in the labeling area were also observed in the FS + ASC group from the 1st to the 14th day, however, with no statistical differences (Figure 3L). Comparison between both groups revealed that the labeling area is more than 10 folds higher in the absence of fibrin sealant on the 1st day ($p < 0.001$; Figure 3L), reinforcing its importance as a scaffold that avoids cell spreading in the tissue. No PKH26-labeled ASC was observed on the 21st day after tendon transection in both ASC and FS + ASC groups (not shown).

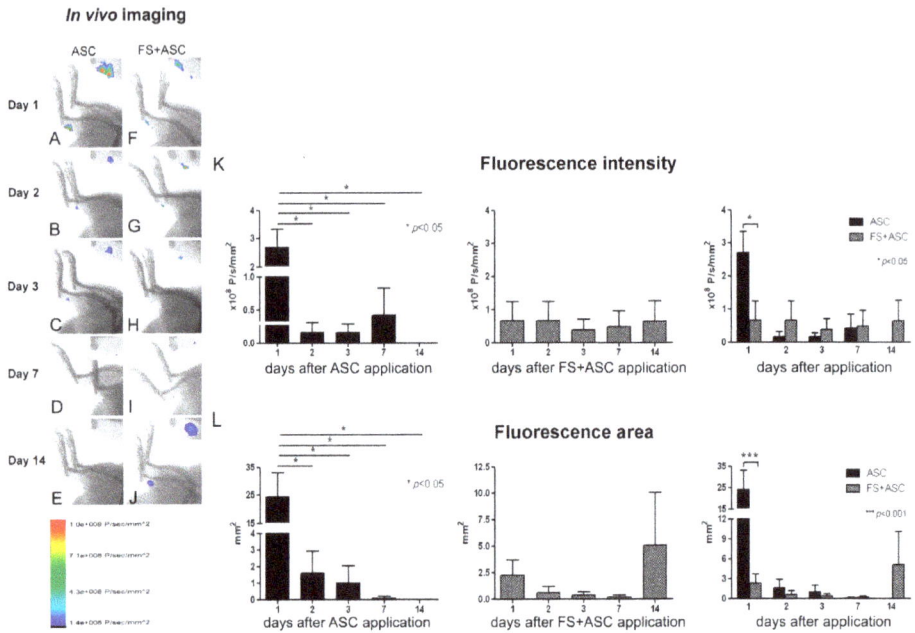

Figure 3. (**A**) In vivo imaging for detection and quantification of PKH26-labeled ASC in the tendon transected region: ASC (**A–E**) and FS + ASC (**F–J**) groups were analyzed on the 1st, 2nd, 3rd, 7th,14th and 21st days after injury ($n = 3$). Observe the fluorescence intensity and area in the detail of each image of tendons (**A–J**). Scale: fluorescence intensity ($\times 10^8$ P/s/mm^2). Fluorescence intensity (**K**) and fluorescence area (**L**) occupied by ASC after quantification of images from in vivo imaging. Significant difference represented by (*) and (***) between the ASC and FS + ASC groups.

3.4. Cell Migration Assay and Macrophages Identification

CD90+CD105+ ASC were visualized in the transected region of tendons of both ASC and FS + ASC groups (Figure 4), proving the ASC migration.

A small amount of macrophages was observed in the normal (Figure 5A,B) as well as in the transected tendons regardless the treatment (Figure 5C–K). Of importance, in tendons that received cells it was possible to observe very few macrophages carrying PKH26 signal (Figure 5G,H,J,K), probably due to phagocytosis of dead ASC.

Immunofluorescence for CD90 and CD105

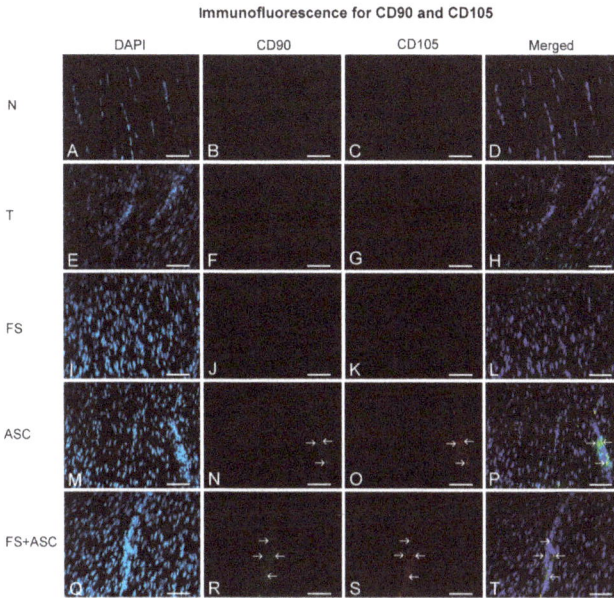

Figure 4. Immunofluorescence for CD90 and CD105 observed in the central portion of the TR of tendons on the 21st day (*n* = 5). Groups N (**A–D**), T (**E–H**), FS (**I–L**), ASC (**M–P**) and FS + ASC (**Q–T**). Note CD90 and CD105 (→) positive marking in the transected region of ASC and FS + ASC groups. Bar = 50 μm.

Immunofluorescence for Iba1

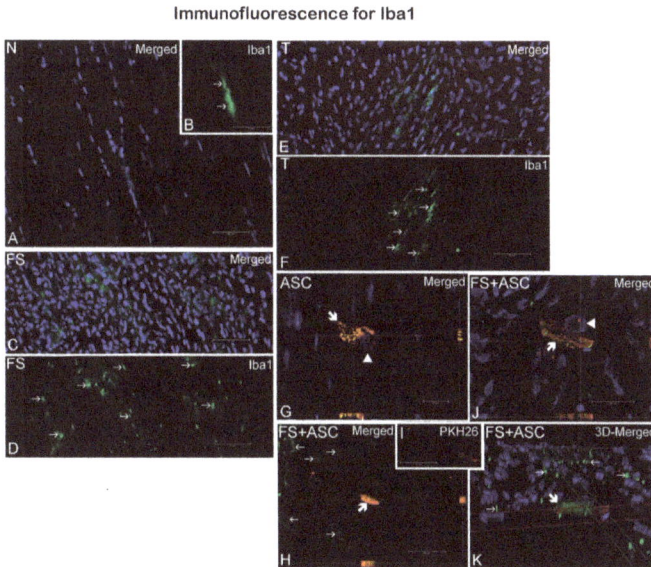

Figure 5. Iba1 labeled macrophages (green, →) were observed in all tendons (*n* = 3) regardless the treatment as follows: N group (**A, B**), FS (**C, D**), T (**E, F**), ASC (**G**) and FS + ASC (**H–K**). Eventually, in groups that received PKH26-labeled ASC (ASC and FS + ASC) fluorescence signal (red) could be seen inside of macrophages (✔). (**B**) Detail of A, showing Iba1 labeled macrophages. (**I**) Detail of H, evidencing PHK26 labeled ASC. (**G, H, J**) orthogonal sectioning; K, 3D deconvolution. Nuclei were stained with DAPI (blue, ▶). Scale bars: **A–F, H, I**, 50μm; **G, J**, 20μm.

3.5. Genes Expression Analysis

A higher expression level of the *Tnmd* gene could be observed in the ASC group in relation to T and FS groups. Trends to a greater expression of *Mmp2, Timp2, Scx* and Ilb1 were observed in the ASC group compared to the other transected groups. In comparison to N group, the transected groups presented higher expression of *Mmp2, Mmp9, Timp2, Gdf5, Scx* and *Tnmd* (Figure 6).

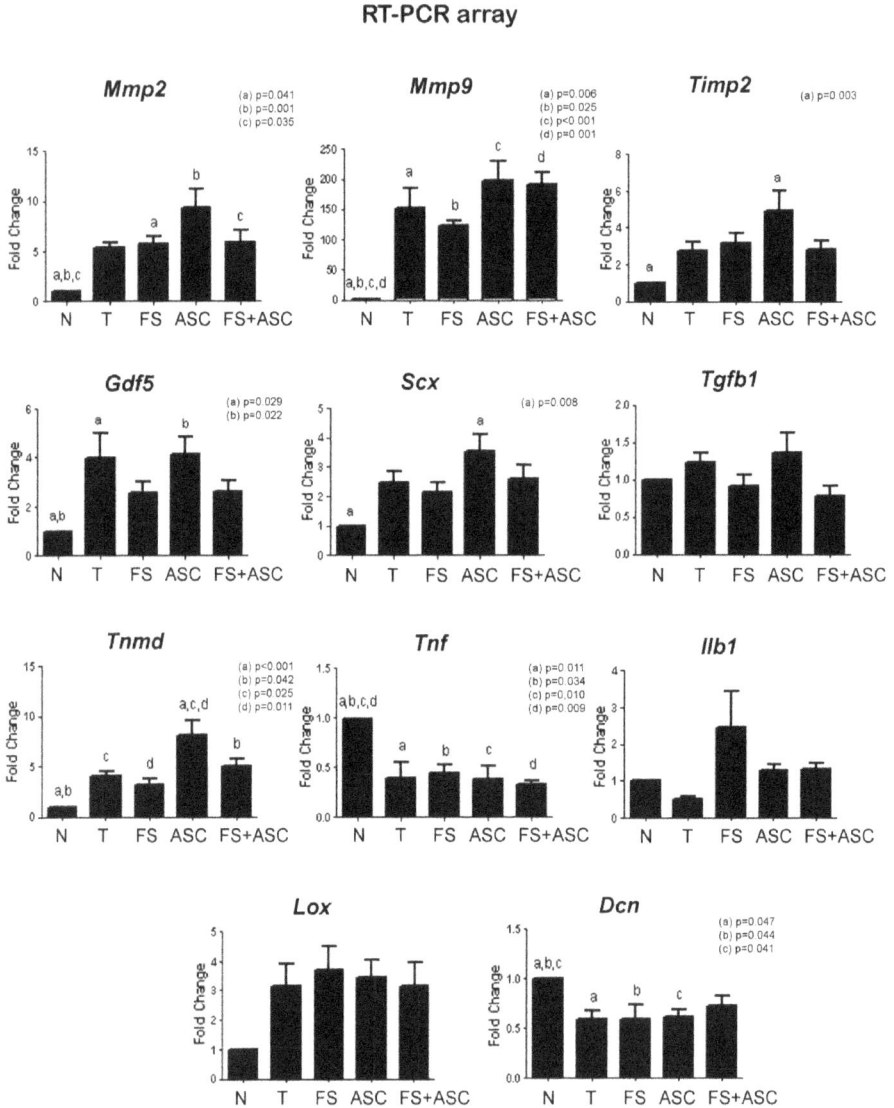

Figure 6. Real-time PCR array data of genes on the 21st day after the tendon transection (*n* = 4). Note the higher *Tnmd* expression in the ASC group in relation to the T and FS. Difference between N group and T, FS, ASC and FS + ASC groups can be observed for genes expression of *Mmp2, Mmp9, Timp2, Gdf5, Scx, Tnmd, Tnf* and *Dcn*. Same letters (a, b, c, d) = significant difference between the groups.

3.6. Total Collagen Quantification

Higher concentration of hydroxyproline was observed in the ASC group compared to the N group (Table 1).

Table 1. Hydroxyproline concentration (mg/g tissue) in the entire tendon (n = 8). (*) Significant difference between N and ASC groups.

Groups	N	T	FS	ASC	FS+ASC
Hydroxyproline (mg/g tissue)	100.8 ± 7.5 *	103.6 ± 9.1	104.9 ± 8.5	107.9 ± 6.5 *	101.6 ± 6.7

$* p < 0.041.$

3.7. Collagen I and Tnmd quantification

The presence of collagen type I and Tnmd was observed in all groups. Densitometry analysis of the bands (pixels) showed a difference only between the normal tendon and the transected tendons, with exception of the ASC group (Figure 7B).

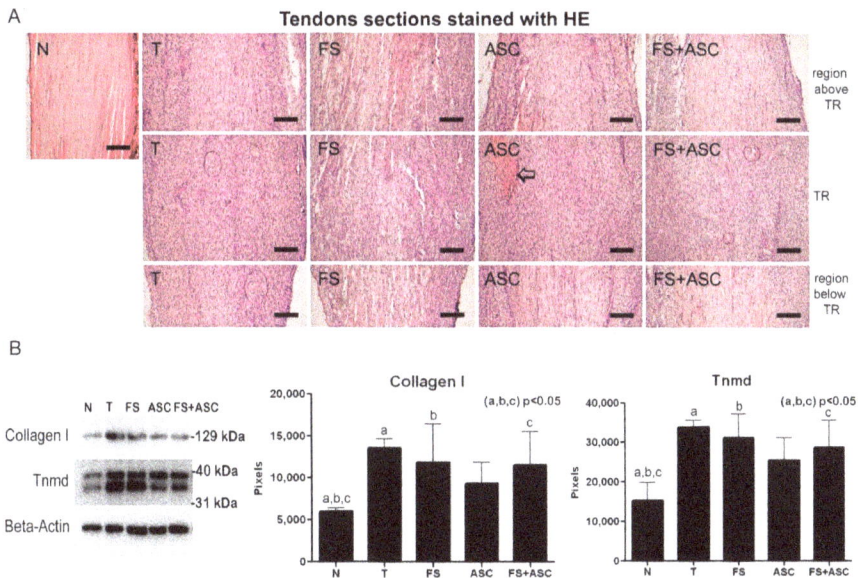

Figure 7. (**A**) Panoramic view of tendons sections stained with HE (*n* = 3). Comparing all transected tendons with normal tendon (proximal region), higher cellularity and light staining of ECM (**A**) can be seen in the TR and in the regions above and below TR. Note ECM more intensely stained in the TR of ASC (⬦). Both regions above TR (located closer to the insertion of tendon in the gastrocnemius muscle) and the region below TR (located closer to the insertion of tendon in the calcaneus bone) are located in the proximal region of the tendon. Bar = 200 μm. (**B**) Western blotting showing collagen type I and Tnmd in the entire tendon (*n* = 4). Beta-actin was utilized as an endogenous control (43 kDa). For the significant differences between the groups, see the band densitometry analysis in the graphics. The same letters (a, b, c) between the groups correspond to a significant difference between them.

3.8. Panoramic View of Tendons Organization and Histomorphometry

The sections stained with HE showed few fibroblasts arranged between dense and organized matrix collagen bundles and characteristics of the proximal region of a normal tendon (Figure 7A). Higher cellularity and light staining of ECM were observed in the TR and in the regions above and below TR of transected tendons in relation to the normal tendon, indicating a strong remodeling process after lesion. The weak staining of collagen fibers by eosin shows that the reorganization of the matrix is not complete on the 21st day. Histomorphometry analysis showed no differences in the number of fibroblasts between the transected tendons. However, the N group presented a smaller number of cells compared to the T, FS, ASC and FS + ASC groups (Table 2).

Table 2. Total number of fibroblasts: no differences were observed between the transected tendons. (*) Significant difference between the N group and the transected groups ($n = 3$).

Groups	N	T	FS	ASC	FS + ASC
Fibroblasts	26.6 ± 3.8 *	88.7 ± 11.7 *	93.9 ± 9.8 *	96.7 ± 12.3 *	90.5 ± 14.2 *

$* p < 0.001$.

3.9. Collagen Fibers Organization Measurements

The analysis of tendons of different groups, under polarizing microscopy, showed higher birefringence values in FS in relation to all groups and in FS + ASC in comparison to the ASC group, indicating difference in the collagen bundles organization (Figure 8 A–G). N group exhibited higher birefringence values in relation to all transected groups. The image analysis of tendons sections from FS groups, showed a typical crimp arrangement in the transected region (Figure 8B). When DIC (Differential Interference Contrast) was used, which increases the contrast in unstained collagen bundles (for details see Vidal et al. [57]), it was possible to detect in the new ECM in formation, a blue color due to the presence of packed and organized collagen bundles and a background red color of unpacked or less organized collagen bundles (Figure 8C). The frequency histograms of birefringence gray average (GA) values showed the heterogeneous data distribution, with marked low birefringence values for group T and higher values especially for the FS group.

3.10. Biomechanical Properties of Tendons

The analysis of mechanical properties of transected tendons for the maximum load (N), maximum displacement (mm), maximum strain (mm), cross-sectional area (mm^2) and maximum stress (MPa), showed some alterations (Figure 9). The T group had higher value of maximum load compared to the groups ASC ($p = 0.011$) and FS + ASC ($p = 0.031$). For the cross-sectional area and maximum stress, N group had lower and higher values, respectively, compared to the transected groups ($p < 0.001$).

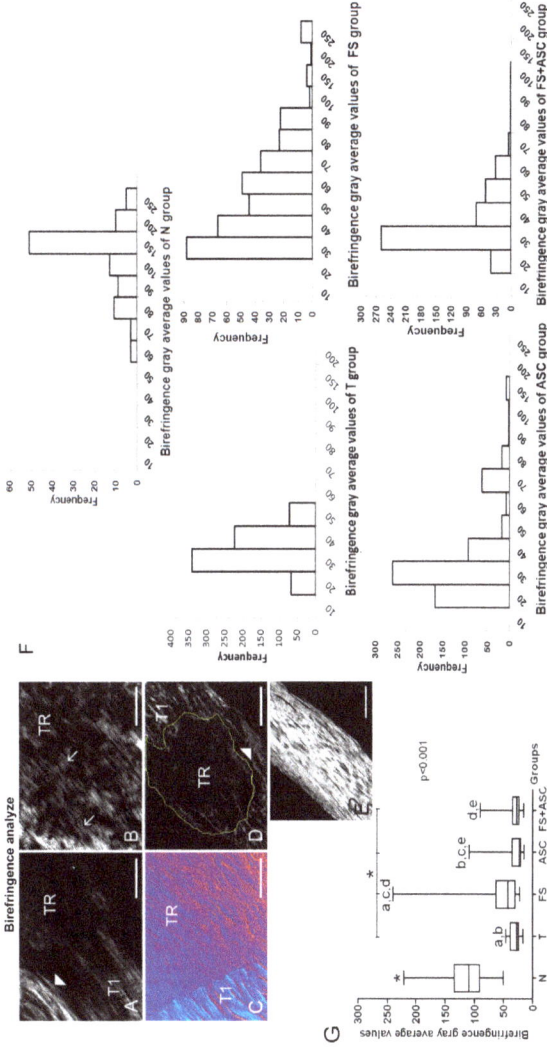

Figure 8. Images of birefringence of tendon longitudinal sections on 21st day using polarization microscopy ($n = 5$). The larger tendon axis was set 45° from the crossed polarizers. The variation of brightness intensity (gray levels) is due the variation of the collagen bundles organization. (**A**) T group: little birefringence brightness is observed in TR (tendon transected region) because of the disorganization of collagen bundles. T1 is the region which border the TR. (▶) remaining portion of the tendon located below the TR. (**B**) FS group: the increase in birefringence brightness was remarkable and a typical well-developed crimp (→) pattern was observed only in this group. (**C**) ASC group: image using DIC (differential interference contrast microscopy), where it is possible to visualize in red a smaller organization of the collagen bundles and in intense blue (according to Michel-Lévy's table) the high degree of compaction of the collagen bundles. (**D**) FS + ASC group: observe a higher birefringence of the collagen fibers compared to the ASC group and observe an imbrication between collagen fibers that were not cut in T1 and fibrils in the TR (delimited by yellow line). (▶) remaining portion of the tendon located below the TR. (**E**) N group: collagen fibers exhibiting strong birefringence. Bar = 100 μm (**A**, **C**, **D**, **E**) and bar = 200 μm (**B**, **F**) Frequency histograms of birefringence gray average (GA) values expressed in pixels in the groups N, T, FS, ASC and FS + ASC, which reflect the variability of the collagen fibers organization on the TR region of the Achilles tendon. (**G**) Birefringence GA (pixels) median between the groups. The measurements data showed in graphics (f and g) were obtained with the larger tendon axis positioned at 45° from the crossed polarizers. (*) Significant differences between the group N and groups with transected tendons. The same letters between the groups correspond to a significant difference between them.

Mechanical properties of tendons

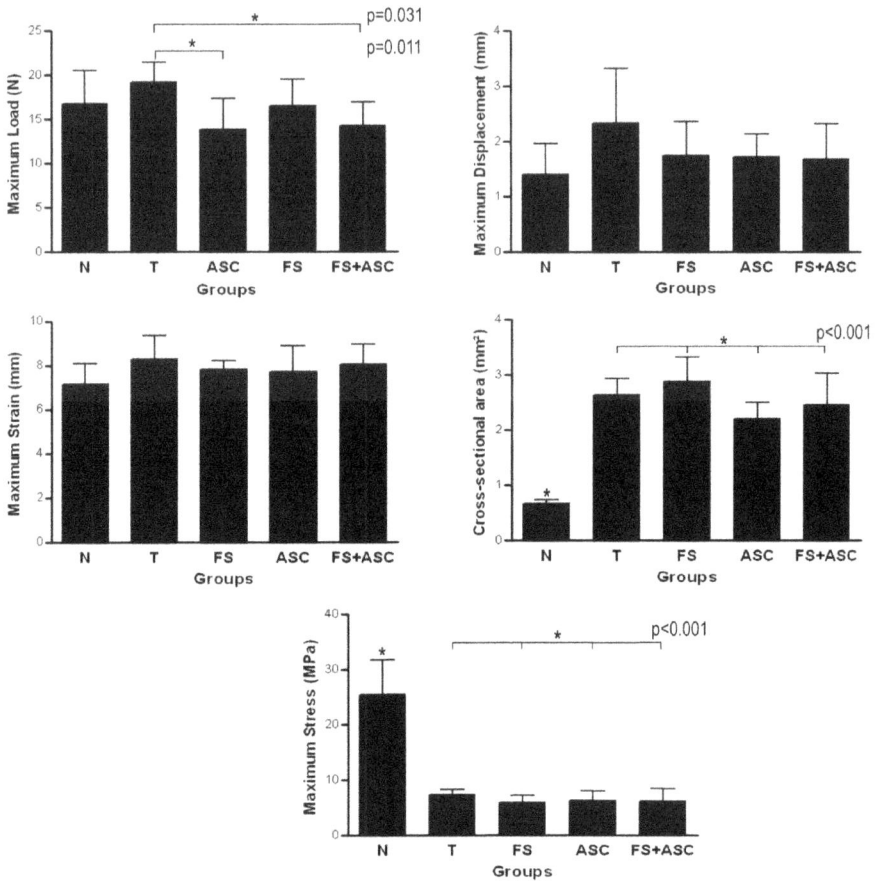

Figure 9. Mechanical properties of transected tendons from groups T, FS, ASC and FS + ASC ($n = 8$). (*) Significant difference between the groups.

For a systematic overview table of results see Table 3.

Table 3. Systematic overview table of results.

	N	T	FS	ASC	FS + ASC
In vivo imaging for ASC detection on tendon	-	-	-	Detection of labeled-ASC until the 7th day	Detection of labeled-ASC until the 14th day
Cell migration assay	-	-	-	Presence of ASC on the 21st day	Presence of ASC on the 21st day
Macrophages identification	Presence of few macrophages	Presence of few macrophages	Presence of few macrophages	Presence of few macrophages carring PKH26 signal	Presence of few macrophages carring PKH26 signal
Genes expression analysis	Lower expression of *Mmp2*, *Mmp9*, *Timp2*, *Gdf5*, *Scx* and *Tnmd*; and higher expression of *Tnf* and *Dcn* in relation to T, FS, ASC and FS + ASC groups	Lower *Tnmd* expression in relation to ASC group	Lower *Tnmd* expression in relation to ASC group	Higher *Tnmd* expression in relation to T and FS groups	No differences between the treatments
Total collagen concentration (entire tendon)	Lower concentration in relation to ASC group	No differences between the treatments and N group	No differences between the treatments and N group	No differences between the treatments and higher concentration in relation to N group	No differences between the treatments and N group
Collagen I and Tnmd quantification (entire tendon)	Lower amount in relation to T, FS and FS+ASC groups	No differences between the treatments	No differences between the treatments	No differences between the treatments and N group	No differences between the treatments
Total number of fibroblasts in the TR	Higher number of cells in relation to T, FS, ASC and FS+ASC groups	No differences between the treatments	No differences between the treatments	No differences between the treatments	No differences between the treatments
Collagen fibers organization measurements	Higher birefringence in relation to T, FS, ASC and FS+ASC groups	Lower birefringence in relation to FS, ASC and FS+ASC groups	Higher birefringence in relation to T, ASC and FS+ASC groups	Higher birefringence in relation to T group	Higher birefringence in relation to T and ASC groups
Biomechanical properties of tendons	*Maximum Load*: no differences in relation to T, FS, ASC and FS+ASC groups; *Cross-sectional area*: lower value in relation to T, FS, ASC and FS+ASC groups; *Maximum Displacement and Strain*: no differences in relation to T, FS, ASC and FS+ASC groups; *Maximum Stress*: higher value in relation to T, FS, ASC and FS+ASC groups	*Maximum Load*: higher value in relation to ASC and FS+ASC groups; *Cross-sectional area*: no differences between the treatments; *Maximum Displacement and Strain*: no differences between the treatments; *Maximum Stress*: no differences between the treatments	*Maximum Load*: no differences between the treatments; *Cross-sectional area*: no differences between the treatments; *Maximum Displacement and Strain*: no differences between the treatments; *Maximum Stress*: no differences between the treatments	*Maximum Load*: lower value in relation to T group; *Cross-sectional area*: no differences between the treatments; *Maximum Displacement and Strain*: no differences between the treatments; *Maximum Stress*: no differences between the treatments	*Maximum Load*: lower value in relation to T group; *Cross-sectional area*: no differences between the treatments; *Maximum Displacement and Strain*: no differences between the treatments; *Maximum Stress*: no differences between the treatments

4. Discussion

In the present study, the 21st day after the transection was chosen because it marks the beginning of the remodeling phase, where there is a marked decrease in cellularity and formation of fibrous tissue [49–51], which is the focus of our study. In order to analyze the presence of ASC in isolation or associated with FS in the transected tendon region, in vivo imaging was used, where the striking result was observed in the ASC group due a high fluorescence intensity and area on the 1st day after injury, in relation to the FS + ASC group. Fibrin sealant works as a 3D scaffold that maintains cells grouped and possibly prevents the quantification of all labeled cells, while in its absence, cells are allowed to spread in the tissue. As a consequence of initial cell spreading (accommodation), increased labeling area was observed on the 1st day in the ASC group. But, as also demonstrated by Qiao et al. [62], our data showed a substantial ASC loss, when only stem cells are transplanted, from the 1st to the 2nd day. From the 2nd to the 7th day, no differences were observed among the groups when considering the fluorescence intensity and area. It is important to observe that only the FS + ASC group showed marking on the 14th day, with similar fluorescence intensity to that observed in previous periods and a larger area of fluorescence, possibly due to the fibrin sealant being degraded and consequently releasing the ASC. Our data supports the in vivo study by Wolbank et al. [63], which subcutaneously implanted in rats, fibrin clots with fluorophore-labeled and monitored their degradation for 21 days by in vivo imaging. The study showed that much of the fibrin clot was degraded by about 14 days and after 16 days, the degradation was total. Bensaïd et al. [64], through an in vitro study, demonstrated that fibrin scaffold together with human mesenchymal stem cells, depending on the concentration of fibrinogen and thrombin, is degraded within 14 to 15 days. Also in an agreement with our data, Spejo et al. [65] showed that the mesenchymal stem cells remain at region with 28 days post injury in glial scar region in the ventral funiculus of rats, evidencing that the sealant (the same as in the present study) kept these cells protected from the phagocytic system of the organism. This sealant is produced from the blood components of buffaloes, mainly fibrinogen, resulting in a fibrin clot and because it is heterologous, its fibrinolysis is delayed, causing it to act with an excellent scaffold for cells [29].

Orsi et al. [66] did not observe cytotoxicity of FS and thrombin-like enzyme (snake venom compound), both the same as in the present study, associated to mesenchymal stem cells. No marking was detected by in vivo imaging from the 15th day. However, the presence of a small number of ASC with CD90 and CD105 positive marking was showed into the newly formed matrix of the transected region in both ASC and FS + ASC groups on the 21st day, confirming the presence of ASC in the tendon. It is important to mention that the marking detected in tendons on the 21st day corresponds to a majority of live ASC since a few macrophages carrying PKH26 signal due to phagocytosis of dead ASC, were observed in the transected region of tendons.

The literature has already reported the paracrine effects of ASC on the modulation of cellular activity [67–69], as well as the effects of fibroblast secreted molecules on ASC activity [70]. In our injury model, the interaction between the transplanted ASC and the cells present in the transected region is evident since the ASC with CD90 e CD105 positive marking were identified in tendon ECM on the 21st day after transection. Some in vitro studies show the capacity of tenogenic differentiation of ASC in the presence of appropriate biological stimuli, such as growth factor, stress forces [71] and when cultivated in co-culture with tenocytes [70]. *Tnmd* is a good phenotype marker for tendon fibroblasts and acts as a regulator of cell proliferation and differentiation [72]. Our results showed significant increase of the *Tnmd* gene expression in the ASC group in relation to the groups without transplanted ASC, indicating a greater differentiation in tenocytes mainly of the transplanted ASC [70,73,74]. Thus, it is concluded that the molecular signaling present in the injury environment possibly affects the differentiation of ASC in tenocytes, only in the transected region of tendons. Considering the entire tendon, no differences were observed in the Tnmd amount, in a protein level, between the transected tendons. In addition to the *Tnmd*, Scx is also a key molecule involved in the process of tendon development. Scx is a transcription factor responsible for the differentiation of stem cells into tenocytes and is specifically detected in precursor populations of tendon cells [70,75]. Our results showed a trend to the increase of

Scx expression in the ASC group in relation to the other transected groups, suggesting its relationship with the significant increase of the *Tnmd* expression.

In order to analyze the effect of the treatments with ASC and FS reflecting in the entire tendon, collagen and tenomodulin, markers of tendon, were analyzed also in the entire tendons considering that the collagen fibers are also responsible for the mechanotransduction during gait, which is important for the cellular signaling in the transected region, affecting the recovery of the ECM at the site of injury. As well as it is important to consider that the molecular events in the transected region can also influence the organization and composition of the not damaged areas of tendons. In addition, collagen is the main constituent of tendons, mainly collagen type I, also being directly related to the recovery of the structure and biomechanical properties of tendons during repair process. Interestingly, the ASC group presented superior total collagen concentration, estimated by the hydroxyproline dosage, in relation to the N group. *Gdf5* and *Tgfb1* are related to the stimulation of collagen synthesis [26,70,76–78], supporting our results in which was observed a higher *Gdf5* expression in transected region of the ASC group, in comparison to N group and higher total collagen concentration in the entire tendon. However, no differences for the collagen I was observed between N and ASC groups. According to Uysal et al. [76], ASC applied on the injured calcaneus tendon of rabbits helped in tendon healing and increased the production of type I collagen.

Data from Peres [79] showed a marked increase of inflammatory infiltrate during the initial stage of healing of the skin of rats after an isolated application of either FS (the same FS used in our study) or after its association with ASC, in relation to the other sutured groups. The trend towards greater expression of *Il1b* observed after the FS application in our study could be related to an inflammatory process with remnants still to the 21st day, considering the presence of fibrin until the 14th day in the injury region and period of its absorption [63]. Yücel et al. [80] showed that although the FS application increases inflammation on the first few days after application, the fibrin matrix helps tissue recovery.

The Yücel et al. [80] study mentioned above supports our structural data. Higher values of birefringence were detected in the FS group in comparison to all groups, followed by the FS+ASC and ASC groups in relation to T, respectively, reflecting greater organization of the collagen bundles. Considering that tissue reorganization after injury is extremely important for the recovery of the tendon functionality, our structural result is highly relevant considering the 21st day of repair, because depending on the extent of the injury, the tendon may take months for complete healing [81]. The use of FS acted as a scaffold for resident tendon cells, as well as for the interaction of ASC transplanted with resident cells of the transected region since both FS and FS + ASC groups had higher birefringence values, showing that fibrin is slowly replaced by connective tissue. A study of Tuan et al. [82] supports our structural results because demonstrated that the cultured fibroblasts together with the fibrin gel, synthesize collagen and actively reorganize the fibrin matrix which is largely replaced by collagen fibrils. Regarding the analysis of genes involved in ECM remodeling, no striking results were observed in relation to the expression of *Lox*, *Dcn*, *Mmp2*, *Mmp9* and *Timp2* in the transected tendons, suggesting the involvement of other genes in the higher collagen fibers organization observed specially in the FS, followed by FS + ASC and ASC groups

In the biomechanical analysis, no differences were observed between the transected groups for the stress parameter on the 21st day, supporting no differences in the amount of collagen type I in the entire tendon, which is directly related to the resistance of tendon. Although, the literature points to the beneficial effects of FS and ASC on the biomechanics of tendons in the healing process [83]. Therefore, it is possible to conclude that the structural data did not directly reflect the biomechanics of the tendons, perhaps due to the recovery time, 21 days, which is marked by the beginning of the remodeling process. A point to be considered is that the number of crosslinks present in tendons treated with ASC and with FS isolated or in combination, reflected in the higher organization of these tendons but was not enough to warrant a more resistant tendon on the 21st day after transection. We believe that long-term time point analysis could demonstrate that transected tendon treated with FS exhibited superior strength in comparison with T group.

According to our data, transplanted ASC migrated to the transected region and consequently the ASC up-regulated the *Tnmd* expression suggesting a differentiation of transplanted cells in tenocytes and the ASC increased the collagen fibers organization. The application of FS alone was able to improve the molecular organization of the collagen fibers but when associated with the ASC, a higher number of cells was kept in the transected region, with the ASC protected from the phagocytic system. In conclusion, our data suggest this FS to be a good scaffold for treatment during tendon repair because it was the most effective one regarding tendon organization recovering, followed by the FS treatment associated with ASC and finally by the transplanted ASC on the 21st day. Further investigations in long-term time points of the tendon repair are needed to analyze if the higher tissue organization found with the FS scaffold will improve the biomechanics of the tendons.

Author Contributions: Conceptualization, A.A.d.A. and E.R.P.; methodology, K.F., L.F.R.T., G.D.C., F.C.d.V., D.L.F., A.L.B., C.P.V., M.E.C.d.A., R.S.F.J., B.B., M.A.M.E., B.d.C.V. and A.A.d.A; formal analysis, K.F., G.D.C., F.C.d.V., D.L.F., A.L.B., P.W.S., L.E.A., A.L.R.d.O., C.P.V., M.E.C.d.A., M.A.M.E., B.d.C.V., E.R.P., A.A.d.A; resources, E.R.P., A.A.d.A. and M.A.M.E.; writing—original draft preparation, K.F. and A.A.d.A.; writing—review and editing, K.F., B.d.C.V., M.A.M.E., R.S.F.J., A.A.d.A. and E.R.P.; supervision, A.A.d.A. and E.R.P.

Funding: The authors thank FAPESP (Fundação de Amparo à Pesquisa do Estado de São Paulo, 2012/14973-8) and CAPES (Coordenação de Aperfeiçoamento de Pessoal de Nível Superior) for the financial support.

Acknowledgments: The authors thank Rafaela Rosa-Ribeiro for her technical assistance during samples preparation for confocal microscopy and Mariana Ozello Baratti for the assistance during the use of the Confocal microscope. We also thank the access to equipment and assistance provided by the National Institute of Science and Technology on Photonics Applied to Cell Biology (INFABIC) at the State University of Campinas.

Conflicts of Interest: The authors declare no conflict of interest.

References

1. Nourissat, G.; Berenbaum, F.; Duprez, D. Tendon injury: From biology to tendon repair. *Nat. Rev. Rheumatol.* **2015**, *11*, 223–233. [CrossRef] [PubMed]
2. Andarawis-Puri, N.; Flatow, E.L.; Soslowsky, L.J. Tendon Basic Science: Development, Repair, Regeneration, and Healing. *J. Orthop. Res.* **2015**, *33*, 780–784. [CrossRef] [PubMed]
3. Christiansen, D.L.; Huang, E.K.; Silver, F.H. Assembly of type I collagen: Fusion of fibril subunits and the influence of fibril diameter on mechanical properties. *Matrix Biol.* **2000**, *19*, 409–420. [CrossRef]
4. Vidal, B.C. Cell and extracellular matrix interaction: A feedback theory based on molecular order recognition-adhesion events. *Rev. Fac. Ciên. Med. Unicamp.* **1994**, *4*, 11–14.
5. Lavagnino, M.; Wall, M.E.; Little, D.; Banes, A.J.; Guilak, F.; Arnoczky, S.P. Tendon Mechanobiology: Current Knowledge and Future Research Opportunities. *J. Orthop. Res.* **2015**, *33*, 813–822. [CrossRef] [PubMed]
6. Arnoczky, S.P.; Lavagnino, M.; Whallon, J.H.; Hoonjan, A. In situ cell nucleus deformation in tendons under tensile load; a morphological analysis using confocal laser microscopy. *J. Orthop. Res.* **2002**, *20*, 29–35. [CrossRef]
7. Arnoczky, S.P.; Lavagnino, M.; Egerbacher, M. The mechanobiological aetiopathogenesis of tendinopathy: Is it the over-stimulation or the under-stimulation of tendon cells? *Int. J. Exp. Pathol.* **2007**, *88*, 217–226. [CrossRef]
8. Archambault, J.M.; Wiley, J.P.; Bray, R.C. Exercise loading of tendons and the development of overuse injuries. A review of current literature. *Sports Med.* **1995**, *20*, 77–89. [CrossRef]
9. Durgam, S.; Stewart, M. Cellular and Molecular Factors Influencing Tendon Repair. *Tissue Eng. Part B Rev.* **2017**, *23*, 307–317. [CrossRef]
10. Docheva, D.; Müller, S.A.; Majewski, M.; Evans, C.H. Biologics for tendon repair. *Adv. Drug Deliv. Rev.* **2015**, *84*, 222–239. [CrossRef]
11. Aro, A.A.; Nishan, U.; Perez, M.O.; Rodrigues, R.A.; Foglio, M.A.; Carvalho, J.E.; Gomes, L.; Vidal, B.C.; Pimentel, E.R. Structural and biochemical alterations during the healing process of tendons treated with Aloe vera. *Life Sci.* **2012**, *91*, 885–893. [CrossRef] [PubMed]
12. Leadbetter, W.B. Cell-matrix response in tendon injury. *Clin. Sports Med.* **1992**, *11*, 533–578. [PubMed]

13. Merceron, C.; Vinatier, C.; Clouet, J.; Colliec-Jouault, S.; Weiss, P.; Guicheux, J. Adipose-derived mesenchymal stem cells and biomaterials for cartilage tissue engineering. *Jt. Bone Spine* **2008**, *75*, 672–674. [CrossRef] [PubMed]

14. Chen, F.H.; Rousche, K.T.; Tuan, R.S. Technology Insight: Adult stem cells in cartilage regeneration and tissue engineering. *Nat. Clin. Pract. Rheumatol.* **2006**, *2*, 373–382. [CrossRef] [PubMed]

15. Gonçalves, A.I.; Rodrigues, M.T.; Lee, S.J.; Atala, A.; Yoo, J.J.; Reis, R.L.; Gomes, M.E. Understanding the Role of Growth Factors in Modulating Stem Cell Tenogenesis. *PLoS ONE* **2013**, *8*, e83734. [CrossRef] [PubMed]

16. Oedayrajsingh-Varma, M.J.; Van Ham, S.M.; Knippenberg, M.; Helder, M.N.; Klein-Nulend, J.; Schouten, T.E.; Ritt, M.J.; van Milligen, F.J. Adipose tissue-derived mesenchymal stem cell yield and growth characteristics are affected by the tissue-harvesting procedure. *Cytotherapy* **2006**, *8*, 166–177. [CrossRef] [PubMed]

17. Katz, A.J.; Tholpady, A.; Tholpady, S.S.; Shang, H.; Ogle, R.C. Cell surface and transcriptional characterization of human adipose-derived adherent stromal (hADAS) cells. *Stem Cells* **2005**, *23*, 412–423. [CrossRef] [PubMed]

18. Hu, R.; Ling, W.; Xu, W.; Han, D. Fibroblast-Like Cells Differentiated from Adipose-Derived Mesenchymal Stem Cells for Vocal Fold Wound Healing. *PLoS ONE* **2014**, *9*, e92676. [CrossRef]

19. Altman, A.M.; Matthias, N.; Yan, Y.; Song, Y.-H.; Bai, X.; Chiu, E.S.; Slakey, D.P.; Alt, E.U. Dermal matrix as a carrier for in vivo delivery of human adipose-derived stem cells. *Biomaterials* **2008**, *29*, 1431–1442. [CrossRef]

20. Kern, S.; Eichler, H.; Stoeve, J.; Klüter, H.; Bieback, K. Comparative Analysis of Mesenchymal Stem Cells from Bone Marrow, Umbilical Cord Blood, or Adipose Tissue. *Stem Cells* **2006**, *24*, 1294–1301. [CrossRef]

21. Rodriguez, A.-M.; Elabd, C.; Amri, E.-Z.; Ailhaud, G.; Dani, C. The human adipose tissue is a source of multipotent stem cells. *Biochimie* **2005**, *87*, 125–128. [CrossRef] [PubMed]

22. De Ugarte, D.A.; Morizono, K.; Elbarbary, A.; Alfonso, Z.; Zuk, P.A.; Zhu, M.; Dragoo, J.L.; Ashjian, P.; Thomas, B.; Benhaim, P.; et al. Comparison of Multi-Lineage Cells from Human Adipose Tissue and Bone Marrow. *Cells Tissues Organs* **2003**, *174*, 101–109. [CrossRef]

23. Hassan, W.U.; Greiser, U.; Wang, W. Role of adipose-derived stem cells in wound healing. *Wound Repair Regen.* **2014**, *22*, 313–325. [CrossRef] [PubMed]

24. Cai, L.; Johnstone, B.H.; Cook, T.G.; Liang, Z.; Traktuev, D.; Cornetta, K.; Ingram, D.A.; Rosen, E.D.; March, K.L. Suppression of hepatocyte growth factor production impairs the ability of adipose derived stem cells to promote ischemic tissue revascularization. *Stem Cells* **2007**, *25*, 3234–3243. [CrossRef] [PubMed]

25. Hattori, H.; Ishihara, M. Altered protein secretions during interactions between adipose tissue- or bone marrow-derived stromal cells and inflammatory cells. *Stem Cell Res. Ther.* **2015**, *6*, 70. [CrossRef] [PubMed]

26. Uysal, C.A.; Tobita, M.; Hyakusoku, H.; Mizuno, H. Adipose-derived stem cells enhance primary tendon repair: Biomechanical and immunohistochemical evaluation. *J. Plast. Reconstr. Aesthet. Surg.* **2012**, *65*, 1712–1719. [CrossRef]

27. Kim, J.H.; Jung, M.; Kim, H.S.; Kim, Y.M.; Choi, E.H. Adipose-derived stem cells as a new therapeutic modality for ageing skin. *Exp. Dermatol.* **2011**, *20*, 383–387. [CrossRef]

28. Shingyochi, Y.; Orbay, H.; Mizuno, H. Adipose-derived stem cells for wound repair and regeneration. *Expert Opin. Biol. Ther.* **2015**, *15*, 1285–1292. [CrossRef]

29. Gasparotto, V.P.O.; Landim-Alvarenga, F.C.; Oliveira, A.L.R.; Simões, G.; Lima-Neto, J.F.; Barraviera, B.; Ferreira, R.S. A new fibrin sealant as a three-dimensional scaffold candidate for mesenchymal stem cells. *Stem Cell Res. Ther.* **2014**, *5*, 78. [CrossRef]

30. James, R.; Kumbar, S.G.; Laurencin, C.T.; Balian, G.; Chhabra, A.B. Tendon tissue engineering: Adipose-derived stem cell and GDF-5 mediated regeneration using electrospun matrix systems. *Biomed. Mater.* **2011**, *6*, 025011. [CrossRef]

31. Biscola, N.P.; Cartarozzi, L.P.; Ulian-Benitez, S.; Barbizan, R.; Castro, M.V.; Spejo, A.B.; Ferreira, R.S.; Barraviera, B.; Oliveira, A.L.R. Multiple uses of fibrin sealant for nervous system treatment following injury and disease. *J. Venom. Anim. Toxins Incl. Trop. Dis.* **2017**, *23*, 13. [CrossRef] [PubMed]

32. Thomazini-Santos, I.A. Fibrin adhesive from snake venom: The effect of adding epsilon-aminocaproic acid, tranexamic acid, and aprotinin for coaptation of wound edges in rat skin incisions. *J. Venom. Anim. Toxins* **2001**, *7*, 148–149. [CrossRef]

33. Thomazini-Santos, I.A.; Barraviera, S.R.C.S.; Mendes-Giannini, M.J.S.; Barraviera, B. Surgical adhesives. *J. Venom. Anim. Toxins* **2001**, *7*, 159–171. [CrossRef]

34. Ferreira, R.S.; de Barros, L.C.; Abbade, L.P.F.; Barraviera, S.R.C.S.; Silvares, M.R.C.; de Pontes, L.G.; dos Santos, L.D.; Barraviera, B. Heterologous fibrin sealant derived from snake venom: From bench to bedside—An overview. *J. Venom. Anim. Toxins Incl. Trop. Dis.* **2017**, *23*, 21. [CrossRef] [PubMed]

35. Barros, L.C.; Ferreira, R.S.; Barraviera, S.R.C.S.; Stolf, H.O.; Thomazini-Santos, I.A.; Mendes-Giannini, M.J.S.; Toscano, E.; Barraviera, B. A new fibrin sealant from Crotalus durissus terrificus venom: Applications in medicine. *J. Toxicol. Environ. Health Part B Crit. Rev.* **2009**, *12*, 553–571. [CrossRef] [PubMed]

36. Gazzeri, R.; Fiore, C.; Galarza, M. Role of EVICEL Fibrin Sealant to Assist Hemostasis in Cranial and Spinal Epidural Space: A Neurosurgical Clinical Study. *Surg. Technol. Int.* **2015**, *26*, 364–369. [PubMed]

37. Scognamiglio, F.; Travan, A.; Rustighi, I.; Tarchi, P.; Palmisano, S.; Marsich, E.; Borgogna, M.; Donati, I.; de Manzini, N.; Paoletti, S. Adhesive and sealant interfaces for general surgery applications. *J. Biomed. Mater. Res. Part B Appl. Biomater.* **2016**, *104*, 626–639. [CrossRef]

38. Noori, A.; Ashrafi, S.J.; Vaez-Ghaemi, R.; Hatamian-Zaremi, A.; Webster, T.J. A review of fibrin and fibrin composites for bone tissue engineering. *Int. J. Nanomed.* **2017**, *12*, 4937–4961. [CrossRef]

39. Barsotti, M.C.; Felice, F.; Balbarini, A.; Di Stefano, R. Fibrin as a scaffold for cardiac tissue engineering. *Biotechnol. Appl. Biochem.* **2011**, *58*, 301–310. [CrossRef]

40. Sameem, M.; Wood, T.J.; Bain, J.R. A systematic review on the use of fibrin glue for peripheral nerve repair. *Plast Reconstr. Surg.* **2011**, *127*, 2381–2390. [CrossRef]

41. Kayaalp, C.; Ertugrul, I.; Tolan, K.; Sumer, F. Fibrin sealant use in pilonidal sinus: Systematic review. *World J. Gastrointest. Surg.* **2016**, *8*, 266–273. [CrossRef] [PubMed]

42. Yang, X.-F.; He, X.; He, J.; Zhang, L.-H.; Su, X.-J.; Dong, Z.-Y.; Xu, Y.-J.; Li, Y.; Li, Y.-L. High efficient isolation and systematic identification of human adipose-derived mesenchymal stem cells. *J. Biomed. Sci.* **2011**, *18*, 59. [CrossRef] [PubMed]

43. Dominici, M.; Le Blanc, K.; Mueller, I.; Slaper-Cortenbach, I.; Marini, F.C.; Krause, D.S.; Deans, R.J.; Keating, A.; Prockop, D.J.; Horwitz, E.M. Minimal criteria for defining multipotent mesenchymal stromal cells. The International Society for Cellular Therapy position statement. *Cytotherapy* **2006**, *8*, 315–317. [CrossRef] [PubMed]

44. Ferreira, R.S. Autologous or heterologous fibrin sealant scaffold: Which is the better choice? *J. Venom. Anim. Toxins Incl. Trop. Dis.* **2014**, *20*, 31. [CrossRef] [PubMed]

45. Barros, L.; Soares, A.; Costa, F.; Rodrigues, V.; Fuly, A.; Giglio, J.; Gallacci, M.; Thomazini-Santos, I.; Barraviera, S.; Barraviera, B.; et al. Biochemical and biological evaluation of gyroxin isolated from Crotalus durissus terrificus venom. *J. Venom. Anim. Toxins Incl. Trop. Dis.* **2011**, *17*, 23–33. [CrossRef]

46. Aro, A.A.; Perez, M.O.; Vieira, C.P.; Esquisatto, M.A.M.; Rodrigues, R.A.F.; Gomes, L.; Pimentel, E.R. Effect of calendula officinalis cream on achilles tendon healing. *Anat. Rec.* **2015**, *298*, 428–435. [CrossRef] [PubMed]

47. Aro, A.A.; Esquisatto, M.A.M.; Nishan, U.; Perez, M.O.; Rodrigues, R.A.F.; Foglio, M.A.; De Carvalho, J.E.; Gomes, L.; Vidal, B.D.C.; Pimentel, E.R. Effect of A loe vera application on the content and molecular arrangement of glycosaminoglycans during calcaneal tendon healing. *Microsc. Res. Tech.* **2014**, *77*, 964–973. [CrossRef] [PubMed]

48. Aro, A.A.; Freitas, K.M.; Foglio, M.A.; Carvalho, J.E.; Dolder, H.; Gomes, L.; Vidal, B.C.; Pimentel, E.R. Effect of the Arrabidaea chica extract on collagen fiber organization during healing of partially transected tendon. *Life Sci.* **2013**, *92*, 799–807. [CrossRef]

49. Aro, A.A.; Simões, G.F.; Esquisatto, M.A.M.; Foglio, M.A.; Carvalho, J.E.; Oliveira, A.L.R.; Gomes, L.; Pimentel, E.R. Arrabidaea chica extract improves gait recovery and changes collagen content during healing of the Achilles tendon. *Injury* **2013**, *44*, 884–892. [CrossRef]

50. Guerra, F.D.R.; Vieira, C.P.; Almeida, M.S.; Oliveira, L.P.; De Aro, A.A.; Pimentel, E.R. LLLT improves tendon healing through increase of MMP activity and collagen synthesis. *Lasers Med. Sci.* **2013**, *28*, 1281–1288. [CrossRef]

51. Tomiosso, T.C.; Nakagaki, W.R.; Gomes, L.; Hyslop, S.; Pimentel, E.R. Organization of collagen bundles during tendon healing in rats treated with L-NAME. *Cell Tissue Res.* **2009**, *337*, 235–242. [CrossRef] [PubMed]

52. Godoy, J.A.P.; Block, D.B.; Tollefsen, D.M.; Werneck, C.C.; Vicente, C.P. Dermatan sulfate and bone marrow mononuclear cells used as a new therapeutic strategy after arterial injury in mice. *Cytotherapy* **2011**, *13*, 695–704. [CrossRef] [PubMed]

53. Marqueti, R.C.; Durigan, J.L.Q.; Oliveira, A.J.S.; Mekaro, M.S.; Guzzoni, V.; Aro, A.A.; Pimentel, E.R.; Selistre-De-Araujo, H.S. Effects of aging and resistance training in rat tendon remodeling. *FASEB J.* **2018**, *32*, 353–368. [CrossRef] [PubMed]

54. Stegemann, H.; Stalder, K. Determination of hydroxyproline. *Clin. Chim. Acta* **1967**, *18*, 267–273. [CrossRef]

55. Jorge, M.P.; Madjarof, C.; Ruiz, A.L.T.G.; Fernandes, A.T.; Rodrigues, R.A.F.; de Oliveira Sousa, I.M.; Foglio, M.A.; de Carvalho, J.E. Evaluation of wound healing properties of Arrabidaea chica Verlot extract. *J. Ethnopharmacol.* **2008**, *118*, 361–366. [CrossRef] [PubMed]

56. de Campos Vidal, B. Fluorescence, aggregation properties and FT-IR microspectroscopy of elastin and collagen fibers. *Acta Histochem.* **2014**, *116*, 1359–1366. [CrossRef] [PubMed]

57. de Campos Vidal, B.; dos Anjos, E.H.M.; Mello, M.L.S. Optical anisotropy reveals molecular order in a mouse enthesis. *Cell Tissue Res.* **2015**, *362*, 177–185. [CrossRef]

58. de Campos Vidal, B. Form birefringence as applied to biopolymer and inorganic material supraorganization. *Biotech. Histochem.* **2010**, *85*, 365–378. [CrossRef]

59. de Campos Vidal, B.; Mello, M.L.S. Optical anisotropy of collagen fibers of rat calcaneal tendons: An approach to spatially resolved supramolecular organization. *Acta Histochem.* **2010**, *112*, 53–61. [CrossRef]

60. Biancalana, A.; Veloso, L.; Gomes, L. Obesity affects collagen fibril diameter and mechanical properties of tendons in Zucker rats. *Connect. Tissue Res.* **2010**, *51*, 171–178. [CrossRef]

61. Sparrow, K.J.; Finucane, S.D.; Owen, J.R.; Wayne, J.S. The Effects of Low-Intensity Ultrasound on Medial Collateral Ligament Healing in the Rabbit Model. *Am. J. Sports Med.* **2005**, *33*, 1048–1056. [CrossRef]

62. Qiao, H.; Surti, S.; Choi, S.R.; Raju, K.; Zhang, H.; Ponde, D.E.; Kung, H.F.; Karp, J.; Zhou, R. Death and proliferation time course of stem cells transplanted in the myocardium. *Mol. Imaging Biol.* **2009**, *11*, 408–414. [CrossRef]

63. Wolbank, S.; Pichler, V.; Ferguson, J.C.; Meinl, A.; Van Griensven, M.; Goppelt, A.; Redl, H. Non-invasive in vivo tracking of fibrin degradation by fluorescence imaging. *J. Tissue Eng. Regen. Med.* **2015**, *9*, 973–976. [CrossRef] [PubMed]

64. Bensaïd, W.; Triffitt, J.T.; Blanchat, C.; Oudina, K.; Sedel, L.; Petite, H. A biodegradable fibrin scaffold for mesenchymal stem cell transplantation. *Biomaterials* **2003**, *24*, 2497–2502. [CrossRef]

65. Spejo, A.B.; Chiarotto, G.B.; Ferreira, A.D.F.; Gomes, D.A.; Ferreira, R.S., Jr.; Barraviera, B.; Oliveira, A.L.R. Neuroprotection and immunomodulation following intraspinal axotomy of motoneurons by treatment with adult mesenchymal stem cells. *J. Neuroinflamm.* **2018**, *15*, 230. [CrossRef] [PubMed]

66. Orsi, P.R.; Landim-Alvarenga, F.C.; Justulin, L.A., Jr.; Kaneno, R.; Golim, M.A.; Santos, D.C.; Creste, C.F.Z.; Oba, E.; Maia, L.; Barraviera, B.; et al. A unique heterologous fibrin sealant (HFS) as a candidate biological scaffold for mesenchymal stem cells in osteoporotic rats. *Stem Cell Res. Ther.* **2017**, *8*, 205. [CrossRef]

67. Wang, Y.; Chen, X.; Cao, W.; Shi, Y. Plasticity of mesenchymal stem cells in immunomodulation: pathological and therapeutic implications. *Nat. Immunol.* **2014**, *15*, 1009–1016. [CrossRef]

68. Peroni, J.F.; Borjesson, D.L. Anti-Inflammatory and Immunomodulatory Activities of Stem Cells. *Vet. Clin. N. Am. Equine Pract.* **2011**, *27*, 351–362. [CrossRef]

69. Lee, R.H.; Oh, J.Y.; Choi, H.; Bazhanov, N. Therapeutic factors secreted by mesenchymal stromal cells and tissue repair. *J. Cell. Biochem.* **2011**, *112*, 3073–3078. [CrossRef]

70. Veronesi, F.; Salamanna, F.; Tschon, M.; Maglio, M.; Nicoli Aldini, N.; Fini, M. Mesenchymal stem cells for tendon healing: What is on the horizon? *J. Tissue Eng. Regen. Med.* **2017**, *11*, 3202–3219. [CrossRef]

71. Raabe, O.; Shell, K.; Fietz, D.; Freitag, C.; Ohrndorf, A.; Christ, H.J.; Wenisch, S.; Arnhold, S. Tenogenic differentiation of equine adipose-tissue-derived stem cells under the influence of tensile strain, growth differentiation factors and various oxygen tensions. *Cell Tissue Res.* **2013**, *352*, 509–521. [CrossRef] [PubMed]

72. Shukunami, C.; Takimoto, A.; Oro, M.; Hiraki, Y. Scleraxis positively regulates the expression of tenomodulin, a differentiation marker of tenocytes. *Dev. Biol.* **2006**, *298*, 234–247. [CrossRef] [PubMed]

73. Lui, P.P.Y. Identity of tendon stem cells—How much do we know? *J. Cell. Mol. Med.* **2013**, *17*, 55–64. [CrossRef] [PubMed]

74. Uysal, A.C.; Mizuno, H. Differentiation of Adipose-Derived Stem Cells for Tendon Repair. In *Methods in Molecular Biology (Clifton, N.J.)*; Humana Press: Totowa, NJ, USA, 2011; Volume 702, pp. 443–451.

75. James, R.; Kesturu, G.; Balian, G.; Chhabra, A.B. Tendon: Biology, Biomechanics, Repair, Growth Factors, and Evolving Treatment Options. *J. Hand Surg. Am.* **2008**, *33*, 102–112. [CrossRef]

76. Kyurkchiev, D.; Bochev, I.; Ivanova-Todorova, E.; Mourdjeva, M.; Oreshkova, T.; Belemezova, K.; Kyurkchiev, S. Secretion of immunoregulatory cytokines by mesenchymal stem cells. *World J. Stem Cells* **2014**, *6*, 552. [CrossRef]

77. Klein, M.B.; Yalamanchi, N.; Pham, H.; Longaker, M.T.; Chang, J. Flexor tendon healing in vitro: Effects of TGF-beta on tendon cell collagen production. *J. Hand Surg. Am.* **2002**, *27*, 615–620. [CrossRef]

78. Rickert, M.; Jung, M.; Adiyaman, M.; Richter, W.; Simank, H.G. A growth and differentiation factor-5 (GDF-5)-coated suture stimulates tendon healing in an Achilles tendon model in rats. *Growth Factors* **2001**, *19*, 115–126. [CrossRef]

79. Peres, V.S. [Unesp] Efeito Do Selante De Fibrina Derivado De Peçonha De Serpente Associado A Células—Tronco Mesenquimais Na Cicatrização De Ferida Cirúrgica em Ratos. Dissertação de Mestrado UNESP, Universidade Estadual Paulista, Sao Paulo, Brazil, 2014.

80. Yücel, E.A.; Oral, O.; Olgaç, V.; Oral, C.K. Effects of fibrin glue on wound healing in oral cavity. *J. Dent.* **2003**, *31*, 569–575. [CrossRef]

81. Chailakhyan, R.K.; Shekhter, A.B.; Ivannikov, S.V.; Tel'pukhov, V.I.; Suslin, D.S.; Gerasimov, Y.V.; Tonenkov, A.M.; Grosheva, A.G.; Panyushkin, P.V.; Moskvina, I.L.; et al. Reconstruction of Ligament and Tendon Defects Using Cell Technologies. *Bull. Exp. Biol. Med.* **2017**, *162*, 563–568. [CrossRef]

82. Tuan, T.-L.; Song, A.; Chang, S.; Younai, S.; Nimni, M.E. In VitroFibroplasia: Matrix Contraction, Cell Growth, and Collagen Production of Fibroblasts Cultured in Fibrin Gels. *Exp. Cell Res.* **1996**, *223*, 127–134. [CrossRef]

83. Ferraro, G.C.; Moraes, J.R.; Shimano, A.C.; Pereira, G.T.; Moraes, F.R.; Bueno de Camargo, M.H. Effect of snake venom derived fibrin glue on the tendon healing in dogs: Clinical and biomechanical study. *J. Venom. Anim. Toxins Incl. Trop. Dis.* **2005**, *11*, 261–274. [CrossRef]

cells

MDPI

Review

Tumor Extracellular Matrix Remodeling: New Perspectives as a Circulating Tool in the Diagnosis and Prognosis of Solid Tumors

Marta Giussani [1], Tiziana Triulzi [1], Gabriella Sozzi [2] and Elda Tagliabue [1,*

[1] Molecular Targeting Unit, Department of Research, Fondazione IRCCS Istituto Nazionale dei Tumori di, 20133 Milano, Italy; marta.giussani@istitutotumori.mi.it (M.G.); tiziana.triulzi@istitutotumori.mi.it (T.T.)
[2] Tumor Genomics Unit, Department of Research, Fondazione IRCCS Istituto Nazionale dei Tumori di, 20133 Milano, Italy; gabriella.sozzi@istitutotumori.mi.it
* Correspondence: elda.tagliabue@istitutotumori.mi.it; Tel.: +39-02-2390-3013; Fax: +39-02-2390-2692

Received: 14 December 2018; Accepted: 21 January 2019; Published: 23 January 2019

Abstract: In recent years, it has become increasingly evident that cancer cells and the local microenvironment are crucial in the development and progression of tumors. One of the major components of the tumor microenvironment is the extracellular matrix (ECM), which comprises a complex mixture of components, including proteins, glycoproteins, proteoglycans, and polysaccharides. In addition to providing structural and biochemical support to tumor tissue, the ECM undergoes remodeling that alters the biochemical and mechanical properties of the tumor microenvironment and contributes to tumor progression and resistance to therapy. A novel concept has emerged, in which tumor-driven ECM remodeling affects the release of ECM components into peripheral blood, the levels of which are potential diagnostic or prognostic markers for tumors. This review discusses the most recent evidence on ECM remodeling-derived signals that are detectable in the bloodstream, as new early diagnostic and risk prediction tools for the most frequent solid cancers.

Keywords: extracellular matrix; circulating biomarkers; diagnosis; prognosis

---

1. Introduction

Solid tumors are complex entities that are characterized by the coexistence of cancer cells and their microenvironment, composed of various cell types, including fibroblasts, adipocytes, endothelial cells, bone marrow-derived immune cells, and the extracellular matrix (ECM) [1]. All of these elements contribute to the development of cancer, which is not an entirely cancer cell-autonomous process but depends on the ability of cellular and noncellular components in the microenvironment to: i) establish a pro-tumor milieu, ii) regulate tumor cell behavior, and iii) coevolve with cancer cells [2,3]. This complex dynamic interaction between tumor cells and their microenvironment has been clearly demonstrated to influence the development, progression, and response to therapy of most solid cancers [4].

The ECM is a complex network of noncellular components, including structural proteins—predominantly collagens—, matricellular proteins—e.g., periostin, thrombospondins, osteopontin and secreted protein acidic and rich in cysteine (SPARC)—, glycoproteins, proteoglycans, and polysaccharides. These molecules contribute in deposition and arrangement of ECM and modulate cell-matrix interaction through their distinct biochemical and physical properties. In addition to its structural function, the ECM is highly dynamic and versatile and is an essential part of the tissue milieu, governing crucial aspects of cell biology. In this context, abnormal ECM dynamics contribute to cancer and lead to the conversion of the stem cell niche into a cancer stem cell niche, promoting the organization of premetastatic/metastatic environments and disrupting tissue polarity, inducing tissue invasion [5].

The composition and mechanical properties of the ECM modulate many of the tumor cell responses that represent the hallmarks of cancer, such as the evasion of apoptosis, insensitivity to growth inhibitors, sustained angiogenesis, self-sufficient growth, limitless replicative potential, and tissue invasion and metastasis [6]. The ECM is highly dynamic, versatile, and constantly remodeled as a result of activities of cells that reside within, which in turn are influenced by the ECM itself [5]. ECM remodeling is characterized by the increased synthesis and deposition of collagens and upregulation of matrix metalloproteinases (MMPs) [7]. These enzymes process matrix components, such as collagens [8], leading to the production and release of bioactive fragments mainly from non-fibrillar collagens [9]. MMP-derived changes in the tumor microenvironment typify tumor prognosis, supporting the function of structural remodeling of the ECM in the progression of many epithelial cancers, including breast, lung, and pancreatic tumors [10–12].

Consequently, interest in reactive stroma and ECM remodeling as potential tissue biomarkers in the management of cancer has grown. Cancer research in the past several decades has focused on identifying stroma/ECM-related characteristics as histological parameters [13] and gene-based signatures with diagnostic and prognostic significance in the most frequent solid tumors [14–17], with the possibility of being applied across several cancer types [18,19].

The tumor stroma is a potential source of new biomarkers, not merely in situ at the tissue level. ECM molecules that are derived from stromal changes are also released into the bloodstream and might represent surrogate markers of tumor development [3,20]. Thus, there is growing interest in studying ECM remodeling-derived molecules as circulating biomarkers. In this review, we discuss the results that have been obtained in the past 10 years of cancer research on the potential function of circulating ECM remodeling-derived molecules in the diagnosis and prognosis of solid tumors.

2. Presence of ECM Remodeling-Derived Molecules in Blood

Using a conditional transgenic HER2/neu-induced mouse model of breast cancer, Pitteri et al. [21] detected proteins that originated from the microenvironment in plasma, demonstrating that the plasma proteome is a sensitive biomonitor of tumor-host interactions. Applying a quantitative proteomic approach to plasma of these mice, the authors identified a set of proteins that changed in relative abundance during tumor induction/progression and regression upon doxycycline administration and withdrawal, respectively. Comparing plasma proteins detected in tumor-bearing mice to proteome, gene expression profiling and immunohistochemistry (IHC) of human breast cancer cell lines and murine induced breast tumors, circulating proteins resulted to originate from the tumor and local microenvironment. In plasma from mice with early-stage breast cancer, they observed higher levels of acute-phase response proteins, complement system proteins, and immune cell proteins, in addition to tumor microenvironment-specific molecules.

When considering the human context, although a direct demonstration of the stromal origin of circulating molecules is not achievable, the tumor stromal compartment was found positive for the same ECM remodeling-derived molecules that were detected in blood in various cancer types by immunohistochemical analysis of tumor tissues. Indeed, collagen 18, the precursor molecule of endostatin, which was upregulated in serum from colorectal cancer patients, was positive in basement membrane structures around invasive tumors and in desmoplastic stromal areas by IHC performed on tumor specimens of the same patients [22]. In non-small-cell lung cancer, higher serum levels of osteopontin, a glycophosphoprotein that has been implicated in tissue remodeling, were noted in patients with osteopontin-positive versus -negative lung cancers, as evaluated by IHC [23]. In breast cancer patients, IHC analysis on disease specimens showed stromal cell positivity of the same ECM molecules that, in blood, discriminated breast cancer patients from those with benign disease [24].

3. ECM Remodeling as Circulating Biomarkers in Cancer Diagnosis

The potential use of molecules that are derived from tumor-microenvironment crosstalk as circulating biomarkers has generated notable results, primarily in the diagnosis of cancer (Table 1). Early detection is a significant need in cancer management, because it influences patient mortality.

Table 1. Extracellular matrix (ECM)-derived molecules in cancer diagnosis.

Solid Cancer Type	ECM Molecules	Diagnostic Setting	Author (Reference)
Breast cancer (BC)	COL4	BC patients vs. healthy donors	Mazzoni [25]
	C1M, C3M, C4M, and C4M12	BC patients vs. healthy donors	Bager [26]
	COMP, COL11, and COL10	BC patients vs. healthy donors	Giussani [24]
		BC vs. benign breast disease	
	Fibronectin	BC patients vs. healthy donors	Moon [27]
		BC vs. benign breast disease	
	MMP-9	BC vs. benign breast disease	Golubnitschaja [28]
	α-1-microglobulin/inter-α-trypsin inhibitor light chain precursor, gelsolin, clusterin, and biotinidase	Infiltrating ductal BC vs. other infiltrating histotype	Cohen [29]
	Fibronectin	Triple-negative vs. hormone-positive BC	Cohen [29]
Lung cancer (LC)	COL10 and SPARC	LC patients vs. healthy donors	Andriani [30]
	Osteopontin	LC patients vs. healthy donors	Kerenidi [31]
	C1M, VCIM	LC patients vs. healthy donors	Willumsen [32]
	Tumstatin (COL18A1)	LC patients vs. healthy donors	Nielsen [33]
		LC vs. COPD [1] or idiopathic pulmonary fibrosis	
Colorectal cancer (CRC)	MMP-9	CRC patients vs. healthy donors	Wilson [34]; Emara [35]; Biasi [36]; Mroczko [37]
	MMP-9	CRC patients vs. healthy donors	Gimeno-García [38]
		CRC vs. nonadvanced adenomas	
	MMP-9 activity	CRC vs. adenomas	Baisi [36]
	TIMP-1	CRC patients vs. healthy donors	Mroczko [37]
		CRC vs. adenomas	
	COL6A3	CRC patients vs. healthy donors	Qiao [39]
	COL10	CRC patients vs. healthy donors	Solé [40]
		Adenoma patients vs. healthy donors	
	C1M, C3M, and pro-C3	CRC patients vs. healthy donors	Kehlet [7]
	C1M and pro-C3	CRC vs. adenomas	Kehlet [7]
	C1M, C3M, C4M, and pro-C3	Stage IV vs. stage I-III CRC	Kehlet [7]
	Endostatin (COL18A1)	Stage I vs. advanced stage CRC	Kantola [22]
		Advanced stage (T3–4) vs. nonadvanced stage (T1–2)	
Urological cancer (Renal cell carcinoma—RCC; Urinary bladder cancer—UBC)	MMP2, MMP9, TIMP-1, and TIMP-2	Metastatic RCC patients vs. healthy donors	Miyake [41]
	MMP7	Metastatic or node-positive RCC vs. localized RCC	Niedworok [42]
	MMP7	Advanced stage vs. nonadvanced stage RCC	Szarvas [43]
		Node-positive vs. node-negative UBC patients	
	MMP9 and TIMP-2	UBC patients vs. healthy donors	Ramón de Fata [44]

Table 1. *Cont.*

Solid Cancer Type	ECM Molecules	Diagnostic Setting	Author (Reference)
Pancreatic cancer (PC)	MMP1, MMP3, MMP7, MMP9, MMP10, and MMP12	PC patients vs. healthy donors	Kahlert [45]
	COL4	PC patients vs. healthy donors	Ohlund [46]
	C1M, C3M and C4M	PDAC [2] vs. healthy donors	Willumsen [20]
	MMP7 and connective tissue growth factor	PDAC [2] vs. healthy donors	Resovi [47]
	MMP7, connective tissue growth factor, plasminogen, fibronectin and COL4	PDAC [2] vs. chronic pancreatitis patients	
	COL4, endostatin (COL18A1), tenascin C, and osteopontin	PDAC patients vs. healthy donors	Franklin [48]
	COL6A3	PDAC patients vs. healthy donors PDAC vs. benign lesions	Kang [49]

COPD: chronic obstructive pulmonary disease; [2] PDAC: pancreatic ductal adenocarcinoma.

3.1. Breast Cancer

Although the use of standard screening tests in the management of breast cancer (BC), such as mammography, ultrasonography, and magnetic resonance imaging, have reduced patient mortality, early diagnosis with noninvasive procedures remains a challenge [50]. Several limitations that require invasive procedures persist, such as discriminating benign from malignant nodules in women with imaging that is merely suggestive of a tumor, which occurs in approximately 10% of women with a nodule detected during early screening.

Collagens are the main proteins of stromal origin that are detected in blood. In particular, collagen 4, as measured by enzyme immunoassay, is significantly higher in serum from patients with BC compared with healthy women [25]. MMP-degradation products of collagen 1 (C1M), collagen 3 (C3M) and collagen 4 (C4M and C4M12) are significantly elevated in serum from BC patients versus healthy donors, all with accuracies of over 75% in discriminating cancer patients [26].

Other ECM-derived molecules have been detected in plasma, such as collagen oligomeric matrix protein (COMP) and fibronectin. In particular, the combination of COMP, collagen 11, and collagen 10 has discriminatory capacity for BC patients versus healthy donors and patients with benign breast disease [24]. Plasma fibronectin levels are significantly higher in BC patients compared with healthy controls, patients with noncancerous diseases (e.g., inflammatory disease), and women with benign breast disease [27]. In addition, MMP-9 serum proteolytic activity is greater in women with malignant versus benign nodules [28].

By liquid chromatography–mass spectrometry (LC–MS/MS) analysis of plasma samples, Cohen and colleagues [29] demonstrated that a subset of peptides, corresponding to α-1-microglobulin/inter-α-trypsin inhibitor light chain precursor, gelsolin, clusterin, and biotinidase, classifies BC patients into two histological types: infiltrating ductal carcinomas and invasive mammary carcinomas with a lobular, tubular, mucinous, or medullary histotype. In addition, they found that circulating levels of a fibronectin-specific peptide differed significantly between triple-negative BC patients and patients who were positive for at least one receptor (estrogen, progesterone, or epidermal growth factor 2), indicating that these ECM-derived circulating biomarkers discriminate BCs of various subtypes during early detection.

3.2. Lung Cancer

For lung cancer, early diagnosis is also urgent, because the disease is often diagnosed at an advanced stage, for which there are no effective clinical options. Thus, the combination of diagnostic imaging by low-dose computed tomography (CT), which is the only evidence-based method for the early detection of lung cancer that reduces mortality, with new potential biomarkers is still a challenge [51]. In this context, many molecules that are derived from ECM remodeling have been analyzed with regard to their discriminatory potential in patients with lung cancer versus healthy controls.

Andriani and colleagues [30] reported that plasma samples from patients with lung cancer had significantly higher levels of collagen 10a1 and the collagen-binding matricellular protein SPARC compared with healthy controls who were matched for clinical parameters, such as sex, age, and smoking status. Levels of collagen 10 differed significantly only in females, whereas SPARC maintained significant differences between gender subgroups, with good overall discrimination of patients from controls.

Osteopontin, another protein that has been implicated in tissue remodeling, is significantly higher in serum from patients with lung cancer compared with healthy volunteers, and moreover, when smoking history was considered, the levels of circulating osteopontin were higher in smokers than in nonsmoking and ex-smoking lung cancer patients [31], suggesting that smoking status is an important parameter that must be taken into account when searching for new lung cancer biomarkers. ECM remodeling is an active process in the pathogenesis of lung cancer and chronic obstructive pulmonary disease (COPD) [52], rendering it difficult to identify microenvironment-derived markers that are specific for neoplastic conditions. Willumsen and colleagues [32] showed that MMP-degraded collagen 1 (C1M) and MMP-degraded citrullinated vimentin (VICM) are significantly elevated in serum from lung cancer patients compared with healthy donors, with excellent performance in detecting subjects with lung cancer. However, the diagnostic power of this two-marker combination fell dramatically when subjects with idiopathic pulmonary fibrosis or COPD were considered as the control group. In contrast, serum tumstatin, a matrikine protein that is derived from collagen 4a3, has tremendous diagnostic potential for non-small-cell lung cancer compared with healthy donors and patients with idiopathic pulmonary fibrosis or COPD as control groups [33].

3.3. Colorectal Cancer

Considering the gradual evolution of most cases of colorectal cancer (CRC) from adenomas, studies for the identification of diagnostic biomarkers in this pathology should also consider their ability to discriminate CRC from precancerous lesions [53]. MMPs and their tissue inhibitors are involved in tumor progression and the invasion of CRCs [54]. One of the most extensively studied MMPs, MMP9, is significantly elevated primarily in serum from CRC patients versus healthy controls [34–38]. Moreover, its enzymatic activity is improved in sera from patients with malignant tumors with respect to controls by gel zymography, reaching significant differences in stage II and III cancer. In contrast, the activity in serum from patients with adenomas remained consistently within the same range as in controls, suggesting the value of this biomarker in discriminating patients with adenomas from those with CRC [36].

Notably, a recent study by Gimeno-García and colleagues [38] analyzed plasma samples that were prospectively collected from patients who were undergoing colonoscopy and included nonadvanced adenomas, advanced adenomas, and CRCs. The levels of MMP9 were significantly higher in CRC patients compared with healthy controls and nonadvanced adenoma patients, whereas no significant differences were observed between CRC and advanced adenoma patients.

Another protein that discriminates between malignant and nonmalignant lesions is tissue inhibitor of metalloproteinase type 1 (TIMP-1). TIMP-1 levels are significantly higher in serum from CRC patients versus colorectal adenoma patients and controls [37], and a recent meta-analysis of 9 published studies, comprising 819 CRC patients and 1067 healthy controls, showed that circulating levels of TIMP-1 have significant clinical value with moderately high sensitivity and specificity in identifying CRC patients [55].

With regard to collagens, the plasma concentration of collagen 6a3 is significantly higher in CRC patients compared with healthy subjects [39], and serum levels of collagen 10 are significantly elevated in CRC and adenoma patients versus controls but fail to distinguish malignant and non-malignant lesions [40]. C1M, C3M, and collagen 3 (pro-C3) identifies CRC subjects with respect to controls, and C1M and pro-C3 are significantly higher in serum from CRC versus adenoma patients, rendering them potential markers in making an early differential diagnosis [7]. These biomarkers have also been considered in relation to the clinicopathological features of CRC patients—specifically, C1M, C3M, C4M, and pro-C3 are significantly higher in stage IV than in stages I, II, and III [7]. Similarly, serum endostatin levels are significantly lower in stage I patients compared with more advanced stages and significantly higher in T3-4 versus T1-2 patients, supporting the increase in remodeling of the tumor microenvironment during progression. Although it is significantly higher in serum from CRC patients compared with healthy controls, endostatin has not shown satisfactory discriminatory diagnostic power [22].

3.4. Urological Cancer

As in CRC, MMPs and their tissue inhibitors are widely considered to be circulating biomarkers in urological cancers, such as renal and bladder tumors. Two independent studies have demonstrated the value of this class of proteins. Serum levels of MMP2, MMP9, TIMP-1, and TIMP-2 are significantly lower in healthy controls compared with patients with metastatic renal cell carcinoma (RCC) [41].

MMP7, a promising tissue marker of poor prognosis [56], is significantly higher in the serum of node-positive and metastatic RCC patients than in those with localized RCC and in the serum of advanced versus nonadvanced stage patients [42].

In addition, this group reported the value of MMP7 in discriminating RCC patients from healthy controls, finding significantly lower levels in the latter. In urinary bladder cancer (UBC), serum MMP7 levels discriminate patients with metastases, wherein MMP7 is 2.9-fold higher in samples of patients with node-positive compared with node-negative tumors [43]. Serum levels of MMP9 and TIMP-2 are significantly greater in UBC patients versus age- and gender-matched healthy individuals [44].

3.5. Pancreatic Cancer

The stroma is the predominant constituent of pancreatic cancer (PC), forming approximately 80% of the tumor mass, especially in pancreatic ductal adenocarcinoma (PDAC), the most frequent pancreatic oncotype [20]. Therefore, based on the urgent clinical need for biomarkers in the early diagnosis of PDAC, stroma-derived molecules are garnering particular interest as circulating biomarkers. A panel of MMPs was analyzed in serum from PC patients—specifically, MMP-1, MMP-3, MMP-7, MMP-9, MMP-10, and MMP-12 were higher in cancer patients compared with healthy donors, showing good diagnostic accuracy, of which MMP-7 and MMP-12 achieved perfect discrimination [45].

Collagen proteins and their fragments are the most frequent biomarkers that are identified in the diagnosis of PC. A pilot study of control subjects and PC patients (n = 8 and 9, respectively) revealed elevations in circulating collagen 4 protein in serum samples from the latter [46]. A subsequent study by Willumsen and colleagues [20] confirmed this evidence in a larger number of samples. Specifically, they observed significantly higher levels of MMP-generated fragments of collagen 1, 3, and 4 in serum from PDAC patients compared with healthy controls, the combination of which attained extraordinary diagnostic power. Other soluble stroma-related molecules were found associated with PDAC, especially when considered in combination with tumor-related CA19.9 biomarker. Indeed Resovi and colleagues [47] demonstrated that both the combinations of CA19.9 with MMP-7 and with connective tissue growth factor (CCN2) display an almost perfect accuracy in discriminating PDAC patients from healthy subjects. Moreover, a panel consisting of CCN2, plasminogen (PGL), fibronectin, collagen 4 and CA19.9 was found able to distinguish PDAC from chronic pancreatitis patients [47]. The possibility of combining more than two biomarkers has also been considered by Franklin and colleagues [48], who evaluated four stroma-derived biomarkers—collagen 4, endostatin, tenascin C, and osteopontin—and 4 conventional markers—cancer antigens CA 19.9 and CA 125, CEA (carcinoembryonic antigen), and TPS (tissue polypeptide-specific antigen). Notably, in addition to the significantly higher levels of all stroma-derived proteins in PDAC patients versus controls, they found a narrower dynamic range of these markers in each group with respect to tumor-derived markers, strengthening their potential in the diagnostic setting. Collagen 6a3 is also differentially expressed at the tissue level between PDAC and adjacent nonmalignant tissue [57]. This protein is significantly higher in serum from PDAC patients compared with patients with benign lesions and healthy controls [49].

Overall, the existing evidence highlights the value of stroma-related circulating proteins as early diagnostic biomarkers, potentially overcoming the molecules that are strictly tumor-related.

4. ECM Remodeling as Circulating Biomarkers in Cancer Prognosis

In addition to early detection, an important priority is the identification of prognostic biomarkers that impact clinical decisions and overall outcomes. Based on the well-established involvement of ECM-derived molecules in the initiation and progression of cancer, they are also being examined as prognostic markers (Table 2).

Table 2. ECM-derived molecules in cancer prognosis.

Solid Cancer Type	ECM Molecules	Patients	Prognostic Setting	Author (Reference)
Breast cancer (BC)	C1M, C3M, C4M, and pro-C3	Hormone receptor + metastatic BC	shorter OS [1]	Lipton [58]
	C1M and C3M	HER2 + metastatic BC	shorter TTP [2]	
	pro-C3		shorter OS	
	C1M, C3M, C4M, and pro-C3		shorter TTP	
	hyaluronic acid	Metastatic BC	shorter PFS [3]	Peng [59]
			shorter OS	
Lung cancer (LC)	osteopontin	advanced non-small-cell lung cancer	shorter OS and PFS	Isa [60]; Mack [61]
	osteopontin	non-small-cell lung cancer	worse OS	Takenaka [62]
	osteopontin	primary lung cancer	worse OS	Kerenidi [31]
	osteopontin	primary non-small-cell lung cancer	shorter DMFS [4] and OS	Rouanne [23]
	Thrombospondin-1		longer OS	
	osteopontin/thrombospondin-1		↑ risk of metastases and death	
Colorectal cancer (CRC)	COL4	Metastatic CRC	at time of disease progression	Nystrom [63]
	COL4 and CEA	Liver metastatic CRC	worse OS	Nystrom [64]
	TIMP-1	CRC	poor prognosis	Lee [65]
	MMP8	CRC	worse DSS [5]	Bockelman [66]
		left-sided CRC and no systemic inflammatory condition	worse DSS	
	TIMP-1	CRC	worse DSS	
		left-sided CRC and no systemic inflammatory condition	worse DSS	
	MMP9/TIMP-1 ratio	CRC	longer DSS	
Urological cancer (Renal cell carcinoma—RCC; Urinary bladder cancer—UBC)	MMP9/TIMP-2 ratio	Metastatic clear-cell RCC	shorter PFS	Miyake [41]
	MMP7	RCC	poor OS, DSS, and MFS [6]	Niedworok [42]
	MMP7	UBC	poor DSS and OS	Szarvas [67]
	endostatin	UBC	poor DSS and MFS	Szarvas [68]
Pancreatic cancer (PC)	COL4	PC	shorter OS	Ohlund [46]
	COL4, endostatin (COL4A3), and osteopontin	PC	shorter OS	Franklin [48]

[1] OS: overall survival; [2] TTP: time to progression; [3] PFS: progression-free survival; [4] DMFS: distant metastasis-free survival; [5] DSS: disease-specific survival; [6] MFS: metastasis-free survival.

4.1. Breast Cancer

MMP-mediated degradation products of collagens, studied extensively as diagnostic biomarkers in BC, were recently associated with patient outcomes—specifically, in two independent cohorts of metastatic BCs: hormone receptor-positive and HER2-positive. Levels of C1M, C3M, C4M, and pro-C3 were measured in serum from metastatic BC patients prior to initiation of therapy and were predictive of a shorter time to progression (TTP) and overall survival (OS) at high levels [58]. Specifically, in hormone receptor-positive tumors, higher levels of all collagen fragments were associated with significantly lower OS, and elevated C1M and C3M were linked to a shorter TTP. Similarly, in the HER2-positive metastatic BC cohort, elevated pro-C3 was significantly linked to shorter OS, whereas higher levels of all 4 molecules correlated significantly with shorter TTP.

The metastatic setting is the main context in BC that is considered with regard to determining the value of ECM-derived molecules as prognostic biomarkers. MMP2 serum levels are significantly higher in advanced HER2 positive BC patients who develop bone and central nervous system metastases compared with those without metastases [69]. Moreover, plasma levels of hyaluronic acid, in addition to being significantly higher in BC patients with versus without metastases, had independent prognostic value in the metastatic group. Considering the two independent metastatic BC cohorts, high levels of hyaluronic acid were significantly associated with shorter progression-free survival (PFS) and OS [59].

4.2. Lung Cancer

Osteopontin has been studied extensively as a circulating prognostic marker in lung cancer, alone and in combination with other molecules.

Two concomitant studies of nested patient samples, the Japan-Multinational Trial Organization (JMTO) Lung Cancer (LC) 0004 in the Japanese population and the SouthWest Oncology Group (SWOG) 0003 in the US population, obtained the same evidence regarding the prognostic significance of circulating osteopontin in advanced non-small-cell lung cancers [60,61]. In both studies, osteopontin was evaluated in samples from patients before treatment, and low levels were significantly associated with a favorable prognosis, in terms of OS and PFS; in the SWOG 0003 study, the prognostic value of circulating osteopontin was significant, regardless of treatment.

Consistent with these reports, the clinical value of circulating osteopontin was recently confirmed in the management of non-small-cell lung cancer [62] and in newly diagnosed primary lung cancer patients, in whom increased levels of osteopontin were significantly linked to worse survival [31]. Elevated osteopontin levels were significantly associated with shorter distant metastasis-free survival (DMFS) and OS; conversely, higher thrombospondin-1 content correlated significantly only with longer OS. The combination of the two proteins, expressed as the osteopontin:thrombospondin-1 ratio, was more significant than each individual molecule in primary non-small-cell lung cancer patients, wherein a rise in osteopontin:thrombospondin-1 ratio was associated with a 30% and 40% increased risk of metastasis and death, respectively [23].

4.3. Colorectal Cancer

In colorectal cancer, the prognostic value of circulating ECM-derived proteins is notable in such patients with metastases. Circulating levels of collagen 4 are significantly higher in CRC patients with versus without liver metastases. Moreover, in the former, these levels increase further at the time of disease progression compared with when metastasis is detected [63]. These data were validated in a second study from the same group in which they also demonstrated the prognostic significance of circulating collagen 4, in combination with carcinoembryonic antigen (CEA)—specifically, patients with CRC metastatic to the liver with low levels of both biomarkers experienced better OS compared with those with high levels of both markers (47% survival three years after surgery) [64].

The prognostic performance of circulating TIMP-1 has been examined extensively and was discussed in a meta-analysis of 10 studies that considered approximately 3000 CRC patients. High levels of TIMP-1 in plasma or serum were significantly associated with a poor prognosis in CRC patients, based on age-, gender-, grade-, and clinical stage-adjusted hazard ratios [65]. Bockelman and colleagues [66] measured MMP-8, MMP-9, and TIMP-1 in serum from CRC patients within 30 days prior to surgery and found that MMP-8 and TIMP-1 alone served as prognostic factors, high levels of which correlated significantly with worse five-year disease-specific survival (DSS). These proteins also had prognostic significance in subgroup analyses: high MMP-8 and TIMP-1 content was linked to poor DSS compared with low concentrations in left-sided CRC patients and patients with no systemic inflammation. Conversely, MMP-9, widely described as a diagnostic marker for CRC patients, reached significance as a predictor of DSS only when combined with TIMP-1 and expressed as the molar ratio MMP-9:TIMP-1.

4.4. Urological Cancer

In urological cancer, MMPs and their tissue inhibitors reflect the clinical course in such patients. In a cohort of metastatic clear cell RCC patients, MMP-9:TIMP-2 ratio was significantly elevated at the time of progression versus diagnosis and was a significant predictor of progression-free survival in univariate and multivariate analyses [41]. Further, Niedworok and colleagues [42] found that high levels of serum MMP-7 are strongly associated with an unfavorable prognosis, in terms of OS, DSS, and DMFS. Moreover, this biomarker is an independent prognostic factor of OS with greater significance than the presence of metastases and tumor grade and stage.

In urinary bladder cancer, circulating MMP-7 is a stage- and grade-independent prognostic factor of DSS and OS, wherein patients with high MMP-7 concentrations have a poor prognosis [67]. In addition, endostatin, the generation of which is strictly related to MMP-7, is a prognostic marker—high serum levels correlate significantly with unfavorable DSS and DMFS in bladder cancer patients [68].

4.5. Pancreatic Cancer

In contrast to the other solid cancer types that we have discussed, ECM-derived circulating molecules have been poorly studied as prognostic biomarkers in PC. A study of a small tumor cohort ($n = 14$) reported that plasma levels of collagen 4, measured after surgery, in addition to having diagnostic value, correlated with patient prognosis; specifically, PC patients with high levels of collagen 4 experienced shorter survival than those with low levels [46]. Similarly, Franklin and colleagues [48] showed that in the postoperative setting, only stroma-derived circulating markers, such as collagen 4, endostatin, and osteopontin, were associated with shorter survival at high levels, whereas conventional tumor markers, such as CA19.9, CA125, and CEA, failed.

Overall, independent of the ECM-derived molecule and cancer type, higher levels of this class of molecules are generally associated with an unfavorable prognosis in cancer patients.

5. Conclusions

Circulating biomarkers in cancer remain a challenging research topic—the possibility of detecting incipient tumor cells significantly increases the potential for treatment and a cure. In contrast to invasive procedures, such as biopsies, which cannot be performed repeatedly and are at times even impractical, tumor-derived molecules in the blood have the advantage of being able to be measured using minimally invasive methods that allow frequent repeat testing during patient follow-up. Thus, in cancer, noninvasive modalities for early detection and risk assessment should become a priority.

Several serum-based tumor markers that are derived solely from neoplastic cells have been described, but none are used for tumor diagnosis—merely to monitor treatment response. Thus, new strategies are continually being developed, one of which is represented by molecules that are derived from the interplay between a tumor and its microenvironment. Such crosstalk fosters and amplifies stromal changes that ultimately manifest in the bloodstream.

Because tissue remodeling commonly occurs in cancers, microenvironment-related molecules might be useful in detecting and monitoring tumors of many oncotypes, as demonstrated for collagen degradation products and MMPs, which are prevalent in the bloodstream of patients with various tumors.

In this context, the C-terminal portion of collagen 8 is increased in the serum of patients who have been diagnosed with breast, lung, prostate, colon, and ovarian carcinoma and melanoma, compared with healthy subjects. MMP-degraded collagen 1 (C1M) and MMP-degraded vimentin (VICM), are significantly higher in serum from lung, gastric, prostate, and melanoma cancer patients versus controls. Similarly, collagen 4 and MMPs and their tissue inhibitors have prognostic value in various solid cancers, such as CRC, urological cancers, and pancreatic cancer. The ability of circulating collagen 1 and 4 to reflect the presence of a tumor is likely attributed to the abundance of collagen 1 in the stroma, where tumors develop, and the finding that collagen 4 is a core component of all basement membranes including those at the tumor and vasculature level in nearly all solid cancers.

Other stroma-derived molecules have been found to be specific for certain cancer oncotypes, such as MMP7 and osteopontin for urological and lung cancer, respectively. Thus, combining ECM molecules could constitute a new tool for detecting incipient tumor cells and, in some cases, determining their origin. A multimarker test based on stroma-related circulating molecules, combined with biomarkers that are related to the characteristics of tumor cells (e.g., CA-125, CEA, CA 15-3 CA19.9, TPS), is a promising platform for tumor diagnosis and to perform patients' follow-up through blood analysis.

The detection of tumor-stroma interplay in the blood might be a feasible strategy for overcoming intratumor heterogeneity, which impedes diagnostic approaches that are based on soluble tumor-related molecules. ECM-related molecules that are detectable in the bloodstream sense a tumor independently of the complexity of its clonal evolution, as supported by the results of cohorts of metastatic cancer patients in whom circulating ECM-related markers were predictive of TTP and OS. This finding suggests that remodeling of the tumor microenvironment also occurs during metastatic dissemination at various sites and that this phenomenon mainly involves the same molecules that are implicated in the development of tumors at primary sites.

In conclusion, these studies provide proof of concept that tissue remodeling is a promising source of novel biomarkers in the management of cancer. Searching for novel tumor biomarkers in circulation beyond the cancer cell itself is a new noninvasive diagnostic and prognostic approach, if we consider that these biomarkers are easily detectable by enzyme-linked immunosorbent assay, alone or combined in multiplex assays.

The next step in this endeavor is to validate the clinical value of circulating stroma-derived molecules by designing large prospective studies in which their diagnostic and prognostic capacity is analyzed with regard to the presence and progression of tumor cells and the existence of non-neoplastic alterations, such as fibrosis and local inflammation that could occur before development of a tumor.

Since molecules deriving from tumor-stroma interplay potentially overcome intra-tumor heterogeneity and are shared by multiple tumor oncotypes, their targeting even more opens the possibility of cancer therapeutic purposes. In this context, strategies aimed at targeting molecules involved in ECM remodeling such as MMP, transforming growth factor-β, enzymes involved in collagen cross-linking (lysyl oxidase and lysyl oxidase-like 2) are being considered as potentially new therapeutic tools [70]. Clinical trials are currently ongoing to demonstrate the efficacy of these approaches in inhibiting cancer progression [71].

Author Contributions: M.G.: systematic review of literature and the design and writing of the manuscript. T.T.: contributed to the supervision of the editing of the manuscript. G.S.: contributed to the supervision of the editing of the manuscript. E.T.: design and writing of the manuscript. All authors participated in revising manuscript and have read and approved the final version.

Funding: This work was supported by Associazione Italiana Ricerca sul Cancro (AIRC), (No 12162) (ET and GS).

Acknowledgments: We thank Mameli L. for secretarial assistance.

Conflicts of Interest: The authors declare no conflicts of interest.

References

1. Poltavets, V.; Kochetkova, M.; Pitson, S.M.; Samuel, M.S. The Role of the Extracellular Matrix and Its Molecular and Cellular Regulators in Cancer Cell Plasticity. *Front. Oncol.* **2018**, *8*, 431. [CrossRef] [PubMed]
2. Werb, Z.; Lu, P. The Role of Stroma in Tumor Development. *Cancer J.* **2015**, *21*, 250–253. [CrossRef]
3. Sund, M.; Kalluri, R. Tumor stroma derived biomarkers in cancer. *Cancer Metastasis Rev.* **2009**, *28*, 177–183. [CrossRef]
4. Joyce, J.A.; Fearon, D.T. T cell exclusion, immune privilege, and the tumor microenvironment. *Science* **2015**, *348*, 74–80. [CrossRef]
5. Lu, P.; Weaver, V.M.; Werb, Z. The extracellular matrix: A dynamic niche in cancer progression. *J. Cell Biol.* **2012**, *196*, 395–406. [CrossRef] [PubMed]
6. Pickup, M.W.; Mouw, J.K.; Weaver, V.M. The extracellular matrix modulates the hallmarks of cancer. *EMBO Rep.* **2014**, *15*, 1243–1253. [CrossRef] [PubMed]
7. Kehlet, S.N.; Sanz-Pamplona, R.; Brix, S.; Leeming, D.J.; Karsdal, M.A.; Moreno, V. Excessive collagen turnover products are released during colorectal cancer progression and elevated in serum from metastatic colorectal cancer patients. *Sci. Rep.* **2016**, *6*, 30599. [CrossRef] [PubMed]
8. Kessenbrock, K.; Plaks, V.; Werb, Z. Matrix metalloproteinases: Regulators of the tumor microenvironment. *Cell* **2010**, *141*, 52–67. [CrossRef] [PubMed]
9. Karamanos, N.K.; Theocharis, A.D.; Neill, T.; Iozzo, R.V. Matrix modeling and remodeling: A biological interplay regulating tissue homeostasis and diseases. *Matrix Biol.* **2018**. [CrossRef] [PubMed]
10. Moriggi, M.; Giussani, M.; Torretta, E.; Capitanio, D.; Sandri, M.; Leone, R.; De, P.S.; Vasso, M.; Vozzi, G.; Tagliabue, E.; et al. ECM Remodelling in Breast Cancer with Different Grade: Contribution of 2D-DIGE Proteomics. *Proteomics* **2018**, *18*, e1800278. [CrossRef]
11. Burgess, J.K.; Mauad, T.; Tjin, G.; Karlsson, J.C.; Westergren-Thorsson, G. The extracellular matrix—The under-recognized element in lung disease? *J. Pathol.* **2016**, *240*, 397–409. [CrossRef] [PubMed]
12. Chaturvedi, P.; Singh, A.P.; Moniaux, N.; Senapati, S.; Chakraborty, S.; Meza, J.L.; Batra, S.K. MUC4 mucin potentiates pancreatic tumor cell proliferation, survival, and invasive properties and interferes with its interaction to extracellular matrix proteins. *Mol. Cancer Res.* **2007**, *5*, 309–320. [CrossRef] [PubMed]
13. Van den Eynden, G.G.; Colpaert, C.G.; Couvelard, A.; Pezzella, F.; Dirix, L.Y.; Vermeulen, P.B.; van Marck, E.A.; Hasebe, T. A fibrotic focus is a prognostic factor and a surrogate marker for hypoxia and (lymph)angiogenesis in breast cancer: Review of the literature and proposal on the criteria of evaluation. *Histopathology* **2007**, *51*, 440–451. [CrossRef] [PubMed]
14. Giussani, M.; Merlino, G.; Cappelletti, V.; Tagliabue, E.; Daidone, M.G. Tumor-Extracellular matrix interactions: Identification of tools associated with breast cancer progression. *Semin. Cancer Biol.* **2015**, *35*, 3–10. [CrossRef] [PubMed]
15. Triulzi, T.; Casalini, P.; Sandri, M.; Ratti, F.; Carcangiu, M.L.; Colombo, M.P.; Balsari, A.; Ménard, S.; Orlandi, R.; Tagliabue, E. Neoplastic and stromal cells contribute to an extracellular matrix gene expression profile defining a breast cancer subtype likely to progress. *PLoS ONE* **2013**, *8*, e56761. [CrossRef] [PubMed]
16. Naba, A.; Clauser, K.R.; Whittaker, C.A.; Carr, S.A.; Tanabe, K.K.; Hynes, R.O. Extracellular matrix signatures of human primary metastatic colon cancers and their metastases to liver. *BMC Cancer* **2014**, *14*, 518. [CrossRef] [PubMed]
17. Lim, S.B.; Tan, S.J.; Lim, W.T.; Lim, C.T. An extracellular matrix-related prognostic and predictive indicator for early-stage non-small cell lung cancer. *Nat. Commun.* **2017**, *8*, 1734. [CrossRef]
18. Yuzhalin, A.E.; Urbonas, T.; Silva, M.A.; Muschel, R.J.; Gordon-Weeks, A.N. A core matrisome gene signature predicts cancer outcome. *Br. J. Cancer* **2018**, *118*, 435–440. [CrossRef]
19. Gandellini, P.; Andriani, F.; Merlino, G.; D'Aiuto, F.; Roz, L.; Callari, M. Complexity in the tumour microenvironment: Cancer associated fibroblast gene expression patterns identify both common and unique features of tumour-stroma crosstalk across cancer types. *Semin. Cancer Biol.* **2015**, *35*, 96–106. [CrossRef]

20. Willumsen, N.; Bager, C.L.; Leeming, D.J.; Smith, V.; Karsdal, M.A.; Dornan, D.; Bay-Jensen, A.C. Extracellular matrix specific protein fingerprints measured in serum can separate pancreatic cancer patients from healthy controls. *BMC Cancer* **2013**, *13*, 554. [CrossRef]

21. Pitteri, S.J.; Kelly-Spratt, K.S.; Gurley, K.E.; Kennedy, J.; Buson, T.B.; Chin, A.; Wang, H.; Zhang, Q.; Wong, C.H.; Chodosh, L.A.; et al. Tumor microenvironment-derived proteins dominate the plasma proteome response during breast cancer induction and progression. *Cancer Res.* **2011**, *71*, 5090–5100. [CrossRef] [PubMed]

22. Kantola, T.; Vayrynen, J.P.; Klintrup, K.; Makela, J.; Karppinen, S.M.; Pihlajaniemi, T.; Autio-Harmainen, H.; Karttunen, T.J.; Makinen, M.J.; Tuomisto, A. Serum endostatin levels are elevated in colorectal cancer and correlate with invasion and systemic inflammatory markers. *Br. J. Cancer* **2014**, *111*, 1605–1613. [CrossRef] [PubMed]

23. Rouanne, M.; Adam, J.; Goubar, A.; Robin, A.; Ohana, C.; Louvet, E.; Cormier, J.; Mercier, O.; Dorfmuller, P.; Fattal, S.; et al. Osteopontin and thrombospondin-1 play opposite roles in promoting tumor aggressiveness of primary resected non-small cell lung cancer. *BMC Cancer* **2016**, *16*, 483. [CrossRef] [PubMed]

24. Giussani, M.; Landoni, E.; Merlino, G.; Turdo, F.; Veneroni, S.; Paolini, B.; Cappelletti, V.; Miceli, R.; Orlandi, R.; Triulzi, T.; et al. Extracellular matrix proteins as dignostic markers of breast carcinoma. *J. Cell Physiol.* **2018**, *233*, 6280–6290. [CrossRef] [PubMed]

25. Mazouni, C.; Arun, B.; Andre, F.; Ayers, M.; Krishnamurthy, S.; Wang, B.; Hortobagyi, G.N.; Buzdar, A.U.; Pusztai, L. Collagen IV levels are elevated in the serum of patients with primary breast cancer compared to healthy volunteers. *Br. J. Cancer* **2008**, *99*, 68–71. [CrossRef]

26. Bager, C.L.; Willumsen, N.; Leeming, D.J.; Smith, V.; Karsdal, M.A.; Dornan, D.; Bay-Jensen, A.C. Collagen degradation products measured in serum can separate ovarian and breast cancer patients from healthy controls: A preliminary study. *Cancer Biomark* **2015**, *15*, 783–788. [CrossRef] [PubMed]

27. Moon, P.G.; Lee, J.E.; Cho, Y.E.; Lee, S.J.; Chae, Y.S.; Jung, J.H.; Kim, I.S.; Park, H.Y.; Baek, M.C. Fibronectin on circulating extracellular vesicles as a liquid biopsy to detect breast cancer. *Oncotarget* **2016**, *7*, 40189–40199. [CrossRef] [PubMed]

28. Golubnitschaja, O.; Yeghiazaryan, K.; Abraham, J.A.; Schild, H.H.; Costigliola, V.; Debald, M.; Kuhn, W. Breast cancer risk assessment: A non-invasive multiparametric approach to stratify patients by MMP-9 serum activity and RhoA expression patterns in circulating leucocytes. *Amino Acids* **2017**, *49*, 273–281. [CrossRef]

29. Cohen, A.; Wang, E.; Chisholm, K.A.; Kostyleva, R.; O'Connor-McCourt, M.; Pinto, D.M. A mass spectrometry-based plasma protein panel targeting the tumor microenvironment in patients with breast cancer. *J. Proteomics* **2013**, *81*, 135–147. [CrossRef]

30. Andriani, F.; Landoni, E.; Mensah, M.; Facchinetti, F.; Miceli, R.; Tagliabue, E.; Giussani, M.; Callari, M.; De Cecco, L.; Colombo, M.P.; et al. Diagnostic role of circulating extracellular matrix-related proteins in non-small cell lung cancer. *BMC Cancer* **2018**, *18*, 899. [CrossRef]

31. Kerenidi, T.; Kazakou, A.P.; Lada, M.; Tsilioni, I.; Daniil, Z.; Gourgoulianis, K.I. Clinical Significance of Circulating Osteopontin Levels in Patients with Lung Cancer and Correlation with VEGF and MMP-9. *Cancer Investig.* **2016**, *34*, 385–392. [CrossRef] [PubMed]

32. Willumsen, N.; Bager, C.L.; Leeming, D.J.; Smith, V.; Christiansen, C.; Karsdal, M.A.; Dornan, D.; Bay-Jensen, A.C. Serum biomarkers reflecting specific tumor tissue remodeling processes are valuable diagnostic tools for lung cancer. *Cancer Med.* **2014**, *3*, 1136–1145. [CrossRef] [PubMed]

33. Nielsen, S.H.; Willumsen, N.; Brix, S.; Sun, S.; Manon-Jensen, T.; Karsdal, M.; Genovese, F. Tumstatin, a Matrikine Derived from Collagen Type IValpha3, is Elevated in Serum from Patients with Non-Small Cell Lung Cancer. *Transl. Oncol.* **2018**, *11*, 528–534. [CrossRef] [PubMed]

34. Wilson, S.; Damery, S.; Stocken, D.D.; Dowswell, G.; Holder, R.; Ward, S.T.; Redman, V.; Wakelam, M.J.; James, J.; Hobbs, F.D.; et al. Serum matrix metalloproteinase 9 and colorectal neoplasia: A community-based evaluation of a potential diagnostic test. *Br. J. Cancer* **2012**, *106*, 1431–1438. [CrossRef]

35. Emara, M.; Cheung, P.Y.; Grabowski, K.; Sawicki, G.; Wozniak, M. Serum levels of matrix metalloproteinase-2 and -9 and conventional tumor markers (CEA and CA 19-9) in patients with colorectal and gastric cancers. *Clin. Chem. Lab. Med.* **2009**, *47*, 993–1000. [CrossRef] [PubMed]

36. Biasi, F.; Guina, T.; Maina, M.; Nano, M.; Falcone, A.; Aroasio, E.; Saracco, G.M.; Papotti, M.; Leonarduzzi, G.; Poli, G. Progressive increase of matrix metalloprotease-9 and interleukin-8 serum levels during carcinogenic process in human colorectal tract. *PLoS ONE* **2012**, *7*, e41839. [CrossRef]

37. Mroczko, B.; Groblewska, M.; Okulczyk, B.; Kedra, B.; Szmitkowski, M. The diagnostic value of matrix metalloproteinase 9 (MMP-9) and tissue inhibitor of matrix metalloproteinases 1 (TIMP-1) determination in the sera of colorectal adenoma and cancer patients. *Int. J. Colorectal. Dis.* **2010**, *25*, 1177–1184. [CrossRef]

38. Gimeno-Garcia, A.Z.; Trinanes, J.; Quintero, E.; Salido, E.; Nicolas-Perez, D.; Adrian-de-Ganzo, Z.; Alarcon-Fernandez, O.; Abrante, B.; Romero, R.; Carrillo, M.; et al. Plasma matrix metalloproteinase 9 as an early surrogate biomarker of advanced colorectal neoplasia. *Gastroenterol. Hepatol.* **2016**, *39*, 433–441. [CrossRef]

39. Qiao, J.; Fang, C.Y.; Chen, S.X.; Wang, X.Q.; Cui, S.J.; Liu, X.H.; Jiang, Y.H.; Wang, J.; Zhang, Y.; Yang, P.Y.; et al. Stroma derived COL6A3 is a potential prognosis marker of colorectal carcinoma revealed by quantitative proteomics. *Oncotarget* **2015**, *6*, 29929–29946. [CrossRef]

40. Sole, X.; Crous-Bou, M.; Cordero, D.; Olivares, D.; Guino, E.; Sanz-Pamplona, R.; Rodriguez-Moranta, F.; Sanjuan, X.; de Oca, J.; Salazar, R.; et al. Discovery and validation of new potential biomarkers for early detection of colon cancer. *PLoS ONE* **2014**, *9*, e106748. [CrossRef]

41. Miyake, H.; Nishikawa, M.; Tei, H.; Furukawa, J.; Harada, K.; Fujisawa, M. Significance of circulating matrix metalloproteinase-9 to tissue inhibitor of metalloproteinases-2 ratio as a predictor of disease progression in patients with metastatic renal cell carcinoma receiving sunitinib. *Urol. Oncol.* **2014**, *32*, 584–588. [CrossRef]

42. Niedworok, C.; vom Dorp, F.; Tschirdewahn, S.; Rubben, H.; Reis, H.; Szucs, M.; Szarvas, T. Validation of the diagnostic and prognostic relevance of serum MMP-7 levels in renal cell cancer by using a novel automated fluorescent immunoassay method. *Int. Urol. Nephrol.* **2016**, *48*, 355–361. [CrossRef] [PubMed]

43. Szarvas, T.; Becker, M.; vom Dorp, F.; Gethmann, C.; Totsch, M.; Bankfalvi, A.; Schmid, K.W.; Romics, I.; Rubben, H.; Ergun, S. Matrix metalloproteinase-7 as a marker of metastasis and predictor of poor survival in bladder cancer. *Cancer Sci.* **2010**, *101*, 1300–1308. [CrossRef] [PubMed]

44. Ramon de, F.F.; Ferruelo, A.; Andres, G.; Gimbernat, H.; Sanchez-Chapado, M.; Angulo, J.C. The role of matrix metalloproteinase MMP-9 and TIMP-2 tissue inhibitor of metalloproteinases as serum markers of bladder cancer. *Actas Urol. Esp.* **2013**, *37*, 480–488.

45. Kahlert, C.; Fiala, M.; Musso, G.; Halama, N.; Keim, S.; Mazzone, M.; Lasitschka, F.; Pecqueux, M.; Klupp, F.; Schmidt, T.; et al. Prognostic impact of a compartment-specific angiogenic marker profile in patients with pancreatic cancer. *Oncotarget* **2014**, *5*, 12978–12989. [CrossRef] [PubMed]

46. Ohlund, D.; Lundin, C.; Ardnor, B.; Oman, M.; Naredi, P.; Sund, M. Type IV collagen is a tumour stroma-derived biomarker for pancreas cancer. *Br. J. Cancer* **2009**, *101*, 91–97. [CrossRef] [PubMed]

47. Resovi, A.; Bani, M.R.; Porcu, L.; Anastasia, A.; Minoli, L.; Allavena, P.; Cappello, P.; Novelli, F.; Scarpa, A.; Morandi, E.; et al. Soluble stroma-related biomarkers of pancreatic cancer. *EMBO Mol. Med.* **2018**, *10*, e8741. [CrossRef] [PubMed]

48. Franklin, O.; Ohlund, D.; Lundin, C.; Oman, M.; Naredi, P.; Wang, W.; Sund, M. Combining conventional and stroma-derived tumour markers in pancreatic ductal adenocarcinoma. *Cancer Biomark* **2015**, *15*, 1–10. [CrossRef]

49. Kang, C.Y.; Wang, J.; Axell-House, D.; Soni, P.; Chu, M.L.; Chipitsyna, G.; Sarosiek, K.; Sendecki, J.; Hyslop, T.; Al-Zoubi, M.; et al. Clinical significance of serum COL6A3 in pancreatic ductal adenocarcinoma. *J. Gastrointest. Surg.* **2014**, *18*, 7–15. [CrossRef]

50. Trecate, G.; Sinues, P.M.; Orlandi, R. Noninvasive strategies for breast cancer early detection. *Future Oncol.* **2016**, *12*, 1395–1411. [CrossRef]

51. Oudkerk, M.; Devaraj, A.; Vliegenthart, R.; Henzler, T.; Prosch, H.; Heussel, C.P.; Bastarrika, G.; Sverzellati, N.; Mascalchi, M.; Delorme, S.; et al. European position statement on lung cancer screening. *Lancet Oncol.* **2017**, *18*, e754–e766. [CrossRef]

52. Berg, J.; Halvorsen, A.R.; Bengtson, M.B.; Tasken, K.A.; Maelandsmo, G.M.; Yndestad, A.; Halvorsen, B.; Brustugun, O.T.; Aukrust, P.; Ueland, T.; et al. Levels and prognostic impact of circulating markers of inflammation, endothelial activation and extracellular matrix remodelling in patients with lung cancer and chronic obstructive pulmonary disease. *BMC Cancer* **2018**, *18*, 739–4659. [CrossRef] [PubMed]

53. Chen, C.D.; Yen, M.F.; Wang, W.M.; Wong, J.M.; Chen, T.H. A case-cohort study for the disease natural history of adenoma-carcinoma and de novo carcinoma and surveillance of colon and rectum after polypectomy: Implication for efficacy of colonoscopy. *Br. J. Cancer* **2003**, *88*, 1866–1873. [CrossRef] [PubMed]

54. Said, A.H.; Raufman, J.P.; Xie, G. The role of matrix metalloproteinases in colorectal cancer. *Cancers (Basel)* **2014**, *6*, 366–375. [CrossRef] [PubMed]

55. Meng, C.; Yin, X.; Liu, J.; Tang, K.; Tang, H.; Liao, J. TIMP-1 is a novel serum biomarker for the diagnosis of colorectal cancer: A meta-analysis. *PLoS ONE* **2018**, *13*, e0207039. [CrossRef] [PubMed]

56. Miyata, Y.; Iwata, T.; Ohba, K.; Kanda, S.; Nishikido, M.; Kanetake, H. Expression of matrix metalloproteinase-7 on cancer cells and tissue endothelial cells in renal cell carcinoma: Prognostic implications and clinical significance for invasion and metastasis. *Clin. Cancer Res.* **2006**, *12*, 6998–7003. [CrossRef] [PubMed]

57. Arafat, H.; Lazar, M.; Salem, K.; Chipitsyna, G.; Gong, Q.; Pan, T.C.; Zhang, R.Z.; Yeo, C.J.; Chu, M.L. Tumor-specific expression and alternative splicing of the COL6A3 gene in pancreatic cancer. *Surgery* **2011**, *150*, 306–315. [CrossRef] [PubMed]

58. Lipton, A.; Leitzel, K.; Ali, S.M.; Polimera, H.V.; Nagabhairu, V.; Marks, E.; Richardson, A.E.; Krecko, L.; Ali, A.; Koestler, W.; et al. High turnover of extracellular matrix reflected by specific protein fragments measured in serum is associated with poor outcomes in two metastatic breast cancer cohorts. *Int. J. Cancer* **2018**, *143*, 3027–3034. [CrossRef]

59. Peng, C.; Wallwiener, M.; Rudolph, A.; Cuk, K.; Eilber, U.; Celik, M.; Modugno, C.; Trumpp, A.; Heil, J.; Marme, F.; et al. Plasma hyaluronic acid level as a prognostic and monitoring marker of metastatic breast cancer. *Int. J. Cancer* **2016**, *138*, 2499–2509. [CrossRef]

60. Isa, S.; Kawaguchi, T.; Teramukai, S.; Minato, K.; Ohsaki, Y.; Shibata, K.; Yonei, T.; Hayashibara, K.; Fukushima, M.; Kawahara, M.; et al. Serum osteopontin levels are highly prognostic for survival in advanced non-small cell lung cancer: Results from JMTO LC 0004. *J. Thorac. Oncol.* **2009**, *4*, 1104–1110. [CrossRef]

61. Mack, P.C.; Redman, M.W.; Chansky, K.; Williamson, S.K.; Farneth, N.C.; Lara, P.N., Jr.; Franklin, W.A.; Le, Q.T.; Crowley, J.J.; Gandara, D.R. Lower osteopontin plasma levels are associated with superior outcomes in advanced non-small-cell lung cancer patients receiving platinum-based chemotherapy: SWOG Study S0003. *J. Clin. Oncol.* **2008**, *26*, 4771–4776. [CrossRef] [PubMed]

62. Takenaka, M.; Hanagiri, T.; Shinohara, S.; Yasuda, M.; Chikaishi, Y.; Oka, S.; Shimokawa, H.; Nagata, Y.; Nakagawa, M.; Uramoto, H.; et al. Serum level of osteopontin as a prognostic factor in patients who underwent surgical resection for non-small-cell lung cancer. *Clin. Lung Cancer* **2013**, *14*, 288–294. [CrossRef] [PubMed]

63. Nystrom, H.; Naredi, P.; Hafstrom, L.; Sund, M. Type IV collagen as a tumour marker for colorectal liver metastases. *Eur. J. Surg. Oncol.* **2011**, *37*, 611–617. [CrossRef] [PubMed]

64. Nystrom, H.; Tavelin, B.; Bjorklund, M.; Naredi, P.; Sund, M. Improved tumour marker sensitivity in detecting colorectal liver metastases by combined type IV collagen and CEA measurement. *Tumour Biol.* **2015**, *36*, 9839–9847. [CrossRef] [PubMed]

65. Lee, J.H.; Choi, J.W.; Kim, Y.S. Plasma or serum TIMP-1 is a predictor of survival outcomes in colorectal cancer: A meta-analysis. *J. Gastrointestin. Liver Dis.* **2011**, *20*, 287–291. [PubMed]

66. Bockelman, C.; Beilmann-Lehtonen, I.; Kaprio, T.; Koskensalo, S.; Tervahartiala, T.; Mustonen, H.; Stenman, U.H.; Sorsa, T.; Haglund, C. Serum MMP-8 and TIMP-1 predict prognosis in colorectal cancer. *BMC Cancer* **2018**, *18*, 679–4589. [CrossRef]

67. Szarvas, T.; Jager, T.; Becker, M.; Tschirdewahn, S.; Niedworok, C.; Kovalszky, I.; Rubben, H.; Ergun, S.; vom Dorp, F. Validation of circulating MMP-7 level as an independent prognostic marker of poor survival in urinary bladder cancer. *Pathol. Oncol. Res.* **2011**, *17*, 325–332. [CrossRef]

68. Szarvas, T.; Laszlo, V.; vom Dorp, F.; Reis, H.; Szendroi, A.; Romics, I.; Tilki, D.; Rubben, H.; Ergun, S. Serum endostatin levels correlate with enhanced extracellular matrix degradation and poor patients' prognosis in bladder cancer. *Int. J. Cancer* **2012**, *130*, 2922–2929. [CrossRef]

69. Skerenova, M.; Mikulova, V.; Capoun, O.; Zima, T.; Tesarova, P. Circulating tumor cells and serum levels of MMP-2, MMP-9 and VEGF as markers of the metastatic process in patients with high risk of metastatic progression. *Biomed. Pap. Med. Fac. Univ. Palacky Olomouc Czech Repub.* **2017**, *161*, 272–280. [CrossRef]

70. Bonnans, C.; Chou, J.; Werb, Z. Remodelling the extracellular matrix in development and disease. *Nat. Rev. Mol. Cell Biol.* **2014**, *15*, 786–801. [CrossRef]
71. Lampi, M.C.; Reinhart-King, C.A. Targeting extracellular matrix stiffness to attenuate disease: From molecular mechanisms to clinical trials. *Sci. Transl. Med.* **2018**, *10*. [CrossRef] [PubMed]

cells

MDPI

Article

Low Molecular Mass Myocardial Hyaluronan in Human Hypertrophic Cardiomyopathy

Christina E. Lorén [1], Christen P. Dahl [2,3,4], Lan Do [1], Vibeke M. Almaas [2], Odd R. Geiran [5,6], Stellan Mörner [1] and Urban Hellman [1,*]

[1] Cardiology, Heart Centre, Department of Public Health and Clinical Medicine, Umeå University, 901 85 Umeå, Sweden; christinae.loren@gmail.com (C.E.L.); lan.do@umu.se (L.D.); stellan.morner@umu.se (S.M.)

[2] Department of Cardiology, Oslo University Hospital Rikshospitalet, 0424 Oslo, Norway; christen.peder.dahl@rr-research.no (C.P.D.); vibalm@ous-hf.no (V.M.A.)

[3] Department of Clinical Medicine, UiT, the Arctic University of Norway, 9019 Tromsø, Norway

[4] Research Institute of Internal Medicine, Oslo University Hospital Rikshospitalet, 0372 Oslo, Norway

[5] Faculty of Medicine, University of Oslo, 0318 Oslo, Norway; ogeiran@medisin.uio.no

[6] Department of Thoracic and Cardiovascular Surgery, Oslo University Hospital Rikshospitalet, 0424 Oslo, Norway

* Correspondence: urban.hellman@umu.se; Tel.: +46-907851720; Fax: +46-90137633

Received: 13 December 2018; Accepted: 23 January 2019; Published: 29 January 2019

Abstract: During the development of hypertrophic cardiomyopathy, the heart returns to fetal energy metabolism where cells utilize more glucose instead of fatty acids as a source of energy. Metabolism of glucose can increase synthesis of the extracellular glycosaminoglycan hyaluronan, which has been shown to be involved in the development of cardiac hypertrophy and fibrosis. The aim of this study was to investigate hyaluronan metabolism in cardiac tissue from patients with hypertrophic cardiomyopathy in relation to cardiac growth. NMR and qRT-PCR analysis of human cardiac tissue from hypertrophic cardiomyopathy patients and healthy control hearts showed dysregulated glucose and hyaluronan metabolism in the patients. Gas phase electrophoresis revealed a higher amount of low molecular mass hyaluronan and larger cardiomyocytes in cardiac tissue from patients with hypertrophic cardiomyopathy. Histochemistry showed high concentrations of hyaluronan around individual cardiomyocytes in hearts from hypertrophic cardiomyopathy patients. Experimentally, we could also observe accumulation of low molecular mass hyaluronan in cardiac hypertrophy in a rat model. In conclusion, the development of hypertrophic cardiomyopathy with increased glucose metabolism affected both hyaluronan molecular mass and amount. The process of regulating cardiomyocyte size seems to involve fragmentation of hyaluronan.

Keywords: hypertrophic cardiomyopathy; hyaluronan; metabolomics; GEMMA; glucose

1. Introduction

Hypertrophic cardiomyopathy (HCM) is a disorder with a prevalence of 1/500 and an annual mortality of approximately 1% [1]. It is a monogenic inherited disease associated with cardiac dysfunction and life threatening arrhythmias [2,3]. HCM exhibits a wide phenotypic variability ranging from asymptomatic to severe symptoms and is an important cause of sudden cardiac death in young adults and athletes. At the cellular level, HCM is characterized by patches of cardiomyocyte hypertrophy, cardiomyocyte disarray, interstitial fibrosis, and small vessel disease.

When subjected to hemodynamic or metabolic stress, the heart returns to fetal metabolism and the fetal gene program where the cells prefer usage of glucose over fatty acid as source of energy [4]. It has been proposed that dysfunctional regulation of the glucose metabolism and cardiac energy metabolism is a prominent feature of the maladapted failing heart and HCM hearts [5,6]. A rise of glucose level results in an increased influx trough of the hexosamine biosynthetic pathway (HBP) [7], resulting in elevated levels of uridine diphosphate *N*-acetyl-D-glucosamine (UDP-GlcNAc).

A significant consequence of higher levels of UDP-GlcNAc is the increased synthesis of the extracellular matrix (ECM) glycosaminoglycan hyaluronan (HA), which has not previously been explored in HCM. HA is a polydisperse unbranched polymer that greatly varies in molecular mass ranging from 5 to 10,000 kDa, which makes it challenging to analyze. It is present in the ECM of all vertebrates, and is highly expressed during development, wound healing, and regeneration. HA is synthesized at the cellular membrane by the linkage of the UDP-sugar precursors D-glucuronic acid (GlcUA) and *N*-acetyl-D-glucosamine (GlcNAc), a reaction catalyzed by the membrane bound enzyme hyaluronan synthases (HAS). The HA synthesis is strongly dependent on the cytoplasmic concentrations of UDP-GlcNAc and UDP-GlcUA, and it has been shown that increased levels of these UDP-sugar precursors enhance HA synthesis [8,9].

Several reports have shown that diverse sizes of HA exert a wide spectrum of functions. In health and tissue homeostasis, HA is present as high molecular mass (MM) HA and has structural and hydrating features as well as an anti-inflammatory effect [10]. On the contrary, low MM HA has been shown to have a pro-inflammatory effect [11]. In addition, many receptors and extracellular proteins have been shown to bind HA, creating a molecular network with a wide range of structural and signaling properties. Degradation of HA is mainly carried out by hyaluronidases (HYAL, CEMIP) or by reactive oxygen species (ROS). Recent work has revealed HA as a driving factor in the development of fibrosis by stimulating both fibroblast proliferation, differentiation, and motility [12]. Furthermore, it has been shown that high MM HA depolarizes the membrane in cell cultures in a concentration dependent manner, which could be reversed by cell surface digestion of HA by hyaluronidase [13]. If this is also true for cardiomyocytes, it could potentially change the cardiac action potential leading to rhythm disturbances and arrhythmias [14].

We have earlier shown elevated gene expression of *HAS 1, 2* and the HA receptor *CD44* as well as increased cardiac levels of HA correlating with pro-hypertrophic gene expression, using a rat model for cardiac hypertrophy [15,16]. We have also identified a crosstalk between cultured cardiomyocytes and fibroblasts resulting in increased HA synthesis in the fibroblasts [17]. In addition, HA staining was stronger in human cardiac tissues from HCM patients compared to autopsy material from previously healthy individuals [18], and in the rat heart HA occurs around myofibrils [19].

In this study we further investigate HA in HCM. We observed an altered metabolism of HA in HCM and changes in molecular mass distribution of HA corresponding with cardiomyocyte size.

2. Materials and Methods

2.1. Human Tissue Samples

Tissue aliquots from human septal myocardium were obtained during surgery with basal septal myectomy from five patients with hypertrophic obstructive cardiomyopathy. Two of the patients were diagnosed with coronary disease. None were diagnosed with diabetes or hypertension. Characteristics are presented in Table 1.

Control (non-failing) human left ventricular, right ventricular, and septal tissue was obtained from five sex- and age-matched subjects whose hearts were rejected as cardiac donors for surgical reasons. The cause of death of donors was cerebrovascular accident or carbon monoxide poisoning, and none had a history of heart disease or known previous medication. Myocardium from these subjects was kept on ice for 1 to 4 h before tissue sampling. All cardiac tissues were snap-frozen in liquid nitrogen and stored at −80 °C until use.

Table 1. Clinical characteristics of the patients with hypertrophic obstructive cardiomyopathy (HOCM).

Characteristics	HOCM Patients (n = 5)
Age (year)	55.4 ± 12.5
Gender (male/female)	2/3
NYHA class (II/III/IV)	1/4/0
IVSd (mm)	19.3 ± 3.8
LVPWd (mm)	12.3 ± 1.9
LVIDd (mm)	43.7 ± 3.5
LVOT (mmHg)	47.4 ± 29.8
Medication	
Calcium antagonist	2
Acetylsalicylic acid	4
β-antagonist	3
Aldosterone antagonist	1
Statins	3

Data are presented as the mean ± SD or number of subjects. NYHA, New York Heart Association Functional Classification; IVSd, interventricular septum in end diastole; LVPWd, left ventricular posterior wall in end diastole; LVIDd, left ventricular internal diameter in end diastole; LVOT, left ventricular outflow tract.

2.2. Rat Model for Hypertrophy

Cardiac tissues were analyzed from a rat model of cardiac hypertrophy from a study published several years ago (2008) [15]. Due to the newly developed GEMMA (gas-phase electrophoretic molecular mobility analyzer) analysis of HA mass distribution, it was now possible to perform HA mass analysis on the small pieces of cardiac tissue saved from the previous study [20]. The surgical procedure of the rats has been described elsewhere [15]. Briefly, rats were anesthetized with 0.2 mL pentobarbital intra-abdominal and after abdominal incision, a titanium clip of 1.15 mm inner diameter was administered on the aorta proximal to the renal arteries on male Wistar rats (n = 3). Age-matched control rats (n = 3) were sham operated, i.e., subjected to the same procedure but without the administration of a clip on the aorta. After being anesthetized with 0.4 mL pentobarbital, the rats were sacrificed at 1 and 42 days after operation and their hearts were harvested.

2.3. Compliance with Ethical Standards

All procedures performed in studies involving human participants were in accordance with the ethical standards of the Regional Health Authorities of South-Eastern Norway and with the 1964 Helsinki declaration and its later amendments. Informed consent was obtained from all individual participants included in the study.

The animal study was performed in Paris at Inserm U689 and all procedures performed in studies involving animals were in accordance with animal welfare regulations, and the study protocol was approved by the Ethical Committee of Inserm. All investigations conformed to guidelines set by the French Ministry of Agriculture and the Guide for the Care and Use of Laboratory Animals published by the US National Institutes of Health.

2.4. [1]H High-Resolution Magic Angle Spinning Nuclear Magnetic Resonance (HR MAS NMR) Spectroscopy

Tissue samples were thawed at room temperature and kept on ice during preparation. Each tissue sample (20–30 mg wet weight) was inserted into disposable 30-μL Teflon NMR inserts followed by the addition of deuterium oxide to the insert to complete the required volume and homogenize insert contents. Inserts were then packed into a 4 mm zirconia MAS rotor (40 μL capacity). All the NMR experiments on tissue samples were carried out on a 500 MHz MAS-NMR spectrometer (Bruker, Billerica, MA, USA) at 300 K.

2.5. NMR Data Processing and Analysis

Spectra were imported into MATLAB (R2015a) (MathWorks Inc., Natick, MA, USA) integrated using in-house developed scripts and normalized by the sum of all intensities. All metabolites in human cardiac tissues were identified using Chenomx NMR Suite 7.7 software (Chenomx Inc., Edmonton, AB, Canada) with full resolution NMR data and a standard two-dimensional (2D) NMR experiment on a selected sample. For identifying metabolites contributing to the discrimination between groups, the normalized ^{1}H-NMR data were uploaded to SIMCA (version 14, Umetrics, Umeå, Sweden) for orthogonal partial least squares-discriminant analysis (OPLS-DA). The spectral variables were scaled to unit variance, and 7-fold internal cross-validation was performed to evaluate the quality of the resulting statistical models by considering the diagnostic measures R2 and Q2. Potential metabolites were selected based on the variable importance in projection (VIP) score > 1.0.

2.6. Immunohistochemistry

Before paraffin embedding, human cardiac tissues were fixed in 4% buffered formaldehyde for 48 h in room temperature. Paraffin sections (4 μm) were then mounted on SuperFrost Plus Slides (Thermo Fisher Scientific Inc., Waltham, MA, USA) and dried overnight at 37 °C. Prior to staining, sections were deparaffinized in xylene, rehydrated in series of graded ethanol, and washed in PBS. To visualize HA only, deparaffinized histological sections were blocked for 30 min using 2% bovine serum in PBS. Samples were incubated with a biotinylated HA binding protein (HABP) probe overnight at 4 °C [21]. HABP was prepared at the Institute of Medical and Physiological Chemistry, University of Uppsala, at a concentration of 100 μg/mL and was diluted 1:40 [22]. The samples were rinsed three times in PBS and then incubated with streptavidin 488 (1:500, Thermo Fisher Scientific Inc., MA, USA) for 1 h at room temperature.

To visualize HA and collagen (I, III, and VI, respectively) in the same section, antigen retrieval was performed by boiling the sections in a citrate buffer in a microwave oven for 9 min followed by cooling for 30 min before washing in water for 5 min. The sections were then blocked for 30 min using 2% bovine serum in PBS. Finally, samples were incubated overnight at 4 °C with the primary reagents HABP and rabbit anti collagen I (ab292 1:200, Abcam, Cambridge, UK), rabbit anti collagen III (ab7778 1:400, Abcam, UK), or rabbit anti collagen VI (ab6588 1:400, Abcam, UK), respectively. After washing in PBS, the samples were incubated with streptavidin 488 (1:500, Thermo Fisher Scientific Inc., MA, USA) and donkey anti rabbit 594 (1:500, Thermo Fisher Scientific Inc., MA, USA). Nuclei were stained with Hoechst 33,343 (1:5000, Thermo Fisher Scientific Inc., MA, USA), and slides were mounted in ProLong Gold Antifade (Thermo Fisher Scientific Inc., MA, USA). Phalloidin-iFluor 594 1:1000 (ab176757, Abcam, UK) was used as a cell marker. All stainings were visualized using the Zeiss LSM 710 laser-scanning confocal microscope (Zeiss, Oberkochen, Germany).

2.7. Cardiomyocyte Area Analysis

Cardiomyocytes were visualized by iFluor 594-labeled phalloidin (1:1000, Abcam, UK) staining actin and ECM was visualized by staining collagen III (ab 7778 1:400, Abcam, UK). Each slide was objected to 10 images. Morphometric analysis was performed using the software Fiji (http://fiji.sc). An average area value from each heart was calculated by the use of the measurements of 30–50 cells containing a central nucleus. Statistical analysis was performed using a two-tailed Student t test assuming unequal variances. Each pixel was calibrated to 0.83 μm according to Zeiss confocal data. Fluorescence images were collected at 20× objective using Zeiss LSM 710 laser-scanning confocal microscope (Zeiss, Oberkochen, Germany).

2.8. HA Molecular Mass Distribution

Cardiac tissue samples, wet weight 27–112 mg, were dried (n = 5 in each group) and homogenized. Proteins and nucleic acids were digested with proteinase K (Sigma-Aldrich, St. Louis, MO, USA), benzonase nuclease (Sigma-Aldrich, MO, USA), and chondroitinase ABC (Sigma-Aldrich, MO, USA) on three consecutive days. At the end of each day, chloroform was added to each sample and the extracted aqueous phase was dialyzed against 0.1 M NaCl using Amicon Ultra 3K concentration units (Millipore, Burlington, MA, USA) followed by overnight precipitation in 99% ethanol (EtOH). Samples were then loaded on anion exchange mini spin columns (Thermo Fisher Scientific, MA, USA) and centrifuged to wash out sulphated glycosaminoglycans and remaining non-HA contaminants, based on NaCl-binding. Finally, to remove salt the sample was dialyzed against 20 mM ammonium acetate (pH 8.0) in Amicon Ultra 3K concentration units. HA molecular mass analyses were performed using a nano-electrospray gas-phase electrophoretic molecular mobility analyzer (GEMMA) (TSI Corp., Shoreview, MN, USA).

Each sample of purified HA (n = 5 in each group) was scanned three times in the GEMMA and the final size distribution spectrum was a sum of the three scans. The raw counts from the GEMMA spectrum were calibrated according to the previously described method [23]. The molecule diameter analyzed in the GEMMA was converted to molecular mass by analyzing HA standards ranging from 30 to 2500 kDa (Hyalose LLC, OC, USA). The relation between area under the curve (AUC) in the GEMMA spectrum to the HA concentration enables an estimation of the relative concentration of different MM of HA. Counts on the Y-axis correspond to the number of detected molecules and the X-axis to the MM of HA. Number of counts were normalized to the dry weight of the sample. Due to the physical properties of HA and the shape dependence of the GEMMA method, the analysis achieves a good separation of low MM HA up to ca. 100 kDa whereas the resolution for higher MM is poorer. The relative amount of HA with an MM less than about 100 kDa cannot be compared with HA of an MM greater than 100 kDa [23].

In this study, we have defined low MM HA as a mass up to 50 kDa. HA extracted from cardiac tissue was degraded with hyaluronidase from *Streptomyces hyalurolyticus* (Sigma-Aldrich, MO, USA) and reanalyzed to test for extraction specificity.

2.9. RNA Extraction and qRT-PCR

To obtain RNA, the cardiac tissues were homogenized in Qiazol lysis agent and with beads using Precellys lysing kit (Bertin Instruments, Montigny-le-Bretonneux, France) and purified using the RNeasy plus Universal Mini Kit (QIAGEN, Waltham, MA, USA). Reverse transcription was performed with 1 µg of total RNA using the High Capacity RNA to DNA kit (Thermo Fisher Scientific, MA, USA). The extracted RNA and cDNA concentration, respectively, were quantified using a NanoDrop Spectrometer ND-1000 (NanoDrop, Thermo Fisher Scientific Inc., Waltham, MA, USA).

The real-time quantitative PCR was performed on a 7900 HT Fast Real-Time PCR system (Thermo Fisher Scientific, MA, USA) using 1 µg cDNA, TaqMan®Gene Expression Assays, and 1 µL Gene Assay Mix for the genes *HAS1-3, HYAL 1* and *2, CEMIP, CD44, VCAN,* and *TSG6* (Thermo Fisher Scientific, MA, USA). *GAPDH* (Thermo Fisher Scientific, MA, USA) was used as an endogenous reference gene. Forty cycles of amplification were performed. The gene of interest was normalized to the reference gene using the ΔCt method [24].

2.10. Statistical Analysis

For metabolomics statistics, non-parametric Mann–Whitney U test with Benjamini–Hochberg correction using in-house software written and compiled in MATLAB (MathWorks Inc., Natick, MA, USA) was used. For analysis of gene expression and relative amount of water, low and high MM HA non-parametric independent Mann–Whitney U test was performed using the SPSS statistic software (version 25, IBM, Armonk, NY, USA). P values of less than 0.05 were considered to be significant. Factor analysis was performed with the principal components method to analyze the correlation matrix and two factors were extracted.

Generation of box plots for cardiomyocyte area was performed using SPSS statistic software.

3. Results

3.1. Metabolomic Analysis of Cardiac Tissue from HCM Patients and Non-Failing Hearts

It has been shown that increase in the synthesis of HA is strongly dependent on the concentration of UDP-GlcNAc and UDP-GlcUA. A higher level of these UDP hexosamines leads to increased synthesis of ECM HA. Therefore, we wanted to analyze the metabolomics and known metabolites in this process in HCM and healthy patients.

NMR was performed using cardiac tissues from non-failing septum and left ventricle (n = 10) and basal septal myectomies from HCM patients (n = 5). UDP hexosamines were identified as a merged multiple peak with an almost 2-fold increase. UDP-GlcUA can be formed from *myo*-Inositol, which was increased 1.5-fold in HCM. Glucose, on the other hand, was decreased 1.5-fold. Lactate was increased about 3.5-fold in HCM. Glutamate and glutamine showed a 3- and 2-fold increase, respectively, in HCM, and glutathione levels were increased almost 2-fold in HCM. Finally, fatty acid levels were decreased 1.5–2-fold in HCM, as shown in Table 2. These results support the proposal that dysfunctional regulation of the glucose metabolism may be involved in the development of HCM. An interesting observation is the increase in glutamate and glutamine, which are substrates for gluthation known to have a protective function in the regulation of ROS.

Table 2. NMR analysis of cardiac tissue from patients with hypertrophic cardiomyopathy (HCM) (n = 5) compared to healthy controls' left wall and septum (n = 10). The variable importance in projection (VIP) score and fold change of metabolites with significant difference between HCM and non-failing left chamber wall.

	VIP Score	pos/neg Correlation	Fold Change	*P*-Value
Sugars and related metabolites				
UDP-glucose, UDP-GlcUA, UDP-galactose, UDP-GlcNAc	1.03	0.13	1.86	0.054 *
glucose	1.44	−0.18	0.63	0.007
Amino acids				
glutamate	1.83	0.22	3.04	0.002
glutamine	1.43	0.18	2.14	0.013
Other metabolites				
taurine	1.58	0.21	2.52	0.009
myo-Inositol	1.55	0.18	1.63	0.005
lactate	1.31	0.17	3.74	0.054 *
acetate	1.34	0.17	3.39	0.023
glutathion	1.53	0.19	1.77	0.007
Fatty acids				
$CH_2CH_2CH_2$– lipid	1.27	−0.16	0.54	0.031
CH_2CO– lipid	1.36	−0.17	0.55	0.017
$CH_2CH_2CH_2CO$– lipid	1.38	−0.17	0.55	0.023
$CH_2CH_2CH=CH$– lipid	1.39	−0.17	0.64	0.017
CH_3CH_2– lipid	1.43	−0.18	0.58	0.017

Variable importance in projection (VIP). (*) *P* values > 0.05. Fold change was calculated by using the means in the included tissues and compared with a non-parametric Mann–Whitney test. For metabolomics statistics, non-parametric Mann–Whitney U test with Benjamini–Hochberg correction using in-house software written and compiled in MATLAB (MathWorks Inc., MA, USA) was used.

3.2. Assessment of HA Morphology in Basal Septal Myectomies and Non-Failing Septum

To better understand the nature of development of HCM, we wanted to investigate the morphology and distribution of HA and collagen in HCM and non-failing myocardial tissues. Therefore, HA and collagen distribution was compared in cardiac tissue from HCM patients and non-failing controls subjected to immunohistochemistry and studied in the confocal microscope.

In non-failing hearts, the natural intermyofibrillar space is aligned with HA forming long fine fibers running parallel to the myofibers, as shown in Figure 1a. These fibers were connected with HA staining strands occasionally seen to run adjacent to collagen strands, as shown in Figure 1e. Both HA fibers and strands were regularly covered with dots of HA, forming an appearance of a string of pearls, as shown in Figure 1b. HA fills the ECM around capillaries and larger vessels. Collagen I could be visualized around individual cardiomyocytes and in the weaves surrounding the myofibrils, as shown in Figure 1e. In contrast, HA staining was weak or absent between individual cardiomyocytes, but located immediately adjacent outside collagen weaves as well as covering collagen higher order structures such as coils, as shown in Figure 1d.

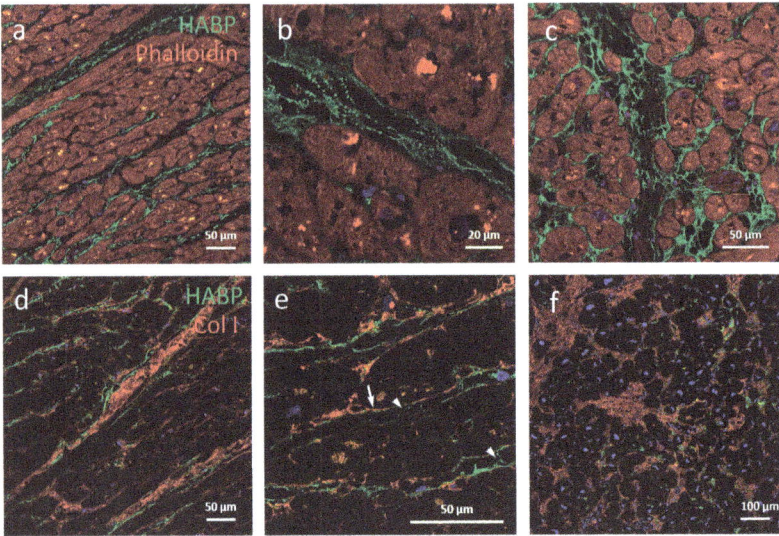

Figure 1. Morphological distribution of HA in cardiac tissue. Cardiomyocytes from non-failing septum (healthy controls) (**a**,**b**,**d**–**e**) n = 5 and HCM (**c**,**f**) n = 5. (**a**) In non-failing hearts, HA form long fine fibers in the perimysium running parallel to myofibers. (**b**) Sometimes these fibers are aligned with stained dots of HA forming an appearance of a string of pearls. (**d**) Collagen coil structures are surrounded by, but exclude, HA. (**e**) HA is located immediately adjacent to collagen weaves (arrow), and form strand-like structures between fibers (arrowheads). (**c**) In HCM, individual cardiomyocytes are separated by HA, which is also more abundant regularly filling the perimysium forming large patches with enlarged ring-like structures. (**f**) In areas with expanded ECM with fibrosis, there were patches where collagen and HA were mutually excluded. Hyaluronan binding protein (HABP, green), cardiomyocytes (phalloidin, red), nuclei (Hoechst, blue) (**a**–**c**). Hyaluronan binding protein (HABP, green), collagen I (Col I, red), nuclei (Hoechst, blue) (**d**–**f**). Large pale red dots inside cells are artefacts due to auto fluorescence.

Individual cardiomyocytes were separated or absent in the HCM tissue in areas with largely expanded and unorganized ECM, as shown in Figure 1f. In addition, these areas were filled with patches of HA. Also, the intermyofibrillar HA staining fibers and strands were thicker than in non-failing tissues, and in some samples, large, non HA staining holes were identified in areas with HA patches, as shown in Figure 1c. In some regions HA and collagen I were localized in the same patches, whereas in some areas within the same sample, regions were observed where HA and collagen I seemed excluded from each other, as shown in Figure 1f. Stained dots of HA were also less frequent in HCM myectomies. No gross difference in staining of collagen I, III, and VI was observed. These data suggest that HA levels, structure, and distribution play a role in the development of HCM.

Since HCM is described as a "patchy" disease, the differential expression pattern of collagen related to HA may reflect different stages of progression in HCM.

3.3. Comparison of Cardiomyocyte Size

Since it is known that HCM is characterized by patches of cardiomyocyte hypertrophy, we wanted to further analyze the distribution of hypertrophic cardiomyocytes in the septum, right and left ventricle wall, respectively. Cardiomyocyte size from HCM patients and non-failing controls were histologically examined with confocal microscopy. Cell size determination showed that cardiomyocytes in right wall, left wall, and septum in controls were of significantly ($P < 0.001$) different sizes, as shown in Figure 2a. In addition, the cells of the right wall were less densely packed. No gross difference in morphology or distribution of HA was identified between the right wall and left wall, as shown in Figure 2c,d. In HCM, very large cardiomyocytes were identified which were significantly larger than cardiomyocytes in healthy left ventricles, as shown in Figure 2e,d, Table 3.

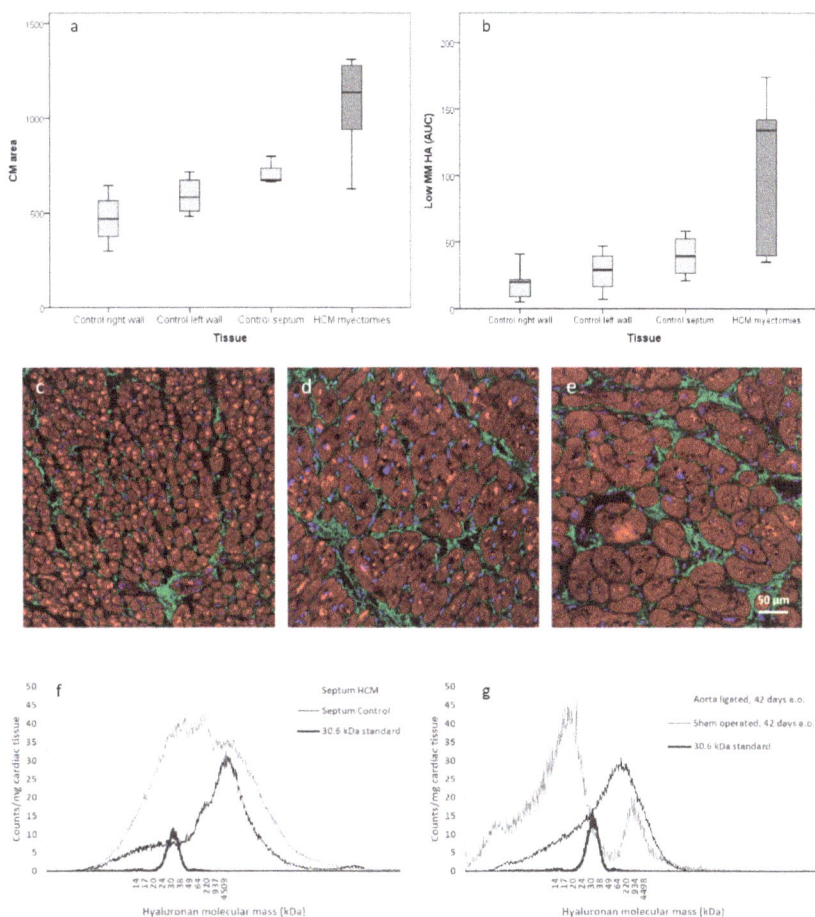

Figure 2. Cardiomyocyte (CM) size and amount of low molecular mass (MM) HA. Box plot illustrating cardiomyocyte size comparison between HCM and non-failing right wall septum and left wall, respectively (**a**). Box plot illustrating amount (corresponds to area under curve, AUC) of low MM HA in HCM and non-failing right wall septum and left wall, respectively (**b**). Immunohistochemistry of human heart tissues from right- (**c**), and left- (**d**) wall and HCM myectomies (**e**). The cardiomyocytes

were larger in left ventricle than right wall. (**a,c,d**) The largest cells could be found in HCM myectomies which also contained the lowest MM HA (**b**). Analysis of HA molecular mass distribution, where HA was extracted from human (n = 5/group) and rat hearts (n = 3/group) and separated by gas-phase electrophoretic molecular mobility analyzer (GEMMA) according to molecular charge (**f,g**). In the myectomies, there was a significant increase of low MM HA ($P < 0.001$) (**f**). There was a clear shift from high MM HA to low MM HA in aorta ligated rats compared to sham operated rats 42 days after surgery (**g**). Cardiomyocytes (phalloidin, red), ECM (collagen III, green), nuclei (Hoechst, blue). Size bar in (**e**) is representative for all immunohistochemistry (**c–e**).

3.4. Analysis of HA Mass Distribution

It has been reported that high MM HA has an anti-inflammatory effect in tissue homeostasis. In contrast, low MM HA has been reported to have a pro-inflammatory effect. Therefore, HA molecular mass in HCM and healthy hearts was investigated. Cardiac tissues from human subjects and from a rat model of induced hypertrophy [15] were subjected to GEMMA to determine HA mass distribution, as shown in Figure 2f,g, Table 3.

In HCM myectomies there was an increased amount of low MM HA compared to left septum from non-failing controls, as shown in Figure 2b,f, Table 3. A difference could also be seen within controls, as shown in Figure 2b. Cardiac tissue with larger cardiomyocytes seems to contain more low MM HA and amounts of low MM HA correlated with the size of cardiomyocytes ($P < 0.001$, Pearson correlation coefficient = 0.792), as shown in Figure 2a,b. Surgically-induced cardiac hypertrophy revealed a clear shift from high MM HA to low MM HA in the aorta-ligated rats 42 days after surgery, as shown in Figure 2g. Samples degraded with hyaluronidase showed no remaining contaminations.

Table 3. Gene characteristics and results of gene expression analysis. Changes in relative amount of low and high MM HA, cardiomyocyte (CM) area, and content of water in basal septum myectomies from HCM patients (n = 5) compared to non-failing septum and left ventricle (LV) (n = 10).

Gene	Description	Applied Biosystems Assay Number	Fold Change in Myectomized vs. Healthy LV	*P*-Value
HYAL2	Hyaluronidase 2	Hs01117343_g1	↓ 1.8	0.003
HAS2	Hyaluronan synthase 2	Hs00193435_m1	↓ 3.1	0.030
HAS3	Hyaluronan synthase 3	Hs00193436_m1	↑ 2.1	0.002
Change in:				
Low MM HA	Relative change in amount of low MM HA		↑ 3.1	0.045
High MM HA	Relative change in amount of high MM HA		↑ 1.6	0.171
CM area	Relative change in CM area		↑ 1.7	0.000
Water content in cardiac tissue	Relative change in water content/mg tissue		↓ 1.3	0.048

Statistical analysis using non-parametric Mann–Whitney U test. Arrows indicate up or down regulated.

3.5. Gene Expression

To further understand the dynamics of HA metabolism, mRNA from HCM and non-failing human cardiac tissue was analyzed for the expression of *HAS1-3, HYAL 1-2, CEMIP, CD44, VCAN*, and *TSG6*. There was a 1.8-fold decrease of *HYAL2* ($P = 0.003$) and a 3.1-fold decrease of *HAS2* in myectomies compared to non-failing septum ($P = 0.016$). *HAS3* showed a 2.1-fold increase in myectomies ($P = 0.008$), as shown in Table 3. The other genes tested were all expressed although no significant statistical difference could be identified (data not shown). No significant difference between left and right ventricle gene expression was observed (data not shown). Levels of *HYAL2* and *HAS3* correlated to amounts of low MM HA in non-failing left ventricles (*P* = 0.009, Pearson correlation coefficient = 0.840 and *P* = 0.012, Pearson correlation coefficient = 0.822, respectively) but not in HCM myectomies.

Factor analysis of gene expression level correlation in left ventricle of non-failing hearts showed that *HYAL2*, *HAS2*, and *HAS3* formed a correlation cluster while the genes *CEMIP*, *CD44*, *HAS1*, *TSG6*, and *VCAN* formed another, as shown in Figure 3a. In HCM, the expression levels of *CEMIP*, *CD44*, and *VCAN* formed a new correlation cluster with *HAS3*. *TSG6*, *HAS2*, and *HYAL1* formed another cluster, as shown in Figure 3b. The levels of *HAS1* and *HYAL2* no longer correlated with any of the other genes investigated.

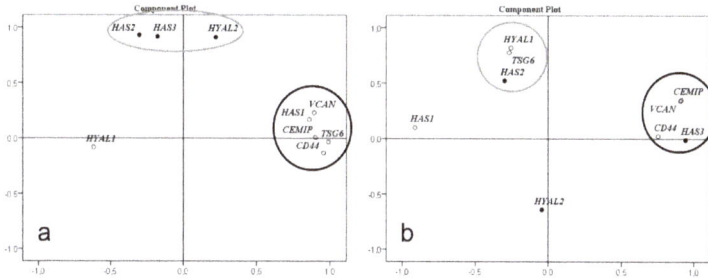

Figure 3. Factor analysis for correlation of gene expression levels. *HAS1-3*, *HYAL1-2*, *CEMIP*, *CD44*, *VCAN*, and *TSG6* in non-failing left wall and septum (n = 9) (**a**) and HCM myectomies (n = 5) (**b**). (**a**) In left wall and septum there are two clusters where *HAS1*, *CEMIP*, *CD44*, *VCAN*, and *TSG6* form one cluster, and *HAS2*, *HAS3*, and *HYAL2* form another. (**b**) In basal septal myectomies from HCM patients the expression levels of *CEMIP*, *CD44*, and *VCAN* formed a new correlation cluster with *HAS3*. *HAS2*, *HYAL1*, and *TSG6* formed another cluster. The levels of *HAS1* and *HYAL2* no longer correlated with any of the other genes investigated. Factor analysis was performed with the principal components method to analyze the correlation matrix and two factors were extracted.

4. Discussion

During development of HCM, the heart is subjected to hemodynamic or metabolic stress and returns to the fetal gene program and to fetal metabolism where cells prefer usage of glucose over fatty acid as a source of energy [4,25]. Our study material is based on basal septal myectomies from HCM patients subjected to surgical intervention due to a symptomatic and advanced disease where fibrosis and disarranged morphology is already established.

NMR analysis of metabolites in cardiac tissue from the HCM patients compared to healthy heart tissues from non-failing controls showed that fatty acids were decreased, and lactate was increased in the heart of HCM patients, indicating altered and less efficient glucose metabolism. UDP-sugar precursors, e.g., UDP-GlcNAc and UDP-GlcUA, the two substrates for HA synthesis, were also increased in the heart of HCM patients, which could be explained by the observed increase of metabolites in the HBP resulting in subsequent increase of UDP-GlcNAc. In addition, *myo*-Inositol and UDP-glucose, which are precursors for UDP-GlcUA, were also increased.

It has previously been shown that elevated levels of HA substrates enhance the synthesis of HA [8] and our metabolic results could be confirmed with an increase of low MM HA in myectomies from HCM patients compared to non-failing hearts, as shown in Figure 2f. An unexpected finding was a small but significant decrease of water content in cardiac tissue from HCM patients. Tissue accumulation of HA normally causes increased water content. Clinically human HCM is not associated with edema and all patients had end stage HCM. Possibly the low MM HA in the hypertrophic heart does not retain water in the tissue.

The factor analysis of expression level of genes involved in synthesis, degradation, and binding of HA clearly demonstrated that expression of genes involved in HA metabolism are altered in HCM.

Interestingly, the expression levels of the three genes *HYAL2*, *HAS2* and *HAS3* correlated closely in the control left ventricle. Furthermore, levels of *HYAL2* and *HAS3* also correlated to amounts of low MM HA in the controls. In cardiac tissue from the HCM patients, these three genes showed significant changes in expression levels, both up and down regulated, but they had lost their mutual correlation and also to low MM HA. These clusters of gene expression levels might indicate genes involved in the same cellular process, governed by the same set of transcription factors. The pathological process of HCM induces another set of transcription factors causing different clusters of gene expression levels. This implies a common transcriptional regulation in the healthy heart which is disrupted during the development of HCM affecting the metabolism of HA, both synthesis and degradation. However, we cannot evaluate the functional significance of these gene expression clusters based on our results.

Downregulation of *HYAL2* suggests that less HA is degraded and eliminated intracellularly. However, oxidative stress and excess production of ROS is a feature of HCM, which also has been implied in degrading HA into low MM HA. Glutathione has a protective function in the attenuation of ROS in HCM. In our metabolomic analysis glutamine and glutamate, substrates for glutathione, as well as glutathione itself were increased in cardiac tissue from the HCM patients, indicating a cellular response of high content of ROS. Thus, ROS might be a possible explanation for the fragmentation and accumulation of low MM HA in the hypertrophic heart.

Based on our results, a hypothetical explanation of the increased levels of low MM HA seen in the cardiac tissue from HCM patients might be increased availability of HA substrates from increased glucose metabolism, upregulated expression of HAS3, downregulation of internal degradation by HYAL2 with subsequent extracellular fragmentation by ROS.

We observed that the size of cardiomyocytes corresponded with the amount of low MM HA in human cardiac tissue, both in non-failing heart tissue as well as in basal septal myectomies from HCM patients, as shown in Figure 2a–e. We experimentally confirmed accumulation of low MM HA in cardiac tissue from a model of induced cardiac hypertrophy in rat, showing a rapid shift from high MM HA to low MM HA after surgery, as shown in Figure 2g,f. This suggests that fragmentation of HA into low MM HA occurs both in different parts of non-failing hearts and during the development of cardiac hypertrophy with levels corresponding to cardiomyocyte size.

An interesting parallel is found in experimental myocardial infarction research [26]. When HA-based hydrogels with HA of different mass were injected in the infarcted area, the gel with the smallest mass HA (50 kDa) showed the most significant regeneration of myocardium and functional recovery. This further supports the hypothesis that low MM HA is not pathogenic in itself but possibly a part of a compensatory process initiated by the need of increased cardiac capacity.

An important feature in HCM is the development of arrhythmias including lethal arrhythmias. Increased fibrosis is known to disrupt the electrical conductivity between cardiomyocytes and act as a substrate for re-entrant arrhythmias [27]. In our study, we observed an increased staining intensity of HA surrounding individual cardiomyocytes in cardiac tissues from HCM patients compared to healthy controls, as shown in Figure 1c,f. Furthermore, it has been shown that high MM HA depolarizes the membrane potential in human fibroblasts, human embryonic kidney (HEK), and neurons in a concentration dependent manner which could be reversed by digestion of HA by hyaluronidase [13]. Our observations and others results [13] could indicate that a local increase of high MM HA in HCM may play a role in membrane polarizing of individual cardiomyocyte cell membranes.

The morphology of hypertrophic myocardium is characterized by regions of seemingly normal tissue, neighboring large disarrayed cardiomyocytes with a remodeled and expanded ECM. We identified regions within expanded ECM areas with stronger HA and weaker collagen staining and vice versa. Accumulation of HA has been shown to precede the development of fibrosis, and low MM HA has been predicted to have an essential role in promoting fibrosis [12]. Low MM HA binds to the Toll-like receptor 2 (TLR2) and it has been shown that inhibition of TLR2 reduces cardiac fibrosis [28,29]. In addition, HA occurs in the myocardial infarction border zone [30] and degradation of HA with hyaluronidases in early treatment of myocardial infarction has been shown to reduce fibrosis and infarct size [31]. Possibly, areas with either more HA or collagen respectively mirrors different stages of progress of the disease and formation of fibrosis.

Both HA's effect on cellular action potential and on the development of fibrosis suggests involvement of HA in development of arrhythmia. In addition, the increased amount of HA contributes to the expanded ECM, thus separating cardiomyocytes within myofibrils and disrupting their cell-to-cell connection and impulse conduction, which also could be a potential risk for arrhythmia.

Cardiac energy metabolism in HCM affects the heart in several ways on a molecular level, e.g., O-GlcNAcylation and mTOR activation. Here we have introduced changes in HA metabolism as another consequence of the dysregulated glucose metabolism.

The study is limited by the availability of human cardiac tissue. Cardiac tissue from five patients and five healthy control hearts are few in a statistical view but valuable material to support results from experimental models.

In conclusion, we have, in various steps, shown that both HA molecular mass and amount changes in the development of HCM. The return to fetal energy metabolism in the HCM heart causes an increased generation of substrates for HA, which together with an altered gene expression changes the metabolism of HA. HA might add to the risk of arrhythmias in HCM and the process of regulating cardiomyocyte size seems to involve fragmentation of HA into low MM HA. The connection of glucose metabolism to HCM and the heart needs further investigation since diet and disease, e.g., diabetes mellitus, can affect the cellular uptake of glucose. This is a novel addition to the underlying mechanisms of hypertrophic cardiomyopathy.

Author Contributions: Conceptualization, U.H., S.M., and C.P.D.; Methodology, U.H., C.E.L., and L.D.; Validation, U.H. and C.E.L.; Visualization, U.H. and C.E.L.; Formal Analysis, U.H., C.E.L., and L.D.; Investigation, U.H., C.E.L., and L.D.; Resources, C.P.D., V.M.A., and O.R.G.; Data Curation, U.H. and C.E.L.; Writing—Original Draft Preparation, C.E.L. and U.H.; Writing—Review & Editing, C.E.L., C.P.D., L.D., V.M.A., O.R.G., S.M., and U.H.; Supervision, U.H. and C.P.D.; Project Administration, U.H. and C.P.D.; Funding Acquisition, U.H., C.P.D., and S.M.

Funding: This study was funded by the Swedish Society of Medicine, the Heart Foundation of Northern Sweden, and Västerbotten County Council.

Acknowledgments: We thank Ann-Britt Lundström for her valuable help with immunohistochemistry and Ilona Dudka at NMR for Life and SciLifeLab for performing NMR analysis. We also thank Anna Engström-Laurent for editorial assistance.

Conflicts of Interest: The authors declare no conflict of interest.

References

1. Maron, B.J.; Gardin, J.M.; Flack, J.M.; Gidding, S.S.; Kurosaki, T.T.; Bild, D.E. Prevalence of hypertrophic cardiomyopathy in a general population of young adults. Echocardiographic analysis of 4111 subjects in the CARDIA Study. Coronary Artery Risk Development in (Young) Adults. *Circulation* **1995**, *92*, 785–789. [CrossRef] [PubMed]

2. Ho, C.Y.; Charron, P.; Richard, P.; Girolami, F.; Van Spaendonck-Zwarts, K.Y.; Pinto, Y. Genetic advances in sarcomeric cardiomyopathies: State of the art. *Cardiovasc. Res.* **2015**, *105*, 397–408. [CrossRef]

3. Olivotto, I.; d'Amati, G.; Basso, C.; Van Rossum, A.; Patten, M.; Emdin, M.; Pinto, Y.; Tomberli, B.; Camici, P.G.; Michels, M. Defining phenotypes and disease progression in sarcomeric cardiomyopathies: Contemporary role of clinical investigations. *Cardiovasc. Res.* **2015**, *105*, 409–423. [CrossRef] [PubMed]

4. Rajabi, M.; Kassiotis, C.; Razeghi, P.; Taegtmeyer, H. Return to the fetal gene program protects the stressed heart: A strong hypothesis. *Heart Fail. Rev.* **2007**, *12*, 331–343. [CrossRef]

5. Kundu, B.K.; Zhong, M.; Sen, S.; Davogustto, G.; Keller, S.R.; Taegtmeyer, H. Remodeling of Glucose Metabolism Precedes Pressure Overload-Induced Left Ventricular Hypertrophy: Review of a Hypothesis. *Cardiology* **2015**, *130*, 211–220. [CrossRef]

6. Crilley, J.G.; Boehm, E.A.; Blair, E.; Rajagopalan, B.; Blamire, A.M.; Styles, P.; McKenna, W.J.; Ostman-Smith, I.; Clarke, K.; Watkins, H. Hypertrophic cardiomyopathy due to sarcomeric gene mutations is characterized by impaired energy metabolism irrespective of the degree of hypertrophy. *J. Am. Coll. Cardiol.* **2003**, *41*, 1776–1782. [CrossRef]

7.   Hebert, L.F., Jr.; Daniels, M.C.; Zhou, J.; Crook, E.D.; Turner, R.L.; Simmons, S.T.; Neidigh, J.L.; Zhu, J.S.; Baron, A.D.; McClain, D.A. Overexpression of glutamine:fructose-6-phosphate amidotransferase in transgenic mice leads to insulin resistance. *J. Clin. Investig.* **1996**, *98*, 930–936. [CrossRef]

8.   Jokela, T.A.; Jauhiainen, M.; Auriola, S.; Kauhanen, M.; Tiihonen, R.; Tammi, M.I.; Tammi, R.H. Mannose inhibits hyaluronan synthesis by down-regulation of the cellular pool of UDP-N-acetylhexosamines. *J. Biol. Chem.* **2008**, *283*, 7666–7673. [CrossRef] [PubMed]

9.   Vigetti, D.; Ori, M.; Viola, M.; Genasetti, A.; Karousou, E.; Rizzi, M.; Pallotti, F.; Nardi, I.; Hascall, V.C.; De Luca, G.; et al. Molecular cloning and characterization of UDP-glucose dehydrogenase from the amphibian Xenopus laevis and its involvement in hyaluronan synthesis. *J. Biol. Chem.* **2006**, *281*, 8254–8263. [CrossRef] [PubMed]

10. Delmage, J.M.; Powars, D.R.; Jaynes, P.K.; Allerton, S.E. The selective suppression of immunogenicity by hyaluronic acid. *Ann. Clin. Lab. Sci.* **1986**, *16*, 303–310. [PubMed]

11. Petrey, A.C.; de la Motte, C.A. Hyaluronan, a crucial regulator of inflammation. *Front. Immunol.* **2014**, *5*, 101. [CrossRef] [PubMed]

12. Albeiroti, S.; Soroosh, A.; de la Motte, C.A. Hyaluronan's Role in Fibrosis: A Pathogenic Factor or a Passive Player? *Biomed. Res. Int.* **2015**, *2015*, 790203. [CrossRef]

13. Hagenfeld, D.; Schulz, T.; Ehling, P.; Budde, T.; Schumacher, U.; Prehm, P. Depolarization of the membrane potential by hyaluronan. *J. Cell. Biochem.* **2010**, *111*, 858–864. [CrossRef]

14. Hellman, U. Hyaluronan, a beneficial glycosaminoglycan that may affect the phenotype of cardiac hypertrophy—A hypothesis. *Int. Cardiovasc. Forum J.* **2014**, *1*, 226–229. [CrossRef]

15. Hellman, U.; Hellstrom, M.; Morner, S.; Engstrom-Laurent, A.; Aberg, A.M.; Oliviero, P.; Samuel, J.L.; Waldenstrom, A. Parallel up-regulation of FGF-2 and hyaluronan during development of cardiac hypertrophy in rat. *Cell Tissue Res.* **2008**, *332*, 49–56. [CrossRef] [PubMed]

16. Hellman, U.; Morner, S.; Engstrom-Laurent, A.; Samuel, J.L.; Waldenstrom, A. Temporal correlation between transcriptional changes and increased synthesis of hyaluronan in experimental cardiac hypertrophy. *Genomics* **2010**, *96*, 73–81. [CrossRef] [PubMed]

17. Hellman, U.; Malm, L.; Ma, L.P.; Larsson, G.; Morner, S.; Fu, M.; Engstrom-Laurent, A.; Waldenstrom, A. Growth factor PDGF-BB stimulates cultured cardiomyocytes to synthesize the extracellular matrix component hyaluronan. *PLoS ONE* **2010**, *5*, e14393. [CrossRef] [PubMed]

18. Hellstrom, M.; Engstrom-Laurent, A.; Morner, S.; Johansson, B. Hyaluronan and collagen in human hypertrophic cardiomyopathy: A morphological analysis. *Cardiol. Res. Pract.* **2012**, *2012*, 545219. [CrossRef] [PubMed]

19. Laurent, C.; Johnson-Wells, G.; Hellstrom, S.; Engstrom-Laurent, A.; Wells, A.F. Localization of hyaluronan in various muscular tissues. A morphological study in the rat. *Cell Tissue Res.* **1991**, *263*, 201–205. [CrossRef]

20. Do, L.; Dahl, C.P.; Kerje, S.; Hansell, P.; Morner, S.; Lindqvist, U.; Engstrom-Laurent, A.; Larsson, G.; Hellman, U. High Sensitivity Method to Estimate Distribution of Hyaluronan Molecular Sizes in Small Biological Samples Using Gas-Phase Electrophoretic Mobility Molecular Analysis. *Int. J. Cell Biol.* **2015**, *2015*, 938013. [CrossRef]

21. Hellstrom, M.; Johansson, B.; Engstrom-Laurent, A. Hyaluronan and its receptor CD44 in the heart of newborn and adult rats. *Anat. Rec. A Discov. Mol. Cell. Evol. Biol.* **2006**, *288*, 587–592. [CrossRef] [PubMed]

22. Tengblad, A. Affinity chromatography on immobilized hyaluronate and its application to the isolation of hyaluronate binding properties from cartilage. *Biochim. Biophys. Acta* **1979**, *578*, 281–289. [CrossRef]

23. Malm, L.; Hellman, U.; Larsson, G. Size determination of hyaluronan using a gas-phase electrophoretic mobility molecular analysis. *Glycobiology* **2012**, *22*, 7–11. [CrossRef] [PubMed]

24. Pfaffl, M.W. A new mathematical model for relative quantification in real-time RT-PCR. *Nucleic Acids Res.* **2001**, *29*, e45. [CrossRef]

25. Oka, T.; Xu, J.; Molkentin, J.D. Re-employment of developmental transcription factors in adult heart disease. *Semin. Cell Dev. Biol.* **2007**, *18*, 117–131. [CrossRef] [PubMed]

26. Yoon, S.J.; Hong, S.; Fang, Y.H.; Song, M.; Son, K.H.; Son, H.S.; Kim, S.K.; Sun, K.; Park, Y. Differential regeneration of myocardial infarction depending on the progression of disease and the composition of biomimetic hydrogel. *J. Biosci. Bioeng.* **2014**, *118*, 461–468. [CrossRef]

27. Long, C.S.; Brown, R.D. The cardiac fibroblast, another therapeutic target for mending the broken heart? *J. Mol. Cell. Cardiol.* **2002**, *34*, 1273–1278. [CrossRef]

28. Scheibner, K.A.; Lutz, M.A.; Boodoo, S.; Fenton, M.J.; Powell, J.D.; Horton, M.R. Hyaluronan fragments act as an endogenous danger signal by engaging TLR2. *J. Immunol.* **2006**, *177*, 1272–1281. [CrossRef]
29. Wang, L.; Li, Y.L.; Zhang, C.C.; Cui, W.; Wang, X.; Xia, Y.; Du, J.; Li, H.H. Inhibition of Toll-like receptor 2 reduces cardiac fibrosis by attenuating macrophage-mediated inflammation. *Cardiovasc. Res.* **2014**, *101*, 383–392. [CrossRef]
30. Dobaczewski, M.; Bujak, M.; Zymek, P.; Ren, G.; Entman, M.L.; Frangogiannis, N.G. Extracellular matrix remodeling in canine and mouse myocardial infarcts. *Cell Tissue Res.* **2006**, *324*, 475–488. [CrossRef]
31. Maclean, D.; Fishbein, M.C.; Maroko, P.R.; Braunwald, E. Hyaluronidase-induced reductions in myocardial infarct size. *Science* **1976**, *194*, 199–200. [CrossRef] [PubMed]

Article

Transcriptome-Wide Analysis of Human Chondrocyte Expansion on Synoviocyte Matrix

Thomas J. Kean [1,*], Zhongqi Ge [2], Yumei Li [2], Rui Chen [2] and James E. Dennis [1]

[1]   Orthopedic Surgery, Baylor College of Medicine, Houston, TX 77030, USA; James.Dennis@bcm.edu
[2]   Department of Molecular and Human Genetics, Baylor College of Medicine, Houston, TX 77030, USA;
       zge@mdanderson.org (Z.G.); yumeil@bcm.edu (Y.L.); ruichen@bcm.edu (R.C.)
*   Correspondence: tjkean@bcm.edu; Tel.: +1-983-986-6025

Received: 9 December 2018; Accepted: 21 January 2019; Published: 24 January 2019

Abstract: Human chondrocytes are expanded and used in autologous chondrocyte implantation techniques and are known to rapidly de-differentiate in culture. These chondrocytes, when cultured on tissue culture plastic (TCP), undergo both phenotypical and morphological changes and quickly lose the ability to re-differentiate to produce hyaline-like matrix. Growth on synoviocyte-derived extracellular matrix (SDECM) reduces this de-differentiation, allowing for more than twice the number of population doublings (PD) whilst retaining chondrogenic capacity. The goal of this study was to apply RNA sequencing (RNA-Seq) analysis to examine the differences between TCP-expanded and SDECM-expanded human chondrocytes. Human chondrocytes from three donors were thawed from primary stocks and cultured on TCP flasks or on SDECM-coated flasks at physiological oxygen tension (5%) for 4 passages. During log expansion, RNA was extracted from the cell layer (70–90% confluence) at passages 1 and 4. Total RNA was column-purified and DNAse-treated before quality control analysis and next-generation RNA sequencing. Significant effects on gene expression were observed due to both culture surface and passage number. These results offer insight into the mechanism of how SDECM provides a more chondrogenesis-preserving environment for cell expansion, the transcriptome-wide changes that occur with culture, and potential mechanisms for further enhancement of chondrogenesis-preserving growth.

Keywords: chondrocyte RNA-Seq; dedifferentiation; chondrogenesis; synoviocyte matrix; physioxia; RNA-Seq; cell senescence

1. Introduction

Arthritis is not only a debilitating disease, but an expensive one, with total arthritis-attributable medical expenditures and lost earnings surpassing $300 billion USD in 2013 [1]. Tissue engineering methods have been applied as a means to treat osteoarthritic lesions and they hold great potential for joint repair. However, human chondrocytes, expanded and used in autologous chondrocyte implantation techniques, are known to rapidly de-differentiate in culture [2], which has a detrimental impact on their utility for tissue engineering applications. The use of the term de-differentiation is distinct from the use of the term in re-programming or stem cell literature, as it indicates that the cells no longer have the ability to form hyaline-like cartilage tissue. Culture-expanded chondrocytes undergo both morphological and phenotypical changes and, eventually, lose the ability to produce hyaline like matrix. By passage 4, chondrogenic potential is essentially absent when tested in re-differentiation culture. This loss of differentiation potential limits their efficacy in the clinic and seriously impedes our ability to produce clinical-scale tissue engineering of human cartilage with suitable biomechanical properties.

Synoviocyte-derived extracellular matrix has been shown to support enhanced chondrocyte expansion whilst retaining re-differentiation potential in human [3] and porcine cells [4].

Next-generation sequencing offers an opportunity to take a global view of the transcriptional changes that occur during in vitro culture both on tissue culture plastic and on devitalized synoviocyte matrix. Relatively few RNA sequencing (RNA-Seq) studies have been conducted on human chondrocytes partially due to the dense extracellular matrix they form, making RNA isolation problematic in terms of both yield and quality. No RNA-Seq reports on the effect of synoviocyte matrix or the effect of culture de-differentiation were found. This study is an expansion of previous work [3] which showed typical chondrocyte-related gene expression changes by RT-qPCR (*COL1A1, COL2A1, SOX9, ACAN, MATN1, MMP13, COL10A1*) the levels of which, by fragments per kilobase of exon model per million fragments mapped reads (FPKM), are well correlated. In addition, RNA-Seq has been shown to be well correlated with RT-qPCR data across over 15,300 genes [5]. Preliminary analysis of the effect of passage on the transcriptome of chondrocytes grown on tissue culture plastic (TCP) was presented [6]. The results from this study give us a deeper understanding of chondrocyte biology, thus providing a better foundation for future therapeutic interventions.

2. Materials and Methods

Human chondrocytes (three donors) were thawed from frozen, end of primary culture stocks collected under an institutional review board (IRB)-approved protocol (H-36683) of the IRB for Baylor College of Medicine and Affiliated Hospitals. Chondrocytes were cultured on TCP flasks and synoviocyte-derived extracellular matrix (SDECM) at physiological oxygen tension (5%) for 4 passages (Figure 1) in growth media (DMEM-LG [HyClone, Pittsburgh, PA, USA] supplemented with 10% FBS [Atlanta Biologicals, Flowery Branch, GA, USA] and 1% penicillin/streptomycin [Gibco, Gaithersburg, MD, USA]). At the end of the first and fourth passage, cells were lysed and RNA extracted from the cell layer during the log expansion (70–90% confluence) phase. Cell lysis was performed using a guanidine chloride-based buffer (TRK Lysis buffer; (E.Z.N.A.® Tissue RNA Kit, Omega Bio-Tek, Norcross, GA, USA) and the lysate was frozen on dry ice and stored (−80 °C, 1–6 weeks). Companion flasks that were not lysed were trypsinized (0.25% trypsin/EDTA; Gibco, Gaithersburg, MD, USA) for 5 min at 37 °C, then trypsin-neutralized with an equal volume of growth media. Cells were collected and centrifuged (500× *g*, 5 min, room temperature). Cell pellets were resuspended and an aliquot counted using a hemocytometer with trypan blue (1:1; Gibco). Cells were seeded at 6000 cells/cm^2 and media were exchanged on day 2–3 and the cells cultured for 5–6 days at each passage.

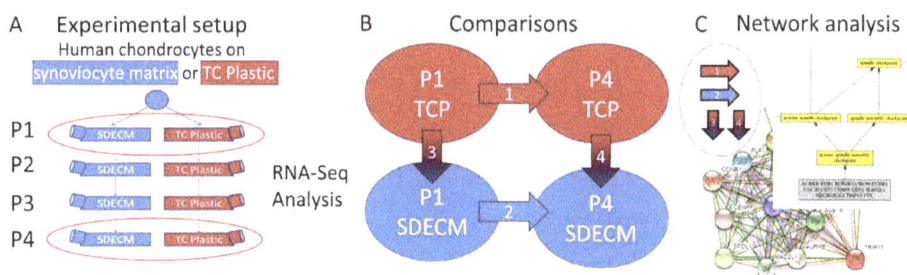

Figure 1. Experimental outline and comparisons. (**A**) Experimental setup: human articular chondrocytes were thawed from frozen stocks and seeded onto both synoviocyte-derived extracellular matrix (SDECM) and tissue culture plastic (TCP) flasks, passaged 4 times and RNA collected at passage (P) P1 and P4 for RNA sequencing (RNA-Seq) analysis. (**B**) Comparisons of gene expression profiles (1) P1 vs. P4 on TCP, (2) P1 vs. P4 on SDECM, (3) TCP vs. SDECM at P1, (4) TCP vs. SDECM at P4. (**C**) Network and gene enrichment analyses of differentially expressed genes from each comparison.

When all flasks had been lysed, total RNA was isolated from the lysate after thawing on ice using column purification (Direct-zol RNA mini-prep, Zymo Research, Irvine, CA, USA), as per manufacturer's instructions, with on-column DNA digest (DNase I, Zymo Research). RNA Purity was analyzed by a 260 nm/280 nm ratio (Tecan Nanoquant, Morrisville, NC, USA) and degradation/quality assessed using an Agilent Bioanalyzer (Genomic and RNA Profiling Core at Baylor College of Medicine, Houston, TX, USA). Samples were then submitted for next-generation RNA sequencing. mRNAs were captured by oligo-dT magnetic beads and fragmented. First-strand cDNA was generated using random primers, and second-strand cDNA was synthesized with deoxyuridine triphosphate (dUTP). The library was generated using the ds cDNA as a template. Briefly, templates were end-repaired, and a 3'A was added before Y-shaped adaptors were added to each end. The strand with dUTP was then digested using uracil-DNA glycosylase prior to PCR. Sequencing was carried out on Illumina HiSeq 2000. RNA-Seq reads were mapped to human genome hg19 and splice junction sites with Bowtie (v0.12.7) [7] and Tophat (v2.0.0) [8]. Read counts mapped to each gene were calculated by HTseq [9], and fragments per kilobase of exon model per million fragments mapped reads (FPKM) values were calculated using Cufflinks (version 2.1.1; http://cole-trapnell-lab.github.io/cufflinks/releases/v2.1.1/) [10]. Differential expression was analyzed with R (http://www.R-project.org) [11] and the Bioconductor R package DESeq [12]. Heatmaps were generated from FPKM values with heatmap3. Genes with an adjusted *p*-value less than 0.01 and greater than a 2log2 (4-fold change) expression were considered to be differentially expressed. Network analysis was performed using STRING (http://string-db.org) [13]. Gene Ontology (GO) term enrichment was queried using a logically accelerated GO term finder (LAGO) [14]. Common genes were determined by combining table queries and uncommon genes by unmatched queries in Microsoft Access (v14.0.7214.5000, Redmond, WA, USA). GO Term lists were downloaded from Jax [15].

RNA-Seq data was deposited in the SRA database under accession number SRP156000.

3. Results

Chondrocytes grown on TCP typically expanded for 1.8 population doublings (PD) in P1 and 1.4 PD at P4, while those grown on SDECM expanded 3.6 PD in P1 and 4.0 PD at P4 [3]. At the end of P4, cells grown on TCP had undergone 7.6 PDs and those on SDECM had undergone 16.0 PDs. RNA was of high quality with RNA integrity numbers (RINs) ≥7.6 (See supplemental data Figure S1 for electropherograms and Table S1 for summary). Principal component analysis showed that donors clustered by both passage (P1 vs. P4) and surface (SDECM vs. TCP; Figure 2).

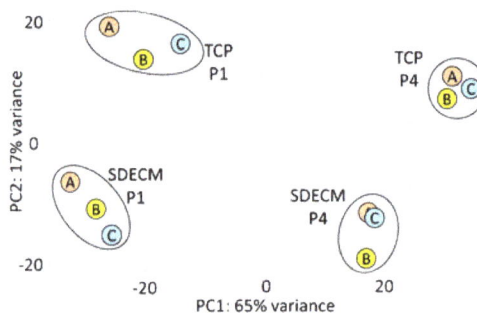

Figure 2. Principal component analysis. Samples are clearly clustered by their passage (P) in principal component (PC) PC1 and surface in PC2; donors are indicated by the letter within the circle (A, B, C).

Many genes were significantly affected by passage and culture surface, an overview of the up and down regulated genes for each comparison is shown in Table 1. Of the genes that were differentially regulated between P1 and P4, 512 were common between the two comparisons with 228 genes being upregulated at P4 and 283 genes downregulated. Only lipase G (*LIPG*) switched direction in cells cultured on the two different surfaces, being downregulated at P4 on SDECM and upregulated on TCP. Looking at the genes that were not common between the two surfaces, 175 were upregulated on TCP and 337 were downregulated. The upregulated genes in this subset were enriched for terms such as "tissue development" and "multicellular development". The downregulated genes were enriched for terms like "skeletal development", "system development" and "extracellular region". Of the 251 genes on SDECM that were not common between the two comparisons, 88 genes were upregulated and 163 downregulated. The upregulated genes in this subset were not enriched for any biological process, cellular component or molecular function. The downregulated genes in this subset were enriched for sterol-, cholesterol-, and lipid-related terms.

Table 1. Summary of differentially expressed genes (Adjusted p-value < 0.01 and log2 fold-change >2).

Comparison	Upregulated (%)	Downregulated (%)	Total	Common [3]
(1) P1 vs. P4 on TCP [1]	404 (39)	620 (61)	1024	
(2) P1 vs. P4 on SDECM [1]	316 (41)	447 (59)	763	512
(3) TCP vs. SDECM P1 [2]	151 (46)	180 (54)	331	
(4) TCP vs. SDECM P4 [2]	162 (48)	177 (52)	339	107

[1] Up/downregulated at P4. [2] Up/downregulated on SDECM. [3] Genes that were differentially expressed in both comparisons 1 and 2 or comparisons 3 and 4.

The top 20 genes for each comparison, in terms of adjusted p-value, are shown in Tables 2 and 3. Full lists of differentially expressed genes, the gene count output from HTseq, the FPKM output from Cufflinks, and the results of DESeq comparisons are shown in supplemental data (S2).

Table 2. Top 20 differentially expressed genes in comparisons 1 and 2.

	Comparison 1			Comparison 2	
Gene Symbol	Gene Name	FC [1]	Gene Symbol	Gene Name	FC [1]
MAFB	MAF BZIP transcription factor B	25.5 D	APLN	Apelin	21.8 D
PLK1	Polo-like kinase 1	23.1 U	MMP1	Matrix metallopeptidase 1	26.6 D
CENPF	Centromere protein F	12.0 U	KIF20A	Kinesin family member 20A	27.6 U
SLC40A1	Solute carrier family 40 member 1	166. D	TOP2A	DNA topoisomerase II alpha	16.3 U
TOP2A	DNA topoisomerase II alpha	13.0 U	CDC20	Cell division cycle 20	21.1 U
MKI67	Marker of proliferation Ki-67	14.5 U	TPX2	TPX2, microtubule nucleation factor	16.0 U
CCNB1	Cyclin B1	15.0 U	FOXM1	Forkhead box M1	12.3 U
FGFR2	Fibroblast growth factor receptor 2	19.0 D	BIRC5	Baculoviral IAP repeat containing 5	19.8 U
ISM1	Isthmin 1	25.5 D	MKI67	Marker of proliferation Ki-67	21.4 U
CDC20	Cell division cycle 20	19.6 U	PRC1	Protein regulator of cytokinesis 1	12.6 U
PRC1	Protein regulator of cytokinesis 1	12.1 U	DLGAP5	DL-associated protein 5	23.6 U
A2M	Alpha-2-macroglobulin	21.9 D	PLK1	Polo-like kinase 1	17.7 U
HMMR	Hyaluronan-mediated motility receptor	20.9 U	SLC40A1	Solute carrier family 40 member 1	25.9 D
DLGAP5	DLG-associated protein 5	17.1 U	ASPM	Abnormal spindle microtubule assembly	10.9 U
AURKA	Aurora kinase A	20.0 U	CCNB1	Cyclin B1	11.7 U
CCNB2	Cyclin B2	17.2 U	PRR11	Proline rich 11	9.43 U
TPX2	TPX2, microtubule nucleation factor	14.2 U	CENPF	Centromere protein F	12.6 U
STEAP4	STEAP4 metalloreductase	14.1 D	KIF23	Kinesin family member 23	13.6 U
ELN	Elastin	7.66 U	CEP55	Centrosomal protein 55	14.6 U
PRR11	Proline rich 11	11.9 U	ANLN	Anillin actin binding protein	12.2 U

[1] Fold change shows the up (U)- or down (D)- regulated fold change of the respective gene in P4. Adjusted p-values were $\leq 2.4 \times 10^{-52}$. Genes highlighted with the same colour are common between the two comparisons.

Table 3. Top 20 differentially expressed genes in comparisons 3 and 4.

	Comparison 3			Comparison 4	
Gene Symbol	Gene Name	FC [1]	Gene Symbol	Gene Name	FC [1]
POSTN	Periostin	16.9 U	COLEC12	Collectin subfamily member 12	32.3 U
COMP	Cartilage oligomeric matrix protein	28.5 D	MME	Membrane metalloendopeptidase	9.11 U
CLEC3B	C-type lectin domain family 3 member B	14.7 D	CRLF1	Cytokine receptor-like factor 1	7.62 D
DCN	Decorin	5.87 D	MFAP5	Tetratricopeptide repeat domain 9	14.5 D
PODN	Podocan	8.89 D	TTC9	Microfibril-associated protein 5	12.1 D
CTGF	Connective tissue growth factor	5.52 D	IGFBP1	Insulin-like growth factor binding protein 1	11.9 D
ADAMTS5	ADAM metallopeptidase thrombospondin type 1 motif 5	9.99 D	CLEC3B	C-type lectin domain family 3 member B	6.48 D
PPAP2B	Phospholipid phosphatase 3	5.57 D	DSP	Desmoplakin	10.1 D
TAGLN	Transgelin	7.23 D	LOX	Lysyl oxidase	4.02 D
SFRP4	Secreted frizzled-related protein 4	9.26 D	IGFBP5	Insulin-like growth factor binding protein 5	4.61 D
CAMK2N1	Calcium/calmodulin-dependent protein kinase II inhibitor 1	11.3 U	C6orf132	Chromosome 6 open reading frame 132	9.53 D
FHL1	Four and a half LIM domains 1	5.66 D	MOXD1	Monooxygenase DBH-like 1	7.35 U
OMD	Osteomodulin	4.92 D	CAMK2N1	Calcium/calmodulin-dependent protein kinase II inhibitor 1	20.1 U
COL18A1	Collagen type XVIII alpha 1 chain	6.22 U	PHLDA1	Pleckstrin homology-like domain family A Member 1	4.67 U
SEMA5A	Semaphorin 5A	4.22 U	EMB	Embigin	6.48 U
SCD	Stearoyl-CoA desaturase	5.78 U	PITPNM3	PITPNM family member 3	4.76 D
MYL9	Myosin light chain 9	6.15 D	INHBB	Inhibin subunit beta B	10.0 D
CPM	Carboxypeptidase M	6.23 U	ACTA2	Actin, alpha 2, smooth muscle, aorta	4.66 D
ITIH5	Inter-alpha-trypsin inhibitor heavy chain family member 5	6.58 D	LOC100505633	Long intergenic non-protein coding RNA 1133	5.19 D
AP1S3	Adaptor-related protein complex 1 subunit sigma 3	11.3 D	FOXQ1	Forkhead box Q1	4.96 U

[1] Fold change shows the up (U)- or down (D)- regulated fold change of the respective gene on SDECM. Adjusted p-values were $\leq 1.8 \times 10^{-50}$. Genes highlighted with the same colour were common between the two comparisons.

In comparisons 1 and 2, the major enriched pathways by GO analysis were cell cycle-associated. This was primarily due to the upregulated genes and resulted in an increase in both the G1/S and G2/M cell cycle checkpoint genes (Figures 3 and 4). Downregulated genes were enriched for GO terms like "system development" and "multicellular organism development". Genes were tallied from GO terms associated with positive or negative regulation of the cell cycle; there is an increase in the number of genes negatively regulating cell proliferation on TCP (140 genes vs. 51 on SDECM). However, caution should be exercised in using these GO term lists alone, as 44 of the 182 genes were present in both positve and negative terms. There were 35 genes associated only with positive regulation of proliferation on SDECM at P4 and 23 on TCP; 25 genes were only associated with negative regulation of proliferation on SDECM at P4 and 95 on TCP (tabulated data are included in S5).

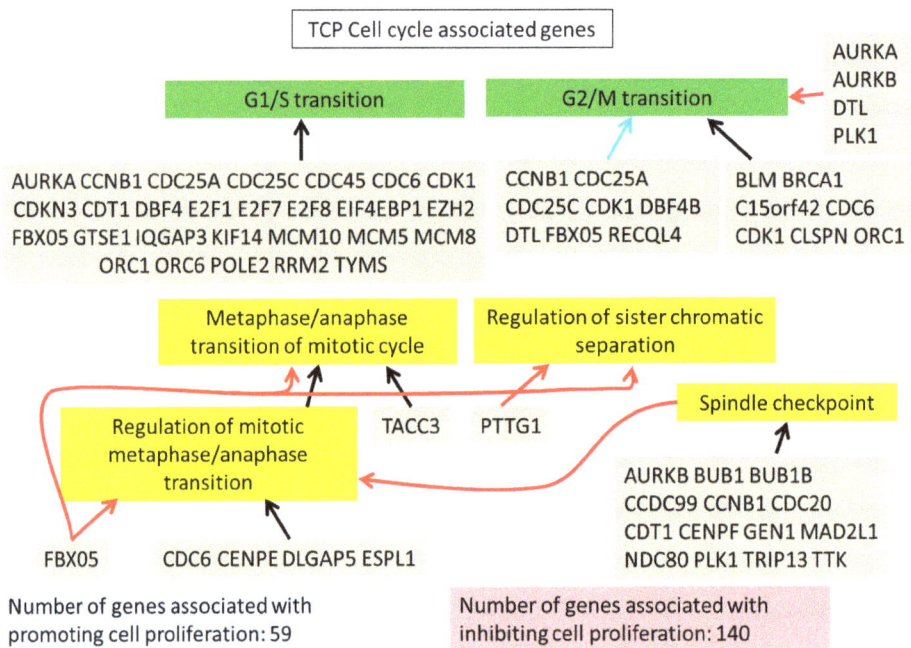

Figure 3. TCP cell cycle-associated genes. A summary of some of the cell cycle-associated genes from LAGO analysis which were upregulated in P4 chondrocytes on TCP. Similar cell cycle enrichment was seen in upregulated genes at P4 on SDECM (Figure 4). Red arrows connecting the genes to a term indicate inhibition of that term, blue arrows = promotion and black arrows = association. For the full interaction chart see S3 (TCP) and for the list of Gene Ontology (GO) terms which were significantly enriched see S5.

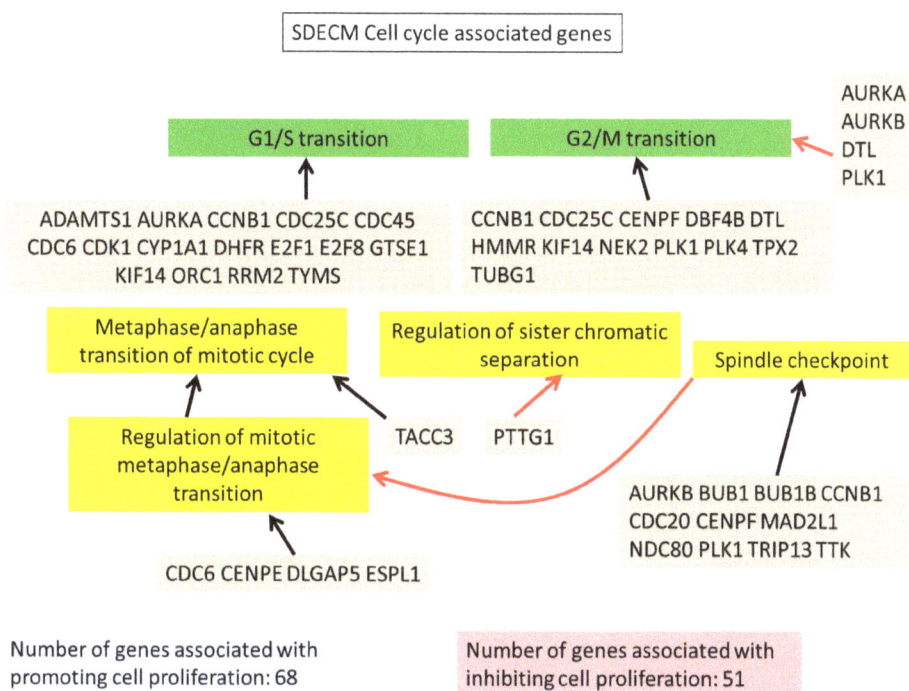

Figure 4. SDECM cell cycle-associated genes. A summary of some of the cell cycle-associated genes from LAGO analysis which were upregulated in P4 chondrocytes on SDECM. Red arrows connecting the genes to a term indicate inhibition of that term, blue arrows = promotion and black arrows = association. For the full interaction chart see S4 (SDECM) and for the list of GO terms which were significantly enriched see S5.

The GO term for "extracellular matrix organization" (GO: 0030198) is associated with 268 genes. When the differentially expressed genes from comparisons 1, 2, 3, and 4 were queried with this list, 50 genes were identified as being differentially expressed in one or more comparison (Figure 5). Comparison 1 showed differential expression of 30 genes, with eight of those genes being upregulated at P4 on TCP, and 22 downregulated at P4 on TCP. Comparison 2 identified 22 differentially expressed genes, 4 of which were upregulated at P4 on SDECM and 18 downregulated at P4 on SDECM. Extracellular matrix disassembly (GO: 0022617) identified 29 genes, of which only three genes were differentially regulated in any of the comparisons: matrix metalloproteinase 13 (*MMP13*), semaphorin 5A (*SEMA5A*) and fibroblast growth factor receptor 4 (*FGFR4*). *MMP13* decreased with passage on both surfaces but was increased at both P1 and P4 by culture on SDECM. *SEMA5A* was increased at P1 by culture on SDECM. *FGFR4* significantly decreased with passage on SDECM but was increased on SDECM in comparison with TCP at P1.

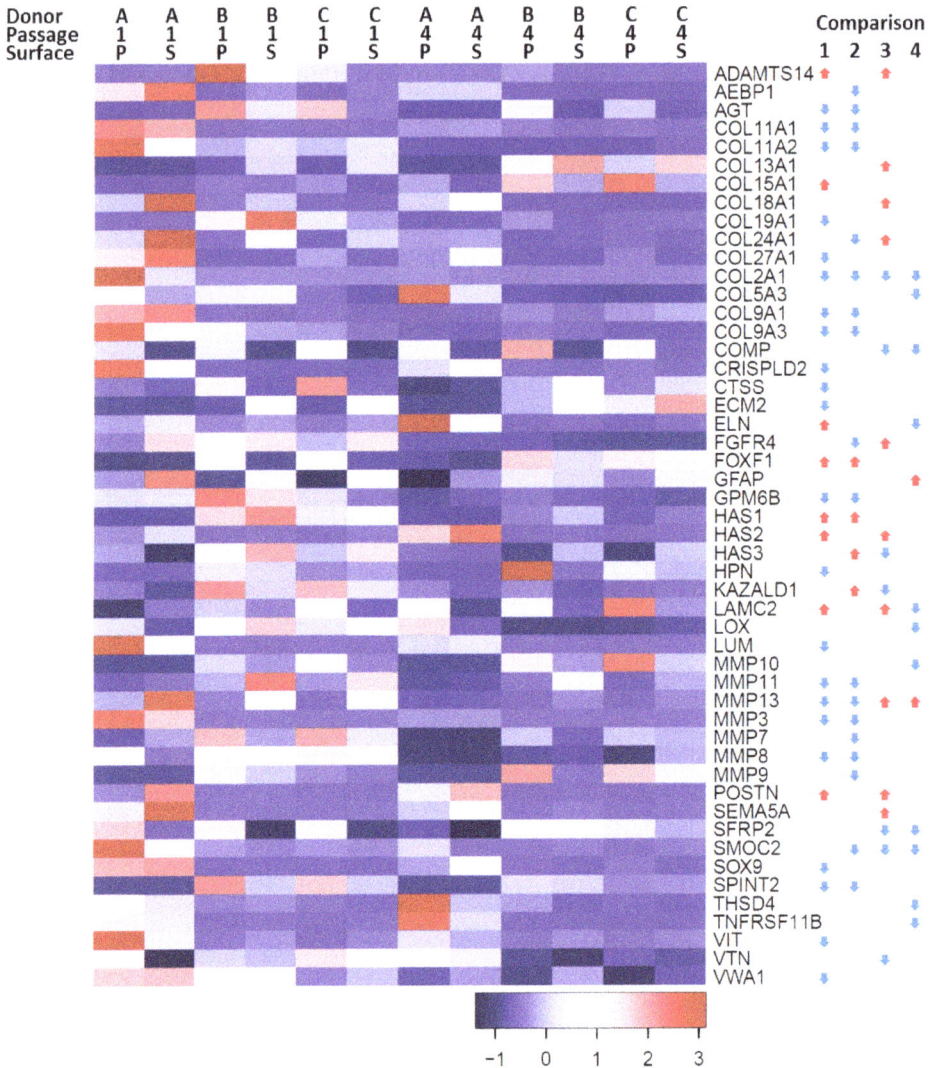

Figure 5. Differentially expressed extracellular matrix organization genes. This heatmap shows differentially expressed genes from the four comparisons which are associated with the GO term "extracellular matrix organization". Donor (A, B, C), passage (1, 4) and surface (P = TCP and S = SDECM) are indicated at the top of the heatmap. The gene symbol is to the right of the heatmap. The arrows indicate which comparisons were significantly different at a ≥4-fold change with $p < 0.01$; no arrow means that the comparison did not meet that threshold (1 = TCP P1 vs. P4, 2 = SDECM P1 vs. P4, 3 = TCP vs. SDECM at P1, 4 = TCP vs. SDECM at P4). A red arrow pointing upwards (↑) indicates that gene expression increased and a blue downward-pointing arrow (↓) that it decreased.

Of the 204 genes that are identified by the gene ontology term "cartilage development" (GO:0051216), 43 genes were differentially expressed in one or more comparisons. At P4 on TCP, 28 genes were significantly decreased vs. 16 on SDECM at P4; only four were upregulated on TCP and three on SDECM (Figure 6).

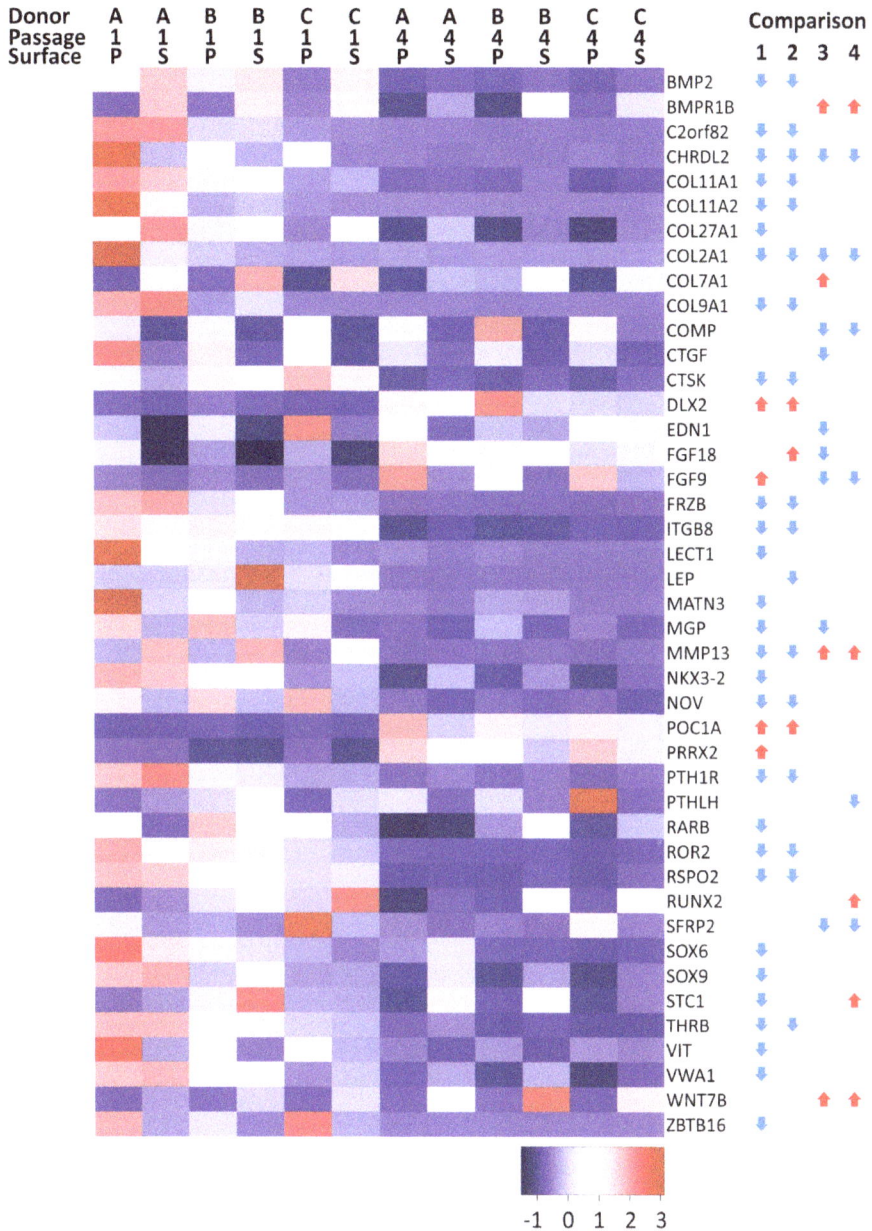

Figure 6. Differentially expressed cartilage development genes. Heatmap summary of differentially expressed genes associated with the GO term "cartilage development". Donor (A, B, C), passage (1, 4), and surface (P = TCP and S = SDECM) are indicated at the top of the heatmap. The gene symbol is to the right of the heatmap. The arrows indicate which comparisons were significantly different at a ≥4-fold change with $p < 0.01$; no arrow means that the comparison did not meet that threshold (1 = TCP P1 vs. P4, 2 = SDECM P1 vs. P4, 3 = TCP vs. SDECM at P1, 4 = TCP vs. SDECM at P4). A red arrow pointing upwards (↑) indicates that gene expression increased and a blue downward-pointing arrow (↓) that it decreased.

Looking at genes associated with the GO Term "cell senescence" (GO:0090398, 63 genes), nine were differentially regulated in comparison 1: *BCL6, C2orf40, NOX4, SERPINE1, BRCA2, FBXO5, SUV39H1, CDK1*, and *FOXM1*. Four of these genes were also differentially regulated in comparison 2: *NOX4, CDK1, FOXM1*, and *C2orf40*.

COL2A1 is a well-known marker for hyaline cartilage, and since it was decreased in all four comparisons, transcription factors regulating its expression were further investigated. Only *BCL6* and *PPARG* were significantly downregulated ≥4-fold on TCP, and *PPARG* alone on SDECM. The conversion from type II collagen expression to type I collagen expression is a distinctive marker for fibrocartilage vs. hyaline cartilage [16,17]. While *COL1A1* is not one of those genes which met the 4-fold increase selection criterion, it was still increased 2.8-fold on TCP with an adjusted *p*-value of 1×10^{-30}; the upregulation on SDECM was 1.2-fold and non-significant. Out of the over 300 transcription factors with an identified (or potential) binding sequence in the *COL1A1* gene [18] only five were significantly upregulated on TCP and 4 on SDECM Table 4.

Table 4. Upregulated [1] transcription factors in the *COL1A1* gene promoter/enhancer.

Gene Symbol	Gene Name	TCP FC	TCP padj	SDECM FC	SDECM padj
BRCA1	BRCA1, DNA repair associated	4.60	3.9×10^{-38}	6.54	2.1×10^{-31}
CEBPA	CCAAT enhancer binding protein alpha	2.11	2.7×10^{-1}	4.82	8.0×10^{-4}
E2F8	E2F transcription factor 8	8.16	1.1×10^{-22}	21.21	3.2×10^{-22}
EZH2	Enhancer of zeste 2 polycomb repressive complex 2 subunit	5.76	1.5×10^{-46}	3.87	1.1×10^{-18}
FOSL1	FOS-like 1, AP-1 transcription factor subunit	4.82	5.0×10^{-43}	1.10	6.5×10^{-1}
MXD3	MAX dimerization protein 3	4.79	1.8×10^{-24}	4.43	2.7×10^{-11}

[1] Upregulated at P4 by ≥4-fold on either TCP or SDECM. Highlighted cells indicate significant data. FC—fold change, padj—adjusted *p*-value.

When considering the effect of culture surface, in comparison 3, the most significantly enriched GO terms related to extracellular matrix (14 of 115 GO terms, *p* < 0.01; S5). Downregulated genes were also significantly enriched for "extracellular matrix" and "cell motility" GO Terms (S6). Upregulated genes were enriched for GO Terms related to development and differentiation (S7).

In comparison 4, the most significantly enriched GO terms related to "development" and "extracellular matrix". "Extracellular matrix" terms dominated the enriched GO term list for downregulated genes (S8) whilst neuron-related terms predominated in the upregulated GO term list (S5, S9). Of the 339 differentially expressed genes at P4, 107 were common with P1; 52 went up and 55 went down. Two genes switched their direction: angiopoietin-like 7 (*ANGPTL7*) went from being downregulated on SDECM at P1 to being upregulated on SDECM at P4; laminin subunit gamma 2 (*LAMC2*) was upregulated on SDECM at P1 and flipped to being downregulated at P4. GO Term analysis of the downregulated genes recapitulated the predominance of terms for "extracellular matrix" as did the upregulated genes for neuronal terms. Differentially expressed genes at P1 that were not common between the two surface comparisons: 224 genes, 99 were upregulated on SDECM and 125 were downregulated. Upregulated genes were enriched in GO terms for extracellular matrix organization and vasculature development. Downregulated genes were enriched in GO terms for "extracellular region". Differentially expressed genes at P4 that were not common between the two surface comparisons: 232 genes, 110 were upregulated on SDECM and 122 were downregulated. Upregulated genes were enriched in GO terms for "multicellular" and "system development". Downregulated genes were enriched in cellular component GO terms for "extracellular region". Selection of genes that were unchanged on SDECM between P1 and P4 (less than a 10% increase or decrease), but that were significantly differentially expressed on TCP gave 15 genes (Table 5).

Table 5. Stable genes on SDECM (P1 vs. P4) that were differentially expressed on TCP [1].

Gene Symbol	Gene Name	Fold Change	padj
FOSL1	FOS Like 1, AP-1 transcription factor subunit	4.81 U	5.0×10^{-43}
STC1	Stanniocalcin 1	4.73 D	2.2×10^{-40}
CDH2	Cadherin 2	11.7 U	1.0×10^{-27}
NXPH3	Neurexophilin 3	4.46 D	7.1×10^{-19}
POM121L9P	POM121 transmembrane nucleoporin-like 9, pseudogene	10.9 D	9.6×10^{-19}
MSC	Musculin	4.62 U	4.3×10^{-12}
POSTN	Periostin	7.64 U	1.2×10^{-11}
CCRL1	Atypical chemokine receptor 4	8.45 D	4.7×10^{-10}
ADAMTS14	ADAM metallopeptidase with thrombospondin type 1 motif 14	4.60 U	8.6×10^{-10}
ACSS1	Acyl-CoA synthetase short chain family member 1	4.24 D	1.5×10^{-8}
SH3TC2	SH3 domain and tetratricopeptide repeats 2	5.20 U	1.1×10^{-6}
NTF3	Neurotrophin 3	4.00 U	2.1×10^{-6}
MYBPH	Myosin binding protein H	6.36 D	1.2×10^{-5}
ELOVL2	ELOVL fatty acid elongase 2	9.81 U	1.2×10^{-4}
GRIK5	Glutamate ionotropic receptor kainate type subunit 5	4.46 D	9.2×10^{-3}

[1] Differentially regulated at P4 by $\geq$4-fold on TCP.

4. Discussion

The earliest RNA-Seq study on chondrocytes appears to be that by Peffers et al. [19] comparing young and old horse cartilage. Peffers et al. found 396 genes changed more than 2.6-fold with a *p*-value <0.05; a relatively smaller differential gene expression than that reported in this study. In fact, the large number of differentially expressed genes prompted the more stringent *p*-value and higher fold-change cut offs. There was a large degree of correlation between comparisons 1 and 2, with half of the top 20 genes being differentially regulated. Overall, half of the genes were common from the TCP comparison and two-thirds of those from the SDECM comparison. The coincidence of the differential regulation of two lipid-related genes—lipase G (*LIPG*) and peroxisome proliferator activated receptor gamma (*PPARG*)—is interesting particularly since *PPARG* stimulation by Rosiglitazone has been shown to be chondroprotective [19]. This chondroprotection potentially acts through a blunting of NF-kappa-B-mediated inflammatory signals [20]. However, in a more recent study, Xu et al. found that *PPARG* stimulation was detrimental to the expression of *COL2A1* and promoted hypertrophy [21]. It is worth noting that the transcription factors *SOX9* and *SOX6*, commonly thought essential for chondrogenesis, were not significantly downregulated on SDECM at P4; this is potentially a major contributor to the perpetuation of chondrogenicity found in cells cultured on synoviocyte matrix. Interestingly, *SOX5*, another putative essential cartilage transcription factor [22,23], is 3.4-fold downregulated at P4 on TCP (Padj 7.9×10^{-11}) vs. 1.7-fold downregulated at P4 on SDECM (Padj 6.4×10^{-4}), and *SOX5* was upregulated in both comparisons 3 and 4 (1.4-fold and 2.8-fold respectively).

Whilst immortalized hepatocytes have been shown to increase hepatocyte markers when cultured on soft substrates [24], this has not been shown for de-differentiated primary human chondrocytes. An increase in chondrogenic markers is often achieved in cell culture expanded cells [25,26]. However, the expression of these markers is commonly deficient when compared to non-expanded chondrocytes. Also, soft substrates do not promote, but inhibit expansion [27]. In the case of hepatocytes and b-cells, de-differentiation has been linked to epithelial–mesenchymal transition [28]; this would not be the case for chondrocytes, as stimulation of transforming growth factor beta (TGFβ) pathways, rather than inhibition, promotes chondrogenesis.

The increase in cell cycle control genes could be indicative of cell culture adaptation, as this was described by Barta et al. [29] who showed increased cyclin E and cyclin A in human embryonic stem cells cultured for over 240 passages; both cyclins were increased at passage 4 (3.2-fold and 9.1-fold in comparisons 1 and 2, respectively). This potentially futile growth, a conversion from reversible cell cycle arrest to irreversible senescence, could be indicative of geroconversion as described by Blagosklonny [30–32]. Interestingly, by blocking mTOR signaling using rapamycin, several studies have shown a positive impact on chondrogenesis and in osteoarthritis [33–35]. When we looked at cell senescence-associated genes several were differentially regulated at higher passage on both surfaces. Those genes, *NOX4, CDK1, FOXM1,* and *C2orf40*, represent interesting targets for modulation to further enhance the retention of chondrogenic capacity found on synoviocyte matrix. F-box protein 5 (*FBXO5*) is notable because of its presence in the cells passaged on TCP (comparison 1, Figure 3) and its absence on SDECM (comparison 2, Figure 4). FBXO5 is part of the Skp, Cullin, F-box containing complex that catalyzes the ubiquitination of proteins marking them for degradation. Increased FBXO5 could result in more cells being held at the G1/S and G2/M checkpoints but warrants more study. The increased number of genes promoting proliferation on SDECM vs. TCP (Figures 3 and 4), and perhaps more importantly, the ratio of genes promoting vs. inhibiting proliferation is thought to be partially responsible for the increased proliferation of chondrocytes on SDECM.

Comparisons 3 and 4 resulted in upregulation of development-related genes and down regulation of matrix-associated genes and those involved in motility. It is postulated that, because there is an extracellular matrix already present, the chondrocytes are less invested in laying down a new matrix. Chondrocytes in their native environment are relatively immobile though in vitro motility has been demonstrated by several studies [36]. The potential reduction in movement on SDECM could indicate that the chondrocytes are in a more "native" state and is another interesting avenue for investigation.

Genes that were stable on SDECM but significantly changed on TCP were identified, as this represents the pool of genes that could be targeted for modulation to recapitulate the effects of growth on SDECM (Table 5). Network and gene enrichment analyses of the genes in Table 4 showed no overall connection between them. This is potentially a consequence of the dearth of studies on the musculoskeletal transcriptome, particularly in cartilage, whereas many cancer studies have contributed to the gene ontology terms and their association. It should also be noted that chondrocyte proliferation, studied here, appears to be a somewhat separate process from chondrogenesis or cartilage development in terms of differentiation markers.

5. Conclusions

Chondrocyte growth on devitalized synoviocyte matrix dramatically changes the transcriptomic signature of the cells, predominantly in extracellular matrix-associated genes and those related to cell motility. De-differentiation due to passage of chondrocytes also dramatically alters the transcriptome, predominantly resulting in cell cycle gene expression changes.

Supplementary Materials: The supplementary materials are available online at http://www.mdpi.com/2073-4409/8/2/85/s1.

Author Contributions: Conceptualization, T.J.K. and J.E.D.; Data curation, T.J.K. and Z.G.; Formal analysis, T.J.K. and Z.G.; Funding acquisition, T.J.K., R.C., and J.E.D.; Investigation, T.J.K. and Y.L.; Methodology, T.J.K., Y.L., and R.C.; Project administration, J.E.D.; Resources, R.C. and J.E.D.; Supervision, R.C. and J.E.D.; Visualization, T.J.K.; Writing—original draft, T.J.K.; Writing—review and editing, T.J.K., R.C., and J.E.D.

Funding: This research was funded by Baylor College of Medicine (J.E.D.) and the Bone Disease Program of Texas (T.J.K.). The sequencing was performed at the functional genomics core which is partially supported by the shared instrument grant S10OD023469 to R.C.

Conflicts of Interest: The authors declare no conflict of interest. The funders had no role in the design of the study; in the collection, analyses, or interpretation of data; in the writing of the manuscript, or in the decision to publish the results.

References

1. Murphy, L.B.; Cisternas, M.G.; Pasta, D.J.; Helmick, C.G.; Yelin, E.H. Medical Expenditures and Earnings Losses Among US Adults With Arthritis in 2013. *Arthritis Care Res.* **2018**, *70*, 869–876. [CrossRef] [PubMed]
2. Giovannini, S.; Diaz-Romero, J.; Aigner, T.; Mainil-Varlet, P.; Nesic, D. Population doublings and percentage of S100-positive cells as predictors of in vitro chondrogenicity of expanded human articular chondrocytes. *J. Cell. Physiol.* **2010**, *222*, 411–420. [CrossRef] [PubMed]

3. Kean, T.J.; Dennis, J.E. Synoviocyte Derived-Extracellular Matrix Enhances Human Articular Chondrocyte Proliferation and Maintains Re-Differentiation Capacity at Both Low and Atmospheric Oxygen Tensions. *PLoS ONE* **2015**, *10*, e0129961.

4. Pei, M.; He, F. Extracellular matrix deposited by synovium-derived stem cells delays replicative senescent chondrocyte dedifferentiation and enhances redifferentiation. *J. Cell. Physiol.* **2012**, *227*, 2163–2174. [CrossRef] [PubMed]

5. Everaert, C.; Luypaert, M.; Maag, J.L.V.; Cheng, Q.X.; Dinger, M.E.; Hellemans, J.; Mestdagh, P. Benchmarking of RNA-sequencing analysis workflows using whole-transcriptome RT-qPCR expression data. *Sci Rep.* **2017**, *7*, 1559. [CrossRef] [PubMed]

6. Kean, T.J.; Koelewyn, S.; Ge, Z.; Li, Y.; Chen, R.; Dennis, J.E. Transcriptome-Wide Analysis of Human Chondrocyte Dedifferentiation on TC Plastic. In Proceedings of the Orthopedic Research Society Annual Meeting, Orlando, FL, USA, 5–8 March 2016.

7. Langmead, B.; Trapnell, C.; Pop, M.; Salzberg, S.L. Ultrafast and memory-efficient alignment of short DNA sequences to the human genome. *Genome Biol.* **2009**, *10*. [CrossRef] [PubMed]

8. Trapnell, C.; Pachter, L.; Salzberg, S.L. TopHat: Discovering splice junctions with RNA-Seq. *Bioinformatics* **2009**, *25*. [CrossRef] [PubMed]

9. Anders, S.; Pyl, P.T.; Huber, W. HTSeq—A Python framework to work with high-throughput sequencing data. *Bioinformatics* **2015**, *31*, 166–169. [CrossRef] [PubMed]

10. Trapnell, C.; Williams, B.A.; Pertea, G.; Mortazavi, A.; Kwan, G.; van Baren, M.J.; Salzberg, S.L.; Wold, B.J.; Pachter, L. Transcript assembly and quantification by RNA-Seq reveals unannotated transcripts and isoform switching during cell differentiation. *Nat. Biotechnol.* **2010**, *28*, 511–515. [CrossRef]

11. R Core Team. R: A Language and Environment for Statistical Computing. R Foundation for Statistical Computing: Vienna, Austria, 2018. Available online: https://www.R-project.org/ (accessed on 9 December 2018).

12. Love, M.I.; Huber, W.; Anders, S. Moderated estimation of fold change and dispersion for RNA-seq data with DESeq2. *Genome Biol.* **2014**, *15*, 550. [CrossRef]

13. Szklarczyk, D.; Morris, J.H.; Cook, H.; Kuhn, M.; Wyder, S.; Simonovic, M.; Santos, A.; Doncheva, N.T.; Roth, A.; Bork, P.; et al. The STRING database in 2017: quality-controlled protein-protein association networks, made broadly accessible. *Nucleic Acids Res.* **2017**, *45*. [CrossRef] [PubMed]

14. Boyle, E.I.; Weng, S.; Gollub, J.; Jin, H.; Botstein, D.; Cherry, J.M.; Sherlock, G. GO::TermFinder—Open source software for accessing Gene Ontology information and finding significantly enriched Gene Ontology terms associated with a list of genes. *Bioinformatics* **2004**, *20*, 3710–3715. [CrossRef] [PubMed]

15. Jax. Gene Ontology Browser. 2018. Available online: http://www.informatics.jax.org/vocab/gene_ontology/ (accessed on 8 November 2018).

16. Marlovits, S.; Hombauer, M.; Truppe, M.; Vecsei, V.; Schlegel, W. Changes in the ratio of type-I and type-II collagen expression during monolayer culture of human chondrocytes. *J. Bone Jt. Surg. J.* **2004**, *86*, 286–295. [CrossRef] [PubMed]

17. Melrose, J.; Smith, S.; Cake, M.; Read, R.; Whitelock, J. Comparative spatial and temporal localisation of perlecan, aggrecan and type I.; II and IV collagen in the ovine meniscus: An ageing study. *Histochem. Cell Biol.* **2005**, *124*, 225–235. [CrossRef] [PubMed]

18. Rosen, N.; Chalifa-Caspi, V.; Shmueli, O.; Adato, A.; Lapidot, M.; Stampnitzky, J.; Safran, M.; Lancet, D. GeneLoc: Exon-based integration of human genome maps. *Bioinformatics* **2003**, *19*, i222–i224. [CrossRef] [PubMed]

19. Peffers, M.; Liu, X.; Clegg, P. Transcriptomic signatures in cartilage ageing. *Arthritis Res. Ther.* **2013**, *15*, R98. [CrossRef] [PubMed]

20. Cuzzocrea, S.; Mazzon, E.; Dugo, L.; Patel, N.S.; Serraino, I.; Di Paola, R.; Genovese, T.; Britti, D.; De Maio, M.; Caputi, A.P.; et al. Reduction in the evolution of murine type II collagen-induced arthritis by treatment with rosiglitazone, a ligand of the peroxisome proliferator-activated receptor gamma. *Arthritis Rheum.* **2003**, *48*, 3544–3556. [CrossRef]

21. Xu, J.; Lv, S.; Hou, Y.; Xu, K.; Sun, D.; Zheng, Y.; Zhang, Z.; Li, X.; Li, Y.; Chi, G. miR-27b promotes type II collagen expression by targetting peroxisome proliferator-activated receptor-gamma2 during rat articular chondrocyte differentiation. *Biosci. Rep.* **2018**, *38*. [CrossRef]

22. Liu, C.F.; Lefebvre, V. The transcription factors SOX9 and SOX5/SOX6 cooperate genome-wide through super-enhancers to drive chondrogenesis. *Nucleic Acids Res.* **2015**, *43*, 8183–8203. [CrossRef]

23. De Crombrugghe, B.; Lefebvre, V.; Behringer, R.R.; Bi, W.; Murakami, S.; Huang, W. Transcriptional mechanisms of chondrocyte differentiation. *Matrix Biol.* **2000**, *19*, 389–394. [CrossRef]

24. Cozzolino, A.M.; Noce, V.; Battistelli, C.; Marchetti, A.; Grassi, G.; Cicchini, C.; Tripodi, M.; Amicone, L. Modulating the Substrate Stiffness to Manipulate Differentiation of Resident Liver Stem Cells and to Improve the Differentiation State of Hepatocytes. *Stem Cells Int.* **2016**, *2016*, 5481493. [CrossRef] [PubMed]

25. Recha-Sancho, L.; Moutos, F.T.; Abella, J.; Guilak, F.; Semino, C.E. Dedifferentiated Human Articular Chondrocytes Redifferentiate to a Cartilage-Like Tissue Phenotype in a Poly(epsilon-Caprolactone)/Self-Assembling Peptide Composite Scaffold. *Materials* **2016**, *9*, 472. [CrossRef] [PubMed]

26. Zhang, T.; Gong, T.; Xie, J.; Lin, S.; Liu, Y.; Zhou, T.; Lin, Y. Softening Substrates Promote Chondrocytes Phenotype via RhoA/ROCK Pathway. *ACS Appl. Mater. Interfaces* **2016**, *8*, 22884–22891. [CrossRef] [PubMed]

27. Callahan, L.A.; Ganios, A.M.; Childers, E.P.; Weiner, S.D.; Becker, M.L. Primary human chondrocyte extracellular matrix formation and phenotype maintenance using RGD-derivatized PEGDM hydrogels possessing a continuous Young's modulus gradient. *Acta Biomater.* **2013**, *9*, 6095–6104. [CrossRef] [PubMed]

28. Efrat, S. Mechanisms of adult human beta-cell in vitro dedifferentiation and redifferentiation. *Diabetes Obes. Metab.* **2016**, *18* (Suppl. 1), 97–101. [CrossRef] [PubMed]

29. Barta, T.; Dolezalova, D.; Holubcova, Z.; Hampl, A. Cell cycle regulation in human embryonic stem cells: Links to adaptation to cell culture. *Exp. Biol. Med.* **2013**, *238*, 271–275. [CrossRef]

30. Blagosklonny, M.V. Geroconversion: Irreversible step to cellular senescence. *Cell Cycle* **2014**, *13*, 3628–3635. [CrossRef]

31. Blagosklonny, M.V. Cell cycle arrest is not senescence. *Aging* **2011**, *3*, 94–101. [CrossRef]

32. Leontieva, O.V.; Blagosklonny, M.V. Gerosuppression in confluent cells. *Aging* **2014**, *6*, 1010–1018. [CrossRef]

33. De Luna-Preitschopf, A.; Zwickl, H.; Nehrer, S.; Hengstschlager, M.; Mikula, M. Rapamycin Maintains the Chondrocytic Phenotype and Interferes with Inflammatory Cytokine Induced Processes. *Int. J. Mol. Sci.* **2017**, *18*, 1494. [CrossRef]

34. Yan, B.; Zhang, Z.; Jin, D.; Cai, C.; Jia, C.; Liu, W.; Wang, T.; Li, S.; Zhang, H.; Huang, B.; et al. mTORC1 regulates PTHrP to coordinate chondrocyte growth, proliferation and differentiation. *Nat. Commun.* **2016**, *7*, 11151. [CrossRef] [PubMed]

35. Preitschopf, A.; Schorghofer, D.; Kinslechner, K.; Schutz, B.; Zwickl, H.; Rosner, M.; Joo, J.G.; Nehrer, S.; Hengstschlager, M.; Mikula, M. Rapamycin-Induced Hypoxia Inducible Factor 2A Is Essential for Chondrogenic Differentiation of Amniotic Fluid Stem Cells. *Stem Cells Transl. Med.* **2016**, *5*, 580–590. [CrossRef] [PubMed]

36. Morales, T.I. Chondrocyte moves: Clever strategies? *Osteoarthr. Cartil.* **2007**, *15*, 861–871. [CrossRef] [PubMed]

cells

MDPI

Article

Wound Healing Fluid Reflects the Inflammatory Nature and Aggressiveness of Breast Tumors

Roberto Agresti [1], Tiziana Triulzi [2], Marianna Sasso [2], Cristina Ghirelli [2], Piera Aiello [2], Ilona Rybinska [2], Manuela Campiglio [2], Lucia Sfondrini [3], Elda Tagliabue [2,*] and Francesca Bianchi [2]

[1]   Division of Surgical Oncology, Breast Unit, Fondazione IRCCS Istituto Nazionale dei Tumori di Milano, 20133 Milan, Italy; Roberto.agresti@istitutotumori.mi.it
[2]   Molecular Targeting Unit, Fondazione IRCCS Istituto Nazionale dei Tumori di Milano, 20133 Milan, Italy; Tiziana.triulzi@istitutotumori.mi.it (T.T.); Marianna.sasso@gmail.com (M.S.); ghirelli.cristina@istitutotumori.mi.it (C.G.); piera.chef@gmail.com (P.A.); ilona.rybinska@istitutotumori.mi.it (I.R.); manuela.campiglio@me.com (M.C.); francesca.bianchi@istitutotumori.mi.it (F.B.)
[3]   Dipartimento di Scienze Biomediche per la Salute, Università degli Studi di Milano, 20133 Milan, Italy; Lucia.sfondrini@unimi.it
*   Correspondence: elda.tagliabue@istitutotumori.mi.it; Tel.: +39-022-390-3013

Received: 14 December 2018; Accepted: 14 February 2019; Published: 19 February 2019

Abstract: Wound healing fluid that originates from breast surgery increases the aggressiveness of cancer cells that remain after the surgery. We determined the effects of the extent of surgery and tumor-driven remodeling of the surrounding microenvironment on the ability of wound-healing to promote breast cancer progression. In our analysis of a panel of 34 cytokines, chemokines, and growth factors in wound healing fluid, obtained from 27 breast carcinoma patients after surgery, the levels of several small molecules were associated with the extent of cellular damage that was induced by surgery. In addition, the composition of the resulting wound healing fluid was associated with molecular features of the removed tumor. Specifically, IP-10, IL-6, G-CSF, osteopontin, MIP-1a, MIP-1b, and MCP1-MCAF were higher in more aggressive tumors. Altogether, our findings indicate that the release of factors that are induced by removal of the primary tumor and subsequent wound healing is influenced by the extent of damage due to surgery and the reactive stroma that is derived from the continuously evolving network of interactions between neoplastic cells and the microenvironment, based on the molecular characteristics of breast carcinoma cells.

Keywords: wound; extracellular matrix; cytokines; breast cancer; surgery; IL-6; G-CSF; osteopontin

1. Introduction

Women with breast cancer undergo breast-conserving surgery or mastectomy as part of their treatment. The relationship between breast cancer surgery and the risk of relapse has been studied extensively. The risk of recurrence after primary breast cancer removal peaks at 10 months, implicating an event at the time of the surgery that accelerates or stimulates the metastatic process [1]. Several theories have been proposed to explain the mechanism that underlies early relapse on surgery. Metastatic processes that are triggered by breast tumor resection can not merely be ascribed to the release of tumor cells from the surgical bed [2]. The relevance of inflammation that is driven by wound healing is well established. Increasing evidence of a reduction in breast cancer relapses with the use of non-steroidal anti-inflammatory drugs (NSAIDs) has demonstrated the involvement of surgery-derived inflammation in metastatic processes [1,3,4].

One of the most obvious effects of wound healing is the stimulation of residual malignant cells or quiescent tumor stem cells by factors that are released in response to inflammation [5,6].

Moreover, angiogenic processes that are induced by the re-epithelization of wound tissue coincides with the generation of the vascular niche that supports tumor stem cell proliferation [7].

Following tissue injury via incision, the first step in the wound healing cascade is hemostasis, during which blood vessels constrict to limit blood flow, after which platelet aggregates and clotted blood impregnate the wound immediately to seal the lesion. During the inflammatory stage, an influx of immune cells ensues to control bleeding and prevent infection. Next, after the first 24 hours, angiogenesis and re-epithelization occur in the proliferation step to set down new cellular and extracellular matrix (ECM) components, in parallel with the release of cellular substances and mediators. Finally, the ECM is remodeled during the maturation phase [8].

Wound healing fluid (WHF) that results from surgical sites might provide a glimpse of the activity of cells that coordinate to release growth factors, cytokines, and chemokines that are fundamental in healing [9,10]. The composition of biological fluids in humans is influenced by their physiological characteristics, but several pathological conditions, such as metabolic disorders, can affect the levels of small molecules in these fluids [11].

During tumor formation and progression, cancer cells participate in dense crosstalk with all of the cell types that form the surrounding tissue and with the ECM that provides structural support [12]. The tumor stroma and ECM are considered essential for sustaining tumor growth because tumor cells foster changes to the surrounding niche through a complex network of signals, creating an environment that favors their proliferation and dissemination [13].

WHF can also force the escape of immune-controlled cancer cells from dormancy and promotes the transformation of already damaged cells [5,14–16]. Accordingly, we have established that growth factors that are released during wound healing following surgery account for the early relapse of the HER2-positive breast cancer subtype [17,18]. Moreover, we recently demonstrated that exposure of triple-negative (TN) breast cancer cells to WHF in vitro increases their expression of CDCP1, a molecule that is associated with a poor prognosis in several types of tumor, including TNBC [19,20].

At the time of tumor surgery, breast cancer cells have already modified the adjacent tissues extensively to create a microenvironment that favors tumor growth and dissemination. In this study, we examined whether the composition of WHF depends on and reflects remodeling of the ECM and the crosstalk with stromal cells provoked by tumor growth and whether such an interaction culminates in the release of molecules that are relevant to tumor aggressiveness into the wound fluid.

2. Materials and Methods

2.1. Collection of Drainage

WHF from breast cancer patients who were treated surgically at Fondazione IRCCS Istituto Nazionale dei Tumori di Milano from 2010–2012 was collected from the first clearance of surgical closed-type breast drains (no abdomen or armpit) under suction during the first 24 h postsurgery. The WHF was centrifuged immediately at 3000 g, and the supernatant was aliquoted and stored at −80 °C until analysis. The protein concentration in the WHF, as determined by Biureto method, ranged from 3.7 to 5.1 g/dL. A human serum (HS) sample comprised a pool of HS from four healthy blood donors. The pathobiological features of the breast cancer patients, from whom WHF was collected, were registered, and a database was created.

The collection did not include WHF from patients with concomitant diseases other than breast cancer that are known to affect the release of small molecules into biological fluids (e.g., hyperglycemia) or patients who received chemotherapy or hormone therapy before surgery.

2.2. Analysis of Small Molecules

The composition of the WHF was analyzed on a Bio-Plex™ 2200 system (Bio-Rad Laboratories, Hercules, CA, USA), testing small-molecule signaling mediators, including cytokines, chemokines, and growth factors, that are involved in the initiation and progression of cancer.

Specifically, the Bio-Plex Pro Human Cytokine 27-Plex Group I assay was used, including assays for PDGF-BB, IL-1b, IP-10, IL-1ra, IL-2, IL-4, IL-5, IL-6, IL-7, IL-8, IL-9, IL-10, IL-12p70, IL-13, IL-15, IL-17, eotaxin, FGF basic, G-CSF, GM-CSF, IFN-g, MCP-1MCAF, MIP-1a, MIP-1b, RANTES, TNF-α, and VEGF. In the indicated experiments, this panel of analytes was integrated with assays for PDGF-AB/BB, sTIE-2, HGF, osteopontin, TGF-beta1, TGF-beta2, and TGF-beta3. The analysis was performed according to the manufacturer's instructions and as described [21–23].

Briefly, samples were incubated in a 96-well plate with polystyrene beads that were coated with small molecule-specific antibodies and then exposed to detection antibodies prior to incubation with streptavidin-PE. Data are presented as concentration (pg/mL). The concentration of each analyte, bound to its specific bead, is proportional to the median fluorescence intensity (MFI) of the reporter signal. All samples were assayed in duplicate.

2.3. Cell Lines, Cultures, and Treatments

The human breast cancer cell lines were maintained at 37 °C in a humidified atmosphere of 5% CO_2 in air as follows: MDA-MB-231, BT-549, SK-BR-3, HCC1937, T-47D and MCF-7 in RPMI 1640 (Life Technologies, Grand Island, NY, USA) and MDA-MB-468 in Dulbecco's modified Eagle's medium (DMEM) (Lonza, Basel, Switzerland). For the stimulation with WHF, cells were starved in serum-free medium for 24 h and then treated with WHF that was diluted in culture medium and passed through a 0.22-µm syringe PVDF filter (Millipore, Burlington, MA, USA) [19,20].

2.4. In Vitro Growth and Migration Assays

Relative 2D cell growth over 4 days was measured by sulforhodamine B (SRB) assay as described [24]. Optical density (OD) was determined on an ELISA microplate reader (Bio-Rad Laboratories). Proliferation was spectrophotometrically assessed by SRB assay after 4 days of treatment (96 h). For each cell line, the optical density (OD) of each experimental condition was normalized on the OD of the same cell line measured immediately before starting the treatments. 0% represents the growth index of cells cultured for 4 days in absence of WHF or fetal bovine serum (FBS) or HS.

To examine the ability of WHF to enhance cell migration, cells were starved in serum-free medium for 48 h, treated with or without a 5% WHF pool for 2 h, and then seeded at the top of a 8-µm Boyden chamber (Sigma-Aldrich) in serum-free medium, with medium that contained 10% FBS placed in the well below as the chemoattractant. After 12 h for MDA-MB-231 cells or 6 h for BT-549 cells, the cells in the upper chamber were removed with cotton swabs, and those that traversed the 8-µm semipermeable membrane were fixed in 100% ethanol, stained with SRB, and imaged under an ECLIPSE TE2000-S inverted microscope (Nikon Instruments, Amstelveen, Netherlands). The results were expressed as the area that was occupied by cells in the bottom of the Transwell, evaluated by digital image analysis using the appropriate software (Image Pro-Plus 7.0 application, Media Cybernetics, Rockville, MD, USA). The mean of 3 independent experiments (± SEM) was calculated.

2.5. Statistical Analysis

Relationships between categorical variables in Table S1 that were related to primary breast cancers from which the WHF was derived were analyzed by Fisher's exact test. To compare the levels of small molecules between 2 independent groups, mean values were compared by nonparametric Mann–Whitney test. The effect of WHF on cell migration was analyzed by student's *t*-test. All analyses were performed using GraphPad Prism, version 5.0 for Windows (GraphPad Software, San Diego, CA, USA). Differences were considered to be significant at $p \leq 0.05$.

Overall survival (OS) was defined as the time elapsed from the date of surgery to the date of death. Distant metastasis free survival (DMFS) was defined as the time elapsed from the date of surgery to the date of the first event. Univariate survival analysis was carried out using Cox proportional hazards regression models, and the effects of explanatory variables on event hazard were quantified by hazard ratios (HR). Small-molecules amount was analyzed after logarithmic transformation (log2). Analysis has been performed by using SAS software (SAS Institute Inc, Cary, NC, USA).

3. Results

3.1. Analysis of the Composition of WHF in Breast Cancer Patients

To determine the feasibility of analyzing the small molecules in WHF at the site of breast cancer surgery, a pool of 5 breast cancer WHF samples, a pool of 4 human serum (HS) samples, and fetal bovine serum (FBS) were evaluated by a Bio-Plex Pro Human Cytokine 27-Plex Group I assay. The WHF pool was enriched in cytokines, chemokines, and growth factors compared with FBS and HS. Overall, the peak concentration of all small molecules in the HS was approximately 100 pg/mL, as expected, based on the literature, whereas that of over half of the analyzed molecules in WHF (17/25) was higher than 100 pg/mL, six of which—MIP-1b, PDGF-bb, IL-1ra, IP-10, IL-6, and IL-8—exceeded 1000 pg/mL. HS and FBS did not differ significantly with regard to any molecule (Figure 1).

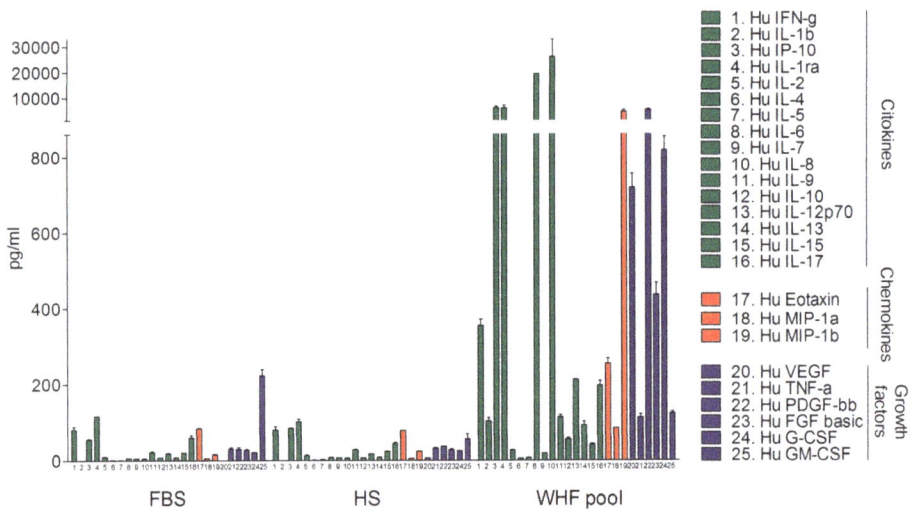

Figure 1. Bio-plex analysis of fetal bovine serum, human serum, and wound healing fluid composition. Concentration (pg/mL) of 25 molecules is shown. Level of cytokines, chemokines, and growth factors, was assessed by Bio-plex assay in fetal bovine serum (FBS), human serum (HS), and a pool of five wound healing fluids from breast cancer surgery (WHF Pool) (mean ± SD).

To dissect the composition of fluids that are released immediately after breast cancer surgery, the levels of 34 small molecules were analyzed in a collection of 27 WHF samples (Table S1) by Bio-Plex Pro Human Cytokine 27-Plex Group I assay, integrated with a panel of nine additional small molecules (see Materials and Methods). Several small molecules were differentially expressed in WHF from breast cancer patients who underwent mastectomy or quadrantectomy. Specifically, the levels of IL-1b, IL-1ra, IL-6, osteopontin, IFN-γ, G-CSF, MIP-1b, and IP-10 were significantly higher in mastectomized patients than in those who underwent quadrantectomy (Figure 2A–D). Frequency analysis showed that the type of surgery was not significantly associated with any pathological variable, despite a near-significant trend for a negative association between mastectomy and the luminal tumor intrinsic molecular subtype and for a positive association between mastectomy and invasive tumor (Table S2).

Considering that all patients with in situ breast cancer in our case study underwent breast-conserving surgery, the composition of WHF regarding tumor histology was analyzed according to quadrantectomy. In patients who underwent quadrantectomy, IL-6, G-CSF, and MCP1-MCAF were significantly enriched in invasive versus in situ breast cancer surgery drainages (Figure 2E), suggesting that the crosstalk between invasive tumor cells and the surrounding microenvironment influences the production and release of these small molecules.

With regard to intrinsic subtype tumors, osteopontin (OPN) was significantly higher in fluids from TN (defined as estrogen receptor (ER) and/or progesterone receptor (PgR) <10% and HER2 0, 1+, 2+ CISH-negative) compared with luminal breast cancer (defined as ER and PgR expression >10% and HER2 0, 1+, 2+ CISH-negative) (Figure 2F). Notably, albeit insignificantly, the level of OPN was higher in fluids from TN breast cancer patients, independent of the type of breast tumor surgery (Figure S1).

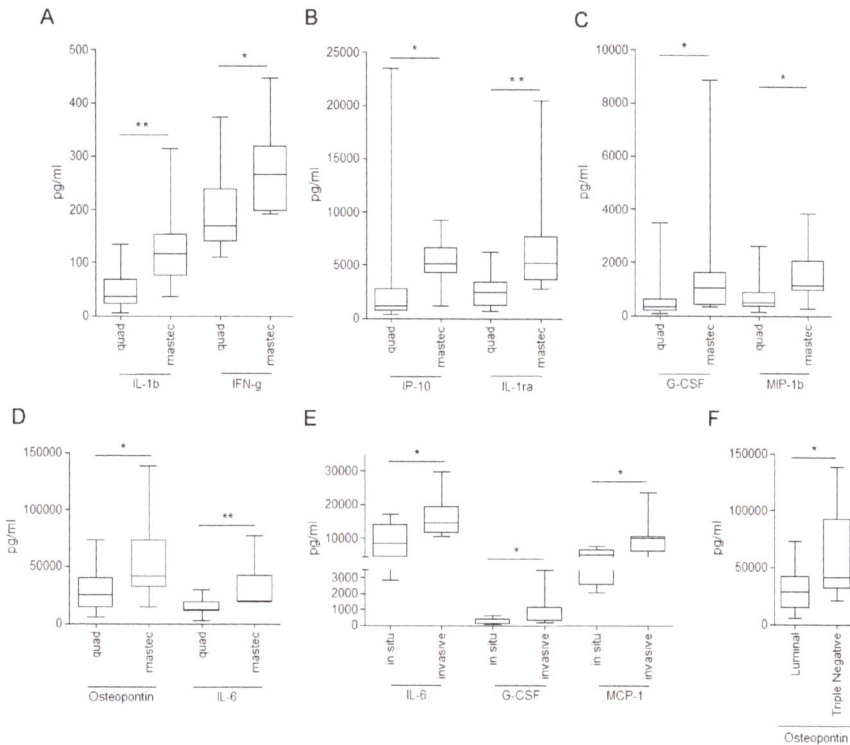

Figure 2. Differences in small-molecule composition of wound healing fluid from breast carcinoma surgery according to type of surgery, tumor histology, and tumor molecular subtype. Concentration (pg/mL) of 34 molecules, including cytokines, chemokines, and growth factors (listed in Material and Methods), was assessed by Bio-plex assay in 27 wound healing fluids from breast cancer patients. Levels of small molecules were differentially enriched by surgery (**A–D**), tumor histology (**E**), and tumor molecular subtype (**F**) (* p-value ≤ 0.05; ** p-value ≤ 0.01; Mann–Whitney test).

IL-6 and OPN were significantly enriched in WHF from surgeries for breast cancers ≥ 2 cm (Figure 3A). G-CSF, MIP-1a and MIP-1b content was significantly higher and TIE-2 was significantly lower in grade III versus grade II breast tumors (Figure 3B). Level of IP-10 was significantly upregulated, whereas TGF-β1 and TGF-β2 were significantly lower, in WHF from N-positive compared with N-negative tumors (Figure 3C).

In our cohort, none of the clinical characteristics of breast cancer patients, from which wound healing fluids have been derived, were associated with prognosis. Nevertheless, a trend towards significance was found for association of IP-10 and OPN amount with both DMFS (IP-10, HR: 1.97,

95% Confident Interval (CI): 0.86–4.55, p = 0.1104; OPN, HR: 2.67, 95% CI: 0.98–7.25, p = 0.0537) and OS (IP-10, HR: 2.15, 95% CI: 0.86–5.34, p = 0.1006; OPN, HR: 2.95, 95% CI: 1.06–8.22, p = 0.0381).

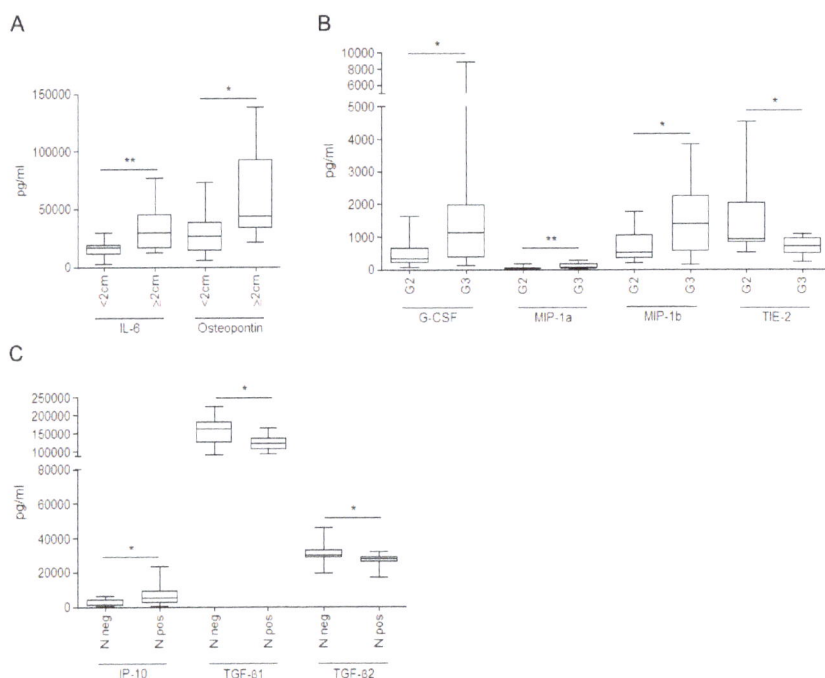

Figure 3. Differences in small-molecule composition of wound healing fluid from breast carcinoma surgery according to tumor size, tumor grade, and lymph node positivity. Concentration (pg/mL) of 34 molecules, including cytokines, chemokines, and growth factors (listed in Material and Methods), was assessed by Bio-plex assay in 27 wound healing fluids from breast cancer patients. Levels of small molecules were differentially enriched by breast cancer size (**A**), tumor grade (grade II (G2); grade III (G3) (**B**), and lymph node (**C**) (* p-value $\leq$ 0.05; ** p-value $\leq$ 0.01; Mann–Whitney test).

3.2. Relationship between WHF and Cancer Cell Aggressiveness

To determine the effects of WHF on cell growth, a panel of seven breast cancer cell lines, representing various breast cancer subtypes—MDA-MB-231 and BT-549 (basal-B TN subtype), MDA-MB-468 and HCC1937 (basal-A TN subtype), SK-BR-3 (HER2 subtype), MCF-7, and T-47D (luminal subtype) [25]—were starved for 24 h, treated with 1% FBS or 1% of pools of human serum from breast cancer patients or of five different WHFs for 96 h, or left untreated. The WHF stimulated robust cell growth in all cell lines (Figure 4A). Similar results were obtained using a wide panel of 45 WHFs tested on four breast cancer cells—MDA-MB-468, MDA-MB-231, BT-549 and MCF-7—strengthening the effect of WHF on cell proliferation (Figure S2).

Because several chemokines were enriched in WHF, its ability to promote migration was also examined in the basal-B TN breast cancer cell lines MDA-MB-231 and BT-549, which have been reported to migrate in vitro. Briefly, cells were starved for 48 h, treated with pools of WHFs for 2 h, or left untreated and seeded in a chamber assay that contained FBS in the lower chamber as the chemoattractant (Figure 4B). In both cell lines, migration was improved on pretreatment with WHF.

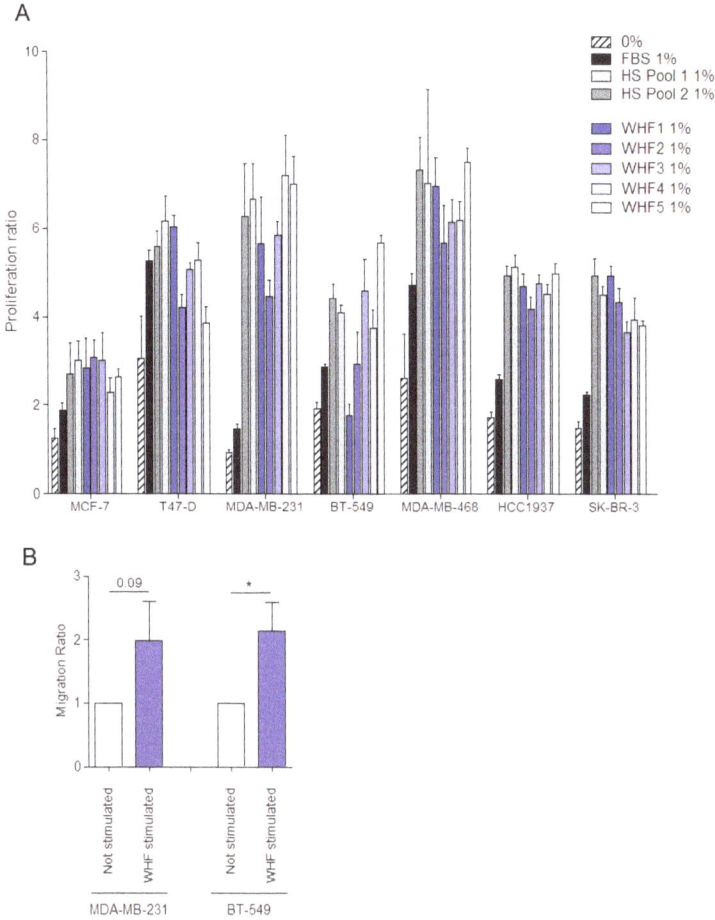

Figure 4. Effect of WHF on breast cancer cell proliferation and migration. (**A**) MCF-7, T-47D, MDA-MB-231, BT-549, MDA-MB-468, HCC1937, and SK-BR-3 breast cancer cell lines were starved for 24 h (0% FBS) and then treated for 96 h with 1% FBS or 1% of pools of human serum from breast cancer patients or 1% of five WHFs. Relative 2D-cell growth was measured by sulforhodamine B (SRB) assay. The ability of WHF to induce proliferation was indicated as the optical density (OD) of each cell line after four days of treatment (96 h) normalized on the OD of the same cell line measured immediately before starting the treatments. 0% represents the growth index of cells cultured for four days in absence of WHF or FBS or serum from breast cancer patients (HS Pool) (mean ± SD). (**B**) MDA-MB-231 and BT-549 cells were starved for 48 h (0% FBS) and then treated for 2 h with a pool of five WHFs and plated into Boyden chambers for migration assay toward FBS 10%. The area occupied by migrated cells in the Transwell was evaluated by digital image analysis (Image ProPlus 7.0 application, Media Cybernetics). Results are expressed as area occupied by cells on the bottom of the Transwell by digital image analysis (mean ± SEM; * *p*-value < 0.05; paired student's *t*-test).

3.3. Relevance of Breast Cancer Subtype in WHF-Driven Stimulation

To determine whether the WHF from breast tumor surgery for various molecular subtypes of tumors preferentially affects the proliferation of cancer cells with the same molecular characteristics, the TN luminal breast cancer cell lines MDA-MB-468 and MCF-7 were stimulated as described with WHF from TN or luminal breast cancer surgery, and proliferation was evaluated by SRB assay (Figure 5).

In addition to the difference in intrinsic proliferation rate between the cell lines—the proliferation rate of MDA-MB-468 cells was higher than that of T-47D cells, as expected—the proliferation of MDA-MB-468 and BT-549 cells increased primarily by WHF from the surgical removal of TN versus luminal breast tumors. Similarly, the proliferation of MCF-7 and T-47D that was induced by luminal-derived WHF was greater than that by TN breast tumors. These data support our hypothesis that tumors per se can modify their environment to favor their progression.

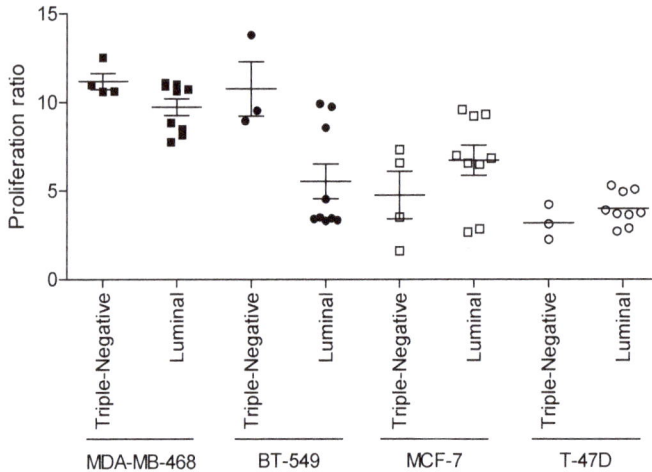

Figure 5. Effect of WHF originating from triple-negative or luminal breast cancer surgery on proliferation of triple-negative- or luminal-subtype breast cancer cells. The TN breast cancer cell lines MDA-MB-468 and BT-549 and the luminal breast cancer cell lines MCF-7 and T-47D were starved for 24 h (0% FBS) and then treated for 96 h with 1% WHFs from patients bearing TN or luminal breast cancer. WHF-induced cell growth was analyzed according to resected breast cancer subtype. The ability of WHF to induce proliferation was indicated as the ratio of the OD of each cell line (WHF-treated at 96 h/OD at 0%, before treatment).

4. Discussion

We have demonstrated that surgery-induced mobilization of factors is related to the biological features of the breast tumor that is removed. The remodeling of the microenvironment by tumor cells is reflected in the small-molecule composition of WHF. Several protumorigenic small molecules were upregulated in WHF from breast cancers with aggressive features versus less aggressive tumors; such molecules can ultimately cooperate to establish conditions that favor relapse. As a support, a higher amount of molecules associated with tumor aggressive features was found in patients with worse event free survival and overall survival.

Recently, the effects of wound healing were recently found to depend primarily on the subsequent inflammatory response, independent of the nature of the resected tumor [16]. By testing surgical wounds regardless of the presence of tumor, wound healing has been shown to activate distant disseminated tumor cells. Clearly, the transient acute inflammation that is provoked by surgery is a systemic event, and factors that are released at the surgical site, such as cytokines and chemokines, can reach, through the circulatory system, and stimulate cancer cells that have already disseminated at the time of the primary tumor resection.

Nevertheless, we highlight the function of the crosstalk between the tumor and its microenvironment, as we previously observed in other tumor types [26,27]. Several cell types, including fibroblasts, endothelial cells, and immune cells, interact with tumor cells, and their activity in shaping the tumor microenvironment is reflected in the composition of the wound healing fluid that is collected during the first 24 h after breast cancer surgery. In our cohort, differences in WHF composition were observed with regard to primary breast tumor histology, intrinsic subtype, size, grade, and lymph node status. Particularly, small molecules with reported functions in tumor inflammation and progression were enriched in WHF from the surgery of tumors with more aggressive features.

In this context, IL-6, G-CSF, and MCP-1 were enriched in WHF from patients with invasive versus in situ breast cancer, independent of the extent of the surgery. Consistent with its function in cancer progression, IL-6 was also upregulated in WHF from larger late-stage tumors. IL-6 regulates the tumor microenvironment, the generation of breast cancer stem cells, and metastasis through the downregulation of E-cadherin [28]. Increased serum IL-6 was correlated with survival [29].

Notably, G-CSF was enriched in WHF from invasive and high-grade breast tumors. G-CSF is one of the chief growth factors that control the maturation of neutrophils. Elevated neutrophil levels have been associated with detrimental outcomes in breast cancer, such that neutrophil count has been proposed as a prognostic and predictive biomarker [30]. Neoplastic cells draw in neutrophils that are recruited to a wound, increasing their proliferation [31].

IP-10/CXCL10, enriched in N-positive tumors, has several functions, such as chemoattraction for immune cells and the promotion of angiogenesis [32]. Notably, the levels of TGF-β1 and TGF-β2, reportedly acting as potent growth inhibitors [33], were reduced in N-positive tumors. With regard to the composition of WHF from the surgery bed following the removal of various molecular subtypes of tumor, the matricellular protein osteopontin (OPN, Spp-1) was the only small molecule that differed in concentration between WHFs from luminal and TN breast cancer. Nevertheless, the small size of WHF cohort, a trend towards significance was found for association of IP-10 and OPN amount with both event free survival and overall survival, supporting the hypothesis of the role of these molecules in tumor aggressiveness. OPN acts as an immune modulator, promoting neutrophil and mast cell recruitment to inflammatory sites. In addition, OPN mediates cell activation and cytokine production and enhances cell survival by regulating apoptosis [34,35]. OPN is overexpressed in breast cancer, particularly the TN subtype [36,37]. Recently, it was discovered that OPN is the basis for one of the mechanisms of breast cancer cell metastasis. OPN that is produced by tumor cells supports their survival in the bloodstream, whereas tumor- and host-derived OPN, particularly from myeloid cells, render the metastatic site more immunosuppressive [38]. The enrichment of OPN in TN tumor fluids could reflect its roles in tumor survival and spreading promotion and explain the high recurrence rate of this breast cancer subtype.

WHF stimulates in vitro proliferation of breast cancer cells of all intrinsic subtypes, consistent with what was reported by Wang and colleagues [39]. Although the increase in proliferation by WHFs was similar between cancer cell lines of various subtypes, the effect of proliferation was greater when the cells were treated with WHF from the surgery of tumors of the same intrinsic subtype. This effect was observed in more aggressive TN cell-WHF pairs and in less proliferating luminal tumors. These data suggest that breast tumors can purposely condition the extracellular microenvironment and, consequently, the factors in WHF.

We compared the levels of small molecules found in our WHF with data reported in literature. D.E. Lyon and colleagues [40], by using the same Multiplex bead array assay we used, analyzed levels of 17 cytokines in serum of women with breast cancer and of women with a suspicious breast mass who were found subsequently to have a negative breast biopsy. They observed a significantly higher cytokine concentration in women with cancer compared to women without cancer for the majority of the analyzed cytokines. Interestingly, comparing the cytokine levels of our WHF with those found in serum by Lyon et al., five cytokines (IL-6; IL-8; G-CSF; IFN-γ; MCP-1) were highly enriched in WHF compared with their serum counterpart (data not shown). Here, we found that these five cytokines were found differently represented in WHFs according to type of surgery, tumor histology, and tumor molecular subtype. Altogether, we can speculate that specific WHF cytokines that are related to tumor aggressiveness are also enriched in serum of BC patients. Consistently, we observed that pools of breast cancer sera increase proliferation of breast cancer cells similarly to what was observed upon WHF treatment. Factors released in the tumor microenvironment, as a consequence of a tumor's own growth and of interaction with stroma and with surrounding cells (immune cells, fibroblasts, etc.), enter the bloodstream and could contribute to promoting tumor growth and dissemination. Moreover, we can assume that the cytokines we found more represented in WHF compared to serum levels reported in the literature presumably act on tumor aggressiveness features other than proliferation, such as immune tumor control.

Finally, our data show that the composition of fluid that is released immediately after the surgery of breast cancer patients mirrors the extent of surgery, and, in particular, all small mediators that we have described were enriched in WHF from patients who underwent mastectomy. Surgery is a key cancer therapy and remains the most effective treatment for breast cancers. The two main types of surgery to remove breast cancer are quadrantectomy and mastectomy. A minimal surgical approach, where applicable, will benefit patients by reducing stimulation of inflammation. Several inflammatory mediators were enriched in the drainage of patients who underwent mastectomy, supporting that a highly destructive surgery increases inflammation [41,42]. Moreover, more reports are indicating that tissue damage due to cancer surgery provides a favorable niche for tumor recurrence [43], facilitating the growth of pre-existing micrometastases [5,18,44,45], enhancing the cancer stem cell population [6], creating a reactive oxygen species (ROS)-rich environment [46], and affecting patient outcomes.

5. Conclusions

Our data, in addition to confirming previous evidence that inflammation that stems from surgical wound healing affects breast cancer cell aggressiveness, highlight the relevance of tumor-induced modifications to the surgical bed. Breast cancers with aggressive features can specifically modify the tumor environment to ultimately favor their growth. Because surgery remains the preferred option for treating cancer, it is essential to improving our understanding of the inflammatory response that occurs as a consequence of local wounding and of the exposure of cancer cells to wounds.

Supplementary Materials: The following are available online at http://www.mdpi.com/2073-4409/8/2/181/s1, Table S1: Supplementary Table 1. Clinical characteristics of breast cancer patients from which wound healing fluids were derived. Table S2: Supplementary Table 2. Clinical characteristics of breast cancer patients from which wound healing fluids were derived according to extent of surgery. Figure S1: Supplementary Figure 1. Osteopontin levels in wound healing fluid from breast carcinoma patients according to breast tumor molecular subtype and type of surgery. Figure S2: Supplementary Figure 2. Effect of WHF on breast cancer cell proliferation.

Author Contributions: Conceptualization: F.B., M.C., E.T., and R.A.; Methodology: M.S., F.B., C.G., P.A., and I.R.; Formal Analysis: F.B., E.T., and T.T.; Resources: R.A.; Writing—Original Draft Preparation: F.B. and E.T.; Writing—Review and Editing: L.S., T.T., and R.A.; Funding Acquisition: M.C. and E.T.

Funding: This work was supported by the Associazione Italiana per la Ricerca sul Cancro (IG 11484 to M.C.; IG 15359 and IG 20264 to E.T.) and by Fondazione Umberto Veronesi (Fellowship 2017 and 2018 to F.B.). The funding sources had no role in the study design; the collection, analysis, and interpretation of the data; the writing of the manuscript; or the decision to submit the manuscript for publication.

Acknowledgments: The authors thank Laura Mameli for secretarial assistance.

Conflicts of Interest: The authors declare no conflicts of interest.

References

1. Retsky, M.; Demicheli, R.; Hrushesky, W.J.; Forget, P.; De, K.M.; Gukas, I.; Rogers, R.A.; Baum, M.; Sukhatme, V.; Vaidya, J.S. Reduction of breast cancer relapses with perioperative non-steroidal anti-inflammatory drugs: New findings and a review. *Curr. Med. Chem.* **2013**, *20*, 4163–4176. [CrossRef] [PubMed]
2. Eschwege, P.; Dumas, F.; Blanchet, P.; Le, V.M.; Benoit, G.; Jardin, A.; Lacour, B.; Loric, S. Haematogenous dissemination of prostatic epithelial cells during radical prostatectomy. *Lancet* **1995**, *346*, 1528–1530. [CrossRef]
3. Forget, P.; Vandenhende, J.; Berliere, M.; Machiels, J.P.; Nussbaum, B.; Legrand, C.; De, K.M. Do intraoperative analgesics influence breast cancer recurrence after mastectomy? A retrospective analysis. *Anesth. Analg.* **2010**, *110*, 1630–1635. [CrossRef] [PubMed]

4. Desmedt, C.; Demicheli, R.; Fornili, M.; Bachir, I.; Duca, M.; Viglietti, G.; Berliere, M.; Piccart, M.; Sotiriou, C.; Sosnowski, M.; et al. Potential Benefit of Intra-operative Administration of Ketorolac on Breast Cancer Recurrence According to the Patient's Body Mass Index. *J. Natl. Cancer Inst.* **2018**, *110*, 1115–1122. [CrossRef] [PubMed]

5. Ceelen, W.; Pattyn, P.; Mareel, M. Surgery, wound healing, and metastasis: Recent insights and clinical implications. *Crit. Rev. Oncol. Hematol.* **2014**, *89*, 16–26. [CrossRef] [PubMed]

6. Arnold, K.M.; Opdenaker, L.M.; Flynn, D.; Sims-Mourtada, J. Wound healing and cancer stem cells: inflammation as a driver of treatment resistance in breast cancer. *Cancer Growth Metastasis* **2015**, *8*, 1–13. [CrossRef] [PubMed]

7. Plaks, V.; Kong, N.; Werb, Z. The cancer stem cell niche: How essential is the niche in regulating stemness of tumor cells? *Cell Stem Cell* **2015**, *16*, 225–238. [CrossRef] [PubMed]

8. Sinno, H.; Prakash, S. Complements and the wound healing cascade: An updated review. *Plast. Surg. Int.* **2013**, *2013*, 146764. [CrossRef] [PubMed]

9. Shah, J.M.; Omar, E.; Pai, D.R.; Sood, S. Cellular events and biomarkers of wound healing. *Indian J. Plast. Surg.* **2012**, *45*, 220–228. [PubMed]

10. Werner, S.; Grose, R. Regulation of wound healing by growth factors and cytokines. *Physiol. Rev.* **2003**, *83*, 835–870. [CrossRef] [PubMed]

11. Anderson, K.; Hamm, R.L. Factors that impair wound healing. *J. Am. Coll. Clin. Wound Spec.* **2014**, *4*, 84–91. [CrossRef] [PubMed]

12. Lu, P.; Weaver, V.M.; Werb, Z. The extracellular matrix: A dynamic niche in cancer progression. *J. Cell Biol.* **2012**, *196*, 395–406. [CrossRef] [PubMed]

13. Carlini, M.J.; De Lorenzo, M.S.; Puricelli, L. Cross-talk between tumor cells and the microenvironment at the metastatic niche. *Curr. Pharm. Biotechnol.* **2011**, *12*, 1900–1908. [CrossRef] [PubMed]

14. Demicheli, R.; Abbattista, A.; Miceli, R.; Valagussa, P.; Bonadonna, G. Time distribution of the recurrence risk for breast cancer patients undergoing mastectomy: Further support about the concept of tumor dormancy. *Breast Cancer Res. Treat.* **1996**, *41*, 177–185. [CrossRef] [PubMed]

15. Demicheli, R.; Retsky, M.W.; Hrushesky, W.J.; Baum, M. Tumor dormancy and surgery-driven interruption of dormancy in breast cancer: Learning from failures. *Nat. Clin. Pr. Oncol.* **2007**, *4*, 699–710. [CrossRef] [PubMed]

16. Krall, J.A.; Reinhardt, F.; Mercury, O.A.; Pattabiraman, D.R.; Brooks, M.W.; Dougan, M.; Lambert, A.W.; Bierie, B.; Ploegh, H.L.; Dougan, S.K.; et al. The systemic response to surgery triggers the outgrowth of distant immune-controlled tumors in mouse models of dormancy. *Sci. Transl. Med.* **2018**, *10*, eaan3464. [CrossRef] [PubMed]

17. Tagliabue, E.; Agresti, R.; Carcangiu, M.L.; Ghirelli, C.; Morelli, D.; Campiglio, M.; Martel, M.; Giovanazzi, R.; Greco, M.; Balsari, A.; et al. Role of HER2 in wound-induced breast carcinoma proliferation. *Lancet* **2003**, *362*, 527–533. [CrossRef]

18. Tagliabue, E.; Agresti, R.; Casalini, P.; Mariani, L.; Carcangiu, M.L.; Balsari, A.; Veronesi, U.; Ménard, S. Linking survival of HER2-positive breast carcinoma patients with surgical invasiveness. *Eur. J. Cancer* **2006**, *42*, 1057–1061. [CrossRef] [PubMed]

19. Turdo, F.; Bianchi, F.; Gasparini, P.; Sandri, M.; Sasso, M.; De Cecco, L.; Forte, L.; Casalini, P.; Aiello, P.; Sfondrini, L.; et al. CDCP1 is a novel marker of the most aggressiveness human triple-negative breast cancer. *Oncotarget* **2016**, *7*, 69649–69665. [CrossRef] [PubMed]

20. Forte, L.; Turdo, F.; Ghirelli, C.; Aiello, P.; Casalini, P.; Iorio, M.V.; D'Ippolito, E.; Gasparini, P.; Agresti, R.; Belmonte, B.; et al. The PDGFRbeta/ERK1/2 pathway regulates CDCP1 expression in triple-negative breast cancer. *BMC Cancer* **2018**, *18*, 586. [CrossRef] [PubMed]

21. De Cesare, M.; Sfondrini, L.; Campiglio, M.; Sommariva, M.; Bianchi, F.; Perego, P.; van Rooijen, N.; Supino, R.; Rumio, C.; Zunino, F.; et al. Ascites regression and survival increase in mice bearing advanced-stage human ovarian carcinomas and repeatedly treated intraperitoneally with CpG-ODN. *J. Immunother.* **2010**, *33*, 8–15. [CrossRef] [PubMed]

22. Martinetti, A.; Miceli, R.; Sottotetti, E.; Di Bartolomeo, M.; de, B.F.; Gevorgyan, A.; Dotti, K.F.; Bajetta, E.; Campiglio, M.; Bianchi, F.; et al. Circulating biomarkers in advanced colorectal cancer patients randomly assigned to three bevacizumab-based regimens. *Cancers (Basel)* **2014**, *6*, 1753–1768. [CrossRef] [PubMed]

23. Plantamura, I.; Casalini, P.; Dugnani, E.; Sasso, M.; D'Ippolito, E.; Tortoreto, M.; Cacciatore, M.; Guarnotta, C.; Ghirelli, C.; Barajon, I.; et al. PDGFRβ and FGFR2 mediate endothelial cell differentiation capability of triple negative breast carcinoma cells. *Mol. Oncol.* **2014**, *8*, 968–981. [CrossRef] [PubMed]

24. Bianchi, F.; Sasso, M.; Turdo, F.; Beretta, G.L.; Casalini, P.; Ghirelli, C.; Sfondrini, L.; Menard, S.; Tagliabue, E.; Campiglio, M. Fhit nuclear import following EGF stimulation sustains proliferation of Breast Cancer cells. *J. Cell Physiol.* **2015**, *230*, 2661–2670. [CrossRef] [PubMed]

25. Neve, R.M.; Chin, K.; Fridlyand, J.; Yeh, J.; Baehner, F.L.; Fevr, T.; Clark, L.; Bayani, N.; Coppe, J.P.; Tong, F.; et al. A collection of breast cancer cell lines for the study of functionally distinct cancer subtypes. *Cancer Cell* **2006**, *10*, 515–527. [CrossRef] [PubMed]

26. De Cesare, A.; Sfondrini, L.; Pennati, M.; De Marco, C.; Motta, V.; Tagliabue, E.; Deraco, M.; Balsari, A.; Zaffaroni, N. CpG-oligodeoxynucleotides exert remarkable antitumor activity against diffuse malignant peritoneal mesothelioma orthotopic xenografts. *J. Transl. Med.* **2016**, *14*, 25. [CrossRef] [PubMed]

27. Le Noci, V.; Sommariva, M.; Tortoreto, M.; Zaffaroni, N.; Campiglio, M.; Tagliabue, E.; Balsari, A.; Sfondrini, L. Reprogramming the lung microenvironment by inhaled immunotherapt forsters immune destruction of tumor. *OncoImmunology* **2016**, *5*, e1234571. [CrossRef] [PubMed]

28. Bussard, K.M.; Mutkus, L.; Stumpf, K.; Gomez-Manzano, C.; Marini, F.C. Tumor-associated stromal cells as key contributors to the tumor microenvironment. *Breast Cancer Res.* **2016**, *18*, 84. [CrossRef] [PubMed]

29. Lippitz, B.E.; Harris, R.A. Cytokine patterns in cancer patients: A review of the correlation between interleukin 6 and prognosis. *OncoImmunology* **2016**, *5*, e1093722. [CrossRef] [PubMed]

30. Ocana, A.; Nieto-Jimenez, C.; Pandiella, A.; Templeton, A.J. Neutrophils in cancer: Prognostic role and therapeutic strategies. *Mol. Cancer* **2017**, *16*, 137. [CrossRef] [PubMed]

31. Antonio, N.; Bonnelykke-Behrndtz, M.L.; Ward, L.C.; Collin, J.; Christensen, I.J.; Steiniche, T.; Schmidt, H.; Feng, Y.; Martin, P. The wound inflammatory response exacerbates growth of pre-neoplastic cells and progression to cancer. *EMBO J.* **2015**, *34*, 2219–2236. [CrossRef] [PubMed]

32. Liu, M.; Guo, S.; Stiles, J.K. The emerging role of CXCL10 in cancer (Review). *Oncol. Lett.* **2011**, *2*, 583–589. [CrossRef] [PubMed]

33. Imamura, T.; Hikita, A.; Inoue, Y. The roles of TGF-beta signaling in carcinogenesis and breast cancer metastasis. *Breast Cancer* **2012**, *19*, 118–124. [CrossRef] [PubMed]

34. Lund, S.A.; Giachelli, C.M.; Scatena, M. The role of osteopontin in inflammatory processes. *J. Cell Commun. Signal.* **2009**, *3*, 311–322. [CrossRef] [PubMed]

35. Zhao, H.; Chen, Q.; Alam, A.; Cui, J.; Suen, K.C.; Soo, A.P.; Eguchi, S.; Gu, J.; Ma, D. The role of osteopontin in the progression of solid organ tumour. *Cell Death Dis.* **2018**, *9*, 356–391. [CrossRef] [PubMed]

36. Wang, X.; Chao, L.; Ma, G.; Chen, L.; Tian, B.; Zang, Y.; Sun, J. Increased expression of osteopontin in patients with triple-negative breast cancer. *Eur. J. Clin. Investig.* **2008**, *38*, 438–446. [CrossRef] [PubMed]

37. Thorat, D.; Sahu, A.; Behera, R.; Lohite, K.; Deshmukh, S.; Mane, A.; Karnik, S.; Doke, S.; Kundu, G.C. Association of osteopontin and cyclooxygenase-2 expression with breast cancer subtypes and their use as potential biomarkers. *Oncol. Lett.* **2013**, *6*, 1559–1564. [CrossRef] [PubMed]

38. Sangaletti, S.; Tripodo, C.; Sandri, S.; Torselli, I.; Vitali, C.; Ratti, C.; Botti, L.; Burocchi, A.; Porcasi, R.; Tomirotti, A.; et al. Osteopontin shapes immunosuppression in the metastatic niche. *Cancer Res.* **2014**, *74*, 4706–4719. [CrossRef] [PubMed]

39. Wang, D.; Hu, K.; Gao, N.; Zhang, H.; Jiang, Y.; Liu, C.; Wang, S.; Zhao, Z. High throughput screening of cytokines, chemokines and matrix metalloproteinases in wound fluid induced by mammary surgery. *Oncotarget* **2015**, *6*, 29296–29310. [CrossRef] [PubMed]

40. Lyon, D.E.; McCain, N.L.; Walter, J.; Schubert, C. Cytokine comparisons between women with breast cancer and women with a negative breast biopsy. *Nurs. Res.* **2008**, *57*, 51–58. [CrossRef] [PubMed]

41. Vilar-Compte, D.; Jacquemin, B.; Robles-Vidal, C.; Volkow, P. Surgical site infections in breast surgery: Case-control study. *World J. Surg.* **2004**, *28*, 242–246. [PubMed]
42. Woodworth, P.A.; McBoyle, M.F.; Helmer, S.D.; Beamer, R.L. Seroma formation after breast cancer surgery: Incidence and predicting factors. *Am. Surg.* **2000**, *66*, 444–450. [PubMed]
43. Hofer, S.O.; Shrayer, D.; Reichner, J.S.; Hoekstra, H.J.; Wanebo, H.J. Wound-induced tumor progression: A probable role in recurrence after tumor resection. *Arch. Surg.* **1998**, *133*, 383–389. [CrossRef] [PubMed]
44. Bogden, A.E.; Moreau, J.P.; Eden, P.A. Proliferative response of human and animal tumours to surgical wounding of normal tissues: Onset, duration and inhibition. *Br. J. Cancer* **1997**, *75*, 1021–1027. [CrossRef] [PubMed]
45. Kuraishy, A.; Karin, M.; Grivennikov, S.I. Tumor promotion via injury- and death-induced inflammation. *Immunity* **2011**, *35*, 467–477. [CrossRef] [PubMed]
46. O'Leary, D.P.; Wang, J.H.; Cotter, T.G.; Redmond, H.P. Less stress, more success? Oncological implications of surgery-induced oxidative stress. *Gut* **2013**, *62*, 461–470. [CrossRef] [PubMed]

cells

MDPI

Article

Extracellular Matrix and Fibrocyte Accumulation in BALB/c Mouse Lung upon Transient Overexpression of Oncostatin M

Fernando M. Botelho, Rebecca Rodrigues, Jessica Guerette, Steven Wong, Dominik K. Fritz and Carl D. Richards *

McMaster Immunology Research Centre, Department of Pathology and Molecular Medicine, McMaster University, Hamilton, ON L8S 4L8, Canada; botelhf@mcmaster.ca (F.M.B.); rebeccamrodrigues@gmail.com (R.R.); jessica.guerete@gmail.com (J.G.); swong2018@meds.uwo.ca (S.W.); fritz.dkp@gmail.com (D.K.F.)
* Correspondence: richards@mcmaster.ca; Tel.:+905-525-9140 (ext. 22391)

Received: 14 December 2018; Accepted: 27 January 2019; Published: 5 February 2019

Abstract: The accumulation of extracellular matrix in lung diseases involves numerous factors, including cytokines and chemokines that participate in cell activation in lung tissues and the circulation of fibrocytes that contribute to local fibrotic responses. The transient overexpression of the gp130 cytokine Oncostatin M can induce extracellular matrix (ECM) accumulation in mouse lungs, and here, we assess a role for IL-13 in this activity using gene deficient mice. The endotracheal administration of an adenovirus vector encoding Oncostatin M (AdOSM) caused increases in parenchymal lung collagen accumulation, neutrophil numbers, and CXCL1/KC chemokine elevation in bronchioalveolar lavage fluids. These effects were similar in IL-13-/- mice at day 7; however, the ECM matrix induced by Oncostatin M (OSM) was reduced at day 14 in the IL-13-/- mice. CD45+col1+ fibrocyte numbers were elevated at day 7 due to AdOSM whereas macrophages were not. Day 14 levels of CD45+col1+ fibrocytes were maintained in the wildtype mice treated with AdOSM but were reduced in IL-13-/- mice. The expression of the fibrocyte chemotactic factor CXCL12/SDF-1 was suppressed marginally by AdOSM in vivo and significantly in vitro in mouse lung fibroblast cell cultures. Thus, Oncostatin M can stimulate inflammation in an IL-13-independent manner in BALB/c lungs; however, the ECM remodeling and fibrocyte accumulation is reduced in IL-13 deficiency.

Keywords: inflammation; fibrocytes; ECM accumulation; cytokines; Oncostatin M; fibrosis

---

1. Introduction

Chronic lung inflammatory diseases such as asthma, chronic obstructive pulmonary disease (COPD), and pulmonary fibrosis collectively affect a significant proportion of patients in North America. At various stages of these conditions, the pathological excess of extracellular matrix (ECM) can compromise lung function, and thus, the effects of key molecules and cytokines including TGF-β on ECM deposition has been keenly investigated [1–3]. TGF-β clearly acts in milieu of other factors in vivo, and although thought to be a central mediator of ECM accumulation and fibrotic mechanisms including epithelial to mesenchymal transition (EMT) [3,4], other cytokines may be involved. As suggested by numerous studies (reviewed in Reference [5]), members of the gp130 cytokine family including IL-6, IL-11, and Oncostatin M (OSM) appear to participate in the mechanisms involved in lung ECM remodeling [6–8]. The OSM-induced matrix accumulation in lung was shown to be independent of the TGF-beta signaling in mouse models [8]. Gp130 cytokines robustly activate STAT3 cell signaling, and in genetic models in mice, the over-activation of STAT3 renders animals much more sensitive to the ECM remodeling effects in the bleomycin model of lung fibrosis [9]. Furthermore, this model is independent of the canonical TGFb signaling pathway SMAD3, emphasizing other pathways can lead to lung fibrosis in vivo [9].

As a member of the gp130 cytokine family, OSM is a multifunctional cytokine that can regulate homeostatic functions as well as disease processes [10–12]. Mouse OSM engages in receptor complexes that include the gp130 signaling molecule and the OSMRβ chain [13–15], both of which are broadly expressed in connective tissue cells. OSM has been shown to induce a number of responses that regulate the remodeling of ECM in articular joints [16–18], skin [19,20], and bones [21,22]. OSM induces the ECM remodelling of lungs in animal models [8,23,24] and may contribute significantly to chronic inflammatory lung disease in humans since it is found at elevated levels locally in sputum samples from severe asthmatics [25] and in the bronchoalveolar lavage fluid (BALF) of idiopathic pulmonary fibrosis patients [8]. OSM can also regulate chemokine release from cells derived from the lung, including fibroblasts [26], airway smooth muscle cells [27], as well as airway epithelial cells [28–30].

There is increasing evidence that conditions that result in ECM remodeling involve the activation of a number of cell types, including resident fibroblasts, myofibroblasts, and circulating fibrocytes [31–33]. Chemokines such as CXCL12/SDF-1 have chemotactic activities for fibrocytes [34,35] and may play a role in generating the fibrocyte response and local fibrogenic sequale. It is currently unclear whether OSM induces chemokine production and subsequent fibrocyte accumulation for these cells.

Th2 cytokines have been implicated in inducing the ECM accumulation in lungs [36]. The overexpression of IL-13 using the adenovirus vector for pulmonary transgene expression induced marked increases in the inflammatory ECM deposition [37]. We have shown that the transient pulmonary transgene expression of OSM induces lung ECM in C57Bl/6 mice [23], although the mechanisms may be quite different in BALB/c mice [24]. In other models of airway inflammation, BALB/c mice have been typically characterized as more biased toward Th2 immune responses. Thus, we have examined here the BALB/c responses to the transient pulmonary overexpression of OSM, their dependence on IL-13, and the accumulation of fibrocytes and CXCL12/SDF-1 in lung tissue.

2. Materials and Methods

2.1. Animals

Female BALB/c mice (8–10 weeks old) were purchased from Charles River Laboratory (Ottawa, ON, Canada). Female BALB/c IL-13-/- mice were a courtesy of Dr. Waliul Khan (McMaster University). All mice were housed in standard conditions with food and water ad libitum. All procedures were approved by the Animals Research Ethics Board at McMaster.

2.2. Administration of Adenovirus Constructs and Sample Collection

Wildtype and IL-13-/- mice were administered 5×10^7 pfu of replication deficient AdDl70 or Ad encoding OSM (AdOSM) through the endotracheal route of administration as previously published [23,38]. Mice were euthanized after 5, 7, or 14 days and bled, and alveolar lavage was performed as previously described [23,38]. Alveolar lavage was centrifuged, and supernatants were stored for future analysis by ELISA. Cell pellets were resuspended, counted, and subjected to cytocentrifugation at 300 rpm for two minutes. Differential counts were determined after staining with protocol Hema3 (Fisher Scientific, Ottawa, ON, Canada). Left lungs were perfused with 10% formalin and fixed for 48 h subsequent to histological preparation and histochemical staining. Right lungs were snap frozen and stored at −80 °C for RNA extraction.

2.3. Reagents and Cell Culture

Mouse lung fibroblast (MLF) cultures were generated from BALB/c or C57Bl/6 mouse lungs as previously described [26]. LA-4 and A549 cells were purchased from ATCC and cultured under recommended conditions. Recombinant E. coli derived cytokines (mouse OSM, IL-1β, and IL-4) were purchased from R&D systems.

2.4. ELISA and Extracellular Matrix Analysis

Duoset ELISA kits were purchased from R&D Systems (Minneapolis, MN, USA) to measure the protein levels of CXCL1/KC, mouse IL-6, mouse CXCL12/SDF-1, mouse CCL11/Eotaxin-1, and mouse VEGF in BALF samples stored at −20 °C. The quantification was completed as per the manufacturer's instructions. For the measurement of collagen accumulation, the intensity of picrosirius-stained tissue sections under polarized light was quantified from images (>20 images per lung) by using the Ashcroft Method [39] as previously described for BALB/c mouse lungs by Wong et al. [24].

2.5. RNA Extraction

Lung tissues or cell culture extracts were homogenized in Trizol (Invitrogen, Life Technologies). RNA was reverse transcribed, and the levels of CXCL12/SDF-1 and IL-6 were assessed by real time Q-PCR (Taqman) using primers with FAM-5′ end-labeled fluorogenic probes for CXCL12/SDF-1 and IL-6 and VIC-5′ end-labeled fluorogenic probes for 18S. mRNA expression was expressed as the ΔΔCt (cycle threshold) values for CXCL12/SDF-1 or IL-6 relative to that of the 18S control RNA. All obtained gene expression assays came from Applied Biosystems, Thermos-Fisher and were performed as previously described [40].

2.6. Isolation of Lung Mononuclear Cells and Flow Cytometric Analysis

Lung mononuclear cell suspensions were generated by mechanical mincing and collagenase digestion. Debris were removed by passage through a 40 micrometer screen size nylon mesh (this results in primarily CD45+ hematopoietic cells introduced to the flow cytometry analysis), and cells were resuspended in phosphate-buffered saline (PBS) containing 0.3% bovine serum albumin (BSA) (Invitrogen, Burlington, ON, Canada) or in RPMI media supplemented with 10% fetal bovine serum (FBS) (Sigma-Aldrich, Oakville, ON, Canada), 1% L-glutamine, and 1% penicillin/streptomycin (Invitrogen, Burlington, ON, Canada). Washed once with PBS/0.3% BSA and stained with primary antibodies directly conjugated to fluorochromes for 30 min at 4 °C were 1×10^6 lung mononuclear cells. Acquired on an LSR II (BD Biosciences, San Jose, CA, USA) flow cytometer were 10^5 live events, and the data were analyzed with FlowJo analysis software (FlowJo, LLC., Ashland, OR, USA). Side scatter and forward scatter parameters were used to define live cell and lymphocyte gates. All antibodies were purchased from BD Biosciences (San Jose, CA, USA) or eBiosciences (San Diego, CA, USA) unless otherwise stated. The following antibodies were used for flow cytometric analysis: APC-cy7-conjugated anti-CD45, PerCP-cy5.5-conjugated anti-CD11c, PE-conjugated anti-CD11b, PE-cy7-conjugated anti-DX5, Pacific Blue-conjugated anti-CD3, and Pacific Orange-conjugated anti-Gr-1. Neutrophils were defined as CD45+, CD11b-hi, and Gr-1+. Macrophage cells were defined as being CD11b+, CD11c-, DX5-, CD3-, and Gr-1-. For the fibrocyte analysis, CD45+ Collagen I (Col1+) fibrocytes were detected first by surface staining the cells with APC-cy7-conjugated anti-CD45 for 30 min at 4 °C and followed by intracellular staining with rabbit anti-collagen I (Rockland, Gilbertsville, PA, USA) and then FITC-conjugated anti-rabbit (Jackson ImmunoResearch, Westgrove, PA, USA) antibodies, each for 30 min at 4 °C in 1× Perm/Wash buffer (BD Biosciences, San Jose, CA, USA) with washes in 1x Perm/Wash between intracellular staining steps. Cells were then washed with 1× PBS/0.3% BSA prior to analysis on an LSR II flow cytometer.

2.7. Statistics

Data were analyzed using GraphPad Prism version 5.1 software and presented as mean +/− standard error of the mean (SEM). For in vivo experiments, five animals per group were utilized. One-way analysis of variance (ANOVA) was used to determine the statistical significance, which was defined as $p < 0.05$ using GraphPad Prism. The p-values are indicated in the individual figures.

3. Results

We have previously shown that intranasal or endotracheal administration of the adenovirus vector expressing mouse OSM in C57Bl/6 and in BALB/c mice causes the transient pulmonary gene transduction of OSM, which leads to a pronounced remodeling of the lung, including relatively rapid increases in both the collagen gene expression and protein [24]. The AdOSM vector induces OSM levels in BALF transiently (250 +/− 25 pg/mL at day 2 and 1540 +/− 150 pg/mL at day 7) whereas levels in naïve or AdDl70 animals were undetectable. In Figure 1A, we examined the histopathology of wild type BALB/c and IL-13-/- BALB/c mice 7 days after delivery of AdOSM. AdOSM-treated mice showed a disruption of the lung architecture, with thicker alveolar walls compared to the AdDl70-treated counterparts. Representative images of the picrosirus red-stained histological sections are shown and indicate a qualitative increase in the staining within the lung parenchyma in AdOSM-treated mice from both the wildtype and IL-13-/- strains. The quantification of PSR-stained sections showed similar increases in the picrosirius red-staining in the parenchyma of both the AdOSM-treated wildtype and IL-13-/- mice (Figure 1A, right panel). We further assessed the histopathology and parenchymal collagen accumulation after a prolonged 14-day time point with AdOSM in the BALB/c wildtype and IL-13-deficient mice. As shown in Figure 1B, the thickening of the lung architecture and the collagen accumulation in AdOSM-treated mice were significantly reduced in IL-13-/- mice as compared to the wildtype mice. Ashcroft scores of the PSR-stained sections also showed a significant reduction in the collagen staining in IL-13-/- mice as compared to the wildtype mice.

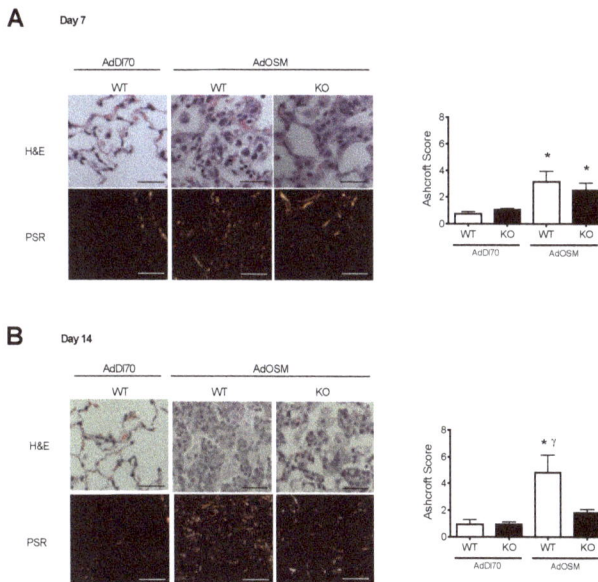

Figure 1. Prolonged induction of interstitial collagen in the lungs of AdOSM-infected BALB/c mice is reduced in IL-13-deficient mice. Wildtype (WT) and IL-13-/- (KO) BALB/c mice were treated with an endotracheal administration of AdDl70 or AdOSM (5 × 10⁷ pfu), culled at day 7 (**A**) or day 14 (**B**), and tissues were prepared for histology. Images of H&E and picrosirius red (PSR) stained tissue sections are shown from mice treated with AdDl70 or AdOSM. Quantification of picrosirius red staining was completed by Ashcroft scores to assess the degree of staining in the lung parenchyma. Data i shown as the mean +/− SEM (*n* = 5 per group). White bars are wildtype and black bars are IL-13-/- data. Scale bars for photomicrographs represent a length of 100um. * indicates significant difference of *p* < 0.05, as compared to AdDl70-treated mice. γ indicates significant difference of *p* < 0.05 as compared to AdOSM-treated IL-13-/- mice.

To determine if the inflammation induced by OSM was affected by the absence of IL-13, the BALF fluid was collected, cell frees supernatants were stored, and cytocentrifuged cells were assessed by differential staining. Figure 2A shows that macrophage and neutrophil numbers and percentages in BALF were significantly elevated in the AdOSM-treated BALB/c mice compared to the control vector AdDl70-treated mice.

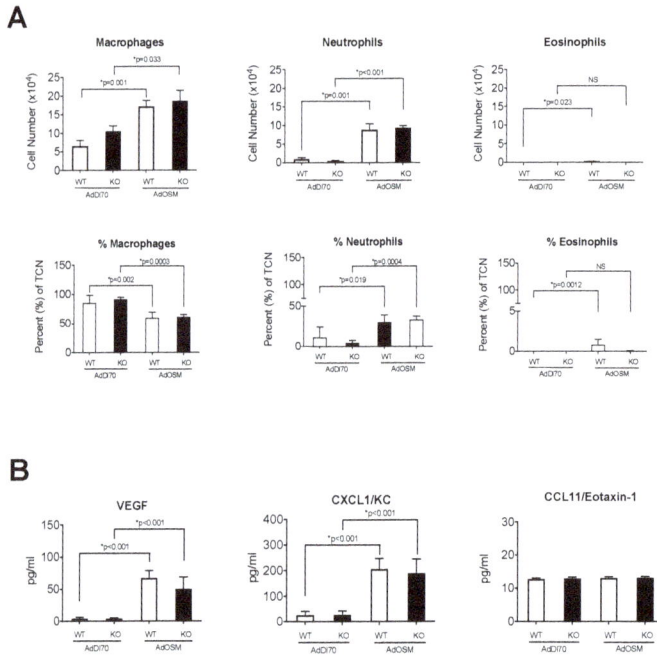

Figure 2. Inflammatory cells and cytokines in BALF at Day 7 in Response to AdOSM. (**A**) Wildtype (WT) and IL-13-/- (KO) BALB/c mice were treated with an endotracheal administration of AdDl70 or AdOSM (5×10^7 pfu), culled at day 7, and bronchoalveolar lavage fluid (BALF) was collected and analyzed for cell numbers and percent of total cell number (TCN) of macrophages, neutrophils, and eosinophils in stained cytocentrifuge smears. (**B**) IL-6, CXCL1/KC and CCL11/Eotaxin-1 cytokine levels in BALF used in (**A**) was measured by ELISA. White bars are wildtype and black bars are IL-13-/- data. Data is shown as the mean +/− SEM (*n* = 5 per group). Statistical significant differences are noted with their p values between the indicated treatment groups.

Similar levels of macrophages and neutrophils were observed in the IL-13-/- in response to the AdOSM treatment in comparison to the wildtype mice. Similar to previous works comparing responses to AdOSM in BALB/c and C57Bl/6 mice [23,24,41], little to no eosinophils were detected in the BALF in either the wildtype BALB/c or IL-13-/- mice in response to AdOSM administration. The chemokine CXCL1/KC (chemoattractant for neutrophils) levels were induced by AdOSM with no difference in the IL-13-/- mice, and the CCL11/Eotaxin-1 levels were low/undetectable and unaltered between the different treatment groups. The vascular endothelial growth factor (VEGF) was also increased in AdOSM-treated mice similarly in both wildtype and IL-13-/- mice (Figure 2B).

To examine the accumulation of CD45 positive cells in this system, flow cytometry was used to assess the different populations of the whole lung upon collagenase digestion and the generation of single cell preparations. Using the strategy shown in Figure 3A, we were able to detect CD45+ collagen1A1+ (CD45+ coll+) cells in the BALB/c lung.

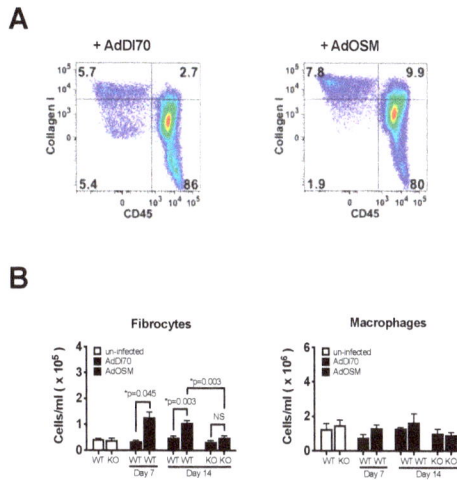

Figure 3. Fibrocyte numbers in AdOSM-treated BALB/c Mouse Lungs. Wildtype BALB/c mice untreated naive mice or wildtype and IL-13-/- (KO) treated mice were endotracheal administered AdDl70 or AdOSM (5×10^7 pfu), culled at day 7 or 14, and mononuclear cell suspensions were analyzed by flow cytometry (as described in methods). (**A**) Representative flow cytometry plots showing CD45+col1+ populations (fibrocytes) at day 7 (**B**) CD45+ Collagen 1A1+ Fibrocytes and CD11c- CD11b+ Gr-1- DX5- Macrophage cells from whole lung tissue, after 7 and 14 day treatment, were also quantitated by flow cytometry. Data is shown as the mean +/− SEM (n = 5 per group). Statistical significant differences are noted with their p values between the indicated treatment groups. No statistical significance was observed between macrophage numbers in 3B.

The numbers of these cells were quantified in Figure 3B. The results indicate that similar numbers of CD45+ col1+ cells were evident in both the uninfected naïve and control AdDl70-treated mice and elevated in cell numbers upon AdOSM administration at day 7. The numbers of CD45+col1+ cells were sustained at day 14 in the wildtype mice and decreased to basal levels in the IL-13-/- mice. As a comparator, tissue macrophage levels (defined as CD11b+, CD11c-, Gr-1-, DX5-, and CD45+) were similar at day 7 or day 14 between the wildtype and IL-13-deficient mice (Figure 3B). We have previously observed significant increases in alternatively activated/M2 macrophage markers in C57Bl/6 mice treated with AdOSM [40]. However, as shown in Supplemental Figure S1, Arginase-1 mRNA was not significantly elevated in AdOSM-treated BALB/c lungs, while C57Bl/6 mice showed marked increases ($p < 0.005$), consistent with previous results [40]. In addition, the mRNA from alveolar macrophages retrieved from AdOSM-treated BALB/c mice did not show any difference in Arginase-1 or CD206 mRNA (both markers of M2 macrophage phenotypes) from those alveolar macrophages retrieved from AdDl70-treated control animals.

We then assessed the expression of CXCL12/SDF-1 (chemotactic for fibrocytes) and IL-6 (an inflammatory cytokine, previously shown to be induced by OSM in C57Bl/6 mice [38]). In BALF, as assessed by ELISA, CXCL12/SDF-1 showed a trend of decreased levels in animals treated with AdOSM in either the wildtype or IL-13-/- mice (Figure 4A).

In contrast, the levels of IL-6 in BALF were elevated in AdOSM-treated mice in both the wildtype and IL-13-/- mice, relative to the AdDl70 controls. Assessing the mRNA expression (Figure 4B) in lung tissue at day 7 following the AdDl70- or AdOSM-treatments of the wildtype show that the levels of CXCL12/SDF-1 mRNA were not increased in response to OSM, while the IL-6 mRNA expression levels were elevated in AdOSM-treated mice.

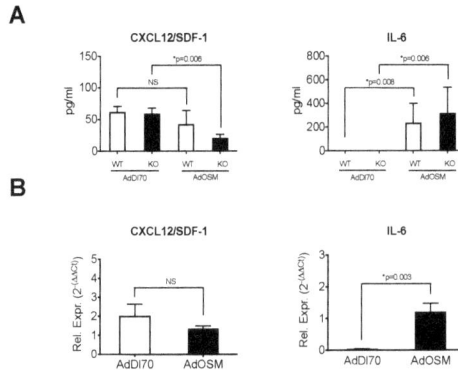

Figure 4. Regulation of fibrocyte chemokine mRNA by OSM. Wildtype BALB/c mice or IL-13-/- (KO) were treated with an endotracheal administration of AdDl70 or AdOSM (5×10^7 pfu) as indicated, culled, BALF retrieved and lung tissues were isolated and prepared for RNA extraction. (**A**) CXCL12/SDF-1 and IL-6 proteins in BALF were analyzed by ELISA. (**B**) RNA was probed for CXCL12/SDF-1 and IL-6 using q-RT-PCR. Rel. Expr.: Relative expression (Relative to 18S RNA). Data is shown as the mean +/− SEM (*n* = 5 per group. Statistical significant differences are noted with their p values between the indicated treatment groups.

To assess the apparent downregulation of CXCL12/SDF-1 by OSM in an in vitro system, MLF cultures from BALB/c mice were isolated and stimulated with varying concentrations of recombinant (E. coli derived) OSM (Figure 5A, left panel).

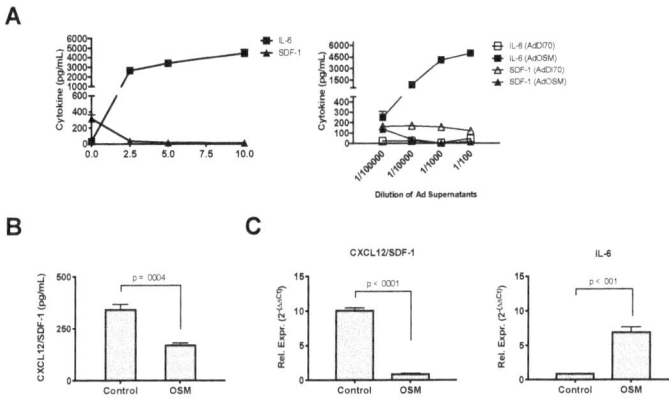

Figure 5. OSM downregulates expression of CXCL12/SDF-1 in mouse lung fibroblasts. BALB/c-derived mouse lung fibroblasts (MLF) were stimulated for 24 h (in triplicate) with increasing dosing of recombinant murine OSM (ng/mL) and supernatants analyzed by ELISA for CXCL12/SDF-1 or IL-6 (**A, left panel**). In (**A right panel**), MLF cells were incubated with dilutions of supernatants from AdDl70 (control)- or AdOSM-infected A549 cells (from 10^{-5} to 10^{-2}), for 24 h, and MLF supernatants analyzed by ELISA for CXCL12/SDF-1 and IL-6. In (**B**), MLF cells derived from C57Bl/6 mice were stimulated for 24 h with recombinant OSM (10 ng/mL) and secreted CXCL12/SDF-1 protein was measured by ELISA. (**C**) CXCL12/SDF-1 and IL-6 mRNA expression in C57Bl/6-derived MLF cells was measured by quantitative PCR. Rel. Expr.: Relative expression (Relative to 18S RNA). Data represent one of at least three separate experiments. Data is shown as the mean +/− SD (*n* = 4 replicate treatments per group). Statistical significant differences are noted with their p values between the indicated treatment groups.

Levels above 300 pg/mL of the CXCL12/SDF-1 protein were detected in the unstimulated cells; however, the levels decreased in cells stimulated with 2.5 ng/mL of OSM or higher. The same supernatants contained increased IL-6 levels, stimulated by OSM in a dose-dependent manner. To assess whether the mammalian-expressed mouse OSM had the same effect, MLF cultures were stimulated with the supernatant from A549 cells infected with either the AdDl70 control or AdOSM vector. Figure 5A (right panel) shows the levels of both thee CXCL12/SDF-1 and IL-6 in MLF supernatants after 24 h of stimulation with various dilutions of the Ad-infected A549 supernatants. Both the CXCL12/SDF-1 and IL-6 levels were unregulated in the MLF cultures treated with Addel-70-infected supernatants, whereas the AdOSM-infected supernatants caused a decrease in the CXCL12/SDF-1 levels and a marked increase in the IL-6 levels in a dose-dependent fashion in the MLF cultures. Thus, both the E. coli-derived OSM and mammalian cell-expressed OSM were able to potently suppress the SDF1 production in MLF cultures.

To determine if the CXCL12/SDF-1 protein decrease in vitro in BALB/c mice was also observed in other mouse strains, the total lung RNA and supernatants were assessed by q-RT-PCR and ELISA, respectively in C57Bl/6-derived MLF cells. OSM induced a reduction in the CXCL12/SDF-1 basal levels of proteins in the supernatants (Figure 5B) and CXCL12/SDF-1 mRNA expression (Figure 5C, left panel) following a 24-h treatment of C57Bl/6 MLF cells. In contrast, we observed elevated levels of IL-6 mRNA expression in these cells (Figure 5C, right panel).

4. Discussion

The regulation of ECM remodeling in lungs involves the participation of various cytokines and cells. In the model system explored here, we have shown that the Ad-vector-mediated overexpression of OSM in BALB/c mice induces neutrophil infiltration and CXCL1/KC chemokine levels into bronchoalveolar spaces in a manner independent of IL-13 at day 7. We observed that AdOSM induces the accumulation of parenchymal collagen that was evident at day 7 (consistent with previous work [24]) and was further elevated at day 14. The collagen increase was evident in IL-13-/- mice at day 7 but was markedly reduced at day 14. We also observed the fibrocyte accumulation at day 14 as defined by CD45+Coll1+ cells in the lungs, which was also reduced in the IL-13-/- mice. The CXCL12/SDF-1 chemotactic factor for fibrocytes was suppressed by AdOSM in vivo and in OSM-stimulated mouse lung fibroblasts in vitro.

The lack of requirement for IL-13 during the early matrix deposition observed here (day 7) is consistent with the studies done on C57Bl/6 mice, where Mozafarrian et al. showed that OSM induced ECM remodeling in the lungs of similar mice that were also treated with IL-13R.Fc (blocks only IL-13 and not IL-4) or the control Fc administered intraperitoneally [8]. IL-13R.Fc did block the eosinophil accumulation. Such results in C57Bl/6 mice were also consistent with subsequent data showing that the AdOSM-induced elevation of ECM was independent of STAT6, a major signaling pathway of both IL-4 and IL-13 but was required for the effects on eosinophil accumulation [23]. However, as we have previously published [24], the AdOSM vector induces significant parenchymal collagen in BALB/c mice but not a detectable increase in the BALF levels of IL-4/IL-13, CCL24/Eotaxin-2, or the eosinophil accumulation. Collectively, the data suggests that eosinophils are not involved in the ECM increases in either BALB/c or C57Bl6 lungs due to OSM overexpression. We have examined the expression of macrophage markers (by mRNA analysis) in BALB/c mouse lungs treated with AdOSM and have not observed any significant changes in alternatively activated "M2" markers (see Supplemental Figure S1). In contrast, C57Bl/6 mice show elevated CD206+ cells and arginase-1 expression (as typical M2 macrophage products) in response to AdOSM [40]. Other macrophage phenotypes may be involved in BALB/c lungs in this system; however, this would require further investigation.

AdOSM expression increased the CXCL1/KC levels and neutrophils numbers in BALF, and the lack of difference in the IL-13-/- mice indicated that the ECM effects and neutrophil presence correlate in the BALB/c system here. There is a possibility that the sources of endogenous OSM such as that from neutrophils as well as macrophages contribute to the load of OSM protein in lung inflammatory conditions. Although activated macrophages and T cells are major sources of OSM, Grenier et al. [42] have found that neutrophils can synthesize OSM and release preformed OSM, and that neutrophils from patients with acute lung injury release OSM. Thus, the effects in this model system in BALB/c mice may reflect the OSM generated by the adenovirus vector and endogenous OSM from inflammatory cells.

Fibrocytes have been implicated in fibrotic mechanisms in both mouse and human systems [32,35,43]. These cells have been shown to be recruited to mouse lungs in models of bleomycin-induced pulmonary fibrosis through CXCL12/SDF-1 [35], which interacts with CXCR-4. IL-4 and IL-13 have been shown to stimulate the fibrocyte production of matrix components [44]. Our results indicate that OSM can engage the recruitment of CD45+col1+ fibrocytes to the lung and may contribute to the remodeling of ECM in BALB/c lungs. The correlation between the reduction of fibrocytes at day 14 in the IL-13-/- mice and the reduction in parenchymal collagen is consistent with this. The accumulation of fibrocytes over the time course of the ECM accumulation in this model may be affected at other time points in IL-13KO mice. The association with the decreased fibrocyte number and decreased ECM at day 14 does not show causality, and further experimentation to determine if fibrocytes are critical to the ECM accumulation at day 14 would be required. The role of IL-13 in the fibrocyte accumulation is not clear. Since we cannot detect the induction of IL-13 protein (BALF ELISA, data not shown), it may be that the required IL-13 concentration locally in the lung is low. It is possible that IL-13 (or CXCL12/SDF-1) is elevated at specific times not captured by the present time-points analysed or indeed in specific populations (such as separate macrophage phenotypes). Assessing specific macrophage populations or other cells over the time course of the model development would assist with this and would be the subject of future experimentation. Alternatively, IL-13 may be required at other organ sites to maximally support fibrocyte generation or recruitment. In addition, IL-13 may act directly on the local matrix synthesizing cells, such as fibroblasts, where low levels may enable maximal OSM regulation of collagen synthesis.

Previous work has shown that human OSM could induce the CXCL12/SDF-1 protein expression in vitro in cardiac myocytes and cardiac fibroblasts [45]. In our system using AdOSM in vivo, we did not observe increases in CXCL12/SDF-1. Furthermore, we observed a suppression of CXCL12/SDF-1 by OSM in the in vitro analysis of the MLF responses. Our contrasting results may be due to the cell type/tissue source studied or to possible differences between species responses. There is a possibility that since the ELISA is specific for CXCL12/SDF-1α and not CXCL12/SDF-1β, we are missing CXCL12/SDF-1β upregulation. However, the RNA probes we used would have detected both the CXCL12/SDF-1 isoforms described. The mechanism by which the suppression of CXCL12/SDF-1 by OSM occurs is not known. Clearly, this is a specific suppression since IL-6 was simultaneously induced strongly. Fibrocyte activation has also been shown through the engagement of CCR2 and its ligands CCL2/MCP-1 and CCL12/MCP-5 [46–48]. Others have shown in the FITC model of lung fibrosis that CCL12/MCP-5 was deemed to be the active ligand for CCR2 involvement in the fibrotic response in vivo [47]. Further analysis is required to assess the role of CCL12/MCP-5, possibly using inhibitors in vivo, to determine if it is central to the mechanism in the system with OSM overexpression.

5. Conclusions

The OSM-induced inflammation in BALB/c mouse lungs does not require IL-13; however, IL-13 is required for maximal extracellular matrix deposition. CD45+ col1+ fibrocytes are induced by OSM but are also reduced in IL-13-/- mice. Chemotactic factors for fibrocytes other than CXCL12/SDF-1 may be involved since OSM selectively suppresses the CXCL12/SDF-1 expression in this system. These results suggest additional mechanisms in the OSM-induced ECM accumulation include the IL-13-dependent fibrocyte accumulation.

Supplementary Materials: The following are available online at http://www.mdpi.com/2073-4409/8/2/126/s1. Reference [49] is cited in the supplementary materials.

Author Contributions: Conceptualization, F.M.B. and C.D.R.; methodology, F.M.B.; formal analysis, F.M.B., R.R., J.G., S.W., D.K.F., and C.D.R.; investigation, F.M.B., R.R., J.G., S.W., and D.K.F.; data curation, F.M.B.; writing—original draft preparation, F.M.B.; writing—review and editing, F.M.B. and C.D.R.; supervision, C.D.R.; funding acquisition, C.D.R.

Funding: The research was funded by the Canadian Institutes for Health through grants #102562 and #137013, as well as the Ontario Thoracic Society (Canadian Lung Association).

Conflicts of Interest: The authors declare no conflict of interest.

References

1. Leask, A.; Abraham, D.J. TGF-beta signaling and the fibrotic response. *FASEB J.* **2004**, *18*, 816–827. [CrossRef] [PubMed]
2. Gauldie, J.; Bonniaud, P.; Sime, P.; Ask, K.; Kolb, M. TGF-beta, Smad3 and the process of progressive fibrosis. *Biochem. Soc. Trans.* **2007**, *35*, 661–664. [CrossRef] [PubMed]
3. Wynn, T.A.; Ramalingam, T.R. Mechanisms of fibrosis: Therapeutic translation for fibrotic disease. *Nat. Med.* **2012**, *18*, 1028–1040. [CrossRef] [PubMed]
4. Kalluri, R.; Neilson, E.G. Epithelial-mesenchymal transition and its implications for fibrosis. *J. Clin. Investig.* **2003**, *112*, 1776–1784. [CrossRef] [PubMed]
5. Silver, J.S.; Hunter, C.A. gp130 at the nexus of inflammation, autoimmunity, and cancer. *J. Leukoc. Biol.* **2010**, *88*, 1145–1156. [CrossRef] [PubMed]
6. Kuhn, C., III; Homer, R.J.; Zhu, Z.; Ward, N.; Flavell, R.A.; Geba, G.P.; Elias, J.A. Airway hyperresponsiveness and airway obstruction in transgenic mice. Morphologic correlates in mice overexpressing interleukin (IL)-11 and IL-6 in the lung. *Am. J. Respir. Cell Mol. Biol.* **2000**, *22*, 289–295. [CrossRef] [PubMed]
7. Tang, W.; Geba, G.P.; Zheng, T.; Ray, P.; Homer, R.J.; Kuhn, C.; Flavell, R.A.; Elias, J.A. Targeted expression of IL-11 in the murine airway causes lymphocytic inflammation, bronchial remodeling, and airways obstruction. *J. Clin. Investig.* **1996**, *98*, 2845–2853. [CrossRef]
8. Mozaffarian, A.; Brewer, A.W.; Trueblood, E.S.; Luzina, I.R.; Todd, N.W.; Atamas, S.P.; Arnett, H.A. Mechanisms of oncostatin M-induced pulmonary inflammation and fibrosis. *J. Immunol.* **2008**, *181*, 7243–7253. [CrossRef]
9. O'Donoghue, R.J.; Knight, D.A.; Richards, C.D.; Prêle, C.M.; Lau, H.L.; Jarnicki, A.G.; Jones, J.; Bozinovski, S.; Vlahos, R.; Thiem, S.; et al. Genetic partitioning of interleukin-6 signalling in mice dissociates Stat3 from Smad3-mediated lung fibrosis. *EMBO Mol. Med.* **2012**, *4*, 939–951. [CrossRef]
10. Tanaka, M.; Miyajima, A. Oncostatin M, A multifunctional cytokine. *Rev. Physiol. Biochem. Pharmacol.* **2003**, *149*, 39–52.
11. Chen, S.H.; Benveniste, E.N. Oncostatin M: A pleiotropic cytokine in the central nervous system. *Cytokine Growth Factor Rev.* **2004**, *15*, 379–391. [CrossRef] [PubMed]
12. Komori, T.; Tanaka, M.; Senba, E.; Miyajima, A.; Morikawa, Y. Lack of Oncostatin M Receptor beta Leads to Adipose Tissue Inflammation and Insulin Resistance by Switching Macrophage Phenotype. *J. Biol. Chem.* **2013**, *288*, 21861–21875. [CrossRef] [PubMed]
13. Ichihara, M.; Hara, T.; Kim, H.; Murate, T.; Miyajima, A. Oncostatin M and leukemia inhibitory factor do not use the same functional receptor in mice. *Blood* **1997**, *90*, 165–173. [PubMed]
14. Lindberg, R.A.; Juan, T.S.; Welcher, A.A.; Sun, Y.; Cupples, R.; Guthrie, B.; Fletcher, F.A. Cloning and characterization of a specific receptor for mouse oncostatin M. *Mol. Cell. Biol.* **1998**, *18*, 3357–3367. [CrossRef] [PubMed]
15. Tanaka, M.; Hara, T.; Copeland, N.G.; Gilbert, D.J.; Jenkins, N.A.; Miyajima, A. Reconstitution of the functional mouse oncostatin M (OSM) receptor: Molecular cloning of the mouse OSM receptor beta subunit. *Blood* **1999**, *93*, 804–815. [PubMed]
16. Cawston, T.; Billington, C.; Cleaver, C.; Elliott, S.; Hul, W.; Koshy, P.; Shingleton, B.; Rowan, A. The regulation of MMPs and TIMPs in cartilage turnover. *Ann. N. Y. Acad. Sci.* **1999**, *878*, 120–129. [CrossRef] [PubMed]
17. Rowan, A.D.; Koshy, P.J.; Shingleton, W.D.; Degnan, B.A.; Heath, J.K.; Vernallis, A.B.; Spaull, J.R.; Life, P.F.; Hudson, K.; Cawston, T.E. Synergistic effects of glycoprotein 130 binding cytokines in combination with interleukin-1 on cartilage collagen breakdown. *Arthritis Rheum.* **2001**, *44*, 1620–1632. [CrossRef]

18. Langdon, C.; Kerr, C.; Hassen, M.; Hara, T.; Arsenault, A.L.; Richards, C.D. Murine oncostatin M stimulates mouse synovial fibroblasts in vitro and induces inflammation and destruction in mouse joints in vivo. *Am. J. Pathol.* **2000**, *157*, 1187–1196. [CrossRef]

19. Finelt, N.; Gazel, A.; Gorelick, S.; Blumenberg, M. Transcriptional responses of human epidermal keratinocytes to Oncostatin-M. *Cytokine* **2005**, *31*, 305–313. [CrossRef]

20. Boniface, K.; Diveu, C.; Morel, F.; Pedretti, N.; Froger, J.; Ravon, E.; Garcia, M.; Venereau, E.; Preisser, L.; Guignouard, E.; et al. Oncostatin M secreted by skin infiltrating T lymphocytes is a potent keratinocyte activator involved in skin inflammation. *J. Immunol.* **2007**, *178*, 4615–4622. [CrossRef]

21. Sims, N.A.; Walsh, N.C. GP130 cytokines and bone remodelling in health and disease. *BMB Rep.* **2010**, *43*, 513–523. [CrossRef] [PubMed]

22. Guihard, P.; Danger, Y.; Brounais, B. Induction of osteogenesis in mesenchymal stem cells by activated monocytes/macrophages depends on oncostatin M signaling. *Stem Cells* **2012**, *30*, 762–772. [CrossRef] [PubMed]

23. Fritz, D.K.; Kerr, C.; Fattouh, R.; Llop-Guevara, A.; Khan, W.I.; Jordana, M.; Richards, C.D. A mouse model of airway disease: Oncostatin M-induced pulmonary eosinophilia, goblet cell hyperplasia, and airway hyperresponsiveness are STAT6 dependent, and interstitial pulmonary fibrosis is STAT6 independent. *J. Immunol.* **2011**, *186*, 1107–1118. [CrossRef] [PubMed]

24. Wong, S.; Botelho, F.M.; Rodrigues, R.M.; Richards, C.D. Oncostatin M overexpression induces matrix deposition, STAT3 activation, and SMAD1 Dysregulation in lungs of fibrosis-resistant BALB/c mice. *Lab. Investig.* **2014**, *94*, 1003–1016. [CrossRef] [PubMed]

25. Simpson, J.L.; Baines, K.J.; Boyle, M.J.; Scott, R.J.; Gibson, P.G. Oncostatin M (OSM) is increased in asthma with incompletely reversible airflow obstruction. *Exp. Lung Res.* **2009**, *35*, 781–794. [CrossRef] [PubMed]

26. Fritz, D.K.; Kerr, C.; Tong, L.; Smyth, D.; Richards, C.D. Oncostatin-M up-regulates VCAM-1 and synergizes with IL-4 in eotaxin expression: Involvement of STAT6. *J. Immunol.* **2006**, *176*, 4352–4360. [CrossRef] [PubMed]

27. Faffe, D.S.; Flynt, L.; Mellema, M.; Moore, P.E.; Silverman, E.S.; Subramaniam, V.; Jones, M.R.; Mizgerd, J.P.; Whitehead, T.; Imrich, A.; et al. Oncostatin M causes eotaxin-1 release from airway smooth muscle: Synergy with IL-4 and IL-13. *J. Allergy Clin. Immunol.* **2005**, *115*, 514–520. [CrossRef] [PubMed]

28. Loewen, G.M.; Tracy, E.; Blanchard, F.; Tan, D.; Yu, J.; Raza, S.; Matsui, S.; Baumann, H. Transformation of human bronchial epithelial cells alters responsiveness to inflammatory cytokines. *BMC Cancer* **2005**, *5*, 145. [CrossRef]

29. Chattopadhyay, S.; Tracy, E.; Liang, P.; Robledo, O.; Rose-John, S.; Baumann, H. Interleukin-31 and oncostatin-M mediate distinct signaling reactions and response patterns in lung epithelial cells. *J. Biol. Chem.* **2007**, *282*, 3014–3026. [CrossRef]

30. Sallenave, J.M.; Tremblay, G.M.; Gauldie, J.; Richards, C.D. Oncostatin M, but not interleukin-6 or leukemia inhibitory factor, stimulates expression of alpha1-proteinase inhibitor in A549 human alveolar epithelial cells. *J. Interferon Cytokine Res.* **1997**, *17*, 337–346. [CrossRef]

31. Quan, T.E.; Cowper, S.; Wu, S.P.; Bockenstedt, L.K.; Bucala, R. Circulating fibrocytes: Collagen-secreting cells of the peripheral blood. *Int. J. Biochem. Cell Biol.* **2004**, *36*, 598–606. [CrossRef] [PubMed]

32. Maharaj, S.S.; Baroke, E.; Gauldie, J.; Kolb, M.R. Fibrocytes in chronic lung disease–facts and controversies. *Pulm. Pharmacol. Ther.* **2012**, *25*, 263–267. [CrossRef]

33. Gomperts, B.N.; Strieter, R.M. Fibrocytes in lung disease. *J. Leukoc. Biol.* **2007**, *82*, 449–456. [CrossRef] [PubMed]

34. Lama, V.N.; Phan, S.H. The extrapulmonary origin of fibroblasts: Stem/progenitor cells and beyond. *Proc. Am. Thorac. Soc.* **2006**, *3*, 373–376. [CrossRef] [PubMed]

35. Phillips, R.J.; Burdick, M.D.; Hong, K. Circulating fibrocytes traffic to the lungs in response to CXCL12 and mediate fibrosis. *J. Clin. Investig.* **2004**, *114*, 438–446. [CrossRef] [PubMed]

36. Chen, Q.; Rabach, L.; Noble, P.; Zheng, T.; Lee, C.G.; Homer, R.J.; Elias, J.A. IL-11 receptor alpha in the pathogenesis of IL-13-induced inflammation and remodeling. *J. Immunol.* **2005**, *174*, 2305–2313. [CrossRef]

37. Therien, A.G.; Bernier, V.; Weicker, S.; Tawa, P.; Falgueyret, J.; Mathieu, M.; Honsberger, J.; Pomerleau, V.; Robichaud, A.; Stocco, R.; et al. Adenovirus IL-13-induced airway disease in mice: A corticosteroid-resistant model of severe asthma. *Am. J. Respir. Cell Mol. Biol.* **2008**, *39*, 26–35. [CrossRef]

38. Botelho, F.M.; Rangel-Moreno, J.; Fritz, D.; Randall, T.D.; Xing, Z.; Richards, C.D. Pulmonary Expression of Oncostatin M (OSM) Promotes Inducible BALT Formation Independently of IL-6, Despite a Role for IL-6 in OSM-Driven Pulmonary Inflammation. *J. Immunol.* **2013**, *191*, 1453–1464. [CrossRef]
39. Ashcroft, T.; Simpson, J.M.; Timbrell, V. Simple method of estimating severity of pulmonary fibrosis on a numerical scale. *J. Clin. Pathol.* **1998**, *41*, 467–470. [CrossRef]
40. Dubey, A.; Izakelian, L.; Ayaub, E.A.; Ho, L.; Stephenson, K.; Wong, S.; Kwofie, K.; Austin, R.C.; Botelho, F.; Ask, K.; et al. Separate roles of IL-6 and oncostatin M in mouse macrophage polarization in vitro and in vivo. *Immunol. Cell Biol.* **2018**, *96*, 257–272. [CrossRef]
41. Langdon, C.; Kerr, C.; Tong, L.; Richards, C.D. Oncostatin M regulates eotaxin expression in fibroblasts and eosinophilic inflammation in C57BL/6 mice. *J. Immunol.* **2003**, *170*, 548–555. [CrossRef] [PubMed]
42. Grenier, A.; Combaux, D.; Chastre, J.; Gougerot-Pocidalo, M.A.; Gibert, C.; Dehoux, M.; Chollet-Martin, S. Oncostatin M production by blood and alveolar neutrophils during acute lung injury. *Lab. Investig.* **2001**, *81*, 133–141. [CrossRef] [PubMed]
43. Moeller, A.; Gilpin, S.E.; Ask, K.; Cox, G.; Cook, D.; Gauldie, J.; Margetts, P.J.; Farkas, L.; Dobranowski, J.; Boylan, C.; et al. Circulating fibrocytes are an indicator of poor prognosis in idiopathic pulmonary fibrosis. *Am. J. Respir. Crit. Care Med.* **2009**, *179*, 588–594. [CrossRef] [PubMed]
44. Bellini, A.; Marini, M.A.; Bianchetti, L.; Barczyk, M.; Schmidt, M.; Mattoli, S. Interleukin (IL)-4, IL-13 and IL-17A differentially affect the profibrotic an dproinflammatory functions of fibrocytes from asthmatic patients. *Mucosal Immunol.* **2012**, *5*, 140–149. [CrossRef] [PubMed]
45. Hohensinner, P.J.; Kaun, C.; Rychli, K.; Nlessner, A.; Pfaffenberger, S.; Rega, G.; Furnkranz, A.; Uhrin, P.; Zaujec, J.; Afonyushkin, T.; et al. The inflammatory mediator oncostatin M induces stromal derived factor-1 in human adult cardiac cells. *FASEB J.* **2009**, *23*, 774–782. [CrossRef]
46. Moore, B.B.; Kolodsick, J.E.; Thannickal, V.J.; Cooke, K.; Moore, T.A.; Hogaboam, C.; Wilke, C.A.; Toews, G.B. CCR2-mediated recruitment of fibrocytes to the alveolar space after fibrotic injury. *Am. J. Pathol.* **2005**, *166*, 675–684. [CrossRef]
47. Moore, B.B.; Murray, L.; Das, A.; Wilke, C.A.; Herrygers, A.B.; Toews, G.B. The role of CCL12 in the recruitment of fibrocytes and lung fibrosis. *Am. J. Respir. Cell Mol. Biol.* **2006**, *35*, 175–181. [CrossRef]
48. Sun, L.; Louie, M.C.; Vannella, K.M.; Wilke, C.A.; LeVine, A.M.; Moore, B.B.; Shanley, T.P. New concepts of IL-10-induced lung fibrosis: Fibrocyte recruitment and M2 activation in a CCL2/CCR2 axis. *Am. J. Physiol. Lung Cell Mol. Physiol.* **2011**, *300*, L341–L353. [CrossRef]
49. Ayaub, E.A.; Dubey, A.; Imani, J.; Botelho, F.; Kolb, M.R.J.; Richards, C.D.; Ask, K. Overexpression of OSM and IL-6 impacts the polarization of pro-fibrotic macrophages and the development of bleomycin-induced lung. *Sci. Rep.* **2017**, *7*, 13281. [CrossRef]

Review

Behavior of Metalloproteinases in Adipose Tissue, Liver and Arterial Wall: An Update of Extracellular Matrix Remodeling

Gabriela Berg [1,2,3,*], Magalí Barchuk [1,2] and Verónica Miksztowicz [1,2,3]

[1] Laboratorio de Lípidos y Aterosclerosis, Departamento de Bioquímica Clínica, Facultad de Farmacia y Bioquímica, Universidad de Buenos Aires, Buenos Aires 1113, Argentina; magalibarchuk@gmail.com (M.B.); veronicajmik@gmail.com (V.M.)

[2] Facultad de Farmacia y Bioquímica, Instituto de Fisiopatología y Bioquímica Clínica (INFIBIOC), Universidad de Buenos Aires, Buenos Aires 1113, Argentina

[3] Facultad de Farmacia y Bioquímica, CONICET, Universidad de Buenos Aires, Buenos Aires C1425FQB, Argentina

* Correspondence: gaberg@ffyb.uba.ar; Tel.: +5411-4964-8297; Fax: +5411-5950-8692

Received: 17 December 2018; Accepted: 1 February 2019; Published: 14 February 2019

Abstract: Extracellular matrix (ECM) remodeling is required for many physiological and pathological processes. Metalloproteinases (MMPs) are endopeptidases which are able to degrade different components of the ECM and nucleus matrix and to cleave numerous non-ECM proteins. Among pathological processes, MMPs are involved in adipose tissue expansion, liver fibrosis, and atherosclerotic plaque development and vulnerability. The expression and the activity of these enzymes are regulated by different hormones and growth factors, such as insulin, leptin, and adiponectin. The controversial results reported up to this moment regarding MMPs behavior in ECM biology could be consequence of the different expression patterns among species and the stage of the studied pathology. The aim of the present review was to update the knowledge of the role of MMPs and its inhibitors in ECM remodeling in high incidence pathologies such as obesity, liver fibrosis, and cardiovascular disease.

Keywords: extracellular matrix; metalloproteinases; adipose tissue; liver; arterial wall

---

1. Introduction

The extracellular matrix (ECM) is a multimolecular complex structure comprising collagen and elastin fibers, structural glycoproteins including fibronectin and laminin, and mucopolysaccharides. It is organized into a three-dimensional (3D) network, and in physiological states, a balance between synthesis, deposit and degradation of ECM components exists [1]. ECM provides the structural framework to cells. The composition of ECM varies among multicellular structures, with the fibroblasts and epithelial cells being the most common cell types in the stromal architecture. The local components and composition of ECM collectively determine the biochemical properties of the connective tissue [2]. Different pathological processes, such as adipose tissue (AT) expansion, fibrosis, and atherosclerotic plaque development and rupture are characterized by ECM remodeling, in which many regulator factors and proteolytic enzymes are involved. Metalloproteinases (MMPs) are the main actors involved in ECM degradation, and the regulation of MMPs expression and activity is crucial for tissues homeostasis. Changes in the MMPs pattern expression or in the balance between MMPs and their tissue-specific inhibitors alters ECM biology. The aim of the present review was to update the knowledge of the role of MMPs and its inhibitors in ECM remodeling in high incidence pathologies such as obesity, liver fibrosis, and cardiovascular disease.

2. Metalloproteinases Characteristics

Metalloproteinases (MMPs) constitute a family of zinc-calcium dependent endopeptidases able to degrade different components of ECM and to cleave numerous non-ECM proteins, such as adhesion molecules, cytokines, protease inhibitors, and membrane receptors. Structurally, MMPs are constituted by a N-terminal propeptide domain, a cysteine-containing switch motif Zn^{2+}-containing conservative catalytic domain, a C-terminal proline-rich hinge region, and a hemopexin domain [3].

So far, 28 MMPs are known which are classified according to their substrate specificity in the following: Collagenases (MMP-1, MMP-8, MMP-13 and MMP-18), which cleave interstitial collagen I, II, and III and also other ECM and non-ECM molecules, such as bradykinin and angiotensin I; gelatinases (MMP-2 and MMP-9), mainly responsible for type IV collagen degradation; stromelysins (MMP-3 and MMP-10), which degrade fibronectin, laminin, gelatins-I, -III, -IV and -V, collagen fibers and proteoglycans; matrilysins (MMP-7 and MMP-26), responsible for fibronectin, gelatins, and human plasminogen hydrolysis [4]. In addition, membrane-type MMPs (MT-MMPs), including transmembrane MMP-14, MMP-15, MMP-16 and MMP-24, and membrane-anchored MMP-17 and MMP-25 have been defined. MT-MMPs can degrade type-I, -II, and -III collagen and other components of ECM, and pro-MMP to active MMP. Finally, there are some non-classified MMPs, such as MMP-12, MMP-19, MMP-20, MMP-21, MMP-23, and MMP-28, in which expression is often tissue-specific [4] (Figure 1).

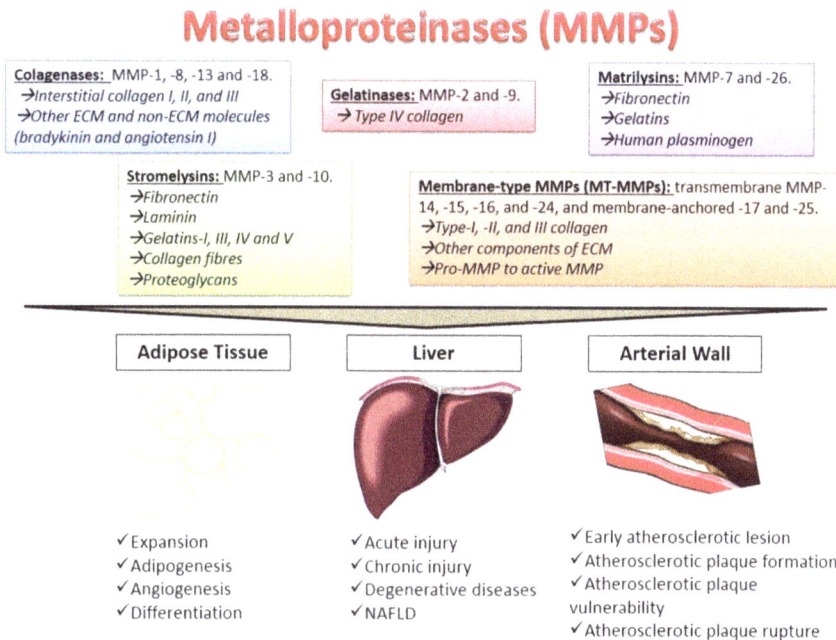

Figure 1. The metalloproteinases family and their main actions in adipose tissue, the liver, and the arterial wall. Arrows and italics represent Metalloproteinases (MMPs) substrates.

According to the MMPs functions mentioned above, these enzymes play important roles in physiological processes such cell proliferation, angiogenesis, and wound healing, among others; However, they are also involved in pathological processes including AT expansion, liver fibrosis and atherosclerosis [5] (Figure 1).

Besides the well-known role of MMPs in ECM remodeling, these enzymes are important in numerous nuclear events. It has been demonstrated that many MMPs have a nuclear localization signaling sequence that allows them to translocate into the nucleus after activation and regulate several mechanisms such as the degradation of nuclear proteins and transcription regulation [6]. Different studies suggest that nuclear MMPs could induce apoptosis in cardiac myocytes, endothelial cells [7], and renal tubular cells [8]. Moreover, some nuclear MMPs can bind directly to DNA promoters and regulate the transcription of multiple genes, such as heat shock family proteins and different growth factors [6].

3. MMPs Regulation

The MMPs are regulated at different levels: Gene expression, proteolytic activation of the proenzymes, inhibition of the catalytic activity by chemical and biological agents, and complexing with specific tissue inhibitors (TIMPs). MMPs are synthesized by different cells types such as endothelial cells, fibroblasts, adipocytes, hepatic stellate cells (HSC), and immune cells (Figure 2). In physiological conditions, they are expressed at some baseline levels but are differentially expressed in response to certain hormones, growth factors, and inflammatory and fibrogenic cytokines [9], via the activation of MAPK, JNK, and NF-κB-dependent signaling pathways [10].

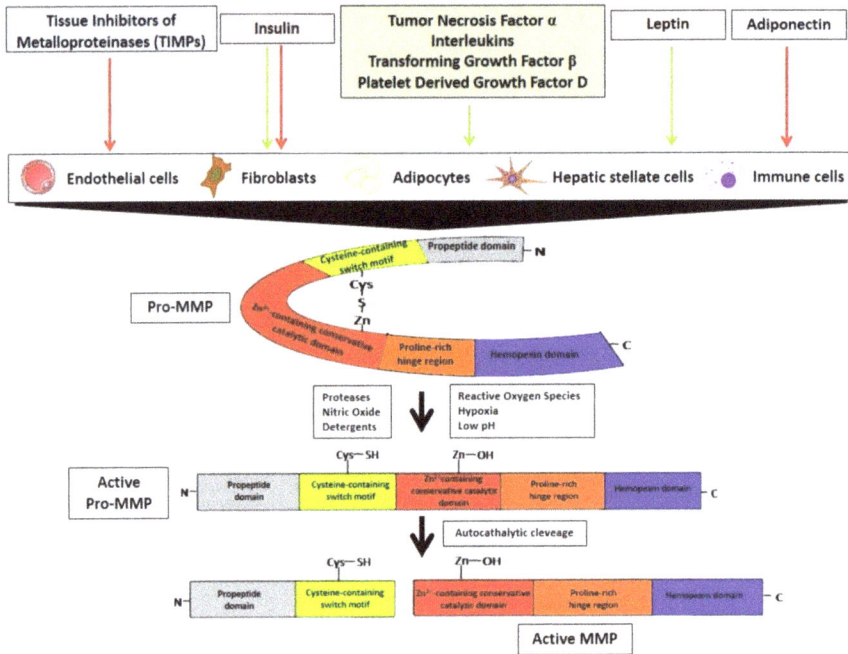

Figure 2. The activation of Metalloproteinases. MMPs are secreted by different cell types as inactive Pro-MMPs. These Pro-MMPs could be activated by different factors via the hydrolysis of the Cys-S-Zn bond. The active Pro-MMPs undergo auto-cleavage to generate the active MMPs. Many hormones regulate MMPs production by inhibiting and/or activating it. The green arrows indicate activation; red arrows indicate inhibition.

MMPs are secreted as latent zymogens (pro-MMPs) which are inactive due to the interaction of the zinc ion in the catalytic domain with the sulfhydryl group of the cysteine residue in the prodomain. Pro-MMPs are activated by the disruption of the cysteine switch by different mechanisms such as the proteolytic action of other proteases, conformational changes generated by nitric oxide (NO), reactive oxygen species (ROS) and hypoxia processes, or by chaotropic agents and denaturants such as sodium dodecyl sulfate and low pH and heat treatment, among others. In all cases the -SH group is replaced by H_2O and the enzyme gains the ability to hydrolyze the propeptide for its complete activation [11] (Figure 2).

Finally, MMPs can be inhibited by TIMPs. Four TIMPs (TIMP-1, TIMP-2, TIMP-3, and TIMP-4) have been reported which form a complex in MMP catalytic domains in a 1:1 stoichiometric ratio. TIMPs have a 2-domain structure in which the N-terminal domain contains the inhibitory residues [12]. TIMPs inhibit most MMPs, except those from TIMP-1, which has a weak inhibitory effect only on MT1-MMP, MT3-MMP, MT5-MMP, and MMP-19 [12]. The local balance between activated MMPs and TIMPs determines the net result of MMPs activity in tissues. In pathological situations, this balance might be disrupted leading to the uncontrolled activation of the MMPs [13].

As mentioned above, MMPs expression is regulated by different factors. Insulin or insulin-like growth factor-1 signaling, through the PI3K/Akt cascade, regulates MMPs in different ways according to the target organ. In animal models, it has been demonstrated that insulin promotes the activation of gelatinases and MT1-MMP through the increase of pro-inflammatory cytokines production [14]. However, hyperinsulinemia showed a contrary effect in the liver, decreasing the expression of the MMP-2, MMP-9, and MT1-MMP [15].

Leptin and adiponectin are adipocytokines synthetized by AT, which also modulate MMPs expression in different manners. Leptin has central and peripheral effects on regulating food intake and energy expenditure, and it is involved in ECM remodeling by regulating the expression of MMPs and TIMPs in many tissues. In vitro studies have reported that leptin promotes MMP-2 secretion by 3T3-L1 preadipocytes [16]. Schram et al. [17] reported that incubation of rat cardiac myofibroblasts with leptin significantly stimulated MT1-MMP expression, resulting in an increase in MMP-2 activity, without changes in protein levels. More recently, studies in vitro and in vivo demonstrated that leptin induces MMP-9 expression in smooth muscle cells [18]. Liberale et al. [19] studied serum levels of leptin and MMP-8 in morbidly obese type 2 diabetes mellitus patients before and after bariatric surgery, showing a significant decrease in both proteins after surgery. The authors concluded that bariatric surgery would be associated with an acute abrogation of leptin affecting MMP-8 levels.

Adiponectin is an anti-inflammatory and anti-atherogenic adipocytokine, inversely associated with expanded AT. Beneficial effects of adiponectin on the vascular wall and atherosclerotic plaque have been further studied. In vitro studies have reported that adiponectin inhibits foam cell formation by increasing NO levels and reducing class A scavenger receptor expression in macrophages [20]. In reference to MMPs, it has been reported that adiponectin would decrease MMPs activity through the increase of TIMPs expression and secretion in human monocyte-derived macrophages via the Syk pathway [21]. In patients, Miksztowicz et al. [22] reported an inverse association between plasma adiponectin levels and circulating activity of MMP-2 in patients with insulin resistance. In accordance, Kou et al. [23] recently demonstrated that adiponectin levels inversely correlate with circulating MMP-9 levels in non-diabetic hypertensive patients.

4. The Role of MMPs in Different Tissues

4.1. Adipose Tissue

The expansion of AT is associated with adipogenesis and angiogenesis [24] and different studies have demonstrated that MMPs are involved in both processes. AT is constituted by different cell types (adipocytes, preadipocytes, fibroblasts, endothelial cells, and immune cells) which can be sources of MMPs. Until now, there are several studies about MMPs behavior in AT, still with controversial results, probably due to the different pattern expression that these enzymes present according to the experimental model and the AT distribution.

In abdominal AT from obese animal models, an increase of MMP-3, MMP-11, MMP-12, MMP-13 and MMP-14 levels with decreased expression of MMP-7, MMP-9, MMP-16, MMP-24 and TIMP-4 [25] has been reported. In our laboratory, we also investigated the behavior of MMPs in visceral AT from an animal model of obesity induced by a high-fat diet. In this model, an increase of MMP-9 activity was observed, inversely associated with PPARγ levels, without differences in MMP-2 [26]. In contrast, a decrease in MMP-2 and MMP-9 activity in AT from an animal model of early insulin-resistance (IR) induced by a sucrose-rich diet, without changes in MMPs plasma activity, has been reported [27]. Furthermore, studies in vitro in 3T3-F442A preadipocytes demonstrated that MMPs and TIMPs are expressed with different patterns during the adipogenesis process, with some discrepancies in comparison to in vivo models [25]. MMP-2, MMP-9, MMP-11, and MMP-16 and TIMP-2 mRNAs were expressed in the in vitro differentiated adipocytes; however, they were not detected in isolated mature adipocytes from fat pads obtained from obese mice. These differences could be due to the uncompleted differentiation observed in vitro. Indeed, although a homogenous population of mature adipocytes was isolated from fat pads, in vitro adipogenesis generated a mixture of cells at different stages of differentiation. Moreover, adipocytes generated in vitro displayed a multilocular morphology that differed from the unilocular adipocytes observed in vivo [25].

In obesity, hypoxia and inflammation occur as a consequence of AT expansion, leading to an increase of proinflammatory adipocytokines secretion. As previously mentioned, adipocytokines are involved in the regulation of MMPs expression. In vitro studies have demonstrated that MMP-2 secretion was significantly promoted by leptin treatment in 3T3-L1 preadipocytes [16]. Little is known about the direct effect of adiponectin on MMPs from AT. It has been suggested that adiponectin could directly affect the balance of MMP/TIMP expression in macrophages from AT [13]; however, further studies are necessary to elucidate the role of this cytokine in MMPs behavior in this tissue.

Obesity is also characterized by collagen deposition in AT. Song et al. [28] studied the association between Toll-like receptor (TLR)-2 and collagen I in AT from TLR-2 knockout mice fed a high-fat diet. The authors reported decreased levels of collagen I and TIMP-1 with increased levels of MMP-1 [28], suggesting that TLR-2 would be involved in collagen I metabolism and would exhibit a role in MMP-1 and TIMP-1 behavior in AT from obese mice. In AT, adipose-derived mesenchymal stem cells (AMSCs) are also responsible for the remodeling of three-dimensional ECM barriers during differentiation. However, the molecular mechanism is still not completely described. Almalki et al. [29] recently studied in vitro the expression of MMPs during the differentiation of AMSCs isolated from porcine abdominal AT to endothelial cells [29]. The authors demonstrated that the up-regulation of MMP-2 and MMP-14 has an inhibitory effect on the differentiation of AMSCs to endothelial cells; the silencing these MMPs inhibits the cleavage of the VEGF-receptor and stimulates the differentiation of AMSCs to endothelial cells and consequently the formation of endothelial tubes. These findings provide a potential mechanism for the regulatory roles of MMP-2 and MMP-14 in the re-endothelialization of coronary arteries.

MMPs have also been studied in human AT. Gummesson et al. [30] studied the gene expression of MMP-9 in subcutaneous AT and plasma in men with and without metabolic syndrome treated with a weight-reducing diet. The authors found a lack of association between AT mRNA and plasma levels of MMP-9, suggesting that this tissue is not a major contributor to circulating MMP-9. Nowadays, one of the main treatments for obesity is bariatric surgery (BS). The impact of BS on MMPs is controversial. It has been reported that in morbidly obese patients, serum MMP-2 and MMP-9 levels significantly decrease after BS [31]. Otherwise, in diabetic obese patients, serum MMP-7 levels remained unchanged after BS [32]. In obese nondiabetic patients, Liu et al. [33] reported an increase in degradation of collagen I and III in subcutaneous AT 1 year after BS, accompanied by increased MMP-2 and MMP-9 activity; however, these differences were not observed in obese diabetic patients. Further studies are needed to explore the behavior and the balance between MMPs and TIMPs in the context of AT remodeling after BS.

In the last few years, attention has been focused on epicardial adipose tissue (EAT), a visceral AT, which surrounds the myocardium and coronary arteries. Given that there are no fascial boundaries between EAT and heart, this AT has a direct impact on coronaries and myocardium. Different studies have demonstrated that expanded EAT would be associated with coronary artery disease (CAD) [34,35]. The expansion of EAT in CAD requires the plasticity of ECM, in which increased MMPs activity could be involved. Recently, we reported an increase in MMP-2 and MMP-9 activity in EAT from CAD patients compared to patients with no CAD, accompanied by an augmented vascular density [36]. MMPs are essential during the angiogenic process, and even more during the ECM degradation [37]. In EAT from CAD patients, we also reported the association between MMPs activity and vascular endothelial growth (VEGF) levels [36]. Qorri et al. [38] demonstrated that MMP-9 releases VEGF bound to proteoglycans within ECM, enhancing the angiogenesis process. In EAT, MMP-2, and MMP-9 were located mainly in perivascular connective stroma and in the basement membrane surrounding the adipocytes [36]. These results suggest that both MMPs would be partially responsible for the ECM remodeling and the major vascular density necessary for EAT expansion in CAD patients.

Different studies have demonstrated that expanded EAT in CAD patients is infiltrated by inflammatory immune cells, mainly macrophages and lymphocytes [36,39]. In our laboratory, to determine the macrophages' polarization to the proinflammatory (M1) or the anti-inflammatory (M2) phenotype, M1 and M2 markers were evaluated. In EAT from CAD patients, an increase of M1 markers was observed, associated with elevated levels of MMP-1, an indicator of activated M1 macrophages (data not published). These results would confirm the observed inflammatory profile of EAT in CAD patients [35,36].

4.2. The Liver

Acute and chronic liver diseases are characterized by ECM remodeling, which is necessary for the fibrogenesis observed in these pathologies. The most abundant protein within ECM is collagen, of which types I, III, and V are predominant in healthy livers, whereas type IV collagen and laminin are essential constituents of basement membranes. The dysregulation of ECM homeostasis is often associated with degenerative diseases [40]. Depending on the nature of the liver injury, the normal composition of ECM in the space of Disse, the Glisson's capsule, around portal tracts, and around the central veins is replaced by up to tenfold the amount of type I and type III collagen during fibrogenesis [41]. In healthy livers, HSCs, localized in the perisinusoidal space, are the most important source of ECM. The maintenance of the dynamic structure and composition of the ECM depends on an exactly regulated, moderate turn-over directed by MMPs [42]. Many MMPs are absent or constitutively expressed in healthy liver, but after hepatic injury, their expression increases and probably mediates the ECM breakdown and control of cellular functions. TIMP-1 and TIMP-2 are secreted by HSCs and are abundantly expressed in the liver. Studies in animal models of liver fibrosis suggest that an imbalance between TIMPs and MMPs leads to fibrogenesis [43].

Until now, the nature of the behavior of MMPs in the liver fibrosis process has been controversial. Different factors could lead to acute liver injuries, such as toxic xenobiotic metabolites and hepatotrophic viruses producing apoptosis and necrosis of hepatocytes. Beside the role of HSC in secreting ECM components, these cells are also the principal source of acute liver injury-associated gene expression and the early state activation of MMPs. Different studies have demonstrated that, directly after liver injury and previous to fibrogenesis, an increase in some MMPs expression occurs [42,44]. Injured hepatocytes release ROS and pro-inflammatory cytokines with a capacity of activated HSCs which proliferate and secrete ECM components contributing to liver fibrosis. It has been suggested that MMPs activation in the space of Disse would be crucial for the fibrotic activation of HSC in ECM and in liver fibrogenesis [45]. Insulin-like growth factor binding protein-related protein 1 (IGFBPrP1) has been proposed as a novel mediator of hepatic fibrogenesis. Recently, Ren et al. [46] reported that hepatic expression of IGFBPrP1 was increased in an animal model with induced hepatic fibrosis, associated with a concomitant MMP2/TIMP2 and MMP9/TIMP1 imbalance. Furthermore, the IGFBPrP1 knockdown attenuated liver fibrosis by re-establishing the MMP2/TIMP2 and MMP9/TIMP1 balance concomitant with the inhibition of HSCs activation and degradation of the ECM [46].

In reference to liver chronic damage, non-alcoholic fatty liver disease (NAFLD) is the most prevalent chronic hepatic disease in the world, and it is considered to be the hepatic manifestation of metabolic syndrome. One-third of patients with NAFLD progress to liver fibrosis within 4–5 years after the first liver biopsy. The increased levels of pro-inflammatory adipocytokines, characteristic of NAFLD, could be partially responsible for liver damage through the alteration of MMPs and TIMPs balance, leading to enhanced matrix deposition by HSCs and diminished breakdown of connective tissue proteins. Toyoda et al. [47] reported that mRNA and serum levels of MMP-2 were up-regulated in hepatic steatosis; however, Miele et al. [48] found only increased TIMP-1 serum levels. In our laboratory, we recently described a decrease in liver MMP-2 activity in severe fibrosis in comparison to non- to moderate fibrosis stages and no differences in MMP-9 activity in patients with NAFLD [49]. In accordance, Hemman et al. [50] previously reported that in activated HSCs, TIMP-1 expression is increased, and concomitantly MMPs activity is reduced, leading to protein accumulation. Recently, an increase of pro-MMP-2 accompanied by increased TIMP-1 and TIMP-2 levels in NAFLD liver fibrosis [51] has been reported. In accordance, Yilmaz et al. [52] described an increase of TIMP-1 serum levels in patients with significant fibrosis, and this was identified as an independent predictor of histological fibrosis.

Different antidiabetic drugs have shown antifibrotic properties in the liver. In animal models, it has been reported that treatment with metformin protects against hepatic fibrosis by decreasing the expression of profibrogenic biomarkers such as TIMP-1 [53]. Furthermore, in an experimental liver fibrosis model, a dipeptidyl peptidase-4 inhibitor (DPP4-I), sitagliptin, attenuated liver fibrosis development along with the suppression of hepatic transforming growth factor (TGF)-β1, total collagen, and TIMP-1 levels in a dose-dependent manner [54]. These suppressive effects occurred almost concurrently with the attenuation of HSCs activation [54]. These results suggest therapeutic potential for antidiabetic drugs in humans with liver fibrosis.

4.3. The Arterial Wall

It has been proposed that MMPs participate in atherosclerotic plaque development and vulnerability [55] due to their role in ECM degradation. Numerous studies in cell culture, animal models, and humans have deepened our understanding of the behavior of these enzymes and inhibitors in the arterial wall. It is well known that the sentinel cause of plaque formation is the lipid deposition in the intima promoting foam cell development. Modified lipoproteins such as small dense LDL, oxidized LDL (ox-LDL), and remnants of lipoproteins are mainly taken up by macrophages via scavenger receptors (SR), with particularly the type A (SR-A) and CD36 contributing to early foam cell formation and atherosclerotic plaque progression [56]. At the same time, activated vascular smooth muscle cells (VSMC) migrate to the intimal layer and proliferate, producing ECM components. Subsequently, intimal growth is mediated by an increase in the number and diversity of cells and by the accumulation of new ECM for the development and maintenance of the fibrous cap. Ox-LDL promotes MMPs expression through different pathways. The exposure of human macrophages to ox-LDL up-regulates MMP-1 and MMP-14 via NF-κB [57]. These MMPs could participate in the early atherosclerotic plaque development because they are found in early fatty streak lesions [58]. Lectin-like oxidized low-density lipoprotein receptor-1 (LOX-1) is the major ox-LDL receptor in ECs [59], VSMC, and macrophages [60].

Studies in HUVEC demonstrated that activated expression of LOX-1 with ox-LDL up-regulates inflammatory proteins and MMP-2 and MMP-9 expression [61]. On the other hand, MMP-2, MMP-9, and MMP-14 promote VSMC migration and proliferation, which could increase fibrous cap thickness and promote plaque stability [62]. Johnson et al. [63] found that stimulation of macrophages with ox-LDL did not affect MMP-14 and TIMP-3 mRNA expression. However, they observed that the later polarization of macrophages to the M1 phenotype increased mRNA expression of MMP-14 and decreased mRNA levels of TIMP-3, leading to heightened invasive capability, increased proliferation, and augmented susceptibility to apoptosis. In contrast, the macrophages' polarization to the anti-inflammatory type M2 produced the opposite effect. The fact that foam cells expressed high MMP-14 and less TIMP-3 protein could implicate epigenetic mechanisms [63].

Within the early atherosclerotic lesions, due to the progressive accumulation of lipids and inflammatory cells, fatty streaks tend to progress to advanced lesions, containing lipid droplets, foam cells, macrophages, and lymphocytes. These inflammatory cells have crucial roles in atherosclerosis progression, expressing and releasing different cytokines, which mediate the chemotaxis of additional macrophages, as well as B and T cells [64], raising aberrant inflammatory responses and fostering an inflammatory milieu. In this environment, the high-level expression of MMPs contributes to plaque instability. TNF-α, IL-1β, and CD40L stimulate MMP expression in both monocytes and macrophages [64]. As described in AT, in isolated human plaque-derived cells, Monaco et al. [65] verified that activation of TLR-2 was directly implicated in MMP-1 and MMP-3 production. These inflammatory mediators share the ability to activate the MAPKs, extracellular signal-related kinases 1/2 (ERK1/2), p38 MAPK, and c-jun N-terminal kinase (JNK), as well as phosphoinositide-3 kinase (PI3 kinase) and the inhibitor of κB kinase-2 (IKK2) that leads to the activation of NF-κB [64], which may account for the upregulation of MMPs [66]. More recently, in an apolipoprotein E (ApoE) knockout model with mice fed a high-fat diet, it was demonstrated that T-box21 (T-bet) knockout has no effect on M1 macrophage polarization and only marginally increases M2 polarization. Moreover, T-bet knockout does not reduce and may even increase the mRNA expression of several MMPs that have been shown to promote plaque development [67]. The chronic state of hypoxia found in advanced atherosclerotic plaque increases MMP-7 expression [68], and transcriptomic data from hypoxic macrophages indicates that MMP-1, MMP-3, MMP-10, and MMP-12 are also significantly upregulated, possibly due to increased production of IL-1α,β [69]. Moreover, hypoxia-inducible factor-1α and hypoxia-inducible factor-2α [70] and JAK2/STAT-3 pathways have been implicated in the overexpression of these MMPs [71].

In contrast, the overexpression of suppressors of cytokine signaling (SOCS) proteins inhibits MMP-1 and MMP-9 expression in monocytes and macrophages. PPARα selectively inhibits MMP-12 production by directly binding to components of the AP-1 complex [72], whereas PPARα and PPARγ inhibit MMP-9 secretion from human macrophages [64]. Herein the role of PPAR agonists as potential therapeutic options must be highlighted.

Nevertheless, given the significant differences in the MMP mRNAs expression pattern and in the morphological characteristics of the atherosclerotic plaques among species, the extrapolation from animal models to humans is discouraging. Mice and human regulate a different spectrum of proteinases, including MMPs, during physiological responses [55]. So far, studies of genome-wide association (GWAS) in humans have been important. These studies provided evidence of the pathogenic role of MMP-12, in which gen was significantly overexpressed in carotid plaques compared to atherosclerosis-free control arteries [73]. GWAS studies also indicated that genetic variation determines a significant portion of circulating MMP-8 concentrations associated with cardiovascular disease [74]. Biobank studies have also demonstrated an association between MMP-8, -9 and -12 with plaque morphologies, suggesting vulnerability to rupture [64]. Further modern studies involving Mendelian randomization should consider MMPs association with plaque rupture and cardiovascular disease.

Statins are known to exert stabilizing effects on atherosclerotic plaque. Studies in vitro demonstrated that atorvastatin reduces the matrix degradation potential of proinflammatory macrophages by reducing intracellular MMP-14 activation [75]. In patients with carotid artery stenosis, Eilenberg et al. [76] reported that circulating neutrophil gelatinase-associated lipocalin (NGAL) and MMP-9/NGAL are associated with plaque vulnerability and that statin treatment could contribute to plaque stabilization by reducing circulating NGAL and MMP-9/NGAL levels. Finally, incretin analog drugs have been also tested for therapeutic properties as modulators of atherosclerosis. Gallego-Colon et al. [77] have recently shown that Exenatide downregulates the expression of MMP-1, MMP-2, and MMP-9 in coronary artery smooth muscle cells under inflammatory conditions, suggesting a possible role of incretin analog drugs in therapy for coronary atherosclerosis.

5. Conclusions

MMPs and TIMPs are the main actors in the ECM remodeling, and their behavior is altered in pathological states. The imbalance between the synthesis and degradation of ECM components underlies the development of many diseases. The controversial results reported up to this moment could be consequences of the pattern of expression and activity of MMPs and their inhibitors, which depend on the stage of the diseases.

Targeting the ECM as a potential therapeutic option should involve MMPs and TIMPs modulation. A deeper understanding of the MMPs behavior in ECM biology must be attained to uncover new targets for future therapies in high incidence pathologies such as obesity and hepatic and cardiovascular diseases.

Author Contributions: G.B. and V.M. were responsible for bibliographic research and for writing the manuscript. M.B. was responsible for figures design and development.

Funding: This research was funded by a grant from University of Buenos Aires [Funder Id: 10.13039/501100005363, Grant Number: N B036 (2016-2018)] Buenos Aires, Argentina and a grant from the National Agency of Promotion of Science and Technology [Grant Number: N PICT 2016-0920].

Conflicts of Interest: The authors declare no conflict of interest.

References

1. Robert, S.; Gicquel, T.; Victoni, T.; Valença, S.; Barreto, E.; Bailly-Maître, B. Involvement of matrix metalloproteinases (MMPs) and inflammasome pathway in molecular mechanisms of fibrosis. *Biosci. Rep.* **2016**, *36*, e00360. [CrossRef]
2. Yue, B. Biology of the extracellular matrix: An overview. *J. Glaucoma* **2014**, *23*, S20–S23. [CrossRef] [PubMed]
3. Benjamin, M.M.; Khalil, R.A. Matrix metalloproteinase inhibitors as investigative tools in the pathogenesis and management of vascular disease. *Exp. Suppl.* **2012**, *103*, 209–279. [PubMed]
4. Chen, Q.; Jin, M.; Yang, F.; Zhu, J.; Xiao, Q.; Zhang, L. Matrix metalloproteinases: Inflammatory regulators of cell behaviors in vascular formation and remodeling. *Mediat. Inflamm.* **2013**, *2013*, 928315. [CrossRef] [PubMed]
5. Berg, G.; Miksztowicz, V.; Schreier, L. Metalloproteinases in metabolic syndrome. *Clin. Chim. Acta* **2011**, *412*, 1731–1739. [CrossRef] [PubMed]
6. Xie, Y.; Mustafa, A.; Yerzhan, A.; Merzhakupova, D.; Yerlan, P.; Orakov, A.; Wang, X.; Huang, Y.; Miao, L. Nuclear matrix metalloproteinases: Functions resemble the evolution from the intracellular to the extracellular compartment. *Cell Death Discov.* **2017**, *3*, 17036. [CrossRef]
7. Aldonyte, R.; Brantly, M.; Block, E.; Patel, J.; Zhang, J. Nuclear localization of active matrix metalloproteinase-2 in cigarette smoke-exposed apoptotic endothelial cells. *Exp. Lung Res.* **2009**, *35*, 59–75. [PubMed]
8. Tsai, J.P.; Liou, J.H.; Kao, W.T.; Wang, S.C.; Lian, J.D.; Chang, H.R. Increased expression of intranuclear matrix metalloproteinase 9 in atrophic renal tubules is associated with renal fibrosis. *PLoS ONE* **2012**, *7*, e48164. [CrossRef] [PubMed]
9. Roderfeld, M. Matrix metalloproteinase functions in hepatic injury and fibrosis. *Matrix Biol.* **2018**, *68–69*, 452–462. [CrossRef]
10. Liang, K.C.; Lee, C.W.; Lin, W.N.; Lin, C.C.; Wu, C.B.; Luo, S.F.; Yang, C.M. Interleukin-1 beta induces MMP-9 expression via p42/p44 MAPK, p38 MAPK, JNK, and nuclear factor-kappaB signaling pathways in human tracheal smooth muscle cells. *J. Cell. Physiol.* **2007**, *211*, 759–770. [CrossRef]
11. Yamamoto, K.; Murphy, G.; Troeberg, L. Extracellular regulation of metalloproteinases. *Matrix Biol.* **2015**, *44–46*, 255–263. [CrossRef] [PubMed]
12. Nagase, H.; Visse, R.; Murphy, G. Structure and function of matrix metalloproteinases and TIMPs. *Cardiovasc. Res.* **2006**, *69*, 562–573. [CrossRef] [PubMed]
13. Berg, G.; Miksztowicz, V. Metalloproteinases in the pathogenesis and progression of metabolic syndrome: Potential targets for improved outcomes. *J. Metalloproteinases in Med.* **2015**, *2*, 51–59. [CrossRef]
14. Boden, G.; Song, W.; Pashko, L.; Kresge, K. In vivo effects of insulin and free fatty acids on matrix metalloproteinases in rat aorta. *Diabetes* **2008**, *57*, 476–483. [CrossRef]

15. Boden, G.; Song, W.; Kresge, K.; Mozzoli, M.; Cheung, P. Effects of hyperinsulinemia on hepatic metalloproteinases and their tissue inhibitors. *Am. J. Physiol. Endocrinol. Metab.* **2008**, *295*, 692–697. [CrossRef]
16. Moon, H.S.; Lee, H.G.; Seo, J.H.; Chung, C.S.; Guo, D.D.; Kim, T.G.; Choi, Y.J.; Cho, C.S. Leptin-induced matrix metalloproteinase-2 secretion is suppressed by trans-10, cis-12 conjugated linoleic acid. *Biochem. Biophys. Res. Commun.* **2007**, *356*, 955–956. [CrossRef]
17. Schram, K.; Wong, M.M.; Palanivel, R.; No, E.K.; Dixon, I.M.; Sweeney, G. Increased expression and cell surface localization of MT1-MMP plays a role in stimulation of MMP-2 activity by leptin in neonatal rat cardiac myofibroblasts. *J. Mol. Cell Cardiol.* **2008**, *44*, 874–881. [CrossRef]
18. Liu, R.; Chen, B.; Chen, J.; Lan, J. Leptin upregulates smooth muscle cell expression of MMP-9 to promote plaque destabilization by activating AP-1 via the leptin receptor/MAPK/ERK signaling pathways. *Exp. Ther. Med.* **2018**, *16*, 5327–5333. [CrossRef]
19. Liberale, L.; Bonaventura, A.; Carbone, F.; Bertolotto, M.; Contini, P.; Scopinaro, N.; Camerini, G.B.; Papadia, F.S.; Cordera, R.; Camici, G.G.; et al. Early reduction of matrix metalloproteinase-8 serum levels is associated with leptin drop and predicts diabetes remission after bariatric surgery. *Int. J. Cardiol.* **2017**, *245*, 257–262. [CrossRef]
20. Matsuzawa, Y. Adiponectin: Identification, physiology and clinical relevance in metabolic and vascular disease. *Atheroscler. Suppl.* **2005**, *6*, 7–14. [CrossRef]
21. Hu, D.; Fukuhara, A.; Miyata, Y.; Yokoyama, C.; Otsuki, M.; Kihara, S.; Shimomura, I. Adiponectin regulates vascular endothelial growth factor-C expression in macrophages via Syk-ERK pathway. *PLoS ONE* **2013**, *8*, e56071. [CrossRef] [PubMed]
22. Miksztowicz, V.; Fernandez Machulsky, N.; Lucero, D.; Fassio, E.; Schreier, L.; Berg, G. Adiponectin predicts MMP-2 activity independently of obesity. *Eur. J. Clin. Investig.* **2014**, *44*, 951–957. [CrossRef] [PubMed]
23. Kou, H.; Deng, J.; Gao, D.; Song, A.; Han, Z.; Wei, J.; Jin, X.; Ma, R.; Zheng, Q. Relationship among adiponectin, insulin resistance and atherosclerosis in non-diabetic hypertensive patients and healthy adults. *Clin. Exp. Hypertens.* **2018**, *40*, 656–663. [CrossRef] [PubMed]
24. Lee, M.J.; Wu, Y.; Fried, S.K. Adipose tissue remodeling in pathophysiology of obesity. *Curr. Opin. Clin. Nutr. Metab. Care* **2010**, *13*, 371–376. [CrossRef] [PubMed]
25. Maquoi, E.; Munaut, C.; Colige, A.; Collen, D.; Lijnen, H.R. Modulation of adipose tissue expression of murine matrix metalloproteinases and their tissue inhibitors with obesity. *Diabetes* **2002**, *51*, 1093–1101. [CrossRef] [PubMed]
26. Barchuk, M.; Morales, C.; Zago, V.; Friedman, S.; Schreier, L.; Miksztowicz, V.; Berg, G. Gelatinases behavior in adipose tissue, heart and liver in a diet induced obesity model. *Medicina* **2016**, *76*, 196.
27. Miksztowicz, V.; Morales, C.; Zago, V.; Friedman, S.; Schreier, L.; Berg, G. Effect of insulin-resistance on circulating and adipose tissue MMP-2 and MMP-9 activity in rats fed a sucrose-rich diet. *Nutr. Metab. Cardiovasc. Dis.* **2014**, *24*, 294–300. [CrossRef]
28. Song, B.; Zhang, H.; Zhang, S. Toll-like receptor 2 mediates deposition of collagen I in adipose tissue of high fat diet-induced obese mice. *Mol. Med. Rep.* **2018**, *17*, 5958–5963. [CrossRef]
29. Almalki, S.G.; Llamas Valle, Y.; Agrawal, D.K. MMP-2 and MMP-14 Silencing Inhibits VEGFR2 Cleavage and Induces the Differentiation of Porcine Adipose-Derived Mesenchymal Stem Cells to Endothelial Cells. *Stem Cells Transl. Med.* **2017**, *6*, 1385–1398. [CrossRef]
30. Gummesson, A.; Hagg, D.; Olson, F.J.; Hulthe, J.; Carlsson, L.M.; Fagerberg, B. Adipose tissue is not an important source for matrix metalloproteinase-9 in the circulation. *Scand. J. Clin. Lab. Investig.* **2009**, *69*, 636–642. [CrossRef]
31. Domienik-Karłowicz, J.; Rymarczyk, Z.; Dzikowska-Diduch, O.; Lisik, W.; Chmura, A.; Demkow, U.; Pruszczyk, P. Emerging markers of atherosclerosis before and after bariatric surgery. *Obes. Surg.* **2015**, *25*, 486–493. [CrossRef] [PubMed]
32. Yang, P.J.; Ser, K.H.; Lin, M.T.; Nien, H.C.; Chen, C.N.; Yang, W.S.; Lee, W.J. Diabetes Associated Markers After Bariatric Surgery: Fetuin-A, but Not Matrix Metalloproteinase-7, Is Reduced. *Obes. Surg.* **2015**, *25*, 2328–2334. [CrossRef] [PubMed]
33. Liu, Y.; Aron-Wisnewsky, J.; Marcelin, G.; Genser, L.; Le Naour, G.; Torcivia, A.; Bauvois, B.; Bouchet, S.; Pelloux, V.; Sasso, M.; et al. Accumulation and Changes in Composition of Collagens in Subcutaneous Adipose Tissue After Bariatric Surgery. *J. Clin. Endocrinol. Metab.* **2016**, *101*, 293–304. [CrossRef] [PubMed]

34. Djaberi, R.; Schuijf, J.D; van Werkhoven, J.M.; Nucifora, G.; Jukema, J.W.; Bax, J.J. Relation of epicardial adipose tissue to coronary atherosclerosis. *Am. J. Cardiol.* **2008**, *102*, 1602–1607. [CrossRef] [PubMed]

35. Iacobellis, G.; Lonn, E.; Lamy, A.; Singh, N.; Sharma, A.M. Epicardial fat thickness and CAD correlate independently of obesity. *Int. J. Cardiol.* **2011**, *146*, 452–454. [CrossRef] [PubMed]

36. Miksztowicz, V.; Morales, C.; Barchuk, M.; López, G.; Póveda, R.; Gelpi, R.; Schreier, L.; Rubio, M.; Berg, G. Metalloproteinase 2 and 9 Activity Increase in Epicardial Adipose Tissue of Patients with Coronary Artery Disease. *Curr. Vasc. Pharmacol.* **2017**, *15*, 135–143. [CrossRef]

37. McKenney, M.L.; Schultz, K.A.; Boyd, J.H.; Byrd, J.P.; Alloosh, M.; Teague, S.D.; Arce-Esquivel, A.A.; Fain, J.N.; Laughlin, M.H.; Sacks, H.S.; et al. Epicardial adipose excision slows the progression of porcine coronary atherosclerosis. *J. Cardiothorac. Surg.* **2014**, *9*, 2. [CrossRef] [PubMed]

38. Qorri, B.; Kalaydina, R.V.; Velickovic, A.; Kaplya, Y.; Decarlo, A.; Szewczuk, M.R. Agonist-Biased Signaling via Matrix Metalloproteinase-9 Promotes Extracellular Matrix Remodeling. *Cells* **2018**, *7*, 117. [CrossRef]

39. Mazurek, T.; Zhang, L.; Zalewski, A.; Mannion, J.D.; Diehl, J.T.; Arafat, H.; Sarov-Blat, L.; O'Brien, S.; Keiper, E.A.; Johnson, A.G.; et al. Human epicardial adipose tissue is a source of inflammatory mediators. *Circulation* **2003**, *108*, 2460–2466. [CrossRef]

40. Friedman, S.L.; Maher, J.J.; Bissell, D.M. Mechanisms and therapy of hepatic fibrosis: Report of the AASLD Single Topic Basic Research Conference. *Hepatology* **2000**, *32*, 1403–1408. [CrossRef]

41. Schuppan, D.; Ruehl, M.; Somasundaram, R.; Hahn, E.G. Matrix as a modulator of hepatic fibrogenesis. *Semin. Liver Dis.* **2001**, *21*, 351–372. [CrossRef] [PubMed]

42. Roderfeld, M.; Geier, A.; Dietrich, C.G.; Siewert, E.; Jansen, B.; Gartung, C.; Roeb, E. Cytokine blockade inhibits hepatic tissue inhibitor of metalloproteinase-1 expression and upregulates matrix metalloproteinase-9 in toxic liver injury. *Liver Int.* **2006**, *26*, 579–586. [CrossRef] [PubMed]

43. Kurzepa, J.; Mądro, A.; Czechowska, G.; Kurzepa, J.; Celiński, K.; Kazmierak, W.; Slomka, M. Role of MMP-2 and MMP-9 and their natural inhibitors in liver fibrosis, chronic pancreatitis and non-specific inflammatory bowel diseases. *Hepatobiliary Pancreat Dis. Int.* **2014**, *13*, 570–579. [CrossRef]

44. Duarte, S.; Shen, X.D.; Fondevila, C.; Busuttil, R.W.; Coito, A.J. Fibronectin-α4β1 interactions in hepatic cold ischemia and reperfusion injury: Regulation of MMP-9 and MT1-MMP via the p38 MAPK pathway. *Am. J. Transplant.* **2012**, *12*, 2689–2699. [CrossRef] [PubMed]

45. Han, Y.P.; Zhou, L.; Wang, J.; Xiong, S.; Garner, W.L.; French, S.W.; Tsukamoto, H. Essential role of matrix metalloproteinases in interleukin-1-induced myofibroblastic activation of hepatic stellate cell in collagen. *J. Biol. Chem.* **2004**, *279*, 4820–4828. [CrossRef] [PubMed]

46. Ren, J.J.; Huang, T.J.; Zhang, Q.Q.; Zhang, H.Y.; Guo, X.H.; Fan, H.Q.; Li, R.K.; Liu, L.X. Insulin-like growth factor binding protein related protein 1 knockdown attenuates hepatic fibrosis via the regulation of MMPs/TIMPs in mice. *Hepatobiliary Pancreat. Dis. Int.* **2018**. [CrossRef] [PubMed]

47. Toyoda, H.; Kumada, T.; Kiriyama, S.; Tanikawa, M.; Hisanaga, Y.; Kanamori, A.; Tada, T.; Murakami, Y. Higher hepatic gene expression and serum levels of matrix metalloproteinase-2 are associated with steatohepatitis in non-alcoholic fatty liver diseases. *Biomarkers* **2013**, *18*, 82–87. [CrossRef] [PubMed]

48. Miele, L.; Forgione, A.; La Torre, G.; Vero, V.; Cefalo, C.; Racco, S.; Vellone, V.G.; Vecchio, F.M.; Gasbarrini, G.; Rapaccini, G.L.; et al. Serum levels of hyaluronic acid and tissue metalloproteinase inhibitor-1 combined with age predict the presence of nonalcoholic steatohepatitis in a pilot cohort of subjects with nonalcoholic fatty liver disease. *Transl. Res.* **2009**, *154*, 194–201. [CrossRef]

49. Barchuk, M.; Schreier, L.; Berg, G.; Miksztowicz, V. Metalloproteinases in non-alcoholic fatty liver disease and their behavior in liver fibrosis. *Horm. Mol. Biol. Clin. Investig.* **2018**. [CrossRef]

50. Hemmann, S.; Graf, J.; Roderfeld, M.; Roeb, E. Expression of MMPs and TIMPs in liver fibrosis a systematic review with special emphasis on anti-fibrotic strategies. *J. Hepatol.* **2007**, *46*, 955–975. [CrossRef]

51. Munsterman, I.D.; Kendall, T.J.; Khelil, N.; Popa, M.; Lomme, R.; Drenth, J.P.H.; Tjwa, E.T.T. Extracellular matrix components indicate remodelling activity in different fibrosis stages of human non-alcoholic fatty liver disease. *Histopathology* **2018**, *73*, 612–621. [CrossRef] [PubMed]

52. Yilmaz, Y.; Eren, F. Serum biomarkers of fibrosis and extracellular matrix remodeling in patients with nonalcoholic fatty liver disease: Association with liver histology. *Eur. J. Gastroenterol. Hepatol.* **2019**, *31*, 43–46. [CrossRef] [PubMed]

53. Al-Hashem, F.; Al-Humayed, S.; Amin, S.N.; Kamar, S.S.; Mansy, S.S.; Hassan, S.; Abdel-Salam, L.O.; Ellatif, M.A.; Alfaifi, M.; Haidara, M.A.; et al. Metformin inhibits mTOR-HIF-1α axis and profibrogenic and inflammatory biomarkers in thioacetamide-induced hepatic tissue alterations. *J. Cell. Physiol.* **2018**. [CrossRef]

54. Kaji, K.; Yoshiji, H.; Ikenaka, Y.; Noguchi, R.; Aihara, Y.; Douhara, A.; Moriya, K.; Kawaratani, H.; Shirai, Y.; Yoshii, J.; et al. Dipeptidyl peptidase-4 inhibitor attenuates hepatic fibrosis via suppression of activated hepatic stellate cell in rats. *J. Gastroenterol.* **2014**, *49*, 481–491. [CrossRef]

55. Newby, A.C. Metalloproteinases promote plaque rupture and myocardial infarction: A persuasive concept waiting for clinical translation. *Matrix Biol.* **2015**, *44–46*, 157–166. [CrossRef]

56. Chistiakov, D.A.; Melnichenko, A.A.; Myasoedova, V.A.; Grechko, A.V.; Orekhov, A.N. Mechanisms of foam cell formation in atherosclerosis. *J. Mol. Med.* **2017**, *95*, 1153–1165. [CrossRef] [PubMed]

57. Chase, A.; Bond, M.; Crook, M.F.; Newby, A.C. Role of nuclear factor-κB activation in metalloproteinase-1, -3 and -9 secretion by human macrophages in vitro and rabbit foam cells produced in vivo. *Arterioscler. Thromb. Vasc. Biol.* **2002**, *22*, 765–771. [CrossRef]

58. Aikawa, M.; Rabkin, E.; Sugiyama, S.; Voglic, S.J.; Fukumoto, Y.; Furukawa, Y.; Shiomi, M.; Schoen, F.J.; Libby, P. An HMG-CoA reductase inhibitor, cerivastatin, suppresses growth of macrophages expressing matrix metalloproteinases and tissue factor in vivo and in vitro. *Circulation* **2001**, *103*, 276–283. [CrossRef]

59. Chen, M.; Masaki, T.; Sawamura, T. LOX-1, the receptor for oxidized low-density lipoprotein identified from endothelial cells: Implications in endothelial dysfunction and atherosclerosis. *Pharmacol. Ther.* **2002**, *95*, 89–100. [CrossRef]

60. Sawamura, T.; Kume, N.; Aoyama, T.; Moriwaki, H.; Hoshikawa, H.; Aiba, Y.; Tanaka, T.; Miwa, S.; Katsura, Y.; Kita, T.; et al. An endothelial receptor for oxidized low-density lipoprotein. *Nature* **1997**, *386*, 73–77. [CrossRef]

61. Sugimoto, K.; Ishibashi, T.; Sawamura, T.; Inoue, N.; Kamioka, M.; Uekita, H.; Ohkawara, H.; Sakamoto, T.; Sakamoto, N.; Okamoto, Y.; et al. LOX-1-MT1-MMP axis is crucial for RhoA and Rac1 activation induced by oxidized low-density lipoprotein in endothelial cells. *Cardiovasc. Res.* **2009**, *84*, 127–136. [CrossRef] [PubMed]

62. Newby, A.C. Metalloproteinases and vulnerable atherosclerotic plaques. *Trends Cardiovasc. Med.* **2007**, *17*, 253–258. [CrossRef] [PubMed]

63. Johnson, J.L.; Jenkins, N.P.; Huang, W.C.; Di Gregoli, K.; Sala-Newby, G.B.; Scholtes, V.P.; Moll, F.L.; Pasterkamp, G.; Newby, A.C. Relationship of MMP-14 and TIMP-3 expression with macrophage activation and human atherosclerotic plaque vulnerability. *Mediators Inflamm.* **2014**, *2014*, 276457. [CrossRef] [PubMed]

64. Newby, A.C. Metalloproteinase production from macrophages—A perfect storm leading to atherosclerotic plaque rupture and myocardial infarction. *Exp. Physiol.* **2016**, *101*, 1327–1337. [CrossRef] [PubMed]

65. Monaco, C.; Gregan, S.M.; Navin, T.J.; Foxwell, B.M.; Davies, A.H.; Feldmann, M. Toll-like receptor-2 mediates inflammation and matrix degradation in human atherosclerosis. *Circulation* **2009**, *120*, 2462–2469. [CrossRef] [PubMed]

66. Schaper, F.; Rose-John, S. Interleukin-6: Biology, signaling and strategies of blockade. *Cytokine Growth Factor* **2015**, *26*, 475–487. [CrossRef]

67. Tsaousi, A.; Hayes, E.M.; Di Gregoli, K.; Bond, A.R.; Bevan, L.; Thomas, A.C.; Newby, A.C. Plaque Size Is Decreased but M1 Macrophage Polarization and Rupture Related Metalloproteinase Expression Are Maintained after Deleting T-Bet in ApoE Null Mice. *PLoS ONE* **2016**, *11*, e0148873. [CrossRef]

68. Sluimer, J.C.; Gasc, J.M; van Wanroij, J.L.; Kisters, N.; Groeneweg, M.; Sollewijn Gelpke, M.D.; Cleutjens, J.P.; van den Akker, L.H.; Corvol, P.; Wouters, B.G.; et al. Hypoxia, hypoxia-inducible transcription factor, and macrophages in human atherosclerotic plaques are correlated with intraplaque angiogenesis. *J. Am. Coll. Cardiol.* **2008**, *51*, 1258–1265. [CrossRef]

69. Fang, H.Y.; Hughes, R.; Murdoch, C.; Coffelt, S.B.; Biswas, S.K.; Harris, A.L.; Johnson, R.S.; Imityaz, H.Z.; Simon, M.C.; Fredlund, E.; et al. Hypoxia-inducible factors 1 and 2 are important transcriptional effectors in primary macrophages experiencing hypoxia. *Blood* **2009**, *114*, 844–859. [CrossRef]

70. Lee, Y.A.; Choi, H.M.; Lee, S.H.; Hong, S.J.; Yang, H.I.; Yoo, M.C.; Kim, K.S. Hypoxia differentially affects IL-1β-stimulated MMP-1 and MMP-13 expression of fibroblast-like synoviocytes in an HIF-1α-dependent manner. *Rheumatology* **2012**, *51*, 443–450. [CrossRef]

71. Gao, W.; McCormick, J.; Connolly, M.; Balogh, E.; Veale, D.J.; Fearon, U. Hypoxia and STAT3 signalling interactions regulate pro-inflammatory pathways in rheumatoid arthritis. *Ann. Rheum Dis.* **2015**, *74*, 1275–1283. [CrossRef] [PubMed]

72. Souissi, I.J.; Billiet, L.; Cuaz-Pérolin, C.; Slimane, M.N.; Rouis, M. Matrix metalloproteinase-12 gene regulation by a PPAR alpha agonist in human monocyte-derived macrophages. *Exp. Cell Res.* **2008**, *314*, 3405–3414. [CrossRef] [PubMed]

73. Traylor, M.; Mäkelä, K.M.; Kilarski, L.L.; Holliday, E.G.; Devan, W.J.; Nalls, M.A.; Wiggins, K.L.; Zhao, W.; Cheng, Y.C.; Achterberg, S.; et al. A novel MMP12 locus is associated with large artery atherosclerotic stroke using a genome-wide age-at-onset informed approach. *PLoS Genet.* **2014**, *10*, e1004469. [CrossRef] [PubMed]

74. Salminen, A.; Vlachopoulou, E.; Havulinna, A.S.; Tervahartiala, T.; Sattler, W.; Lokki, M.L.; Nieminen, M.S.; Perola, M.; Salomaa, V.; Sinisalo, J.; et al. Genetic Variants Contributing to Circulating Matrix Metalloproteinase 8 Levels and Their Association with Cardiovascular Diseases: A Genome-Wide Analysis. *Circ. Cardiovasc. Genet.* **2017**, *10*, e001731. [CrossRef] [PubMed]

75. Hohensinner, P.J.; Baumgartner, J.; Ebenbauer, B.; Thaler, B.; Fischer, M.B.; Huber, K.; Speidl, W.S.; Wojta, J. Statin treatment reduces matrix degradation capacity of proinflammatory polarized macrophages. *Vascul. Pharmacol.* **2018**, *110*, 49–54. [CrossRef]

76. Eilenberg, W.; Stojkovic, S.; Kaider, A.; Kozakowski, N.; Domenig, C.M.; Burghuber, C.; Nanobachvili, J.; Huber, K.; Klinger, M.; Neumayer, C.; et al. NGAL and MMP-9/NGAL as biomarkers of plaque vulnerability and targets of statins in patients with carotid atherosclerosis. *Clin. Chem Lab. Med.* **2017**, *56*, 147–156. [CrossRef] [PubMed]

77. Gallego-Colon, E.; Klych-Ratuszny, A.; Kosowska, A.; Garczorz, W.; Aghdam, M.; Wozniak, M.; Francuz, T. Exenatide modulates metalloproteinase expression in human cardiac smooth muscle cells via the inhibition of Akt signaling pathway. *Pharmacol. Rep.* **2018**, *70*, 178–183.

Article

Nuclear Progestin Receptor Phosphorylation by Cdk9 Is Required for the Expression of Mmp15, a Protease Indispensable for Ovulation in Medaka

Katsueki Ogiwara and Takayuki Takahashi *

Laboratory of Reproductive and Developmental Biology, Faculty of Science, Hokkaido University,
Sapporo 060–0810, Japan; kogi@sci.hokudai.ac.jp
* Correspondence: ttakaha@sci.hokudai.ac.jp

Received: 13 December 2018; Accepted: 26 February 2019; Published: 4 March 2019

Abstract: Ovulation denotes the discharge of fertilizable oocytes from ovarian follicles. Follicle rupture during ovulation requires extracellular matrix (ECM) degradation at the apex of the follicle. In the teleost medaka, an excellent model for vertebrate ovulation studies, LH-inducible matrix metalloproteinase 15 (Mmp15) plays a critical role during rupture. In this study, we found that follicle ovulation was inhibited not only by roscovitine, the cyclin-dependent protein kinase (CDK) inhibitor, but also by CDK9-inhibitor II, a specific CDK9 inhibitor. Inhibition of follicle ovulation by the inhibitors was accompanied by the suppression of Mmp15 expression in the follicle. In follicles treated with the inhibitors, the formation of the phosphorylated nuclear progestin receptor (Pgr) was inhibited. Roscovitine treatment caused a reduction in the binding of Pgr to the promoter region of *mmp15*. The expression of Cdk9 and cyclin I (Ccni), and their association in the follicle was demonstrated, suggesting that Cdk9 and Ccni may be involved in the phosphorylation of Pgr in vivo. LH-induced follicular expression of *ccni*/Ccni was also shown. This study is the first to report the involvement of CDK in ECM degradation during ovulation in a vertebrate species.

Keywords: medaka; ovulation; extracellular matrix degradation; cyclin-dependent kinase 9; Ccni; matrix metalloproteinase 15 expression; nuclear progestin receptor phosphorylation

---

1. Introduction

Vertebrate ovaries represent a vigorously dynamic structure due to the constant change in follicles over time, and such structural changes in the ovaries are closely associated with remodeling of the extracellular matrix (ECM). In particular, ECM remodeling underlies major changes during ovulation, which denotes the shedding of one or more viable oocytes from fully grown ovarian follicles into the reproductive tract. Ovulation is triggered by a surge in gonadotropin luteinizing hormone (LH) in all vertebrates [1–4]. It is widely recognized that follicle rupture during vertebrate ovulation involves proteolytic degradation at the apex of ovulating follicles. Using the teleost medaka, a suitable model for ovulation studies, proteolytic processes accompanying follicle rupture during ovulation have been explored [5]. Our previous studies demonstrated that activation of the plasminogen activator/plasmin (Plau/Plasmin) system and matrix metalloproteinase (Mmp) system is required for the hydrolysis of ECM proteins present in the follicle layers of ovulating follicles [6–9]. Upon activation of the Plau/Plasmin system, active plasmin is generated by proteolytic processing of the liver-derived precursor plasminogen by the follicle-produced active urokinase-type plasminogen activator-1 (Plau1). Plasmin, thus generated, is responsible for hydrolyzing laminin, a major ECM component of the basement membrane [7,8]. After the Plau/Plasmin system is kept active for several hours, another proteolytic system involving three distinct Mmp enzymes, membrane type 1-Mmp (Mt1-mmp/Mmp14), membrane type 2-Mmp (Mt2-mmp/Mmp15), and gelatinase A (Mmp2), becomes active. Mmp2 activated proteolytically by Mmp14 hydrolyzes type IV collagen, another principal component of the basement membrane, while Mmp15 degrades the type I collagen most abundantly present in the theca cell layer [6,9]. During these hydrolytic processes, plasminogen activator inhibitor-1 [8] and the tissue inhibitor of metalloproteinase-2b (Timp2b) [6] are involved in regulating the Plau/Plasmin system and Mmp system, respectively. Among the proteases involved in follicle layer ECM hydrolysis, Mmp15 is the only LH-inducible enzyme. Our recent study demonstrated that LH-induced expression of the *mmp15* gene is accomplished in two steps. In the first step, nuclear progestin receptor Pgr is induced by the LH surge, and the resulting Pgr is then complexed with 17α, 20β-dihydroxy-4-pregnen-3-one (17,20βP)—the physiological progestin ligand for medaka Pgr [10–12]—to become an active transcription factor. In the second step, activated Pgr, together with the transcription factor CCAAT/enhancer-binding protein β (Cebpb), contributes to the expression of *mmp15* mRNA [13].

In our attempt to search for genes/proteins involved in the expression of *mmp15* mRNA, we found that the cyclin-dependent protein kinase (CDK) inhibitor roscovitine inhibited not only follicle ovulation, but also the follicular expression of *mmp15* mRNA in the medaka, implicating CDK in the expression of the protease gene in the follicle that is destined to ovulate. In this study, we suggest that after phosphorylation, Pgr becomes a functional transcription factor for *mmp15* gene expression, and that Cdk9 and cyclin I (Ccni) are involved in the process of Pgr phosphorylation.

2. Materials and Methods

2.1. Animals and Tissues

Adult orange-red variety of medaka, *Oryzias latipes* (himedaka) were purchased from a local dealer and used for the experiments. The fish were maintained in aquariums under an illumination cycle, 14 h of light and 10 h of dark, at 26–27 °C [14], and were fed 3–4 times a day with commercial fish diet (Otohime, Nisshin Co. Tokyo, Japan). Fish under artificial lighting conditions ovulated every day around the transition time from dark to light period. Ovulation hour 0 was set to the start of the light period. Ovaries, ovarian follicles, follicle layers of the follicles, and oocytes were isolated from spawning female fish as previously described [14]. Animal cultures and experimentation were conducted in accordance with the guidelines for animal experiments of Hokkaido University and were approved by the Committee of Experimental Plants and Animals, Hokkaido University (16-0072).

2.2. In Vitro Culture of Isolated Follicles

In female medaka with a 24-h spawning cycle, postvitellogenic follicles undergo an LH surge approximately 18 h before ovulation and germinal vesicle breakdown (GVBD), an important milestone for oocyte maturation, which occurs 6 h before ovulation in vivo. A procedure for in vitro ovulation using postvitellogenic follicles was established [5]. In the current study, follicle cultures were conducted in 4 mL of 90% M199 medium containing 50 μM gentamycin (pH 7.4), using follicles isolated either 22 h before ovulation (designated as the −22 h-follicle) or 14 h before ovulation (designated as the −14 h-follicle). For the −22 h-follicles, which had not yet been exposed to the in vivo surge of LH, 100 μg/mL medaka recombinant LH (rLH) was included in the culture medium to initiate a series of ovulatory reactions in the follicles. The −14 h-follicles were incubated without medaka rLH, but with various other chemicals, because they had already been exposed to the ovulatory LH surge in vivo. The chemicals used were roscovitine (Merck Millipore, Billerica, MA, USA), MEK inhibitor PD98059 (Merck Millipore), CDK9 inhibitor II (Merck Millipore), and RU486 (also known as mifepristone, Sigma-Aldrich, St. Louis, MO, USA). Compared to the in vivo situation, GVBD and follicle ovulation take more hours under the in vitro culture. Rates of GVBD and ovulation in the incubated follicles were assessed. In addition, expression levels of various genes/proteins in follicles or follicle layers of the follicles were determined. An outline of the in vitro follicle culture used in this study is shown in Figure 1. Follicles were obtained from two to three fish ovaries, pooled, and then divided into control and test groups. The number of follicles per group was approximately 20–25. The duration of incubation and time points at which follicles and/or follicle layers were collected for target gene expression analysis are indicated in the text. Medaka rLH was produced in Chinese hamster ovary k-1 cells as previously described [14].

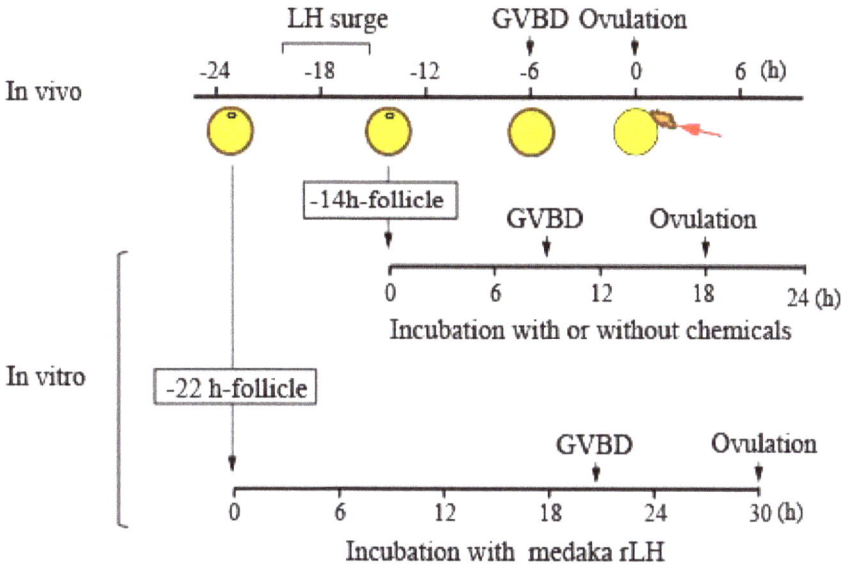

Figure 1. An outline of the in vitro culture experiments using medaka preovulatory follicles. Large follicles destined to ovulate were isolated 22 h before ovulation (−22 h-follicle) or 14 h before ovulation (−14 h-follicle). The follicles were incubated in medium containing medaka recombinant luteinizing hormone (rLH) (for the −22 h-follicle) or without LH (for the −14 h-follicle). Various chemicals were tested in culture using the −14 h-follicles to assess their effects on ovulation, oocyte maturation, as well as gene and protein expression.

2.3. cDNA Cloning for ccni

cDNA cloning and the sequencing were conducted for medaka *ccni* because the cDNA sequence was not available from public databases. It was determined by PCR using a KOD FX Neo-DNA polymerase (Toyobo, Osaka, Japan) and medaka ovary cDNA. The primers used were Cyclin I 5′-SS and Cyclin I 3′-AS (Table S1). The amplified products were phosphorylated, gel-purified, inserted into the pBluescript II vector (Agilent Technologies, Santa Clara, CA, USA), and sequenced. The sequence determined was deposited into the DDBJ/GenBank/NCBI database (accession number LC435346).

2.4. RNA Isolation, Reverse Transcription (RT), and Real-Time Polymerase Chain Reaction (PCR)

Total RNA isolation, RT, and real-time RT-PCR were conducted as previously described [13]. Eukaryotic translation elongation factor 1 alpha 1 (*eef1a1*) was used as a reference gene to normalize the expression of the target genes examined. Relative expression levels were expressed compared to the appropriate controls. The primers used in this study are listed in Table S1.

2.5. Preparation of Antigens and Antibodies

RT-PCR was performed to amplify cDNAs encoding medaka Cdk1 (303 residues), Cdk9 (393 residues), Ccni (342 residues), and Ribosomal protein L7 (Rpl7; 245 residues) using KOD Neo DNA polymerase (Toyobo, Tokyo, Japan) with ovary cDNA. The primers used are listed in Supplementary

Table S1. The amplified products were phosphorylated and inserted into the prokaryotic expression vector pET30a (Novagen, Madison, WI, USA), which had been previously digested by *Eco*RV. Expression, purification, and dialysis of the protein were performed according to previously published methods [15]. Anti-medaka Cdk1, Cdk9, Ccni, and Rpl7 antibodies were generated using mice according to methods described previously [14]. Mouse anti-medaka Pgr antibody [16], rat anti-medaka Pgr antibody [16], and rabbit anti-medaka Mmp15 antibody [6] were prepared as described previously. Antibodies were purified as previously described [7], and the resulting purified antibodies were used in the experiments.

2.6. Immunoprecipitation and Western Blot Analysis

Immunoprecipitation/western blot analysis for Pgr was performed as previously described [12], except that an anti-mouse IgG, HRP-Linked Whole Ab Sheep, was used as a secondary antibody. Immunoprecipitation/western blot analysis for Cdk1, Cdk9, and Ccni was performed in a manner similar to that for Pgr, as described above. Detection of medaka Mmp15 [6], Pgr [16], and Cebpb [13] proteins in follicle layer extracts was performed as previously described. For the detection of Cdk1 and Cdk9 protein, western blot analysis was performed according to the previous method described in Reference [6]. Rpl7 was used as loading control.

2.7. Phosphatase Treatment

The materials immunoprecipitated using rat anti-medaka Pgr antibody were incubated with or without Lambda Protein Phosphatase (New England BioLabs Inc., Ipswich, MA, USA) for 60 min at 30 °C in 1 × NEB Buffer for Protein Metallo Phosphatase supplemented with 1 mM $MnCl_2$ and gentle agitation. Incubation was terminated by adding 1 × SDS sample buffer and boiling for 10 min.

2.8. Primary Culture of Medaka Granulosa Cells (GC)

Isolation and primary culture of medaka GCs were performed as previously described [17], except that GCs were isolated from preovulatory follicles, 7 h before ovulation. After culturing for 24 h, the cells were harvested, and the expression of target mRNAs was examined by real-time RT-PCR.

2.9. Digestion of ECM Proteins by Medaka Recombinant Mmp15

Five micrograms of bovine type I collagen (Sigma-aldrich, St. Loui, MO, USA), bovine type IV collagen, human fibronectin (Sigma), human laminin (Sigma), porcine gelatin (Difco Laboratories, Inc., Detroit, MI, USA), and medaka collagen type I were incubated in 50 mM Tris-HCl buffer (pH 7.5) containing 5 mM $CaCl_2$ and 50 µM $ZnSO_4$ with medaka recombinant Mmp15 (100 ng) for 16 h at 27 °C. After incubation, reactions were terminated by adding 1 × SDS sample buffer and boiling for 10 min. The resulting samples were subjected to sodium dodecyl sulfate-polyacrylamide gel electrophoresis (SDS-PAGE). After electrophoresis, the gel was visualized by Coomassie brilliant blue R-250 (CBB) staining. Medaka recombinant Mmp15 preparation, active site titration of the enzyme, and purification of medaka collagen type I were performed as previously described [6].

2.10. Culturing Cells Stably Expressing Medaka Pgr

Establishment of a cell line stably expressing medaka Pgr [13] using OLHNI-2 cells [16] was performed according to the previous method. The cells were treated for 24 h with or without 17,20βP (100 nM) and/or roscovitine (50 µM). After the treatment, cells were harvested, and the expression of *mmp15* mRNA was examined by real-time RT-PCR.

2.11. Chromatin Immunoprecipitation (ChIP)

ChIP assays were performed according to the methods previously described [13], except that cultured follicles were analyzed in this study. Briefly, the −14 h-follicles were incubated with or without roscovitine and the resulting follicles were used. Preparation of follicle cell layer fractions, sonication, immunoprecipitation with protein G-Sepharose-coupled medaka Pgr antibody, elution, reverse-crosslinking, and real time PCR were carried out as previously described [13]. Eight sets of primers were used, and are listed in Supplementary Table S1. Putative binding sites of Pgr were identified using a free program TFBIND (http://tfbind.hgc.jp/).

2.12. Detection of Phosphorylated Pgr

Phos-tag SDS-PAGE using a 7% polyacrylamide gel containing 50 µM Phos-tag® acrylamide (Wako Chemicals, Osaka, Japan) and 50 µM MnCl$_2$ was carried out according to the manufacturer's instructions. Briefly, materials immunoprecipitated with rat anti-medaka Pgr antibody were subjected to Phos-tag SDS-PAGE. After electrophoresis, the gel was incubated twice in a buffer (0.2 M glycine, 20 mM Tris, 0.1% SDS, and 20% methanol) containing 10 mM EDTA, followed by a buffer without EDTA. Transfer to the Immobilon PVDF membranes (Merck Millipore), blocking, and detection of Pgr were performed.

2.13. Immunohistochemistry

Immunohistochemistry was performed according to the method previously described in [8], except that the signal was detected using an ImmPACT AMEC Red Peroxidase (HRP) Substrate (Vector Laboratories, Burlingame, CA, USA), according to the manufacturer's instructions.

2.14. Knockout Experiments for cdk9

The Cas9 nuclease expression vector was generated as follows. The hygromycin B resistance gene cassette was prepared by PCR amplification using primer pair Hyg SS and Hyg AS (Table S1). pGloSensor™-22F cAMP Plasmid (Promega corporation, Madison, Germany) was used as the template. The PCR product was phosphorylated and inserted into pCS2+hSpCas9 (Addgene Plasmid 51815), which had been digested with Mfe I, filled in by Klenow fragment, and dephosphorylated. The resultant vector, pCS2+hSpCas9/Hyg, was used for knockout experiments. The sgRNA expression vector was generated according to the previous method described in [18]. Briefly, a pair of oligonucleotides was annealed and ligated into pDR274 (Addgene Plasmid 42250), which had been digested with Bsa I. The resultant vector, pDR274-CDK9, was used for KO experiments. A combination of pDR274-CDK9 and pCS2+hSpCas9/Hyg, or pDR274-Mock and pCS2+hSpCas9/Hyg, was co-transfected into OLHNI-2 cells stably expressing medaka Pgr using ScreenFect A (Wako) according to the manufacturer's instructions. After culture for 48 h, the medium was changed into fresh medium containing 100 µg/mL hygromycin (Wako), and cells were cultured for another 48 h. The cells were harvested and used for the immunoprecipitation/Phos-tag SDS-PAGE analysis.

2.15. Coimmunoprecipitation

Preovulatory follicles isolated from ovaries 7 h before ovulation were sonicated for a few seconds in 50 mM Tris-HCl (pH 8.0) containing a 1 × protease inhibitor cocktail (Wako) and 1× phosphatase inhibitor cocktail (Wako), and then centrifuged at 13,000× *g* for 10 min. The resulting supernatants were used for coimmunoprecipitation analysis. The samples were treated with Protein G-Sepharose (GE Healthcare) that had been previously coupled with anti-medaka Cdk9 antibody. After incubation at 4 °C for 16 h, they were washed four times using 50 mM Tris-HCl (pH 8.0). The precipitant materials were boiled in 1× SDS sample buffer for 10 min and used for western blot analyses.

2.16. Statistical Analysis

The experiments performed in this study were repeated independently four to seven times, and all values are presented as the mean ± S.E.M. Statistical significance was verified using Student's *t*-test or one-way ANOVA followed by Dunnett's post hoc test, or Kruskal-Wallis test, as appropriate. Equal variation was confirmed by the F-test or Bartlett's test, as appropriate. The minimum level of statistical significance was set at $p < 0.05$. The experiments for western blot, immunoprecipitation/western blot, and immunohistochemistry analyses were repeated three to five times to confirm their reproducibility, and the results of one experiment are presented.

3. Results

3.1. Inhibitory Effect of Roscovitine on Medaka Follicle Ovulation

The −14 h-follicles spontaneously ovulated in the culture medium in vitro after approximately 20 h, without externally added medaka rLH. Under these conditions, virtually all follicles underwent GVBD within 14 h. However, the follicles failed to ovulate in the medium containing 5 and 50 µM of the CDK inhibitor roscovitine (Figure 2A). Roscovitine also inhibited GVBD (Figure 2B). To determine the viability of the follicles that had been incubated with the inhibitor at 50 µM for 14 h, a trypan blue exclusion test was conducted. The oocyte and follicle cells of the follicles both showed a clear cytoplasm, indicating that the inhibition of ovulation and GVBD in roscovitine-treated follicles was not due to the CDK inhibitor's toxic effect. Concerning the inhibitory effect of roscovitine on follicle ovulation, we hypothesized that the inhibitor might affect the expression and/or activation of ECM-degrading enzyme(s). Therefore, we examined the effect of roscovitine treatment on the follicular expression of MMPs and their intrinsic inhibitor genes, which are involved in follicle layer ECM degradation at medaka ovulation [6,13]. Transcript levels of *mmp2*, *mmp14*, *mmp15*, and *timp2b* were compared in follicles that had been incubated with or without roscovitine. Among them, *mmp15* expression was drastically and selectively reduced by roscovitine treatment (Figure 2C). Consistent with the result, inhibition of Mmp15 protein expression was observed in follicle layer extracts of roscovitine-treated follicles (Figure 2D). We next examined the in vitro proteolytic activities of medaka recombinant Mmp15 toward several ECM proteins. The enzyme exhibited degrading activities toward bovine and medaka collagen type I, human fibronectin, and porcine gelatin, but not bovine collagen type IV or human laminin (Figure 2E), confirming our previous finding that Mmp15 selectively hydrolyzed collagen type I among the three major ECM proteins (collagen type I, collagen type IV, and laminin) [6,8,19,20]. We further tried to detect the reduction Mmp15 activity using extracts from follicular layers of follicles that had been previously treated with roscovitine. However, this attempt was not successful because the enzyme activity was too low to detect, even with follicle layer extracts of untreated follicles (data not shown).

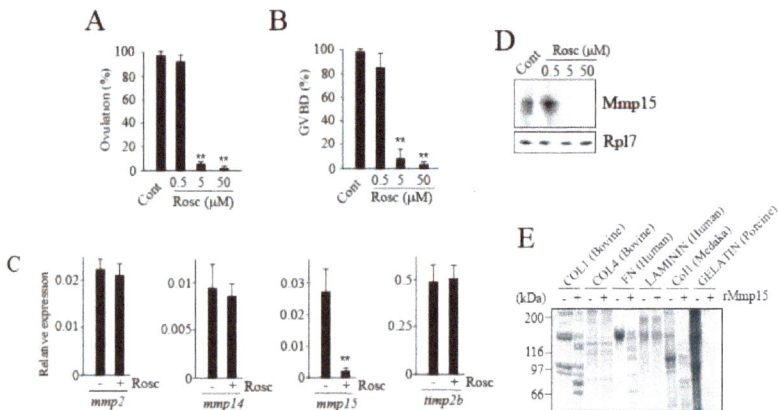

Figure 2. Effects of roscovitine on ovulation, germinal vesicle breakdown (GVBD), and gene expression of ovulation-related Mmps and its intrinsic inhibitor. (**A**) The −14 h-follicles were incubated in vitro with roscovitine (Rosc) at 0.5, 5, and 50 µM for 18 h, and the rate of ovulation was determined. Asterisks indicate significance at $p < 0.01$ (**) compared to follicles incubated without any additives (ANOVA and Dunnett's post hoc test, N = 5). (**B**) GVBD rate in the follicles was determined. Follicle incubation

was conducted as in (**A**), except that the duration of incubation was 14 h. Asterisks indicate a significance at $p < 0.01$ (**) compared to follicles incubated without any additives (ANOVA and Dunnett's post hoc test, N = 5). (**C**) The −14 h-follicles were incubated with or without Rosc for 18 h, and the expression levels of various Mmps and timp2b were determined by real time RT-PCR. Asterisks indicate significance at $p < 0.01$ (**) (*t*-test, N = 5). (**D**) Follicle layers of follicles that had been incubated with or without Rosc for 18 h were analyzed for Mmp15 expression by western blotting. (**E**) The hydrolyzing activity of medaka recombinant Mmp15 was tested in vitro using various ECM proteins.

In summary, the suppression of follicle ovulation by roscovitine treatment appears to be associated with the reduced expression of *mmp15*/Mmp15.

3.2. Effects of Roscovitine on Transcription Factors Involved in mmp15 Gene Expression

We examined the mechanism by which roscovitine inhibited the expression of the *mmp15* gene in preovulatory follicles destined for ovulation. The effects of roscovitine on the follicular expression of two transcription factors, CCAAT/enhancer-binding protein β (Cebpb) and classical nuclear progestin receptor (Pgr), were examined, based on previous knowledge demonstrating that both factors play critical roles in the expression of *mmp15* [13]. Treating the −14 h follicles with roscovitine had no effect on the transcription factors' expression at the protein level (Figure 3A). Next, we examined the effect of roscovitine treatment on the phosphorylation status of Pgr. In regular SDS-PAGE of untreated follicles, Pgr synthesized in follicles was detected as a single polypeptide throughout the 24-h spawning cycle (Figure 3B, lower panel). This result is consistent with our previous observation on the temporal expression profile of Pgr in follicles destined to ovulate [13]. Pgr was present at low levels in postvitellogenic follicles, but its robust expression occurred several hours after the LH surge. On the other hand, in Phos-tag SDS-PAGE analysis, a single band corresponding to unphosphorylated Pgr was detected in the first half (-23 to -13 h) of the spawning cycle (arrowhead in Figure 3B, upper panel), but multiple slow-migrating bands appeared at later times in the cycle (arrow in Figure 3B, upper panel). Almost all bands with reduced electrophoretic mobility disappeared when the sample was treated with protein phosphatase prior to electrophoresis (Figure 3C), demonstrating that they are phosphorylated forms of Pgr. Roscovitine treatment of the follicle strongly inhibited the formation of phosphorylated Pgr, with the exception of a few bands (Figure 3D). Incubating the follicles with the MEK inhibitor resulted in no significant change compared to the control. In a separate experiment, we confirmed that the phosphorylation of a medaka MAPK protein (Erk), which corresponds to ERK42/44 in mammalian species, was almost completely blocked when the follicle incubated with the MEK inhibitor was analyzed (data not shown), indicating that the MEK inhibitor was effective in inhibiting MEK.

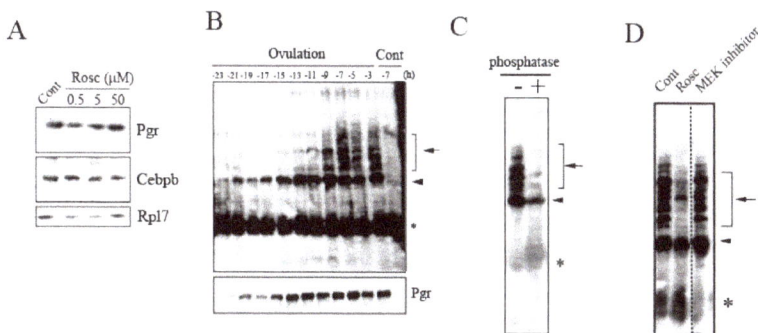

Figure 3. Roscovitine inhibition of Pgr phosphorylation in preovulatory follicles. (**A**) The −14 h-follicles were incubated with roscovitine (Rosc) at 0.5, 5, or 50 µM for 14 h, and the follicle layer extracts were analyzed by western blotting. (**B**) Preovulatory follicles were isolated at various time points, and their

extracts were analyzed by immunoprecipitation/Phos-tag SDS-PAGE/western blotting (upper panel) and immunoprecipitation/regular SDS-PAGE/western blotting using an antibody for medaka Pgr (lower panel). Phosphorylated (indicated by arrow) and unphosphorylated forms of Pgr (indicated by arrowhead) are shown in the upper panel. An asterisk indicates the bands corresponding to the antibody used for immunoprecipitation. As controls, the −7 h-follicle extracts were immunoprecipitated using normal IgG. (**C**) The materials immunoprecipitated from follicle extracts obtained from the −7 h-follicles were incubated with or without phosphatase, and the samples were analyzed by Phos-tag SDS-PAGE/western blotting using the antibody for medaka Pgr. Phosphorylated (indicated by arrow) and unphosphorylated forms of Pgr (indicated by arrowhead) are shown. An asterisk indicates the bands corresponding to the antibody used for immunoprecipitation. (**D**) The −14 h-follicles were incubated for 14 h with Rosc (50 μM), with MEK inhibitor (10 μM), or without any additives, and the follicle extracts of the treated follicles were analyzed by immunoprecipitation/Phos-tag SDS-PAGE/western blotting using the antibody for medaka Pgr. Phosphorylated (indicate by arrow) and unphosphorylated forms of Pgr (indicated by arrowhead) are shown. An asterisk indicates the bands corresponding to the antibody used for immunoprecipitation.

The above results strongly suggest that Pgr is phosphorylated during the last 10 h of the 24-h spawning cycle in follicles destined to ovulate, and that Pgr phosphorylation could be blocked by roscovitine treatment.

3.3. Loss of Pgr Binding Ability to the Promoter Region of the mmp15 Gene in Preovulatory Follicles under the Effect of Roscovitine Treatment

We recently reported that Pgr is an important transcription factor necessary for the expression of *mmp15*, and that it binds to the promoter region of the Mmp gene in granulosa cells of follicles undergoing ovulation [13]. Therefore, the effect of roscovitine treatment on Pgr binding to the *mmp15* promoter in ovulating follicles was examined by performing the ChIP assay (Figure 4A) and using eight primer pairs, as previously reported [12]. First, −14 h-follicles were incubated for 14 h without roscovitine (DMSO only as control) and then analyzed. Among them, primer pair-1, which was designed to generate 101-bp nucleotides corresponding to the sequence between −101 and −1 upstream of the transcription start site of the *mmp15* gene, was only effective for amplifying an expected size of nucleotides (Figure 4B). In this experiment, significant enrichment was not observed with primer pair-2, which—as reported in our previous study [13]—was also effective for the amplification of nucleotides, although the extent of amplification was not as great as that of primer pair-1. The inconsistency in the results of primer pair-2 between our previous and present studies is not clear. In the previous study, we performed ChIP assays using intact follicles isolated from ovaries, whereas in the present study, the assay was conducted using cultured follicles, which might be a reason for the inconsistency. Furthermore, -14 h follicles were incubated with DMSO alone for 7 or 14 h and analyzed using primer pair-1. Pgr recruitment to the promoter was observed for follicles incubated for 14 h (Figure 4C). Next, we examined the effect of roscovitine treatment of follicles on Pgr binding to the promoter region of the *mmp15* gene using eight primer pairs. No significant amplification of nucleotides was found (Figure 4D). Finally, we examined the effect of roscovitine on the 17,20βP-stimulated expression of *mmp15* using OLHNI-2 cells, a cell line established using cells that originated from the medaka fin (Figure 4E). When the cells stably expressing Pgr were treated with 17,20βP for 24 h, *mmp15* expression was significantly increased, confirming that the transcription factor Pgr becomes active by associating with 17,20βP. However, the 17,20βP-induced expression was nullified by the addition of roscovitine.

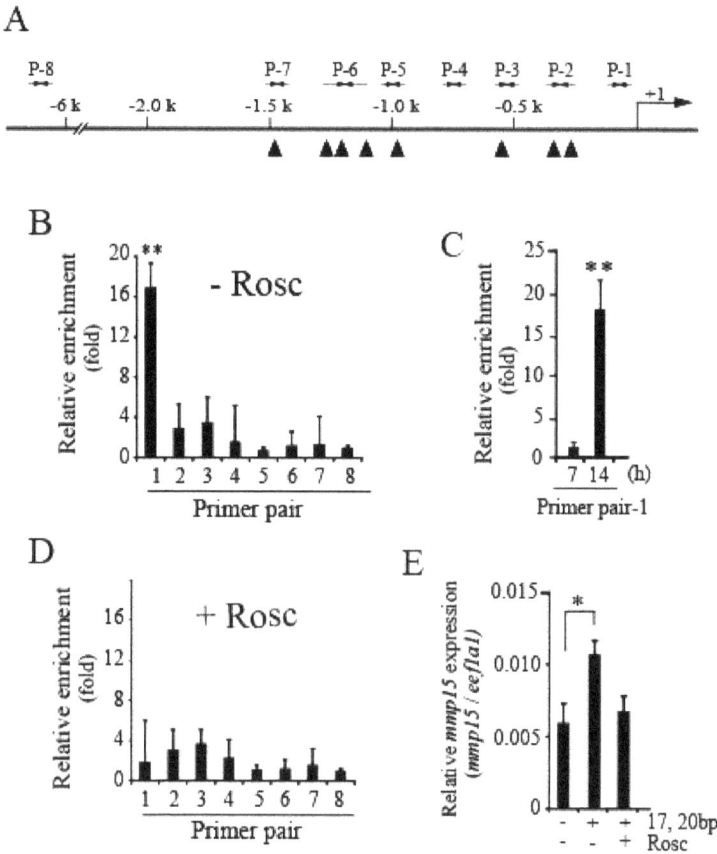

Figure 4. Effect of roscovitine treatment of follicles on Pgr binding to the *mmp15* promoter region. (**A**) For ChIP assays to examine the binding of Pgr to *mmp15* promoter region, seven ChIP primer pairs (P1–P7) in the 1.5 kb upstream region of the transcription start site (indicated as +1) of the *mmp15* gene and another primer pair (P8 as a negative control) were prepared. Putative progestin receptor elements (PREs) in the region are indicated by boxes. (**B**) The −14 h-follicles were incubated in vitro without roscovitine (Rosc) for 14 h, and the resulting follicles were used for ChIP assays and amplified with primer pairs P1 to P8. The sheared DNA immunoprecipitated with anti-medaka Pgr antibody was analyzed by real-time RT-PCR. Asterisks indicate a significant difference at $p < 0.01$ (**) compared to the negative control (ANOVA and Dunnett's post hoc test, N = 4). (**C**) The −14 h-follicles were incubated in vitro for 7 h or 14 h, and the resulting follicles were used for ChIP assays using primer pair P1. The sheared DNA immunoprecipitated with anti-medaka Pgr antibody was analyzed by real-time RT-PCR. Asterisks indicate a significant difference at $p < 0.01$ (**) compared with the 7 h-incubated follicles (*t*-test, N = 4). (**D**) The −14 h-follicles were incubated in vitro with Rosc (50 µM) for 14 h, and the resulting follicles were used in ChIP assays as in (**B**). (**E**) OLHNI-2 cells stably expressing medaka Pgr were cultured alone, in the presence of 17,20βP (100nM), or in the presence of both 17,20βP and Rosc. After culturing for 24 h, the expression levels of *mmp15* were examined by real-time RT-PCR. Asterisks indicate a significant difference at $p < 0.05$ (*) (ANOVA and Dunnett's post hoc test, N = 4).

The results described above indicate that roscovitine treatment of preovulatory follicles causes the inhibition of Pgr binding to the *mmp15* promoter region.

3.4. Expression of CDKs in Preovulatory Follicles of the Medaka Ovary

The results described in the preceding section suggest an involvement of CDK in Pgr phosphorylation in follicles. We therefore examined the expression of CDKs in ovarian follicles of fish ovaries. A computer search for *cdk* genes using the draft medaka genome database (Ensembl genome database: https://asia.ensembl.org/index.html.) revealed that medaka contains a total of 34 *cdk* genes, including *cdk*-like genes. Our recent screening for genes associated with medaka ovulation using next-generation sequencing identified nine *cdk* genes that are presumed to be activated during the ovarian follicle 6 h before ovulation, the time at which Pgr is phosphorylated by Cdk(s). The nine *cdk* genes were indeed expressed in the follicle layer of the -6 h-follicle, as confirmed by real-time RT-PCR; among them, *cdk1*, *cdk9*, and *cdk11b* exhibited relatively high expression (Figure 5A). The levels of the three *cdk* transcripts were not altered in the follicle layers during the 24-h spawning cycle (Figure S1). To examine whether these *cdk* genes are expressed in granulosa cells of preovulatory follicles, real-time RT-PCR analysis was conducted using total RNAs prepared from granulosa cells derived from −6 h-follicles. Amplified products for *cdk1* and *cdk9* were detected (Figure 5B). The expression levels of *cdk11b* mRNA were extremely low, arguing against the idea that Cdk11b might be involved in Pgr phosphorylation in the granulosa cells of follicles.

We next raised specific antibodies for fish Cdk1 and Cdk9 (Figure S2A,B). Single polypeptides of 34 kDa (for Cdk1) and 43 kDa (for Cdk9) were detected by immunoprecipitation/western blot analysis with whole follicle extracts (Figure 5C). The staining intensities of the Cdk1 and Cdk9 bands did not change significantly in the follicles throughout the 24-h spawning cycle. However, follicle layer extracts of follicles predicted to ovulate in 7, 5, and 3 h were analyzed, and Cdk9, but not Cdk1, was detected (Figure 5D). In immunohistochemical analysis, strong signals associated with Cdk1 were observed in the oocyte cytoplasm of small-sized growing follicles (Figure 5E; left panels). Weak signals for Cdk1 were also detected in association with the follicle layers of medium- and large-sized follicles. In addition, a signal was detectable in the oocyte cytoplasm of large-sized follicles. On the other hand, strong signals for Cdk9 were found in the follicle layers of all growing follicles (Figure 5E; right panels).

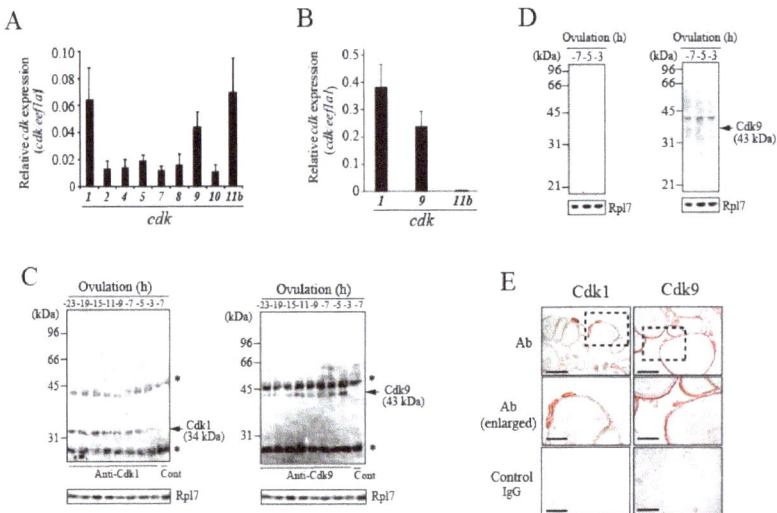

Figure 5. Expression of Cdks in the medaka ovary. (**A**) Relative expression levels of *cdk* transcripts were determined by real-time RT-PCR using total RNAs isolated from follicle layers of preovulatory follicles isolated 6 h before ovulation. (**B**) Relative expression levels of *cdk1*, *cdk9*, and *cdk11b* transcripts were determined by real-time RT-PCR using total RNAs from granulosa cells isolated from the −6 h-follicles.

(**C**) Expression of Cdk1 (left panel) and Cdk9 proteins (right panel) in the preovulatory follicles were analyzed. Extracts from follicles were prepared at various time points in the 24 h spawning cycle and immunoprecipitated with specific Cdk1 and Cdk9 antibodies. The immunoprecipitated materials were then analyzed by western blotting using the same antibodies. Asterisks indicate the bands corresponding to the antibody used for immunoprecipitation. As control, the −7 h-follicle extracts were immunoprecipitated with normal IgG. (**D**) Expression of Cdk1 (left panel) and Cdk9 proteins (right panel) in follicle layer extracts were analyzed. The extracts were prepared using the follicle layers of follicles isolated at 7, 5, or 3 h before ovulation and examined by western blotting using specific antibodies. (**E**) Immunohistochemical analyses were performed in the sections of spawning fish ovaries isolated 15 h before ovulation. Cdk1 (left, upper and middle panels) and Cdk9 (right, upper and middle panels) were stained using the purified antibodies. The area indicated by a box in the upper panels is shown at a higher magnification in the middle panels. Normal IgG was used as a control (lower panels). Bars represent 200 μm in the upper and lower four panels, and 100 μm in the middle two panels.

In vitro treatment of preovulatory follicles with roscovitine did not affect the expression levels of *cdk1*, *cdk9*, and *cdk11b* transcripts in the follicle layer of follicles (Figure S3).

The above results indicate that Cdk9 is the only Cdk species that is constitutively expressed in granulosa cells of ovulating follicles when Pgr undergoes phosphorylation.

3.5. Expression of Cyclins in Preovulatory Follicles of the Medaka Ovary

Cyclins (Ccns) that partner with Cdk for Pgr phosphorylation during medaka ovulation were searched. An Ensembl database search suggested that medaka contains 29 *cyclin* genes, including *cyclin*-like genes. In our attempt to search for ovulation-related genes by next-generation sequencing using preovulatory follicles, three *cyclin* genes, *cyclin G2* (*ccng2*), *cyclin I* (*ccni*), and *cyclin E2* (*ccne2*), were identified as candidates on the basis of FPKM values. Real-time RT-PCR analysis indicated that *ccni* mRNA was most abundantly expressed in −6 h-follicles (Figure 6A). *ccng2* transcripts were also detected in the follicle, but levels were less than one-third those of *ccni*. When granulosa cells were prepared from the −6 h follicle and used for real-time RT-PCR analysis, only *ccni* transcripts were detected (Figure 6B). Therefore, subsequent analysis was conducted with a focus on the *ccni* gene. Time course analysis of *ccni* mRNA levels was investigated using preovulatory follicles isolated from fish ovaries at various time points of the 24-h spawning cycle. The levels of *ccni* mRNA remained low between 23 h and 15 h before ovulation, but *ccni* expression levels increased as ovulation approached (Figure 6C). When the −22 h-preovulatory follicles were cultured in vitro with medaka rLH, the levels of *ccni* mRNA increased (Figure 6D). This rLH-induced *ccni* mRNA expression was significantly reduced by the presence of RU486 (Figure 6E), which was previously shown to act as an antagonist for medaka Pgr [16]. These results suggest that *ccni* mRNA expression in the follicle is regulated by the transcription factor Pgr. To detect Ccni protein in follicles, a specific antibody for medaka Ccni was prepared (Figure S2C). Using the antibody, immunoprecipitation/western blot analysis was performed on whole follicle extracts at various time points. Two bands of approximately 40 and 41 kDa for Ccni were detected between −7 h and −3 h of ovulation (Figure 6F). Similarly, the synthesis of Ccni in the follicle was confirmed between 18 h and 30 h of incubation when the −22 h follicles were cultured with medaka rLH (Figure 6G). These results are consistent with those of the mRNA expression studies described above.

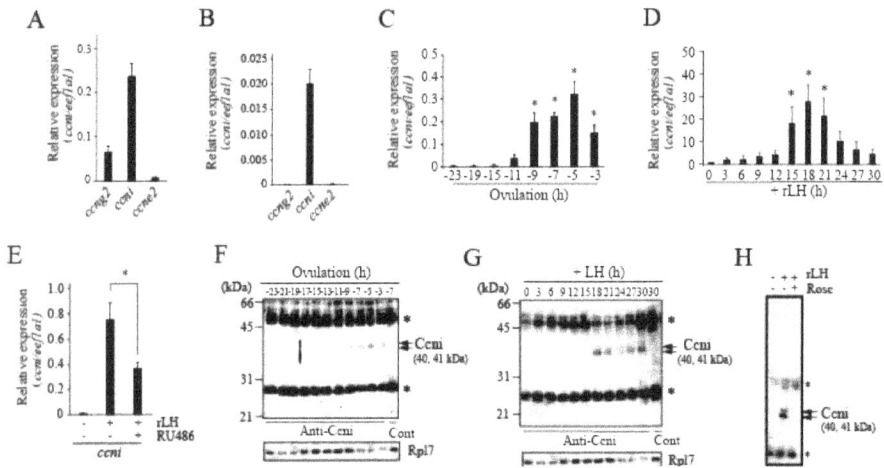

Figure 6. Expression of cyclins in the medaka ovary. (**A**) Relative expression levels of *ccn* transcripts were determined by real-time RT-PCR using total RNAs derived from follicle layers of preovulatory follicles isolated 6 h before ovulation. (**B**) Relative expression levels of *ccng2*, *ccni*, and *ccne2* transcripts were determined by real-time RT-PCR using total RNAs of granulosa cells isolated from the −6 h-follicles. (**C**) Total RNAs were prepared from the follicles at various time points in the 24 h spawning cycle and were used for real-time RT-PCR of *ccni* mRNA. The expression levels were normalized to those of *eef1a1*. Asterisks indicate a significant difference at $p < 0.05$ (*) (Kruskal–Wallis test, N = 5–7). (**D**) The −22 h-follicles were incubated in vitro with or without recombinant medaka LH (100 μg/mL). At various time points of incubation, total RNAs were prepared from follicles and used for real-time RT-PCR of *ccni* mRNA. The expression levels wre normalized to those of *eef1a1* and expressed as the fold change compared to levels of 0 h follicles. Asterisks indicate a significant difference at $p < 0.05$ (*) (Kruskal–Wallis test, N = 4–5). (**E**) The −22 h-follicles were incubated with recombinant medaka LH (100 μg/mL) with or without RU-486 (100 μM). After 18 h of incubation, total RNA was extracted from the follicles and used for real-time RT-PCR analysis of *ccni* mRNA. The expression levels were normalized to those of *eef1a1*. An asterisk indicates a significant difference at $p < 0.05$ (*) (*t*-test, N = 4). (**F**) Expression of Ccni protein in the preovulatory follicles was analyzed. Extracts of the follicles were prepared at various time points in the 24 h spawning cycle and immunoprecipitated using the specific Ccni antibody. The immunoprecipitated materials were then analyzed by western blotting using the same antibody. Asterisks indicate the band corresponding to the antibody used for immunoprecipitation. As a control, the −7 h-follicle extracts were immunoprecipitated with normal IgG. (**G**) The −22 h-follicles were incubated with medaka rLH (100 μg/mL). At various time points of incubation, follicle extracts were prepared and immunoprecipitated with the Ccni antibody. The immunoprecipitated materials were then analyzed by western blotting using the same antibody. Asterisks indicate the band corresponding to the antibody used for immunoprecipitation. As a control, the follicles that had been incubated with medaka rLH for 30 h were immunoprecipitated with normal IgG. (**H**) The −22 h-follicles were incubated with medaka rLH (100 μg/mL) with or without roscovitine (Rosc; 50 μM). After 18 h of incubation, the follicular extracts were immunoprecipitated using the Ccni antibody, and the resultant materials were analyzed by western blotting using the same antibody. An asterisk indicates the bands corresponding to the antibody used for immunoprecipitation.

The effect of roscovitine on the follicular expression of Ccni was examined. In vitro roscovitine treatment of −14 h follicles strongly inhibited Ccni expression (Figure 6H), indicating that Cdk may be involved in *ccni*/Ccni expression in the follicle.

The above results showed that follicle cells of follicles destined for ovulation express Ccni, and that its expression is induced by LH. The results also indicate that LH-induced *ccni* expression in the follicle is suppressed by roscovitine treatment.

3.6. Possible Involvement of Ccni/Cdk9 in Pgr Phosphorylation in the Preovulatory Follicle

We examined the effect of CDK9-inhibitor II, a specific inhibitor of Cdk9 [21], on the rate of ovulation using an in vitro follicle culture system. The inhibitor completely suppressed ovulation at 10 µM (Figure 7A). The rate of GVBD was not affected by the inhibitor (Figure 7B), suggesting that Cdk9 is not involved in oocyte maturation in the medaka ovulating follicle. In CDK9 inhibitor II-treated follicles, Mmp15 expression in the follicle layer was strongly suppressed (Figure 7C). To evaluate the possible role of Cdk9 in Pgr phosphorylation, we conducted experiments using follicles cultured with or without CDK9 inhibitor II. In control follicles, phosphorylated Pgr was detected in multiple forms (Figure 7D; Control), but some of the phosphorylated Pgr disappeared or its staining intensity was noticeably reduced after treating the follicles with CDK9-inhibitor II (Figure 7D). A similar result was observed when Cdk9 knockout was generated using the Crispr/Cas9 system in OLHNI-2 cells stably expressing Pgr. In cells transfected with the mock vector, bands for phosphorylated Pgr were detected (Figure 7E; left lane), whereas Pgr phosphorylation was significantly suppressed following Cdk9 knockout (Figure 7E; middle lane). To confirm the reproducibility of the finding, we also conducted experiments using another sgRNA targeting a different Cdk9 sequence, and a similar result was obtained (data not shown), suggesting that the system achieved site-specific DNA recognition and cleavage of the target specifically, without off-target effects. Finally, we examined whether Cdk9 interacts with Ccni in follicles destined to ovulate. Immunoprecipitation was conducted using a Cdk9 antibody with follicle layer extracts of -6 h follicles, and the resulting precipitated materials were further subjected to western blot analysis. The materials contained not only Cdk9 but also Ccni protein (Figure 7F). These results indicate that Ccni and Cdk9 are complexed in follicle cells of follicles undergoing ovulation.

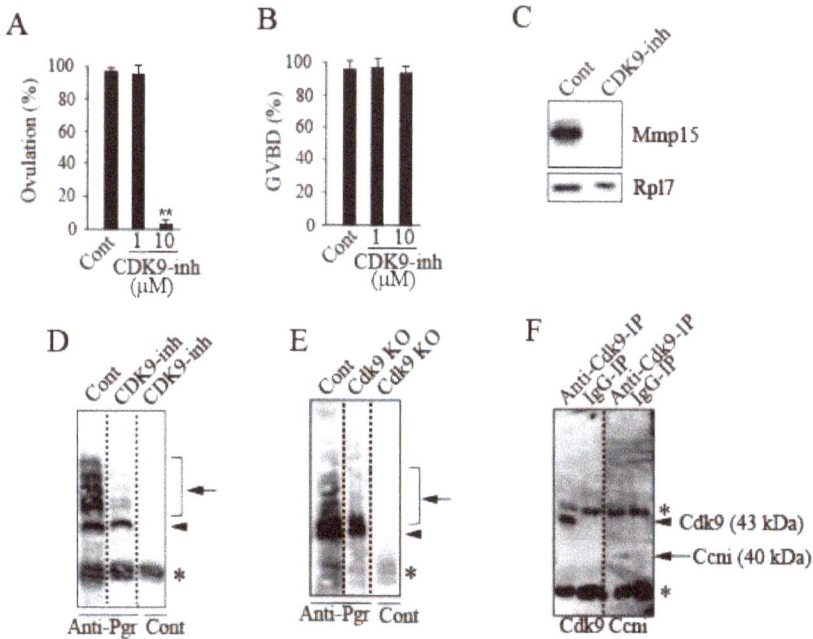

Figure 7. Further evidence for the role of Cdk9/Ccni in the follicles that are destined to ovulate. (**A**) The −14 h-follicles were incubated in vitro with the specific CDK9 inhibitor, CDK9-inhibitor II, at 1 and 10 µM for 18 h, and the rate of ovulation was determined. Asterisks indicate significance at $p < 0.01$ (**) compared to follicles incubated without any additives (ANOVA and Dunnett's post hoc test, N = 5).

(**B**) Incubation of follicles with CDK9-inhibitor II was conducted as in (**A**) except the duration of incubation was 14 h. Note that the inhibitor had no effect on the GVBD of follicles (N = 5). (**C**) The -14 h-follicles were incubated in vitro with or without the specific CDK9 inhibitor CDK9-inhibitor II (10 μM) for 18 h, and the expression of Mmp15 in the follicle layer of the follicles was analyzed by western blotting. (**D**) The −14 h-follicles were incubated in vitro with or without the specific CDK9 inhibitor CDK9-inhibitor II (10 μM) for 14 h, and Pgr phosphorylation in the follicles was examined by immunoprecipitation/Phos-tag SDS-PAGE/western blot analysis. Positions of phosphorylated (indicated by arrow) and unphosphorylated Pgr (indicated by arrowhead) are shown. As control, extracts of the follicles treated with CDK9-inhibitor II were immunoprecipitated with normal IgG. An asterisk indicates the bands corresponding to the antibody used for immunoprecipitation. (**E**) OLHNI-2 cells stably expressing medaka Pgr (Cont) and Cdk9-deficient OLHNI-2 cells (Cdk9 KO), which were generated from the above cells with CRISPR/Cas9 technology, were immunoprecipitated with medaka Pgr antibody, and the resulting precipitated materials were analyzed by Phos-tag SDS-PAGE/western blot analysis. Positions of phosphorylated (indicated by arrow) and unphosphorylated Pgr (indicated by arrowhead) are shown. As a control, extracts of Cdk9-deficient OLHNI-2 cells were immunoprecipitated with normal IgG. An asterisk indicates the bands corresponding to the antibody used for immunoprecipitation. (**F**) Extracts of the −6 h-follicles were immunoprecipitated with medaka Cdk9 antibody or normal IgG, and the resulting precipitated materials were then analyzed by western blotting using the same Cdk9 antibody (left two lanes) or medaka Ccni antibody (right two lanes). Asterisks indicate the bands corresponding to the antibody used for immunoprecipitation.

4. Discussion

Follicle rupture during ovulation in vertebrates involves well-regulated ECM degradation at the apical region of ovulating follicles. Our previous studies using the teleost medaka showed that two distinct proteolytic enzyme systems, the Plau/plasmin system and the MMP system, contribute to follicle rupture. Among the proteolytic enzymes involved in the process, the role of Mmp15 is of particular interest, in that it serves as a protease that hydrolyzes collagen type I, a major ECM protein in the follicle layer of ovulating follicles, and that its expression is drastically induced in the granulosa cells of follicles at ovulation [6,14]. We recently reported the involvement of at least two transcription factors, Pgr and Cebpb, in the expression of the *mmp15* gene [13]. The current study was initiated following the finding that the CDK inhibitor roscovitine suppressed in vitro follicle ovulation, and the results of this study provide additional information on the regulatory mechanism for *mmp15* expression occurring in the granulosa cells of ovulating follicles. Our new findings are as follows: (i) Pgr is phosphorylated, (ii) the Ccni–Cdk9 complex may be involved in Pgr phosphorylation, and (iii) Ccni is induced by the ovulatory LH surge.

In the present study, we demonstrated that medaka Pgr undergoes phosphorylation prior to its binding to the promoter region of the *mmp15* gene. Electrophoretic analyses of Pgr revealed that the protein is phosphorylated at various sites, as indicated by the appearance of multiple polypeptide bands with a molecular mass greater than that of unphosphorylated Pgr. Our data also suggest that Pgr phosphorylation is largely due to the action of Cdk, because almost all phosphorylated Pgr bands disappeared upon roscovitine treatment of the follicle. Treatment with phosphatase, rescovitine, or CDK9-inhibitor II resulted in no apparent increase in unphosphorylated Pgr. In a separate experiment, we found that the phosphorylated form of Pgr was estimated to be approximately 20% of the total Pgr. Since the proportion of phosphorylated Pgr to total Pgr is low, we presume that changes in the Pgr protein level could not be clearly observed after phosphatase or inhibitor treatment. Previous studies documented that, like medaka Pgr, the human counterpart is also phosphorylated [22,23]. Three PGR isoforms (generally known as PR-A, PR-B, and PR-C) are known in humans, and two major protein kinases involved in human PGR phosphorylation are reported to be MAPK and CDK2. Furthermore, serine residues that are phosphorylated by these protein kinases have been identified. Sequence homologies in fish Pgr and human PGR are very low, hence, a prediction of possible phosphorylation sites in medaka Pgr by searching conserved serine residues between the fish and human proteins was fruitless. Therefore, phosphorylation sites of medaka Pgr were predicted using the free program KinasePhos (http://kinasephos.mbc.nctu.edu.tw/). With this prediction tool, Ser208 and Ser250 were shown to be phosphorylated by Cdk. In addition, Thr14, Thr143, Ser208, and Thr212 could be possible phosphorylation sites for MAPK, although our current data indicated that MAPK has little, if any, effect on Pgr phosphorylation in the fish ovulating follicle. Further studies are needed to establish the validity of such predictions in the future.

We showed that Cdk9 may be responsible for the phosphorylation of Pgr. This idea is supported by the fact that (i) Cdk9 is expressed in the granulosa cells of ovulating follicles, as demonstrated by both immunohistochemical and western blot analyses; (ii) similar to roscovitine, treatment of follicles with the CDK9 inhibitor consistently inhibited follicle ovulation, Pgr phosphorylation, and Mmp15 expression; (iii) Cdk9 knockout experiments using OLHNI-2 cells stably expressing Pgr resulted in a reduction of Pgr phosphorylation; and more importantly, (iv) Cdk9 forms a complex with Ccni in the follicle layers of ovulating follicles. In the context of Pgr phosphorylation, the involvement of Cdk1 is less likely, because the Cdk1 protein is lacking in granulosa cells of ovulating follicles despite the presence of a large abundance of the corresponding mRNA in these cells. Considering that Cdk1 is abundantly detected immunohistochemically in small-growing follicles, Cdk1 may play a role in the early stage of follicle growth. We previously found that, similar to *cdk1*, transcripts of enteropeptidase [15] and melatonin receptor subtype 1c [17] were expressed in the fish ovary, but their translation products were not detectable. Direct evidence for the involvement of CDK9 in Pgr phosphorylation in the granulosa cells of ovulating follicles is needed in future studies.

Interestingly, CDK9-inhibitor II treatment of preovulatory follicles inhibited ovulation, while roscovitine showed inhibitory effects on both ovulation and GVBD of the follicles. This finding indicates that CDKs other than CDK9 may be involved in GVBD, an LH-induced event occurring in the oocyte of medaka.

Ccni was the only cyclin expressed in the granulosa cells of ovulating follicles. Therefore, we tentatively assume that Ccni serves as the regulatory subunit of Cdk9 in the follicle, although Cdk9 activation by Ccni must be experimentally verified. Indeed, we found that Ccni was present in granulosa cells of the follicle in association with Cdk9 protein. Another important finding of the present study is that Ccni is an LH-inducible protein in cells of ovulating follicles. Interestingly, treatment of follicles with RU486, a Pgr antagonist, significantly reduced LH-induced *ccni* expression. This observation strongly suggests that Ccni is expressed in follicles in an LH- and Pgr-dependent manner.

Based on the present observations together with our previous findings [6,13], we propose a schematic model for the role of Ccni/Cdk9 in the pathway leading to *mmp15*/Mmp15 expression after the LH surge in follicles (Figure 8). In this model, we focus on the events occurring in granulosa cells, because these cells have been documented to play a major role in fish ovulation [24]. Roscovitine's effects are also shown in this model. A variety of ovulatory reactions, including the activation of many ovulation-related genes, are evoked in cells expressing the LH receptor in response to the surge of LH, which occurs approximately 18 h before ovulation in the 24-h spawning cycle [14]. Activation of the LH receptor by the LH surge immediately causes an increase in the intracellular cAMP concentration, which leads to the activation of Pgr. Pgr, thus synthesized, would then undergo phosphorylation at multiple sites, mainly via Cdk9. Cdk9, which is constitutively expressed in granulosa cells, should be activated by Ccni. Ccni is also induced in granulosa cells in response to LH stimulation. Our current data suggest that the expression of Ccni is driven by the 17,20βP-activated transcription factor Pgr, and that Pgr-dependent Ccni expression is initiated approximately 9 h before ovulation. Activated Cdk9 complexed with Ccni is capable of phosphorylating Pgr at various sites. The increase in phosphorylated Pgr levels begins at 9 h before ovulation in the granulosa cells of ovulating follicles, and such a situation lasts thereafter until the time of ovulation. Another transcription factor, Cebpb, which is also induced in an LH-dependent manner, is synthesized in granulosa cells of the follicle [13]. At several hours preceding follicle rupture, both phosphorylated Pgr and Cebpb bind to the promoter region of the *mmp15* gene for activation. The resulting translation product, Mmp15, is eventually expressed on the surface of granulosa cells. To substantiate the above hypothesis, further analyses using siRNA and knockout experiments targeting the *ccni* and *cdk9* genes are necessary.

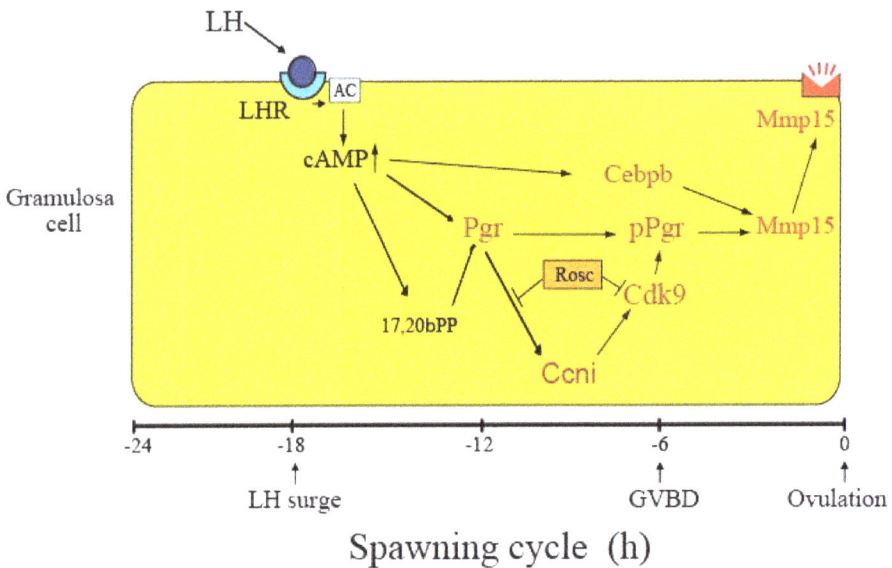

Figure 8. A model for the roles of Ccni and Cdk9 in the process of LH-induced expression of Mmp15 in medaka ovulation. For details, see the text. Note that transcription of *pgr*, *cebpb*, *ccni*, *cdk9*, and *mmp15* gene occurs in the nucleus of the granulosa cell.

The present study points to the role of Ccni/Cdk9 in the generation of phosphorylated Pgr, which serves as a functional transcription factor for the expression of *mmp15*/Mmp15. However, as documented in studies using mammalian models, alterations in the transcriptional activities of target genes are some of the roles known for phosphorylated Pgr [22]. This posttranslational modification of Pgr may also influence promoter specificity [25], receptor turnover [26], and nuclear association [27]. Further studies are needed to comprehensively clarify the roles of phosphorylated Pgr in the follicle rupture process involving Mmp15 in medaka ovulation.

Another intriguing observation of the present study is that LH-induced Ccni synthesis itself was inhibited by roscovitine. This suggests the involvement of Cdk in the expression of Ccni in granulosa cells of ovulating follicles. Cdk9 is the only Cdk detectable in granulosa cells throughout the 24-h spawning cycle, therefore, we speculate that Cdk9 may contribute to the expression of Ccni. However, there are two important questions relevant to the LH-induced expression of Ccni: i) what molecular species of Ccn could be an activator of Cdk9? and ii) in which process is activated Cdk involved, i.e., transcriptional or translational events of the *ccni* gene? To solve these questions, further studies are required.

In summary, we showed that Pgr, a transcription factor required for the expression of *mmp15*, undergoes considerable phosphorylation in granulosa cells of ovulating follicles, and that only the phosphorylated form of Pgr likely functions as a potent transcription factor for *mmp15* expression. Our data indicated that Ccni/Cdk9 may be responsible for Pgr phosphorylation. To our knowledge, this is the first study to demonstrate the involvement of Cdk in the process of follicle rupture during ovulation in vertebrates. The findings provided in this study help elucidate the whole process of ECM degradation that occurs in follicles during ovulation after the ovulatory surge of LH.

Supplementary Materials: The following are available online at http://www.mdpi.com/2073-4409/8/3/215/s1. Figure S1: Expression of *cdk1*, *cdk9*, and *cdk11b* mRNA in medaka preovulatory follicles during 24 h-spawning cycle, Figure S2: Characterization of antibodies prepared in this study, Figure S3: Expression of cdk1, cdk9, and cdk11b mRNA in preovulatory follicles cultured with or without roscovitine, Table S1: PCR primers used in this study.

Author Contributions: K.O. and T.T. designed the study; K.O. performed all of the experiments; K.O. and T.T. analyzed the data and wrote the manuscript.

Funding: This work was supported by Grants-in-Aid for Scientific Research 16H04810 from the Ministry of Education, Culture, Sports, Science and Technology of Japan.

Acknowledgments: We acknowledge financial support from the Ministry of Education, Culture, Sports, Science and Technology of Japan.

Conflicts of Interest: The authors declare no conflict of interest.

References

1. Espey, L.L.; Richards, J.S. *Knobil and Neill's Physiology of Reproduction*, 3rd ed.; Neill, J.D., Ed.; Academic Press: Amsterdam, The Netherlands, 2006; Volume 1, pp. 425–474. ISBN 0-12-515401-1.
2. Lubzens, E.; Young, G.; Bobe, J.; Cerdà, J. Oogenesis in teleosts: How fish eggs are formed. *Gen. Comp. Endocrinol.* **2010**, *165*, 367–389. [CrossRef] [PubMed]
3. Zhang, Z.; Zhu, B.; Ge, W. Genetic analysis of zebrafish gonadotropin (FSH and LH) functions by TALEN-mediated gene disruption. *Mol. Endocrinol.* **2015**, *29*, 76–98. [CrossRef] [PubMed]
4. Takahashi, A.; Kanda, S.; Abe, T.; Oka, Y. Evolution of the hypothalamic-pituitary-gonadal axis regulation in vertebrates revealed by knockout medaka. *Endocrinology* **2016**, *157*, 3994–4002. [CrossRef] [PubMed]
5. Takahashi, T.; Fujimori, C.; Hagiwara, A.; Ogiwara, K. Recent advances in the understanding of teleost medaka ovulation: The role of proteases and prostaglandins. *Zool. Sci.* **2013**, *30*, 239–247. [CrossRef] [PubMed]
6. Ogiwara, K.; Takano, N.; Shinohara, M.; Murakami, M.; Takahashi, T. Gelatinase A and membrane-type matrix metalloproteinases 1 and 2 are responsible for follicle rupture during ovulation in the medaka. *Proc. Natl. Acad. Sci. USA* **2005**, *102*, 8442–8447. [CrossRef] [PubMed]
7. Ogiwara, K.; Minagawa, K.; Takano, N.; Kageyama, T.; Takahashi, T. Apparent involvement of plasmin in early-stage of follicle rupture during ovulation in medaka. *Biol. Reprod.* **2012**, *86*, 1–10. [CrossRef] [PubMed]
8. Ogiwara, K.; Hagiwara, A.; Rajapakse, S.; Takahashi, T. The role of urokinase plasminogen activator and plasminogen activator inhibitor-1 in follicle rupture during ovulation in the teleost medaka. *Biol. Reprod.* **2015**, *92*, 1–17. [CrossRef] [PubMed]
9. Takahashi, T.; Hagiwara, A.; Ogiwara, K. Follicle rupture during ovulation with an emphasis on recent progress in fish models. *Reproduction* **2018**, REP-18-0251. [CrossRef] [PubMed]
10. Sakai, N.; Iwamatsu, T.; Yamauchi, K.; Nagahama, Y. Development of thesteroidogenic capacity of medaka (*Oryzias latipes*) ovarian follicles during vitellogenesis and oocyte maturation. *Gen. Comp. Endocrinol.* **1987**, *66*, 333–342. [CrossRef]
11. Fukada, S.; Sakai, N.; Adachi, S.; Nagahama, Y. Steroidogenesis in the ovarian follicle of medaka (*Oryzias latipes*, a daily spawner) during oocyte maturation. *Dev. Growth Differ.* **1994**, *36*, 81–88. [CrossRef]
12. Nagahama, Y.; Yamashita, M. Regulation of oocyte maturation in fish. *Dev. Growth Differ.* **2008**, *50*, S195–S219. [CrossRef] [PubMed]
13. Ogiwara, K.; Takahashi, T. Involvement of the nuclear progestin receptor in LH-induced expression of membrane type 2-matrix metalloproteinase required for follicle rupture during ovulation in the medaka, *Oryzias latipes*. *Mol. Cell. Endocrinol.* **2017**, *450*, 54–63. [CrossRef] [PubMed]
14. Ogiwara, K.; Fujimori, C.; Rajapakse, S.; Takahashi, T. Characterization of luteinizing hormone and luteinizing hormone receptor and their indispensable role in the ovulatory process of the medaka. *PLoS ONE* **2013**, *8*, e54482. [CrossRef] [PubMed]
15. Ogiwara, K.; Takahashi, T. Specificity of the medaka enteropeptidase serine protease and its usefulness as a biotechnological tool for fusion-protein cleavage. *Proc. Natl. Acad. Sci. USA* **2007**, *104*, 7021–7026. [CrossRef] [PubMed]

16. Hagiwara, A.; Ogiwara, K.; Katsu, Y.; Takahashi, T. Luteinizing hormone-induced expression of Ptge4b, a prostaglandin E$_2$ receptor indispensable for ovulation of the medaka *Oryzias latipes*, is regulated by a genomic mechanism involving nuclear protestin receptor. *Biol. Reprod.* **2014**, *90*, 1–14. [CrossRef] [PubMed]
17. Ogiwara, K.; Takahashi, T. A dual role for melatonin in medaka ovulation: Ensuring prostaglandin synthesis and actin cytoskeleton rearrangement in follicular cells. *Biol. Reprod.* **2016**, *94*, 1–15. [CrossRef] [PubMed]
18. Ansai, S.; Kinoshita, M. Targeted mutagenesis using CRISPR/Cas system in medaka. *Biol. Open* **2014**, *3*, 362–371. [CrossRef] [PubMed]
19. Horiguchi, M.; Fujimori, C.; Ogiwara, K.; Moriyama, A.; Takahashi, T. Collagen type I α1 chain mRNA is expressed in the follicle cells of the medaka ovary. *Zool. Sci.* **2008**, 937–945. [CrossRef] [PubMed]
20. Kato, Y.; Ogiwara, K.; Fujimori, C.; Kimura, A.; Takahashi, T. Expression and localization of collagen type IV α1 chain in medaka ovary. *Cell Tissue Res.* **2010**, *340*, 595–605. [CrossRef] [PubMed]
21. Lu, H.; Xue, Y.; Yu, G.K.; Arias, C.; Lin, J.; Fong, S.; Faure, M.; Weisburd, B.; Ji, X.; Mercier, A.; et al. Compensatory induction of MYC expression by sustained CDK9 inhibition via a BRD4-dependent mechanism. *eLife* **2015**, *4*, e06535. [CrossRef] [PubMed]
22. Hagen, C.R.; Daniel, A.R.; Dressing, G.E.; Lange, C.A. Role of phosphorylation in progesterone receptor signaling and specificity. *Mol. Cell. Endocrinol.* **2012**, *357*, 43–49. [CrossRef] [PubMed]
23. Abdel-Hafiz, H.A.; Horwitz, K.B. Post-translational modifications of the progesterone receptors. *J. Steroid Biochem. Mol. Biol.* **2014**, *140*, 80–89. [CrossRef] [PubMed]
24. Takahashi, T.; Hagiwara, A.; Ogiwara, K. Prostaglandins in teleost ovulation: A review of the roles with a view to comparison with prostaglandins in mammalian ovulation. *Mol. Cell. Endocrinol.* **2017**, *461*, 236–247. [CrossRef] [PubMed]
25. Qiu, M.; Lange, C.A. MAP kinases couple multiple functions of human progesterone receptors: Degradation, transcriptional synergy, and nuclear association. *J. Steroid Biochem. Mol. Biol.* **2003**, *85*, 147–157. [CrossRef]
26. Lange, C.A.; Shen, T.; Horwitz, K.B. Phosphorylation of human progesterone receptors at serine-294 by mitogen-activated protein kinase signals their degradation by the 26S proteasome. *Proc. Natl. Acad. Sci. USA* **2000**, *97*, 1032–1037. [CrossRef] [PubMed]
27. Qiu, M.; Olsen, A.; Faivere, E.; Horwitz, K.B.; Lange, C.A. Mitogen-activated protein kinase regulates nuclear association of human progesterone receptors. *Mol. Endocrinol.* **2003**, *17*, 628–642. [CrossRef] [PubMed]

cells

MDPI

Review

Cysteine Cathepsins and Their Extracellular Roles: Shaping the Microenvironment

Eva Vidak [1,2,†], Urban Javoršek [1,2,†], Matej Vizovišek [1,3,*] and Boris Turk [1,4,*]

1   Jozef Stefan Institute, Department of Biochemistry and Molecular and Structural Biology, Jamova 39, SI-1000 Ljubljana, Slovenia; eva.vidak@ijs.si (E.V.); urban.javorsek@ijs.si (U.J.)
2   International Postgraduate School Jozef Stefan, Jamova 39, SI-1000 Ljubljana, Slovenia
3   Department of Biology, Institute of Molecular Systems Biology, ETH Zürich Otto-Stern-Weg 3, 8093 Zürich, Switzerland
4   Faculty of Chemistry and Chemical Technology, University of Ljubljana, Vecna pot 113, SI-1000 Ljubljana, Slovenia
*   Correspondence: vizovisek@imsb.biol.ethz.ch (M.V.); boris.turk@ijs.si (B.T.)
†   These authors have contribution equally.

Received: 7 February 2019; Accepted: 15 March 2019; Published: 20 March 2019

Abstract: For a long time, cysteine cathepsins were considered primarily as proteases crucial for nonspecific bulk proteolysis in the endolysosomal system. However, this view has dramatically changed, and cathepsins are now considered key players in many important physiological processes, including in diseases like cancer, rheumatoid arthritis, and various inflammatory diseases. Cathepsins are emerging as important players in the extracellular space, and the paradigm is shifting from the degrading enzymes to the enzymes that can also specifically modify extracellular proteins. In pathological conditions, the activity of cathepsins is often dysregulated, resulting in their overexpression and secretion into the extracellular space. This is typically observed in cancer and inflammation, and cathepsins are therefore considered valuable diagnostic and therapeutic targets. In particular, the investigation of limited proteolysis by cathepsins in the extracellular space is opening numerous possibilities for future break-through discoveries. In this review, we highlight the most important findings that establish cysteine cathepsins as important players in the extracellular space and discuss their roles that reach beyond processing and degradation of extracellular matrix (ECM) components. In addition, we discuss the recent developments in cathepsin research and the new possibilities that are opening in translational medicine.

Keywords: cathepsin; inflammation associated disease; cancer; osteoporosis; extracellular matrix

---

1. Introduction

Cysteine cathepsins are an important group of proteases that regulate numerous physiological processes and are normally found in high concentrations in endosomes and lysosomes where they are crucial for protein breakdown and major histocompatibility complex (MHC) class II-mediated immune responses [1,2]. There are 11 cathepsins encoded in the human genome (B, C, F, H, K, L, O, S, V, W, and X) [3], and different studies have shown that a number of them have largely overlapping specificities [4–7]. Nevertheless, there are several examples of specific proteolytic functions of cathepsins demonstrating that their roles are not limited to the endolysosomal system. Cathepsins have thus been found in the cytoplasm, cell nucleus, and the extracellular space [8,9], and their extra-lysosomal localization and activity is frequently associated with ongoing pathological processes [10]. Moreover, high cathepsin activities, especially in extracellular spaces, are a hallmark of inflammation that often accompanies different diseases including cancer, arthritis, cardiovascular disease, and bone and joint disorders as a consequence of dysregulated localization, activation or transcription, as well as inhibitor imbalance [10–12].

Extracellular cathepsins have been shown to participate in extracellular matrix remodeling by degrading abundant structural components of the extracellular matrix (ECM) (e.g., collagen or elastin), but their extracellular functions go beyond simple proteolysis [13–15]. Accordingly, cathepsins have been found to be involved in the processing of cytokines and chemokines, thereby representing an important bridge between inflammation and diseases like cancer and psoriasis [16–20]. Moreover, other more specific functions of cathepsins were reported recently, and cathepsins were thus found to shed a group of extracellular receptors and cell adhesion molecules, demonstrating that their limited extracellular proteolysis could not only directly impact the cell surface but also influence intracellular signaling pathways (e.g., kinase receptor signaling), thereby contributing to the disease progression [21]. In addition, cathepsin S was identified as the critical sheddase of the membrane-anchored chemokine fractalkine, thereby critically contributing to the neuropathic pain [22,23].

With the emerging novel roles of cathepsins, new possibilities are opening for the development of diagnostic and therapeutic tools that will further improve our understanding of their extracellular roles and support of translational medicine. In this review, we focus on cysteine cathepsins and their roles in limited proteolysis in the extracellular space. We describe their roles in pathologies, highlight their most important disease-specific extracellular substrates, and discuss how the new findings can translate into improved diagnostic and therapeutic tools. We also discuss the future perspectives of cathepsin research that will benefit from the emerging systems biology approaches.

2. Cysteine Cathepsins: Structure, Function, and Regulation

Cysteine cathepsins belong to the papain family of cysteine proteases sharing the typical papain-like fold, which is composed of two domains (L—left domain, R—right domain) that form the active protease with the catalytic Cys-His ion pair located in the active site cleft on opposite sides (Cys25-His159; papain numbering) [24] (Figure 1A). In general, cysteine cathepsins are monomers with a MW in the 20–35 kDa range, with variations being the consequence of different posttranslational modifications (e.g., glycosylation). The only exception is the tetrameric cathepsin C with a MW of 200 kDa. Several cysteine cathepsins, including cathepsins B, C, F, H, L, O, and X, are ubiquitously expressed in human tissues and cells, whereas cathepsins K, S, V, and W have more specific localization due to their more specific functions (Table 1). Cathepsin K is expressed in osteoclasts and synovial fibroblasts, cathepsin S is expressed predominantly in immune cells, cathepsin V in thymus and testes, and cathepsin W in CD 8+ lymphocytes and natural killer (NK) cells [1,3,12]. All cathepsins are synthesized as proenzymes, and, after activation, their activity is kept under tight control by pH, compartmentalization, and by their endogenous protein inhibitors stefins, cystatins, kininogens, thyropins, and serpins, which are important for the fine-tuning of their proteolytic activity [25,26].

The majority of cathepsins are potent endopeptidases; nevertheless, some cathepsins have exopeptidase activities due to loops and propeptide regions that limit the accessibility of the active site. Accordingly, cathepsins B and X are carboxypeptidases and cathepsins C and H are aminopeptidases, although cathepsin B can also have endopeptidase activity at a neutral pH [2]. Moreover, cathepsins are not very specific enzymes with highly similar substrate specificities and a moderate preference for cleavage after basic and hydrophobic residues. The only substrate recognition site that actually forms a defined pocket is the S2 site, which, together with the S1 and S1′ sites, seems to be the major substrate recognition site. Several studies using combinatorial peptide libraries and proteomic approaches have demonstrated a strong preference for small hydrophobic amino acid residues (Leu, Val, Ile) in the P2 position, although aromatic amino acid residues (Phe, Tyr) are also accepted (Figure 1B). However, there are a few exceptions. The first is the acceptance of Pro in the P2 position of cathepsin K, which is important for the collagenolytic activity of the latter. The second is the acceptance of Arg in the P2 position of cathepsin B [4–7,24]. The broad specificity of cathepsins is in good agreement with their roles in protein turnover and degradation, including in antigen processing. On the other hand, such broad substrate specificity combined with the high proteolytic efficiency ensures that even in unfavorable conditions, cathepsins can have major roles not only in protein degradation but also in more subtle protein processing, thereby having more signaling roles. This is supported by numerous examples where several cathepsins were found to cleave their protein substrates at the same or similar cleavage sites, although there were also examples of substrates like collagen, osteocalcin, cytokines, and chemokines, which were only cleaved by a subset of cathepsins [10,14].

Table 1. Cysteine cathepsins. Overview of cysteine cathepsins, their peptidase activities, expression, and gene names according to the HUGO Gene Nomenclature Committee.

Cysteine Cathepsins.	Gene Name	Peptidase Activity	Expression
Cathepsin B	CTSB	Carboxydipeptidase, Endopeptidase	Ubiquitous
Cathepsin C	CTSC	Aminodipeptidase	Ubiquitous
Cathepsin F	CTSF	Endopeptidase	Ubiquitous
Cathepsin H	CTSH	Aminopeptidase, Endopeptidase	Ubiquitous
Cathepsin K	CTSK	Endopeptidase	Osteoclasts [27]
Cathepsin L	CTSL	Endopeptidase	Ubiquitous
Cathepsin O	CTSO	Unknown	Ubiquitous
Cathepsin S	CTSS	Endopeptidase	Antigen-presenting cells (e.g., dendritic cells, B-cells) [28,29]
Cathepsin V	CTSV	Endopeptidase	Thymus, testis [30,31]
Cathepsin W	CTSW	Unknown	Natural killer cells, cytotoxic T cells [32]
Cathepsin Z (Cathepsin X)	CTSZ	Carboxymonopeptidase	Ubiquitous

If not otherwise stated, expression profiles are from ref. [3].

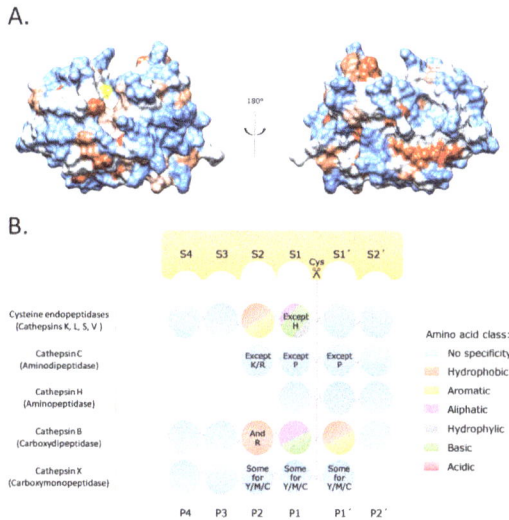

A.

B.

Figure 1. Cysteine cathepsin structures and specificities. (**A**) Crystal structure of the two-chain form of cathepsin L (PDB 1icf [33]) in standard orientation colored according to surface hydrophobicity (red: most hydrophobic, blue: most hydrophilic). Active site Cys25 is colored in yellow. (**B**) Substrate specificity of different cysteine cathepsins relative to the cleavage site, which is between the P1 and P1' residues. S4–S1 and S1'–S2' represent the substrate binding sites into which the corresponding substrate residues P4–P2' bind, with P4–P1 designating the substrate residue N-terminals of the cleavage site and P1'–P2' designating residue C-terminals of the cleavage site, respectively. Amino acid residues of the substrate, which bind to their designated sites, are shown with colored circles, with each color representing a different amino acid class required for the binding of the substrate to the active site.

For their optimal activity, cysteine cathepsins require reducing and mildly acidic conditions and, except cathepsin S, all are irreversibly inactivated at a neutral pH with cathepsin L being the most unstable [34–37]. While these findings originate mostly from in vitro assays, different components in the extracellular milieu and ECM building blocks can stabilize or alter cathepsin activity in the extracellular milieu. A good example is the glycosaminoglycans (GAGs) that can stabilize cathepsins at a neutral pH [38]. Moreover, experiments have shown that GAGs and other negatively charged surfaces enable autocatalytic activation of cathepsins even at neutral pH and possibly contribute to the cathepsin activity in the extracellular space [39]. Nevertheless, effects can be substantially different. Accordingly, the collagenolytic activity of cathepsin K is reduced in case of dermatan sulfate, heparan sulfate, and heparin, but GAGs like keratan and chondroitin sulfates can potentiate it [40]. Another good example is chondroitin sulfate, which inhibits the elastolytic activity of cathepsins V, K, and S [41]. In addition, very potent regulators of cysteine cathepsins are their endogenous inhibitors: cystatins (stefins, cystatins, and kininogens), thyropins, and serpins. They inhibit most of the cathepsins with very high affinities in the nM to pM range. While stefins are essentially cytosolic, the others are primarily extracellular, and their main function is to block the cathepsins escaped into the extracellular milieu, thereby acting as emergency-type inhibitors [1,26,42].

3. Extracellular Cysteine Cathepsin Origins

There is long-lasting evidence that cysteine cathepsins can be present in the extracellular milieu. Under physiological conditions, they are commonly involved in the processes of wound healing, bone remodeling, and prohormone activation [9,43]; however, their extracellular localization is far more common in different pathological conditions [10,12,44]. Prolonged extracellular cathepsin activity upon their secretion is unusual since they have optimal stability under acidic conditions. Nevertheless, the loss of activity at neutral pH can at least partly be delayed by secretion in zymogen form in which they remain relatively stable until their activation [45,46]. Apart from that, cysteine cathepsins can be secreted in an active form and their concentration is sufficient for ECM degradation [47]. An important factor for their secretion is also the type of cell. High levels of cathepsins are most commonly secreted by different immune cells [48], which is in agreement with the fact that cathepsins are elevated in inflammation [49]. In addition, keratinocytes, osteoclasts, smooth muscle cells, and thyroid cells also secrete cathepsins [10]. Nevertheless, the concentration is much lower than in the case of immune cells, where it can reach up to 100 nM in the macrophage secretome [21].

Diverse cellular mechanisms and pathways were shown to be involved in the secretion of cysteine cathepsins (Figure 2A), which is often accompanied by acidification of the extracellular milieu. Experiments performed with macrophages showed that acidification can be achieved either by vacuolar-type H^+ ATPase, which undergoes activation by protein kinase C or serotonin [50–52] or by Na^+/H^+ exchanger I, activated following the binding of immunoglobulin E to high-affinity immunoglobulin-ε receptor [53]. Acidification is especially pronounced in the tumor microenvironment, where tumor-associated immune cells secrete large amounts of cathepsins, and their extracellular presence is connected with more aggressive cancers and inflammation [54]. Apart from cancer, acidification was also observed to be present in advanced osteoarthritis [55] and in atherosclerotic plaques [56]. Usually, secretion of cysteine cathepsins is linked with their overexpression, a common consequence of activation of transcription factor EB [57], or signal transducer and activator of transcription (STAT) signaling pathways by activation of STAT3 or STAT6 [58,59]. Elevated expression can also be the result of extracellular stimuli provided by different cytokines and interleukins [60,61]. The main examples of cytokine-triggered overexpression and secretion are cathepsins S and K. Cathepsin S is overexpressed and secreted in its active form by human chondrocytes upon stimulation from pro-inflammatory cytokines interleukin 1α (IL-1α) and tumor necrosis factor α (TNF α) [60], while regulation of cathepsin K expression is controlled by RANKL (receptor activator of NF-κB ligand; NF-κB, nuclear factor kappa-light-chain-enhancer of activated B cells) [62]. Another factor that can act as an initiator of the positive feedback loops that drive the cathepsin secretion are proteolytic products of ECM degradation [63], whereas elevation of intracellular levels of Ca^{2+}, which triggers the fusion of lysosomes with the plasma membrane, enables secretion of cathepsins by vesicular exocytosis [64,65]. In addition, secretion of cathepsins may even result from increased concentrations of reactive oxidative species, which may lead to the permeabilization of the lysosomal membrane and the release of lysosomal proteases into the cytoplasm and further into the extracellular milieu [66,67].

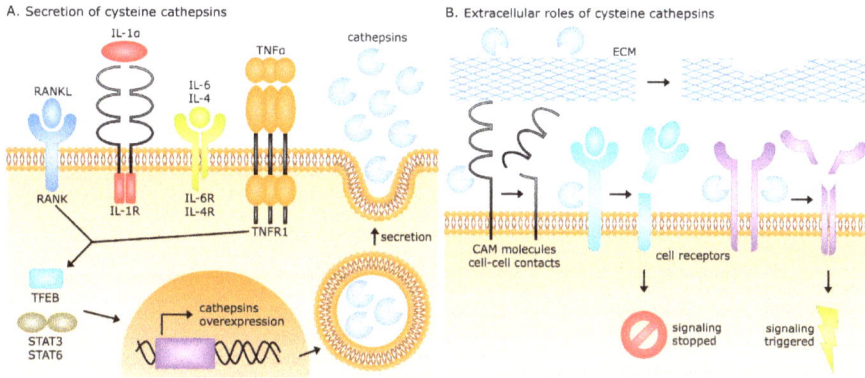

Figure 2. Cysteine cathepsin secretion and their extracellular roles. (**A**) Secretion of extracellular cathepsin is often tightly connected with their overexpression and can be triggered by diverse cell signaling pathways. Overexpressed cysteine cathepsins are usually secreted with vesicular exocytosis. (**B**) In the extracellular milieu, cysteine cathepsins cleave different targets. Cleavages of cell adhesion molecules (CAM), cell-cell contacts, and proteins of ECM mainly influence cell adhesion and migration. Additionally, proteolytic products of these cleavages can act as signaling molecules and have an impact on cell growth, invasion, and angiogenesis. Other main target of cysteine cathepsins are cell receptors, and their cleavage can result in either constantly triggered signaling, in the case of partial trimming of the receptor, or inhibited signaling, in the case of a complete removal of the extracellular domain. CAM, cell adhesion molecules; ECM, extracellular matrix; IL, interleukin; RANK, receptor activator of NF-κB; RANKL, receptor activator of NF-κB ligand; STAT, signal transducer and activator of transcription; TFEB, transcription factor EB; TNFα, tumor necrosis factor alpha; TNFR1, tumor necrosis factor receptor 1.

Once secreted, cathepsins can either remain bound to the plasma membrane or interact with molecules from the extracellular milieu [15]. In particular, the former may be of special importance as such membrane association may also protect the cathepsins against inactivation in otherwise unfavorable conditions of the extracellular milieu. Cathepsin S was thus shown to associate with the plasma membrane and co-localizes with $\alpha_v\beta_3$ integrin on the surface of smooth muscle cells present in the vasculature [68], but its exact binding partners remain unknown. Another cathepsin that translocates to the plasma membrane is cathepsin X, which acts as an activator of β_2 integrins that are crucial for cell adhesion of dendritic cells and lymphocytes during their maturation [69–71]. Translocation of cysteine cathepsins occurs also in pathological conditions as shown for cathepsin B, which localizes to the cell membrane in cancer cells [72] either by binding to annexin II tetramers [73] or by association to the caveolae site [74]. Moreover, its membrane-bound localization was correlated with shortened survival in the case of colorectal cancer [75,76].

4. ECM Proteolysis and the Cathepsins

The ECM that surrounds the cells is composed of the proteins these cells secrete [77]. The main components of ECM are structural proteins, like collagen, elastin, fibronectin and laminins, non-structural matricellular proteins, polysaccharides, mainly glycosaminoglycans and hyaluronan, and proteoglycans [78,79]. The structure of the ECM is dynamic and depends on the equilibrium between synthesis and degradation [80,81]. Modifications of the ECM are achieved through both enzymatic and non-enzymatic processes, which can influence the stability and function of the ECM [82]. Among these processes, proteolytic processing is one of the most important mechanisms involved in the regulation of the ECM. It is therefore not surprising that a vast number of proteases were found to be involved, with major players being various metalloproteases such as a disintegrin and metalloprotease (ADAMs), matrix metalloprotease (MMPs), meprins and bone morphogenetic protein (BMP)/tolloid metalloproteases [83] and various cysteine proteases [15].

Cysteine cathepsins are the main cysteine proteases participating in the reorganization of ECM and are associated with non-specific degradation of abundant ECM proteins, which can take place both extracellularly and intracellularly following endocytosis [82]. Proteolysis of ECM proteins occurs under normal conditions, as in the case of collagen degradation in bone resorption or elastin degradation in the vascular system, as well as under pathological conditions, since ECM degradation is an influential factor in many diseases including cancer, cardiovascular diseases, and arthritis [14,15,84]. In particular, extracellular proteolysis seems to be of crucial importance, since the extracellular cathepsins can cleave a plethora of structural and functional proteins, thereby affecting not only the structural aspects of ECM but also the associated signaling pathways (Figure 2B). While non-specific and specific proteolysis by cysteine cathepsins are important for ECM remodeling in disease, one has to keep in mind that extracellular substrates of cysteine cathepsins were mostly identified in in vitro studies. Nevertheless, much evidence for their physiological significance and substrates confirming their involvement in disease development and progression has emerged just in the last decade (Table 2), and the next sections will provide a detailed overview of the extracellular roles of cathepsins in different pathologies.

Table 2. Cathepsins in disease.

Disease	Cathepsins Involved	Cleaved Targets	Selected References
Angiogenesis/Leukocyte recruitment	B, K, L, S	ELR (glutamate-leucin-arginin motif) chemokines/non-ELR chemokines, CD18	[16,85]
Cancer	B, K, L, S, X	Tenascin-C, nidogen-1, fibronectin, osteonectin, laminin, periostin, collagen IV, general degradation	[86–91]
Cardiovascular and kidney diseases (e.g., atherosclerosis, abdominal aortic aneurysm, chronic kidney disease)	K, L, S, V	Elastin, CX$_3$CL, heparanase, collagen I (catK: *Gly61-Lys62, Arg144-Gly145, Gln189-Gly190*)	[41,92,93]
Lung fibrosis	S	Decorin	[94]
Neuroinflammation/hyperalgesia	S	CX$_3$CL1, PAR2	[95–97]
Osteoarthritis and rheumatoid arthritis	K, B, L, S	Collagen II (catK: *Gly61-Lys62, Arg144-Gly145, Gln189-Gly190*), aggrecan (catB: *Asn341-Phe342, Gly344-Val345*, catL: *Gly344-Val345*)	[98–100]
Osteoporosis	K, B, L, S, H	Collagen I (catK: *Gly61-Lys62, Arg144-Gly145, Gln189-Gly190*), osteonectin, osteocalcin (catB: *Arg44-Phe45*, catL: *Gly7-Ala8*, catS: *Arg43-Arg44*, catS: *Gly7-Ala8*)	[101–104]
Tuberculosis	K	Collagen I (casK: *Gly61-Lys62, Arg144-Gly145, Gln189-Gly190*)	[105]

Diseases where cysteine cathepsins are involved in the development and progression of the pathology together with their extracellular substrates. Known cleavage sites of extracellular substrates for each cathepsin are shown in parentheses.

4.1. Cathepsins in Cancer

Cysteine cathepsins were historically first linked to extracellular proteolysis in cancer, as first demonstrated for cathepsin B almost 40 years ago [106]. However, despite the use of various in vivo and in vitro cancer progression models, the exact roles of individual cysteine cathepsins are not completely understood [107–109]. A comprehensive understanding of the functions of cathepsins B, C, H, K, L, S, and X in the extracellular milieu has proved complicated because of their broad substrate specificity [1], their endogenous inhibitors [110], compensatory effects [111,112], effects that are not associated with their proteolytic function [113], their ability to act as tumor suppressors [114], and the various different cell types that comprise many tumors [115].

Cathepsins are released into the tumor microenvironment by different cells including tumor cells, endothelial cells, tumor-associated macrophages (TAM), myoepithelial cells, fibroblasts, and other cells, which infiltrate the tumor site [48,54]. Among these, TAMs are considered to release the largest amount of cathepsins primarily due to stimulation by the IL-4, IL-6, and IL-10 cytokines in the tumor microenvironment [10,48]. Interestingly, the source and role of secreted cathepsins responsible for cancer progression are not universal among different types of cancer, as evident primarily from different mouse *in vivo* cancer models. In the RIP1-Tag2 model of the pancreatic neuroendocrine cancer, cathepsins B, H, and S, derived from the TAMs, had a predominant, if not exclusive, effect on cancer progression through the reduction of tumor burden, increase in apoptosis, and decrease of angiogenesis [48]. On the other hand, cathepsin L-mediated cancer progression was largely due to its secretion from cancer cells, whereas the effect of cathepsin X was due to its release from both sources. Moreover, cathepsin X was found to partially compensate for the loss of cathepsins B and S, as revealed by simultaneous deletion of the two cathepsin genes. In addition, simultaneous deletion of cathepsins B and S revealed additive effects in early stages, but at late stages, several differences were restored to the wild-type level, although the mechanisms are not known [111,113]. Interestingly, expression of cathepsin C was also increased in RIP1-Tag2 and MMTV-PyMT mammary gland models of cancer but had no functional role in their progression [109,116]. Cathepsin S was also found to have a major role in a syngeneic colorectal carcinoma murine model, where its release from cancer cells, endothelial cells, and TAMs was found to be responsible for the progression of cancer through the promotion of neovascularization and tumor growth [117]. An analogous effect was also observed *in vivo* in the MMTV-PyMT mammary gland cancer model, where cathepsin B, originating from both tumor cells and macrophages, was suggested to have a major role in the tumor progression and metastasis spread [107,118], which could have been partially compensated by cathepsin X [112]. However, using an orthotopic transplantation of primary mouse PyMT cancer cells overexpressing cathepsin B showed that the enzyme expressed and secreted from tumor, but not stromal, cells increased invasiveness into adjacent tissues by excessive extracellular matrix degradation [119]. Another *in vivo* mouse model, where the roles of individual cathepsins were systematically investigated, was the K14-HPV16 model of squamous cell carcinoma. In this model, cathepsin C, but not cathepsin B, released from stromal cells was shown to be responsible for cancer development and angiogenesis. Moreover, cathepsin L was found to have a tumor suppressor role in this model, further showing the complexity of cathepsins involvement in tumorigenesis [109].

There are also numerous cellular studies available where secretion of cathepsins was shown to be an important factor in changing cellular properties possibly leading to tumor progression, although these studies were mostly focused on tumor cell lines but not on macrophages or related immune cells. Among the cathepsins, cathepsin B was most often associated with changes of cellular properties, although there were also numerous reports about cathepsins L, X, and K. Cathepsins B and X were thus shown to be important in the epithelial to mesenchymal transition in two breast cancer cell lines, MCF7 and MDA-MB-231, which express low and high amounts of cathepsins, respectively. Upregulation of cathepsins B and X in MCF7 cells decreased E-cadherin, a marker for epithelial cells, which was mainly attributed to cathepsin B [120]. Similar observations involving cathepsin X were also reported in hepatocellular cancer where its overexpression increased invasiveness and motility in a matrix-coated transwell model and a wound healing assay [121], although the enzyme seems to be primarily expressed in the immune cells [70]. Release of cathepsins B and L was recently reported in different melanoma cell lines, where Abl/Arg nonreceptor tyrosine kinases were shown to play an important role in cathepsin B and L expression and their release from the cells [122]. Tumor progression and increased cell invasiveness following cathepsin B release are also characteristic for pancreatic ductal adenocarcinoma [123], esophageal adenocarcinoma [124], and glioma, where cathepsins K and X were also found to be involved [125–127]. Recently, a mechanism of cathepsin B release followed by enhanced lung cancer cell migration was proposed [128], indicating a possible role in the progression of lung cancer. However, the role of cathepsins in cancer is not only extracellular but also has an important intracellular component [48,129]. An example of this is the leakage of cathepsin B to the cytoplasm following chemotherapy-induced lysosomal membrane permeabilization in myeloid-derived suppressor cells (MDSC). Upon release, cathepsin B was found to interact with NLRP3, leading to a non-proteolytic activation of the inflammasome and pro-IL-1β processing. The released IL-1β enhanced IL-17 production in CD 4⁺ T cells, which resulted in angiogenesis and tumor relapse [130].

Unfortunately, there are almost no in vivo studies on the role of cysteine cathepsin inhibitors that could shed some more light on the possible imbalance between the endogenous inhibitors and the target proteases as one of the reasons for tumor progression, as suggested some time ago [54]. The first study revealed that genetic ablation of cystatin C, major extracellular cathepsin inhibitor, in the pancreatic neurocrine tumor model (Rip-Tag2) resulted in an increased size of islet cell carcinomas and angiogenic islets, which was linked to deregulated cathepsin S activity leading to increased endostatin generation [91]. The other two studies available included genetic ablation of stefin B, the major intracellular cathepsin inhibitor [131], and of cystatin C [132], in the mammary gland PyMT mouse model. However, contrary to all expectation, tumors were smaller in both inhibitor knock-out models despite the increased cathepsin activity in the tumors. While in the case of stefin B this was linked to increased sensitivity to lysosomal cathepsin-mediated cell death and oxidative stress, the potential link in case of cystatin C were the 14-3-3 proteins, but the evidence is not entirely conclusive. This further supports the idea of differential and context-dependent roles of cathepsins and their inhibitors in cancer.

Nevertheless, it seems that the major role of cathepsins in cancer is in extracellular matrix degradation [87], shedding of receptors and adhesion molecules [21,85], activation of cytokines and growth factors [48], and cleavage of proteins forming cell-cell junctions [133]. In addition, cathepsins were also suggested to be involved in the activation of other tumor-associated proteases, although the evidence is primarily based on in vitro studies. An example is cathepsin B, which was suggested to be involved in the activation of urokinase-type plasminogen activator (uPA), which influences uPA/uPAR signaling, thereby possibly affecting cell migration, whereas cathepsin L was suggested to be involved in the activation of MMP1 and MMP3 [13,54].

Despite many phenotypic changes following cathepsin ablation or inhibition in in vivo and in vitro tumor models, the exact roles, besides the degradation of extracellular matrix, are less-well known. This is supported by a recent proteomic study where individual genetic ablation of cathepsins B, H, L, S, and X in the Rip-Tag2 pancreatic tumor model was shown to have a predominant effect on the degradation of the extracellular matrix, with very few limited proteolysis events detected [87]. Nevertheless, limited proteolysis of E-cadherin, an important adhesion molecule and marker for epithelial phenotype, by cathepsins B, L, and S was shown to potentially drive invasiveness of cancer cells in the same RIP1-Tag2 model [116]. There are several other substrates of individual cathepsins identified in different cancer models, but their importance for cancer progression is unclear. Among these targets is CD18, which was found to be cleaved by cathepsin B released from adhering leukocytes following physiological levels of shear stress [134] and may have a role in leukocyte recruitment in angiogenic vessels [85]. Another set of substrates of extracellular cathepsins associated with angiogenesis are basement membrane proteins, which are cleaved during basement membrane degradation, in mother vessels formation, following vascular endothelial growth factor A (VEGF-A) stimulation [135]. One of these is nidogen-1, which was found to be degraded by cathepsin S in patients with non-small cell lung cancer [136]. In addition, cathepsins B, L, and S were reported to cleave laminin, with cathepsin S generating a fragment with pro-angiogenic effects [91]. Cathepsins B, L, and S can also cleave fibronectin, the results of which are still poorly understood [15,88]. Another such substrate is tenascin-C, which was found to be cleaved by cathepsin B, resulting in a pro-angiogenic effect in glioma [86]. An important cathepsin substrate also seems to be collagen XVIII, which can serve as a source of endostatin that can be generated by cathepsins L and S, thereby affecting angiogenesis [137,138]. Finally, cathepsin K was demonstrated to cleave periostin, which may be linked to breast cancer bone metastasis. The decrease of C-terminal intact periostin was namely shown to be a marker for osteolytic lesions in breast cancer bone metastasis and could be used to detect bone relapse in patients [89]. Cathepsin K cleavage of osteonectin was also implicated in experimental prostate to bone tumor metastasis, where both the protease and substrate expression levels were higher. Additionally, high levels of released inflammatory cytokines suggested that cathepsin K and/or osteonectin may regulate their release [90]. Some additional cathepsin ECM substrates linked to cancer have also been described elsewhere [15,82]. A more detailed overview on the roles of cysteine cathepsins in cancer development and progression can be found elsewhere [48].

4.2. Cysteine Cathepsins and Tissue Remodeling

The role of cathepsins in bone and cartilage processing is extremely well described, with cathepsin K being the most studied. The main source of cathepsin K in these processes are osteoclasts. Cathepsin K is one of the few proteases capable of cleaving the collagen triple helix in the polyproline region, and its activity differs from metalloproteases. It has multiple cleavage sites in the polyproline region of both collagens I and II [102,139] and can also cleave collagen in the telopeptide region [101]. Collagenolytic activity of cathepsin K is strong in the acidic environment at the site where osteoclasts attach to the bone surface and is crucial for normal bone homeostasis [14]. It has been shown that inhibition of cathepsin K can result in an altered structure of the bone, leading to changed crystallinity and crystal structure [140]. Another factor that influences cathepsin K activity is the presence of GAGs, since its collagenolytic activity is completely dependent on the formation of an oligomeric complex between cathepsin K and GAGs [141,142]. Aging-associated changes of collagen fibers were also suggested to influence the collagenolytic activity of cathepsin K. Accumulation of advanced glycation end-products (AGEs) and mineralization were thus shown to reduce cathepsin K collagenolytic activity, while removal of GAGs completely blocked it [143]. Pathological collagen degradation is connected with cathepsin K overexpression and unbalanced osteoclast/osteoblast activation, leading to collagen I degradation and bone loss as seen in osteoporosis [144]. While cathepsin K is the most important cathepsin in bone and cartilage remodeling, one must keep in mind that in vitro studies have identified cathepsins B, L, H, and S as potentially being involved in bone remodeling since they can cleave osteocalcin [103], a biomarker of bone degradation in osteoporosis [145].

Other pathological conditions with similar causes are osteoarthritis and rheumatoid arthritis, where cathepsin K degrades collagen II from the N-terminus, which leads to cartilage erosion [99,102]. Moreover, the low pH, as observed in advanced osteoarthritis, favors cathepsin K as a major collagenase in these diseases [55]. In addition, cathepsin K can cleave the proteoglycan aggrecan, which is also cleaved by cathepsins B, L, and S [98,100], and ECM bone protein osteonectin [104], further supporting its crucial role in osteoarthritis. However, in vivo studies in mouse models showed that disease progression was substantially diminished in cathepsin S-deficient animals in collagen-induced arthritis and in cathepsin L-deficient animals in antigen-induced arthritis [29], suggesting that other cathepsins are also involved in the arthritis progression. Moreover, substantial levels of cathepsins B, L, and S were identified in synovial fluids of patients with rheumatoid arthritis and osteoarthritis, with higher levels detected in rheumatoid arthritis patients, suggesting their involvement in inflammation and cartilage destruction [146–150]. However, despite high serum levels of cathepsins S and L in rheumatoid arthritis patients, they did not correlate with the severity of the disease, arguing against their use as disease biomarkers [151]. Anyhow, substantial differences in the serum levels of these two cathepsins were observed between the rheumatoid arthritis patients with and without autoantibodies, suggesting specific roles of cathepsins S and L in disease development and progression in seropositive patients [150].

Cysteine cathepsin-mediated ECM remodeling is also important in the cardiovascular system, where cathepsins K, S, and V can cleave elastin, with cathepsin V displaying the highest elastolytic activity [84]. This is a result of two unique exosites present in its structure, which stabilizes the elastin-cathepsin V complex [92]. Since elastin is responsible for tissue stability against stretching forces, its fragmentation decreases the elasticity of the blood vessels and can lead to their rupture [152]. Excessive elastin degradation was observed in cardiovascular diseases, atherosclerosis and abdominal aortic aneurism, kidney diseases, and during aging [15,84], as upon reaching adulthood, elastin is not synthesized de novo anymore, and consequently its degradation is irreversible [153]. Besides cathepsin S, cathepsin K and V also exhibit elastinolytic activity important in cardiovascular pathology, and their ablation in mice showed decreased signs of atherosclerotic pathology [154–156]. Elastin degradation by cysteine cathepsins occurs mainly extracellularly, with only one-third of elastin being degraded intracellularly [41]. However, not all functions of extracellular cysteine cathepsins in the cardiovascular system are harmful. A recent study has shown that cathepsin K has a pivotal role in cardiac remodeling following myocardial infarction, since sufficient collagen degradation is needed to reduce cardiac fibrosis [93]. However, processing of membrane-bound chemokine fractalkine (CX$_3$CL1 or (C-X3-C motif) ligand 1) by cathepsin S homes additional inflammatory cells to the site of atherogenesis, thereby sustaining the inflammation and progressing the pathology [157]. The ECM degradation capability of cathepsin B is also considered to be a marker of the disease, whereas cathepsin X has a role in homing the inflammatory cells, especially T-cells, to the atheroma site [158].

Cysteine cathepsins B and L have an important role in neural tissue remodeling, especially in axon growth, but exact mechanisms remain mostly unknown [159,160]. Nevertheless, both can cleave perlecan, and this cleavage generates a C-terminal fragment with neuroprotective roles [161]. Another role of cathepsin B is the degradation of chondroitin sulfate proteoglycans (CSPGs) [160], which have inhibitory effects on axon growth and regeneration transmitted by receptor protein tyrosine phosphatase σ (RPTKσ) [162]. A recent study found that regulation of cathepsin B expression and secretion by intracellular sigma peptide (ISP)-modified RPTKσ signaling can possibly lead to axon overgrowth in a CSPG-rich environment [160]. Another study also identified cathepsin B as a myokine that is systemically secreted during running and can cross the blood-brain barrier. This running-induced, systemically secreted cathepsin B induced neurogenesis, improved spatial memory, and also influenced plasticity, cell survival, differentiation, and neuronal migration [163]. In addition, in vitro experiments have also identified extracellular cathepsin L as a potential stimulus of axonal growth [159].

Cysteine cathepsins also participate in lung pathologies [164]. In silicosis, cysteine cathepsins B, H, K, L, and S were found in large excess compared to their inhibitors in the patient's bronchoalveolar lavage fluid (BALF) and are likely involved in the breakdown and remodeling of the ECM [165]. Recently, cathepsin K overexpression and release was also found in lymphangioleiomyomatosis (LAM), a rare nodule-forming disease. Interestingly, cathepsin K was overexpressed and released from fibroblasts associated with LAM cells similar to tumor cell–stroma interaction and could be the driving factor behind matrix degradation and cytokine processing [166]. In the lung, cysteine cathepsin activity is involved in tuberculosis where cathepsin K is one of the collagenolytic proteases that degrade the ECM and cause the formation of lung cavities [105,167]. On the other hand, cathepsin K overexpression and extracellular activity warranted protection against bleomycin-induced pulmonary fibrosis. Additionally, cathepsin K knock-out mice deposited more extracellular matrix and had decreased collagenolytic activity, which points to an important role of cathepsin K in lung collagenolytic activity [168,169]. The same model also provided evidence for cathepsin B overexpression and presence in BALF [170]. Furthermore, cathepsin S was shown to cleave decorin and produce a fragment, which can be robustly detected in the serum of fibrosis and cancer patients [94]. Finally, pharmacological inhibition of cathepsin S was shown to substantially reduce a cystic fibrosis-like disease in a mouse model, possibly via a link to protease-activated receptor 2 (PAR2) [171], which was previously reported to be a cathepsin S substrate [172].

More examples of cathepsin-mediated tissue remodeling are scattered throughout tissue types. In the case of pre-adipocyte cells, cleavage of fibronectin by cathepsin S was suggested as a possible mechanism for their differentiation [173]. Recent in silico and in vitro studies reported that cathepsins K, L, and S could cleave fibrinogen and fibrin, resulting in fragments that differ from the ones produced by plasmin, which opened the door to further investigate their roles in vascular homeostasis, especially in coagulation-associated diseases [174,175]. Moreover, cathepsin K was recently demonstrated to have an essential role in skeletal muscle remodeling, dysfunction, and fibrosis following injury [176]. The roles of cysteine cathepsins in tissue remodeling are diverse and can be both detrimental and beneficial; therefore, cysteine cathepsins are emerging as possible targets in therapeutic strategies.

4.3. Cysteine Cathepsins in Inflammation

Cysteine cathepsin secretion often accompanies different inflammation-driven pathologies where they are released from recruited immune cells or aberrantly expressed or processed in the inflamed tissue. Because of this general mechanism, there are numerous organ systems or tissues where cathepsins can be detected extracellularly during inflammation, and their extracellular localization has been implicated in different aspects of the immune cell physiology [10]. Extracellular cathepsins, in particular cathepsins L, S, and K, were shown to process and activate the glutamate-leucin-arginin motif (ELR) and inactivate the non-ELR CXC (N-terminal Cys-X-Cys motif) chemokines, thereby regulating chemotaxis and angiogenesis [16]. In the case of cathepsin S, its extracellular localization can influence macrophage and monocyte migration through the basal membrane [154]. The arginin-glycin-aspartate (RGD) motif in procathepsin X is responsible for its binding to integrin $\alpha_v\beta_3$, thus modulating the binding of cells to the ECM components [177]. In intestinal goblet cells, cathepsin K is highly expressed and released and provides an antimicrobial effect. As a result, cathepsin K-deficient mice exhibit more severe colitis and have an altered microbial community [178]. During inflammation, a crosstalk is established with the coagulation cascade where fibrin, which can be cleaved by cathepsins K, L, and S in vitro, plays an important role [179].

Moreover, it seems there is also a connection between the nervous and immune systems, where cathepsins were suggested to play a role [180]. At this neuroimmune interface, cathepsin S has been shown to cleave PAR2 in a different pattern as previously described for serine proteases, inducing hyperalgesia [96]. Additionally, cathepsin S cleavage of CX_3CL1 from neurons results in microglial stimulation, which is critical in chronic pain maintenance [181]. Furthermore, microglia cells were shown to secrete cathepsin B, S, and X, which are considered important in inflammation-induced neurodegeneration [182].

Cysteine cathepsins also play a role in both acute and chronic phases of kidney disease, albeit there are currently more known intracellular functions compared to extracellular ones [183]. Nevertheless, cathepsin L-mediated heparanase activation was shown to play a role in the pathogenesis of diabetic nephropathy, causing proteinuria and renal damage [97]. Finally, cathepsin S elastolytic activity was also implicated in the occurrence of calcifications during the pathogenesis of chronic kidney disease [184].

5. Extracellular Cathepsins and Their Translation into Clinical Applications

Elevated activity of cysteine cathepsins in the extracellular space is now widely recognized as an important hallmark of developing or ongoing disease. Therefore, cathepsins are getting increasing attention in the development of novel therapeutic and diagnostic tools (Figure 3). In particular, therapeutic inhibition of cathepsins was the driving force in the field since the discovery that cathepsin K has a crucial role in bone resorption and thus in osteoporosis [37,104,185,186], and that cathepsin S is the key enzyme in the MHC II-mediated immune response [29,31]. A number of small-molecule inhibitors of cathepsins were developed, and despite several inhibitors of cathepsin K and S showing good initial results for treatment of osteoporosis, aortic aneurysm, arthritis, and neuropathic pain, none has entered clinical use so far, and very few are in clinical trials at the moment [10,84]. There are several reasons for this, with perhaps the best-known being the on-target toxicity revealed in the case of cathepsin K inhibitors, which became evident after long-term treatment. This was demonstrated in the case of Odanacatib (Merck), a non-basic nitrile, which was very successful in preclinical stages and even successfully concluded phase III clinical trials, but prolonged investigations of the stroke-related side effects lead to its discontinuation [84,187,188]. However, the problem was already at least partially raised with Balicatib (Novartis), the first cathepsin K inhibitors that entered clinical trials for osteoporosis treatment and was discontinued after Phase II [188]. In order to overcome the problem, research went in the direction of exosite inhibitors of cathepsin K that would only block collagen degradation, whereas cytokine processing, such as that of TGF-β, which is active site-driven, would be unchanged. A good progress in this area was demonstrated with tanshinones, a group of so-called ectosteric inhibitors targeting the cathepsin K exosite originating from plants, which already showed good results in preclinical in vivo studies [189,190]. Currently, a lot of hope is also in the new inhibitors of cathepsin S for neuropathic pain treatment, but the outcome of the clinical studies remains to be seen [10]. Another strategy is the use of inhibitory antibodies, which demonstrated good potential for cathepsin S inhibition, resulting in reduced tumor growth and improved chemotherapeutic efficiency [191,192], and further developments are expected to be seen in this area as well.

However, the high extracellular cathepsin levels secreted from the immune cells in various inflammation-associated diseases including many cancers opened the door for in vivo diagnostic applications. Different activity-based probes (ABPs) with fluorescent tags or internally quenched substrates that start to emit the reporter signals after cathepsin cleavage were used to detect pathologic cathepsin activity [193]. There are several examples where probes were used to detect inflammation [194], visualization of cancer cells [195], and lung fibrosis [196]. Good results were also achieved with fluorogenic substrates in preclinical imaging of cathepsins, especially with the application of the reverse-design principle where medicinal chemistry-optimized small molecule inhibitors were converted to fluorescent activatable substrates [197]. A cathepsin S-selective lipidated fluorescent substrate based on this principle was successfully used for in vivo imaging of mammary gland mouse tumors [198]. Another set of tools that can be used for labeling cathepsins are designed ankyrin repeat proteins (DARPins). A cathepsin B-selective DARPin with high affinity was successfully used for in vivo imaging in two mouse models of mammary gland cancer [199]. While selective tools perform well at the preclinical level, pan-cathepsin probes were much more successful in image-guided surgery applications to visualize the tumor tissue and thus increase the chances of its complete removal [200], leading to the first compounds being evaluated in clinical trials [10].

Concepts for targeted drug delivery can also largely benefit from the elevated activity of cysteine cathepsins in disease. First, cathepsins can be used as drug activators. Accordingly, several drugs are synthesized as prodrugs or antibody-drug-conjugates (ADCs) and become active only after cathepsin cleavage. This concept has been successfully used in oncology with a good example being ADCETRIS®, which is already clinically approved [10], while several other prodrugs are at different stages of development [201,202]. Second, targeting extracellular or membrane-associated cathepsins also emerged as a promising drug delivery strategy. The power of this concept was demonstrated when cancer cell membrane-associated cathepsin B was targeted by liposomes with a selective cathepsin B inhibitor as a targeting moiety, and the system demonstrated improved selectivity and targeting efficiency [203]. However, while the majority of research has been focused on how to exploit the extracellular presence of cysteine cathepsins for medical applications, the extracellular substrate pool itself represents a major source of potential future therapeutic targets, targeting moieties, and biomarkers that will likely result in the development of new diagnostic and therapeutic strategies with major potential for future clinical applications [204].

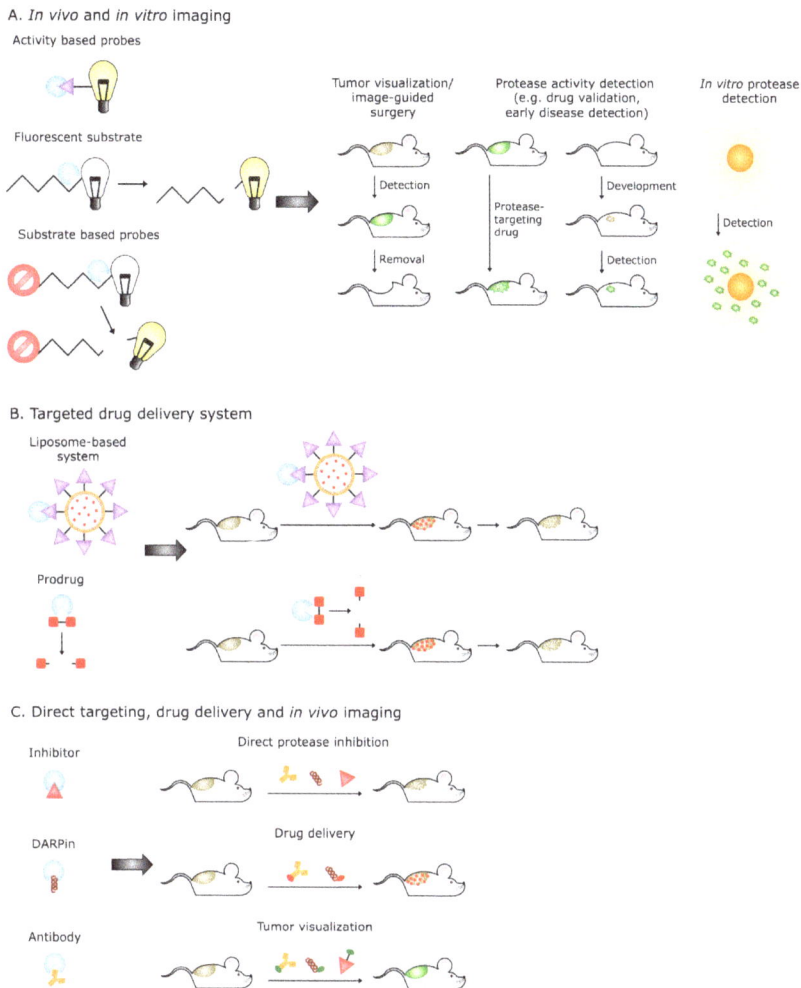

Figure 3. Extracellular cathepsins as diagnostic targets, prodrug activators, and targets for targeted drug delivery. The high levels of cathepsins in the ECM can be utilized in different imaging and targeting techniques. (**A**) Fluorescent substrates, substrate-based probes, and activity-based probes are the most commonly used tools for in vivo and in vitro imaging. (**B**) Extracellular cathepsins can be also used for targeted drug delivery and for prodrug activation. (**C**) Since many cathepsins are also overexpressed and active in different cancers, they can be targeted directly using their inhibitors, designed ankyrin repeat proteins (DARPins), or antibodies. These molecules can also be conjugated and, as such, used for drug delivery and tumor visualization.

6. Concluding Remarks and Future Perspectives

Recent findings on the roles of cysteine cathepsins in the extracellular space have substantially improved our understanding of these important proteases and of their normal and pathological proteolysis in the ECM. While cathepsins are widely recognized as important diagnostic and therapeutic targets largely for the diseases that involve ECM remodeling, their multifunctional roles pose a problem in the design of successful tools for their therapeutic targeting as demonstrated by the systematic failure of a number of cathepsin inhibitors in clinical trials. Therefore, we believe that in the future, the clinical paradigm on cysteine cathepsins will shift from targets for therapeutic intervention to diagnostic targets and targets for image-guided surgery, targets for targeted drug delivery, modifiers of the cancer cell surfaceome, and generators of cancer cell-specific fingerprints with biomarker potential, thereby leading to the development of new cathepsin-based diagnostic and therapeutic applications.

Author Contributions: All the authors wrote the manuscript.

Acknowledgments: This work was supported by the research grants from Slovene Research Agency (P1-0140) and ICGEB (CRP/SVN 16-01) to B.T.

Conflicts of Interest: The authors declare no conflict of interest.

References

1. Turk, V.; Stoka, V.; Vasiljeva, O.; Renko, M.; Sun, T.; Turk, B.; Turk, D. Cysteine cathepsins: From structure, function and regulation to new frontiers. *Biochim. Biophys. Acta* **2012**, *1824*, 68–88. [CrossRef]
2. Turk, V.; Turk, B.; Turk, D. Lysosomal cysteine proteases: Facts and opportunities. *EMBO J.* **2001**, *20*, 4629–4633. [CrossRef]
3. Rossi, A.; Deveraux, Q.; Turk, B.; Sali, A. Comprehensive search for cysteine cathepsins in the human genome. *Biol. Chem.* **2004**, *385*, 363–372. [CrossRef] [PubMed]
4. Vidmar, R.; Vizovisek, M.; Turk, D.; Turk, B.; Fonovic, M. Protease cleavage site *fingerprinting* by label-free in-gel degradomics reveals pH-dependent specificity switch of legumain. *EMBO J.* **2017**, *36*, 2455–2465. [CrossRef]
5. Vizovisek, M.; Vidmar, R.; Van Quickelberghe, E.; Impens, F.; Andjelkovic, U.; Sobotic, B.; Stoka, V.; Gevaert, K.; Turk, B.; Fonovic, M. Fast profiling of protease specificity reveals similar substrate specificities for cathepsins K, L and S. *Proteomics* **2015**, *15*, 2479–2490. [CrossRef]
6. Biniossek, M.L.; Nagler, D.K.; Becker-Pauly, C.; Schilling, O. Proteomic identification of protease cleavage sites characterizes prime and non-prime specificity of cysteine cathepsins B, L, and S. *J. Proteome Res.* **2011**, *10*, 5363–5373. [CrossRef]
7. Choe, Y.; Leonetti, F.; Greenbaum, D.C.; Lecaille, F.; Bogyo, M.; Bromme, D.; Ellman, J.A.; Craik, C.S. Substrate profiling of cysteine proteases using a combinatorial peptide library identifies functionally unique specificities. *J. Biol. Chem.* **2006**, *281*, 12824–12832. [CrossRef] [PubMed]
8. Brix, K.; Dunkhorst, A.; Mayer, K.; Jordans, S. Cysteine cathepsins: Cellular roadmap to different functions. *Biochimie* **2008**, *90*, 194–207. [CrossRef]
9. Brix, K.; Linke, M.; Tepel, C.; Herzog, V. Cysteine proteinases mediate extracellular prohormone processing in the thyroid. *Biol. Chem.* **2001**, *382*, 717–725. [CrossRef] [PubMed]
10. Kramer, L.; Turk, D.; Turk, B. The Future of Cysteine Cathepsins in Disease Management. *Trends Pharm. Sci.* **2017**, *38*, 873–898. [CrossRef]
11. Vasiljeva, O.; Reinheckel, T.; Peters, C.; Turk, D.; Turk, V.; Turk, B. Emerging roles of cysteine cathepsins in disease and their potential as drug targets. *Curr. Pharm. Des.* **2007**, *13*, 387–403. [CrossRef]
12. Reiser, J.; Adair, B.; Reinheckel, T. Specialized roles for cysteine cathepsins in health and disease. *J. Clin. Investig.* **2010**, *120*, 3421–3431. [CrossRef]
13. Gocheva, V.; Joyce, J.A. Cysteine cathepsins and the cutting edge of cancer invasion. *Cell Cycle* **2007**, *6*, 60–64. [CrossRef] [PubMed]
14. Bromme, D.; Wilson, S. Role of Cysteine Cathepsins in Extracellular Proteolysis. In *Extracellular Matrix Degradation*; Parks, W.C., Mecham, R.P., Eds.; Springer: Berlin/Heidelberg, Germany, 2011; pp. 23–51. [CrossRef]
15. Vizovisek, M.; Fonovic, M.; Turk, B. Cysteine cathepsins in extracellular matrix remodeling: Extracellular matrix degradation and beyond. *Matrix Biol.* **2019**, *75–76*, 141–159. [CrossRef]
16. Repnik, U.; Starr, A.E.; Overall, C.M.; Turk, B. Cysteine Cathepsins Activate ELR Chemokines and Inactivate Non-ELR Chemokines. *J. Biol. Chem.* **2015**, *290*, 13800–13811. [CrossRef] [PubMed]
17. Ainscough, J.S.; Macleod, T.; McGonagle, D.; Brakefield, R.; Baron, J.M.; Alase, A.; Wittmann, M.; Stacey, M. Cathepsin S is the major activator of the psoriasis-associated proinflammatory cytokine IL-36γ. *Proc. Natl. Acad. Sci. USA* **2017**, *114*, E2748–E2757. [CrossRef]

18. Breznik, B.; Motaln, H.; Lah Turnsek, T. Proteases and cytokines as mediators of interactions between cancer and stromal cells in tumours. *Biol. Chem.* **2017**, *398*, 709–719. [CrossRef]
19. Ohashi, K.; Naruto, M.; Nakaki, T.; Sano, E. Identification of interleukin-8 converting enzyme as cathepsin L. *Biochim. Biophys. Acta* **2003**, *1649*, 30–39. [CrossRef]
20. Hira, V.V.; Verbovsek, U.; Breznik, B.; Srdic, M.; Novinec, M.; Kakar, H.; Wormer, J.; der Swaan, B.V.; Lenarcic, B.; Juliano, L.; et al. Cathepsin K cleavage of SDF-1alpha inhibits its chemotactic activity towards glioblastoma stem-like cells. *Biochim. Biophys. Acta Mol. Cell Res.* **2017**, *1864*, 594–603. [CrossRef] [PubMed]
21. Sobotic, B.; Vizovisek, M.; Vidmar, R.; Van Damme, P.; Gocheva, V.; Joyce, J.A.; Gevaert, K.; Turk, V.; Turk, B.; Fonovic, M. Proteomic Identification of Cysteine Cathepsin Substrates Shed from the Surface of Cancer Cells. *Mol. Cell. Proteom.* **2015**, *14*, 2213–2228. [CrossRef]
22. Clark, A.K.; Grist, J.; Al-Kashi, A.; Perretti, M.; Malcangio, M. Spinal cathepsin S and fractalkine contribute to chronic pain in the collagen-induced arthritis model. *Arthritis Rheum.* **2012**, *64*, 2038–2047. [CrossRef] [PubMed]
23. Clark, A.K.; Yip, P.K.; Malcangio, M. The liberation of fractalkine in the dorsal horn requires microglial cathepsin S. *J. Neurosci.* **2009**, *29*, 6945–6954. [CrossRef] [PubMed]
24. Turk, D.; Guncar, G.; Podobnik, M.; Turk, B. Revised definition of substrate binding sites of papain-like cysteine proteases. *Biol. Chem.* **1998**, *379*, 137–147. [CrossRef]
25. Turk, V.; Stoka, V.; Turk, D. Cystatins: Biochemical and structural properties, and medical relevance. *Front. Biosci.* **2008**, *13*, 5406–5420. [CrossRef] [PubMed]
26. Turk, B.; Turk, D.; Salvesen, G.S. Regulating cysteine protease activity: Essential role of protease inhibitors as guardians and regulators. *Curr. Pharm. Des.* **2002**, *8*, 1623–1637. [CrossRef] [PubMed]
27. Drake, F.H.; Dodds, R.A.; James, I.E.; Connor, J.R.; Debouck, C.; Richardson, S.; Lee-Rykaczewski, E.; Coleman, L.; Rieman, D.; Barthlow, R.; et al. Cathepsin K, but not cathepsins B, L, or S, is abundantly expressed in human osteoclasts. *J. Biol. Chem.* **1996**, *271*, 12511–12516. [CrossRef] [PubMed]
28. Riese, R.J.; Mitchell, R.N.; Villadangos, J.A.; Shi, G.P.; Palmer, J.T.; Karp, E.R.; De Sanctis, G.T.; Ploegh, H.L.; Chapman, H.A. Cathepsin S activity regulates antigen presentation and immunity. *J. Clin. Investig.* **1998**, *101*, 2351–2363. [CrossRef]
29. Nakagawa, T.Y.; Brissette, W.H.; Lira, P.D.; Griffiths, R.J.; Petrushova, N.; Stock, J.; McNeish, J.D.; Eastman, S.E.; Howard, E.D.; Clarke, S.R.; et al. Impaired invariant chain degradation and antigen presentation and diminished collagen-induced arthritis in cathepsin S null mice. *Immunity* **1999**, *10*, 207–217. [CrossRef]
30. Bromme, D.; Li, Z.; Barnes, M.; Mehler, E. Human cathepsin V functional expression, tissue distribution, electrostatic surface potential, enzymatic characterization, and chromosomal localization. *Biochemistry* **1999**, *38*, 2377–2385. [CrossRef]
31. Shi, G.P.; Villadangos, J.A.; Dranoff, G.; Small, C.; Gu, L.; Haley, K.J.; Riese, R.; Ploegh, H.L.; Chapman, H.A. Cathepsin S required for normal MHC class II peptide loading and germinal center development. *Immunity* **1999**, *10*, 197–206. [CrossRef]
32. Wex, T.; Buhling, F.; Wex, H.; Gunther, D.; Malfertheiner, P.; Weber, E.; Bromme, D. Human cathepsin W, a cysteine protease predominantly expressed in NK cells, is mainly localized in the endoplasmic reticulum. *J. Immunol.* **2001**, *167*, 2172–2178. [CrossRef] [PubMed]
33. Guncar, G.; Pungercic, G.; Klemencic, I.; Turk, V.; Turk, D. Crystal structure of MHC class II-associated p41 Ii fragment bound to cathepsin L reveals the structural basis for differentiation between cathepsins L and S. *EMBO J.* **1999**, *18*, 793–803. [CrossRef] [PubMed]
34. Turk, B.; Dolenc, I.; Turk, V.; Bieth, J.G. Kinetics of the pH-induced inactivation of human cathepsin L. *Biochemistry* **1993**, *32*, 375–380. [CrossRef]
35. Turk, B.; Bieth, J.G.; Bjork, I.; Dolenc, I.; Turk, D.; Cimerman, N.; Kos, J.; Colic, A.; Stoka, V.; Turk, V. Regulation of the activity of lysosomal cysteine proteinases by pH-induced inactivation and/or endogenous protein inhibitors, cystatins. *Biol. Chem. Hoppe Seyler* **1995**, *376*, 225–230. [CrossRef]
36. Kirschke, H.; Wiederanders, B.; Bromme, D.; Rinne, A. Cathepsin S from bovine spleen. Purification, distribution, intracellular localization and action on proteins. *Biochem. J.* **1989**, *264*, 467–473. [CrossRef] [PubMed]

37. Bromme, D.; Okamoto, K.; Wang, B.B.; Biroc, S. Human cathepsin O2, a matrix protein-degrading cysteine protease expressed in osteoclasts. Functional expression of human cathepsin O2 in Spodoptera frugiperda and characterization of the enzyme. *J. Biol. Chem.* **1996**, *271*, 2126–2132. [CrossRef] [PubMed]
38. Almeida, P.C.; Nantes, I.L.; Rizzi, C.C.; Júdice, W.A.; Chagas, J.R.; Juliano, L.; Nader, H.B.; Tersariol, I.L. Cysteine proteinase activity regulation. A possible role of heparin and heparin-like glycosaminoglycans. *J. Biol. Chem.* **1999**, *274*, 30433–30438. [CrossRef]
39. Caglic, D.; Pungercar, J.R.; Pejler, G.; Turk, V.; Turk, B. Glycosaminoglycans facilitate procathepsin B activation through disruption of propeptide-mature enzyme interactions. *J. Biol. Chem.* **2007**, *282*, 33076–33085. [CrossRef] [PubMed]
40. Li, Z.; Yasuda, Y.; Li, W.; Bogyo, M.; Katz, N.; Gordon, R.E.; Fields, G.B.; Bromme, D. Regulation of collagenase activities of human cathepsins by glycosaminoglycans. *J. Biol. Chem.* **2004**, *279*, 5470–5479. [CrossRef] [PubMed]
41. Yasuda, Y.; Li, Z.; Greenbaum, D.; Bogyo, M.; Weber, E.; Brömme, D. Cathepsin V, a novel and potent elastolytic activity expressed in activated macrophages. *J. Biol. Chem.* **2004**, *279*, 36761–36770. [CrossRef]
42. Turk, B. Targeting proteases: Successes, failures and future prospects. *Nat. Rev. Drug Discov.* **2006**, *5*, 785–799. [CrossRef]
43. Friedrichs, B.; Tepel, C.; Reinheckel, T.; Deussing, J.; von Figura, K.; Herzog, V.; Peters, C.; Saftig, P.; Brix, K. Thyroid functions of mouse cathepsins B, K, and L. *J. Clin. Investig.* **2003**, *111*, 1733–1745. [CrossRef]
44. Sukhova, G.K.; Shi, G.P.; Simon, D.I.; Chapman, H.A.; Libby, P. Expression of the elastolytic cathepsins S and K in human atheroma and regulation of their production in smooth muscle cells. *J. Clin. Investig.* **1998**, *102*, 576–583. [CrossRef] [PubMed]
45. Jordans, S.; Jenko-Kokalj, S.; Kühl, N.M.; Tedelind, S.; Sendt, W.; Brömme, D.; Turk, D.; Brix, K. Monitoring compartment-specific substrate cleavage by cathepsins B, K, L, and S at physiological pH and redox conditions. *BMC Biochem.* **2009**, *10*, 23. [CrossRef]
46. Godat, E.; Hervé-Grvépinet, V.; Veillard, F.; Lecaille, F.; Belghazi, M.; Brömme, D.; Lalmanach, G. Regulation of cathepsin K activity by hydrogen peroxide. *Biol. Chem.* **2008**, *389*, 1123–1126. [CrossRef]
47. Rozhin, J.; Sameni, M.; Ziegler, G.; Sloane, B.F. Pericellular pH affects distribution and secretion of cathepsin B in malignant cells. *Cancer Res.* **1994**, *54*, 6517–6525. [PubMed]
48. Olson, O.C.; Joyce, J.A. Cysteine cathepsin proteases: Regulators of cancer progression and therapeutic response. *Nat. Rev. Cancer* **2015**, *15*, 712–729. [CrossRef]
49. Wu, H.; Du, Q.; Dai, Q.; Ge, J.; Cheng, X. Cysteine Protease Cathepsins in Atherosclerotic Cardiovascular Diseases. *J. Atheroscler. Thromb.* **2017**, *25*, 111–123. [CrossRef] [PubMed]
50. Nanda, A.; Gukovskaya, A.; Tseng, J.; Grinstein, S. Activation of vacuolar-type proton pumps by protein kinase C. Role in neutrophil pH regulation. *J. Biol. Chem.* **1992**, *267*, 22740–22746.
51. Reddy, V.Y.; Zhang, Q.Y.; Weiss, S.J. Pericellular mobilization of the tissue-destructive cysteine proteinases, cathepsins B, L, and S, by human monocyte-derived macrophages. *Proc. Natl. Acad. Sci. USA* **1995**, *92*, 3849–3853. [CrossRef]
52. Dames, P.; Zimmermann, B.; Schmidt, R.; Rein, J.; Voss, M.; Schewe, B.; Walz, B.; Baumann, O. cAMP regulates plasma membrane vacuolar-type H+-ATPase assembly and activity in blowfly salivary glands. *Proc. Natl. Acad. Sci. USA* **2006**, *103*, 3926–3931. [CrossRef] [PubMed]
53. Wang, J.; Cheng, X.; Xiang, M.X.; Alanne-Kinnunen, M.; Wang, J.A.; Chen, H.; He, A.; Sun, X.; Lin, Y.; Tang, T.T.; et al. IgE stimulates human and mouse arterial cell apoptosis and cytokine expression and promotes atherogenesis in Apoe$^-$/$^-$ mice. *J. Clin. Investig.* **2011**, *121*, 3564–3577. [CrossRef]
54. Mohamed, M.M.; Sloane, B.F. Cysteine cathepsins: Multifunctional enzymes in cancer. *Nat. Rev. Cancer* **2006**, *6*, 764–775. [CrossRef]
55. Konttinen, Y.T.; Mandelin, J.; Li, T.F.; Salo, J.; Lassus, J.; Liljeström, M.; Hukkanen, M.; Takagi, M.; Virtanen, I.; Santavirta, S. Acidic cysteine endoproteinase cathepsin K in the degeneration of the superficial articular hyaline cartilage in osteoarthritis. *Arthritis Rheum.* **2002**, *46*, 953–960. [CrossRef]
56. Naghavi, M.; John, R.; Naguib, S.; Siadaty, M.S.; Grasu, R.; Kurian, K.C.; van Winkle, W.B.; Soller, B.; Litovsky, S.; Madjid, M.; et al. pH Heterogeneity of human and rabbit atherosclerotic plaques; a new insight into detection of vulnerable plaque. *Atherosclerosis* **2002**, *164*, 27–35. [CrossRef]

57. Settembre, C.; Di Malta, C.; Polito, V.A.; Garcia Arencibia, M.; Vetrini, F.; Erdin, S.; Erdin, S.U.; Huynh, T.; Medina, D.; Colella, P.; et al. TFEB links autophagy to lysosomal biogenesis. *Science* **2011**, *332*, 1429–1433. [CrossRef]

58. Yan, D.; Wang, H.W.; Bowman, R.L.; Joyce, J.A. STAT3 and STAT6 Signaling Pathways Synergize to Promote Cathepsin Secretion from Macrophages via IRE1α Activation. *Cell Rep.* **2016**, *16*, 2914–2927. [CrossRef] [PubMed]

59. Kreuzaler, P.A.; Staniszewska, A.D.; Li, W.; Omidvar, N.; Kedjouar, B.; Turkson, J.; Poli, V.; Flavell, R.A.; Clarkson, R.W.; Watson, C.J. Stat3 controls lysosomal-mediated cell death in vivo. *Nat. Cell Biol.* **2011**, *13*, 303–309. [CrossRef] [PubMed]

60. Caglič, D.; Repnik, U.; Jedeszko, C.; Kosec, G.; Miniejew, C.; Kindermann, M.; Vasiljeva, O.; Turk, V.; Wendt, K.U.; Sloane, B.F.; et al. The proinflammatory cytokines interleukin-1α and tumor necrosis factor α promote the expression and secretion of proteolytically active cathepsin S from human chondrocytes. *Biol. Chem.* **2013**, *394*, 307–316. [CrossRef] [PubMed]

61. Mohamed, M.M.; Cavallo-Medved, D.; Rudy, D.; Anbalagan, A.; Moin, K.; Sloane, B.F. Interleukin-6 increases expression and secretion of cathepsin B by breast tumor-associated monocytes. *Cell. Physiol. Biochem.* **2010**, *25*, 315–324. [CrossRef]

62. Troen, B.R. The regulation of cathepsin K gene expression. *Ann. N. Y. Acad. Sci.* **2006**, *1068*, 165–172. [CrossRef] [PubMed]

63. Ruettger, A.; Schueler, S.; Mollenhauer, J.A.; Wiederanders, B. Cathepsins B, K, and L are regulated by a defined collagen type II peptide via activation of classical protein kinase C and p38 MAP kinase in articular chondrocytes. *J. Biol. Chem.* **2007**, *283*, 1043–1051. [CrossRef] [PubMed]

64. Hashimoto, Y.; Kondo, C.; Katunuma, N. An Active 32-kDa Cathepsin L Is Secreted Directly from HT 1080 Fibrosarcoma Cells and Not via Lysosomal Exocytosis. *PLoS ONE* **2015**, *10*, e0145067. [CrossRef] [PubMed]

65. Rodríguez, A.; Webster, P.; Ortego, J.; Andrews, N.W. Lysosomes behave as Ca^{2+}-regulated exocytic vesicles in fibroblasts and epithelial cells. *J. Cell Biol.* **1997**, *137*, 93–104. [CrossRef]

66. Ichinose, S.; Usuda, J.; Hirata, T.; Inoue, T.; Ohtani, K.; Maehara, S.; Kubota, M.; Imai, K.; Tsunoda, Y.; Kuroiwa, Y.; et al. Lysosomal cathepsin initiates apoptosis, which is regulated by photodamage to Bcl-2 at mitochondria in photodynamic therapy using a novel photosensitizer, ATX-s10 (Na). *Int. J. Oncol.* **2006**, *29*, 349–355. [CrossRef]

67. Chwieralski, C.E.; Welte, T.; Bühling, F. Cathepsin-regulated apoptosis. *Apoptosis* **2006**, *11*, 143–149. [CrossRef] [PubMed]

68. Cheng, X.W.; Kuzuya, M.; Nakamura, K.; Di, Q.; Liu, Z.; Sasaki, T.; Kanda, S.; Jin, H.; Shi, G.P.; Murohara, T.; et al. Localization of cysteine protease, cathepsin S, to the surface of vascular smooth muscle cells by association with integrin alphanubeta3. *Am. J. Pathol.* **2006**, *168*, 685–694. [CrossRef]

69. Obermajer, N.; Svajger, U.; Bogyo, M.; Jeras, M.; Kos, J. Maturation of dendritic cells depends on proteolytic cleavage by cathepsin X. *J. Leukoc. Biol.* **2008**, *84*, 1306–1315. [CrossRef]

70. Obermajer, N.; Premzl, A.; Zavasnik Bergant, T.; Turk, B.; Kos, J. Carboxypeptidase cathepsin X mediates beta2-integrin-dependent adhesion of differentiated U-937 cells. *Exp. Cell Res.* **2006**, *312*, 2515–2527. [CrossRef] [PubMed]

71. Nascimento, F.D.; Rizzi, C.C.; Nantes, I.L.; Stefe, I.; Turk, B.; Carmona, A.K.; Nader, H.B.; Juliano, L.; Tersariol, I.L. Cathepsin X binds to cell surface heparan sulfate proteoglycans. *Arch. Biochem. Biophys.* **2005**, *436*, 323–332. [CrossRef]

72. Sloane, B.F.; Rozhin, J.; Johnson, K.; Taylor, H.; Crissman, J.D.; Honn, K.V. Cathepsin B: Association with plasma membrane in metastatic tumors. *Proc. Natl. Acad. Sci. USA* **1986**, *83*, 2483–2487. [CrossRef] [PubMed]

73. Mai, J.; Finley, R.L.; Waisman, D.M.; Sloane, B.F. Human procathepsin B interacts with the annexin II tetramer on the surface of tumor cells. *J. Biol. Chem.* **2000**, *275*, 12806–12812. [CrossRef] [PubMed]

74. Cavallo-Medved, D.; Dosescu, J.; Linebaugh, B.E.; Sameni, M.; Rudy, D.; Sloane, B.F. Mutant K-ras regulates cathepsin B localization on the surface of human colorectal carcinoma cells. *Neoplasia* **2003**, *5*, 507–519. [CrossRef]

75. Campo, E.; Muñoz, J.; Miquel, R.; Palacín, A.; Cardesa, A.; Sloane, B.F.; Emmert-Buck, M.R. Cathepsin B expression in colorectal carcinomas correlates with tumor progression and shortened patient survival. *Am. J. Pathol.* **1994**, *145*, 301–309.

76. Hazen, L.G.; Bleeker, F.E.; Lauritzen, B.; Bahns, S.; Song, J.; Jonker, A.; Van Driel, B.E.; Lyon, H.; Hansen, U.; Köhler, A.; et al. Comparative localization of cathepsin B protein and activity in colorectal cancer. *J. Histochem. Cytochem.* **2000**, *48*, 1421–1430. [CrossRef] [PubMed]
77. Rozario, T.; DeSimone, D.W. The extracellular matrix in development and morphogenesis: A dynamic view. *Dev. Biol.* **2009**, *341*, 126–140. [CrossRef]
78. Ozbek, S.; Balasubramanian, P.G.; Chiquet-Ehrismann, R.; Tucker, R.P.; Adams, J.C. The evolution of extracellular matrix. *Mol. Biol. Cell* **2010**, *21*, 4300–4305. [CrossRef]
79. Murphy-Ullrich, J.E.; Sage, E.H. Revisiting the matricellular concept. *Matrix Biol.* **2014**, *37*, 1–14. [CrossRef] [PubMed]
80. Bonnans, C.; Chou, J.; Werb, Z. Remodelling the extracellular matrix in development and disease. *Nat. Rev. Mol. Cell Biol.* **2014**, *15*, 786–801. [CrossRef] [PubMed]
81. Werb, Z. ECM and cell surface proteolysis: Regulating cellular ecology. *Cell* **1997**, *91*, 439–442. [CrossRef]
82. Fonovic, M.; Turk, B. Cysteine cathepsins and extracellular matrix degradation. *Biochim. Biophys. Acta* **2014**, *1840*, 2560–2570. [CrossRef] [PubMed]
83. Apte, S.S.; Parks, W.C. Metalloproteinases: A parade of functions in matrix biology and an outlook for the future. *Matrix Biol.* **2015**, *44–46*, 1–6. [CrossRef]
84. Liu, C.L.; Guo, J.; Zhang, X.; Sukhova, G.K.; Libby, P.; Shi, G.P. Cysteine protease cathepsins in cardiovascular disease: From basic research to clinical trials. *Nat. Rev. Cardiol.* **2018**, *15*, 351–370. [CrossRef]
85. Nakao, S.; Zandi, S.; Sun, D.; Hafezi-Moghadam, A. Cathepsin B-mediated CD18 shedding regulates leukocyte recruitment from angiogenic vessels. *FASEB J.* **2018**, *32*, 143–154. [CrossRef]
86. Mai, J.; Sameni, M.; Mikkelsen, T.; Sloane, B.F. Degradation of extracellular matrix protein tenascin-C by cathepsin B: An interaction involved in the progression of gliomas. *Biol. Chem.* **2002**, *383*, 1407–1413. [CrossRef] [PubMed]
87. Prudova, A.; Gocheva, V.; Auf dem Keller, U.; Eckhard, U.; Olson, O.C.; Akkari, L.; Butler, G.S.; Fortelny, N.; Lange, P.F.; Mark, J.C.; et al. TAILS N-Terminomics and Proteomics Show Protein Degradation Dominates over Proteolytic Processing by Cathepsins in Pancreatic Tumors. *Cell Rep.* **2016**, *16*, 1762–1773. [CrossRef]
88. Guinec, N.; Dalet-Fumeron, V.; Pagano, M. "In vitro" study of basement membrane degradation by the cysteine proteinases, cathepsins B, B-like and L. Digestion of collagen IV, laminin, fibronectin, and release of gelatinase activities from basement membrane fibronectin. *Biol. Chem. Hoppe Seyler* **1993**, *374*, 1135–1146.
89. Gineyts, E.; Bonnet, N.; Bertholon, C.; Millet, M.; Pagnon-Minot, A.; Borel, O.; Geraci, S.; Bonnelye, E.; Croset, M.; Suhail, A.; et al. The C-Terminal Intact Forms of Periostin (iPTN) Are Surrogate Markers for Osteolytic Lesions in Experimental Breast Cancer Bone Metastasis. *Calcif. Tissue Int.* **2018**, *103*, 567–580. [CrossRef] [PubMed]
90. Podgorski, I.; Linebaugh, B.E.; Koblinski, J.E.; Rudy, D.L.; Herroon, M.K.; Olive, M.B.; Sloane, B.F. Bone marrow-derived cathepsin K cleaves SPARC in bone metastasis. *Am. J. Pathol.* **2009**, *175*, 1255–1269. [CrossRef] [PubMed]
91. Wang, B.; Sun, J.; Kitamoto, S.; Yang, M.; Grubb, A.; Chapman, H.A.; Kalluri, R.; Shi, G.P. Cathepsin S controls angiogenesis and tumor growth via matrix-derived angiogenic factors. *J. Biol. Chem.* **2006**, *281*, 6020–6029. [CrossRef]
92. Du, X.; Chen, N.L.; Wong, A.; Craik, C.S.; Brömme, D. Elastin degradation by cathepsin V requires two exosites. *J. Biol. Chem.* **2013**, *288*, 34871–34881. [CrossRef]
93. Fang, W.; He, A.; Xiang, M.X.; Lin, Y.; Wang, Y.; Li, J.; Yang, C.; Zhang, X.; Liu, C.L.; Sukhova, G.K.; et al. Cathepsin K-deficiency impairs mouse cardiac function after myocardial infarction. *J. Mol. Cell. Cardiol.* **2018**, *127*, 44–56. [CrossRef]
94. Kehlet, S.N.; Bager, C.L.; Willumsen, N.; Dasgupta, B.; Brodmerkel, C.; Curran, M.; Brix, S.; Leeming, D.J.; Karsdal, M.A. Cathepsin-S degraded decorin are elevated in fibrotic lung disorders—Development and biological validation of a new serum biomarker. *BMC Pulm. Med.* **2017**, *17*, 110. [CrossRef]
95. Clark, A.K.; Yip, P.K.; Grist, J.; Gentry, C.; Staniland, A.A.; Marchand, F.; Dehvari, M.; Wotherspoon, G.; Winter, J.; Ullah, J.; et al. Inhibition of spinal microglial cathepsin S for the reversal of neuropathic pain. *Proc. Natl. Acad. Sci. USA* **2007**, *104*, 10655–10660. [CrossRef] [PubMed]
96. Zhao, P.; Lieu, T.; Barlow, N.; Metcalf, M.; Veldhuis, N.A.; Jensen, D.D.; Kocan, M.; Sostegni, S.; Haerteis, S.; Baraznenok, V.; et al. Cathepsin S causes inflammatory pain via biased agonism of PAR2 and TRPV4. *J. Biol. Chem.* **2014**, *289*, 27215–27234. [CrossRef] [PubMed]

97. Garsen, M.; Rops, A.L.; Dijkman, H.; Willemsen, B.; van Kuppevelt, T.H.; Russel, F.G.; Rabelink, T.J.; Berden, J.H.; Reinheckel, T.; van der Vlag, J. Cathepsin L is crucial for the development of early experimental diabetic nephropathy. *Kidney Int.* **2016**, *90*, 1012–1022. [CrossRef] [PubMed]

98. Mort, J.S.; Magny, M.C.; Lee, E.R. Cathepsin B: An alternative protease for the generation of an aggrecan 'metalloproteinase' cleavage neoepitope. *Biochem. J.* **1998**, *335 Pt 3*, 491–494. [CrossRef]

99. Mort, J.S.; Beaudry, F.; Théroux, K.; Emmott, A.A.; Richard, H.; Fisher, W.D.; Lee, E.R.; Poole, A.R.; Laverty, S. Early cathepsin K degradation of type II collagen in vitro and in vivo in articular cartilage. *Osteoarthr. Cartil.* **2016**, *24*, 1461–1469. [CrossRef]

100. Hou, W.S.; Li, Z.; Büttner, F.H.; Bartnik, E.; Brömme, D. Cleavage site specificity of cathepsin K toward cartilage proteoglycans and protease complex formation. *Biol. Chem.* **2003**, *384*, 891–897. [CrossRef]

101. Atley, L.M.; Mort, J.S.; Lalumiere, M.; Eyre, D.R. Proteolysis of human bone collagen by cathepsin K: Characterization of the cleavage sites generating by cross-linked N-telopeptide neoepitope. *Bone* **2000**, *26*, 241–247. [CrossRef]

102. Kafienah, W.; Brömme, D.; Buttle, D.J.; Croucher, L.J.; Hollander, A.P. Human cathepsin K cleaves native type I and II collagens at the N-terminal end of the triple helix. *Biochem. J.* **1998**, *331 Pt 3*, 727–732. [CrossRef]

103. Baumgrass, R.; Williamson, M.K.; Price, P.A. Identification of peptide fragments generated by digestion of bovine and human osteocalcin with the lysosomal proteinases cathepsin B, D, L, H, and S. *J. Bone Min. Res.* **1997**, *12*, 447–455. [CrossRef]

104. Bossard, M.J.; Tomaszek, T.A.; Thompson, S.K.; Amegadzie, B.Y.; Hanning, C.R.; Jones, C.; Kurdyla, J.T.; McNulty, D.E.; Drake, F.H.; Gowen, M.; et al. Proteolytic activity of human osteoclast cathepsin K. Expression, purification, activation, and substrate identification. *J. Biol. Chem.* **1996**, *271*, 12517–12524.

105. Kubler, A.; Larsson, C.; Luna, B.; Andrade, B.B.; Amaral, E.P.; Urbanowski, M.; Orandle, M.; Bock, K.; Ammerman, N.C.; Cheung, L.S.; et al. Cathepsin K Contributes to Cavitation and Collagen Turnover in Pulmonary Tuberculosis. *J. Infect. Dis.* **2016**, *213*, 618–627. [CrossRef] [PubMed]

106. Sloane, B.F.; Dunn, J.R.; Honn, K.V. Lysosomal cathepsin B: Correlation with metastatic potential. *Science* **1981**, *212*, 1151–1153. [CrossRef]

107. Vasiljeva, O.; Papazoglou, A.; Kruger, A.; Brodoefel, H.; Korovin, M.; Deussing, J.; Augustin, N.; Nielsen, B.S.; Almholt, K.; Bogyo, M.; et al. Tumor cell-derived and macrophage-derived cathepsin B promotes progression and lung metastasis of mammary cancer. *Cancer Res.* **2006**, *66*, 5242–5250. [CrossRef] [PubMed]

108. Joyce, J.A.; Baruch, A.; Chehade, K.; Meyer-Morse, N.; Giraudo, E.; Tsai, F.Y.; Greenbaum, D.C.; Hager, J.H.; Bogyo, M.; Hanahan, D. Cathepsin cysteine proteases are effectors of invasive growth and angiogenesis during multistage tumorigenesis. *Cancer Cell* **2004**, *5*, 443–453. [CrossRef]

109. Ruffell, B.; Affara, N.I.; Cottone, L.; Junankar, S.; Johansson, M.; DeNardo, D.G.; Korets, L.; Reinheckel, T.; Sloane, B.F.; Bogyo, M.; et al. Cathepsin C is a tissue-specific regulator of squamous carcinogenesis. *Genes Dev.* **2013**, *27*, 2086–2098. [CrossRef] [PubMed]

110. Cox, J.L. Cystatins and cancer. *Front. Biosci.* **2009**, *14*, 463–474. [CrossRef]

111. Akkari, L.; Gocheva, V.; Quick, M.L.; Kester, J.C.; Spencer, A.K.; Garfall, A.L.; Bowman, R.L.; Joyce, J.A. Combined deletion of cathepsin protease family members reveals compensatory mechanisms in cancer. *Genes Dev.* **2016**, *30*, 220–232. [CrossRef] [PubMed]

112. Sevenich, L.; Schurigt, U.; Sachse, K.; Gajda, M.; Werner, F.; Muller, S.; Vasiljeva, O.; Schwinde, A.; Klemm, N.; Deussing, J.; et al. Synergistic antitumor effects of combined cathepsin B and cathepsin Z deficiencies on breast cancer progression and metastasis in mice. *Proc. Natl. Acad. Sci. USA* **2010**, *107*, 2497–2502. [CrossRef]

113. Akkari, L.; Gocheva, V.; Kester, J.C.; Hunter, K.E.; Quick, M.L.; Sevenich, L.; Wang, H.W.; Peters, C.; Tang, L.H.; Klimstra, D.S.; et al. Distinct functions of macrophage-derived and cancer cell-derived cathepsin Z combine to promote tumor malignancy via interactions with the extracellular matrix. *Genes Dev.* **2014**, *28*, 2134–2150. [CrossRef]

114. Dennemarker, J.; Lohmuller, T.; Mayerle, J.; Tacke, M.; Lerch, M.M.; Coussens, L.M.; Peters, C.; Reinheckel, T. Deficiency for the cysteine protease cathepsin L promotes tumor progression in mouse epidermis. *Oncogene* **2010**, *29*, 1611–1621. [CrossRef] [PubMed]

115. Quail, D.F.; Joyce, J.A. Microenvironmental regulation of tumor progression and metastasis. *Nat. Med.* **2013**, *19*, 1423–1437. [CrossRef] [PubMed]

116. Gocheva, V.; Zeng, W.; Ke, D.; Klimstra, D.; Reinheckel, T.; Peters, C.; Hanahan, D.; Joyce, J.A. Distinct roles for cysteine cathepsin genes in multistage tumorigenesis. *Genes Dev.* **2006**, *20*, 543–556. [CrossRef]

117. Small, D.M.; Burden, R.E.; Jaworski, J.; Hegarty, S.M.; Spence, S.; Burrows, J.F.; McFarlane, C.; Kissenpfennig, A.; McCarthy, H.O.; Johnston, J.A.; et al. Cathepsin S from both tumor and tumor-associated cells promote cancer growth and neovascularization. *Int. J. Cancer* **2013**, *133*, 2102–2112. [CrossRef] [PubMed]

118. Vasiljeva, O.; Korovin, M.; Gajda, M.; Brodoefel, H.; Bojic, L.; Kruger, A.; Schurigt, U.; Sevenich, L.; Turk, B.; Peters, C.; et al. Reduced tumour cell proliferation and delayed development of high-grade mammary carcinomas in cathepsin B-deficient mice. *Oncogene* **2008**, *27*, 4191–4199. [CrossRef] [PubMed]

119. Bengsch, F.; Buck, A.; Gunther, S.C.; Seiz, J.R.; Tacke, M.; Pfeifer, D.; von Elverfeldt, D.; Sevenich, L.; Hillebrand, L.E.; Kern, U.; et al. Cell type-dependent pathogenic functions of overexpressed human cathepsin B in murine breast cancer progression. *Oncogene* **2014**, *33*, 4474–4484. [CrossRef]

120. Mitrovic, A.; Pecar Fonovic, U.; Kos, J. Cysteine cathepsins B and X promote epithelial-mesenchymal transition of tumor cells. *Eur. J. Cell Biol.* **2017**, *96*, 622–631. [CrossRef]

121. Wang, J.; Chen, L.; Li, Y.; Guan, X.Y. Overexpression of cathepsin Z contributes to tumor metastasis by inducing epithelial-mesenchymal transition in hepatocellular carcinoma. *PLoS ONE* **2011**, *6*, e24967. [CrossRef]

122. Tripathi, R.; Fiore, L.S.; Richards, D.L.; Yang, Y.; Liu, J.; Wang, C.; Plattner, R. Abl and Arg mediate cysteine cathepsin secretion to facilitate melanoma invasion and metastasis. *Sci. Signal.* **2018**, *11*, eaao0422. [CrossRef]

123. Gopinathan, A.; Denicola, G.M.; Frese, K.K.; Cook, N.; Karreth, F.A.; Mayerle, J.; Lerch, M.M.; Reinheckel, T.; Tuveson, D.A. Cathepsin B promotes the progression of pancreatic ductal adenocarcinoma in mice. *Gut* **2012**, *61*, 877–884. [CrossRef]

124. Maacha, S.; Hong, J.; von Lersner, A.; Zijlstra, A.; Belkhiri, A. AXL Mediates Esophageal Adenocarcinoma Cell Invasion through Regulation of Extracellular Acidification and Lysosome Trafficking. *Neoplasia* **2018**, *20*, 1008–1022. [CrossRef]

125. Rempel, S.A.; Rosenblum, M.L.; Mikkelsen, T.; Yan, P.S.; Ellis, K.D.; Golembieski, W.A.; Sameni, M.; Rozhin, J.; Ziegler, G.; Sloane, B.F. Cathepsin B expression and localization in glioma progression and invasion. *Cancer Res.* **1994**, *54*, 6027–6031. [PubMed]

126. Breznik, B.; Limback, C.; Porcnik, A.; Blejec, A.; Krajnc, M.K.; Bosnjak, R.; Kos, J.; Van Noorden, C.J.F.; Lah, T.T. Localization patterns of cathepsins K and X and their predictive value in glioblastoma. *Radiol. Oncol.* **2018**, *52*, 433–442. [CrossRef]

127. Breznik, B.; Limbaeck Stokin, C.; Kos, J.; Khurshed, M.; Hira, V.V.V.; Bosnjak, R.; Lah, T.T.; Van Noorden, C.J.F. Cysteine cathepsins B, X and K expression in peri-arteriolar glioblastoma stem cell niches. *J. Mol. Histol.* **2018**, *49*, 481–497. [CrossRef] [PubMed]

128. Shao, G.; Wang, R.; Sun, A.; Wei, J.; Peng, K.; Dai, Q.; Yang, W.; Lin, Q. The E3 ubiquitin ligase NEDD4 mediates cell migration signaling of EGFR in lung cancer cells. *Mol. Cancer* **2018**, *17*, 24. [CrossRef] [PubMed]

129. Vasiljeva, O.; Turk, B. Dual contrasting roles of cysteine cathepsins in cancer progression: Apoptosis versus tumour invasion. *Biochimie* **2008**, *90*, 380–386. [CrossRef]

130. Bruchard, M.; Mignot, G.; Derangere, V.; Chalmin, F.; Chevriaux, A.; Vegran, F.; Boireau, W.; Simon, B.; Ryffel, B.; Connat, J.L.; et al. Chemotherapy-triggered cathepsin B release in myeloid-derived suppressor cells activates the Nlrp3 inflammasome and promotes tumor growth. *Nat. Med.* **2013**, *19*, 57–64. [CrossRef] [PubMed]

131. Butinar, M.; Prebanda, M.T.; Rajkovic, J.; Jeric, B.; Stoka, V.; Peters, C.; Reinheckel, T.; Kruger, A.; Turk, V.; Turk, B.; et al. Stefin B deficiency reduces tumor growth via sensitization of tumor cells to oxidative stress in a breast cancer model. *Oncogene* **2014**, *33*, 3392–3400. [CrossRef]

132. Zavrsnik, J.; Butinar, M.; Prebanda, M.T.; Krajnc, A.; Vidmar, R.; Fonovic, M.; Grubb, A.; Turk, V.; Turk, B.; Vasiljeva, O. Cystatin C deficiency suppresses tumor growth in a breast cancer model through decreased proliferation of tumor cells. *Oncotarget* **2017**, *8*, 73793–73809. [CrossRef]

133. Sevenich, L.; Bowman, R.L.; Mason, S.D.; Quail, D.F.; Rapaport, F.; Elie, B.T.; Brogi, E.; Brastianos, P.K.; Hahn, W.C.; Holsinger, L.J.; et al. Analysis of tumour- and stroma-supplied proteolytic networks reveals a brain-metastasis-promoting role for cathepsin S. *Nat. Cell Biol.* **2014**, *16*, 876–888. [CrossRef]

134. Fukuda, S.; Schmid-Schonbein, G.W. Regulation of CD18 expression on neutrophils in response to fluid shear stress. *Proc. Natl. Acad. Sci. USA* **2003**, *100*, 13152–13157. [CrossRef] [PubMed]

135. Chang, S.H.; Kanasaki, K.; Gocheva, V.; Blum, G.; Harper, J.; Moses, M.A.; Shih, S.C.; Nagy, J.A.; Joyce, J.; Bogyo, M.; et al. VEGF-A induces angiogenesis by perturbing the cathepsin-cysteine protease inhibitor balance in venules, causing basement membrane degradation and mother vessel formation. *Cancer Res.* **2009**, *69*, 4537–4544. [CrossRef]
136. Willumsen, N.; Bager, C.L.; Leeming, D.J.; Bay-Jensen, A.C.; Karsdal, M.A. Nidogen-1 Degraded by Cathepsin S can be Quantified in Serum and is Associated with Non-Small Cell Lung Cancer. *Neoplasia* **2017**, *19*, 271–278. [CrossRef] [PubMed]
137. Felbor, U.; Dreier, L.; Bryant, R.A.; Ploegh, H.L.; Olsen, B.R.; Mothes, W. Secreted cathepsin L generates endostatin from collagen XVIII. *EMBO J.* **2000**, *19*, 1187–1194. [CrossRef] [PubMed]
138. Veillard, F.; Saidi, A.; Burden, R.E.; Scott, C.J.; Gillet, L.; Lecaille, F.; Lalmanach, G. Cysteine cathepsins S and L modulate anti-angiogenic activities of human endostatin. *J. Biol. Chem.* **2011**, *286*, 37158–37167. [CrossRef] [PubMed]
139. Garnero, P.; Borel, O.; Byrjalsen, I.; Ferreras, M.; Drake, F.H.; McQueney, M.S.; Foged, N.T.; Delmas, P.D.; Delaissé, J.M. The collagenolytic activity of cathepsin K is unique among mammalian proteinases. *J. Biol. Chem.* **1998**, *273*, 32347–32352. [CrossRef] [PubMed]
140. Yoshioka, Y.; Yamachika, E.; Nakanishi, M.; Ninomiya, T.; Nakatsuji, K.; Kobayashi, Y.; Fujii, T.; Iida, S. Cathepsin K inhibitor causes changes in crystallinity and crystal structure of newly-formed mandibular bone in rats. *Br. J. Oral Maxillofac. Surg.* **2018**, *56*, 732–738. [CrossRef]
141. Sharma, V.; Panwar, P.; O'Donoghue, A.J.; Cui, H.; Guido, R.V.; Craik, C.S.; Brömme, D. Structural requirements for the collagenase and elastase activity of cathepsin K and its selective inhibition by an exosite inhibitor. *Biochem. J.* **2015**, *465*, 163–173. [CrossRef] [PubMed]
142. Aguda, A.H.; Panwar, P.; Du, X.; Nguyen, N.T.; Brayer, G.D.; Brömme, D. Structural basis of collagen fiber degradation by cathepsin K. *Proc. Natl. Acad. Sci. USA* **2014**, *111*, 17474–17479. [CrossRef]
143. Panwar, P.; Butler, G.S.; Jamroz, A.; Azizi, P.; Overall, C.M.; Brömme, D. Aging-associated modifications of collagen affect its degradation by matrix metalloproteinases. *Matrix Biol.* **2017**, *65*, 30–44. [CrossRef]
144. Yasuda, Y.; Kaleta, J.; Brömme, D. The role of cathepsins in osteoporosis and arthritis: Rationale for the design of new therapeutics. *Adv. Drug Deliv. Rev.* **2005**, *57*, 973–993. [CrossRef]
145. Wheater, G.; Elshahaly, M.; Tuck, S.P.; Datta, H.K.; van Laar, J.M. The clinical utility of bone marker measurements in osteoporosis. *J. Transl. Med.* **2013**, *11*, 201. [CrossRef]
146. Mort, J.S.; Recklies, A.D.; Poole, A.R. Extracellular presence of the lysosomal proteinase cathepsin B in rheumatoid synovium and its activity at neutral pH. *Arthritis Rheum.* **1984**, *27*, 509–515. [CrossRef]
147. Hashimoto, Y.; Kakegawa, H.; Narita, Y.; Hachiya, Y.; Hayakawa, T.; Kos, J.; Turk, V.; Katunuma, N. Significance of cathepsin B accumulation in synovial fluid of rheumatoid arthritis. *Biochem. Biophys. Res. Commun.* **2001**, *283*, 334–339. [CrossRef]
148. Pozgan, U.; Caglic, D.; Rozman, B.; Nagase, H.; Turk, V.; Turk, B. Expression and activity profiling of selected cysteine cathepsins and matrix metalloproteinases in synovial fluids from patients with rheumatoid arthritis and osteoarthritis. *Biol. Chem.* **2010**, *391*, 571–579. [CrossRef]
149. Ben-Aderet, L.; Merquiol, E.; Fahham, D.; Kumar, A.; Reich, E.; Ben-Nun, Y.; Kandel, L.; Haze, A.; Liebergall, M.; Kosinska, M.K.; et al. Detecting cathepsin activity in human osteoarthritis via activity-based probes. *Arthritis Res. Ther.* **2015**, *17*, 69. [CrossRef]
150. Weitoft, T.; Larsson, A.; Manivel, V.A.; Lysholm, J.; Knight, A.; Ronnelid, J. Cathepsin S and cathepsin L in serum and synovial fluid in rheumatoid arthritis with and without autoantibodies. *Rheumatology* **2015**, *54*, 1923–1928. [CrossRef]
151. Ruge, T.; Sodergren, A.; Wallberg-Jonsson, S.; Larsson, A.; Arnlov, J. Circulating plasma levels of cathepsin S and L are not associated with disease severity in patients with rheumatoid arthritis. *Scand. J. Rheumatol.* **2014**, *43*, 371–373. [CrossRef]
152. Robert, L.; Robert, A.M.; Jacotot, B. Elastin-elastase-atherosclerosis revisited. *Atherosclerosis* **1998**, *140*, 281–295. [CrossRef]
153. Robert, L.; Robert, A.M.; Fülöp, T. Rapid increase in human life expectancy: Will it soon be limited by the aging of elastin? *Biogerontology* **2008**, *9*, 119–133. [CrossRef]
154. Sukhova, G.K.; Zhang, Y.; Pan, J.H.; Wada, Y.; Yamamoto, T.; Naito, M.; Kodama, T.; Tsimikas, S.; Witztum, J.L.; Lu, M.L.; et al. Deficiency of cathepsin S reduces atherosclerosis in LDL receptor-deficient mice. *J. Clin. Investig.* **2003**, *111*, 897–906. [CrossRef]

155. Samokhin, A.O.; Wong, A.; Saftig, P.; Bromme, D. Role of cathepsin K in structural changes in brachiocephalic artery during progression of atherosclerosis in apoE-deficient mice. *Atherosclerosis* **2008**, *200*, 58–68. [CrossRef]

156. Kitamoto, S.; Sukhova, G.K.; Sun, J.; Yang, M.; Libby, P.; Love, V.; Duramad, P.; Sun, C.; Zhang, Y.; Yang, X.; et al. Cathepsin L deficiency reduces diet-induced atherosclerosis in low-density lipoprotein receptor-knockout mice. *Circulation* **2007**, *115*, 2065–2075. [CrossRef]

157. Fonovic, U.P.; Jevnikar, Z.; Kos, J. Cathepsin S generates soluble CX3CL1 (fractalkine) in vascular smooth muscle cells. *Biol. Chem.* **2013**, *394*, 1349–1352. [CrossRef]

158. Zhao, C.F.; Herrington, D.M. The function of cathepsins B, D, and X in atherosclerosis. *Am. J. Cardiovasc. Dis.* **2016**, *6*, 163–170.

159. Tohda, C.; Tohda, M. Extracellular cathepsin L stimulates axonal growth in neurons. *BMC Res. Notes* **2017**, *10*, 613. [CrossRef]

160. Tran, A.P.; Sundar, S.; Yu, M.; Lang, B.T.; Silver, J. Modulation of Receptor Protein Tyrosine Phosphatase Sigma Increases Chondroitin Sulfate Proteoglycan Degradation through Cathepsin B Secretion to Enhance Axon Outgrowth. *J. Neurosci.* **2018**, *38*, 5399–5414. [CrossRef]

161. Saini, M.G.; Bix, G.J. Oxygen-glucose deprivation (OGD) and interleukin-1 (IL-1) differentially modulate cathepsin B/L mediated generation of neuroprotective perlecan LG3 by neurons. *Brain Res.* **2012**, *1438*, 65–74. [CrossRef]

162. Shen, Y.; Tenney, A.P.; Busch, S.A.; Horn, K.P.; Cuascut, F.X.; Liu, K.; He, Z.; Silver, J.; Flanagan, J.G. PTPsigma is a receptor for chondroitin sulfate proteoglycan, an inhibitor of neural regeneration. *Science* **2009**, *326*, 592–596. [CrossRef]

163. Moon, H.Y.; Becke, A.; Berron, D.; Becker, B.; Sah, N.; Benoni, G.; Janke, E.; Lubejko, S.T.; Greig, N.H.; Mattison, J.A.; et al. Running-Induced Systemic Cathepsin B Secretion Is Associated with Memory Function. *Cell Metab.* **2016**, *24*, 332–340. [CrossRef]

164. Taggart, C.; Mall, M.A.; Lalmanach, G.; Cataldo, D.; Ludwig, A.; Janciauskiene, S.; Heath, N.; Meiners, S.; Overall, C.M.; Schultz, C.; et al. Protean proteases: At the cutting edge of lung diseases. *Eur. Respir. J.* **2017**, *49*. [CrossRef]

165. Perdereau, C.; Godat, E.; Maurel, M.C.; Hazouard, E.; Diot, E.; Lalmanach, G. Cysteine cathepsins in human silicotic bronchoalveolar lavage fluids. *Biochim. Biophys. Acta* **2006**, *1762*, 351–356. [CrossRef]

166. Dongre, A.; Clements, D.; Fisher, A.J.; Johnson, S.R. Cathepsin K in Lymphangioleiomyomatosis: LAM Cell-Fibroblast Interactions Enhance Protease Activity by Extracellular Acidification. *Am. J. Pathol.* **2017**, *187*, 1750–1762. [CrossRef]

167. Squeglia, F.; Ruggiero, A.; Berisio, R. Collagen degradation in tuberculosis pathogenesis: The biochemical consequences of hosting an undesired guest. *Biochem. J.* **2018**, *475*, 3123–3140. [CrossRef]

168. Srivastava, M.; Steinwede, K.; Kiviranta, R.; Morko, J.; Hoymann, H.G.; Langer, F.; Buhling, F.; Welte, T.; Maus, U.A. Overexpression of cathepsin K in mice decreases collagen deposition and lung resistance in response to bleomycin-induced pulmonary fibrosis. *Respir. Res.* **2008**, *9*, 54. [CrossRef]

169. Buhling, F.; Rocken, C.; Brasch, F.; Hartig, R.; Yasuda, Y.; Saftig, P.; Bromme, D.; Welte, T. Pivotal role of cathepsin K in lung fibrosis. *Am. J. Pathol.* **2004**, *164*, 2203–2216. [CrossRef]

170. Kasabova, M.; Villeret, B.; Gombault, A.; Lecaille, F.; Reinheckel, T.; Marchand-Adam, S.; Couillin, I.; Lalmanach, G. Discordance in cathepsin B and cystatin C expressions in bronchoalveolar fluids between murine bleomycin-induced fibrosis and human idiopathic fibrosis. *Respir. Res.* **2016**, *17*, 118. [CrossRef]

171. Small, D.M.; Brown, R.R.; Doherty, D.F.; Abladey, A.; Zhou-Suckow, Z.; Delaney, R.J.; Kerrigan, L.; Dougan, C.M.; Borensztajn, K.S.; Holsinger, L.; et al. Targeting of Cathepsin S Reduces Cystic Fibrosis-like Lung Disease. *Eur. Respir. J.* **2019**. [CrossRef]

172. Elmariah, S.B.; Reddy, V.B.; Lerner, E.A. Cathepsin S signals via PAR2 and generates a novel tethered ligand receptor agonist. *PLoS ONE* **2014**, *9*, e99702. [CrossRef]

173. Taleb, S.; Cancello, R.; Clément, K.; Lacasa, D. Cathepsin s promotes human preadipocyte differentiation: Possible involvement of fibronectin degradation. *Endocrinology* **2006**, *147*, 4950–4959. [CrossRef]

174. Douglas, S.A.; Lamothe, S.E.; Singleton, T.S.; Averett, R.D.; Platt, M.O. Human cathepsins K, L, and S: Related proteases, but unique fibrinolytic activity. *Biochim. Biophys. Acta Gen. Subj.* **2018**, *1862*, 1925–1932. [CrossRef]

175. Ferrall-Fairbanks, M.C.; West, D.M.; Douglas, S.A.; Averett, R.D.; Platt, M.O. Computational predictions of cysteine cathepsin-mediated fibrinogen proteolysis. *Protein Sci.* **2018**, *27*, 714–724. [CrossRef]

176. Ogasawara, S.; Cheng, X.W.; Inoue, A.; Hu, L.; Piao, L.; Yu, C.; Goto, H.; Xu, W.; Zhao, G.; Lei, Y.; et al. Cathepsin K activity controls cardiotoxin-induced skeletal muscle repair in mice. *J. Cachexia Sarcopenia Muscle* **2017**, *9*, 160–175. [CrossRef]

177. Lechner, A.M.; Assfalg-Machleidt, I.; Zahler, S.; Stoeckelhuber, M.; Machleidt, W.; Jochum, M.; Nagler, D.K. RGD-dependent binding of procathepsin X to integrin alphavbeta3 mediates cell-adhesive properties. *J. Biol. Chem.* **2006**, *281*, 39588–39597. [CrossRef]

178. Sina, C.; Lipinski, S.; Gavrilova, O.; Aden, K.; Rehman, A.; Till, A.; Rittger, A.; Podschun, R.; Meyer-Hoffert, U.; Haesler, R.; et al. Extracellular cathepsin K exerts antimicrobial activity and is protective against chronic intestinal inflammation in mice. *Gut* **2013**, *62*, 520–530. [CrossRef] [PubMed]

179. Jennewein, C.; Tran, N.; Paulus, P.; Ellinghaus, P.; Eble, J.A.; Zacharowski, K. Novel aspects of fibrin(ogen) fragments during inflammation. *Mol. Med.* **2011**, *17*, 568–573. [CrossRef] [PubMed]

180. Libert, C. Inflammation: A nervous connection. *Nature* **2003**, *421*, 328–329. [CrossRef] [PubMed]

181. Wolf, Y.; Yona, S.; Kim, K.W.; Jung, S. Microglia, seen from the CX3CR1 angle. *Front. Cell. Neurosci.* **2013**, *7*, 26. [CrossRef] [PubMed]

182. Pislar, A.; Kos, J. Cysteine cathepsins in neurological disorders. *Mol. Neurobiol.* **2014**, *49*, 1017–1030. [CrossRef]

183. Cocchiaro, P.; De Pasquale, V.; Della Morte, R.; Tafuri, S.; Avallone, L.; Pizard, A.; Moles, A.; Pavone, L.M. The Multifaceted Role of the Lysosomal Protease Cathepsins in Kidney Disease. *Front. Cell Dev. Biol.* **2017**, *5*, 114. [CrossRef] [PubMed]

184. Sena, B.F.; Figueiredo, J.L.; Aikawa, E. Cathepsin S As an Inhibitor of Cardiovascular Inflammation and Calcification in Chronic Kidney Disease. *Front. Cardiovasc. Med.* **2017**, *4*, 88. [CrossRef] [PubMed]

185. Kiviranta, R.; Morko, J.; Uusitalo, H.; Aro, H.T.; Vuorio, E.; Rantakokko, J. Accelerated turnover of metaphyseal trabecular bone in mice overexpressing cathepsin K. *J. Bone Miner. Res.* **2001**, *16*, 1444–1452. [CrossRef]

186. Gowen, M.; Lazner, F.; Dodds, R.; Kapadia, R.; Feild, J.; Tavaria, M.; Bertoncello, I.; Drake, F.; Zavarselk, S.; Tellis, I. Cathepsin K knockout mice develop osteopetrosis due to a deficit in matrix degradation but not demineralization. *J. Bone Miner. Res.* **1999**, *14*, 1654–1663. [CrossRef]

187. Mullard, A. Merck &Co. drops osteoporosis drug odanacatib. *Nat. Rev. Drug Discov.* **2016**, *15*, 669. [CrossRef]

188. Bromme, D.; Panwar, P.; Turan, S. Cathepsin K osteoporosis trials, pycnodysostosis and mouse deficiency models: Commonalities and differences. *Expert Opin. Drug Discov.* **2016**, *11*, 457–472. [CrossRef]

189. Panwar, P.; Soe, K.; Guido, R.V.; Bueno, R.V.; Delaisse, J.M.; Bromme, D. A novel approach to inhibit bone resorption: Exosite inhibitors against cathepsin K. *Br. J. Pharmacol.* **2016**, *173*, 396–410. [CrossRef] [PubMed]

190. Panwar, P.; Xue, L.; Soe, K.; Srivastava, K.; Law, S.; Delaisse, J.M.; Bromme, D. An Ectosteric Inhibitor of Cathepsin K Inhibits Bone Resorption in Ovariectomized Mice. *J. Bone Miner. Res.* **2017**, *32*, 2415–2430. [CrossRef]

191. Burden, R.E.; Gormley, J.A.; Jaquin, T.J.; Small, D.M.; Quinn, D.J.; Hegarty, S.M.; Ward, C.; Walker, B.; Johnston, J.A.; Olwill, S.A.; et al. Antibody-mediated inhibition of cathepsin S blocks colorectal tumor invasion and angiogenesis. *Clin. Cancer Res.* **2009**, *15*, 6042–6051. [CrossRef] [PubMed]

192. Burden, R.E.; Gormley, J.A.; Kuehn, D.; Ward, C.; Kwok, H.F.; Gazdoiu, M.; McClurg, A.; Jaquin, T.J.; Johnston, J.A.; Scott, C.J.; et al. Inhibition of Cathepsin S by Fsn0503 enhances the efficacy of chemotherapy in colorectal carcinomas. *Biochimie* **2012**, *94*, 487–493. [CrossRef] [PubMed]

193. Sanman, L.E.; Bogyo, M. Activity-based profiling of proteases. *Annu. Rev. Biochem.* **2014**, *83*, 249–273. [CrossRef]

194. Withana, N.P.; Saito, T.; Ma, X.; Garland, M.; Liu, C.; Kosuge, H.; Amsallem, M.; Verdoes, M.; Ofori, L.O.; Fischbein, M.; et al. Dual-Modality Activity-Based Probes as Molecular Imaging Agents for Vascular Inflammation. *J. Nucl. Med.* **2016**, *57*, 1583–1590. [CrossRef]

195. Walker, E.; Mann, M.; Honda, K.; Vidimos, A.; Schluchter, M.D.; Straight, B.; Bogyo, M.; Popkin, D.; Basilion, J.P. Rapid visualization of nonmelanoma skin cancer. *J. Am. Acad. Derm.* **2017**, *76*, 209–216.e9. [CrossRef]

196. Withana, N.P.; Ma, X.; McGuire, H.M.; Verdoes, M.; van der Linden, W.A.; Ofori, L.O.; Zhang, R.; Li, H.; Sanman, L.E.; Wei, K.; et al. Non-invasive Imaging of Idiopathic Pulmonary Fibrosis Using Cathepsin Protease Probes. *Sci. Rep.* **2016**, *6*, 19755. [CrossRef]

197. Watzke, A.; Kosec, G.; Kindermann, M.; Jeske, V.; Nestler, H.P.; Turk, V.; Turk, B.; Wendt, K.U. Selective activity-based probes for cysteine cathepsins. *Angew. Chem. Int. Ed. Engl.* **2008**, 47, 406–409. [CrossRef]
198. Hu, H.Y.; Vats, D.; Vizovisek, M.; Kramer, L.; Germanier, C.; Wendt, K.U.; Rudin, M.; Turk, B.; Plettenburg, O.; Schultz, C. In vivo imaging of mouse tumors by a lipidated cathepsin S substrate. *Angew. Chem. Int. Ed. Engl.* **2014**, 53, 7669–7673. [CrossRef]
199. Kramer, L.; Renko, M.; Zavrsnik, J.; Turk, D.; Seeger, M.A.; Vasiljeva, O.; Grutter, M.G.; Turk, V.; Turk, B. Non-invasive in vivo imaging of tumour-associated cathepsin B by a highly selective inhibitory DARPin. *Theranostics* **2017**, 7, 2806–2821. [CrossRef]
200. Garland, M.; Yim, J.J.; Bogyo, M. A Bright Future for Precision Medicine: Advances in Fluorescent Chemical Probe Design and Their Clinical Application. *Cell Chem. Biol.* **2016**, 23, 122–136. [CrossRef]
201. Fang, Y.; Du, F.; Xu, Y.; Meng, H.; Huang, J.; Zhang, X.; Lu, W.; Liu, S.; Yu, J. Enhanced cellular uptake and intracellular drug controlled release of VESylated gemcitabine prodrug nanocapsules. *Colloids Surf. B Biointerfaces* **2015**, 128, 357–362. [CrossRef]
202. Zhang, X.; Tang, K.; Wang, H.; Liu, Y.; Bao, B.; Fang, Y.; Zhang, X.; Lu, W. Design, Synthesis, and Biological Evaluation of New Cathepsin B-Sensitive Camptothecin Nanoparticles Equipped with a Novel Multifuctional Linker. *Bioconj. Chem.* **2016**, 27, 1267–1275. [CrossRef]
203. Mikhaylov, G.; Klimpel, D.; Schaschke, N.; Mikac, U.; Vizovisek, M.; Fonovic, M.; Turk, V.; Turk, B.; Vasiljeva, O. Selective targeting of tumor and stromal cells by a nanocarrier system displaying lipidated cathepsin B inhibitor. *Angew. Chem. Int. Ed. Engl.* **2014**, 53, 10077–10081. [CrossRef]
204. Vizovisek, M.; Vidmar, R.; Drag, M.; Fonovic, M.; Salvesen, G.S.; Turk, B. Protease Specificity: Towards In Vivo Imaging Applications and Biomarker Discovery. *Trends Biochem. Sci.* **2018**, 43, 829–844. [CrossRef]

Review

Extracellular Matrix Component Remodeling in Respiratory Diseases: What Has Been Found in Clinical and Experimental Studies?

Juliana T. Ito [1], Juliana D. Lourenço [1], Renato F. Righetti [1,2], Iolanda F.L.C. Tibério [1], Carla M. Prado [3] and Fernanda D.T.Q.S. Lopes [1,*]

[1] Department of Clinical Medicine, Laboratory of Experimental Therapeutics/LIM-20, School of Medicine of University of Sao Paulo, Sao Paulo 01246-903, Brazil; jutiyakito@hotmail.com (J.T.I.); juliana.dl31@gmail.com (J.D.L.); refragar@gmail.com (R.F.R.); iocalvo@uol.com.br (I.F.L.C.T.)

[2] Rehabilitation service, Sírio-Libanês Hospital, Sao Paulo 01308-050, Brazil

[3] Department of Bioscience, Laboratory of Studies in Pulmonary Inflammation, Federal University of Sao Paulo, Santos 11015-020, Brazil; cmaximoprado@gmail.com

* Correspondence: fernandadtqsl@gmail.com; Tel.: +55-(11)-3061-7180

Received: 7 March 2019; Accepted: 9 April 2019; Published: 11 April 2019

Abstract: Changes in extracellular matrix (ECM) components in the lungs are associated with the progression of respiratory diseases, such as asthma, chronic obstructive pulmonary disease (COPD), and acute respiratory distress syndrome (ARDS). Experimental and clinical studies have revealed that structural changes in ECM components occur under chronic inflammatory conditions, and these changes are associated with impaired lung function. In bronchial asthma, elastic and collagen fiber remodeling, mostly in the airway walls, is associated with an increase in mucus secretion, leading to airway hyperreactivity. In COPD, changes in collagen subtypes I and III and elastin, interfere with the mechanical properties of the lungs, and are believed to play a pivotal role in decreased lung elasticity, during emphysema progression. In ARDS, interstitial edema is often accompanied by excessive deposition of fibronectin and collagen subtypes I and III, which can lead to respiratory failure in the intensive care unit. This review uses experimental models and human studies to describe how inflammatory conditions and ECM remodeling contribute to the loss of lung function in these respiratory diseases.

Keywords: extracellular matrix; lung function; asthma; chronic obstructive pulmonary disease; acute respiratory distress syndrome

1. The Connective Tissue of the Lung

Extracellular matrix (ECM) components are constituents of the connective tissue that play essential structural roles in maintaining organ functionality. The composition of the connective tissue is determined by a hierarchical molecular organization; under inflammatory conditions, this organization depends on the balance between the injury and remodeling of these components, which can lead to chemical and structural changes and reduced tissue functionality [1].

In lung tissues, the main components of the ECM are elastic and collagen fibers, proteoglycans, glycoproteins, and metalloproteinases (MMPs), and their tissue inhibitors (TIMPs). Among these components, collagen is the most abundant [1–4], and the number of collagen fibers is, thus, the primary determinant for the mechanical properties of the lungs [1].

There are more than twenty different subtypes of collagen molecules, and in the lung parenchyma, subtypes I and III mostly constitute the structural framework of the alveolar walls, whereas subtype IV is present in the basement membranes [5]. The fibers of collagen subtype I are stiffer than those of collagen subtype III (as indicated by a comparative histology [6]), and the subtype I/III ratio determines the resistance of these fibers to breakdown under mechanical forces during stretching [1].

Elastic fibers are considered the main component responsible for the elastic recoil properties of the lungs, and these fibers are composed of at least two morphologically distinguishable components—elastin and microfibrils [7,8]. Elastic fibers are mechanically connected to collagen [9] by microfibrils or proteoglycans [10,11], with elastin acting predominantly in pulmonary elasticity, in the presence of normal breathing lung volumes [12], while collagen acts progressively in the presence of volumes that approach the total lung capacity [13].

Microfibrils are composed of fibrillins, microfibril-associated glycoproteins, and transforming growth factor-beta (TGF-β) binding proteins, among which fibrillins are the major components, providing a scaffold for elastin polymer aggregation [7].

Proteoglycans are composed of a protein core and side chains of glycosaminoglycans [2], which are a family of linear polysaccharides that constitute the ECM and basement membrane [14,15]. Glycosaminoglycans play a pivotal role in the maintenance of the collagen fiber assembly, as well as water balance, cell adhesion and migration [5].

The major cells responsible for the ECM production and normal ECM turnover are fibroblasts [16–19]. Different molecules can influence the activity of these cells, such as cytokines, growth factors, and components of the ECM itself. Among the relevant cytokines, TGF-β plays a pivotal role in inducing the production of the ECM components by fibroblasts [16].

Additionally, ECM components usually interact with epithelial cells, serving as ligands to transmit signals to regulate adhesion, migration, proliferation, apoptosis, survival, or differentiation. In addition, they can release growth factors and other signaling molecules that regulate cells behavior [20].

Under pathological conditions, ECM turnover is altered [18,21,22], leading to an altered architecture and consequent failure of the mechanical properties of this tissue (Figure 1) [16]. Considering the importance of the altered composition of the ECM components, under chronic inflammatory conditions for lung function impairment due to respiratory diseases, in this review, we summarized how structural changes in some ECM components can impact the worsening of lung function in three lung diseases—asthma, chronic obstructive pulmonary disease (COPD), and acute respiratory distress syndrome (ARDS).

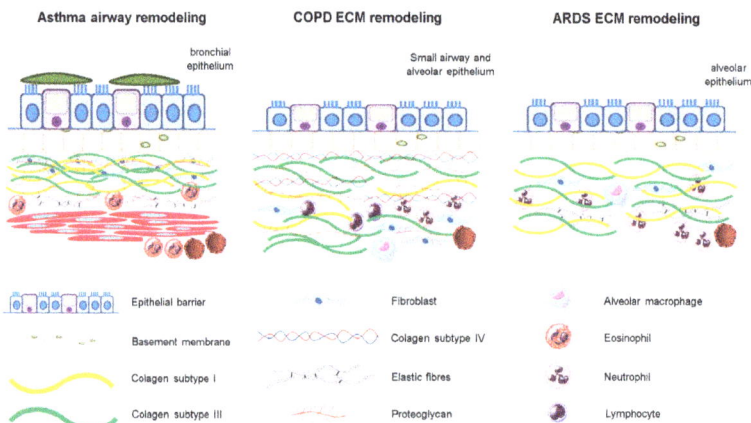

Figure 1. The structural changes of ECM components in respiratory diseases. There is an inflammatory process associated with different fibers rearrangement. In asthma, structural changes are mainly in the bronchial epithelium, whereas in ARDS, these are observed near the alveolar epithelium. In COPD, the ECM remodeling is observed predominantly in small airways and distal areas of parenchyma.

2. Airway Remodeling in Asthma

A recent study confirmed that asthma rates are increasing in countries outside of the United States; approximately 300 million people or 4.3% of the world's population suffer from asthma [23]. The prevalence of asthma increased 2.9% each year from 2001 to 2010 (20.3 million people in the United States had asthma in 2001, compared with 25.7 million in 2010) [24]. The Global Initiative for Asthma estimates that there will be an additional 100 million people with asthma, by the year 2025 [25].

Asthma is a common and potentially severe chronic disease that can be controlled but not cured. It presents symptoms such as, wheezing, shortness of breath, tightness in the chest, and coughing that vary over time, in terms of their occurrence, frequency, and intensity. Asthma is associated with variable expiratory airflow, due to bronchoconstriction, thickening of the airway wall, and mucus. Symptoms can be triggered or worsened by factors such as viral infections, allergens, smoking, exercise, and stress [25].

Inhaled allergens come into contact with the respiratory mucosa and are captured by dendritic cells present in the bronchial epithelium. These cells recognize and process the antigen and present it to T helper (Th) lymphocytes [26]. These cells release mediators and recruit other inflammatory cells into the lung. After these events, the transformation of CD4+ T cells into different profiles occurs. Interleukins (IL)-4, IL-5, and IL-13, induce the proliferation of Th2 cells, interferon-gamma (IFN-γ), and IL-2 induce the proliferation of Th1 cells, and transforming growth factor-beta (TGF-β), and IL-6 induce the proliferation of Th17 cells [27,28].

The inflammatory process in asthma, results in chronic inflammation of the airway walls and lung tissue, eventually triggering bronchoconstriction and structural changes, called airway remodeling [29]. However, several lines of evidence have shown the role of mechanical forces that occur during bronchoconstriction in inflammation-independent airway remodeling [30–34]. Grainge et al. suggest that repeated bronchoconstriction in asthma induces epithelial stress and initiates a tissue response that leads to structural airway changes [35].

In addition, a few studies have addressed the initial occurrence of airway wall remodeling in young to very young children. Pohunek et al. showed evidence for ECM remodeling very early in childhood. In a bronchial biopsy study of 27 children, aged 1.2 to 11.7 years, with chronic respiratory symptoms, the thickness of the subepithelial lamina reticularis was observed to be greater in children with bronchial asthma diagnosed at follow-up, compared with children who did not progress to asthma. This suggests that remodeling might be present even before asthma becomes symptomatic [36]. Corroborating this study, a biopsy study in pre-school children with severe wheezing, reported a number of characteristics of airway remodeling, such as increased basement membrane thickness, increase in airway smooth muscle (ASM), vascularity, and mucus gland area, without any relationship with the inflammatory cell count [37]. These continuous tissue adaptations trigger changes in lung structure, geometry, and tissue properties [38], and they are considered the main causes of the symptoms associated with a decreased lung function [39]. Hill et al. showed in a theoretical model that the inflammatory conditions lead to mechanical stresses, leading to the release of contractile agonists, exacerbating the fiber remodeling in ASM [20]. This process explains why some asthmatic patients present partial and irreversible loss of respiratory function, over time, especially severe asthmatic patients who experience an accelerated decline in pulmonary function, with the disease progression [40]. The changes that occur during airway remodeling consists of subepithelial reticular basement membrane (RBM) thickening, increased ASM thickness, angiogenesis, and goblet cell hyperplasia associated with irreversible loss of lung function [41]. Airway remodeling and subepithelial fibrosis are not inhibited in severe asthma, despite corticosteroid treatment and might lead to worsening of symptoms as the disease progresses [42,43].

The hyperresponsiveness of airways leads to exaggerated airway narrowing which is result of the ASM contraction. Thus, it is important to understand the changes that occur in the ECM that surround the airways, as well as the interactions between the ECM and the ASM [44,45]. Khan et al. demonstrated a significant loss of tethering (interaction of ASM to its surrounding parenchyma) forces in mouse lung microsections incubated overnight with proteolytic enzymes, through two different in vitro experiments [46,47]. In the first study, the authors exposed slices from control animals to porcine pancreatic elastase (PPE) and evaluated the cholinergic responses, in order to verify the hyperresponsiveness, by treating with acetylcholine. The elastase exposure resulted in increased magnitudes and velocities of airway narrowing, impaired relaxation, and increased rupture of the airways, from the surrounding parenchyma [46]. In the second study, the authors treated mouse lung slices with PPE, collagenase, or both, and assessed the hyperresponsiveness, by adding acetylcholine. The treatment with PPE or collagenase increased the lumen narrowing induced by acetylcholine even more. When treated with both proteases together, there was an increase in the velocity of contraction, as well as a decrease in the velocity of relaxation, resulting in a retraction of the airway and a reduction in the tethering forces [47].

2.1. Clinical Studies

In asthmatic patients, smooth muscle mass is increased, due to a coordinated increase in hyperplasia, hypertrophy, and ASM cells. Smooth muscle cells actively participate in inflammatory and remodeling processes, through the release of pro-inflammatory cytokines, chemokines, and ECM proteins; therefore, these cells might contribute to the pathogenesis of asthma [48].

The migration of smooth muscle cells is a recently described feature of airway remodeling. Joubert et al. showed that chemokines have the ability to induce human ASM cell migration and to increase their contractility, revealing another process that could significantly contribute to the overall airflow obstruction in these patients [49]. James et al. showed that human lung smooth muscle in the airways of patients with asthma, must shorten by only 40% of its resting length, in order to completely occlude the airway lumen. These results are consistent with observations made during bronchial challenge [50]. While in normal subjects, the decrease in forced expiratory volume in one second (FEV_1) reaches a plateau, in patients with asthma, the FEV_1 continues to decrease without reaching a plateau [51]. Smooth muscle mass has been correlated with asthma severity [52].

Yick et al. evaluated the spirometry and methacholine responsiveness associated with positive staining of elastin, collagen I, III, and IV, decorin, versican, fibronectin, laminin, and tenascin, in ASM performed in atopic mild asthma and healthy subjects. In this study, ECM in ASM was related to the dynamics of airway function in the absence of differences in ECM expression between asthma and the controls. This indicates that the ASM layer in its full composition, is a major structural component, in determining variable airways obstruction in asthma [45]. In addition, Slats et al. showed that hyperresponsiveness is associated with the level of expression of α-smooth muscle-actin, desmin, and elastin, within the bronchial wall, but not with myosin, calponin, vimentin, type III collagen, or fibronectin. This suggests that expression of each of the contractile and structural smooth muscle proteins, as well as components of the ECM, distinctly influences the dynamic airway function [44].

Airway mucus contains approximately 2% mucins, but some exogenous factors (antigen contact) and endogenous mediators, produced by inflammatory and structural cells, can contribute to submucosal and goblet cell hyperplasia, airway mucus hypersecretion, and upregulation of the MUC protein expression, to varying degrees. However, submucosal gland hyperplasia and goblet cell hyperplasia are observed in the airways of asthmatic patients and are a feature that is particularly evident in fatal asthma [53].

Asthmatic lung biopsies have shown that airway fibroblasts are morphologically distinct from lung parenchymal fibroblasts. Distal lung tissue fibroblasts are broader with more cytoplasmic projections, and airway fibroblasts synthesize more procollagen type I, after TGF-β stimulation [54]. These structural differences might explain, at least partially, why a variety of repair responses are observed in the proximal and distal airways, in response to a damaging stimulus in the lung parenchyma of asthmatic patients.

Produced in response to the cytokine TGF-β, collagen is a potential marker of remodeling in asthma [55]. It has been documented that in moderate (FEV_1 60–80% predicted) and severe asthma ($FEV_1 \leq 60\%$ predicted), there is an increase in the deposition of fibroblasts, collagen I, and collagen III, in bronchial biopsies [42,52]. In addition to its involvement in the inflammatory process, TGF-β is also involved in the changes observed in the ECM, stimulating the production of fibroblasts, collagens type I and III, fibronectin, and proteoglycans [56].

Lung ECM remodeling in asthma is determined by the rate of ongoing deposition and degradation of proteins, including collagens I, III, and V, fibronectin, tenascin, lumican, and biglycan. These components are likely secreted due to the activation of fibroblasts and myofibroblasts, mainly because of the TGF-β signaling [57,58]. Collagen fibers are the most abundant elements of the ECM in the lung, constituting approximately 70% of the lung tissue, and changing the structure, quantity, or geometry of their distribution can trigger changes in lung functioning [58–60].

Considered important in the remodeling process, proteoglycans (decorin, biglycan, and lumican) play roles in the interaction of fibrils and collagen fibrogenesis in the tissue, with other components of the ECM [61]. The increase in decorin expression in patients with severe asthma, could be a protective mechanism for modulating pulmonary remodeling. Conversely, an increased amount of decorin could regulate and stabilize collagen spacing and create a more rigid collagen matrix, which might affect the overall elasticity of the lung tissue [62].

Proteoglycans are formed by the linkage of glycosaminoglycans, such as heparin and heparan sulphate, which play important roles in allergic and inflammatory processes, such as asthma. This molecule binds to chemokines, interleukins, growth factor, and other proteins [14,63], and promotes the recruitment of leukocytes, playing a critical role in airway hyperresponsiveness and inflammation [64]. In asthma, heparin can regulate bronchial hyperresponsiveness, influence the inflammatory process, by inhibiting the recruitment of inflammatory cells, and attenuate tissue damage by binding and neutralizing chemokines and cytokines that are released from inflammatory cells [14,65].

In fatal asthma, which is characterized as clinically severe asthma, fibronectin levels, and elastic fibers are increased in the smooth muscle of the airway wall [66]. Other abnormalities in airway matrix structure in asthma, include a specific increase in the lumican and biglycan isoforms, which is also associated with tissue injury and worsening of the lung function ($FEV_1 \leq 80\%$ predicted) [67]. MMPs play an important role during the remodeling process, by degrading the components of the ECM (especially MMP-9 and MMP-12). The inhibition of the activated MMPs might be rapidly conducted by TIMPs that are produced by most mesenchymal cells [68]. Righetti et al. and Pigati et al. showed that in a model of chronic allergic inflammation, increases in MMP-9 and TIMP-1 were associated with increases in the volume fraction of actin, collagen, and elastic fibers in the airway and the distal parenchyma, revealing the importance of the parenchyma, during the remodeling process [60,69].

Although the remodeling process is potentially harmful to lung function, this response is thought to be an attempt to protect against aggressive pulmonary inflammation. Previous evidence has shown that without tissue remodeling, the patient can experience worsening of symptoms and a faster decline in lung function [70].

As the main inhaled medication used in asthma, several studies have reported the effects of glucocorticoids on airway remodeling, including the reduction of RBM thickness [71,72], improvements in the number of ciliated epithelial cells [73], and decreases in both vessel numbers and percent vascularity in the submucosa [72,74]. However, it is important to note that although some patients under corticosteroids treatment show improvement in symptoms and inflammatory cell numbers, within the sputum of the airways, the improvement in airway hyperresponsiveness to nonspecific stimuli, occurs only after much longer periods of treatment [71,75]. Elliot et al. showed in post-mortem airway sections from asthmatic patients that, some structural changes that lead the hyperresponsiveness might be partly independent of inflammation and, therefore, are not reversible by anti-inflammatory treatment [76].

Alternatively, bronchial thermoplasty is a novel non-drug therapy that targets airway remodeling [77]. Evidence has shown that this technique reduces the ASM area and that this response is correlated with improvements in asthma control and quality of life, and decreases severe exacerbations and hospitalizations [78,79].

2.2. Experimental Models

Experimental allergic asthma in mice is a reliable and clinically relevant model of human disease, because clinical studies of asthma are not able to clarify all aspects of disease pathophysiology [80]. Several experimental protocols and animal species have been used in experimental models of asthma, such as cats, dogs, rats, guinea pigs, pigs, equines, and primates [81,82]. However, the most common species studied in the recent decades have been guinea pigs and mice (particularly BALB/c mice) [83]. The experimental protocols for inducing asthma include two phases—(i) sensitization is achieved by the intraperitoneal or subcutaneous route in mice and by inhalation in guinea pigs (associated with intranasal instillation of antigen), which has been increasingly used because human asthma is induced by inhalation of antigen; and (ii) antigen challenges are performed through intratracheal and intranasal instillation and aerosol inhalation [84]. The classic antigens used in experimental models of asthma are ovalbumin (OVA) and house dust mites [57,58,81].

Several studies using these experimental models of asthma have shown functional alterations in the resistance and elasticity of the respiratory system associated with inflammatory eosinophilic infiltrates; expression of Th2 and Th17 cytokines, MMP-9-, MMP-12-, TIMP-1-, and TGF-β-positive cells; increased deposition of actin, collagen, and elastic fibers, and increased mucus production in the airways and lung tissues [57,59,60,69,85].

Airway resistance is resistance to the in- and outflow of air, exerted by the airway walls and lung elasticity. Also known as elastic resistance, the reciprocal of lung compliance is the pressure change required to elicit a unit of volume change; both parameters represent lung function [86]. Hyperresponsiveness is a feature of asthmatic rats that indicates functional changes in airway resistance and lung elasticity [87]. Measures of airway resistance and lung elasticity are commonly used in experiments, to evaluate hyperresponsiveness in respiratory disease models, including asthma models [85,88,89].

In a guinea pig asthma model sensitized with OVA, Possa et al. showed positive correlations between the volume fractions of actin, collagen, and elastic fibers, and airway resistance and elastance [48]. A decrease in the actin deposition in the airways, reduces the resistance and elastance of the respiratory system, as a result of antigen sensitization. Corroborating this study, Vasconcelos et al. demonstrated in guinea pigs with chronic allergic inflammation that the concomitant reduction of airway hyperresponsiveness and smooth muscle mass are correlated, suggesting that such structural changes could explain the functional change in smooth muscle contractile responsiveness [90]. Additionally, Righetti et al., in the same experimental model, showed that increased actin, collagen, and elastic fibers in the lung tissue are associated with functional alterations in the alveolar lung tissue mechanics [60]. Therefore, airway remodeling in experimental models resembles the pathophysiological features of human asthma [76].

Other components of the ECM appear to participate in airway remodeling in human and experimental models. In a mouse model of asthma exacerbated with lipopolysaccharide (LPS), Camargo et al. showed an increased number of cells positive for MMP-9, MMP-12, TIMP-1, and TGF-β, as well as an increased volume fraction of collagen fibers I and III, decorin, actin, biglycan, lumican, and fibronectin in the lung tissue [57]. These animals were also treated with an anti-IL-17 antibody and showed a decreased pulmonary inflammation, edema, and airway remodeling, compared to the non-treated animals [57]. In a murine asthma model, Dos Santos et al. showed that the presence of the IL-17 and Rho-kinase (ROCK) proteins, enhances the percentage of maximal increase in the respiratory system resistance and elastance, after being challenged with methacholine. Additionally, there were increases in the number of cells positive for ROCK1, ROCK2, TGF-β, MMP-9, MMP-12, and TIMP-1, and the percentages of isoprostane, biglycan, decorin, fibronectin, and collagen fibers, in the asthma group. However, all these changes were attenuated after treatment with an anti-IL-17 antibody or a ROCK inhibitor, and the combination of these treatments potentiated this protective effect [58].

Some of the mediators mentioned above also elicit early mucus hypersecretion [91]. Pardo–Saganta et al. showed increases in the expression of mucous cell-specific genes and the number of ciliated cells in the murine pseudostratified airway epithelium, after the OVA challenge [92]. Asthmatic patients showed higher levels of MUC5AC in the airways and more total mucus, with consequences for pulmonary function [93,94].

These associations in humans were supported in the OVA-sensitized guinea pigs, which exhibited increases in lung tissue resistance and elastance, eosinophilic infiltration in the airways and parenchyma, a significant increase in collagen density, and a concurrent parenchymal contractile response [95]. Almeida–Reis et al. showed in an experimental model of asthma that chronic allergic lung inflammation reduces mucociliary clearance, due to alterations in the rheological properties of mucus, increasing acidity, wettability, and adhesiveness of the mucus [85]. The functional consequences of these abnormalities, mostly result in increased airway wall thickness, sputum production, and airway narrowing, due to sputum secretion [25].

Experimental models have been developed to better understand these mechanisms, to evaluate both the safety and efficacy of therapies, before clinical trials, and to mimic the pathophysiology of human disease [96]. Table 1 summarizes lung function results and markers of airway remodeling from clinical and experimental studies of asthma.

Table 1. Lung function changes and markers of airway remodeling in asthma.

	Clinical Studies	Experimental Studies *
Lung Function	Moderate to severe stages of asthma (FEV1 < 80%) [42,52,66,67]	↑ in lung tissue resistance and elastance [48,58–60,69,88,89]
Remodeling Markers		
Fibroblast	↑ in airways (severe stage) [52]	—
Collagen fibers	—	↑ in airways [48,58,59,69,88] ↑ in lung parenchyma [58–60,69,88,89]
Collagen subtype I	↑ in airways (moderate and severe stages) [42]	↑ in lung parenchyma [57]
Collagen subtype III	↑ in airways (severe stage) [52] ↑ in airways (moderate and severe stages) [42]	↑ in lung parenchyma [57]
Elastic fibers	↑ in large ASM in fatal asthma (severe stage) [66]	↑ in airways [48,59,69] ↑ in lung parenchyma [59,60]
Decorin	—	↑ in airways [58] ↑ in lung parenchyma [57,58]
Lumican	↑ in subepithelial layer (severe stage) [67] ↑ in ASM (moderate stage) [67]	↑ in lung parenchyma [57]
Actin	—	↑ in airways [69] ↑ in lung parenchyma [57,60,69]
Biglycan	↑ in ASM (moderate stage) [67]	↑ in airways [58] ↑ in lung parenchyma [57,58]
Fibronectin	↑ in large ASM in fatal asthma (severe stage) [66]	↑ in airways [58] ↑ in lung parenchyma [57,58]

FEV$_1$: Forced expiratory volume in 1 s; ASM: Airway smooth muscle. * OVA-induced asthma model.

3. Extracellular Matrix Remodeling in COPD

COPD is a common, preventable, and treatable disease, characterized by persistent respiratory symptoms and airflow limitations, caused by a mixture of small airway disease (e.g., obstructive bronchiolitis) and parenchymal destruction (emphysema), mainly induced by smoke exposure [97]. These changes do not always occur together, and there are some variations in the degree of airway disease and emphysema in COPD patients, which might explain the heterogeneity of the response to treatment [98]. Although long-acting bronchodilators have been used in the management of COPD, these drugs are not efficient to control the inflammatory process, as well as the structural changes [99].

Persistence of the inflammatory process leads to structural changes, such as parenchymal tissue destruction (resulting in emphysema) and disruption of normal repair and defense mechanisms (resulting in small airway fibrosis), culminating in a decrease in lung elastic recoil, gas trapping, and progressive airflow limitation [97].

The level of obstruction in COPD patients is determined by spirometry, in which a post-bronchodilator value of FEV_1/forced vital capacity (FVC) < 0.70 confirms the presence of persistent airflow limitation. Table 2 shows the severity of lung function impairment in COPD patients, based on post-bronchodilator FEV_1 [97].

Table 2. Classification of airflow limitation severity in chronic obstructive pulmonary disease (COPD).

In Patients with FEV_1/FVC < 0.70:		
GOLD 1	Mild	$FEV_1 \geq 80\%$ predicted
GOLD 2	Moderate	$50\% \leq FEV_1 < 80\%$ predicted
GOLD 3	Severe	$30\% \leq FEV_1 < 50\%$ predicted
GOLD 4	Very Severe	$FEV_1 < 30\%$ predicted

GOLD: Global Initiative for Chronic Obstructive Lung Disease; FEV_1: Forced expiratory volume in 1 s; FVC: Forced vital capacity.

The chronic inflammatory response in COPD patients is associated with increased numbers of inflammatory cells, such as macrophages, neutrophils, and CD4+ and CD8+ T lymphocytes [100], and fibroblasts in the airways, which play a pivotal role in the upregulation of proteases, such as MMPs, resulting in the destruction and remodeling of ECM components, in the small airways [101] and in the parenchyma [102]. The role of collagenases, such as MMP-1, MMP-8, and MMP-13, in ECM fiber destruction in COPD patients, has been described [103,104], and MMP-12 has been the most commonly described collagenase in experimental models [105–107]. In response to fiber destruction by MMPs, there is a structural reorganization of parenchymal fibers, constituting a dynamic process of repair and remodeling [108]. It is believed that changes in major lung ECM components, such as collagen subtypes I and III, and elastin, are involved in the loss of elasticity, during emphysema progression [1,109–111].

Although emphysema is defined by the destruction of distal air spaces, with or without fibrosis [97], the majority of clinical studies have described an increase in the amount of ECM fiber deposition in the airways and the lung parenchyma [112–115]. However, it is important to emphasize that these evaluations were usually performed in patients who were in advanced stages of the disease and in experimental models, after a few days of disease induction [104,116–118].

3.1. Clinical Studies

In COPD patients, the majority of studies have shown that structural changes in the airway walls are associated with disease progression. The ECM composition and the amount of different constituents are altered in these patients [22,119,120]. Kranenburg et al. showed that these changes occurred mainly in the surface epithelial basement membrane and were characterized by increased deposition of collagen subtypes I, III, and IV, associated with high levels of collagen subtypes I and III, in both the bronchial lamina propria and adventitia, as well as enhanced expression of fibronectin in the vascular intima [22]. Additionally, in this study, a significant, direct correlation was demonstrated between the severity of COPD (moderate and severe stages) and an enhanced expression of these different ECM components.

Several observations about COPD patients have pointed out the importance of examining the airway smooth muscle and its interaction with the surrounding parenchyma (tethering), since the loss of elastic tissue observed in the ECM of COPD patients [121,122] can reduce tethering forces around the airway, resulting in a higher propensity airway narrowing [123,124]. Chen et al. demonstrated in vitro that ASM cells from COPD patients stimulated with cigarette smoke (CS) extract, have higher deposition of collagen type VIII alpha I, but no differences on the deposition of collagen V and fibronectin [125].

Hogg et al. [126] demonstrated that thickening of the airway walls, by the remodeling process, was strongly associated with the progression of obstruction. In addition, the authors showed that the accumulation of inflammatory cells (polymorpho-nuclear leukocytes, macrophages, CD4+ and CD8+ T lymphocytes and B cells) in the lumen of the airways, leads to a malfunction of the mucociliary clearance apparatus [108,126].

It is worth noting that there is no consensus on the number of ECM components in the airways and parenchyma in COPD patients. In moderate COPD patients, Annoni et al. showed a decrease in elastic fibers, collagen subtype I, and versican, in small and large airways, associated with a higher fibronectin fractional area [5]. Additionally, these authors suggested that a decrease in elastic fibers, leads to a loss of airway parenchyma, resulting in airway collapse and gas trapping [5]. Such findings are consistent with those of previous studies, in which the authors showed a decrease in elastic fibers, in both the small airways and the alveolar septa, in lung samples from COPD patients, who were in moderate stages, and showed severe lung function impairment [122,127].

In contrast, in lung samples from surgically resected lobes, Vlahovic et al. previously demonstrated an increase in the volume of the alveolar septum, with a parallel increase in elastic fibers in the COPD patients, with mild to moderate emphysema. Additionally, they observed increased numbers of interstitial fibroblasts and macrophages [112].

Although there has been divergence among studies regarding the different ECM components, there is a consensus about the structural changes in these components, which usually involve fragmentation [5,122]. Abraham and Hogg showed that severe disruption and remodeling of the elastic and collagen fibers, in alveolar airspaces of emphysematous human lung samples and collagen, spreads to alveolar airspaces, indicating extensive alterations in the collagen fiber structures, in the alveolar region [108].

In vitro studies have addressed some of the potential mechanisms that drive the ECM component remodeling. In this context, Sun et al. showed an increased immunoreactivity of LL-37, a protein of the human cathelicidin family, which is involved in the tissue remodeling processes, in the small airway epithelium of COPD patients, compared to healthy smokers. They showed that the expression of LL-37 in the airway epithelium, was correlated with the airway wall thickness, as well as a deposition of collagen in the airway walls. Additionally, the authors showed in vitro, that exposure to CS, induced an increase in LL-37 and augmented the fibroblast collagen production [128].

Milara et al. showed that CS exposure induces chronic lung remodeling in differentiated bronchial epithelial cells, from smokers and moderate COPD patients. These cells undergo mesenchymal transition, as a result of the release of TGF-β1, by enhancing oxidative stress, the phosphorylation of ERK1/2, and SMAD3, and the downregulation of cyclic monophosphate (cAMP) [129]. A similar response was found in all airway wall compartments of smokers and patients with COPD, but mostly in actively-smoking COPD subjects [130].

Anti-TGF-β treatments can attenuate CS-induced lung injury in COPD. Both, in vitro and in vivo studies indicate that, inhibition of TGF-β signaling can protect the lungs from altered lung morphology, impaired lung function, and lung injury [131,132].

Brandsma et al. demonstrated differential effects of fluticasone treatment on different lung compartments, in severe COPD patients [98]. This inhaled steroid stimulated the production of decorin by airway fibroblasts, inducing the restoration of decorin around the airways; while in contrast, in parenchymal fibroblasts, fluticasone inhibited the production of biglycan and procollagen, indicating inhibition of tissue repair in emphysematous areas. The authors attribute this response to the phenotypic differences between lung fibroblasts, including ECM production and the response to TGF-β [19,98]. Since not all treatments are able to reverse tissue damage in COPD, some ECM protein markers have been used in determining disease prognosis. Some findings have demonstrated that serological markers can reflect the extent of structural changes in COPD patients. Sand et al. showed that serological biomarkers of collagen subtypes I, III, IV, and VI were associated with an increased mortality [133]. The increased serum levels of procollagen type I, associated with high levels of IL-6 and IL-8, in COPD patients, might indicate the airway remodeling condition, as the inflammatory process plays an important role in stimulating collagen synthesis [134]. Vignola et al. demonstrated that increased levels of active elastase and overproduction of TIMP-1, relative to MMP-9, were associated with the magnitude of lung changes on high-resolution computed tomography [135]. Papakonstantinou et al. showed that hyaluronic acid levels in the serum of COPD patients are associated with COPD severity and airflow limitation, pointing out this molecule as a potent target to control airway inflammation and remodeling in COPD [136].

3.2. Experimental Models

Since the majority of studies in COPD patients have been restricted to lung samples obtained from pulmonary biopsy or resection, experimental models have been used to understand how abnormal fiber repair, under COPD conditions, could interfere with lung functionality [137].

There are many ways to induce COPD in rodents. The administration of proteases, such as PPE, and exposure to CS, remain the most commonly used strategies to induce lung structural changes that resemble those observed in COPD patients [138].

The CS-induced model is considered to best represent human COPD, since CS is the main risk factor for this disease in humans [139,140]. Several studies have demonstrated parenchymal destruction and remodeling, worsening lung function, and inflammatory processes, over a long period of time, after CS exposure [140–143]. In addition to these characteristic features of COPD, Beckett et al. showed systemic effects on the skeletal muscle and the heart, in a short-term model of COPD [144].

Conversely, the elastase-induced model requires a short time to induce drastic structural changes, compared to the CS-induced model; for this reason, it is the most commonly used model to study how changes in the ECM fiber deposition in the parenchyma, interfere with respiratory mechanics [141,145].

It is important to note that both models show an inflammatory process characterized mainly by increased macrophages [143,146], but neutrophils and the presence of CD4+ and CD8+ T lymphocytes have also been observed. MMP-12 is the metalloprotease most often described in animal models of COPD, and along with TGF-β, it acts by modulating increases in the amounts of elastic and collagen fibers [143,147]. Structural changes are observed mainly in the lung parenchyma, where increased alveolar enlargement is observed, reflecting alveolar wall destruction, and the presence of fragmented elastic and collagen fibers [145].

The use of proteinase inhibitors represents a potential therapeutic treatment for emphysema in animal models. Lourenço et al. showed that treatment with a serine protease inhibitor from *Rhipicephalus (B.) microplus* (rBmTI-A), decreased the MMP-12+ cells, and the collagen fibers in the lung parenchyma, and reversed the loss of elastic recoil and alveolar enlargement, in an emphysema model induced by PPE instillation and CS exposure [143,146]. Similarly, the use of other proteinase inhibitors, also reduced the elastase-induced pulmonary inflammation, remodeling, oxidative stress, and mechanical alterations [148,149].

There are differences in the ECM fiber remodeling patterns between experimental models. Although all newly deposited fibers showed a fragmented appearance, there was an increase in collagen subtype III, with no differences in the collagen subtype I, in the lung parenchyma, and there were differences between the experimental models, when we analyzed the elastic fiber components. In addition, in the CS model, we observed increased fibrillin amounts, while in the PPE-induced model, there was an increase in elastin [145].

Robertoni et al. [150] showed that a reduction in ECM fibers preceded increased deposition of these fibers in the distal parenchyma in a PPE-induced model, suggesting that in this animal model, the destruction and repair processes did not occur simultaneously. These authors showed an initial decrease in the volume proportions of collagen subtype I (on the 3rd day) and subtype III (from the sixth hour until the third day), after the detection of increased MMP-8 and MMP-13 gene expression. On the twenty-first day, there was an increase in the volume proportion of collagen subtype III, and collagen subtype I returned to levels similar to those in the control groups. Additionally, MMP-12 gene expression was increased (from the third hour to the sixth hour) before the decrease in the volume proportion of elastin on the third day, with a subsequent increase in the proportion of this fiber on the twenty-first day. An increase in polymorphonuclear cells was observed beginning, in the first hour, after the PPE instillation, which remained until the third day [150].

The majority of studies describing the impact of ECM fiber remodeling on lung function impairment have emphasized the importance of collagen fibers [1,151,152]. Previous findings have demonstrated that fiber stiffness depends on the relative amounts of subtype I and subtype III collagens, since subtype I collagen is stiffer than subtype III [1,6]. Such findings might explain why, in animal models of emphysema, collagen fibers break at tensions that correspond to those recorded with normal breathing [152]. The stimulation of ECM components, such as collagen subtype I and subtype III, induced the proliferation, migration, and adhesion of ASM cells in rat models of COPD. The concomitant increase in TGF-β expression in these cells, induces overproduction of multiple ECM proteins, which might result in ASM cell hyperplasia [153].

The newly synthetized collagen fibers had altered configurations and were likely to be weaker, compromising the strength of the alveolar walls [152]. To determine how these structural changes impact the lung function, Kononov et al. used an elastase-induced model, to show that, the mechanical forces generated during normal breathing were sufficient to promote tissue damage and stress failure in the remodeled alveolar walls, with increased collagen and elastin [110]. Additionally, Ito et al. showed that parenchymal fibers failed at lower stress, after remodeling in mice, due to four weeks of PPE instillation, and the authors attributed these changes to the newly synthesized collagen fibers [152].

Recently, structural changes have been detected, prior to functional changes. In a PPE-induced model, although structural changes were detected in earlier stages of emphysema development (6 h after PPE instillation), significant decreases in tissue elastance and tissue resistance were observed only twenty-one days after elastase instillation [146,150]. In the CS-induced model, exposure required three months to induce a decrease in tissue elastance and tissue resistance, whereas alveolar enlargement could be detected after one month, with 30 min of exposure, repeated two times per day, for 5 days per week [142,154].

It is interesting that lung function parameters do not reflect the structural changes in COPD animal models, mainly due to technical difficulties in performing evaluations in small animals [86,155]. To detect changes in lung function in animal models, it is necessary to show significant changes in lung structure, which explains why many studies have not shown modifications in lung function but have conducted morphometric analysis [115,155,156]. Therefore, it is important to understand how these structural and functional changes occur at different time points, during disease development in experimental models, to facilitate the choice of the best model, according to the approach and goals. Table 3 summarizes lung function changes and markers of ECM remodeling in clinical and experimental studies of COPD.

Table 3. Lung function changes and markers of extracellular matrix (ECM) remodeling in COPD.

	Clinical Studies	Experimental Studies *
Lung Function		
	Mild to severe stages of COPD [5,22,112,122,128,129] (FEV1/FVC < 70%, FEV1 ≥ 80%)	↓ in lung tissue elastance and resistance [115,143,146,154]
Remodeling Markers		
Collagen fibers	↑ in SEBM (moderate and severe stages) [10] ↑ in small airways (moderate and severe stages) [128] ↑ in interstitial matrix (mild to moderate stages) [112]	↑ in lung parenchyma [115,143,146,154]
Collagen subtype I	↑ in SEBM, lamina propria and bronchial adventitia (moderate and severe stages) [22] ↑ in airway (moderate stage) [129] ↓ in small and large airways (moderate stage) [5]	↓ in lung parenchyma during COPD development [150]
Collagen subtype Ill	↑ in SEBM, lamina propria and bronchial adventitia (moderate and severe stages) [22]	↑ in lung parenchyma [145,150] ↓ in lung parenchyma during COPD development [150]
Collagen subtype IV	↑ in SEBM (moderate and severe stages) [22]	—
Elastic fibers	↓ in alveoli and small airways (moderate stage) [122]	↑ in lung parenchyma [115,143,146] ↓ in lung parenchyma and basement membrane [117]
Elastin	↑ in interstitial matrix (mild to moderate stages) [112]	↑ in lung parenchyma [145,150] ↓ in lung parenchyma during COPD development [150]
Fibrillin	—	↑ in lung parenchyma [145]
Fibronectin	↑ in SEBM (moderate and severe stages) [22] ↑ in small airways (moderate stage) [5]	—

ECM: Extracellular matrix; COPD: Chronic obstructive pulmonary disease; FEV$_1$: Forced expiratory volume in 1 s; FVC: Forced vital capacity; SEBM: Surface epithelial basement membrane; PPE: Porcine pancreatic elastase. * COPD model induced by PPE [117,145,146,150] or papain instillation [112], or cigarette smoke exposure [143,145,154].

4. Extracellular Matrix Remodeling in ARDS

ARDS was defined in 1994, but in 2011, after an initiative of the European Respiratory Society of Intensive Care Medicine endorsed by the American Thoracic Society, this disease was redefined by the Berlin definition. Three categories of ARDS were proposed, based on the degree of hypoxemia—mild (200 mm Hg < PaO$_2$/FIO$_2$ ≤ 300 mm Hg); moderate (100 mm Hg < PaO$_2$/FIO$_2$ ≤ 200 mm Hg); and severe (PaO$_2$/FIO$_2$ ≤ 100 mmHg).

ARDS remains an important cause of death within intensive care units, and approximately 30% of patients die due to ARDS, despite advances in therapeutic strategies [157,158]. Pulmonary fibroproliferation has been associated with higher mortality and ventilator dependence, and it remains an observable clinical feature, in a subset of patients [159].

In this context, there is increasing interest in better understanding the basic and pathophysiological mechanisms that drive the fibroproliferative response in ARDS. The initial site of the lesion is the alveolar epithelium or the endothelium [160,161].

The acute phase of ARDS is characterized by local and systemic inflammatory responses [162], and involves the release of several pro-inflammatory cytokines, such as tumor necrosis factor-alpha (TNF-α), IL-1β, and IL-8 [162–165]. Additionally, neutrophil infiltration, interstitial edema and hypoxemia, are often accompanied by aggressive ECM remodeling [166]. Chen et al. used in vitro models of acid-induced lung epithelial cell injury, to show that the interaction of these cells with monocytes, accelerates the epithelial remodeling process through EMT signaling [167].

Although the physiopathological mechanisms of ECM remodeling among asthma, COPD and ARDS are completely different, the ECM remodeling also requires the action of mechanical forces generated by the migration or contraction of myofibroblasts, by themselves, and the presence of fibronectin, initially produced by macrophages, which is responsible for the adhesion of cells to the matrix. At the end of the acute phase of ARDS, fibronectin is already being produced by myofibroblasts [168].

The fibroproliferative phase, which is mainly characterized by thickening of the alveolar wall associated with interstitial edema and large cellularity, occurs between 7 and 15 days after the primary injury. The cells most involved in this phase are neutrophils, macrophages, myofibroblasts, and type II pneumocytes [169]. The hyaline membrane that arises during the acute phase plays an important role in the fibroproliferative phase of ARDS, since it attaches fibronectin produced by alveolar macrophages to its surface.

Myofibroblasts deposit elements of the collagen and elastic systems, both, in the lumen and inside the alveolar septum, as well as in the walls of the blood vessels. Initially, there is an increase in the deposition of thin fibers of subtype III collagen. Many patients present resolution of the process at this stage, but some progress to the phase of fibrotic remodeling [170–172].

In later stages of this disease, thickening of the vessel wall is present, making gas exchange and local metabolism even more difficult. The main characteristic of this phase is a change in the gene expression of subtype I collagen, which is synthesized in increasing amounts. At the same time, there is an increase in collagenase-digested subtype III collagen (secreted in the previous phases), leading to a tendency towards the accumulation of fibrous tissue, in later stages of ARDS.

Tissue repair includes a variety of mechanisms, as well as edema reabsorption, resolution of inflammation, and cell proliferation, with the aim of repairing the alveolar epithelium [173].

During the fibrotic remodeling phase of ARDS (late phase), there is a trend towards increased deposition of elastic fibers in the alveolar septa, leading to a progressive fibroelastosis [172]. During this stage, alveoli obliterated by fibrosis are adjacent to the ectatic alveoli, with irregular, thickened walls covered by stratified epithelium, or simple columnar epithelium, likely derived from the bronchioles. In the alveolar spaces, there is a large number of pulmonary surfactants produced by numerous type II pneumocytes, which remain active after differentiating from type I pneumocytes, promoting alveolar re-epithelization.

The increase in the number of elastic fibers in the late stages of ARDS might be a compensatory response to the fragmentation and degradation of pre-existing fibers, in the early stages of the process. However, the deposition of large numbers of elastic fibers, leads to progressive elastosis, which is partially responsible for the loss of the normal architecture of the alveolar wall, contributing to the tendency to collapse [172,174].

Since the deposition of microfibrils precedes the appearance of elastin, it should be considered that, during the "de novo" synthesis process of elastic fibers, there will be a stage during which the areas undergoing remodeling will be rich in bundles of microfibrils, with very little or no elastin. Thus, in addition to the absence of the elastic component, the mechanical properties of these inextensible microfibrils, add to those of collagen I, yielding even more tissue resistance to the physical adaptations necessary for a good respiratory performance.

These structural alterations in the ECM have repercussions for the compliance of the pulmonary parenchyma, with impacts on the respiratory mechanics. Evidence has suggested that intrinsic factors, such as genetic patterns of inflammatory response modulation, can influence the production of interleukins involved in the disease progression [175]. Additionally, external factors, such as alcoholism, have been identified as promoters of inflammation and fibrogenesis in ARDS [176–178].

4.1. Clinical Studies

Pioneering studies performed by different groups in the 1990s have showed that, 72 h after the diagnosis of ARDS, patients already showed important increases in collagen synthesis, as detected by high levels of the N-terminal peptide of type III procollagen [179–181]. Additionally, these elevated levels are associated with histological lung fibroproliferation and mortality, in ARDS patients [169].

Although the alveolar level of N-terminal peptide of type III procollagen is considered the best surrogate marker for the diagnosis of lung fibroproliferation, Hamon et al. demonstrated that patients with active lung proliferation have higher fibrosis score, as evaluated by a chest CT scan, which allows an alternative use of this radiological tool, which is less invasive than fibroscopic bronchoalveolar lavage [182]. Thille et al. have showed that histological features of the lungs are related to the duration of ARDS. The authors analyzed 159 patients and found a reduction in the prevalence of exudative changes over time, with greater changes in patients with ARDS, for less than one week, and smaller changes in patients with a disease duration, between one and three weeks. However, the prevalence of proliferative changes increased over time and was greater in patients with a long duration of disease. These authors have also showed that fibrosis was more common in patients whose ARDS origin was pulmonary [183].

Interestingly, an almost complete recovery of lung function has been demonstrated in survivors of ARDS, after 6 to 12 months of evolution [184,185]. However, approximately 30% to 40% of ARDS patients in the late phase, evolve to an exacerbated and progressive remodeling process, culminating in the destruction of the pulmonary architecture and death [172]. In ARDS survivors, a negative correlation has been described between fibroproliferation and quality of life. Burnham et al. studied 82 patients with ARDS and showed that reduced lung compliance measured at the bedside, was associated with radiologic fibroproliferation, 14 days post-diagnosis [166]. These data were interesting since they could be helpful in identifying patients with ARDS who are at risk for complications in clinical conditions.

Zheng et al. showed the protective effects of ResolvinD1, a lipid mediator which attenuates the excessive polymorphonuclear infiltration and transmigration, in the fibroproliferative phase of ARDS. The authors demonstrated that ResolvinD1 inhibited primary human lung fibroblast proliferation, collagen production, and myofibroblast differentiation induced by TGF-β, from ARDS patients [186].

Although lung remodeling is an important feature in ARDS patients and is related to the deterioration of lung function, the assessment of lung repair in patients, remains limited. However, such assessments could be of great interest because of the prognostic relevance of lung repair in ARDS patients.

Unfortunately, no pharmacological agents that focus specifically on fibroproliferation are available at this time for the treatment of ARDS [166]. In this context, basic and experimental studies are relevant and could contribute not only to a better understanding of the fibroproliferative process in ARDS, but also to the development of new therapeutic strategies for ARDS patients.

4.2. Experimental Models

Most experimental models used to study ARDS investigated the acute phase, although at this time, most authors, including our group, have shown deposition of collagen fibers in the alveolar septa.

Once the direct or indirect etiological stimulus has ceased, the behavior of tissue remodeling is completely different. In animals subjected to direct lesions, we observed continuous deposition of collagen that remained stable, until the eighth week, followed by a deposition of elastic fibers, with significant differences after the first week. In animals subjected to an indirect initial insult, the levels of collagen deposition fell to basal levels, during the first week, after the insult, and elastosis was not observed [174].

We used animals challenged with LPS, and evaluated them 6 and 24 h after injury. We found that at 6 h, there was intense inflammation in the lung with high levels of pro-inflammatory cytokines; however, no signs of lung remodeling and no deterioration of lung function were detected at this time. Only at 24 h after LPS instillation did we observe an intense deposition of collagen fibers in the alveolar septa, and a reduction in the respiratory system and lung tissue compliance [187].

Costa et al. developed an experimental model of ARDS, induced by nebulized LPS, and they found that, 24 h after LPS, the animals showed increased pro-inflammatory cytokine levels, increased total septal volume, and a thickening associated with reduced surface density of the alveolar septa. However, after five weeks, the animals showed an increased total lung volume and accentuated collagen deposition, particularly collagen subtype I, associated with reduced MMP-2 protein expression [188]. This model could contribute to a better understanding of the remodeling process in ARDS, and the development of preventive or therapeutic strategies, to counteract lung remodeling in ARDS.

LPS is a widely used model to mimic ARDS alterations in experimental animals. In this regard, Oliveira et al. using in vitro techniques, showed that, LPS increased lung epithelial cell stiffness and is associated to cytoskeletal remodeling [189].

Although no studies in patients have investigated the effects of drugs in lung remodeling, in animals, we showed the effects of different pathways involved in remodeling. An extensive body of literature shows that natural substances can reduce acute lung injury (ALI) in animal models; however, few studies have focused on lung remodeling. In this regard, Mernak et al. showed that sakuranetin, a flavonoid that can significantly reduce lung inflammation, reduced collagen deposition in the parenchyma, when it was administered 6 h after LPS, the point at which animals have intense inflammation. Moreover, this compound also improved the lung tissue elastance of these animals [187]. Park et al. also demonstrated that the human tripeptide glycyl-l-histidyl-l-lysine reduced the reactive oxygen species, TNF-α, and IL-6 production, in murine macrophages stimulated with LPS [190]. These data suggest that this tripeptide is relevant for controlling inflammation and preventing lung remodeling, at least in animals.

Although lung remodeling in ARDS is still not fully understood, lung repair and remodeling, including all the alterations discussed above, is necessary, to allow for the recovery of ARDS. In this regard, Pinheiro et al. clearly showed that pharmacological stimulation of nicotinic receptors by PNU changed macrophage profile from M1 through M2 subtypes. This treatment also attenuated collagen deposition, suggesting that, this change in macrophage profile can explain the resolution of lung inflammation and the improvement in lung function observed in this study [191]. Table 4 summarizes lung function changes and markers of ECM remodeling, in clinical and experimental studies of ARDS.

Table 4. Lung function changes and markers of ECM remodeling in acute respiratory distress syndrome (ARDS).

	Clinical Studies	Experimental Studies *
Lung Function		
	Moderate to severe stage of ARDS (PaO2/FIO2 ≤ 200 mmHg) [179–181]	↑ in respiratory system resistance after 24 h [174,191]
		↑ in respiratory system elastance after 24 h [174,187,191]
		↑ in lung tissue resistance after 24 h [191]
		↑ in lung tissue elastance after 24 h [187,191]
Remodeling Markers		
Collagen fibers	↑ in lung parenchyma (late phase) [172]	↑ in lung parenchyma after 24 h [174,187,191]
		↑ in lung parenchyma after 5 weeks [188]
Collagen subtype I	↑ in BALF and serum (severe stage) [180]	↑ in lung parenchyma after 5 weeks [188]
Collagen subtype III	↑ in BALF and serum (severe stage) [180]	—
Elastic fibers	↑ in lung parenchyma (late phase) [172]	↑ in lung parenchyma after 1 week [174]
Procollagen type I	↑ in plasma (moderate and severe stages) [181]	—
Procollagen type III	↑ in BALF (severe stage) [179,180]	—
	↑ in serum (severe stage) [180]	
	↑ in plasma (moderate and severe stages) [181]	

ECM: Extracellular matrix; ARDS: Acute respiratory distress syndrome; PaO2/FIO2: Arterial partial pressure of oxygen/fraction of inspired oxygen; ALI: Acute lung injury; BALF: Bronchoalveolar lavage fluid; LPS: Lipopolysaccharide. * ALI induced by Escherichia coli LPS intratracheal or intraperitoneal administration [174]. ALI induced by LPS intranasal [187] or intratracheal administration [191] or LPS nebulization [188].

4.3. Final Considerations

Structural changes in ECM components are associated with the worsening of lung function and the progression of asthma, COPD, and ARDS. In this context, clinical studies have been useful for characterizing which ECM components are present in the lung samples of patients, mainly those in advanced stages of respiratory diseases, and for investigating the associations between these changes and the progression of these diseases. In addition, some ECM proteins and inflammatory mediator markers in the serum of patients, have been used as important features for elucidating the extent of structural changes in the lungs, to avoid invasive procedures, facilitating a prognostic evaluation of these respiratory diseases.

In animal models, temporal analyses of inflammatory profiles and respiratory function have been performed, to elucidate the different mechanisms involved in disease progression. Moreover, the opportunity to evaluate in vivo responses to treatment, with inhibitors of specific inflammatory mediators, has better elucidated the different mechanisms involved in the pathogenesis of these diseases and highlighted some possible therapeutic targets, since most of these inhibitors have been shown to attenuate fiber remodeling and improve lung function. It is important to note that although many experimental models have been described in the literature for inducing different respiratory diseases, none of them recreates all of the physiological changes observed in humans. Therefore, before choosing an experimental model, it is very important to consider which inflammatory events and structural changes can be evaluated with different approaches.

In vitro studies [46,47,87,92,125,128,129,153,167,186,189,190,192] have improved our understanding of which ECM components change and elucidate their effects on the impairment of respiratory parameters; they have also made it possible to analyze how different inflammatory mediators impact inflammatory cell activity and recruitment.

However, these studies have some limitations. Despite these advances in our understanding of the mechanisms involved in ECM structural changes in asthma, COPD, and ARDS, there are no clinical studies that have showed an effective treatment to reverse all structural changes, in order to totally restore the lung function. Bronchodilators and corticosteroids have been used to relieve the symptoms of these respiratory diseases, but these approaches cannot control disease progression.

Further investigations are necessary to distinguish how the dysregulation of the different ECM components drive these structural changes progression, as well as how the interactions between cells and the ECM components, during these disease progressions, could impact the lung function.

Author Contributions: Conceptualization, F.D.T.Q.S.L., I.F.L.C.T., C.M.P., and J.T.I.; Methodology, F.D.T.Q.S.L., J.T.I, J.D.L., and R.F.R.; Investigation, F.D.T.Q.S.L., J.T.I., R.F.R., and C.M.P.; Resources, F.D.T.Q.S.L. and J.T.I.; Writing–Original Draft Preparation, F.D.T.Q.S.L., J.T.I., J.D.L., R.F.R., and C.M.P.; Writing–Review & Editing, F.D.T.Q.S.L., J.T.I., J.D.L., R.F.R., and C.M.P.; Visualization, F.D.T.Q.S.L., I.F.L.C.T., C.M.P., and J.T.I.; Supervision, F.D.T.Q.S.L. and J.T.I.; Funding Acquisition, F.D.T.Q.S.L.

Funding: This research received no external funding and the APC was funded by Fundação Faculdade de Medicina through the Programa de Fomento às Atividades de Pesquisa.

Acknowledgments: This study was supported by the Hospital das Clínicas da Faculdade de Medicina da Universidade de São Paulo, Brazil (HC/FMUSP).

Conflicts of Interest: The authors declare no conflicts of interest.

References

1. Suki, B.; Bates, J.H. Extracellular matrix mechanics in lung parenchymal diseases. *Respir. Physiol. Neurobiol.* **2008**, *163*, 33–43. [CrossRef]
2. Burgstaller, G.; Oehrle, B.; Gerckens, M.; White, E.S.; Schiller, H.B.; Eickelberg, O. The instructive extracellular matrix of the lung: Basic composition and alterations in chronic lung disease. *Eur. Respir. J.* **2017**, *50*, 1601805. [CrossRef]
3. Manuyakorn, W.; Howarth, P.H.; Holgate, S.T. Airway remodelling in asthma and novel therapy. *Asian Pac. J. Allergy Immunol.* **2013**, *31*, 3–10.
4. Yue, B. Biology of the extracellular matrix: An overview. *J. Glaucoma* **2014**, *23*, S20–S23. [CrossRef]
5. Annoni, R.; Lancas, T.; Yukimatsu Tanigawa, R.; de Medeiros Matsushita, M.; de Morais Fernezlian, S.; Bruno, A.; Fernando Ferraz da Silva, L.; Roughley, P.J.; Battaglia, S.; Dolhnikoff, M.; et al. Extracellular matrix composition in COPD. *Eur. Respir. J.* **2012**, *40*, 1362–1373. [CrossRef]
6. Silver, F.H.; Birk, D.E. Molecular structure of collagen in solution: Comparison of types I, II, III and V. *Int. J. Biol. Macromol.* **1984**, *6*, 125–132. [CrossRef]
7. Shifren, A.; Mecham, R.P. The stumbling block in lung repair of emphysema: Elastic fiber assembly. *Proc. Am. Thorac. Soc.* **2006**, *3*, 428–433. [CrossRef]
8. Robbesom, A.A.; Koenders, M.M.; Smits, N.C.; Hafmans, T.; Versteeg, E.M.; Bulten, J.; Veerkamp, J.H.; Dekhuijzen, P.N.; van Kuppevelt, T.H. Aberrant fibrillin-1 expression in early emphysematous human lung: A proposed predisposition for emphysema. *Mod. Pathol.* **2008**, *21*, 297–307. [CrossRef] [PubMed]
9. Brown, R.E.; Butler, J.P.; Rogers, R.A.; Leith, D.E. Mechanical connections between elastin and collagen. *Connect. Tissue Res.* **1994**, *30*, 295–308. [CrossRef] [PubMed]
10. Raspanti, M.; Alessandrini, A.; Ottani, V.; Ruggeri, A. Direct visualization of collagen-bound proteoglycans by tapping-mode atomic force microscopy. *J. Struct. Biol.* **1997**, *119*, 118–122. [CrossRef]

11. Kielty, C.M.; Sherratt, M.J.; Shuttleworth, C.A. Elastic fibres. *J. Cell Sci.* **2002**, *115*, 2817–2828.
12. Setnikar, I. Origin and significance of the mechanical property of the lung. *Arch. Fisiol.* **1955**, *55*, 349–374.
13. Yuan, H.; Kononov, S.; Cavalcante, F.S.; Lutchen, K.R.; Ingenito, E.P.; Suki, B. Effects of collagenase and elastase on the mechanical properties of lung tissue strips. *J. Appl. Physiol.* **2000**, *89*, 3–14. [CrossRef] [PubMed]
14. Lever, R.; Page, C. Glycosaminoglycans, airways inflammation and bronchial hyperresponsiveness. *Pulm. Pharmacol. Ther.* **2001**, *14*, 249–254. [CrossRef]
15. Tyrrell, D.J.; Horne, A.P.; Holme, K.R.; Preuss, J.M.; Page, C.P. Heparin in inflammation: Potential therapeutic applications beyond anticoagulation. *Adv. Pharmacol.* **1999**, *46*, 151–208. [PubMed]
16. Salazar, L.M.; Herrera, A.M. Fibrotic response of tissue remodeling in COPD. *Lung* **2011**, *189*, 101–109. [CrossRef]
17. Malmstrom, J.; Larsen, K.; Malmstrom, L.; Tufvesson, E.; Parker, K.; Marchese, J.; Williamson, B.; Hattan, S.; Patterson, D.; Martin, S.; et al. Proteome annotations and identifications of the human pulmonary fibroblast. *J. Proteome Res.* **2004**, *3*, 525–537. [CrossRef]
18. Zandvoort, A.; Postma, D.S.; Jonker, M.R.; Noordhoek, J.A.; Vos, J.T.; Timens, W. Smad gene expression in pulmonary fibroblasts: Indications for defective ECM repair in COPD. *Respir. Res.* **2008**, *9*, 83. [CrossRef]
19. Hallgren, O.; Nihlberg, K.; Dahlback, M.; Bjermer, L.; Eriksson, L.T.; Erjefalt, J.S.; Lofdahl, C.G.; Westergren-Thorsson, G. Altered fibroblast proteoglycan production in COPD. *Respir. Res.* **2010**, *11*, 55. [CrossRef]
20. Hill, M.R.; Philp, C.J.; Billington, C.K.; Tatler, A.L.; Johnson, S.R.; O'Dea, R.D.; Brook, B.S. A theoretical model of inflammation- and mechanotransduction-driven asthmatic airway remodelling. *Biomech. Model. Mechanobiol.* **2018**, *17*, 1451–1470. [CrossRef]
21. Zandvoort, A.; Postma, D.S.; Jonker, M.R.; Noordhoek, J.A.; Vos, J.T.; van der Geld, Y.M.; Timens, W. Altered expression of the Smad signalling pathway: Implications for COPD pathogenesis. *Eur. Respir. J.* **2006**, *28*, 533–541. [CrossRef]
22. Kranenburg, A.R.; Willems-Widyastuti, A.; Moori, W.J.; Sterk, P.J.; Alagappan, V.K.; de Boer, W.I.; Sharma, H.S. Enhanced bronchial expression of extracellular matrix proteins in chronic obstructive pulmonary disease. *Am. J. Clin. Pathol.* **2006**, *126*, 725–735. [CrossRef]
23. Loftus, P.A.; Wise, S.K. Epidemiology of asthma. *Curr. Opin. Otolaryngol. Head Neck Surg.* **2016**, *24*, 245–249. [CrossRef]
24. Moorman, J.E.; Akinbami, L.J.; Bailey, C.M.; Johnson, C.A.; King, M.E.; Liu, X.; Zahran, H.S. National surveillance of asthma: United States, 2001–2010. *Vital Health Stat. Ser. Anal. Epidemiol. Stud.* **2012**, *35*, 1–58.
25. Global Initiative for Asthma (GINA). Global Strategy for Asthma Management and Prevention. 2018. Available online: https://ginasthma.org/ (accessed on 30 March 2019).
26. Branchett, W.J.; Lloyd, C.M. Regulatory cytokine function in the respiratory tract. *Mucosal Immunol.* **2019**. [CrossRef]
27. Ubel, C.; Graser, A.; Koch, S.; Rieker, R.J.; Lehr, H.A.; Muller, M.; Finotto, S. Role of Tyk-2 in Th9 and Th17 cells in allergic asthma. *Sci. Rep.* **2014**, *4*, 5865. [CrossRef]
28. Peng, J.; Li, X.M.; Zhang, G.R.; Cheng, Y.; Chen, X.; Gu, W.; Guo, X.J. TNF-TNFR2 Signaling Inhibits Th2 and Th17 Polarization and Alleviates Allergic Airway Inflammation. *Int. Arch. Allergy Immunol.* **2019**, *178*, 281–290. [CrossRef]
29. Al-Muhsen, S.; Johnson, J.R.; Hamid, Q. Remodeling in asthma. *J. Allergy Clin. Immunol.* **2011**, *128*, 451–462, quiz 463–454. [CrossRef]
30. Nishimura, Y.; Inoue, T.; Morooka, T.; Node, K. Mechanical stretch and angiotensin II increase interleukin-13 production and interleukin-13 receptor alpha2 expression in rat neonatal cardiomyocytes. *Circ. J.* **2008**, *72*, 647–653. [CrossRef]
31. Tschumperlin, D.J.; Drazen, J.M. Mechanical stimuli to airway remodeling. *Am. J. Respir. Crit. Care Med.* **2001**, *164*, S90–S94. [CrossRef]
32. Miyagawa, A.; Chiba, M.; Hayashi, H.; Igarashi, K. Compressive force induces VEGF production in periodontal tissues. *J. Dent. Res.* **2009**, *88*, 752–756. [CrossRef]
33. Park, J.A.; Tschumperlin, D.J. Chronic intermittent mechanical stress increases MUC5AC protein expression. *Am. J. Respir. Cell Mol. Biol.* **2009**, *41*, 459–466. [CrossRef]
34. Park, J.A.; Drazen, J.M.; Tschumperlin, D.J. The chitinase-like protein YKL-40 is secreted by airway epithelial cells at base line and in response to compressive mechanical stress. *J. Biol. Chem.* **2010**, *285*, 29817–29825. [CrossRef]

35. Grainge, C.L.; Lau, L.C.; Ward, J.A.; Dulay, V.; Lahiff, G.; Wilson, S.; Holgate, S.; Davies, D.E.; Howarth, P.H. Effect of bronchoconstriction on airway remodeling in asthma. *N. Engl. J. Med.* **2011**, *364*, 2006–2015. [CrossRef]

36. Pohunek, P.; Warner, J.O.; Turzikova, J.; Kudrmann, J.; Roche, W.R. Markers of eosinophilic inflammation and tissue re-modelling in children before clinically diagnosed bronchial asthma. *Pediatr. Allergy Immunol.* **2005**, *16*, 43–51. [CrossRef]

37. Lezmi, G.; Gosset, P.; Deschildre, A.; Abou-Taam, R.; Mahut, B.; Beydon, N.; de Blic, J. Airway Remodeling in Preschool Children with Severe Recurrent Wheeze. *Am. J. Respir. Crit. Care Med.* **2015**, *192*, 164–171. [CrossRef]

38. James, A. Airway remodeling in asthma. *Curr. Opin. Pulm. Med.* **2005**, *11*, 1–6. [CrossRef]

39. Lazaar, A.L.; Panettieri, R.A., Jr. Is airway remodeling clinically relevant in asthma? *Am. J. Med.* **2003**, *115*, 652–659. [CrossRef]

40. Cohn, L.; Elias, J.A.; Chupp, G.L. Asthma: Mechanisms of disease persistence and progression. *Annu. Rev. Immunol.* **2004**, *22*, 789–815. [CrossRef]

41. James, A. Airway Remodeling in Asthma: Is it Fixed or Variable? *Am. J. Respir. Crit. Care Med.* **2017**, *195*, 968–970. [CrossRef]

42. Chakir, J.; Shannon, J.; Molet, S.; Fukakusa, M.; Elias, J.; Laviolette, M.; Boulet, L.P.; Hamid, Q. Airway remodeling-associated mediators in moderate to severe asthma: Effect of steroids on TGF-beta, IL-11, IL-17, and type I and type III collagen expression. *J. Allergy Clin. Immunol.* **2003**, *111*, 1293–1298. [CrossRef]

43. Kaminska, M.; Foley, S.; Maghni, K.; Storness-Bliss, C.; Coxson, H.; Ghezzo, H.; Lemiere, C.; Olivenstein, R.; Ernst, P.; Hamid, Q.; et al. Airway remodeling in subjects with severe asthma with or without chronic persistent airflow obstruction. *J. Allergy Clin. Immunol.* **2009**, *124*, 45–51.e1-4. [CrossRef]

44. Slats, A.M.; Janssen, K.; van Schadewijk, A.; van der Plas, D.T.; Schot, R.; van den Aardweg, J.G.; de Jongste, J.C.; Hiemstra, P.S.; Mauad, T.; Rabe, K.F.; et al. Expression of smooth muscle and extracellular matrix proteins in relation to airway function in asthma. *J. Allergy Clin. Immunol.* **2008**, *121*, 1196–1202. [CrossRef]

45. Yick, C.Y.; Ferreira, D.S.; Annoni, R.; von der Thusen, J.H.; Kunst, P.W.; Bel, E.H.; Lutter, R.; Mauad, T.; Sterk, P.J. Extracellular matrix in airway smooth muscle is associated with dynamics of airway function in asthma. *Allergy* **2012**, *67*, 552–559. [CrossRef]

46. Khan, M.A.; Kianpour, S.; Stampfli, M.R.; Janssen, L.J. Kinetics of in vitro bronchoconstriction in an elastolytic mouse model of emphysema. *Eur. Respir. J.* **2007**, *30*, 691–700. [CrossRef]

47. Khan, M.A.; Ellis, R.; Inman, M.D.; Bates, J.H.; Sanderson, M.J.; Janssen, L.J. Influence of airway wall stiffness and parenchymal tethering on the dynamics of bronchoconstriction. *Am. J. Physiol. Lung Cell. Mol. Physiol.* **2010**, *299*, L98–L108. [CrossRef]

48. Possa, S.S.; Charafeddine, H.T.; Righetti, R.F.; da Silva, P.A.; Almeida-Reis, R.; Saraiva-Romanholo, B.M.; Perini, A.; Prado, C.M.; Leick-Maldonado, E.A.; Martins, M.A.; et al. Rho-kinase inhibition attenuates airway responsiveness, inflammation, matrix remodeling, and oxidative stress activation induced by chronic inflammation. *Am. J. Physiol. Lung Cell. Mol. Physiol.* **2012**, *303*, L939–L952. [CrossRef]

49. Joubert, P.; Hamid, Q. Role of airway smooth muscle in airway remodeling. *J. Allergy Clin. Immunol.* **2005**, *116*, 713–716. [CrossRef]

50. James, A.L.; Pare, P.D.; Hogg, J.C. The mechanics of airway narrowing in asthma. *Am. Rev. Respir. Dis.* **1989**, *139*, 242–246. [CrossRef]

51. Holloway, L.; Beasley, R.; Roche, W. The pathology of bronchial asthma. In *Asthma and Rhinitis*; Busse, W., Holgate, S., Eds.; Blackwell Scientific Publications: Oxford, UK, 1995.

52. Benayoun, L.; Druilhe, A.; Dombret, M.C.; Aubier, M.; Pretolani, M. Airway structural alterations selectively associated with severe asthma. *Am. J. Respir. Crit. Care Med.* **2003**, *167*, 1360–1368. [CrossRef]

53. Keglowich, L.F.; Borger, P. The Three A's in Asthma—Airway Smooth Muscle, Airway Remodeling & Angiogenesis. *Open Respir. Med. J.* **2015**, *9*, 70–80. [CrossRef]

54. Kotaru, C.; Schoonover, K.J.; Trudeau, J.B.; Huynh, M.L.; Zhou, X.; Hu, H.; Wenzel, S.E. Regional fibroblast heterogeneity in the lung: Implications for remodeling. *Am. J. Respir. Crit. Care Med.* **2006**, *173*, 1208–1215. [CrossRef]

55. Ojiaku, C.A.; Yoo, E.J.; Panettieri, R.A., Jr. Transforming Growth Factor beta1 Function in Airway Remodeling and Hyperresponsiveness. The Missing Link? *Am. J. Respir. Cell Mol. Biol.* **2017**, *56*, 432–442. [CrossRef]

56. Burgess, J.K.; Mauad, T.; Tjin, G.; Karlsson, J.C.; Westergren-Thorsson, G. The extracellular matrix—The under-recognized element in lung disease? *J. Pathol.* **2016**, *240*, 397–409. [CrossRef]

57. Camargo, L.D.N.; Righetti, R.F.; Aristoteles, L.; Dos Santos, T.M.; de Souza, F.C.R.; Fukuzaki, S.; Cruz, M.M.; Alonso-Vale, M.I.C.; Saraiva-Romanholo, B.M.; Prado, C.M.; et al. Effects of Anti-IL-17 on Inflammation, Remodeling, and Oxidative Stress in an Experimental Model of Asthma Exacerbated by LPS. *Front. Immunol.* **2017**, *8*, 1835. [CrossRef]

58. Dos Santos, T.M.; Righetti, R.F.; Camargo, L.D.N.; Saraiva-Romanholo, B.M.; Aristoteles, L.; de Souza, F.C.R.; Fukuzaki, S.; Alonso-Vale, M.I.C.; Cruz, M.M.; Prado, C.M.; et al. Effect of Anti-IL17 Antibody Treatment Alone and in Combination With Rho-Kinase Inhibitor in a Murine Model of Asthma. *Front. Physiol.* **2018**, *9*, 1183. [CrossRef]

59. Bortolozzo, A.S.S.; Rodrigues, A.P.D.; Arantes-Costa, F.M.; Saraiva-Romanholo, B.M.; de Souza, F.C.R.; Bruggemann, T.R.; de Brito, M.V.; Ferreira, R.D.S.; Correia, M.; Paiva, P.M.G.; et al. The Plant Proteinase Inhibitor CrataBL Plays a Role in Controlling Asthma Response in Mice. *BioMed Res. Int.* **2018**, *2018*, 9274817. [CrossRef]

60. Righetti, R.F.; Pigati, P.A.; Possa, S.S.; Habrum, F.C.; Xisto, D.G.; Antunes, M.A.; Leick, E.A.; Prado, C.M.; Martins Mde, A.; Rocco, P.R.; et al. Effects of Rho-kinase inhibition in lung tissue with chronic inflammation. *Respir. Physiol. Neurobiol.* **2014**, *192*, 134–146. [CrossRef]

61. Reese, S.P.; Underwood, C.J.; Weiss, J.A. Effects of decorin proteoglycan on fibrillogenesis, ultrastructure, and mechanics of type I collagen gels. *Matrix Biol.* **2013**, *32*, 414–423. [CrossRef]

62. Kalamajski, S.; Oldberg, A. The role of small leucine-rich proteoglycans in collagen fibrillogenesis. *Matrix Biol.* **2010**, *29*, 248–253. [CrossRef]

63. Sarrazin, S.; Lamanna, W.C.; Esko, J.D. Heparan sulfate proteoglycans. *Cold Spring Harb. Perspect. Biol.* **2011**, *3*, a004952. [CrossRef] [PubMed]

64. Chen, H.C.; Chang, H.T.; Huang, P.H.; Chang, M.D.; Liu, R.S.; Lin, Y.J.; Hsieh, C.H. Molecular imaging of heparan sulfate expression with radiolabeled recombinant eosinophil cationic protein predicts allergic lung inflammation in a mouse model for asthma. *J. Nucl. Med.* **2013**, *54*, 793–800. [CrossRef]

65. Tanaka, Y.; Adams, D.H.; Shaw, S. Proteoglycans on endothelial cells present adhesion-inducing cytokines to leukocytes. *Immunol. Today* **1993**, *14*, 111–115. [CrossRef]

66. Araujo, B.B.; Dolhnikoff, M.; Silva, L.F.; Elliot, J.; Lindeman, J.H.; Ferreira, D.S.; Mulder, A.; Gomes, H.A.; Fernezlian, S.M.; James, A.; et al. Extracellular matrix components and regulators in the airway smooth muscle in asthma. *Eur. Respir. J.* **2008**, *32*, 61–69. [CrossRef] [PubMed]

67. Pini, L.; Hamid, Q.; Shannon, J.; Lemelin, L.; Olivenstein, R.; Ernst, P.; Lemiere, C.; Martin, J.G.; Ludwig, M.S. Differences in proteoglycan deposition in the airways of moderate and severe asthmatics. *Eur. Respir. J.* **2007**, *29*, 71–77. [CrossRef]

68. Murphy, G. Tissue inhibitors of metalloproteinases. *Genome Biol.* **2011**, *12*, 233. [CrossRef] [PubMed]

69. Pigati, P.A.; Righetti, R.F.; Possa, S.S.; Romanholo, B.S.; Rodrigues, A.P.; dos Santos, A.S.; Xisto, D.G.; Antunes, M.A.; Prado, C.M.; Leick, E.A.; et al. Y-27632 is associated with corticosteroid-potentiated control of pulmonary remodeling and inflammation in guinea pigs with chronic allergic inflammation. *BMC Pulm. Med.* **2015**, *15*, 85. [CrossRef] [PubMed]

70. James, A.L.; Wenzel, S. Clinical relevance of airway remodelling in airway diseases. *Eur. Respir. J.* **2007**, *30*, 134–155. [CrossRef]

71. Ward, C.; Pais, M.; Bish, R.; Reid, D.; Feltis, B.; Johns, D.; Walters, E.H. Airway inflammation, basement membrane thickening and bronchial hyperresponsiveness in asthma. *Thorax* **2002**, *57*, 309–316. [CrossRef] [PubMed]

72. Chetta, A.; Zanini, A.; Foresi, A.; Del Donno, M.; Castagnaro, A.; D'Ippolito, R.; Baraldo, S.; Testi, R.; Saetta, M.; Olivieri, D. Vascular component of airway remodeling in asthma is reduced by high dose of fluticasone. *Am. J. Respir. Crit. Care Med.* **2003**, *167*, 751–757. [CrossRef]

73. Laitinen, L.A.; Laitinen, A.; Haahtela, T. A comparative study of the effects of an inhaled corticosteroid, budesonide, and a beta 2-agonist, terbutaline, on airway inflammation in newly diagnosed asthma: A randomized, double-blind, parallel-group controlled trial. *J. Allergy Clin. Immunol.* **1992**, *90*, 32–42. [CrossRef]

74. Hoshino, M.; Takahashi, M.; Takai, Y.; Sim, J.; Aoike, N. Inhaled corticosteroids decrease vascularity of the bronchial mucosa in patients with asthma. *Clin. Exp. Allergy* **2001**, *31*, 722–730. [CrossRef] [PubMed]

75. Sont, J.K.; Han, J.; van Krieken, J.M.; Evertse, C.E.; Hooijer, R.; Willems, L.N.; Sterk, P.J. Relationship between the inflammatory infiltrate in bronchial biopsy specimens and clinical severity of asthma in patients treated with inhaled steroids. *Thorax* **1996**, *51*, 496–502. [CrossRef] [PubMed]

76. Elliot, J.G.; Noble, P.B.; Mauad, T.; Bai, T.R.; Abramson, M.J.; McKay, K.O.; Green, F.H.Y.; James, A.L. Inflammation-dependent and independent airway remodelling in asthma. *Respirology* **2018**, *23*, 1138–1145. [CrossRef]

77. Berair, R.; Brightling, C.E. Asthma therapy and its effect on airway remodelling. *Drugs* **2014**, *74*, 1345–1369. [CrossRef]

78. Pretolani, M.; Dombret, M.C.; Thabut, G.; Knap, D.; Hamidi, F.; Debray, M.P.; Taille, C.; Chanez, P.; Aubier, M. Reduction of airway smooth muscle mass by bronchial thermoplasty in patients with severe asthma. *Am. J. Respir. Crit. Care Med.* **2014**, *190*, 1452–1454. [CrossRef]

79. Chakir, J.; Haj-Salem, I.; Gras, D.; Joubert, P.; Beaudoin, E.L.; Biardel, S.; Lampron, N.; Martel, S.; Chanez, P.; Boulet, L.P.; et al. Effects of Bronchial Thermoplasty on Airway Smooth Muscle and Collagen Deposition in Asthma. *Ann. Am. Thorac. Soc.* **2015**, *12*, 1612–1618. [CrossRef]

80. Williams, K.; Roman, J. Studying human respiratory disease in animals–role of induced and naturally occurring models. *J. Pathol.* **2016**, *238*, 220–232. [CrossRef]

81. Zosky, G.R.; Sly, P.D. Animal models of asthma. *Clin. Exp. Allergy* **2007**, *37*, 973–988. [CrossRef] [PubMed]

82. Mullane, K.; Williams, M. Animal models of asthma: Reprise or reboot? *Biochem. Pharmacol.* **2014**, *87*, 131–139. [CrossRef] [PubMed]

83. Shin, Y.S.; Takeda, K.; Gelfand, E.W. Understanding asthma using animal models. *Allergy Asthma Immunol. Res.* **2009**, *1*, 10–18. [CrossRef] [PubMed]

84. Aun, M.V.; Bonamichi-Santos, R.; Arantes-Costa, F.M.; Kalil, J.; Giavina-Bianchi, P. Animal models of asthma: Utility and limitations. *J. Asthma Allergy* **2017**, *10*, 293–301. [CrossRef] [PubMed]

85. Almeida-Reis, R.; Toledo, A.C.; Reis, F.G.; Marques, R.H.; Prado, C.M.; Dolhnikoff, M.; Martins, M.A.; Leick-Maldonado, E.A.; Tiberio, I.F. Repeated stress reduces mucociliary clearance in animals with chronic allergic airway inflammation. *Respir. Physiol. Neurobiol.* **2010**, *173*, 79–85. [CrossRef] [PubMed]

86. Bates, J.H.; Davis, G.S.; Majumdar, A.; Butnor, K.J.; Suki, B. Linking parenchymal disease progression to changes in lung mechanical function by percolation. *Am. J. Respir. Crit. Care Med.* **2007**, *176*, 617–623. [CrossRef]

87. Park, G.M.; Han, H.W.; Kim, J.Y.; Lee, E.; Cho, H.J.; Yoon, J.; Hong, S.J.; Yang, S.I.; Yang, H.J.; Yu, J. Association of symptom control with changes in lung function, bronchial hyperresponsiveness, and exhaled nitric oxide after inhaled corticosteroid treatment in children with asthma. *Allergol. Int.* **2016**, *65*, 439–443. [CrossRef]

88. Abreu, S.C.; Antunes, M.A.; Mendonca, L.; Branco, V.C.; de Melo, E.B.; Olsen, P.C.; Diaz, B.L.; Weiss, D.J.; Paredes, B.D.; Xisto, D.G.; et al. Effects of bone marrow mononuclear cells from healthy or ovalbumin-induced lung inflammation donors on recipient allergic asthma mice. *Stem Cell Res. Ther.* **2014**, *5*, 108. [CrossRef] [PubMed]

89. Marques, R.H.; Reis, F.G.; Starling, C.M.; Cabido, C.; de Almeida-Reis, R.; Dohlnikoff, M.; Prado, C.M.; Leick, E.A.; Martins, M.A.; Tiberio, I.F. Inducible nitric oxide synthase inhibition attenuates physical stress-induced lung hyper-responsiveness and oxidative stress in animals with lung inflammation. *Neuroimmunomodulation* **2012**, *19*, 158–170. [CrossRef]

90. Vasconcelos, L.H.C.; Silva, M.; Costa, A.C.; de Oliveira, G.A.; de Souza, I.L.L.; Queiroga, F.R.; Araujo, L.; Cardoso, G.A.; Righetti, R.F.; Silva, A.S.; et al. A Guinea Pig Model of Airway Smooth Muscle Hyperreactivity Induced by Chronic Allergic Lung Inflammation: Contribution of Epithelium and Oxidative Stress. *Front. Pharmacol.* **2018**, *9*, 1547. [CrossRef]

91. Hallstrand, T.S.; Henderson, W.R., Jr. An update on the role of leukotrienes in asthma. *Curr. Opin. Allergy Clin. Immunol.* **2010**, *10*, 60–66. [CrossRef]

92. Pardo-Saganta, A.; Law, B.M.; Gonzalez-Celeiro, M.; Vinarsky, V.; Rajagopal, J. Ciliated cells of pseudostratified airway epithelium do not become mucous cells after ovalbumin challenge. *Am. J. Respir. Cell Mol. Biol.* **2013**, *48*, 364–373. [CrossRef]

93. Evans, C.M.; Koo, J.S. Airway mucus: The good, the bad, the sticky. *Pharmacol. Ther.* **2009**, *121*, 332–348. [CrossRef] [PubMed]

94. Bonser, L.R.; Erle, D.J. Airway Mucus and Asthma: The Role of MUC5AC and MUC5B. *J. Clin. Med.* **2017**, *6*, 112. [CrossRef] [PubMed]

95. Lancas, T.; Kasahara, D.I.; Prado, C.M.; Tiberio, I.F.; Martins, M.A.; Dolhnikoff, M. Comparison of early and late responses to antigen of sensitized guinea pig parenchymal lung strips. *J. Appl. Physiol.* **2006**, *100*, 1610–1616. [CrossRef] [PubMed]

96. van der Worp, H.B.; Howells, D.W.; Sena, E.S.; Porritt, M.J.; Rewell, S.; O'Collins, V.; Macleod, M.R. Can animal models of disease reliably inform human studies? *PLoS Med.* **2010**, *7*, e1000245. [CrossRef]

97. Global Initiative for Chronic Obstructive Lung Disease. *Global Strategy for the Diagnosis, Management and Prevention of COPD 2019*; Global Initiative for Chronic Obstructive Lung Disease: Fontana, WI, USA, 2019.

98. Brandsma, C.A.; Timens, W.; Jonker, M.R.; Rutgers, B.; Noordhoek, J.A.; Postma, D.S. Differential effects of fluticasone on extracellular matrix production by airway and parenchymal fibroblasts in severe COPD. *Am. J. Physiol. Lung Cell. Mol. Physiol.* **2013**, *305*, L582–L589. [CrossRef] [PubMed]

99. Barnes, P.J.; Stockley, R.A. COPD: Current therapeutic interventions and future approaches. *Eur. Respir. J.* **2005**, *25*, 1084–1106. [CrossRef]

100. Brusselle, G.G.; Joos, G.F.; Bracke, K.R. New insights into the immunology of chronic obstructive pulmonary disease. *Lancet* **2011**, *378*, 1015–1026. [CrossRef]

101. Higham, A.; Quinn, A.M.; Cancado, J.E.D.; Singh, D. The pathology of small airways disease in COPD: Historical aspects and future directions. *Respir. Res.* **2019**, *20*, 49. [CrossRef]

102. Hogg, J.C.; Timens, W. The pathology of chronic obstructive pulmonary disease. *Annu. Rev. Pathol.* **2009**, *4*, 435–459. [CrossRef]

103. Imai, K.; Dalal, S.S.; Chen, E.S.; Downey, R.; Schulman, L.L.; Ginsburg, M.; D'Armiento, J. Human collagenase (matrix metalloproteinase-1) expression in the lungs of patients with emphysema. *Am. J. Respir. Crit. Care Med.* **2001**, *163*, 786–791. [CrossRef]

104. Segura-Valdez, L.; Pardo, A.; Gaxiola, M.; Uhal, B.D.; Becerril, C.; Selman, M. Upregulation of gelatinases A and B, collagenases 1 and 2, and increased parenchymal cell death in COPD. *Chest* **2000**, *117*, 684–694. [CrossRef] [PubMed]

105. Churg, A.; Wang, R.; Wang, X.; Onnervik, P.O.; Thim, K.; Wright, J.L. Effect of an MMP-9/MMP-12 inhibitor on smoke-induced emphysema and airway remodelling in guinea pigs. *Thorax* **2007**, *62*, 706–713. [CrossRef] [PubMed]

106. Churg, A.; Zhou, S.; Wright, J.L. Series "matrix metalloproteinases in lung health and disease": Matrix metalloproteinases in COPD. *Eur. Respir. J.* **2012**, *39*, 197–209. [CrossRef] [PubMed]

107. Hautamaki, R.D.; Kobayashi, D.K.; Senior, R.M.; Shapiro, S.D. Requirement for macrophage elastase for cigarette smoke-induced emphysema in mice. *Science* **1997**, *277*, 2002–2004. [CrossRef]

108. Abraham, T.; Hogg, J. Extracellular matrix remodeling of lung alveolar walls in three dimensional space identified using second harmonic generation and multiphoton excitation fluorescence. *J. Struct. Biol.* **2010**, *171*, 189–196. [CrossRef] [PubMed]

109. Shifren, A.; Durmowicz, A.G.; Knutsen, R.H.; Hirano, E.; Mecham, R.P. Elastin protein levels are a vital modifier affecting normal lung development and susceptibility to emphysema. *Am. J. Physiol. Lung Cell. Mol. Physiol.* **2007**, *292*, L778–L787. [CrossRef]

110. Kononov, S.; Brewer, K.; Sakai, H.; Cavalcante, F.S.; Sabayanagam, C.R.; Ingenito, E.P.; Suki, B. Roles of mechanical forces and collagen failure in the development of elastase-induced emphysema. *Am. J. Respir. Crit. Care Med.* **2001**, *164*, 1920–1926. [CrossRef]

111. Koenders, M.M.; Wismans, R.G.; Starcher, B.; Hamel, B.C.; Dekhuijzen, R.P.; van Kuppevelt, T.H. Fibrillin-1 staining anomalies are associated with increased staining for TGF-beta and elastic fibre degradation; new clues to the pathogenesis of emphysema. *J. Pathol.* **2009**, *218*, 446–457. [CrossRef] [PubMed]

112. Vlahovic, G.; Russell, M.L.; Mercer, R.R.; Crapo, J.D. Cellular and connective tissue changes in alveolar septal walls in emphysema. *Am. J. Respir. Crit. Care Med.* **1999**, *160*, 2086–2092. [CrossRef]

113. Rubio, M.L.; Martin-Mosquero, M.C.; Ortega, M.; Peces-Barba, G.; Gonzalez-Mangado, N. Oral N-acetylcysteine attenuates elastase-induced pulmonary emphysema in rats. *Chest* **2004**, *125*, 1500–1506. [CrossRef]

114. Churg, A.; Wang, R.D.; Tai, H.; Wang, X.; Xie, C.; Wright, J.L. Tumor necrosis factor-alpha drives 70% of cigarette smoke-induced emphysema in the mouse. *Am. J. Respir. Crit. Care Med.* **2004**, *170*, 492–498. [CrossRef] [PubMed]

115. Anciaes, A.M.; Olivo, C.R.; Prado, C.M.; Kagohara, K.H.; Pinto Tda, S.; Moriya, H.T.; Mauad, T.; Martins Mde, A.; Lopes, F.D. Respiratory mechanics do not always mirror pulmonary histological changes in emphysema. *Clinics* **2011**, *66*, 1797–1803. [PubMed]

116. Stockley, R.A. Proteases and antiproteases. *Novartis Found. Symp.* **2001**, *234*, 189–189; discussion 199–204. [PubMed]

117. Lucey, E.C.; Goldstein, R.H.; Stone, P.J.; Snider, G.L. Remodeling of alveolar walls after elastase treatment of hamsters. Results of elastin and collagen mRNA in situ hybridization. *Am. J. Respir. Crit. Care Med.* **1998**, *158*, 555–564. [CrossRef]

118. Kawakami, M.; Matsuo, Y.; Yoshiura, K.; Nagase, T.; Yamashita, N. Sequential and quantitative analysis of a murine model of elastase-induced emphysema. *Biol. Pharm. Bull.* **2008**, *31*, 1434–1438. [CrossRef]

119. Parameswaran, K.; Willems-Widyastuti, A.; Alagappan, V.K.; Radford, K.; Kranenburg, A.R.; Sharma, H.S. Role of extracellular matrix and its regulators in human airway smooth muscle biology. *Cell Biochem. Biophys.* **2006**, *44*, 139–146. [CrossRef]

120. Li, H.; Cui, D.; Ma, N.; Lu, L.; Gao, Y.; Cui, X.; Wang, D. The effect of extracellular matrix remodeling on airflow obstruction in a rat model of chronic obstructive pulmonary disease. *Zhonghua Jie He He Hu Xi Za Zhi* **2002**, *25*, 403–407.

121. Bidan, C.M.; Veldsink, A.C.; Meurs, H.; Gosens, R. Airway and Extracellular Matrix Mechanics in COPD. *Front. Physiol.* **2015**, *6*, 346. [CrossRef] [PubMed]

122. Black, P.N.; Ching, P.S.; Beaumont, B.; Ranasinghe, S.; Taylor, G.; Merrilees, M.J. Changes in elastic fibres in the small airways and alveoli in COPD. *Eur. Respir. J.* **2008**, *31*, 998–1004. [CrossRef]

123. Gladysheva, E.S.; Malhotra, A.; Owens, R.L. Influencing the decline of lung function in COPD: Use of pharmacotherapy. *Int. J. Chronic Obstr. Pulm. Dis.* **2010**, *5*, 153–164.

124. Pare, P.D.; Mitzner, W. Airway-parenchymal interdependence. *Compr. Physiol.* **2012**, *2*, 1921–1935. [CrossRef] [PubMed]

125. Chen, L.; Ge, Q.; Tjin, G.; Alkhouri, H.; Deng, L.; Brandsma, C.A.; Adcock, I.; Timens, W.; Postma, D.; Burgess, J.K.; et al. Effects of cigarette smoke extract on human airway smooth muscle cells in COPD. *Eur. Respir. J.* **2014**, *44*, 634–646. [CrossRef] [PubMed]

126. Hogg, J.C.; Chu, F.; Utokaparch, S.; Woods, R.; Elliott, W.M.; Buzatu, L.; Cherniack, R.M.; Rogers, R.M.; Sciurba, F.C.; Coxson, H.O.; et al. The nature of small-airway obstruction in chronic obstructive pulmonary disease. *N. Engl. J. Med.* **2004**, *350*, 2645–2653. [CrossRef]

127. Eurlings, I.M.; Dentener, M.A.; Cleutjens, J.P.; Peutz, C.J.; Rohde, G.G.; Wouters, E.F.; Reynaert, N.L. Similar matrix alterations in alveolar and small airway walls of COPD patients. *BMC Pulm. Med.* **2014**, *14*, 90. [CrossRef] [PubMed]

128. Sun, C.; Zhu, M.; Yang, Z.; Pan, X.; Zhang, Y.; Wang, Q.; Xiao, W. LL-37 secreted by epithelium promotes fibroblast collagen production: A potential mechanism of small airway remodeling in chronic obstructive pulmonary disease. *Lab. Investig.* **2014**, *94*, 991–1002. [CrossRef]

129. Milara, J.; Peiro, T.; Serrano, A.; Cortijo, J. Epithelial to mesenchymal transition is increased in patients with COPD and induced by cigarette smoke. *Thorax* **2013**, *68*, 410–420. [CrossRef] [PubMed]

130. Mahmood, M.Q.; Reid, D.; Ward, C.; Muller, H.K.; Knight, D.A.; Sohal, S.S.; Walters, E.H. Transforming growth factor (TGF) beta1 and Smad signalling pathways: A likely key to EMT-associated COPD pathogenesis. *Respirology* **2017**, *22*, 133–140. [CrossRef] [PubMed]

131. Podowski, M.; Calvi, C.; Metzger, S.; Misono, K.; Poonyagariyagorn, H.; Lopez-Mercado, A.; Ku, T.; Lauer, T.; McGrath-Morrow, S.; Berger, A.; et al. Angiotensin receptor blockade attenuates cigarette smoke-induced lung injury and rescues lung architecture in mice. *J. Clin. Investig.* **2012**, *122*, 229–240. [CrossRef]

132. Wang, Z.; Fang, K.; Wang, G.; Guan, X.; Pang, Z.; Guo, Y.; Yuan, Y.; Ran, N.; Liu, Y.; Wang, F. Protective effect of amygdalin on epithelial-mesenchymal transformation in experimental chronic obstructive pulmonary disease mice. *Phytother. Res.* **2019**, *33*, 808–817. [CrossRef] [PubMed]

133. Sand, J.M.; Leeming, D.J.; Byrjalsen, I.; Bihlet, A.R.; Lange, P.; Tal-Singer, R.; Miller, B.E.; Karsdal, M.A.; Vestbo, J. High levels of biomarkers of collagen remodeling are associated with increased mortality in COPD—Results from the ECLIPSE study. *Respir. Res.* **2016**, *17*, 125. [CrossRef]

134. Zeng, Y.Y.; Hu, W.P.; Zuo, Y.H.; Wang, X.R.; Zhang, J. Altered serum levels of type I collagen turnover indicators accompanied by IL-6 and IL-8 release in stable COPD. *Int. J. Chronic Obstr. Pulm. Dis.* **2019**, *14*, 163–168. [CrossRef] [PubMed]

135. Vignola, A.M.; Paganin, F.; Capieu, L.; Scichilone, N.; Bellia, M.; Maakel, L.; Bellia, V.; Godard, P.; Bousquet, J.; Chanez, P. Airway remodelling assessed by sputum and high-resolution computed tomography in asthma and COPD. *Eur. Respir. J.* **2004**, *24*, 910–917. [CrossRef] [PubMed]

136. Papakonstantinou, E.; Bonovolias, I.; Roth, M.; Tamm, M.; Schumann, D.; Baty, F.; Louis, R.; Milenkovic, B.; Boersma, W.; Stieltjes, B.; et al. Serum levels of hyaluronic acid are associated with COPD severity and predict survival. *Eur. Respir. J.* **2019**, *53*, 1801183. [CrossRef] [PubMed]

137. Suki, B.; Lutchen, K.R.; Ingenito, E.P. On the progressive nature of emphysema: Roles of proteases, inflammation, and mechanical forces. *Am. J. Respir. Crit. Care Med.* **2003**, *168*, 516–521. [CrossRef] [PubMed]

138. Fricker, M.; Deane, A.; Hansbro, P.M. Animal models of chronic obstructive pulmonary disease. *Expert Opin. Drug Discov.* **2014**, *9*, 629–645. [CrossRef] [PubMed]

139. Churg, A.; Cosio, M.; Wright, J.L. Mechanisms of cigarette smoke-induced COPD: Insights from animal models. *Am. J. Physiol. Lung Cell. Mol. Physiol.* **2008**, *294*, L612–L631. [CrossRef]

140. Wright, J.L.; Churg, A. Animal models of COPD: Barriers, successes, and challenges. *Pulm. Pharmacol. Ther.* **2008**, *21*, 696–698. [CrossRef] [PubMed]

141. Wright, J.L.; Cosio, M.; Churg, A. Animal models of chronic obstructive pulmonary disease. *Am. J. Physiol. Lung Cell. Mol. Physiol.* **2008**, *295*, L1–L15. [CrossRef]

142. Ito, J.T.; Cervilha, D.A.B.; Lourenco, J.D.; Goncalves, N.G.; Volpini, R.A.; Caldini, E.G.; Landman, G.; Lin, C.J.; Velosa, A.P.P.; Teodoro, W.P.R.; et al. Th17/Treg imbalance in COPD progression: A temporal analysis using a CS-induced model. *PLoS ONE* **2019**, *14*, e0209351. [CrossRef] [PubMed]

143. Lourenco, J.D.; Ito, J.T.; Cervilha, D.A.B.; Sales, D.S.; Riani, A.; Suehiro, C.L.; Genaro, I.S.; Duran, A.; Puzer, L.; Martins, M.A.; et al. The tick-derived rBmTI-A protease inhibitor attenuates the histological and functional changes induced by cigarette smoke exposure. *Histol. Histopathol.* **2018**, *33*, 289–298. [CrossRef] [PubMed]

144. Beckett, E.L.; Stevens, R.L.; Jarnicki, A.G.; Kim, R.Y.; Hanish, I.; Hansbro, N.G.; Deane, A.; Keely, S.; Horvat, J.C.; Yang, M.; et al. A new short-term mouse model of chronic obstructive pulmonary disease identifies a role for mast cell tryptase in pathogenesis. *J. Allergy Clin. Immunol.* **2013**, *131*, 752–762. [CrossRef]

145. Lopes, F.D.; Toledo, A.C.; Olivo, C.R.; Prado, C.M.; Leick, E.A.; Medeiros, M.C.; Santos, A.B.; Garippo, A.; Martins, M.A.; Mauad, T. A comparative study of extracellular matrix remodeling in two murine models of emphysema. *Histol. Histopathol.* **2013**, *28*, 269–276. [CrossRef]

146. Lourenco, J.D.; Neves, L.P.; Olivo, C.R.; Duran, A.; Almeida, F.M.; Arantes, P.M.; Prado, C.M.; Leick, E.A.; Tanaka, A.S.; Martins, M.A.; et al. A treatment with a protease inhibitor recombinant from the cattle tick (Rhipicephalus Boophilus microplus) ameliorates emphysema in mice. *PLoS ONE* **2014**, *9*, e98216. [CrossRef]

147. Rodrigues, R.; Olivo, C.R.; Lourenco, J.D.; Riane, A.; Cervilha, D.A.B.; Ito, J.T.; Martins, M.A.; Lopes, F. A murine model of elastase- and cigarette smoke-induced emphysema. *J. Bras. Pneumol.* **2017**, *43*, 95–100. [CrossRef]

148. Almeida-Reis, R.; Theodoro-Junior, O.A.; Oliveira, B.T.M.; Oliva, L.V.; Toledo-Arruda, A.C.; Bonturi, C.R.; Brito, M.V.; Lopes, F.; Prado, C.M.; Florencio, A.C.; et al. Plant Proteinase Inhibitor BbCI Modulates Lung Inflammatory Responses and Mechanic and Remodeling Alterations Induced by Elastase in Mice. *BioMed Res. Int.* **2017**, *2017*, 8287125. [CrossRef]

149. Theodoro-Junior, O.A.; Righetti, R.F.; Almeida-Reis, R.; Martins-Oliveira, B.T.; Oliva, L.V.; Prado, C.M.; Saraiva-Romanholo, B.M.; Leick, E.A.; Pinheiro, N.M.; Lobo, Y.A.; et al. A Plant Proteinase Inhibitor from Enterolobium contortisiliquum Attenuates Pulmonary Mechanics, Inflammation and Remodeling Induced by Elastase in Mice. *Int. J. Mol. Sci.* **2017**, *18*, 403. [CrossRef]

150. Robertoni, F.S.; Olivo, C.R.; Lourenco, J.D.; Goncalves, N.G.; Velosa, A.P.; Lin, C.J.; Flo, C.M.; Saraiva-Romanholo, B.M.; Sasaki, S.D.; Martins, M.A.; et al. Collagenase mRNA Overexpression and Decreased Extracellular Matrix Components Are Early Events in the Pathogenesis of Emphysema. *PLoS ONE* **2015**, *10*, e0129590. [CrossRef]

151. Silver, F.H.; Horvath, I.; Foran, D.J. Mechanical implications of the domain structure of fiber-forming collagens: Comparison of the molecular and fibrillar flexibilities of the alpha1-chains found in types I-III collagen. *J. Theor. Biol.* **2002**, *216*, 243–254. [CrossRef]

152. Ito, S.; Ingenito, E.P.; Brewer, K.K.; Black, L.D.; Parameswaran, H.; Lutchen, K.R.; Suki, B. Mechanics, nonlinearity, and failure strength of lung tissue in a mouse model of emphysema: Possible role of collagen remodeling. *J. Appl. Physiol.* **2005**, *98*, 503–511. [CrossRef]

153. Wang, Z.; Li, R.; Zhong, R. Extracellular matrix promotes proliferation, migration and adhesion of airway smooth muscle cells in a rat model of chronic obstructive pulmonary disease via upregulation of the PI3K/AKT signaling pathway. *Mol. Med. Rep.* **2018**, *18*, 3143–3152. [CrossRef]

154. Toledo, A.C.; Magalhaes, R.M.; Hizume, D.C.; Vieira, R.P.; Biselli, P.J.; Moriya, H.T.; Mauad, T.; Lopes, F.D.; Martins, M.A. Aerobic exercise attenuates pulmonary injury induced by exposure to cigarette smoke. *Eur. Respir. J.* **2012**, *39*, 254–264. [CrossRef] [PubMed]

155. Gomes, R.F.; Shen, X.; Ramchandani, R.; Tepper, R.S.; Bates, J.H. Comparative respiratory system mechanics in rodents. *J. Appl. Physiol.* **2000**, *89*, 908–916. [CrossRef] [PubMed]

156. Foronjy, R.F.; Mercer, B.A.; Maxfield, M.W.; Powell, C.A.; D'Armiento, J.; Okada, Y. Structural emphysema does not correlate with lung compliance: Lessons from the mouse smoking model. *Exp. Lung Res.* **2005**, *31*, 547–562. [CrossRef] [PubMed]

157. Bellani, G.; Laffey, J.G.; Pham, T.; Fan, E.; Brochard, L.; Esteban, A.; Gattinoni, L.; van Haren, F.; Larsson, A.; McAuley, D.F.; et al. Epidemiology, Patterns of Care, and Mortality for Patients With Acute Respiratory Distress Syndrome in Intensive Care Units in 50 Countries. *JAMA* **2016**, *315*, 788–800. [CrossRef] [PubMed]

158. Amato, M.B.; Meade, M.O.; Slutsky, A.S.; Brochard, L.; Costa, E.L.; Schoenfeld, D.A.; Stewart, T.E.; Briel, M.; Talmor, D.; Mercat, A.; et al. Driving pressure and survival in the acute respiratory distress syndrome. *N. Engl. J. Med.* **2015**, *372*, 747–755. [CrossRef] [PubMed]

159. Papazian, L.; Doddoli, C.; Chetaille, B.; Gernez, Y.; Thirion, X.; Roch, A.; Donati, Y.; Bonnety, M.; Zandotti, C.; Thomas, P. A contributive result of open-lung biopsy improves survival in acute respiratory distress syndrome patients. *Crit. Care Med.* **2007**, *35*, 755–762. [CrossRef] [PubMed]

160. Pelosi, P.; Caironi, P.; Gattinoni, L. Pulmonary and extrapulmonary forms of acute respiratory distress syndrome. *Semin. Respir. Crit. Care Med.* **2001**, *22*, 259–268. [CrossRef]

161. Hoelz, C.; Negri, E.M.; Lichtenfels, A.J.; Concecao, G.M.; Barbas, C.S.; Saldiva, P.H.; Capelozzi, V.L. Morphometric differences in pulmonary lesions in primary and secondary ARDS. A preliminary study in autopsies. *Pathol. Res. Pract.* **2001**, *197*, 521–530. [PubMed]

162. Petersen, A.M.; Pedersen, B.K. The anti-inflammatory effect of exercise. *J. Appl. Physiol.* **2005**, *98*, 1154–1162. [CrossRef]

163. Meduri, G.U.; Kohler, G.; Headley, S.; Tolley, E.; Stentz, F.; Postlethwaite, A. Inflammatory cytokines in the BAL of patients with ARDS. Persistent elevation over time predicts poor outcome. *Chest* **1995**, *108*, 1303–1314. [CrossRef]

164. Zhou, X.; Dai, Q.; Huang, X. Neutrophils in acute lung injury. *Front. Biosci.* **2012**, *17*, 2278–2283. [CrossRef]

165. Huang, W.; McCaig, L.A.; Veldhuizen, R.A.; Yao, L.J.; Lewis, J.F. Mechanisms responsible for surfactant changes in sepsis-induced lung injury. *Eur. Respir. J.* **2005**, *26*, 1074–1079. [CrossRef] [PubMed]

166. Burnham, E.L.; Janssen, W.J.; Riches, D.W.; Moss, M.; Downey, G.P. The fibroproliferative response in acute respiratory distress syndrome: Mechanisms and clinical significance. *Eur. Respir. J.* **2014**, *43*, 276–285. [CrossRef] [PubMed]

167. Chen, Q.; Luo, A.A.; Qiu, H.; Han, B.; Ko, B.H.; Slutsky, A.S.; Zhang, H. Monocyte interaction accelerates HCl-induced lung epithelial remodeling. *BMC Pulm. Med.* **2014**, *14*, 135. [CrossRef]

168. Tomashefski, J.F., Jr. Pulmonary pathology of acute respiratory distress syndrome. *Clin. Chest Med.* **2000**, *21*, 435–466. [CrossRef]

169. Forel, J.M.; Guervilly, C.; Farnarier, C.; Donati, S.Y.; Hraiech, S.; Persico, N.; Allardet-Servent, J.; Coiffard, B.; Gainnier, M.; Loundou, A.; et al. Transforming Growth Factor-beta1 in predicting early lung fibroproliferation in patients with acute respiratory distress syndrome. *PLoS ONE* **2018**, *13*, e0206105. [CrossRef] [PubMed]

170. Rubenfeld, G.D.; Herridge, M.S. Epidemiology and outcomes of acute lung injury. *Chest* **2007**, *131*, 554–562. [CrossRef]

171. Herridge, M.S.; Tansey, C.M.; Matte, A.; Tomlinson, G.; Diaz-Granados, N.; Cooper, A.; Guest, C.B.; Mazer, C.D.; Mehta, S.; Stewart, T.E.; et al. Functional disability 5 years after acute respiratory distress syndrome. *N. Engl. J. Med.* **2011**, *364*, 1293–1304. [CrossRef]

172. Negri, E.M.; Montes, G.S.; Saldiva, P.H.; Capelozzi, V.L. Architectural remodelling in acute and chronic interstitial lung disease: Fibrosis or fibroelastosis? *Histopathology* **2000**, *37*, 393–401. [CrossRef]

173. Gonzalez-Lopez, A.; Albaiceta, G.M. Repair after acute lung injury: Molecular mechanisms and therapeutic opportunities. *Crit. Care* **2012**, *16*, 209. [CrossRef]

174. Santos, F.B.; Nagato, L.K.; Boechem, N.M.; Negri, E.M.; Guimaraes, A.; Capelozzi, V.L.; Faffe, D.S.; Zin, W.A.; Rocco, P.R. Time course of lung parenchyma remodeling in pulmonary and extrapulmonary acute lung injury. *J. Appl. Physiol.* **2006**, *100*, 98–106. [CrossRef]

175. Kamp, R.; Sun, X.; Garcia, J.G. Making genomics functional: Deciphering the genetics of acute lung injury. *Proc. Am. Thorac. Soc.* **2008**, *5*, 348–353. [CrossRef]

176. Burnham, E.L.; Moss, M.; Ritzenthaler, J.D.; Roman, J. Increased fibronectin expression in lung in the setting of chronic alcohol abuse. *Alcohol. Clin. Exp. Res.* **2007**, *31*, 675–683. [CrossRef]

177. Esper, A.; Burnham, E.L.; Moss, M. The effect of alcohol abuse on ARDS and multiple organ dysfunction. *Minerva Anestesiol.* **2006**, *72*, 375–381. [PubMed]

178. Aytacoglu, B.N.; Calikoglu, M.; Tamer, L.; Coskun, B.; Sucu, N.; Kose, N.; Aktas, S.; Dikmengil, M. Alcohol-induced lung damage and increased oxidative stress. *Respiration* **2006**, *73*, 100–104. [CrossRef]
179. Clark, J.G.; Milberg, J.A.; Steinberg, K.P.; Hudson, L.D. Type III procollagen peptide in the adult respiratory distress syndrome. Association of increased peptide levels in bronchoalveolar lavage fluid with increased risk for death. *Ann. Intern. Med.* **1995**, *122*, 17–23. [CrossRef]
180. Farjanel, J.; Hartmann, D.J.; Guidet, B.; Luquel, L.; Offenstadt, G. Four markers of collagen metabolism as possible indicators of disease in the adult respiratory distress syndrome. *Am. Rev. Respir. Dis.* **1993**, *147*, 1091–1099. [CrossRef]
181. Meduri, G.U.; Tolley, E.A.; Chinn, A.; Stentz, F.; Postlethwaite, A. Procollagen types I and III aminoterminal propeptide levels during acute respiratory distress syndrome and in response to methylprednisolone treatment. *Am. J. Respir. Crit. Care Med.* **1998**, *158*, 1432–1441. [CrossRef]
182. Hamon, A.; Scemama, U.; Bourenne, J.; Daviet, F.; Coiffard, B.; Persico, N.; Adda, M.; Guervilly, C.; Hraiech, S.; Chaumoitre, K.; et al. Chest CT scan and alveolar procollagen III to predict lung fibroproliferation in acute respiratory distress syndrome. *Ann. Intensive Care* **2019**, *9*, 42. [CrossRef]
183. Thille, A.W.; Esteban, A.; Fernandez-Segoviano, P.; Rodriguez, J.M.; Aramburu, J.A.; Vargas-Errazuriz, P.; Martin-Pellicer, A.; Lorente, J.A.; Frutos-Vivar, F. Chronology of histological lesions in acute respiratory distress syndrome with diffuse alveolar damage: A prospective cohort study of clinical autopsies. *Lancet Respir. Med.* **2013**, *1*, 395–401. [CrossRef]
184. Heyland, D.K.; Groll, D.; Caeser, M. Survivors of acute respiratory distress syndrome: Relationship between pulmonary dysfunction and long-term health-related quality of life. *Crit. Care Med.* **2005**, *33*, 1549–1556. [CrossRef]
185. McHugh, L.G.; Milberg, J.A.; Whitcomb, M.E.; Schoene, R.B.; Maunder, R.J.; Hudson, L.D. Recovery of function in survivors of the acute respiratory distress syndrome. *Am. J. Respir. Crit. Care Med.* **1994**, *150*, 90–94. [CrossRef] [PubMed]
186. Zheng, S.; Wang, Q.; D'Souza, V.; Bartis, D.; Dancer, R.; Parekh, D.; Gao, F.; Lian, Q.; Jin, S.; Thickett, D.R. ResolvinD1 stimulates epithelial wound repair and inhibits TGF-beta-induced EMT whilst reducing fibroproliferation and collagen production. *Lab. Investig.* **2018**, *98*, 130–140. [CrossRef]
187. Bittencourt-Mernak, M.I.; Pinheiro, N.M.; Santana, F.P.; Guerreiro, M.P.; Saraiva-Romanholo, B.M.; Grecco, S.S.; Caperuto, L.C.; Felizardo, R.J.; Camara, N.O.; Tiberio, I.F.; et al. Prophylactic and therapeutic treatment with the flavonone sakuranetin ameliorates LPS-induced acute lung injury. *Am. J. Physiol. Lung Cell. Mol. Physiol.* **2017**, *312*, L217–L230. [CrossRef] [PubMed]
188. de Souza Xavier Costa, N.; Ribeiro Junior, G.; Dos Santos Alemany, A.A.; Belotti, L.; Zati, D.H.; Frota Cavalcante, M.; Matera Veras, M.; Ribeiro, S.; Kallas, E.G.; Nascimento Saldiva, P.H.; et al. Early and late pulmonary effects of nebulized LPS in mice: An acute lung injury model. *PLoS ONE* **2017**, *12*, e0185474. [CrossRef]
189. Oliveira, V.R.; Uriarte, J.J.; Falcones, B.; Zin, W.A.; Navajas, D.; Farre, R.; Almendros, I. Escherichia coli lipopolysaccharide induces alveolar epithelial cell stiffening. *J. Biomech.* **2019**, *83*, 315–318. [CrossRef] [PubMed]
190. Park, J.R.; Lee, H.; Kim, S.I.; Yang, S.R. The tri-peptide GHK-Cu complex ameliorates lipopolysaccharide-induced acute lung injury in mice. *Oncotarget* **2016**, *7*, 58405–58417. [CrossRef]
191. Pinheiro, N.M.; Santana, F.P.; Almeida, R.R.; Guerreiro, M.; Martins, M.A.; Caperuto, L.C.; Camara, N.O.; Wensing, L.A.; Prado, V.F.; Tiberio, I.F.; et al. Acute lung injury is reduced by the alpha7nAChR agonist PNU-282987 through changes in the macrophage profile. *FASEB J.* **2017**, *31*, 320–332. [CrossRef]
192. Simcock, D.E.; Kanabar, V.; Clarke, G.W.; Mahn, K.; Karner, C.; O'Connor, B.J.; Lee, T.H.; Hirst, S.J. Induction of angiogenesis by airway smooth muscle from patients with asthma. *Am. J. Respir. Crit. Care Med.* **2008**, *178*, 460–468. [CrossRef] [PubMed]

MDPI

St. Alban-Anlage 66

4052 Basel

Switzerland

Tel. +41 61 683 77 34

Fax +41 61 302 89 18

www.mdpi.com

Cells Editorial Office

E-mail: cells@mdpi.com

www.mdpi.com/journal/cells

www.ingramcontent.com/pod-product-compliance
Lightning Source LLC
Chambersburg PA
CBHW051705210326
41597CB00032B/5377